Advancing Frontiers in Mycology & Mycotechnology

Tulasi Satyanarayana
Sunil Kumar Deshmukh
Mukund V. Deshpande
Editors

Advancing Frontiers in Mycology & Mycotechnology

Basic and Applied Aspects of Fungi

Springer

Editors
Tulasi Satyanarayana
Division of Biological Science and Engineering
Netaji Subhas University of Technology
New Delhi, Delhi, India

Sunil Kumar Deshmukh
Biotech & Management of Bioresources Div
The Energy and Resources Institute
New Delhi, Delhi, India

Mukund V. Deshpande
Division of Biological Sciences
CSIR-National Chemical Laboratory
Pune, Maharashtra, India

ISBN 978-981-13-9348-8 ISBN 978-981-13-9349-5 (eBook)
https://doi.org/10.1007/978-981-13-9349-5

This Springer imprint is published by the registered company Springer Nature Singapore Pte Ltd.
The registered company address is: 152 Beach Road, #21-01/04 Gateway East, Singapore 189721, Singapore

Foreword

The term 'mycology' dates back to the eighteenth century, and in its formative years, the study of fungi was guided by reputed botanists like C. H. Persoon, E.M. Fries and A. deBary. During the nineteenth and twentieth centuries, phenomenal progress in classical mycology took place, including descriptive morphology, identification, classification and taxonomy of fungi primarily, due to the efforts of botanists, despite the fact that fungi are not phylogenetically related to plants. The reason for this essentially stems from the devastating fungal diseases in plants, notably the crop plants, causing serious problems in agriculture. There was an urgent need to understand the aetiology of the destructive fungal diseases and devise suitable control measures to save the crops. As an example may be mentioned the studies in India by Prof. M.J. Narasimhan (1930) who collected and studied diverse heterothallic strains of the destructive *Phytophthora*[1] infection on areca nut palms and devising fungicidal treatment schedules for controlling the disease in the rainforest areas of Karnataka (Mysore) state.

During the early years, traditional mycological knowledge facilitated the (a) development of various oriental fermented foods, such as miso, tempeh and sake, based on strains of fungi such as *Aspergillus oryzae*, *Rhizopus oligosporus* and others (Hesseltine 1983), (b) differentiation of edible and poisonous mushrooms and (c) understanding the aetiology of ergot poisoning that caused serious convulsions due to consumption of the ergot sclerotia contaminating the grain which is consumed as food.

During the twentieth century, several milestone advances in biological sciences based on mycological studies can be recognized. As examples, we may mention the following:

(a) The Nobel Prize-winning (1958) studies on *Neurospora* by G.W. Beadle and E.L. Tatum which resulted in their discovery of how genes act by regulating definite chemical events that in turn affected the development in organisms. Based on ingenious selection of mutants deficient in the synthesis of single growth factors, Beadle and Tatum concluded that mutations to genes affected

[1] Recently, the genera *Phytophthora* and *Pythium* have been placed in Kingdom Chromista or Kingdom Straminipila, distinct from Kingdom Fungi (Ho 2018).

the enzymes of organisms. It is noteworthy that these studies provided the early link between genetics and the upcoming new scientific field of molecular biology.

(b) The discovery of truly anaerobic chytrids present in the rumen of herbivores and playing a significant role in the digestion of the cellulosic feed is a milestone discovery of the twentieth century. Classified under *Neocallimastigaceae*, these chytrids with chitinous cell walls are truly anaerobes lacking mitochondria, cytochromes and other biochemical features of the oxidative phosphorylation pathway. The group possesses organelles (hydrogenosomes) for a major part of anaerobic energy metabolism (Nagpal et al. 2009).

Taxonomy and classification of fungi have been central to the mycological studies, and this has led to a well-established database for understanding fungal biodiversity. Classical taxonomy relied upon the morphological features of the spore forms. With the advent of biochemical and molecular data becoming available, several conceptual changes in fungal taxonomy have arisen. While giving an opportunity to gain newer knowledge, these developments also pose several challenges and opinion divides in their interpretation and implementation.

Classical taxonomy classified all the zoospore-forming fungi under Phycomycetes considering their monophyletic origin from algae. Recent advances in biochemical knowledge have established that the *Chytridiomycetes*[2] with chitinous cell walls, despite being zoospore-forming, are distinct from the rest of the zoospore-forming fungi with β-glucan in their cell walls, and these are grouped together under the Oomycota (as opposed to the *Eomycota* under which the *Zygomycetes*, *Ascomycetes* and *Basidiomycetes* are classified). The term Phycomycetes as originally adapted is no longer valid and, hence, abandoned from present-day mycological literature.

DNA sequencing and bar-coding of fungal species are major developments which offer challenges to accepted morphology-based taxonomy. The nuclear ribosomal internal transcribed spacer (ITS) is widely used as a DNA bar-coding marker to characterise the diversity and composition of fungal communities. Since the 1990s, the ITS region has been extensively used in molecular methods as well as ecological studies on fungi.

Estimates of fungal biodiversity have undergone sharp changes following environmental DNA sequencing of fungal ITS reads. Currently, there are around 81,000 accepted fungal names out of a total of up to 5.1 million species (Blackwell 2011). Environmental DNA sequencing has accumulated more than one million fungal ITS reads. Whether ‘sequence-based voucher-less fungi’ (in the absence of actual specimens) should be given a formal nomenclature is widely debated, and the opinions of expert mycologists are sharply divided on this point with reference to the estimation of biodiversity of fungi in the natural environment. Several informative reviews on DNA bar-coding as well as sequence-based fungal nomenclature are published and may be consulted for a more complete understanding of the problem and its

---

[2] Now chytrids are grouped under Chytridiomycota (N. P. Money, in The Fungi (Third Edition), 2016)

perspectives (Shenoy et al. 2007; Begerow et al. 2010; Das and Deb 2015; Hibbett et al. 2016; Hongsanen et al. 2018).

Conventionally, taxonomic principles related to fungi are debated and decisions taken under the authority of the International Code of Botanical Nomenclature (ICBN) held once in 6 years during the International Botanical Congress. Recently, the ICBN has been renamed as International Code of Nomenclature for algae, fungi and plants. In the Melbourne ICBN held in 2011, a momentous decision was taken to abolish Article 59 which permitted the use of dual nomenclature for pleomorphic fungi. In the morphology-based taxonomy, anamorphs (asexual forms) and teleomorphs (sexual forms) of fungi discovered and described independently could receive different but valid names as ruled by Article 59 of the ICBN. In the present era of phylogenetic molecular analysis, co-ordination of anamorphic and teleomorphic elements of a fungus can, in principle, be unequivocally established. Molecular mycologists are strongly in favour of the unification of the fungal nomenclature and achieve the 'one fungus-one name' ideal. However, there are several practical hurdles, and several specialist committees are in serious debates to elucidate the practical feasibility of implementing the same (Gams 2014).

Simultaneous with the focused attention on fungi as troublesome pathogens of plants, animals and humans, beneficial attributes of fungi as sources of valuable metabolites were also realized. Taka-Diastase from *A. oryzae* and citric acid from *Aspergillus niger* are examples of fungal-based processes of manufacture in the late nineteenth and early twentieth centuries. With the discovery of penicillin and the dawn of the antibiotic era, the decades of the twentieth century became an era for screening diverse microorganisms including fungi for discovering novel bioactive molecules and producing them by fermentation. Several industrially useful enzymes, hormones, vitamins and growth factors of fungal origin spearheaded the fermentation technology development. Fungi and their metabolites were also employed in the biotransformation of organic molecules (e.g. steroids).

The beneficial attributes of fungi far outweighed their earlier negative image, and this led to a very healthy and positive attitude towards fungi which deserved to be studied for deriving beneficial products from their metabolism. The term 'biodiversity' acquired a newer dimension in terms of its relevance to applied microbiological and biotechnological research. An immense focus on the physiological, biochemical and molecular aspects of mycodiversity became imminent for bioprocess development based on novel fungal strains.

Heterologous gene expression and production of mammalian proteins in fungal hosts have been successfully carried out for chymosin (cheese manufacture enzyme) in *A. niger* var. *awamori* and insulin in *Komagataella (Pichia) pastoris.*

Spectacular advances in both mycology and mycotechnology during the twentieth century have added new dimensions to mycological research, necessitating newer approaches to achieve success in this field at present. A different mindset and training have become necessary in order to carry out meaningful research projects in the twenty-first century. Some of the salient features of this new preparation may be defined as follows:

(a) Knowledge-based exploration of fungal biodiversity in different ecosystems by mycologists is an important aspect. Understanding the type of habitats in which specific types of fungal populations would be naturally enriched will be very helpful. For example, screening for thermophiles in self-heating composts, bagasse piles or sun-heated soils would be very useful. The utilization of special traits in specific groups of fungi is also very beneficial, such as the overlaying of agar plates with moistened detritus to allow colonies to develop from forcibly 'shot' spores in the case of saprophytic Entomophthorales like *Conidiobolus* and *Basidiobolus* as well as yeasts like *Sporobolomyces.* Adopting selective isolation techniques including the use of specific inhibitory chemicals or antibiotics has significantly enabled the slower growing and rarer populations to be isolated on agar plates overcoming competition from the rapidly growing fungal populations. The topic of selective isolation is a vast one and has been reviewed (Srinivasan 2004, 2008) which may be consulted.
(b) Apart from taxonomic studies, in this biotechnology era, mycologists have the responsibility of conserving their discoveries in pure culture through optimization of culture preservation methods suited for their specific strains. They have to ensure both morphological and genetic stability of their cultures under the laboratory conditions for prolonged periods. Repeated and frequent subcultures on sugar-rich media, like potato dextrose agar, often adversely affect sporulation, particularly in strains associated with decomposing plant litter. While freeze-drying (lyophilisation) is widely accepted as a safe method of long-term conservation, it must be verified in regard to its suitability for the strains under study. It is experienced that some fungi (e.g. *Conidiobolus*) failed to survive lyophilisation, while many of the lignocellulose-degrading hyphomycetes showed reduced sporulation upon revival from lyophilized cultures. After in-depth studies to standardize the optimum conservation method(s), mycologists in India should deposit their cultures in the national fungal germplasm banks such as the National Fungal Culture Collection of India (NFCCI), Agharkar Research Institute, Pune; Microbial Type Culture Collection (MTCC), CSIR-Institute of Microbial Technology, Chandigarh; National Centre for Microbial Resource (NCMR), National Centre for Cell Science, Pune; or the National Collection of Industrial Microorganisms (NCIM), CSIR-National Chemical Laboratory, Pune. While depositing the cultures, it is mandatory that the full information on the optimal methods of conservation for the strain(s) is given to the curators of the culture collections to follow for successful conservation.
(c) When exploring fungal diversity for novel metabolites, it may be necessary to pay attention right from the isolation stage, keeping in mind the purpose of the strain screening. I shall illustrate this point taking the case of 'endophytic fungi' and exploration of their potential to produce the plant metabolite by fermentation. Widespread interest in exploring endophyte fungal populations of diverse medicinal plants has arisen after the report from Prof. Gary Strobel's laboratory

in the USA that an endophyte designated *Taxomyces andreanae* produced the valuable anticancer metabolite taxol by fermentation (Stierle et al. 1993). Presently, diverse endophytic fungi have been isolated and maintained in pure culture in several laboratories with the purpose of studying their potential to produce the valuable plant metabolites by fermentation. A point that deserves to be seriously given consideration is as follows:

The endophytic fungus has imbibed the genes from the medicinal plant through constant association. In the laboratory, culturing it on routine sugar-rich media while facilitating satisfactory growth, there is no selection pressure for the growing fungus to conserve these imbibed genes, and over several subcultures on routine media may even be lost. An alternate suggestion which may be worth considering is to plan the culture isolation of the endophyte on a relatively simple nutrient medium fortified with some of the key chemical intermediate components involved in the biosynthesis of the specific plant metabolite. It is possible on such a medium, the metabolic genes for the plant metabolite are conserved better, and these may be subjected to mutation and gene amplification through molecular techniques, leading to strains which will be able to produce the metabolite at commercially viable levels. Presently, the levels of production observed are very low and insignificant, and much more studies are warranted before practical and viable technologies for endophyte-based plant metabolites would become a reality.

Recent years have witnessed increased emphasis on the beneficial attributes of fungi, recognizing their role in the natural ecosystems and as the source of several metabolites of value for humankind. Rambold et al. (2013) in an article, suggesting that mycology should be recognized as a major field in biology, stated 'Given the ecological and economic relevance of fungi...Mycology is insufficiently recognized as a major field of science'. In the theme for the International Mycological Congress held at Puerto Rico in July 2018, mycological discoveries for a better world and how fungi contributed to the health of society and of ecosystem were given emphasis. Clearly, one can perceive the changing face of mycology from 'fungi as destructive agents to be despised to fungi as true benefactors of humanity to be adored'.

In order to harness the full potential of fungi for mankind's benefit, mycologists must identify themselves as part of multidisciplinary teams exploring fungi for novel metabolites. They should broaden their horizons of interest and acquire knowledge in physiology, biochemistry and molecular biology of fungi. Familiarity with natural product chemistry and bioprocess engineering for mould metabolites including mould morphogenesis in fermentors related to growth conditions for maximum yields are desirable.

Multidisciplinary collaboration to achieve success in technology is a must, and the status of mycology in the twenty-first century is meaningfully reflected in the following quote from Kreger (2003):

> *We all realise that most scientists in the future will be part of multi-disciplinary research teams...the shift is causing a way scientists need to train ... although they must still be an expert in their specialty, they must also become conversant with techniques that once seemed beyond their domain ... they also need to recognize where their knowledge ends and where they should seek the help from others....*

It gives me pleasure in writing the foreword for this book, *Advancing Frontiers in Mycology and Mycotechnology: Basic and Applied Aspects of Fungi.* The invited chapters included in this book cover most of the aspects of fungal biology and applied aspects of fungi. I wish to congratulate the editors and all the contributors for their efforts in bringing out this book. I sincerely hope and wish that the book will be useful for students, scholars, scientists and teachers of biology, microbiology and biotechnology.

Ex-Head, Biochemical Sciences Division
CSIR-NCL
Pune, Maharashtra, India
R.H.17, Planet Millennium
Pune, Maharashtra, India

M. C. Srinivasan

References

Begerow D, Nilsson H, Unterseher M, Maier W (2010) Current state and perspectives of fungal DNA bar coding and rapid identification procedures. Appl Microbiol Biotechnol 87:99–108
Blackwell M (2011) The fungi: 1, 2, 3 … 5.1 million species ? Am J Bot 98: 426–438
Das S, Deb B (2015) DNA bar coding of fungi using ribosomal ITS marker for genetic diversity analysis: a review. Int J Pure Appl Biosci 3:160–167
Gams W (2014) A new nomenclature for fungi. Mycol Iranica 1:1–5
Hesseltine CW (1983) Microbiology of oriental fermented foods. Annu Rev Microbiol 37: 575–601
Hibbett D, Abarenkov K, Kõljalg U, Öpik M, Vhai B, Cole J, Wang Q, Crous P, Robert V, Helgason T, Herr JR, Kirk P, Lueschow S, O'Donnell K, Nilsson RH, Oono R, Schoch C, Smyth C, Walker DM, Porras-Alfaro A, Taylor JW, Geiser DM (2016) Sequence based classification and identification of fungi. *Mycologia* 108:1049–1068
Ho HH (2018) The taxonomy and biology of *Phytophthora* and *Pythium*. J Bacteriol Mycol Open Access 6(1):40–45. https://doi.org/10.15406/jbmoa.2018.06.00174
Hongsanen S, Jeewon R, Purahong W, Xie N, Liu J-K, Jayawardana RS, Ekanayaka AH, Dissanayake A, Raspe O, Hyde KD, Stadler M, Pers D (2018) Can we use environmental DNA as holotypes? Fungal Divers 92:1–30.
Kreeger C (2003) The learning curve. Nat Biotechnol 21:951–952
Nagpal R, Puniya AK, Griffith GW, Goel G, Puniya M, Sehgal JP, Singh K (2009) Anaerobic rumen fungi: potential and applications. Chapter 17, 375–393. In: Khachatourians GG, Arora DK, Rajendran TP, Srivastava AK (eds), Agriculturally important micro-organisms, volume 1, Academic World International
Narasimhan MJ (1930) Studies in the genus *Phytophthora* X Heterothallic strains of *Phytophthora*. Phytopathology 20:201–214
Rambold G, Stadler M, Begerow D (2013) Mycology should be recognised as a field in biology at an eye level with other major disciplines – a memorandum. Mycol Prog 12:455–463
Shenoy BD, Jeewon R, Hyde KD (2007) Impact of DNA sequence-data on the taxonomy of anamorphic fungi. Fungal Divers 26:1–54
Srinivasan MC (2004) Selective isolation of fungi, Chapter 2. In: Practical mycology for industrial biotechnologists. Tata McGraw-Hill Publishing co. Ltd., New Delhi, p 26–46

Srinivasan MC (2008) Unconventional techniques to explore fungal diversity for industrial enzyme technology. In: Sridhar et al (ed) Novel techniques and ideas in mycology. Fungal Diversity Research Series 26:261–270

Srinivasan MC (2014) Role of mycology and mycologists in an era of industrial biotechnology. Kavaka 43:3–5

Stierle A, Strobel G, Stierle D (1993) Taxol and taxane production by *Taxomyces andreanae*, an endophytic fungus of Pacific yew. Science 260:214–216

Preface

Fungi are members of the group of eukaryotic organisms, which include microbes like yeasts, moulds and mushrooms. They are classified as kingdom Fungi and more closely related to animals than to plants, therefore, placed with the animals in the monophyletic group. Analyses using molecular phylogenetics support a monophyletic origin of fungi. They are heterotrophs, as they acquire their food by absorbing dissolved molecules, typically by secreting digestive enzymes into their environment. The origin of fungi can be traced to single-celled marine ancestors in the Mesoproterozoic era more than one billion years ago. Fungi have since then conquered not only land but also almost every potential habitat and substrate. Although all fungi are heterotrophs, the fungal kingdom comprises a wide range of life strategies ranging from saprotrophy through mutualism to parasitism. The fungal kingdom encompasses an enormous diversity of taxa with varied ecologies, life cycle strategies and morphologies ranging from unicellular aquatic chytrids and yeasts to large mushrooms. However, very little is known about the true diversity of kingdom Fungi, which had been estimated at five million species. Fungi are immensely diverse, with 144,000 species named and classified so far at a current rate of around 2,000 per year. Of these, over 8,000 species are known to be detrimental to plants, and at least 300 can be pathogenic to human beings.

Our fascination for the fungal kingdom is a natural and ancient one based on the following: (i) the roles of fungi in the production of a variety of foods and beverages and even as a source of food themselves; (ii) their global ecological impact, especially as the cause of devastating infections of humans and other animals and of plants, including many crops grown around the world; and (iii) their roles as fundamental model systems in genetics and biological research.

Since the pioneering eighteenth- and nineteenth-century taxonomical works of Carl Linnaeus, Christiaan Hendrik Persoon and Elias Magnus Fries, fungi have been classified based on their morphological, physiological and biochemical characteristics. Advances in molecular genetics have opened the way for DNA analysis to be incorporated into taxonomy, which has sometimes challenged the historical groupings based on morphology and other traits. Phylogenetic studies published in the last decade have helped to reshape the classification within kingdom Fungi, which has been divided into seven phyla (*Microsporidia*, *Chytridiomycota*, *Blastocladiomycota*, *Neocallimastigomycota*, *Glomeromycota*, *Ascomycota* and *Basidiomycota*).

Along with bacteria, fungi are the major decomposers in most terrestrial and some aquatic ecosystems; therefore, they play an essential role in nutrient cycling, especially as saprobes and symbionts, degrading organic matter to inorganic molecules, which can then re-enter anabolic metabolic pathways in plants and other organisms. Certain fungi, in particular white-rot fungi, can degrade insecticides, herbicides, pentachlorophenol, creosote, coal tars and heavy fuels and turn them into carbon dioxide, water and basic elements. Fungi have been shown to mineralize uranium oxides, suggesting that they may have application in the bioremediation of radioactively polluted sites.

The human use of fungi for food preparation and other purposes is extensive and has a long history. Mushroom farming and mushroom gathering are large industries in several countries. The global commercial mushroom market was ~US$35 billion in 2015 and is anticipated to grow to as much as ~US$60 billion by 2021. Ethnomycology is the study of the historical uses and sociological impact of fungi. Fungi have the capacity to produce an enormous range of natural products with antimicrobial or other biological activities; therefore, many species have long been used or are being developed for industrial production of antibiotics, vitamins and anticancer and cholesterol-lowering drugs. Very recently, methods have been developed for genetic engineering of fungi that enable metabolic engineering of fungal species. For example, genetic modification of yeast species, which are easy to grow at fast rates in large fermentation vessels, has opened up ways of pharmaceutical production that are potentially more efficient than production by the original source organisms.

Several pivotal discoveries in biology were made by researchers using fungi as model organisms. Indeed, seven Nobel prizes have been awarded to scientists studying yeasts and moulds as model organisms that explain fundamental aspects of cell biology: from Alexander Fleming, Ernst Chain and Howard Florey in 1945 for the discovery of penicillin from *Penicillium notatum*; to George Beadle and Edward Tatum in 1958 for their 'one gene-one enzyme' hypothesis in *Neurospora crassa*; to Paul Nurse and Leland Hartwell in 2001 for cell division and cancer in *Schizosaccharomyces pombe* and *Saccharomyces cerevisiae*; to Roger Kornberg in 2006 for eukaryotic gene transcription in *S. cerevisiae*; to Jack W. Szostak (shared) in 2009 for chromosome telomeres in *S. cerevisiae*; to Randy Schekman in 2013 (shared) for machinery regulating vesicle traffic in *S. cerevisiae*; and to Yoshinori Ohsumi in 2016 for autophagy in *S. cerevisiae*. Other important model fungi, which includes *Aspergillus nidulans*; *Candida albicans*, a dimorphic, opportunistic human pathogen; *Magnaporthe grisea*, a plant pathogen; and *Pichia pastoris*, a yeast widely used for eukaryotic protein production, have more recently emerged, which address specific biological questions relevant to medicine, plant pathology and industrial uses.

In fact, a fungal species was the first eukaryotic organism to have its genome completely sequenced (the model budding yeast *S. cerevisiae*). Advances in genetics and cell biology have contributed to provide a detailed view of how the genome contributes to the functions of the cell and of the organism. Together, these advances in genomics and genetics provide a 'blueprint' for how these species operate and have evolved at a cellular level, and consequently, they offer a wealth of knowledge about how representative species in the fungal kingdom function and the diversity that lies within. This diversity spans from the most basic way that a fungal cell is organized, either as a yeast or as a filamentous hypha, to the myriad ways these species interact with their environment, from aquatic basal fungi (*Chytridiomycota*, *Cryptomycota*) to fungi that are associated with plants and were critical for their emergence from the oceans and colonization of the planet to fungi that are pathogens of plants or animals. This diversity also extends to the biological behaviour and cell biology of fungi, including fungi that can sense light and those that have evolved to be insensitive to light (blind), the modes of sexual reproduction including heterothallism and homothallism, the loss and retention of RNAi pathways, the replacement of regional centromeres by point centromeres and the retention of flagella in basal fungi versus their loss in fungal branches that evolved the ability to be aerially dispersed.

This book entitled *Advancing Frontiers in Mycology and Mycotechnology: Basic and Applied Aspects of Fungi* is aimed at reviewing major recent developments in understanding the biology and potential biotechnological applications of fungi. Parts I, II, III and IV focus on the basic aspects of fungi; their role in environmental sustainability; their interactions with humans, plants and animals; and bioprospects, respectively. Part I includes chapters on the diversity of fungi from different environments, growth, morphogenesis and genetics and conservation and taxonomy. Part II deals with the role of fungi in environmental sustainability (biodegradation and bioremediation) and in nanobiotechnology. The interactions of fungi with plants, animals and humans, such as mycorrhizal association, endophytism, human mycoses and mycotoxins, are covered in Part III. The bioprospecting aspects of entomo- and myco-pathogens, industrial enzymes and secondary metabolites in healthcare are dealt with in Part IV.

We wish to thank all the contributors for readily accepting our invitation and submitting their well-written chapters in their areas of specialization within the stipulated period. We sincerely hope and wish that this book will be useful to students, scholars, teachers and scientists in the broad areas of microbiology, life sciences and biotechnology. Thanks are also due to Springer Nature for publishing the book.

New Delhi, Delhi, India — Tulasi Satyanarayana
New Delhi, Delhi, India — Sunil K. Deshmukh
Pune, Maharashtra, India — Mukund V. Deshpande

Contents

Part IV Bioprospects

About the Editors and Contributors

Editors

Prof. T. Satyanarayana is a UGC-BSR Faculty Fellow at the Division of Biological Sciences & Engineering, Netaji Subhas University of Technology, New Delhi, after superannuating from the Department of Microbiology, University of Delhi South Campus, New Delhi, as Professor and Head, in June 2016. He has over 270 scientific papers and reviews, 8 edited books and 2 patents to his credit. He is a fellow of National Academy of Agricultural Sciences (NAAS), Association of Microbiologists of India (AMI), Biotech Research Society (I), Mycological Society of India (MSI) and Telengana Academy of Sciences. He is a recipient of Dr. Manjrekar award of AMI, Dr. Agnihotrudu award of MSI and Malaviya Memorial award of BRSI. He has over 40 years of research and teaching experience and has mentored 30 scholars for Ph.D. He was the President of AMI and MSI. His research efforts have been focused on understanding the diversity and applications of yeasts, thermophilic fungi and bacteria and their enzymes, metagenomics, carbon sequestration employing extremophilic bacterial carbonic anhydrases and bioethanol production from lignocellulosic substrates using enzyme cocktails.

Dr. Sunil Kumar Deshmukh received his Ph.D. in Mycology from Dr. H.S. Gour University, Sagar (M.P.) in 1983. The veteran industrial mycologist spent a substantial part of his career at Hoechst Marion Roussel Limited [now Sanofi India Limited], Mumbai and Piramal Enterprises Limited, Mumbai in drug discovery. He has to his credit, 8 patents, 120 publications and 9 books on various aspects of Fungi and natural products of microbial origin. He is the past president of the Mycological Society of India. He is a fellow of Mycological Society of India (MSI), the Association of Biotechnology and Pharmacy and the Society for Applied Biotechnology. He is currently Fellow at Nano-Biotechnology Centre, TERI, New Delhi, and Adjunct Associate Professor in Deakin University, Australia, working towards the development of natural food colors, antioxidants and biostimulants through nanotechnology intervention.

Dr. Mukund V. Deshpande obtained his PhD in 1982 in Biochemistry and D.Sc. in Microbiology of the University of Pune in 1994. His extensive work in the area of fungal biology, especially, fungal differentiation earned him D.Sc. He has worked extensively on the use of fungi and fungal products in Biotechnology. Dr. Deshpande successfully completed more than 35 research projects funded by national and international funding agencies, such as Indo-Swiss Collaboration in Biotechnology (ISCB) programme of Department of Biotechnology (DBT), New Delhi and Swiss Development Cooperation (SDC), Berne, Switzerland on development of mycoinsecticide, Indo-Belarus programme of DBT on biopesticides, Indo-Mexico programme of Department of Science and Technology, New Delhi and CONACYT on fungal dimorphism. Dr. Deshpande is an elected fellow of the Maharashtra Academy of Sciences (FMASc, 1994) and the Society for Biocontrol Advancement (FSBA, 2010). He is also a recipient of the Department of Biotechnology Overseas (Short-term) Associateship (1995), and Commonwealth Science Council Fellowship (1998). He has to his credit more than 140 research papers, reviews and chapters, 8 patents, 6 books and a number of popular articles. He has his own start-up Greenvention Biotech located in Urli-Kanchan, Pune for the translational activities in Agricultural Biotechnology.

Contributors

Krishnendu Acharya Molecular and Applied Mycology and Plant Pathology Laboratory, Department of Botany, University of Calcutta, Kolkata, West Bengal, India

Nitin Adhapure Department of Biotechnology and Microbiology, Vivekanand Arts, Sardar Dalip Singh Commerce and Science College, Aurangabad, Maharashtra, India

Shivankar Agrawal TERI-Deakin Nano Biotechnology Centre, Biotechnology and Management of Bioresources Division, The Energy and Resources Institute, New Delhi, India

Centre for Chemistry and Biotechnology (CCB), School of Life and Environmental Sciences, Deakin University, Waurn Ponds, VIC, Australia

Indian Council of Medical Research (ICMR), Delhi, India

Shruti Ahlawat Department of Microbiology, Maharshi Dayanand University, Rohtak, Haryana, India

Colin J. Barrow Centre for Chemistry and Biotechnology (CCB), School of Life and Environmental Sciences, Deakin University, Waurn Ponds, VIC, Australia

Jadson Diogo Pereira Bezerra Departamento de Micologia Prof. Chaves Batista, Programa de Pós-Graduação em Biologia de Fungos (PPG-BF), CB, Universidade Federal de Pernambuco, Recife, PE, Brazil

V. K. Bhalerao AICRP on Fruits, Mahatma Phule Krishi Vidyapeeth Rahuri, Ahmednagar, Maharashtra, India

José L. Cabrera-Ponce Departamento de Ingeniería Genética, Unidad Irapuato, Centro de Investigación y de Estudios Avanzados del IPN, Irapuato, Guanajuato, México

Arunaloke Chakrabarti Department of Medical Microbiology, PGIMER, Chandigarh, India

Bhushan P. Chaudhari Biochemical Sciences Division, CSIR-National Chemical Laboratory, Pune, Maharashtra, India

Upashna Chettri Microbiology Laboratory, Department of Biotechnology and Bioinformatics, North-Eastern Hill University, Shillong, Meghalaya, India

Hemraj Chhipa College of Horticulture and Forestry, Agriculture University Kota, Jhalawar, Rajasthan, India

Priyanka Choudhari Nanobioscience, Agharkar Research Institute, Pune, Maharashtra, India

Ritumbhara Choukade Department of Microbiology, Dr. Harisingh Gour Vishwavidyalaya, Sagar, Madhya Pradesh, India

Bruna de Almeida Martins Department of Chemistry, Exact Sciences Institute, Universidade Federal de Minas Gerais, Belo Horizonte-MG, Brazil

Geane Pereira de Oliveira Department of Chemistry, Exact Sciences Institute, Universidade Federal de Minas Gerais, Belo Horizonte-MG, Brazil

Cristina Maria de Souza-Motta Departamento de Micologia Prof. Chaves Batista, Programa de Pós-Graduação em Biologia de Fungos (PPG-BF), CB, Universidade Federal de Pernambuco, Recife, PE, Brazil

C. D. Deokar Department of Plant Pathology, Mahatma Phule Krishi Vidyapeeth, Ahmednagar, India

Sunil K. Deshmukh Biotech & Management of Bioresources Div, The Energy and Resources Institute, New Delhi, Delhi, India

Mukund V. Deshpande Division of Biological Sciences, CSIR-National Chemical Laboratory, Pune, Maharashtra, India

Thokchom Sarda Devi Department of Botany, University of Delhi, New Delhi, India

A. P. Gaikwad AICRP on Mushroom, College of Agriculture, Pune, India

Vandana Ghormade Nanobioscience, Agharkar Research Institute, Pune, Maharashtra, India

Samta Gupta Department of Botany, University of Delhi, New Delhi, Delhi, India

Akshaya Gupte Department of Microbiology, N V Patel College of Pure & Applied Sciences, Anand, Gujarat, India

Shilpa Gupte Department of Microbiology, Ashok & Rita Patel Institute of Integrated Study & Research in Biotechnology and Allied Sciences, Anand, Gujarat, India

Uttam Kumar Jana Department of Microbiology, Dr. Harisingh Gour Vishwavidyalaya, Sagar, Madhya Pradesh, India

Rajesh Jeewon Department of Health Sciences, Faculty of Science, University of Mauritius, Reduit, Mauritius

S. R. Joshi Microbiology Laboratory, Department of Biotechnology and Bioinformatics, North-Eastern Hill University, Shillong, Meghalaya, India

D. N. Kamra Animal Nutrition Division, ICAR-Indian Veterinary Research Institute, Izatnagar, Uttar Pradesh, India

Naveen Kango Department of Microbiology, Dr. Harisingh Gour Vishwavidyalaya, Sagar, Madhya Pradesh, India

Rupam Kapoor Department of Botany, University of Delhi, New Delhi, India

Durgadas P. Kasbekar Centre for DNA Fingerprinting and Diagnostics, Hyderabad, Telangana, India

Tanveer Kaur Department of Biotechnology, Thapar Institute of Engineering & Technology, Patiala, Punjab, India

Claudia León-Ramírez Departamento de Ingeniería Genética, Unidad Irapuato, Centro de Investigación y de Estudios Avanzados del IPN, Irapuato, Guanajuato, México

Matheus Thomaz Nogueira Silva Lima Department of Food Science, Faculty of Pharmacy, Universidade Federal de Minas Gerais, Belo Horizonte-MG, Brazil

David Lloyd School of Biosciences, Cardiff University, Cardiff, Wales, UK

C. Manoharachary Mycology and Molecular Plant Pathology Laboratory, Department of Botany, Osmania University, Hyderabad, Telangana, India

Aliesha Moudgil Biochemical Sciences Division, CSIR-National Chemical Laboratory, Pune, Maharashtra, India

D. Nagaraju Department of Botany, Government Degree College, Warangal, Telangana, India

Sudeshna Nandi Molecular and Applied Mycology and Plant Pathology Laboratory, Department of Botany, University of Calcutta, Kolkata, West Bengal, India

Redeemson Panmei Biochemical Sciences Division, CSIR-National Chemical Laboratory, Pune, Maharashtra, India

Ejaj K. Pathan Biochemical Sciences Division, CSIR-National Chemical Laboratory, Pune, Maharashtra, India

A. V. Patil Biochemical Sciences Division, CSIR-National Chemical Laboratory, Pune, Maharashtra, India

Fernando Pérez-Rodríguez Departamento de Ingeniería Genética, Unidad Irapuato, Centro de Investigación y de Estudios Avanzados del IPN, Irapuato, Guanajuato, México

K. S. Raghuwanshi Department of Plant Pathology, Mahatma Phule Krishi Vidyapeeth, Ahmednagar, India

Shraddha Rahi Nanobioscience, Agharkar Research Institute, Pune, Maharashtra, India

M. Sudhakara Reddy Department of Biotechnology, Thapar Institute of Engineering & Technology, Patiala, Punjab, India

Darshan M. Rudakiya Department of Microbiology, N V Patel College of Pure & Applied Sciences, Anand, Gujarat, India

José Ruiz-Herrera Departamento de Ingeniería Genética, Unidad Irapuato, Centro de Investigación y de Estudios Avanzados del IPN, Irapuato, Guanajuato, México

Mayela Salazar-Chávez Departamento de Ingeniería Genética, Unidad Irapuato, Centro de Investigación y de Estudios Avanzados del IPN, Irapuato, Guanajuato, México

Alejandro Sánchez-Arreguín Departamento de Ingeniería Genética, Unidad Irapuato, Centro de Investigación y de Estudios Avanzados del IPN, Irapuato, Guanajuato, México

V. Venkateswara Sarma Department of Biotechnology, Pondicherry University, Kalapet, Pondicherry, India

Akshay Shankar Department of Microbiology, Maharshi Dayanand University, Rohtak, Haryana, India

Krishna Kant Sharma Department of Microbiology, Maharshi Dayanand University, Rohtak, Haryana, India

Rohit Sharma National Centre for Microbial Resource (NCMR), National Centre for Cell Science (NCCS), Pune, Maharashtra, India

Rimpa Sikder Molecular and Applied Mycology and Plant Pathology Laboratory, Department of Botany, University of Calcutta, Kolkata, West Bengal, India

Leticia Francisca da Silva Departamento de Micologia Prof. Chaves Batista, Programa de Pós-Graduação em Biologia de Fungos (PPG-BF), CB, Universidade Federal de Pernambuco, Recife, PE, Brazil

B. Singh ICAR-Indian Veterinary Research Institute, Palampur, Himachal Pradesh, India

Shiv Mohan Singh Banaras Hindu University (BHU), Varanasi, Uttar Pradesh, India

Shreya Singh Department of Medical Microbiology, PGIMER, Chandigarh, India

Jacqueline Aparecida Takahashi Department of Chemistry, Exact Sciences Institute, Universidade Federal de Minas Gerais, Belo Horizonte-MG, Brazil

Archana Tripathi Department of Microbiology, N V Patel College of Pure & Applied Sciences, Anand, Gujarat, India

John Vélez-Haro Departamento de Ingeniería Genética, Unidad Irapuato, Centro de Investigación y de Estudios Avanzados del IPN, Irapuato, Guanajuato, México

Tapani Yli-Mattila Molecular Plant Biology, Department of Biochemistry, University of Turku, Turku, Finland

Emre Yörük Department of Molecular Biology and Genetics, Faculty of Arts and Sciences, Istanbul Yeni Yuzyil University, Istanbul, Turkey

Part I

Basic Aspects of Fungi

The Mystical World of Mushrooms

1

V. K. Bhalerao, A. P. Gaikwad, C. D. Deokar, and K. S. Raghuwanshi

Abstract

The mushrooms have existed approximately 130 million years ago, i.e., long before human beings evolved on this planet as per the fossil records. Earlier in Sanskrit mushroom is known as "Ksuonpa." In Hindi, mushroom is known as "*Khumbi*." Since long mushrooms are worshiped and also considered as divine. In nature, mushrooms have not only been a source of food for man and other animals but also have contributed an important role in the cycling of carbon and other elements through the breakdown of lignocellulolytic plant residues and animal dung which serves as the substrates for these saprophytic fungi. The historical records of intentionally cultivated mushrooms estimated that the first mushroom cultivation was started in 600 AD. In the last 20 years, much progress has been made in the field of mechanization of mushroom cultivation, i.e., manure turners, spawning, filling and casing of trays, mechanical harvesting, and polythene bag method of cultivation. In India, cultivation of edible mushrooms is of very recent origin, though methods of cultivation of them were known for many years. The research on different aspects, *viz.*, production, productivity, spawn production, strain improvement, post-harvest technology disease, and pest management, was attempted by several scientists. All mushrooms belong to the group of fungi, a group very distinct from plants, animals, and bacteria. Most fungi have plant-like cells but miss the most important features of plants. The known number species of fungi was about 69,000 till 1990, while it was conservatively estimated that 1.5 million species of fungi actually existed in nature.

V. K. Bhalerao (✉)
AICRP on Fruits, Mahatma Phule Krishi Vidyapeeth, Rahuri, Ahmednagar, Maharashtra, India

A. P. Gaikwad
AICRP on Mushroom, College of Agriculture, Pune, India

C. D. Deokar · K. S. Raghuwanshi
Department of Plant Pathology, Mahatma Phule Krishi Vidyapeeth, Ahmednagar, India

T. Satyanarayana et al. (eds.), *Advancing Frontiers in Mycology & Mycotechnology*,
https://doi.org/10.1007/978-981-13-9349-5_1

Mushrooms are cultivated throughout the world. About 200 species of more than 2000 edible fungi are widely adopted for human consumption. Out of these, only 10–12 species are commercially cultivated since the past few decades in India due to technical advancement. Mushrooms have the capacity to produce highest proteins per unit area and time by utilizing vertical space which is hundred times more than the traditional agriculture and animal husbandry. This hi-tech horticulture venture can reduce the pressure on cultivated land to meet the food shortages all over the world. More than 100 countries are engaged in mushroom farming today which is increasing at an annual rate of 6–7% per annum. The very high levels of mechanization and automation were achieved in mushroom farming in developed countries of Europe and America. As per FAO Stat, the present world production of mushrooms is around 3.5 million tonnes, which is more than 25 million tonnes (estimated) as per claims of Chinese Association of Edible Fungi.The mushrooms can be cultivated under varied climatic conditions. Some of the economically important mushrooms cultivated all over the world under temperate, subtropical, and tropical conditions are *Agaricus bisporus, Lentinula edodes, Flammulina velutipes, Agaricus bitorquis, Pleurotus* spp., *Auricularia* spp., *Agrocybe aegerita, Volvariella* spp., *Calocybe indica, Ganoderma lucidum,* etc.

Mushrooms are rich source of proteins, carbohydrates, valuable salts, and vitamins in diet of human being. Mushroom fungus has the ability to secrete a wide variety of hydrolyzing and oxidizing enzymes which have potential for biotechnological applications. More than 100 medicinal edible mushrooms have been identified. However, important medicinal mushrooms are *Lentinus edodes* (shiitake mushroom), *Ganoderma lucidum* (reishi mushroom), *Grifola frondosa* (maitake mushroom), *Pleurotus ostreatus* (oyster mushroom), *Agaricus bisporus* (button mushroom), *Coriolus versicolor* (PSK), *Boletus edulis, Tremella fuciformis, Auricularia polytricha, Hericium erinaceus,* and *Cordyceps sinensis.*

Keywords
Mushrooms · Oyster mushroom · *Agaricus* sp. · Medicinal values · Cultivation · Commercial production · Nutritional resource

1.1 Introduction

The mushrooms have existed approximately 130 million years ago, i.e., long before human beings evolved on this planet as per the fossil records. The origin of the term "Mushroom" has different views. In Latin, "Fungo" means to flourish. The Greek term mushroom was derived from the word "Sphonggos" or "Sphoggos" which means "Sponge" and referred to the sponge-like structure of some species. In French, the term mushroom was derived from the word "Mousseron" (muceron), "Mousse," or "Moss."

About three and half millennia ago, the Greek hero Perseus founded Mycenae city after a legendary mushroom. "Mycenae" is derived from Greek word "mycology" (i.e., mykes – mushroom + logos – discourse). Earlier in Sanskrit, mushroom is known as "Ksuonpa." In Hindi, mushroom is known as "Khumbi." The other words for fleshy capped fungi are Chatra, Kukurmutta, Kavaka, Bhumi kavak, and Bhustrana. Aryans during migration into the Indian subcontinent around in 1500 B.C. carried with them an intoxicating drink "Soma" which they used in their religious rites. According to Wasson) (1969), the "Soma" in Rig Veda refers to *Amanita muscaria.*

Since long mushrooms are worshiped and also considered as divine. There are lots of superstitions about the mushrooms. According to the ancient Indian, Greek, and Roman myths, mushrooms sprang from a stroke of lightning. The Indians in Mexico believed that mushrooms are sacred because they are borne of the sexual intercourse between a bolt of lightning and the earth. The mushroom-shaped stone carvings have also been found in Central America and the highlands of Guatemala. These objects resemble very closely to the fruit body of *Amanita muscaria.*

In nature, mushrooms have not only been a resource of food for man and other animals but also have an important role in the cycling of carbon and other elements through the breakdown of lignocellulolytic plant residues and animal dung which serves as the substrates for these saprophytic fungi. In this way, mushroom species as agents of decay helps in keeping the environment clean.

The mushrooms were long appreciated because of their flavor, texture, and medicinal or tonic attributes. The recognition, that mushrooms are nutritionally a very good food is much more recent.

The popularity of mushroom is desired from the following highly preferred characteristics of food:

1. Significant taste and flavor.
2. Nutritious, not only because they contain high level of protein containing significant amounts of lysine and methionine (which are generally low in plants), fibers, minerals, and vitamins but also for what they do not have (high calories, sodium, fat, and cholesterol).
3. Can be easily stored by processing, drying, pickling, and canning which allow maximum storage. Mushrooms appeal to different people in different ways. Mushrooms are objects of beauty for artists. Mushrooms are possible source of new drugs for medical people. Architects have constructed minerals, temples, and cupola column in its shape. Jewelers have made expensive pieces of mushroom design. Designers have reproduced the mushroom design on fabrics.

1.2 History of Mushroom Research and Cultivation

1.2.1 International Developments

The historical records of intentionally cultivated mushrooms are shown in Table 1.1. The first artificial cultivation of *Auricularia auricula* mushroom was estimated around 600 AD in China on wood logs and the same time other wood rotting mushrooms, such as *Flammulina velutipes* (800 AD) and *Lentinula edodes* (1000 AD). In Europe, the first cultivated fungi, the mushroom, was introduced in the seventeenth century. *Agaricus bisporus* was not cultivated until 1600 AD which is the leading mushrooms. The *Pleurotus ostreatus* was first cultivated in the USA, and several other species of *Pleurotus* were initially cultivated in India.

The first record of the cultivation of mushrooms was during 1638–1715, i.e., the reign of Louis XIV. Treschow shortly before 1700 suggested a method of growing *A. bisporus* in hotbeds under heat. Tourneforte, a French man in 1707, published description and know-how of growing mushrooms which is similar to the present methods.

Van Griensven (1988) reviewed the historical events in mushroom cultivation, whereas a detailed history was elaborated by Shu-ting and Miles Philip. The cultivation of mushrooms aboveground is originated in Sweden. Bahl (2002) reviewed that Lundberg described mushroom growing in greenhouses in year 1754 whereas, the French started growing mushrooms underground in the quarries around Paris on horse manure in 1800 and Callow reported shelf bed method in a peculiar type of house for mushroom cultivation in 1831.

By the end of the nineteenth century, French mycologists Matrochot and Constantin (1894) discovered the cause of mushroom disease *La mole (Mycogone perniciosa* Magnus) and started sulfur fumigation. They were able to germinate spores for obtaining sterile spawn. Miss Fergussen (1902) of Corne described detailed method for germinating spores.

Duggar (1905), an American, perfected a method of making pure culture spawn from mushroom tissue. Lambert (1929) introduced and marketed pure spore culture bottled spawn in the USA. The cultivation of mushroom aboveground is originated in Sweden. Sinden (1938) patented his grain spawn process. Thomas, Austin, and Jary worked on pests of mushrooms. Ware, Glasscock, and Bewley in the UK and Lambert and Beach in the USA worked on diseases and competitive molds of mushrooms.

In 1945, MGA (Mushroom Growers Association of England and Ireland), MRA (Mushroom Research Association Ltd.), and MGP (Midlands Group of Publication) were formed and established. In the same year, an organization called Mushroom Growers Association of Great Britain and North Ireland was established. This association publishes a mushroom journal, which is supplied to the 50 nations.

In 1946, the first mushroom research station was founded in England with Dr. R.L. Edwards as its first director. The mushroom research at this station is mainly devoted to *Agaricus bisporus* and involves all aspects of this single species. In 1954,

Table 1.1 Historical evidence of commonly cultivated mushrooms (Miles and Chang 1997)

Species	Record first cultivated (est.)
Agaricus bisporus	1600 (Atkins 1979)
Agaricus bitorquis	1961
Agrocybe cylindracea	1950
Amanita caesarea	1984
Armillaria mellea	1983
Auricularia auricula	600
Coprinus comatus	1984
Dictyophora duplicata	1982
Flammulina velutipes	800
Ganoderma spp.	1621
Gloestereum incarnatum	1989
Grifola frondosus	1983
Hericium coralloides	1984
Hericium erinaceus	1960
Hohenbuehelia serotina	1982
Hypsizygus marmoreus	1973
Lentinus edodes	1000
Lentinus tigrinus	1988
Lyophyllum ulmarium	1987 (Wang and Zhang 1987)
Morchella spp.	1986
Oudemansiella radicata	1982
Pholiota nameko	1958
Pleurotus citrinopileatus	1981
Pleurotus cystidiosus	1969
Pleurotus ferulae	1958
Pleurotus flabellatus	1962 (Bano and Srinvatava 1962)
Pleurotus Florida	1958
Pleurotus ostreatus	1900
Pleurotus sajor-caju	1974 (Jandaik 1974)
Poria cocos	1232
Sparassis crispa	1985
Tremella fuciformis	1800
Tremella mesenterica	1985
Tricholoma gambosum	1991
Tricholoma lobayense	1990
Tricholoma mongolicum	1991
Volvariella volvacea	1700

the mushroom research station was transferred along with three staff members and government grants to the new glasshouse crop research station at Littlehampton, Sussex.

In 1959 at Horst in Holland, a fine center on mushroom was opened. In the later years of the eighteenth century, French growers discovered the usefulness of gypsum powder to prevent greasiness in compost. Bewley and Lambert worked on relationship between compost and yield. MRS (Mushroom Research Station) at yarely produced formulae for synthetic compost made recommendation for control of truffle disease and suggested use of peat in casing.

At the GCRI at Little Hampton, much work is being done on the control of cecids, phorids, mites, virus, and bacterial diseases, compost, casing and environment, genetics, nutrition, and nematode control. In Denmark, Rasmussen used pig manure and got good results. Arnold in East Germany, Bukowskii in Poland, Hetlay in Hungary, and Willanis in Belgium were also engaged in state-sponsored mushroom programs. Sinden and Tschierpe have done useful work on composting and environmental control. Bel has done lot of works for the benefit of Dutch mushroom industry. Some of the important developments in the cultivation of edible mushroom at global level are as below.

In the last 20 years, much progress has been made in the field of mechanization of mushroom cultivation, i.e., manure turners, spawning, filling and casing of trays, mechanical harvesting, and polythene bag method of cultivation.

1.2.2 Indian Scenario of Mushroom Research and Cultivation

The cultivation of edible mushrooms in India is of very recent origin, although methods of cultivation for some of them were known for many years. Some important historical advancements in the cultivation of edible mushroom are as below.

1886: N.W. Newton grown and exhibited some specimens of mushrooms at the Annual show of Agric. Hort. Soc. of India.

1896–1897: B.C. Roy carried out chemical analysis of the local mushrooms prevalent in caves or mines.

1908: Sir David Prain was initiated thorough search of edible mushrooms.

1981: Some mushrooms from Calcutta were recorded by Lt. Col. Kirtikar of Indian Medical Services.

1921: S.R. Bose was successful in culturing two agarics on a sterilized dung media and details of which were published in the Indian Science Congress held at Nagpur during 1926.

1939–1945: The Dept. of Agriculture, Madras attempted experimental cultivation of paddy straw mushroom (*Volvariella*).

1940: The method of spawn production and cultivation of paddy straw mushroom (*Volvariella* Sp.) was reported by Su and Seth.

1941: Padwick reported successful cultivation of *Agaricus bisporus* from various countries.

1945: Thomas et al. demonstrated cultivation of paddy straw mushrooms (*V. displasia*) in Madras.

1947: The better yield of paddy straw mushroom by adding red-powdered dal to the beds was reported by R.P. Asthana and suggested that April–June is the most suitable period for cultivating this mushroom in Central Provinces and also carried out the chemical analysis of this mushroom.

1961: A scheme entitled "Development of mushroom cultivation in H.P." was stated at Solan by H.P.S.G. in collaboration with ICAR. This was the first attempt on cultivation of *A. bisporus* in the country.
1962: Bano et al. obtained increased yield of *Pleurotus flabellatus* on paddy straw.
1965: E.F.K. Mantal, F.A.P., Mushroom Export, guided and assisted Dept. of Agril. H.P. for the construction of modern spawn lab and fully A.C. mushroom house. The research work on the evaluation of different strains, use of various agricultural wastes, organic manures and fertilizers for preparing synthetic compost was initiated. Dr. Mantel's consultancy concluded after a period of 7 years.
1971: An ICAR-sponsored coordinated research scheme on mushroom was started in various zones of the country, viz., Solan (H.P.), Ludhiana (Punjab), Bangalore (Karnataka), and New Delhi. This scheme was later converted into AICMIP with its H.Q. at NCMRT, Solan in 1983.
1974: W.A. Hayes, FAO mushroom expert, guided in improving the method of compost preparation, pasteurization, and management of important parameters in the mushroom house such as, new compost formulations and casing material. The important parameters like N-content, moisture in compost and casing soil, air movement, maintenance of proper environmental factors were also standardized which increased the mushroom yields from 7 to 14 kg/m^2.
1977: The 1.27 crore of MDP was launched by the Dept. of Hort. (H.P.) with the financial support from UNDP, wherein the services of Mr. James Tunney were made available. He got bulk pasteurization chambers constructed which enabled ready-made compost and casing soil to the growers of H.P. The UNDP Project concluded in 1982. Since then, the Dept. of Hort. (HP) has been running the project.
1983: The NCMRT was sanctioned by ICAR on October 23, 1982, during VIth plan. The center, however, started functioning w.e.f. June 8, 1983. It was formally inaugurated on June 21, 1986.

The research on different aspects, viz., production, productivity, spawn production, strain improvement, post-harvest technology disease, and pest management, was attempted by several scientists. The successful method of spawn making using cereal grains, millets, and other farm wastes such as straws and bajra husk was reported by Jandaik and Kapoor (1975).

A lot of work has been done on suitability of various substrates for *Pleurotus* production. Gaikwad (1983) studied the yield potential of *Pleurotus* spp. on selected agricultural wastes using wheat grain spawn and wheat straw spawn. Patil and Jadhav (1989) assessed different 14 substrates separately and in 3 combinations of selected substrates on productivity of *P. sajor-caju.* Thilagavathy *et al.* (1991) observed maximum yield of *P. sajor-caju* from banana pseudostem. Similarly, various workers have suggested different substrates for *Pleurotus* cultivation. Some of them are Jowar straw and groundnut pods for *P. florida* (Khandar *et al.* 1991), wheat straw (Gupta and Langar 1988), and old palm monocarp waste (Babu and Nair 1991).

Polybags were used by several research workers (Bano *et al.* 1976; Baskaran *et al.* 1978; Sivaprakasam and Kandaswamy 1980; Bano and Rajarathnam 1982). Gunny bags were used in place of polythene bags (Pal and Thapa 1979). *P. sajor-caju* was successfully cultivated in a modified version of the plastic sac by introducing a PVC tube or bamboo in the center of sac (Bano *et al.* 1979).

In some species, selective hybridization has been carried out between the crossing of monosporous isolates, and variations have been recorded in yield, morphological growth, and fruit body characters in *Agaricus bisporus* (Bhandal and Mehta 1989), *Pleurotus sajor-caju* (Ghosh and Chakravarty 1991), *P. sapidus* (Thakur and Bhandal 1993), and *Calocybe* (Doshi *et al.* 1993). Bahukhandi and Munjal (1989) improved the commonly cultivated strains of *Pleurotus sajor-caju* in respect of better yield by inducing mutation by chemicals. Bahukhandi and Sharma (2002) obtained a specific hybrid by mating of *P. sajor-caju* and *P. cornucopiae.* Kaur (2007) developed funnel-shaped fruit bodies, lateral fruit bodies with wavy margins, and gray color fruit bodies by intraspecific hybridization of *P. florida* PAU-5. Bhalerao *et al.* (2017) attempted intraspecific hybridization, for developing temperature and mold-tolerant strains of oyster mushroom.

The studies on phylogenetic relationships, genetic variability, and use of molecular markers were studied by Swarnendu Chandra *et al.* (2010), Bhavna Gupta et al. (2011), Manjit Singh and Shwet Kamal (2011), Mishra *et al.* (2012), Kaur and Sodhi (2012), and Agarwal *et al.* (2013).

1.3 Taxonomy of Mushroom

All mushrooms belong to the group of fungi, a group very distinct from plants, animals, and bacteria. Most fungi have plant-like cells but miss the most important features of plants, the ability to use energy from the sun directly through the use of chlorophyll. Thus, fungi depend on other organisms for food as does mankind and in fact all animals. All fungi, with the exception of yeasts, form so-called hyphae, tiny threads that originate from the spores. These hyphae will branch out and form the mycelium. After some time, they will enter a sexual phase and form spores. In nature, this is the most striking part of the organism, but in fact it is just the fruiting body.

In nature, fungi multiply by producing millions and billions of spores. When they land in a suitable environment, they can germinate and form a mycelium. In the meantime, the mycelium will colonize the substrate and use the available nutrient. When some nutrients were exhausted or when the weather conditions are unfavorable, the mycelium will reach a different phase: the reproductive sexual stage.

The taxonomic hierarchy of some of the important mushroom based on Ainsworth and Bisby's (2008) *Dictionary of the Fungi* (10th. edition) is as follows:

1.3.1 Phylum: Ascomycota

Subphylum: Pezizomycotina
  Class: Leotiomycetes
    Order: Helotiales
      Family: Hemiphacidiaceae
        Genera: *Chlorencoelia* (*C. torta*)
      Family: Heliotiaceae
        Genera: *Hymenoscyphus fructigenus;* possibly also *Ascocoryne* (*Chlorociboria aeruginascens*), and *lonomidotis* (*I. irregularis*)
      Family: Dermateaceae
        Genera: *Chlorosplenium* (*C. chlora*)
        Genera: *Bulgaria* (*B. inguinans*)
    Order: Rhytismatales
      Family: Rhytismataceae
        Genera: *Colpoma* (*C. guercinum*)
      Family: Cudoniaceae
        Genera: *Cudonia* (*C. circinans*), *Spathularia Spathulariopsis* (*S. velutipes*)
    Order: Leotiales
      Family: Leotiaceae
        Genera: *Leotia* (*L. lubrica*)
  Class: Pezizomycetes
    Order: Pezizales
      Family: Helvellaceae
        Genera: *Helvella*
      Family: Chorioactidaceae
        Genera: *Chorioactis* (*C. geaster*), *Wolfina* (*W. aurantiopsis*)
      Family: Pezizaceae
        Genera: *Sarcosphaera* (S. *comnaria*), *Pachyella* (*P. clypeata*), *Peziza*
      Family: Morchellaceae
        Genera: *Disciotis* (*D. venosa*), *Morchella, Verpa* (*V. bohemica*)
      Family: Pyronemataceae
        Genera: *Tarzetta* (*T. bronca*) *Aleuria* (*A. aurantia*), *Cheilymenia Jafnea* (*J. semitosta*), *Otidea* (*O. onotica*), *Scutellinia* (*S. scutellata*), *Sphaerosporella* (*S. brunnea*),
      Family: Sarcosomataceae
        Genera: *Galiella* (*G. rufa*), *Urnula*
      Family: Sarcoscyphaceae
        Genera: *Microstoma* (*M. floccosum*),
  Class: Sordariomycetes
    (Most "Pyrenomycetes," in 15 orders, 64 families, and over 1000 genera)
        Genera: *Akanthomyces aculeatus*), *Camarops Xylaria.*

1.3.2 Phylum: Basidiomycota

Subphylum: Agaricomycotina
Class: Dacrymycetes
Order: Dacrymycetales
Family: Dacrymycetaceae
Genera: *Calocera* (*C. cornea*), *Dacryopinax* (*D. elegans*)
Class: Tremellomycetes
Order: Tremellales
Family: Phaeotremellaceae
Genera: *Phaeotremella* (*P. frondosa*)
Family: Tremellaceae
Genera treated: *Tremella* (*T. mesenterica*)
Family: Carcinomycetaceae
Genera: *Syzygospora* (*Sjmycetophjlaa*)
Class: Agaricomycetes
Order: Agaricales
Family: Clavariaceae
Genera: *Clavaria* (*C. vermicularis*), *Clavulinopsis* (*CJaeticolor*), *Ramariopsis* (*R. kunzei*)
Family: Amanitaceae
Genera: *Amanita, Limacella*
Family: Entolomataceae
Genera: *Clitopilus* (*C. prunulus*), *Entoloma, Rhodocybe*(*R. mundula*)
Family: Fistulinacea
Genera: *Fistulina* (see *F. hepatica*)
Family: Bolbitiaceae
Genera: *Bolbitius* (*B. vitellinus*), *Conocybe* (*C. jalbbipes*), *Galeropsis* (*Gastrocybe lateritia*)
Family: Hydnangiaceae
Genera: *Laccaria*
Family: Inocybaceae
Genera: *Crepidotus, Flammulaster* (*F. erinaceella*), *Inocybe, Simocybe* (*S. centunculus*);
Family: Lyophyllaceae
Genera: *Asterophora* (*A. lycoperdoides*), *Calocybe* (*C. carnea*), *Hypsizygus* (*H. tessulatus*), *Lyophyllum* (*L. decastes*)
Family: Hygrophoraceae
Genera: *Chrysomphalina* (*C. chrysophylla*), *Cuphophyllus* (*C. pratensis*), *Gliophorus* (*G. psittacinus*),*Hygrocybe* (*H. conica*), *Hygrophorus* (*H. russula*), *Ampulloclitocybe* (*A. clavipes*),
Family: Marasmiaceae
Genera: *Megacollybia*, *Micromphale* (*M. perforans*), *Mycetinis* (*M. scorodonius*), *Omphalotus* (*O. illudens*), *Tetrapyrgos* (*T. nigripes*)*Baeospora* (*B. myosura*), *Clitocybula* (*C. abundans*),

Connopus (*C. acervatus*), *Crinipellis* (*C. zonata*), *Gerronema* (*G. strombodes*), *Gymnopus* (*G. dryophilus*), *Macrocystidia* (*see M. cucumis*), *Marasmiellus* (*M. candidus*), *Marasmius* (*see M. rotula*),

Family: Agaricaceae

Genera: *Leucoagaricus* (*L. naucinus*), *Leucocoprinus* (*L. birnbaumii*), *Lycoperdon* (*L. pulcherrimum*), *Cyathus* (*C. striatus*), *Cystoderma, Cystolepiota* (*C. seminuda*), *Floccularia, Macrolepiota* (*M. procera*), *Battarrea* (*B. phalloides*), *Morganella* (*M. pyriformis*), *Mycenastrum* (*M. corium*), *Nidularia* (*N. pulvinata*), *Podaxis* (*P. longii*), *Ripartitella* (*R. brasiliensis*), *Tulostoma* (*T. lloydii*), *Vascellum* (*V. curtisii*), *Calvatia* (*C. craniiformis*), *Agaricus* (*Agaricus bisporus*), *Lepiota* (*L. cristata*), *Arachnion* (*A. album*), *Bovista* (*B. longispora*), *Chlorophyllum* (*C. molybdites*), *Coprinus* (see *C. comatus*), *Crucibulum* (*C. laeve*)

Family: Omphalotaceae

Genera: *Rhodocollybia* (*R. maculata*).

Family: Physalacriaceae

Genera: *Armillaria, Cyptotrama* (*C. asprata*), *Flammulina, Rhizomarasmius, Rhodotus* (*R. palmatus*), Xeruloid Mushrooms (including *Hymenopellis* and *Paraxerula*)

Family: Mycenaceae

Genera: *Mycena, Panellus* (*P. stipticus*)

Family: Pleurotaceae

Genera: *Hohenbuehelia, Pleurotus* (*P. ostreatus*)

Family: Pluteaceae

Genera: *Pluteus, Volvariella*

Family: Psathyrellaceae

Genera: *Coprinellus* (*C. disseminatus*), *Coprinopsis* (*C. atramentaria*), *Lacrymaria* (*L. velutina*), *Parasola* (*P._plicatilis*), *Psathyrella*

Family: Schizophyllaceae

Genera: *Schizophyllum* (*S. commune*)

Family: Strophariaceae

Genera: *Leratiomyces* (*L. ceres*), *Pholiota, Psilocybe; Stropharia* (*S. rugosoannulata*);*Agrocybe, Deconica* (*D. argentina*), *Galerina* (*G. marginata*), *Hebeloma, Hemipholiota* (*H. populnea*), *Hypholoma, Kuehneromyces* (*K. mutabilis*),

Family: Tapinellaceae

Genera: *Tapinella* (*T. panuoides*)

Family: Tricholomataceae

Genera: *Leucopaxillus, Leucopholiota, Macrocybe* (*M. titans*), *Melanoleuca, Omphalina* (*O. epichysium*), *Resupinatus* (*R. alboniger*), *Tricholoma, Clitocybe, Collybia* (*C. cirrhata*), *Dendrocollybia* (*D. racemosa*), *Infundibulicybe, Callistosporium* (*C. luteo-olivaceum*), *Catathelasma, Caulorhiza* (*C. umbonata*),

Family: Typhulaceae

Genera treated: *Macrotyphula* (*M. juncea*).

Family: Uncertain

Genera: *Phyllotopsis* (see *P. nidulans*), *Rickenella* (*R. fibula;*

Order: Auriculariales

Family: Auriculariaceae

Genera: *Auricularia* (*A. auricula*), *Exidia* (*E. glandulosa*)

Order: Boletales

Family: Hygrophoropsidaceae

Genera: *Hygrophoropsis* (*H. aurantiaca*)

Family: Paxillaceae

Genera: *Paragyrodon* (*P. sphaerosporus*),

Paxillus (*P. vernalis*)

Family: Sclerodermataceae

Genera: *Pisolithus* (*P. tinctorius*), *Scleroderma*

Family: Suillaceae

Genera treated: *Suillus*

Family: Boletaceae

Genera: *Leccinum, Phylloporus, Pulveroboletus* (*P. ravenelii*), *Retiboletus* (*R. ornatipes*), *Rubroboletus* (*R. dupainii*), *Strobilomyces, Tylopilus, Xanthoconium* (*X._purpureum*), *Xerocomellus* (*X. chrysenteron*), *Xerocomus* (*X. subtomentosus*), *Austroboletus, Boletellus, Boletus, Bothia* (*B. castanella*), *Chalciporus* (*C._piperatus*),*Chamonixia* (*C. caespitosa*), *Harrya* (*H. chromapes*), *Heimioporus,*

Family: Boletinellaceae

Genera: *Boletinellus* (*B. merulioides*)

Family: Calostomataceae

Genera: *Calostoma* (*C. cinnabarinum*)

Family: Diplocystidiaceae

Genera: *Astraeus* (*A. hygrometricus*)

Family: Gomphidiaceae

Genera: *Chroogomphus, Gomphidius*

Family: Gyroporaceae

Genera: *Gyroporus*

Order: Cantharellales

Family: Cantharellaceae

Genera: *Cantharellus* (*C. cibarius*), *Craterellus* (*C. fallax*)

Family: Clavulinaceae

Genera: *Clavulina* (*C. cristata*)

Family: Hydnaceae

Genera: *Hydnum* (*H. repandum*)

Order: Geastrales

Family: Geastraceae

Genera: *Geastrum* (*G. saccatum*)

Order: Gloeophyllales

Family: Gloeophyllaceae

Genera: *Gloeophyllum* (*G. sepiarium*)
Order: Gomphales
Family: Clavariadelphaceae
Genera: *Clavariadelphus* (*C. unicolor*)
Family: Gomphaceae
Genera: *Gomphus* (*G. clavatus*), *Ramaria* (*R. botrytis*), *Turbinellus* (*T. floccosus*)
Family: Lentariaceae
Genera: *Lentaria* (*L. micheneri*)
Order: Hymenochaetales
Family: Hymenochaetaceae
Genera: *Coltricia* (*C. cinnamomea*), *Inonotus* (*I. radiatus*), *Phellinus* (*P. gilvus*), Porodaedalea (*P. pini*)
Family: Uncertain
Genera: *Rickenella* (*R. fibula*)
Order: Polyporales
Family: Cerrenaceae
Genera: *Cerrena* (*C. unicolor*), *"Spongipellis"* (*S. unicolor*)
Family: Fomitopsidaceae
Genera: *Daedalea* (*D. guercina*), *Fomitopsis* (*F_pinicola*), *Ischnoderma* (*I. resinosum*), *Piptoporus* (*P. betulinus*), *Pycnoporellus* (*P alboluteus*)
Family: Ganodermataceae
Genera: *Ganoderma* (*G. lucidum*)
Family: Grifolaceae
Genera: *Grifola* (*G. frondosa*)
Family: Laetiporaceae
Genera: *Laetiporus, Phaeolus* (*P. schweinitzii*)
Family: Meripilaceae
Genera: *Abortiporus* (*A. biennis*), *Meripilus* (*M. giganteus*) Family: Meruliaceae
Genera: *Bjerkandera* (*B. adusta*), *Gloeoporus* (*G. dichrous*), *Irpex* (*I. lacteus*), *Mycorrhaphium* (*M. adustum*), *Phlebia* (*P. incarnata*), *Podoscypha* (*P. aculeata*), *Steccherinum* (*S. ochraceum*)
Family: Panaceae
Genera: *Panus* (*P. conchatus*)
Family: Phanerochaetaceae
Genera: *Climacodon* (*C. septentrionale*), *Hapalopilus* (*H. nidulans*),
Family: Polyporaceae
Genera: *Polyporus* (*P. sguamosus*), *Poronidulus, Pycnoporus* (*P cinnabarinus*), *Trametes* (*T. versicolor*), *Trichaptum* (*T. biforme*), *Tyromyces* (*T. chioneus*), *Coriolopsis* (*C._gallica*), *Cryptoporus* (*C. volvatus*), *Daedaleopsis* (*D. confragosa*), *Fomes* (*F. fomentarius*), *Lentinus* (*L. tigrinus*), *Lenzites* (*L. betulina*), *Neofavolus* (*N. alveolaris*), *Nigroporus* (*N. vinosus*), *Perenniporia* (*P. ohiensis*),

Family: Sparassidaceae
Genera treated: *Sparassis* (*S.crispa*)
Order: Russulales
Family: Auriscalpiaceae
Genera: Artomyces (*Artomyces pyxidatus*), *Auriscalpium* (*A. vulgare*), *Lentinellus*
Family: Hericiaceae
Genera: *Hericium*
Family: Peniophoraceae
Genera: *Peniophora* (*P. rufa*)
Family: Russulaceae
Genera: *Arcangeliella* (*A. desjardinii*), *Lactarius, Russula*
Family: Bondarzewiaceae
Genera: *Bondarzewia* (*B. berkeleyi*), *Heterobasidion* (*H. annosum*)
Family: Stereaceae
Genera: *Aleurodiscus* (*A. oakesii*), *Stereum* (*S.ostrea*), *Xylobolus* (*X. rustulatus*)
Order: Sebacinales
Family: Sebacinaceae
Genera: *Tremellodendron* (*T. schweinitzii*)
Order: Thelephorales
Family: Bankeraceae
Genera: *Boletopsis* (*B. leucomelaena*), *Hydnellum, Phellodon* (*P. confluens*), *Sarcodon* (S. *imbricatus*)
Family: Thelephoraceae
Genera: *Polyozellus* (*P. multiplex*), *Thelephora T. multipartita*)
Order: Tremellodendropsidales
Family: Tremellodendropsidaceae
Genera: *Tremellodendropsis* (*T. tuberosa*)

1.4 Magnitude of Mushroom Species

The known number of species of fungi till 1990 was 69,000 (Hawksworth 1991). On average, 700 species were described as new to science each year from 1920 to 1950. However, the annual total catalogued fungi reached around 1400 in 1961 and 1500 in 1968 and averaged 1700 each year from 1986 to 1990.

Fungi are considered as the second largest group of organisms in the biosphere after the insects. Known fungal species constitute only about 5% of their species in the world (Table 1.2). The large number of fungal species is still unknown. Out of about 70,000 described species of fungi, it has been suggested that around 14,000–15,000 species produce fruiting bodies of sufficient size and suitable structure to be considered as macrofungi (mushrooms) (Plate 1.1). About 5000 of the species are measured to possess varying degrees of edibility. More than 2000 species from 31 genera are regarded as prime edible mushrooms. But only 100 of them are

Table 1.2 Comparison of the number known and estimated species in the world of a selected group of organisms

Group	Known species	Total species	% known species
Vascular plants	220,000	270,000	81
Bryophytes	17,000	25,000	68
Algae	40,000	60,000	67
Fungi	69,000	1,500,000	5
Bacteria	3000	30,000	10
Viruses	5000	130,000	4

experimentally grown, 50 economically cultivated, and 30 commercially cultivated. Only about six have reached an industrial scale of production in many countries. Furthermore, about 1800 are medicinal ones. The number of poisonous mushrooms is relatively small (approximately 10%), of these some 30 species are considered to be lethal (Miles and Chang 1997).

1.4.1 Edible Mushroom Types

Mushrooms are cultivated throughout the world. About 200 species of more than 2000 edible fungi are widely adopted for human consumption. Out of these, only 10–12 species are commercially cultivated since past few decades in India due to technical advancement. The most commonly cultivated edible types are:

1. Button mushroom – *Agaricus bisporus, A. bitorquis*
2. Paddy straw mushroom – *Volvariella volvacea*
3. Oyster mushroom – *Pleurotus* spp.
4. Milky mushroom – *Calocybe indica*
5. Shitake mushroom – *Lentinula edodes*
6. Winter mushroom – *Flammulina velutipes*
7. Nameko – *Pholiota nameka*
8. Black ear mushroom (Japanese) – *Auricularia* spp.

In India, many edible types of mushroom have been reported (Table 1.3 and Plate 1.2), but the commercially cultivated types are button, paddy straw, oyster, and milky mushrooms.

1.5 Cultivation Potential of Mushroom

The mushroom is a fleshy, spore-bearing organ of fungi. The fleshy nature of the mushroom is responsible for its attraction to man as a source of food. The biological role of the mushroom for the fungus is the production and dissemination of spores in numbers sufficient to assure the propagation of the species under a diversity of environmental conditions.

Plate 1.1 Wild species of mushroom reported in India. (**a**) *Agaricus* sp. (**b**) *Pleurotus* sp. (**c**). *Termitomyces* sp. (**d**) *Tricholoma* sp. (**e**). *Lactarius* sp. (**f**) *Boletus* sp. (**g**) *Dasyscyphus* sp. H. *Hygrocybe* sp. (**i**) *Calocera* sp. (**j**) *Ramaria* sp.

Plate 1.1 (continued)

Table 1.3 Common edible mushrooms of India

Sr. no.	Name of mushrooms	Sr. no.	Name of mushrooms
1	*Agaricus arvensis schaeff ex seer*	20	*Laccaria laccata*
2	*Agaricus bisporus (Lange) Sing.*	21	*Laetiporus sulphureus*
3	*Agaricus campestris L. exfr.*	22	*Leucocoprinus cepaestipes*
4	*Agrocybe praecox*	23	*Lycoperdon perlatum*
5	*Amanita caesarea*	24	*Lycoperdon pyriformin*
6	*Amanita vaginata*	25	*Macrolepiota mostoidea*
7	*Armillaria mellea*	26	*Macrolepiota procera*
8	*Bovista plumbea*	27	*Morchella conica*
9	*Calvatia cyathiformis*	28	*Morchella deliciosa*
10	*Calvatia utriformis*	29	*Morchella esculenta*
11	*Calvatia bovista*	30	*Pleurotus flabellatus*
12	*Cantharellus cibarius*	31	*Pleurotus ostreatus*
13	*Coprinus atramentarius*	32	*Pleurotus sajor-caju*
14	*Coprinus comatus*	33	*Podabrella microcarpa*
15	*Coprinus micaceus*	34	*Psathyrella hydrophilum*
16	*Flammulina velutipes*	35	*Termitomyces hemili*
17	*Heterobasidion annosum*	36	*Rigidoporus ulmarius*
18	*Hirneola auricula-judae (Auricularia spp.)*	37	*Rusulla emetica*
19	*Hydnum repandum*	38	*Rusulla lepida*

There are many different species of mushrooms that have been cultivated successfully in various parts of the world, but from the worldwide standpoint, a few species account for all but a few percent of the total production. In recent years, there have been five genera whose species are responsible for this cultivation. The genera are *Agaricus*, *Lentinus*, *Flammulina*, *Volvariella*, and *Pleurotus*.

The commercial production of mushrooms by using the agricultural, industrial, forestry, and household wastes into nutritious food (mushrooms). Indoor cultivation of mushrooms uses the vertical space and is regarded as the highest protein producer per unit area and time – almost 100 times more than the conventional agriculture and animal husbandry. This hi-tech horticulture venture has a promising scope

Plate 1.2 Cultivated mushrooms species. (**a**) Oyster mushroom (*Pleurotus sajor-caju).* (**b**) Oyster mushroom (*Pleurotus florida*). (**c**) Oyster mushroom (*Pleurotus eous*). (**d**) Button mushroom (*Agaricus bisporus*). (**e**) Shiitake mushroom (*Lentinula edodes*). (**f**) Paddy straw mushroom (*Volvariella volvacea*). (**g**) Milky mushroom (*Calocybe indica*). (**h**) Black ear Mushroom (*Auricularia polytricha*). (**i**) Winter mushroom (*Flammulina velutipes*)

to meet the food shortages without undue pressure on land. Mushroom farming today is being practiced in more than 100 countries, and its production is increasing at an annual rate of 6–7%. In some developed countries of Europe and America, mushroom farming has attained the status of a high-tech industry with very high levels of mechanization and automation. Present world production of mushrooms is around 3.5 million tonnes as per FAO Stat and is over 25 million tonnes (estimated) as per claims of Chinese Association of Edible Fungi. The wide variation in world production data in FAO Stat and CAEF is partly due to the fact that in FAO Stat, mushroom means button mushroom (*Agaricus* spp.) along with the boletes, morels, and tuber, whereas CAEF data covers all types of mushrooms. China alone is reported to grow more than 20 different types of mushroom at commercial scale, and mushroom cultivation has become China's sixth largest industry. Presently, three geographical regions– Europe, America, and East Asia –contribute to about 96% of world mushroom production. With the rise in the income level, the demand for mushrooms is bound to increase in other parts of the world as well. China has been producing mushrooms at very low costs with the help of seasonal growing,

Table 1.4 The ecological habitats and level of fruiting by artificial cultivation in mushrooms

Ecological habitats			Important genera	Fruiting by artificial cultivation
Non- Ectomycorrhizal mushrooms	Above ground	Wood	*Pleurotus*	Easy ↑
			Lentinus	
			Auricularia	
			Tremella	
			Hericium	
			Pholiota	
			Kuehneromyces	
		Straw	*Volvariella*	
	On earth	Dung	*Coprinus*	
			Agaricus	
			Agrocybe	
			Stropharia	
		Soil	*Lepiota*	
			Dictyophora	
			Lepista	
			Melanoleuca	
			Morchella	
		Insects	*Cordyceps*	
			Termitomyces	
Ectomycorrhizal mushrooms	With roots	Mycorrhizal	*Tricholoma*	Difficult

(continued)

Table 1.4 (continued)

Ecological habitats			Important genera	Fruiting by artificial cultivation
			Ramaria	
			Cantharellus	
			Boletus	
			Suillus	
			Gomphidius	
			Lactarius	
			Russula	
			Amanita	
			Cortinarius	
			Rhizopogon	
			Terfezia	

state subsidies, and capturing the potential markets in the world with processed mushrooms at costs not remunerative to the growers in other mushroom-producing countries.

It is possible to cultivate mushrooms under diverse climatic conditions (Table 1.4). Some of the important mushrooms for temperate, subtropical, and tropical conditions are as below.

1.5.1 Temperate Mushrooms

1.5.1.1 Button Mushroom: *Agaricus bisporus*

It is commonly known as "White button, European mushroom, or temperate mushroom." Sporophores are usually centrally stipulated. Pileus is usually 3.5 cm to 10 cm in diameter. Gills are crowded, and flesh is white; it has excellent flavor, is attractive and widely accepted through the world, and hence is most popular in common people and cultivated under controlled conditions. It requires 16–18 °C temperature with relative humidity more than 90% in the growing rooms. The compost is an essential requirement for cultivation. However, seasonally it can be found in Himachal Pradesh, J&K, Central India, Orissa, Nilgiri, and Kumaon hills of India during cool season.

1.5.1.2 Shiitake Mushroom: *Lentinula edodes*

It is commonly known as shiitake mushroom. Sporophores usually grow on woods of dead deciduous trees. Pileus is up to 11 cm in diameter. It is brown in color and centrally stipulated. Gills are crowded. In India, its cultivation is negligible. However, experiments show that shiitake mushroom can be successfully grown on sawdust when temperature is about 20 °C. There is good potential for the cultivation in the country. This may become a popular mushroom in domestic market and has good potential for export.

1.5.1.3 Flammulina Velutipes

The complete technology for its cultivation has been standardized. It is commonly called as winter mushroom in East Asian countries and is well known for its nutritional and medicinal value. It can be cultivated on sawdust of broad leaves supplemented with 10% wheat bran in the temperature range of 10–14 °C. This mushroom can be grown in variety of containers.

1.5.2 Subtropical Mushrooms

1.5.2.1 Agaricus Bitorquis

It is commonly known as town or "street mushroom." Sporophores are observed solitary or in clusters. It grows in grassy places and sometimes at the roadsides. Pileus is up to 2 cm in diameter. Gill is crowded. Short stipe tapers toward the base. It can grow at quite higher temperature ranging from 22 to 24 °C temperature with relative humidity more than 90% in the growing flesh white. In India, it is found in South West of India and Punjab.

1.5.2.2 Oyster Mushrooms: *Pleurotus* Spp.

The oyster mushroom is also known as wood fungus. In India, it is more popular in the local name as Dhingri mushroom. The sporophore is tongue- or shell-shaped, hence the name oyster. The oyster mushroom grows under natural conditions on trees or dead woody branches of trees as saprophytes and primary decomposers. Pileus is usually 1.5–20 cm in diameter. Stipe may be eccentric or centrally stipulate. Gills are decurrent. It requires a wide range of temperature from 20 to 30 °C. However, 24 ± 2 °C is said to be best with relative humidity more than 90% in the growing rooms. The widely cultivated regions in India where oyster mushroom is cultivated are Maharashtra, West Bengal, Tamil Nadu, Punjab, and Madhya Pradesh. The other common species of oyster are *Pleurotus sajor-caju* (gray color), *P. florida* (white color), *P. eous* (pink color), and *P. flabellatus* (white or tanning red color).

1.5.2.3 Black Ear Mushroom: *Auricularia* Spp.

Sporophore grows on dead branches of *Ficus benghalensis*. The fruiting bodoes are leathery or rubbery and gelatinous when fresh and cartilaginous on drying. It is sessile or sometimes slightly stalked. Pileus is up to 10 cm in diameter. It is red-brown in color when fresh and turns gray or tan on drying. It is ear shaped. It requires a wide range of temperature from 20 to 30 °C. However, 24 ± 2 °C is said to be best with relative humidity more than 90% in the growing rooms.

1.5.2.4 Black Poplar Mushroom *(Agrocybe aegerita)*

Agrocybe aegerita is commonly called as black poplar mushroom *which can be* grown on willow wood. Its cultivation on wheat straw has been standardized. It fruits at temperature around 25 °C.

1.5.3 Subtropical Mushrooms

1.5.3.1 Paddy Straw Mushroom: *Volvariella* Spp.

It is commonly known as paddy straw mushroom or Chinese or tropical mushroom. It is a fast-growing mushroom. Sporophore usually grows solitarily or gregarious on rotten paddy straw heaps. Pileus is usually 5–12 cm in diameter and centrally stipulated. Gills are crowded and grayish in color. It is cultivated easily at 30–35 °C temperature with relative humidity more than 80–90% in the growing rooms. It is cultivated in different parts of India. Commonly cultivated species is *Volvariella volvacea*.

1.5.3.2 Milky Mushroom: Calocybe Indica

It is commonly known as "Dudh chatta." It is bright white in color. Sporophores usually grow solitary in soil. Pileus is 10–14 cm in diameter and milky white-colored. Stipe is long and stout, white, and solid with bulbous base. It is centrally stipulated. Gills are distinctly formed and crowded. It is cultivated at 30–35 °C temperature with relative humidity more than 80%.

1.6 Nutritional and Medicinal Properties of Mushrooms

Mushrooms have been reported as a special kind of food since earlier days. Romans regarded mushrooms as the "food of the Gods," which were served only on festive occasions. The Greeks believed that mushrooms provided strength for warriors in battle. The Chinese considered mushrooms as a healthy food. The mushrooms, being a good source of delicious food with high nutritional attributes and medicinal values, received a remarkable amount of interest in recent periods.

Mushroom provides rich addition to the diet in the form of proteins, carbohydrates, valuable salts, and vitamins. Proximate analysis of different species of edible mushrooms and comparison with other food items are listed in Table 1.5. As a food, the nutritional value of mushrooms lies between meat and vegetables. Mushrooms are well suited to supplement diets which lack protein and in the sense they have rightly been called "Vegetable meat" (Table 1.6). Mushroom protein is a high-protein, low-calorie diet and hence good for heart patients. About 100–200 pounds of mushrooms (dry weight) are required to maintain nutritional balance in a normal human being weighing 70 kg (Table 1.7).

Table 1.5 Proximate analysis of mushrooms and vegetables (percent fresh weight)

Name	Moisture	Protein	Fat	Carbohydrate	Crude fiber	Ash	Calories
Mushrooms	90.1	2.5	0.5	4.9	1.1	0.8	35.0
Cabbage	91.1	1.8	0.1	4.6	1.0	0.6	27.0
Cauliflower	90.8	2.6	0.4	4.0	1.2	1.0	30.0
Potato	74.7	1.6	0.1	22.6	0.4	0.6	97.0

Table 1.6 Yield of dry protein per unit area utilized for farming beef, fish, and *A. bisporus*

Source	Approx. annual yield dry protein (kg/ha)
Beef, cattle, conventional agriculture	78
Fish-intensive pond rearing	675
Agaricus bisporus	65,000

Table 1.7 Proximate composition of cultivated species of edible mushrooms

Species	Moisture[a]	Crude protein[b]	Crude fat[b]	Carbohydrate[b]	Crude fiber[b]	Ash[b]	Energy value[c]
Agaricus bisporus	78.3–90.5	23.9–34.8	1.7–8.0	51.3–62.5	8.0–10.4	7.7–12.0	328–368
Auricularia spp.	89.1	4.2	8.3	82.8	19.8	4.7	351
Flammulina velutipes	89.2	17.6	1.9	73.1	3.7	7.4	378
Lentinula edodes	90.0–91.8	13.4–17.5	4.9–8.0	67.5–78.0	7.3–8.0	3.7–7.0	387–392
Pleurotus eous	92.2	25.0	1.1	59.2	12.0	9.1	261
Pleurotus Florida	91.5	27.0	1.6	58.0	11.5	9.3	265
Pleurotus ostreatus	73.7–90.8	10.5–30.4	1.6–2.2	57.6–81.8	7.5–8.7	6.1–9.8	345–367
Pleurotus sajor-caju	90.1	26.6	2.0	50.7	13.3	6.5	300
Volvariella volvacea	89.1	25.9	2.4	57.0	9.3	8.8	276

[a]% fresh weight; [b]% dry weight; [c]K cal/100 g dry weight

1.6.1 Proteins

The quality of protein is nutritionally more important than its quantity. As single source, vegetable proteins are of poorer quality. The quality of mushroom protein is far superior to the vegetable proteins and is as good as or just inferior to animal proteins. It is because all the essential amino acids are present in mushrooms. Mushrooms are rich in lysine and tryptophan as compared to cysteine and methionine, so they can effectively supplement the cereals in terms of protein quality. Mushrooms protein contains high amount of free amino acids. The digestibility of mushroom protein is as high as 72–83%. The amount of crude protein in mushroom ranks below most animal meats but well above most other foods, including milk, which is an animal product.

Table 1.8 Vitamin content of some edible mushrooms (mg/100 g dry weight)

Species	Vitamins			
	Thiamine	Riboflavin	Niacin	Ascorbic acid
Agaricus bisporus	1.1	5.0	55.7	81.9
Lentinula edodes	7.8	4.9	54.9	00
Pleurotus ostreatus	4.8	4.7	108.7	00
Volvariella volvacea	1.2	3.3	91.9	20.2

Table 1.9 Mineral content of some edible mushrooms (mg/100 g dry weight)

Species	Minerals				
	Ca	P	Fe	Na	K
Agaricus bisporus	23	1429	0.2	N.D.	4762
Lentinula edodes	33	1348	15.2	837	3793
Pleurotus ostreatus	98	476	8.5	61	N.D.
Volvariella volvacea	71	677	17.1	374	3455

N.D. Not determined

The produce of an acre of land can be transformed into ten times as much fungus protein as meat protein (Table 1.8). Thus on an area basis, mushrooms are a more valuable source of protein than either cattle or fish.

1.6.2 Vitamins

It has been reported that edible mushrooms are a good source for several vitamins, viz., thiamine (vit. B1), riboflavin (vit. B2), niacin, biotin, and ascorbic acid (vit. C) (Table 1.8). In general, mushrooms are good source of many vitamins especially those of B-complex group but poor in vitamins A, D, E, and K. Folic acid and vitamin B12 which are generally absent in plant goods are present in mushrooms in small quantities. As little as 3 g of fresh mushrooms may provide the recommended daily intake of vitamin B12.

1.6.3 Minerals

Like most vegetables, mushrooms are rich in minerals (Table 1.9), viz., potassium (45% of total ash content), followed by phosphorus, sodium, magnesium, and calcium which together constitute about 56–70% of total ash content. The copper, zinc, iron, manganese, cadmium, and lead make up for the rest of ash components.

Table 1.10 Lipid, fatty acid, and ergosterol content of some edible mushrooms

Species	Total lipid/fatty acid	Linoleic acid[a]	Ergosterol[b]
Agaricus bisporus	3.1	69.22	0.23
Auricularia auricula	1.3	40.39	0.07
Lentinula edodes	2.1	67.79	0.21
Pleurotus sajor-caju	1.6	32.94	0.13
Volvariella volvacea	3.0	69.91	0.47

[a]% of total fatty acids; [b]% of dry material

1.6.4 Carbohydrate and Fiber

Carbohydrate constitutes the greatest fraction of mushroom dry matter (Table 1.10). Chitin constitutes the major fraction of fiber content. The fiber content in almost all the mushrooms is very high. The fiber absence of starch and high fiber in mushrooms makes it an ideal food for diabetic patients. The constituents of mushroom carbohydrates are pentoses, methylpentoses, hexoses, disaccharides, amino sugars, sugar alcohols, and sugar acids. Fresh mushrooms contain 0.95% mannitol, 0.28% reducing sugars, 0.59% glycogen, and 0.91% hemicellulose. The water-soluble polysaccharides as well as acidic polysaccharides obtained from the fruiting bodies of mushrooms have strong antitumor activity. *Pleurotus* species contain carbohydrates ranging from 46.6% to 81.8% as compared to 60% in *Agaricus bisporus*, *Volvariella volvacea* (40–50%), and *Lentinula edodes* (67.5%) on a dry weight basis (Table 1.7). The fiber content ranges from 7.4% to 27.6% in *Pleurotus* species as compared to 10.4% in *Agaricus bisporus* and 4% to 2% in *Volvariella volvacea*.

1.6.5 Fat

Mushrooms are low-fat food with 2–8% crude fat on dry weight basis. Desirability of the unsaturated fatty acids in human diet is compensated by mushrooms with high proposition of unsaturated fatty acids especially linoleic acid (about 70%) with no cholesterol. The total lipids in mushrooms ranged between 0.6% and 3.1% on dry weight basis. The ergosterol, provitamin D2, is most abundant in mushrooms. Among the different species of mushrooms, *Volvariella volvacea* has the highest provitamin D2 content on a dry weight basis (0.47%), followed by *L. edodes* (0.27%) and *A. bisporus* (0.23%) (Table 1.8).

1.6.6 Energy Value

Mushrooms are a good source of energy. Reports show that one pound (454 g) of fresh mushroom provides 120 K calories. The mushrooms can be classed in the category of low-calorie food due to their low dry matter content and fit in well in this era of healthy eating by cutting down the calories.

1.6.7 Moisture

Moisture content may not be of any nutritional significance but it influences the nutritional value of any food item. The moisture content of fresh cultivated mushrooms varies from 90 to 94% (6–10% dry matter), and 90% may be considered as an average moisture of the mushrooms.

1.7 Medicinal Values of Mushroom

Mushrooms exhibit a wide range of biochemical reactions both biosynthetic and biodegradative in nature. They are endowed with the ability to secrete a wide variety of hydrolyzing and oxidizing enzymes which have potential for biotechnological applications.

More than 100 medicinal edible mushrooms have been identified. However, important medicinal mushrooms are *Ganoderma lucidum* (reishi mushroom), *Lentinus edodes* (shiitake mushroom), *Grifola frondosa* (maitake mushroom), *Pleurotus ostreatus* (oyster mushroom), *Agaricus bisporus* (button mushroom), *Coriolus versicolor* (PSK), *Boletus edulis, Tremella fuciformis, Auricularia polytricha, Hericium erinaceus,* and *Cordyceps sinensis.*

Many mushrooms have been traditionally used as medicine and tonic in China, Korea, and Japan. The importance and uses of mushrooms in Chinese medicines are summarized in Table 1.11.

1.7.1 Health Effects of Medicinal Mushroom

Mushroom is not only an ultimate health food but of late has also proved to be a begin but effective drug because of its medicinal attributes, which is due to its two bioactive components, i.e., (1) nutraceutical and (2) pharmaceutical.

Table 1.11 Mushrooms in Chinese traditional and herbal drugs

Mushrooms	Indication
Agaricus bisporus	Stimulating digestion, curing hypertension
Auricularia polytricha	Strengthening health, helping blood circulation
Ganoderma lucidum	Rejuvenating effect, neurasthenia
Lentinus edodes	Strengthening health and resisting diseases
Pleurotus ostreatus	Causing muscles and joints to relax
Boletus edulis	Causing muscles and joints to relax
Coriolus versicolor	Curing chronic diseases
Grifola frondosa	Strengthening health and resisting diseases
Hericium erinaceus	Beneficial to heart, kidney, liver, lungs, spleen
Tremella fuciformis	Strengthening health and resisting diseases
Tricholoma mongolium	Beneficial to stomach and intestines

1. *Nutraceutical*

   Nutritional constituents of medicinal mushrooms are proteins, vitamins (B-complex, folic acid, B12), minerals (potassium, phosphorus, copper, iron), and small quantities of utilizable sugar and fat (rich in linoleic acid, devoid of cholesterol). These are mushroom extracts which are consumed as capsule or tablet as dietary supplement (not as food). These extracts have potential therapeutic application, i.e., either they increase the body resistance or cause regression in the disease state. Nutritional value of mushroom lies between high-grade vegetable and low-grade meat.
2. *Pharmaceutical*

Mushroom is a nonchemical drug. There are more than 50 types of polysaccharides and 30 types of terpenoids which are the main bioactive components. Mushrooms have little scope of overdose and toxicity and hence does not require commercial license for selling (over-the-counter medicines). Important pharmacological ingredients of mushroom, e.g., *Ganoderma lucidum*, are polysaccharides, triterpenes, lentinan, adenosine, lingzhi-8, PSK (anticancer drug in Japan), and eritadenine butyric acid. Both the cellular components of mushrooms and their secondary metabolites have been shown to have biological activity (Tables 1.12 and 1.13).

Table 1.12 Pharmacological action of bioactive mushroom

Sr. no.	Component	Action	Mechanism of action
1	Polysaccharide	Antitumor Anti-inflammatory	Immune potentiation
2	Triterpenes	Antiviral (HIV, Hep. B)	Alters cell membrane, inhibiting entry of virus in cell
		Antithrombotic Anti-atherosclerotic	Alters platelet aggregation and plaque formation
3	Adenosine	Antidiabetic	Lowers blood sugar by stimulating insulin receptors
4	LZ-8	Antitumor	Immune modulation
5	PSK	General tonic	Scavenging free oxygen radicals
6	Eritadenine butyric acid	Anti-cholelithic	Alteration of lipid profile

Table 1.13 Clinical applications of medicinal mushrooms

Sr. no.	Mushrooms	Indications
1	*Ganoderma lucidum*	Anticancer, antiviral-hepatitis-B, HIV
2	*Agaricus bisporus*	Antihypertensive
3	*Auricularia polytricha*	Cure of piles, anti-atherosclerotic
4	Maitake mushroom	Antidiabetic
5	*Pleurotus sajor-caju*	Antihypertensive, Reno-protective
6	*Lentinula edodes*	Hypocholesterolemic, antihypertensive

1.7.2 Health and Nutritional Benefits of Eating Mushrooms

- Mushrooms contain low sodium, carbohydrate, fat, and calories content and high fiber content. This is the reason why mushrooms are considered good for those aiming for weight loss.
- Mushrooms are an excellent source of potassium. Since potassium helps to lower blood pressure and diminished the risk of stroke, mushrooms are recommended to people suffering from hypertension.
- Mushrooms are rich in copper, a mineral that has cardioprotective properties. A single serving of mushrooms is said to provide about 20–40% of the daily needs of copper.
- Mushrooms are believed to help fight against cancer. They are an excellent source of selenium, an antioxidant that works with vitamin E to protect cells from the damaging effects of free radicals.
- White button mushrooms have been found to restrain the activity of aromatase, an enzyme involved in estrogen production, and 5-alpha-reductase, an enzyme that converts testosterone to DHT.
- Researches have suggested that white button mushrooms can reduce the risk of breast cancer and prostate cancer. In fact, extract of white button mushrooms has been found to help in diminishing cell proliferation as well as tumor size.
- Shiitake mushrooms comprise lentinan, a beta-glucan that has been associated with stimulation of the immune system and thus is believed to be helpful in fighting against AIDS. It also helps fight infection and exhibits antitumor activity.
- Being rich in fiber, protein, and vitamin B, mushrooms help to maintain a healthy metabolism.
- It has been found that mushroom extract helps to stop migraine headaches and is beneficial for people suffering from mental illnesses, like obsessive-compulsive disorder.
- Oyster mushrooms are said to be useful in strengthening of veins and relaxation of the tendons.

1.8 Cultivation Methods of Different Mushrooms

1.8.1 Oyster Mushroom Cultivation

The *Pleurotus* mushroom is generally referred to as "oyster mushroom" in India. It is a basidiomycete and belongs to the genus "*Pleurotus*." It is a lignocellulolytic fungus and grows naturally in the temperate and tropical forest on dead and decaying wooden logs or sometimes on dryings trunks of deciduous or coniferous woods. It can also grow on decaying organic matter.

1.8.1.1 Advantages of Growing Oyster Mushroom

Difference of Substrates

Pleurotus mushroom can degrade any kind of agricultural or forest waste which contains lignin, cellulose, and hemicellulose.

Choice of Species

The *Pleurotus* mushroom has maximum number of commercially cultivated species suitable for round the year cultivation. Moreover, variations in shape color, texture, and aroma are also available as per consumers' choice.

Simple Cultivation Technologies

Pleurotus mycelium can grow on dried straw, and it does not require selective compost for growth. Media preparation for oyster mushroom is very simple. Moreover, it does not require controlled environmental conditions because most of the species have a very wide temperature, relative humidity, and CO_2 tolerance.

Longer Shelf Life

The oyster mushroom fruit bodies can be easily dried and stored. Dried oyster mushrooms can be instantaneously used after soaking in hot water for 5–10 min or it can be used in powdered form for several preparations. Fresh mushrooms have a shelf life of 24–48 h even at room temperature.

High Productivity

The productivity of oyster mushrooms is very high as compared to all other cultivated mushrooms. The biological efficiency of oyster mushroom is 50–70% on dry wheat or paddy straw in 45–60 days.

1.8.1.2 Steps in Oyster Mushroom Cultivation

1. Spawn preparation
2. Preparation of substrate
3. Sterilization of substrate
4. Bed preparation
5. Crop management and harvesting

Spawn Preparation

Grain spawn is prepared as per standard method for this mushroom. Pure culture of *desired species of Pleurotus* is used to inoculate the master spawn and incubate it at 23–25 °C. After completion of mycelial growth in master spawn, commercial spawn can be multiplied and prepared (Plate 1.3).

Preparation of Substrate

Oyster mushroom can be grown on various substrates like wheat, cotton waste, paddy and rye straw, soybean husk, sugarcane thrash, banana leaves, etc. Cultivation on wheat/paddy straw is more economical as these are easily available. Chop up the

Plate 1.3 Steps in cultivation of oyster mushroom. (**a**) Preparation of substrate. (**b**). Soaking of substrate. (**c**) Sterilization of substrate. (**d**) Bed filling. (**e**) Complete spawn run after 15–18 days of bed filling. (**f**) Pinhead formation. (**g**) Mushrooms ready for harvesting

straw into small pieces of 2–3 cm size, and then soak in tap water for 8–10 h for presoaking. Drain out the excess water.

Sterilization of Substrate

Sterilization of the soaked straw is the most important step in mushroom cultivation to remove the unwanted natural contaminating microorganisms from the straw collected from field. The various methods of substrate sterilization are adopted for *Pleurotus* cultivation. They are as follows:

(i) Steam sterilization
(ii) Hot water treatment sterilization
(iii) Chemical sterilization

Steam Sterilization

In steam sterilization, 1 day before, the straw is kept for soaking in cold tap water. On the next day, the excess water is removed from the bag and the straw is exposed to hot steam at 80 °C temperature for 1 h in either autoclave or closed room. The separate boiler unit may be installed on the basis of capacity of your project to generate enough steam for sterilization. Majority of the growers use autoclave of required capacity to sterilize straw by this technique. This is the best method of straw sterilization in which complete disinfection of straw is done due to high pressure of steam.

Hot Water Sterilization

Before exposing the straw for hot water sterilization, 1 day before, the straw is kept (dipped) in cold tap water for overnight for soaking. On the next day, excess water is removed. This bag containing straw is then dipped in water of 80 °C temperature for 1 h.

Chemical Sterilization

In chemical sterilization, it is not necessary to dip the straw for presoaking in cold water 1 day before treatment. However, you can soak the straw for overnight in the solution of formalin (125 ml) and carbendazim (7.5 gm) mixed in 100 l of water for nearly 18 h.

Bed Preparation

The bed filling is carried out with sterilized straw in aseptic environment, disinfected room. Use standard size of the polythene bags of 35 × 55 cm size and disinfect in 5% formalin solution. Follow the most convenient method of layer spawning for filling these bags. Put the first layer of straw measuring about 2–3 cm at the bottom. Then, spread the spawn uniformly over the surface of straw. Likewise, spread the spawn in 3–4 layers by pressing the straw lightly after every layer. The rate of spawning should be at 2% of the wet weight basis of straw. Tie the neck of bags tightly with thread. Prepare small pinhole on the surface of bag to remove the excess moisture and to release the gases during spawn run. Keep these inoculated bags for incubation (spawn run).

Crop Management and Harvesting

Incubation

The spawned bags are kept in the incubation room for fungal mycelial growth. Spawn bags can be kept on a raised platform or shelves or can be hanged in cropping room for mycelial colonization of the substrate. Although mycelium can grow from 10 to 30 °C, the optimum temperature lies between 22 and 26 °C. Higher temperature (more than 30 °C) in the cropping room will inhibit the growth and kill the mycelium. The bed temperature is generally 2–4 °C higher from the room temperature. The mushroom mycelium can tolerate very high CO_2 conc of 15–20%. During mycelial growth, the bags are not to be opened or no ventilation is needed. Moreover, there is no need for any high relative humidity, so no water should be sprayed.

Fruit Body Induction

Once the mycelium has fully colonized the substrate and forms thick mycelial mat, it is ready for fruiting. Contaminated bags with mold may be discarded, while bags with patchy mycelial growth may be left for few more days to complete the mycelial growth. In no case, bags should be opened before 16–18 days except in the case of *P. membranaceus* and *P. djamor* var. *roseus* which forms fruit bodies within 10 days even in closed bags from small holes. There is no need for casing the substrate. All the bundles, cubes, or blocks are arranged on wooden platforms or shelves with a minimum distance of 15–20 cm between each bag in the tier. Various cultural conditions required for fruiting are as follows.

Temperature

The different species have different temperature requirements for fruiting. However, mycelial growth of all the *Pleurotus* spp. can take place between 20 and 30 °C. On the basis of temperature requirement of a species, oyster species can be categorized into two groups – winter or low temperature requiring species (10–2 °C) and summer or moderate temperature requiring species (16–30 °C). Commercial varieties which can be cultivated during summer are *P. flabellatus, P. sapidus, P. citrinopileatus, and P. sajor-caju*. Low temperature requiring species are *P. ostreatus, P. florida, P. eryngii, P. fossulatus,* and *P. cornucopiae*. The growing temperature not only affects the yield but also the quality of produce. The temperature requirement of different *Pleurotus* sp. is given in Table 1.14.

Table 1.14 Temperature requirement of different *Pleurotus* sp. and their yield performance

Sr. no	Species	Optimum temp. of mycelium grown on substrate (°C)	Optimum temp for fruiting (°C)	Temp. range where it can grow (°C)	Yield performance (% B.E.)
A	Summer species				
1	*P. flabellatus*	25–28	22 ± 2	16–28	60–90
2	*P. Sajor-caju*	25–28	24 ± 2	17–30	50–70
3	*P. sapidus*	25–28	24 ± 2	17–30	40–70
4	*P. membranaceus*	25–28	27 ± 2	20–30	40–70
5	*P. Citrinopileatus*	25–28	26 ± 2	20–30	30–60
6	*P. eous*	25–28	24 ± 2	16–28	30–50

Relative Humidity

The high relative humidity (70–80%) during fruiting is the pre-requisite for all the *Pleurotus* species. To maintain relative humidity, water spaying is to be done in the cropping rooms. During hot and dry weather conditions, spraying has to be done 2–3 times, while in hot and humid conditions (monsoon) light spraying will be sufficient. The judgment of spraying can be made by touching the surface of the substrate. Spraying should be done with a fine nozzle to create a mist or fog in the cropping room. Ventilators and exhausts fans should be operated for air circulation so that the excess moisture from the pileus surface evaporates. Humidifiers can be used to keep the relative humidity of cropping rooms between 75 and 80%.

Oxygen and Carbon Dioxide Requirements

The oyster mushroom can tolerate high carbon dioxide concentration during spawn run (up to 20,000 ppm or 20–22%), while it should be less than 600 ppm or 0.6% during cropping. Therefore, sufficient ventilation should be provided during fructification. If the CO_2 concentration is high, the mushrooms will have long stipe and small pileus. Mushrooms will appear like a mouth of trumpet.

Harvesting

The pinheads will start to develop from all side of beds within 3–5 days and will be ready for harvest in a week. Harvest full-size mushroom by twisting clockwise or anticlockwise with thumb and forefinger. After harvesting, scrap the beds lightly with hands so as to remove the upper decomposed layer of straw. Likewise, three flushes can be obtained from a single bed. The average yield of fresh mushroom is 750–800 g/kg of dry weight of substrate.

1.8.2 Cultivation of Milky Mushroom (*Calocybe indica)*

Milky mushroom is also called as white summer mushroom because this can be grown during hot summer months. It can be grown during March to November months and can be best fitted in relay cropping pattern. Its milky white attractive color is the reason behind its name "milky mushroom." It has very good shelf life at room temperature (2–3 days), and in refrigerator it can be kept for 10–12 days. Its taste is different and appealing than other edible mushrooms. It can be grown in the bags and on the shelves prepared for button mushroom cultivation also. The biological efficiency is 80–90%.

Following are the major steps in cultivation technology:

1. Spawn production
2. Substrate preparation and treatment
3. Bed filling and spawning of the substrate
4. Care and maintenance after spawning
5. Casing
6. Aftercare and management during production
7. Harvesting

1.8.2.1 Spawn Preparation for *Calocybe indica*

Grain spawn is prepared as per standard method for this mushroom. Pure culture of *Calocybe indica* is used to inoculate the master spawn and incubate it at 23–25 °C. After completion of mycelial growth in master spawn, commercial spawn can be multiplied and prepared by it.

1.8.2.2 Substrate Preparation and Treatment

Wheat straw, paddy straw, soybean straw, sugarcane trash, etc. are most suitable substrate for cultivation of *Calocybe indica*. The length of straw should be between 1 and 1.5 inches, and it should not be fine; otherwise, there will be problems of aeration in the beds after spawning. The treatment of substrate can be done by two methods:

(a) Chemical steeping treatment
(b) Hot water treatment

Chemical Steeping Treatment

This treatment is less time-consuming and economically cheaper and efficient than other treatments. For this treatment, in 100 liters of water, fungicide like carbendazim (7.5 gm) + formaldehyde (125 ml) + insecticide dichlorvos (20 ml) is mixed in a bucket full of water and poured over wheat straw filled in a plastic drum or cement tank or in a rust-free container. The substrate should be soaked for 14–16 h and after this remove wheat straw and keep it on sloppy surface or wire gage for 1½–2 h so that excess water is decanted.

Hot Water Treatment of the Substrate

Wheat straw is soaked for 10–12 h in potable clean water and then dipped in water of 80 °C for 1 hr. After cooling this immersed straw to room temperature, this should be placed on a sloppy surface for 1–2 h to remove excess water. After complete drain off the excess water, the substrate is ready for spawning.

1.8.2.3 Bed Filling and Spawning of the Substrate

(a) *Thorough Mixing:* After weighing wet substrate say 100 kg, mix spawn in desired quantity (4%) to ensure the equal distribution of spawn. This method of spawning is more suitable for bed cultivation method. In these cases, thc bcd depth should not be more than 7–8 in.
(b) *Layered Spawning:* For bag cultivation method, the substrate should be spawned in layers. The first layer of the spawn is at 4-inch layer of substrate, and subsequent layers should be 6 inches apart. At the top layer, 3-inch straw should be added and the mouth of the bags should be closed by rubber band or thread.

1.8.2.4 Care and Maintenance after Spawning

The spawned bags should be kept at 25–30 °C temperature. If spawning is done on the beds, they should be covered by newspaper and frequent water sprays (3–4 times a day) may be done to keep the upper layer moist. Fresh air circulation through cross-ventilation should be there. Spawn run should be observed every day, and if there is any contamination of molds or *Coprinus*, they should be removed immediately.

1.8.2.5 Casing

At suitable temperature and humidity (temp. 25–27 °C and relative humidity 70–80 °C), spawn run in wheat straw takes 15 days to complete. After this, it needs casing which is a 2.5–3 cm layer spread on the top of the bag or bed. The best suitable casing material is 2-year-old cow dung, 2-year-old farmyard manure, or 2-year-old biogas spent slurry or mixture of these in 1:1 ratio. Following are the importance of casing in the bags of *Calocybe indica*. Casing is necessary for pinhead initiation in the bags, and in its absence either there will be no pinheads, or few or scarce pinheads will appear which will not develop further. Casing material should not contain any nutrients which inhibits further mycelial growth and promotes fruiting. Completed spawn run bags are opened from the top and folded in such a way that it holds the casing material on the top. After opening from the top, it should be leveled for uniform spread of casing and water is sprinkled on it. After spraying water on it, casing layer 2 cm should be spread.

1.8.2.6 Aftercare and Management During Production

Water should be sprayed 2–3 times on bags so that the casing remains moist. The floor of cropping room should be covered with sand so that waterlogging does not occur and excess water is retained in sand. Foggers or humidifier may be used to maintain the humidity. Water should also be sprayed on walls of the cropping room.

After casing, water should be sprayed thrice a day so that it does not dry out. Relative humidity 80% should be maintained inside the cropping room. Temperature of the cropping room should not exceed 30 °C; otherwise, it will affect adversely. Proper light arrangement must be three; for this, two fluorescent tube lights should be fitted in 25′ × 25′ cropping room. This mushroom (*Calocybe indica*) is extremely sensitive to light. For good aeration, cross-ventilation must be there; otherwise, higher CO_2 concentration will adversely affect the fruiting. These windows and ventilations should be opened for 2 h a day. Insect-proof net should be there on the windows and ventilations to prevent insect entry. Pinheads will be initiated after 10–12 days post-casing.

1.8.2.7 Harvesting

Pinheads will be mature after 6–7 days which may be harvested by twisting clockwise. These should not be harvesting by cutting or pulling. Water should be sprayed after harvesting of mushrooms and spraying should also be done on floor and walls of the cropping room. After harvesting, the gap of casing should be filled with sterilized casing material. Pinheads will again initiate after 7–10 days internal, and they should be harvested in the same manner.

Production/Yield

From one bag or bed, mushroom can be harvested for 5–6 times, and total production, if we take 1000 kg dry straw, would yield 600–700 kg of fresh *Calocybe indica*. Total cropping period is 60 days on an average. From spawning to last harvest, total cropping period is 90 days.

1.8.3 Cultivation of Button Mushroom (*Agaricus bisporus*)

Button mushroom is a leading mushroom variety in the world. Cultivation of the common white button mushroom is a complex process and requires special technical skill for raising a successful crop. It requires two different temperatures, 22–28 °C for spawn run and 16–18 °C for fruit body formation. Besides specific temperature, it requires proper humidity (85–95%) and enough ventilation during fructification. Cultivation of white button mushroom is accomplished in three basic steps (Plate 1.4).

1. Production or procurement of spawn
2. Preparation of growth medium (compost)
3. Cultivation

Flowchart

Compost preparation 10–30 days
↓
Pasteurization in short method 5–7 days
↓
Spawning (1 %) of compost
↓
Spawn run 12–15 days
↓
Casing material (soil, 2-years-old cow dung and old farmyard manure) preparation
↓
Casing material's pasteurization (steam/chemical)
↓
Casing(1-2″ thickness) and case run 10 days
↓
Pinning 2–4 days
↓
Cropping 30–60 days
↓
Spent compost can be used as manure after 6 months

Button mushroom (*Agaricus bisporus*) being coprophilous (dung-inhabiting) fungus requires compost for its growth. It requires carbon, nitrogen, phosphorus, sulfur, iron, potassium, and vitamins such as thiamin and biotin for its growth. The ingredients containing these items are fermented in a set pattern to produce selective compost suitable for button mushroom production and that does not favor the growth of other harmful and competing microbes. Button mushroom is the most popular cultivated variety and fetches higher price. For its successful cultivation, careful attention must be paid to four aspects, i.e., good compost, pure and productive spawn, proper environmental conditions [temperature, relative humidity, and fresh air], and hygiene of the farm. The details of cultivation process are given below.

1.8.3.1 Compost Preparation

The substrate in which mushroom mycelium grows is a specially prepared material called compost. The purpose of composting is to convert various raw materials into a selective homogeneous substrate suitable for mushroom mycelium. The compost ingredients can be classified in the following categories based on their functional role.

Vegetative-Based Material

Agricultural residues like wheat and paddy straw are being widely used throughout the world. However, other agri-residues, viz., barley straw, maize stalks, sugarcane bagasse, sugarcane trashes and leaves, soybean stalks, mustard stalks, etc., are also used. These materials act as reservoir of cellulose, hemicellulose, and lignin which

Plate 1.4 Different operations in button mushroom cultivation. (**a**). and (**b**) Long method compost preparation. (**c**) Short method of compost preparation Phase I: pre-wetting. (**d**). Compost ready for Phase II in short method. (**e**) Mushrooms ready for harvest

are utilized by microbes during composting and later on by mushroom mycelium. They also provide nitrogen and being in bulk provides suitable physical structure for fermentation during composting process. Paddy straw is soft and absorbs moisture very fast; hence, care should be taken to adjust watering during wetting and turnings.

Animal Manure

Horse dung undoubtedly is most suitable. Nowadays, chicken manure is being widely used to produce compost. It provides slow releasing nitrogen and also increases bulk of compost. Chicken manure is preferred for short method of compost as chances of disease-causing organisms are more in long method of composting. However, some farmers prefer chicken manure and get better yield through long method of composting.

Nitrogen Fertilizer

In this category, fertilizers urea, ammonium sulfate, and calcium ammonium nitrate are used. Nitrogen content of these fertilizers is very high and is released very fast during composting, thus accelerating the fermentation.

Carbohydrate Sources

Molasses, malt sprout, potato waste apple, and grape pomace are good source of carbohydrate. These items hasten the composting process, balance the C/N ratio, and also help to establish bacterial flora in compost.

Concentrated Meals

In this category, generally, animal feeds are included, viz., wheat/rice bran, cotton, soybean caster, sunflower meal, and neem cake. They provide both nitrogen and carbohydrates, and nitrogen is released slowly.

Supplementation to Rectify Mineral Deficiencies

In addition to C and N, button mushroom requires other elements for its growth. Fertilizers like muriate of potash and superphosphate are used to meet the requirement.

A large number of compost formulae have been developed. However, these can be modified based on availability of raw materials and their cost in the region. Initial nitrogen content of the total raw material should range between 1.5 and 1.75% on dry weight basis and C/N ratio 25–30:1. Once the compost is ready, the total nitrogen content will be around 2% and C/N ratio 16–18:1.

Moisture and Nitrogen Contents (Approximately) of some Raw Materials

Material	Moisture content %	Total nitrogen %
Paddy/wheat straw	10–15	0.4–0.6
Horse manure light	20–30	0.8–1.0
Horse manure heavy	40–50	1.0–1.5
Deep litter chicken manure	20–30	2.5–3.5
Wheat bran	5–10	2.0–2.5
Soybean meal	5–10	6.5–7.2
Cottonseed meal	7–10	7.0–7.2

Some Compost Formulae Are Given below

I

Wheat/paddy straw or	1000 kg
Paddy straw + maize stalk	500 + 500 kg
Ammonium sulfate or calcium ammonium nitrate	30 kg
Superphosphate	10 kg

Urea	15 kg
Muriate of potash	10Kg
Wheat bran or rice bran	145 kg
Gypsum	40 kg
Chalk powder	30 kg
Neem cake/ cottonseed meal	40 kg

II

Paddy straw	1000 kg
Wheat bran or rice bran	166 kg
Urea	16 kg
Gypsum	40 kg
Neem cake/ cottonseed meal	30 kg

III

Paddy straw /wheat straw	1000 kg
Chicken manure	500 kg
Wheat bran or rice bran	30 kg
Urea	15 kg
Gypsum	40 kg

1.8.3.2 Methods of Composting

Long Method of Composting

It takes about 26–30 days to prepare the compost, and 7–8 turnings (2–3 days interval) are given without pasteurization. The schedule of turning is given below.

Day-2: This is pre-wetting period done 2 days in advance of 0 day (the day standard stack is prepared for composting). The straw/manure is spread and watered in layers making a stack of 1 M height. Next day [D-1], it is turned and again watered to saturate it.

Day-0: This day standard stack is prepared for the fermentation of substrate. At this stage, moisture content of straw should be 75–77%. Mixture of fertilizers and bran is mixed with wet straw thoroughly or in layers (20–30 cm) and stacked in a dimension of 1.5–1.8 M [width] × 1.5–1.8 M [height] and length depending on quantity used. Height should be at least 6″ more than width for better air exchange and fermentation. Dry spots are watered, and stack is gently pressed to have firm pile.

Day +4: In the first turning, a portion of stack is broken, thoroughly shaken, mixed, and watered if required and stacked in layers. Subsequently, remaining portions should also be turned and stacked as done above.

Day+7: Second turning is given as above and gypsum is added in layers. Watering is done if required,

Day +10, third; Day +13, fourth; Day+16, fifth; Day+19, sixth; Day +22, seventh; and Day+25 eighth turnings are given.

By this time, ammonia smell should disappear. However, in case ammonia smell is still noticed, it should be further turned for one or two turnings to get compost free from ammonia. At this stage, compost will be dark brown in color and moisture content should be 62–68%. In case cultivation is to be done in bags, moisture content of 62–65% should be preferred. Moisture content can be measured by putting 100 g compost in oven (such five replicates should be used) at 100 °C for 24–48 h to get constant weight. The moisture content % = fresh wt. (100 g)–dry wt.

This is the most popular method for commercial production of button mushroom. It is completed in two phases.

Phase I

Day-4: Straw/manure [horse dung, chicken manure] is spread in layers and thoroughly watered to make a stack of 1 M height. Seepage water should be stored in a pit in the corner of composting platform and reused to water.

Day-2: Stack is turned and watered thoroughly, and seepage water collected in pit should also be used to water substrate.

Day-0: At this stage, moisture content of substrate [straw/manure] should be 75–77%. Fertilizer, bran, is mixed thoroughly or in layers (20–30 cm), and watering is done if required particularly on layers after spreading mixture (fertilizer, etc.). Standard stack [1.5–1.8 M × 1.5–1.8 M × length] is prepared.

Day +4: First turning, Day +7: Second turning, gypsum is added. Watering is done if required.

Day+10: Third turning. Stack is broken and thoroughly shaken and filled in bulk chamber for Phase II. At this stage, compost will be having strong smell of ammonia, moisture content 68–72% and light to dark brown in color.

Phase II

Phase II helps to complete fermentation of the compost and pasteurization to eliminate harmful microbes, insect pests. It is done in bulk chamber.

1.8.3.3 Bulk Pasteurization

In bulk pasteurization system, compost is filled in tunnel which is provided with slanting floor and perforated false floor [25–30% opening] to facilitate airflow from the bottom. Compost is filled on perforated floor to a height of about 1.8–2.2 M. The height of tunnel should be 4 M. The size of the tunnel depends on the quantity of compost. The tunnel of 10 x 3x 4 M will be suitable for about 20–25 tonnes of compost. One side fitted with blower and duct steam outlet, gap between floor and false floor should be 90 cm at this end and on opposite side 15 cm. Blower is connected to upper portion of tunnel and fresh air supply system through duct. Fresh air supply system should be provided with filter and damper to control air supply. After loading compost and placing thermometers at different levels, blower is switched on and 10–15% fresh air is provided for 2–3 h and fresh air supply is stopped for temperature to rise; however, blower should be continuously on to circulate inside air. Next day air temperature may rise to 57 °C on its own; otherwise, steam is introduced. This raises compost temperature 58–60 °C which is maintained for 6–8 h to achieve pasteurization. Temperature should not be allowed to go beyond this limit and can be checked by stopping steam supply and increasing fresh air. On completion of this stage, compost temperature is slowly decreased by introducing fresh air (10–20%) to 55–48 °C and fresh air supply is continued to complete fermentation also known as conditioning for 2–3 days. At this stage, compost should be free from ammonia and dark brown in color with moisture content 62–68%. Once temperature is below 28 °C, compost is filled in trays/polythene bags/shelves and spawned.

Bulk chamber: Infrastructure and machinery required for setting up compost production facility producing 10–12 tonnes of compost in a single operation are given below.

Infrastructure

	Quantity	Size (I × W × H)
i. Composting yard	1 no.	18 × 12 × 4.2 M
ii. Bulk chamber	1 no.	6 × 2.7 × 3.6 M
(slope between floor and false floor 85–12 cm)		
iii. Casing soil chamber	1 no.	3.6 × 2.1 × 2.7 M
iv. Ancillary room	3 nos.	3.6 × 3 × 3 M
v. Service room	1 no.	3.6 × 2.4 × 2.4M
vi. Spawning area	1 no.	6 × 4.5 × 2.4 M

Machinery Required

	Quantity
i. Grating for the tunnel and casing soil chamber	2 sets
ii. Tunnel ventilation system complete with blower and ducting	1 set
iii. Blower for casing chamber	1 no.
iv. Boiler 100–150 kg/h steam generation capacity	1 no.
v. Insulated doors for tunnel and casing chamber	3 nos.
vi. Digital thermometers	4 nos.
vii. Compost making boards	1 set
viii. Electrical fittings and water pipelines	
ix. Misc items like forks, shovels, weighing balance, etc.	

Spawning

Spawning means putting spawn in the compost. The spawning is done by mixing spawn evenly and throughout the compost, and 500 g of spawn is used to spawn 100 kg compost. It takes about 12–14 days to complete spawn run in compost.

After spawning, compost should be covered with newspaper and water is sprayed to keep papers moist. In case polythene bags are used, bags can be loosely folded and watering can be avoided. In case it is not possible to fold, newspaper can be used to cover and light watering should be done to maintain moisture. Temperature 25 ± 2°C and relative humidity [R.H] 85–90% should be maintained by spraying water on shelves and walls. During first 24–48 h, after spawning one should be careful about compost temperature; if temperature goes beyond 30 °C, water is sprayed on newspaper and walls and ventilators are opened as temperature more than 34^{0}c is injurious to mycelium. Mushroom mycelium starts spreading from the spawn and permeates through the compost. It is called "spawn running." It covers the entire compost as whitish gray thread-like structure with fine fluffy appearance of the mycelium. Once spawn running is over, cover is removed from the shelves/beds or bag's top is rolled back to facilitate casing.

Casing

The term casing means covering the compost after spawn run is over, with a layer [2.5–5 cm] of casing material which may retain moisture and help in gaseous exchange. It also provides physical support to growing mushrooms. Casing material can be prepared by mixing equal part of soil, old cow dung, and FYM (more than 2 years old) and adding chalk powder to adjust pH to 7.5.

Casing Pasteurization and Application

Casing material is chemically pasteurized by drenching casing material with formaldehyde solution [4%] 2 weeks before casing so that it should be free from formalin fumes. However, steam pasteurization [65 °C for 6 h] is most effective. Casing material [moist] is filled in trays (6–8″ height and 5–10 cm legs) and stacked inside pasteurization chamber one above the other in rows and 50 cm above the ground. There should be gap between trays and wall [about 30 cm] to facilitate steam movement. Steam is released at ground level and inside air is circulated with the help of blower connected to duct opening at ceiling level and 20–30 cm from ground level to maintain uniform temperature. Casing material's temperature should be maintained 65 °C for 6 h. Temperature should be monitored from outside with the help of dial/digital thermometer with probes/sensors inserted in casing material at top-, middle-, and bottom-level trays. On cooling to room temperature, casing material is spread in about 2.5–5 cm thickness on compost after removing the cover. Temperature of 23 ± 2 °C and R.H. 80–85% are maintained, and watering is done to keep casing layer moist. Over watering should be avoided.

Cropping

It takes about 10 days when mycelium covers casing layer, and this is the time for pinning and cropping. Temperature is lowered to 16–18 °C, and fresh air by opening ventilators or through air handling unit [200 cu ft./hr./ton of compost] is introduced. R.H 80–85% is maintained. These changes result in initiation of pinheads. During this period, mist watering is done. Once mushrooms are pea size, little heavy watering is done. It takes another 3–5 days to attain 2.5–4 cm when mushrooms are harvested by holding cap and twisting gently. After harvesting, lower portions with casing material and root like mycelium are cut with knife, cleaned, and packed in perforated (2–4 mm, 4–6 holes) polythene/pp. bags for fresh market or canned in brine solution. Mushrooms appear in flushes of 3–5 days duration each at an interval of 7–10 days. Generally, harvesting is done for 4–8 weeks. An average yield of 10–15% of compost can be obtained in 6 weeks. However, modern farms with bulk pasteurization facility and fully air-conditioned cropping rooms are producing 20–25% of the compost.

Crop Management

To get better production and quality of mushroom, the following steps should be followed.

(i) Composting should be done on cemented, raised, and sloppy platform with pit for the collection of seepage water. Once composting is over, clean and wash the platform and surrounding area.
(ii) Thermometers/probes, blower, boiler, and other machinery should be checked and made functional.
(iii) Bag filling and spawning area should be cleaned and sprayed with formaldehyde (2%) one day before and after spawning.
(iv) Regular checking for temperature and weed molds incidence during spawn run is important. Once in a week, spray nuvan (0.1%) and bavistin (0.1%) + formaldehyde (0.5%) separately. Spot treatment with cotton soaked in formaldehyde (4%) is advised if any patch of mold is noticed.
(v) Before casing and after casing, spray carbendazim and formaldehyde solution and diclorovos separately and repeat after a week. The spot treatment with formaldehyde is given if weed mold is noticed. During cropping, avoid these sprays. For control of flies use light traps. This can be prepared by coating polythene sheet with sticky materials like mustard oil and attached to a 15 W bulb. Keep it on. This is very useful to control flies.
(vi) Removal of stumps and pieces from cropping room should be carried out immediately after harvest. If bacterial blotch/spot is noticed, spray bleaching powder 200 ppm (2 g in 10 liters of water) and reduce R.H. 80%.
(vii) Once the crop is over, spray formaldehyde (2%) before and after removal of spent mushroom substrate (SMS). It should be disposed of from the farm and can be used as manure after 6-month exposure.

References

Agarwal K, Prasad MP, Rindhe G (2013) Genomic discrimination of eleven commercial mushrooms by DNA fingerprinting using RAPD marker. Int Res J Biol Sci 2(10):1–5

Atkins (1979) Research and the mushroom growers. Mushroom Sci 10:7–13

Babu KM, Nair RK (1991) Mushroom cultivation on oil palm factory wastes. Indian Mushrooms. Proceedings of the national symposium on mushrooms, Thiruvananthapuram, pp 104–108

Bahl N (2002) Handbook of mushroom, New Delhi, 4th edn. Vijay Primlani for oxfor & IBH Publishing Co. Pvt. Ltd

Bahukhandi D, Munjal RL (1989) Effect of chemical mutagens on colony development & crop yield of *Pleurotus sajor-caju*. Indian Phytopathol 42:459–462
Bahukhandi D, Sharma RK (2002) Interspecific hybridization in *Pleurotus* species. Indian Phytopathol 55(1):61–66
Bano Z, Rajarathnam S (1982) Studies on cultivation of *Pleurotus sajor-caju*. Mushroom J 115:243–245
Bano Z, Rajarathnam S, Nagarajan N (1979) Some aspects on cultivation of *Pleurotus flabellatus* in India. Mushroom Sci 10(2):597–608
Bano Z, Srivastava HC (1962) Studies in the cultivation of Pleurotus flabellatus on paddy straw. Food Sci 12:363–368
Bano Z, Srivastava HC, Nagarajan N (1976) The cultivation of mushroom (*Pleurotus flabellatus*) on paddy straw packed in polyethylene bags with vents at room temperature. Indian Food Packers 30(4):52–54
Baskaran TL, Sivaprakasam K, Kandaswamy TK (1978) Compact bag method – a new method of increasing the yield of *P. sajor-caju*. Indian J. Mushrooms 4(2):10–12
Bhalerao VK, Raghuwanshi KS, Jadhav AC, Shinde DB, Deokar CD (2017) Genetics and hybrid breeding of *Pleurotus* sp. for biotic and abiotic stress. In: National symposium on mushroom: trends and innovations in mushroom science, Solan, HP, India, pp 351–354
Bhandal MS, Mehta KB (1989) Evaluation and improvement of strains in Agaricus bisporus. Mushroom Sci 12(1):25–35
Chandra S, Ghosh K, Acharya K (2010) Comparative studies on the Indian cultivated *Pleurotus* species by RAPD fingerprinting. Nat Sci 8(7):90–94
Doshi A, Sharma SS, Trivedi A (1993) Strain selection and development in *Calocybe indica* by means of mycelia anastomosis. In: First international conference on mushroom biology and mushroom products, Hong-Kong, p 87 (abs)
Duggar BM (1905) The principles of mushroom growing and mushroom and mushroom spawn making. Bulletin of the United States Bureau of Plant Industries No. 55
Ferguson MA (1902) Preliminary study of the germination of the spores of *Agaricus campestris* and other basidiomycetes. U S Dept Agr Bur Pl Ind Bull 16:1–40
Gaikwad B M (1983) Comparative study of yield potential of different *Pleurotus* spp. Using selected agricultural wastes. MSc Thesis submitted to M.P.K.V. Rahuri
Ghosh N, Chakravarty DK (1991) Studies on evolving new strains of *Pleurotus sajor-caju* by selective dikaryotization. Indian Mushrooms 9:69–72
Gupta VK, Langar PN (1988) Pleurotus Florida for up-grading the nutritive value of wheat straw. Biological Waste 23:57–64
Gupta B, Reddy N, Kotasthane AS (2011) Molecular characterization and mating type analysis of oyster mushroom (*Pleurotus* spp.) using single basidiospores for strain improvement. World J Microbiol Biotechnol 27(1):1–9
Hawksworth DL (1991) The fungal dimension of biodiversity: magnitude, significance and conservation. Mycol Res 95:641–655
Jandaik CL (1974) Artificial cultivation of Pleurotus sajor-caju(Fr.) Singer. Mushroom J 22:405
Jandaik CL, Kapoor JN (1975) Cultural studies on some edible fungi. Indian J Mushrooms 1:22–26
Kaur J (2007) Selection and breeding for the improvement of oyster mushroom, *Pleurotus florida*. MSc thesis, Punjab Agricultural University, Ludhiana, India, pp 55–56
Kaur J, Sodhi HS (2012) Molecular characterization of mutant strains of *Calocybe indica* using RAPD-PCR. Bioscan 7(3):527–532
Khandar RR, Vaishnav MV, Akhari LF, Andhania JH (1991) Effect of various plant substrates on sporophores production of *P. sajor-caju*. Indian Mushrooms. Proceedings of the national symposium on mushrooms, Triruvananthapuram, pp 112–113
Kirk PM, Cannon PF, Minter DW, Stalpers JA (2008) Ainsworth & Bisby's dictionary of the fungi. CAB International. Wallingford, 771pp
Lambert EB (1929) The production of normal sporophores in monosporous cultures ofAgaricus campestris. Mycologia 21:333–335
Matrochot L, Constantin J (1894) Culture d'un champignon lignicole. CR Acad Paris 119:752–753

Miles PG, Chang S-T (1997) Mushroom biology: concise basics and current developments. World Scientific, River Edge, p 197
Mishra RP, Pandey M, Tripathi UK, Singh M (2012) Genetic diversity within the genus *Pleurotus* determined by random amplified polymorphism DNA (RAPD) analysis. Arch Phytopathol Plant Protect 45(7):873–880
Pal J, Thapa CD (1979) Cultivation of Dhingri (*Pleuroius sajor-caju*) made easy. Indian J Mushrooms 5:17–20
Patil BD, Jadhav SW (1989) Studies on cultivation of *Pleurotus sajor-caju* on different substrates. J Maharashtra Agric Univ 14(2):156–158
Sinden JW (1938) New methods of mushroom culture. Ann Rep Veg Grow Assoc Amer 27:181–188
Singh M, Kamal S (2011) Conventional and molecular approaches for breeding button mushroom. In: Proceedings of the 7th international conference on mushroom biology and mushroom products (ICMBMP7), pp 35–42
Sivaprakasam K, Kandaswamy TK (1980) Effect of cultivation methods on sporophore production of *P. sajor-caju* (F.) Singer. Indian J Mushrooms 6:13–15
Thakur K, Bhandal M (1993) Monosporus isolates and their inter-mating in *Pleurotus sapidus* and *Pleurotus sajor-caju*. Mushroom Research 2:41–44
Thilagavathy D, Kumutha Kavally R, Shanmugam S (1991) Study of oyster mushroom cultivation in various substrates. Indian mushrooms. Proceedings of the national symposium on mushroom, Thiruvananthapuram, pp 86–88
Van Griensven JLD (1988) The cultivation of mushrooms. Darlington Mushroom Laboratories Ltd, Rustington, p 846
Wang, Zhang (1987) Experiment of domestication of Lyophyllum ulmarium. Edible Fungi China 10:417–418
Wasson RG (1969) Soma Devine mushroom of immortality. Harcourt Brace Jovanovich, New York

The Developmental History of *Ustilago maydis*: A Saprophytic Yeast, a Mycelial Fungus, Mushroom-Like, and a Smut

2

José Ruiz-Herrera, José L. Cabrera-Ponce, Claudia León-Ramírez, Fernando Pérez-Rodríguez, Mayela Salazar-Chávez, Alejandro Sánchez-Arreguín, and John Vélez-Haro

Abstract

Ustilago maydis is a basidiomycete fungus of the subphylum Ustilaginomycotina. *U. maydis* is a biotrophic phytopathogen that causes common smut in maize (*Zea mays* L) and its ancestor teozintle (*Zea mays* ssp. *parviglumis* and ssp. *mexicana*). The economical importance of *U. maydis* is not as significant as other smuts because it does not cause heavy losses in agriculture but is considered a classic fungal model for studying the mechanism of DNA recombination and mating, other important aspects of fungal development, and the molecular mechanisms of fungal pathogenesis. The life cycle of *U. maydis* involves one saprophytic phase of haploid yeast form (sporidia) and a hyphal virulent dikaryon formed by mating of two sexually compatible sporidia. This invades the host growing in the hyphal form and finally forms tumors full of diploid teliospores that germinate with the formation of a phragmobasidium and four basidiospores. This cycle has been analyzed at the molecular level. Importantly, it was found that *U. maydis* may be pathogenic under axenic conditions for plants unrelated to maize, but does not complete the sexual cycle. The dimorphic yeast-to-hypha transition occurs also in vitro induced by growth with fatty acids or acetate and at acidic pH, developing into multicellular individuals, and unexpectedly forms basidiocarps in vitro completing the sexual cycle with the formation of holobasi-

J. Ruiz-Herrera (✉) · J. L. Cabrera-Ponce · C. León-Ramírez · F. Pérez-Rodríguez
M. Salazar-Chávez · A. Sánchez-Arreguín · J. Vélez-Haro
Departamento de Ingeniería Genética, Unidad Irapuato, Centro de Investigación y de Estudios Avanzados del IPN, Irapuato, Guanajuato, México
e-mail: jruiz@ira.cinvwestav.mx; jose.ruiz@cinvestav.mx

T. Satyanarayana et al. (eds.), *Advancing Frontiers in Mycology & Mycotechnology*,
https://doi.org/10.1007/978-981-13-9349-5_2

dia instead of phragmobasidia, septal pores, and fibulae. Interestingly, laboratory and natural strains of *U. maydis* harbor a bacterial symbiont that fixes N_2.

Keywords
Ustilago maydis · Fungal virulence · Fungal development · Dimorphism · Basidiocarps · Fungal evolution

2.1 Introduction

Ustilaginales comprise approximately 35 genera and more than 1000 species of phytopathogenic fungi (Stoll et al. 2003). Commonly known as "cereal smuts," they are biotrophic pathogens of a large number of economically important crop plants. Among them, *Ustilago maydis* is possibly the most representative species, and the most used model for the analysis of a number of physiological characteristics of fungi. *U. maydis* belongs to the family *Ustilaginaceae* of the order *Ustilaginales* (Kües et al. 2011), and it was the first basidiomycete parasite of plants whose genome was fully sequenced (Begerow et al. 2006). The investigations of *U. maydis* were originally focused on the characterization of the genes involved in the recombination of DNA and in the determination of types of mating (*loci*). Later on the study of the regulation of genes, signaling, and virulence followed (Kämper et al. 2006; Pérez-Nadales et al. 2014), making the fungus an excellent model organism for studying molecular phytopathology and many aspects of the biology of eukaryotic cell (Steinberg 2007).

U. maydis is a biotrophic pathogen specific to maize (*Zea mays* L) and its possible ancestor, the teozintle (*Zea mays* ssp. *parviglumis* and ssp. *mexicana*). In contrast to other *Ustilaginales* that are responsible for major losses of crops of economic importance, *U. maydis* parasitizes only a reduced number of maize plants. Despite this, it produces a loss of the corn harvest of about 2% per year, equivalent to 1 billion US dollars annually; therefore, it is necessary to study this pathogen (Saville et al. 2012). In addition, *U. maydis* is edible and highly appreciated, not only in the Mexican cuisine, but also internationally.

Although *U. maydis* has been subjected to an exhaustive analysis, some hidden characteristics of it still remain to be studied. Among them is the demonstration that the fungus is able to form fruiting bodies (Cabrera-Ponce et al. 2012), and that it harbors a bacterial endosymbiont that endows the fungus the capacity to fix N2 (Ruiz-Herrera et al. 2015). In this chapter, we describe the most important aspects of the physiology, development, pathogenicity, and evolution of this important phytopathogen.

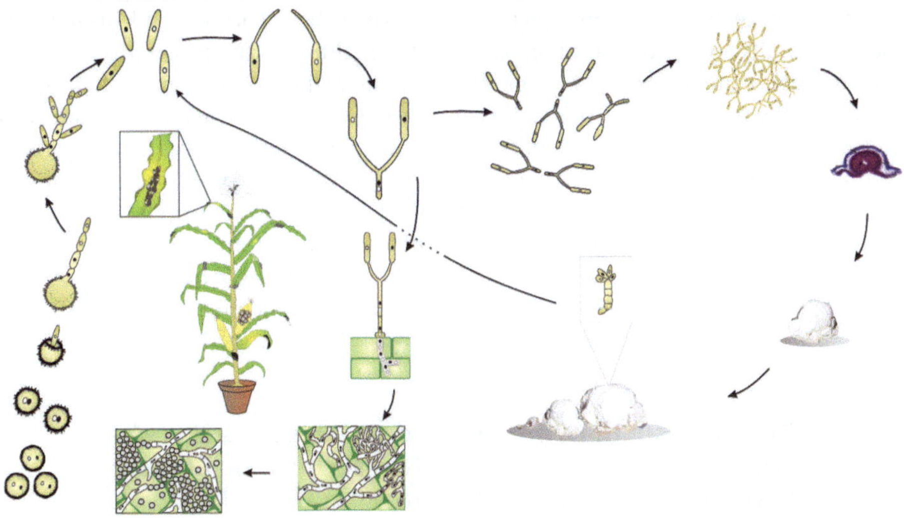

Fig. 2.1 Life cycle of *Ustilago maydis*. The pathogenic cycle as well as the stage of basidiocarp formation is included

2.1.1 Life Cycle of *Ustilago maydis*

Under natural conditions, the life cycle of *U. maydis* involves two separate stages: (i) the saprophytic phase where the fungus grows in a yeast-like form that reproduces by budding and (ii) the pathogenic phase where it grows in a hyphal form. The first phase involves an asexual development, whereas the second one involves the sexual part of the life cycle of the fungus. To these two stages, we must now add the stage that involves the formation of fruiting bodies (Fig. 2.1). This aspect is described below.

It occurs only under specific in vitro conditions (Cabrera-Ponce et al. 2012), and has not yet been observed under natural conditions.

The sexual phase of the cycle is initiated by mating of two sexually compatible yeast cells (sporidia) by means of conjugation tubes that fuse at the tip with formation of a dikaryotic cell that grows in a hyphal form. *U. maydis* is a heterothallic fungus and mating is controlled by two unlinked mating-type *loci*: *a* and *b* (Rowell 1955a, b), the first one with two alleles, more strictly considered to be idiomorphs: *a1* and *a2*, and the second one with about 25 idiomorphs (Puhalla 1970). These idiomorphs contain genes whose function involves a recognition system through the action of pheromones and their opposite recognition receptors (Froeliger and Leong 1991; Bölker et al. 1992). The *a* idiomorphs control the formation of conjugation tubes and cell fusion (Trueheart and Herskowitz 1992), and *b* idiomorphs contain two genes *bE* and *bW* which encode proteins, if they are different; after conjugation form a heterodimer that serves as the master regulator for the mycelial growth and pathogenesis (Gillissen et al. 1992; Kämper et al. 1995). It is important to stress that mating occurs only between sporidia carrying different *a* and *b* idiomorphs.

Initiation of the pathogenic phase occurs when the dikaryotic cell invades the host, normally by formation of an appressorium, although the possibility of infection through wounds also may occur. The invading pathogen then grows in the plant tissues in the hyphal form. At a later stage, karyogamy occurs, and the diploid cells suffer from a series of morphological alterations, round up, and form characteristic spores, named teliospores, which have a thick pigmented cell wall (see Banuett and Herskowitz 1996 for details). Out of the plant, the mature teliospores germinate with the formation of the promycelium (a phragmobasidium) that septates giving rise to four basidiospores. These germinate starting again the life cycle (see reviews by Banuett (1995), Ruiz-Herrera et al. (2000), and León-Ramírez et al. (2014)). The life cycle involving the formation of fruiting bodies is described below.

2.2 Dimorphism of *U. maydis*: Induction of the Mycelial Phase by Different Stimuli, Transcriptomic Control of the Process, and Transition from Unicellular to Pluricellular Form

Dimorphism is the capacity of fungus to grow in hyphal or yeast-like forms depending on the environmental conditions. This process is normally reversible. *U. maydis* is a fungus that displays a dimorphic capacity under different conditions. In nature, the yeast-to-mycelium dimorphic transition of *U. maydis* takes place during the sexual process. It grows in a yeast-like haploid saprophytic form (sporidia) that reproduces by budding. When two sexually compatible sporidia meet, a mating process among them takes place, with the formation of a pathogenic dikaryotic hypha that invades the plant host (see Sect. 2.4, the pathogenic phase section of the chapter).

2.2.1 Stimuli for *U. maydis* Dimorphism

Under in vitro condition, different stimuli can trigger the dimorphic transition in *U. maydis*. Thus, in a medium of neutral pH, the fungus grows in a yeast-like form, whereas in an acid medium with an optimum of pH 3, the fungus grows in a mycelial form (Ruiz-Herrera et al. 1995; see Fig. 2.2) This phenomenon takes place in haploids as well as in diploids and mutants deficient in the mating genes *bE* and *bW*, which encode the heterodimer complex that is a key element for pathogenic development of the fungus (see the corresponding section). This result indicates that this dimorphic phenomenon is independent of the mating process (Ruiz-Herrera et al. 1995). Interestingly, under these conditions, there occurs an active process of H^+ expulsion by the action of a diethylstilbestrol-sensitive, but vanadate-resistant pump. It was suggested that this mechanism was related to the dimorphic phenomenon. A transcriptomic study by a microarray analysis during the yeast-to-mycelium

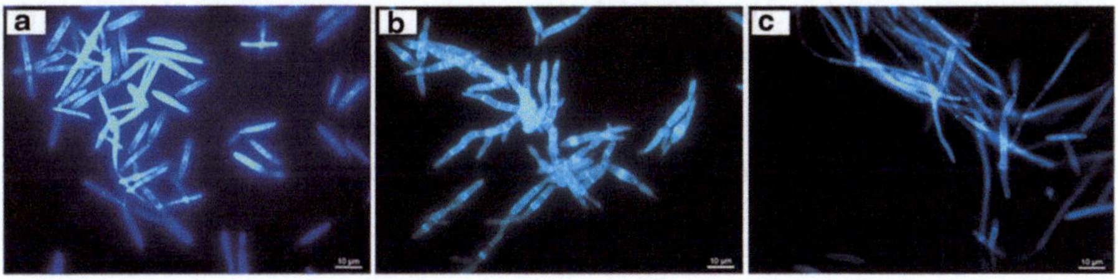

Fig. 2.2 Dimorphic and multicellular transition of *U. maydis* (**a**). Yeast-like cells grown in synthetic glucose medium, pH 7 (**b**). Mycelial cells grown in synthetic glucose medium, pH 3.0 (**c**). Mycelial cells grown in synthetic acetate medium, pH 7.0. All cultures incubated for 48 h. Notice in B and C, the septa of hyphal cells. Magnification bar: 10 μm

dimorphism suggested that 154 genes are specifically involved in the dimorphic phenomenon (Martínez-Soto and Ruiz Herrera 2013). Although a number of the differentially expressed genes were classified as encoding unclassified proteins (59 genes), one of the most representative gene classes identified in this study was metabolism, where interesting genes were found to be involved in nitrogen, fatty acid, and polyamine metabolism. This is significant because of its relationship with dimorphism involving nitrogen starvation (Banuett and Herskowitz 1994a), fatty acids (Klose et al. (2004), and polyamine requirement for dimorphic transition in *U. maydis* (Guevara-Olvera et al. 1997; Valdes-Santiago et al. 2012a, b).

Nitrogen starvation also induces the dimorphic transition of *U. maydis* but occurs only in diploid cells. When diploids are transferred from a rich nitrogen media to minimal media with low nitrogen concentration, they form very thin and long filaments. This process is dependent on *a* and *b* genes (Banuett and Herskowitz 1994a). Another factor that induces the yeast-to-mycelium dimorphic transition in *U. maydis* is the nature of the carbon source. When the fungus grows with a fatty acid as the carbon source, it adopts a mycelial form (Klose et al. 2004). This process is inhibited if glucose is added to the medium together with the fatty acid, indicating that the process is subjected to catabolite repression. Authors suggested that the presence of lipids triggered mycelial growth by a mechanism similar to that existing in the plant surface during the pathogenic phase of *U. maydis*. This was supported by morphological similarities of the in vivo and in vitro filaments that shared the same signaling components (Klose et al. 2004).

More recently we investigated the effect of the probable product of fatty acid metabolism, acetate, used as the sole carbon source for growth of *U. maydis*. It was observed that at neutral or slightly alkaline conditions, *U. maydis* developed in the form of very long and septate hyphae (Fig. 2.2), while at acidic pH, acid inhibited the growth of the fungus.

This result suggests that acetate is the active mycelial-inducing factor of fatty acids (Klose et al. 2004; Salazar-Chávez and Ruiz-Herrera unpublished). In contrast, Kretschmer et al. (2018) studied the effect of acetate as only carbon source and reported that it promotes the formation of reactive species that caused cell death, reduced virulence, also mitochondrial stress, but not mycelium development; the reason for this has not yet been understood.

2.2.2 The Phenomenon of Multicellular Development by *U. maydis*

The observation that incubation of *U. maydis* at acidic pH with glucose or acetate as the carbon source induces unicellular cells of *U. maydis* to grow in the form of septate mycelium (see Fig. 2.2) is interesting. This phenomenon is related to the process of experimental transition from a unicellular to a multicellular form. The phenomenon of transition from a unicellular to a pluricellular stage is important during the evolution of living organisms (Nagy et al. 2018). Fungi have, therefore, been considered suitable models to study the development of multicellularity (Ratcliff et al. 2012).

2.2.3 Signaling Pathways Involved in *U. maydis* Dimorphism

It has been demonstrated that the two principal pathways involved in signal transduction in *U. maydis* are i) the pathway involving the protein kinase dependent on cAMP (PKA) and ii) the pathway involving mitogen-activated protein kinases (MAPK). It was demonstrated that mutants in adenylate cyclase (*uac1*) grow only in the mycelial form (Gold et al. 1994). This result suggested that the PKA pathway was involved in the growth of the yeast form. Later on, a series of genes, designated as *Ubc* (Ustilago Bypass of Cyclase,), has been identified to be involved in the mycelial growth of the fungus (Martínez-Espinoza et al. 2004). Among them was the regulatory subunit of PKA (*ubc1*). In further studies, it was demonstrated that the addition of cAMP prevented the mycelial growth by acidic pH, and that all *ubc* mutants were unable to grow in the mycelial form. These mutants are deficient in several MAPKs (Martínez-Espinoza et al. 2004). Among them, *U. maydis* mutants in *FUZ7*, a putative gene which codes for a serine/threonine tyrosine kinase of the MAP kinase kinase (MAPKK/MEK) family, showed reduced capacity to form conjugation tubes and dikaryotic mycelium in sexual differentiation (Banuett and Herskowitz 1994b). Moreover, different mutants affected in the MAPK pathways were demonstrated to be involved in the dimorphic transition induced by acidic pH (Martínez-Espinosa et al. 1997, 2002). All these observations revealed that the PKA and MAPK pathways are antagonistic; PKA is involved in the growth in yeast-like form, whereas the MAPK pathway is involved in mycelial growth.

2.2.4 The Role of Polyamines in Dimorphism

Polyamines are essential for all living organisms, because they are involved in many cellular processes, especially in proliferation and cellular differentiation. In fungi, the addition of diaminobutanone (DAB), an inhibitor in the biosynthesis of polyamines, in concentrations that do not suppress vegetative growth, arrests specifically several developmental phenomena such as sporulation, spore germination, and dimorphism. Exogenous supplementation of polyamines reverts the inhibitor effect (Calvo-Méndez et al. 1987; Martínez-Pacheco et al. 1988; Ruiz-Herrera 1994; Guevara-Olvera et al. 1993, 1997; Valdés-Santiago et al. 2009; for a review see Valdés-Santiago and Ruiz-Herrera 2015).

The three most important polyamines are putrescine, spermidine, and spermine. *U. maydis* is able to synthesize only putrescine and spermidine. Mutants deficient in ornithine decarboxylase, the first enzyme in the polyamine biosynthetic pathway (Δ*odc*), are unable to synthesize both polyamines, but grow in low concentration of putrescine (0.1 mM), and are unable to perform the yeast-to-mycelium dimorphic transition induced by acidic pH, requiring high concentration of putrescine (5 mM) to grow in the hyphal form (Guevara-Olvera et al. 1997). However, mutants unable to form putrescine by the biosynthetic pathway and by the degradation pathway from spermidine (Δ*spe-sdh*) lose their capacity to grow in the mycelial form, even when supplemented with high concentration of spermidine (0.5 mM) (Valdes-Santiago et al. 2009). These observations make clear the requirement of polyamines for the dimorphic transition of *U. maydis*.

To understand these phenomena, we have obtained the transcriptome of mutants affected in the polyamine biosynthetic pathway, and grown in the presence of high and low concentration of polyamines at an acidic and neutral pH (Pérez-Rodríguez and Ruiz-Herrera unpublished). In our preliminary analysis of data of these microarrays, we identified 2959 regulated genes by putrescine at pH 7.0 (1531 downregulated and 1428 upregulated); 1426 genes regulated by putrescine at pH 3.0; 712 downregulated and 714 upregulated by spermidine at pH 7.0; and only 11 genes regulated by spermidine at pH 3.0 (4 downregulated, 7 upregulated) (unpublished data); analysis of these data is in progress.

As already described during *U. maydis* mating, the formation of conjugation tubes takes place followed by the growth of dikaryotic mycelium. We have observed that polyamines are also involved in these events. Accordingly, we have observed that the addition of a low concentration of polyamines (0.1 mM) reduces significantly these processes in polyamine less mutants, in comparison to mating experiments when high concentrations of polyamines are used (Pérez-Rodríguez and Ruiz-Herrera unpublished).

2.3 Breaking a Rule: Formation of Fruiting Bodies – Morphological Development and Transcription Factors Involved in the Process

According to the characteristics of the members of the phylum Ustilaginomycotina, and in contrast to Agaricomycotina (mushrooms), it might be expected that *U. maydis* is unable to form fruiting bodies (e.g., Alexopoulos 1962). Interestingly, Cabrera-Ponce et al. (2012) showed that under special culture conditions, *U. maydis* had the ability to differentiate and develop fruiting bodies. This occurred when mixtures of sexually compatible *U. maydis* strains were grown on solid media in dual cultures of with embryogenic maize calli. The parameters that control the formation of the fruiting bodies were, besides the presence of the maize embryogenic calli, the addition to the medium of synthetic auxins such as adenine, and most important, dicamba, and an illumination between 10 and 75 mol $m^{-2}s^{-1}$. The fruiting bodies (basidiocarps) developed constantly, reaching large sizes, and when transferred to fresh media, the presence of the calli was unnecessary to form secondary fruiting bodies. Accordingly, once *U. maydis* compromised to form basidiocarps, it acquired the characteristic to live for long periods of time under controlled conditions and to reach considerable macroscopic sizes.

The structure of basidiocarp shows interesting characteristics. In sections, they appear to be formed by three well-delineated layers, the outer one shows an enrichment of not branched skeletal hyphae. This type of hyphae is characteristic of the so-called "woody" mushrooms such as those of the polyporal order (Roy 1972), whose main function is to support the fruiting body. The middle layer is formed by generative hyphae with walls thicker than the ones of the outer layer. The most internal layer is formed by a poorly differentiated tissue where the reproductive structures are accumulated. This layer of mucilaginous consistency that corresponds to the hymenium also brings further surprises. Instead of the usual phragmobasidia characteristics of the germination of the teliospores, non-septate typical holobasidia with their characteristic vessel-like body from which the four basidiospores emerge, each with a nucleus, are formed. In addition the mycelium giving rise to the basidiospores contains a number of clamp connections that are rare in the pathogenic cycle of the fungus and also shows septal pores, not seen normally in the reproductive mycelium. All these data show the extreme plasticity of *U. maydis* that permits its adaptation to the changing conditions of the environment, giving rise to fruiting bodies, which in contrast to normal yeast or mycelial cells can be long lived.

It is important to stress that during this alternative life cycle of *U. maydis*, where fruiting bodies are formed, mating occurs followed by meiosis with the formation of spores according to a Mendelian process. Whereas the basidia are morphologically different to the ones formed in the pathogenic cycle (see above), the basidiospores formed in this process are morphologically undistinguished from the ones formed during the pathogenic cycle, and are as virulent to maize plants as the ones formed in this later process (Cabrera-Ponce et al. (2012).

In order to understand the genetic control of fruiting body formation, we proceed to analyze the regulation of the total genome at different stages of basidiocarp formation. Accordingly, we obtained the transcriptomes of *U. maydis* during the formation of young basidiocarps and during their maturation. A total of 2002 and 1064 genes were found to be regulated, respectively, at each stage. Among them we identified homologs of genes that previously have been found to be regulated during the synthesis of fruiting bodies by typical basidiocarp-forming fungi: *ELN*2-1 CYP502 (Muraguchi and Kamada 2000), *ICH1*, an O-methyltransferase (Muraguchi and Kamada, 1998), *CLP*1 (clampless 1-1, A mating) reported by Inada et al. (2001), and *CFS*1 (cyclopropane fatty acid synthase) described by Liu et al. (2006).

It is known that virulence of *U. maydis* involves PKA and MAPK pathways (Muller et al. 2003; Martínez-Soto and Ruiz-Herrera 2017; see the corresponding section). In addition, they are involved in differentiating processes such as plant invasion (Brefort et al. 2009), mating, and filamentation (Garrido et al. 2004; Brefort et al. 2009). Although little is known about the role of MAPK in the development of fruiting bodies of *Basidiomycota*, in the last 15 years the important role played by these genes in the development of different species like *Lentinula edodes* (Leung et al. 2000), *Coprinopsis cinerea* (Cheng et al. 2013), and *Hypsizygus marmoreus* (Zhang et al. 2015) has been demonstrated.

We, therefore, analyzed to find out whether the same MAPK pathway recognized during the virulent phase of *U. maydis* is involved or not during the process of basidiocarp formation. To this end, we analyzed the capacity to form basidiocarps by two sexually competitive strains with a mutation in the gene *FUZ7* (um01514) that encodes a MAPKK, a member of the MAPK pathway. Our observations suggested that the mixture of these mutant strains is unable to form basidiocarps.

We pointed above that light is an absolute requirement for the development of fruiting bodies (Cabrera-Ponce et al. 2012), and that the quantity and quality of luminous impulse determined the efficiency of the development of the basidiocarps. In other fructifying fungi, the reception of the light stimulus has been previously discussed in relation to different light-sensitive phenomena, specifically the initial steps such as the formation of primordia (Wösten and Wessels 2006). These authors contemplated the role of gene homologs of *WC* involved in blue light reception, and increased expression of genes in the mycelium prior to the formation of fruiting bodies in *L. edodes* and in *H. marmoreus*. Knowing that *U. maydis* contains homologs of these *WC* genes, we have made preliminary analysis of the role of these genes in fruiting body formation. It was observed that in mutants of *WC*1-*a* and *WC*-2, genes of *U. maydis* significantly affected the formation of basidiocarps (Sánchez-Arreguín et al. unpublished). The transcriptional network elaborated from the genes described as transcription factors that are regulated positively in the stages of development of the fruiting bodies shows us the pathway as to how they interact with each other to regulate the genes involved in several pathways of differentiation of the fungus. This is the case of *TEC*1 that self-regulates with *PRO*1b the filamentation path and the formation of the biofilm; *FOXO*3 interacts with various genes for the regulation of pathways related to the cell cycle, morphogenesis, and regulation of oxidative stress. The induction of the transcription factor *TEC*1 could induce the formation of specialized hyphae with structures essential for the development of the formation of fruiting bodies (preliminary observations).

In conclusion, it can be said that the initiation of fruiting body formation requires the regulation of a large number of genes (2002) and that almost half of them are no longer necessary for the maturation stage of the same (1064). With the abovementioned data, we propose an alternative life cycle for the fungus *U. maydis* which includes a stage where after sporidia mating, the promycelium is formed and can follow two ways depending on the existing stimuli: (i) activation of the mechanisms necessary to penetrate the plant tissue and follow the virulent cycle to finally induce the formation of tumors full of teliospores whose germination leads to reinitiate the cycle or (ii) an alternative cycle that leads to the formation of basidiocarps. Unfortunately, whether this alternative cycle occurs in nature or not has not been confirmed.

2.4 The Pathogenic Phase: Natural and Experimental Hosts, Virulence Factors, Effectors, the "Priming" Phenomenon

(a) *U. maydis* virulence. All phytopathogenic fungi have a variety of processes that lead them to recognize their respective hosts, penetrating their physical barriers and leading to overcoming their defenses and injuring substances, finally ending up proliferating within the invaded tissues, multiplying, and dispersing (Martínez-Espinoza et al. 2000). Approximately, of the 30,000 described species of *Basidiomycota* fungi, only 8000 species of rusts and smuts are pathogens of plants, and at least 40 cause diseases in mammals. The genus *Ustilago* spp., *Puccinia* spp., and *Cryptococcus* spp. are the most used models for the study of *Basidiomycota* pathogenesis (Bakkeren et al. 2012). Among these genera, there are species responsible for some of the most important plant diseases in economic terms, including those of cereals. *Ustilaginales* comprise approximately more than 35 genus and more than 1000 species of phytopathogenic fungi (Stoll et al. 2003). Commonly known as "cereal smuts," they are biotrophic pathogens that cause disease in a large number of economically important crop plants. *Ustilago maydis* is a pathogen specific to maize (*Zea mays* L.) and its possible ancestor, teozintle (*Zea mays* ssp. *parviglumis* and ssp. *mexicana*). *U. maydis* is a dimorphic fungus, which grows in the form of haploid yeast-like cells. Its sexual development is initiated by the fusion of two haploid cells (see detailed below). This dikaryon is filamentous and invades the plant cells, establishing a close with the plant host (see reviews by Schirawski et al. 2010; Kämper et al. 2006).

U. maydis presents a pathogenic variability in different maize cultivars with partial resistance. The losses that *U. maydis* causes to the agriculture, although may be significant under some special conditions, are not as devastating as compared with other smut fungi. Infection of corn (*Zea mays* L.), the natural host, has had a great interest in the area of biology and agriculture. Taking into consideration that its natural host, maize, is native to Mexico, it has been concluded that *U. maydis* is native to this country and was taken to southern Europe by the invading Spaniards.

It was not until 1836 that it was identified as a fungus and named as *Ustilago zeae*. However, it was not until 1842 that its correct binomial denomination of *U. maydis* DC.Cda was established, according to the International Standards of Botanical Nomenclature (Christensen 1963).

Until the nineteenth century, it was thought that the tumors formed by the fungus in the maize plant were a physiological alteration of the plant. Between 1893 and 1895, the experimental inoculation of maize with *U. maydis* teliospores was achieved, and in 1927 its sexual cycle was discovered. Studies on its physiology were carried out in the years 1930–1940 of the last century by Christensen and Stackman, and in 1949, Perkins (1949) generated auxotrophic mutants initiating genetic studies of the fungus (see Garcia-Pedrajas et al. 2004; Pérez-Nadales et al. 2014). *U. maydis* virulence has been studied with its natural hosts (maize and teozintle), observing a great variability to the response of the plant resistance according to their variety (Chavan and Smith 2014). In general it has been observed that sweet corn hybrids are the most susceptible varieties to infection (Juárez-Montiel et al. 2011).

Ustilago maydis is a biotrophic pathogen; accordingly, it depends on the living tissue of its host for proliferation and development. Its genome has revealed that it lacks some signatures found in the genomes of aggressive pathogenic fungi. It also may be indicated that the genome of *U. maydis* harbors only a few genes that encode enzymes that hydrolyze the cell wall of plants, such as polysaccharide hydrolases and pectin esterases, which is also in contrast to the genomes of more aggressive pathogenic fungi. These aspects correlate with their biotrophic lifestyle, where damage to the host is minimized to avoid the defense responses of the plant (Kämper et al. 2006).

However, genomic characteristics responsible for the pathogenicity of this organism have been found. Specifically, 12 clusters of genes encode small secreted proteins. Analysis of these genes revealed that most of them exist in groups (clusters) that are commonly regulated in the infected tissue. Mutation of these clusters alters the virulence of *U. maydis* as a complete lack of symptoms of hypervirulence. This infection process of *U. maydis* demonstrated new mechanisms of previously unknown pathogenicity that operate biotrophic fungi (Kämper et al. 2006).

Among the symptoms induced by *U. maydis*, we may cite chlorosis, increased production of anthocyanins, but the development of tumors where huge amounts of black-pigmented diploid teliospores is the most characteristic symptom. Outside of the host, these spores are released and germinate forming a promycelium (phragmobasidium) which in turn produces four haploid sporidia (basidiospores). The life cycle of the fungus, which as was previously described involves a saprophytic stage during which the fungus grows in the form of haploid budding yeasts (sporidia), ends with the formation of a dikaryotic hypha by the mating of two sexually compatible sporidia. This process is controlled by two *loci, a* and *b* (more adequately termed idiomorphs; Banuett et al. 2008). The mating process requires both the cAMP and mitogen-activated protein kinase (MAPK) signaling and the pheromone response to signaling pathways. The idiomorph *a* contains genes that encode the components of the signal transduction pathway, pheromones *mfa1* and *mfa2*, and the corresponding receptors *pra1* and *pra2*, whereas the *b* idiomorph harbors two genes that encode two proteins: bE and bW with characteristic homeodomains (HD). After mating of two sexually compatible strains, there occurs the interaction of the bE protein of one partner with the bW protein of the other forming a bE/bW heterodimer that serves as a master regulator of the pathogenic process (Brachmann et al. 2001; Kahmann and Schirawski, 2007). After pheromone-induced activation, the heterodimer controls the important transcription factor *Prf1*, which in turn induces transcription of a large set of genes (Brefort et al. 2009) and the transcription factor *Biz1* that plays an important role during the colonization of the plant (Flor-Parra et al. 2006).

The dikaryotic hypha invades the host plant through the formation of specialized appresoria in a process that requires the operation of both signal transmission pathways PKA and MAPK and the *Gpa3* heterotrimeric G protein (see Brefort et al. 2009). Mutants in any of these components are avirulent. It should be noted that *U. maydis* does not form haustoria (a form of specialized feeding structures), a characteristic that suggested that exchange of signals and nutrition of the parasite occurs via the biotrophic development of the fungus (Banuett and Herskowitz 1996; Banuett 2002).

The functional classification of the genes induced by *b* genes revealed that several cellular processes, such as cell wall regeneration, cell cycle control, lipid metabolism, DNA replication, and mitosis, are regulated by the bE/bW heterodimer (Kahmann and Schirawski 2007). In fact, the total genes controlled by the heterodimer are 212 upregulated and 135 repressed. Among them, the following have been identified: *dik1* and *dik6*; *egl1* that encodes an endoglucanase; *rep1* and *hum2* encoding a repellent and a hydrophobin, respectively; and *lga2* that encodes a putative mitochondrial protein of unknown function. Another activated gene is *frb52* that encodes a DNA polymerase. Due to the fact that the promoters of almost all these genes lack putative binding sites of bE/bW, the control by the heterodimer is indirect (Basse and Steinberg 2004).

Other genes related to *U. maydis* pathogenesis are encoding effectors (see below), genes related to dimorphism, genes involved in wall biosynthesis, different kinds of cellular transporters and metal-dependent enzymes, and enzymes involved in polyamine metabolism (see Ruiz-Herrera et al. 1995; Banuett 2002; Banuett et al. 2008; Brefort et al. 2009; Valdés-Santiago et al. 2009, 2012a, b; Redkar et al. 2015a, b; Lanver et al. 2018).

The effector proteins secreted by the pathogens are key factors in the infection process (Cristancho et al. 2014). *U. maydis* is the first eukaryotic pathogen where new effectors with functions for pathogenic development were discovered, observing clusters of genes being regulated transcriptionally in the tumor tissue (Kämper et al. 2006). Elimination of some of these clusters reduced the virulence, but surprisingly others led to an increase in virulence (Kämper et al. 2006). It has been suggested that this later observation is related with the biotrophic behavior of *U. maydis* (Brefort et al. 2009). It has been described that the genome of *U. maydis* encodes approximately 550 effector proteins of which the function of a few, e.g., See1, Pep1, Cmu1, Pit2, Tin2, and Cce1, has been elucidated (see Redkar et al. 2015a; Seitner et al. 2018). In summary, transcriptome of *U. maydis* infecting maize has been published permitting the identification of all the possible genes involved in virulence of this fungus (Lanver et al. 2018).

(b) *Alternative hosts.* Many pathogens show a great plasticity of their genes to the change of response imposed by a new host. Its analysis in the pathogenic processes within the host, using comparative genomics, shows that both the gain and loss of genes together with their expansion and contraction are the most likely mechanisms among the different pathosystems (Benevenuto et al. 2018).

We described that *U. maydis* can infect under axenic conditions, not only its natural hosts but also other species of plants, taxonomically unrelated to the natural hosts such as papaya, asparagus, garlic, potato, tobacco, African violets, sorghum, rice, beans (León-Ramírez et al. 2004), and also the model plant, *Arabidopsis thaliana* (Méndez-Morán et al. 2005). The symptoms of the disease were mainly growth of mycelium of fungus on the surface of the leaves, intracellular invasion of the fungus, alteration in root growth, stunting, and in some cases death of the plant. In papaya plants and *Arabidopsis*, tumor-like bodies were formed at the stalks and in beans. Nevertheless no teliospore formation or sexual cycle development was observed.

Analysis of the transcriptome of the infection of *Arabidopsis* by *U. maydis* (Martínez-Soto et al. 2013) revealed that a series of regulated genes coincided with genes known to be involved in maize infection (J. Velez-Haro unpublished data).

More recently, some over-regulated genes identified during maize infection by *U. maydis* have been deleted, and their virulence to *Arabidopsis* and maize has been determined. It was found that the mutants displayed reduced virulence to both *Arabidopsis* and maize (J. Velez-Haro et al. preliminary observations). These observations revealed that *U. maydis* employs similar virulent mechanisms in the infection of maize and *A. thaliana*, suggesting the importance of the analysis of experimental hosts infection to understand the mechanisms of *U. maydis* virulence.

(c) *The "priming" phenomenon.* Plants have an efficient immune system, which allows them to overcome many biological and non-biological threats. This is due to the fact that they develop mechanisms of defense when they have been previously subjected to a specific insult. This phenomenon is known as "priming" and is defined as an adaptive characteristic of induced resistance. This phenomenon starts by the earlier establishment of a previous environmental, biological, or chemical stimulus by which the plants establish a "priming phase" where molecular and biochemical changes occur without a direct activation of the mechanisms of defense. It has been observed that this phase is of long duration in the plant that can be even transmitted to its descendants (Conrath et al. 2006, 2015; Pétriacq et al. 2018).

Recently, this phenomenon was analyzed in the infection of *U. maydis* in corn, comparing the pathogenicity that occurs in plants inoculated under axenic conditions and in non-sterile soil. It was observed that plants grown under axenic conditions were more susceptible to *U. maydis* infection and developed exaggerated symptoms of the disease, mainly formation of tumors and necrosis. Added to this, the production of reactive oxygen species (ROS) as well as cell death and ethylene production was presented in plants grown under axenic conditions. These results suggest not only that the plants are more susceptible to the infection by *U. maydis* under axenic conditions but also that when grown in non-sterile soil, they acquired resistance against the fungus through the above-described mechanism of resistance induced or priming (Martínez-Soto and Ruiz-Herrera 2016).

2.5 Recapitulating the Possible Evolution of *U. maydis*, According to Its Actual Characteristics

Fungi are one of the largest eukaryotic kingdoms, with an estimated 1.5-5 million species. They form a diverse group with a wide variety of life cycles and life style, metabolism, morphogenesis, and ecologies, including mutualism, parasitism, and commensalism with many living organisms. They are found at all temperature zones of the earth with diverse fauna and flora and have a broad and profound impact on the earth ecosystem through their biological activities (Taylor et al. 2004; Petersen 2013).

As described in the previous sections of this review, the characteristics of the species *U. maydis* reveal that it is a fungus with a great plasticity. Thus, to our knowledge it has (i) the capacity to grow as a saprophyte or as a parasite of either a wide spectrum of hosts or to a very specific host; (ii) the characteristic to grow alternatively in hyphal or yeast forms, (iii) the capacity to form characteristic basidiocarps; (iv) the capacity to grow in the form of unicellular or multicellular individuals; and (v) the ability to host bacterial symbionts with the capacity to fix N_2. These characteristics are displayed in response to a number of stimuli and involve the differential expression of selected sets of genes, indicating that these were not lost during the evolution, but only remained silent.

A brief recollection of the evolution of *U. maydis* and its specific host (maize) may help to understand at which stages its present characteristics could have been acquired (Fig. 2.3).

To set the times of appearance of the possible ancestors of *U. maydis*, we may start by indication of the probable appearance and separation of animals, fungi, and plants. It has been considered that separation during evolution of the group of fungi and animals (Opisthokonta) from plants occurred about 1000 mya, and that separation of fungi and animals occurred *ca.* about 800 mya. The appearance of plants has been set to about 500 mya ago. According to these data, before plant appearance, fungal ancestors were either saprobiotic, they preyed on bacteria, or established symbiotic associations with them. It was not until the appearance of plants that phytopathogenic appeared, but Poaceae, the characteristics host of smuts, did not

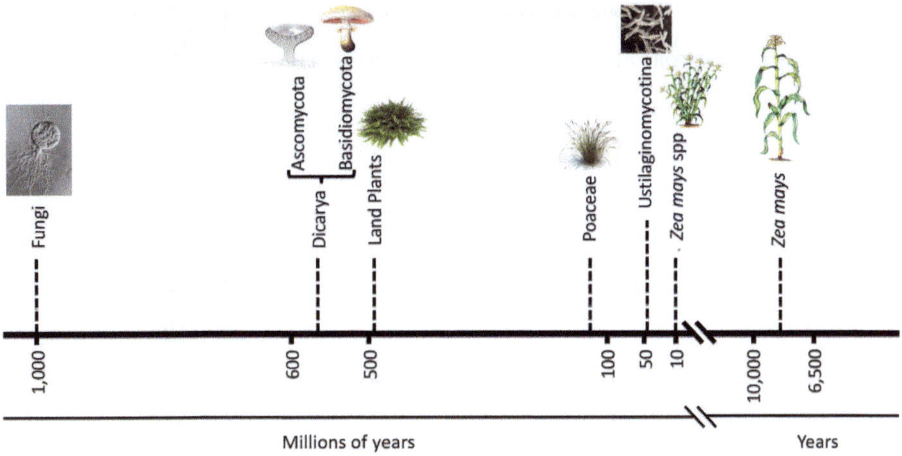

Fig. 2.3 Scheme of evolution of fungi and plants related to the evolution of the *Ustilago maydis* ancestors

appear until about 120 mya. In possible parallel evolution, *Basidiomycota* separated from *Ascomycota ca.* 560 mya, and Ustilaginomycotina separated from the rest of *Basidiomycota* about 49 mya (Oberwinkler 2012).

Regarding the natural hosts of *U. maydis*, teozintle and maize, it has been considered that teozintle is the antecessor of maize. The evolutionary history of maize (*Z. mays L.* ssp. *mays*) has been clarified with genomic data from modern landraces and wild teozintle grasses (Matsuoka et al. 2002; Chia et al. 2012), including archaeological findings that suggest that its domestication occurred between 10,000 and 6500 years ago in southern Mexico (Piperno and Flannery 2001; Piperno et al. 2007).

Taking into consideration the characteristics of the ancestors of *U. maydis* along the evolution, it is suggested that "relics" of them are exhibited by the modern fungus (see Fig. 2.4).

2.6 Concluding Remarks

In the preceding pages, we have tried to describe the main characteristics of *U. maydis* from its life cycle to mating, differentiation, and pathogenesis. It is evident that *U. maydis* has an extreme plasticity that we have tried to exemplify by the title of the chapter. Accordingly, the fungus can thrive as a saprophytic yeast-like unicellular organism but has the property to change its morphology in response to different stimuli, including mating of sexually compatible yeast cells, and grow in a hyphal form. This process may be accompanied by the transformation to a multicellular form. In addition, it can go from a saprophytic mode of life to a pathogenic stage. Thus, it parasitizes members of the *Z. mays* species where it completes its sexual cycle with the advantage to give rise to genetic exchange in its descendants.

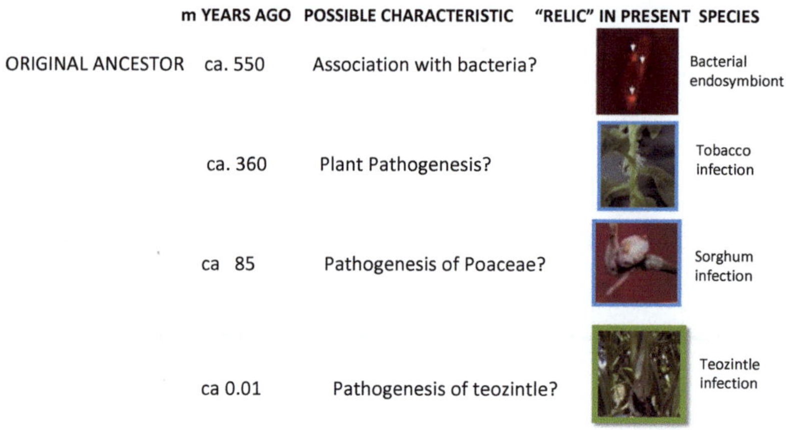

Fig. 2.4 Theoretical scheme of some characteristics acquired by *U. maydis* ancestors whose "relics" remain in the present species

In addition, *U. maydis* has some hidden properties that recall characteristics of its evolution. Accordingly, under special conditions it can form basidiocarps, an uncommon feature to its class, something that may be considered an approaching characteristic with the mushrooms (perhaps an exaggeration, we accept it beforehand, but we want to emphasize that both of them form basidiocarps), and it harbors a bacterium that endows the fungus the capacity to fix N_2. This property may be useful for the growth of the fungus in its hosts, making it independent of the nitrogen compounds of the plant. In addition, this capacity may be an advantage for the biotechnological applications of *U. maydis*, since the fungus can be grown in media devoid of a nitrogen source.

In conclusion, we can state that based on the data presented here, it is clear that *Ustilago maydis* is an ideal subject of study for scholars interested in agronomic and biotechnological aspects.

Acknowledgments The experimental work of the authors described in the text was partially supported by Consejo Nacional de Ciencia y Tecnología (CONACYT) México.

References

Alexopoulos CJ, Mims CW (1962) Introductory mycology. Wiley, New York, p 613

Bakkeren G, Kruzel EK, Hull CM (2012) Sex and virulence in basidiomycete pathogens. In: Sibley LD, Howlett BJ, Heitman J (eds) Evolution of virulence in eukaryotic microbes. Wiley-Blackwell & Sons Inc., Hoboken, pp 437–460

Banuett F (1995) Genetics of *Ustilago maydis* a fungal pathogen that induces tumors in maize. Genetics 29:179–208

Banuett F (2002) Pathogenic development in *Ustilago maydis*. In: Osiewacz HD (ed) Molecular biology of fungal development. CRC Press, Boca Raton, pp 349–398

Banuett F, Herskowitz I (1994a) Morphological transition in the life cycle of *Ustilagomaydis* and their genetic control by the *a* and *bloci*. Exp Mycol 18:247–266

Banuett F, Herskowitz I (1994b) Identification of Fuz7, a *Ustilago maydis* MEK/MAPKK homolog required for *a*-locus-dependent and -independent steps in the fungal life cycle. Genes Dev 8:1367–1378

Banuett F, Herskowitz I (1996) Discrete developmental stages during teliospore formation in the corn smut fungus *Ustilago maydis*. Development 122:2965–2976

Banuett F, Quintanilla RH Jr, Reynaga-Peña CG (2008) The machinery for cell polarity, cell morphogenesis, and the cytoskeleton in the Basidiomycete fungus *Ustilago maydis*-a survey of the genome sequence. Fungal Genet Biol 45:3–14

Basse CW, Steinberg G (2004) *Ustilago maydis*, model system for analysis of the molecular basis of fungal pathogenicity. Mol Plant Pathol 5:83–92

Begerow D, Stoll M, Bauer R (2006) A phylogenetic hypothesis of Ustilaginomycotina based on multiple gene analyses and morphological data. Mycologia 98:906–916

Benevenuto J, Teixeira-Silva NS, Kuramae EE, Croll D, Monteiro-Vitorello CB (2018) Comparative genomics of smut pathogens: insights from orphans and positively selected genes into host specialization. Front Microbiol 9:660–677

Bölker M, Urban M, Kahmann R (1992) The *a* mating type *locus* of *U. maydis* specifies cell signaling components. Cell 68:441–450

Brachmann A, Weinzierl G, Kamper J, Kahmann R (2001) Identification of genes in the bW/bE regulatory cascade in *Ustilago maydis*. Mol Microbiol 42:1047–1063

Brefort T, Doehlemann G, Mendoza-Mendoza A, Reissmann S, Djamei A, Kahmann R (2009) *Ustilago maydis* as a pathogen. Annu Rev Phytopathol 47:423–445

Cabrera-Ponce JL, León-Ramírez CG, Verver-Vargas A, Palma-Tirado L, Ruiz-Herrera J (2012) Metamorphosis of the Basidiomycota*Ustilago maydis*: transformation of yeast-like cells into basidiocarps. Fungal Genet Biol 10:765–771

Calvo-Méndez C, Martínez-Pacheco M, Ruiz-Herrera J (1987) Regulation of ornithine decarboxylase activity in *Mucor bacilliformis* and *Mucor rouxii*. Exp Mycol 11:270–277

Chavan S, Smith SM (2014) A rapid and efficient method for assessing pathogenicity of *Ustilago maydis* on maize and teosinte lines. J Vis Exp 83:1–7

Cheng CK et al (2013) 5′-Serial analysis of gene expression studies reveal a transcriptomic switch during fruiting body development in *Copriniopsis cinerea*. BMC Genomics 14:1–17

Chia JM, Song C, Bradbury PJ, Costich D, de Leon N, Doebley J, Elshire RJ, Gaut B, Geller L, Glaubitz JC et al (2012) Maize HapMap2 identifies extant variation from a genome in flux. Nat Genet 44:803–807

Christenssen JJ (1963) Corn smut caused by *Ustilago maydis*. Am Phytopathol Soc Monogr 2:1–41

Conrath U, Beckers GJ, Flors V et al (2006) Priming: getting ready for battle. Mol Plant-Microbe Interact 19:1062–1071

Conrath U, Beckers GJM, Langenbach CJG, Jaskiewicz MR (2015) Priming for enhanced defense. Annu Rev Phytopathol 53:97–119

Cristancho M, Giraldo W, Botero D, Tabima J, Ortiz D, Peralta A, Gaitán A, Restrepo S, Riaño D (2014) Application of Genome Studies of Coffee Rust. In: Castillo LF, Gustavo-Isaza MC, Andrés Pinzón JM, Corchado R (eds) Advances in Intelligent Systems and Computing volume 232, Advances in Computational Biology Proceedings of the 2nd Colombian Congress on Computational Biology and Bioinformatics (CCBCOL). Springer, New York, pp 133–139

Flor-Parra I, Vranes M, Kämper J, Pérez-Martin J (2006) Biz1, a zinc finger protein required for plant invasion by *Ustilago maydis*, regulates the levels of a mitotic cyclin. Plant Cell 18:2369–2387

Froeliger EH, Leong SA (1991) The *a* mating type genes of *Ustilago maydis* are idiomorphs. Gene 100:113–122

Garcia-Pedrajas MD, Steven JK, Andrews DL, Gold SE (2004) The *Ustilago maydis*–maize interaction. In: Talbot NJ (ed) Plant-pathogen interactions. Blackwell Publishing Ltd, Oxford, pp 166–201

Garrido E, Voss U, Müller P, Castillo-Lluva S, Kahmann R, Pérez-Martín J (2004) The induction of sexual development and virulence in the smut fungus *Ustilago maydis* depends on Crk1, a novel MAPK protein. Genes Dev 18:3117–3130
Gillissen B, Bergmann J, Sandman C, Schoeer M, Bölker M, Kahmann R (1992) A two component regulatory system for self nonself recognition in *U. maydis*. Cell 68:647–657
Gold S, Duncan G, Barrett K, Kronstad J (1994) cAMP regulates morphogenesis in the fungal pathogen *Ustilago maydis*. Genes Dev 8:2805–2816
Guevara-Olvera L, Calvo-Méndez C, Ruiz-Herrera J (1993) The role of polyamine metabolism in dimorphism of *Yarrowia lipolytica*. J Gen Microbiol 139:485–493
Guevara-Olvera L, Xoconostle-Cázares B, Ruiz-Herrera J (1997) Cloning and disruption of the ornithine decarboxylase gene of *Ustilago maydis*: evidence for a role of polyamines in its dimorphic transition. Microbiology 143:2237–2245
Inada K, Morimoto Y, Arina T, Murata Y, Kamada T (2001) The clp1 gene of the mushroom *Coprinus cinereus* is essential for A-regulated sexual development. Genetics 157:133–140
Juárez-Montiel M, Ruiloba de León S, Chávez-Camarillo G, Hernández-Rodríguez C, Villa-Tanaca L (2011) Huitlacoche (corn smut), caused by the phytopathogenic fungus *Ustilago maydis*, as a functional food. Rev Iberoam Micol 28:69–73
Kahmann R, Schirawski J (2007) Mating in the smut fungi: from a to b to the downstream cascades. In: Heitman J, Kronstad JW, Taylor JW, Casselton LA (eds) Sex in fungi: molecular determination and evolutionary implications. ASM Press, Washington, DC, pp 377–387
Kämper J, Reichman M, Romeis T, Bölker M, Kahmann R (1995) Multiallelic recognition: Nonself-dependent dimerization of the bE and bW homeodomain proteins in *Ustilago maydis*. Cell 81:73–83
Kämper J, Kahmann R et al (2006) Insights from the genome of the biotrophic fungal plant pathogen *Ustilago maydis*. Nature 444:97–101
Klose J, Moniz de Sá MM, Kronstad JW (2004) Lipid-induced filamentous growth in *Ustilagomaydis*. Mol Microbiol 52:823–835
Kretschmer M, Lambie S, Croll D, Kronstad JW (2018) Acetate provokes mitochondrial stress and cell death in *Ustilago maydis*. Mol Microbiol 107:488–507
Kües U, Timothy YJ, Heitman J (2011) Mating type in basidiomycetes: unipolar, bipolar, and tetrapolar patterns of sexuality. In: Pöggeler S, Wöstemeyer J (eds) Evolution of fungi and fungal-like organisms, The mycota XIV. Springer, Berlin Heidelberg, pp 97–160
Lanver D, Müller AN, Happe P, Schweizer G, Haas FB, Franitza M, Pellegrin C, Reissmann S, Altmüller J, Rensing SA, Kahmann R (2018) The biotrophic development of *Ustilago maydis* studied by RNA-Seq Analysis. Plant Cell 30:300–323
León-Ramírez CG, Cabrera-Ponce JL, Martínez-Espinoza AD, Herrera-Estrella L, Méndez L, Reynaga-Peña CG, Ruiz-Herrera J (2004) Infection of alternative host plant species by *Ustilago maydis*. New Phytol 164:337–346
León-Ramírez CG, Sánchez-Arreguín JA, Ruiz-Herrera J (2014) *Ustilago maydis*, a delicacy of the aztec cuisine and a model for research. Nat Resour 5:256–267
Leung GSW, Zhang M, Xie WJ, Kwan HS (2000) Identification by RNA fingerprinting of genes differentially expressed during the development of the basidiomycete *Lentinula edodes*. Mol Gen Genet 262:977–990
Liu Y, Srivilai P, Loos S, Aebi M, Kües U (2006) An essential gene for fruiting body initiation in the basidiomycete *Coprinopsis cinerea* is homologous to bacterial cyclopropane fatty acid synthase genes. Genetics 172:873–884
Martínez-Espinosa AD, Ruiz-Herrera J, Gold SE (2000) Las vías de transducción de señales en la patogénesis y la morfogénesis de hongos: los casos de *Ustilago maydis y Magnaporthe grisea*. Rev Mex Fitopatol 18:55–60
Martínez-Espinoza AD, León C, Elizarraraz G, Ruiz-Herrera J (1997) Monomorphic nonpathogenic mutants of *Ustilago maydis*. Phytopathology 87:259–265
Martínez-Espinoza AD, García-Pedrajas MD, Gold SE (2002) The Ustilaginales as plant pests and model systems. Fungal Genet Biol 35:1–20

Martínez-Espinoza AD, Ruiz-Herrera J, León-Ramírez CG, Gold SE (2004) MAP Kinase and cAMP signaling pathways modulate the pH-induced yeast-to-mycelium dimorphic transition in the corn smut fungus *Ustilago maydis*. Curr Microbiol 49:274–281
Martínez-Pacheco M, Rodríguez G, Reyna G, Calvo-Méndez C, Ruiz-Herrera J (1988) Inhibition of the yeast-mycelial transition and the phorogenesis of Mucorales by diaminobutanone. Arch Microbiol 151:10–14
Martínez-Soto D, Ruiz-Herrera J (2013) Transcriptomic analysis of the dimorphic transition of *Ustilago maydis* induced *in vitro* by a change in pH. Fungal Genet Biol 58:116–125
Martínez-Soto D, Ruiz-Herrera J (2016) Induced resistance to *Ustilago maydis* in *Zea mays* inoculated in non-sterile conditions. Int J Biotech Well Ind 5:51–59
Martínez-Soto D, Ruiz-Herrera J (2017) Functional analysis of the MAPK pathways in fungi. Rev Iberoam Micol 34:192–202
Martínez-Soto D, Robledo-Briones AM, Estrada-Luna A, Ruiz-Herrera J (2013) Transcriptomic analysis of *Ustilago maydis* infecting *Arabidopsis* reveals important aspects of the fungus pathogenic mechanisms. Plant Signal Behav 8:1–13
Matsuoka Y, Vigouroux Y, Goodman MM, Sanchez GJ, Buckler E, Doebley J (2002) A single domestication for maize shown by multilocus microsatellite genotyping. Proc Natl Acad Sci U S A 99:6080–6084
Méndez-Morán L, Reynaga-Peña CG, Springer PS, Ruiz-Herrera J (2005) *Ustilagomaydis* infection of the non-natural host *Arabidopsis thaliana*. Phytopathology 95:480–488
Müller P, Weinzierl G, Brachmann A, Feldbrügge M, Kahmann R (2003) Mating and pathogenic development of the Smut fungus *Ustilago maydis* are regulated by one mitogen-activated protein kinase cascade. Eukaryot Cell 2:1187–1199
Muraguchi H, Kamada T (1998) The ich1 gene of the mushroom *Coprinus cinereus* is essential for pileus formation in fruiting. Development 125:3133–3141
Muraguchi H, Kamada T (2000) A mutation in the eln2 gene encoding a cytochrome P450 of *Coprinus cinereus* affects mushroom morphogenesis. Fungal Genet Biol 29:49–59
Nagy LG, Kovács GM, Krizsán K (2018) Complex multicellularity in fungi: evolutionary convergence, single origin, or both? Biol Rev Camb Philos Soc 93:1778–1794
Oberwinkler F (2012) Evolutionary trends in Basidiomycota. Stapfia 96:45–104
Pérez-Nadales E, Almeida-Nogueira MF, Baldin C et al (2014) Fungal model systems and the elucidation of pathogenicity determinants. Fungal Genet Biol 7:42–67
Perkins DD (1949) Biochemical mutants in the smut fungus, *Ustilago maydis*. Genetics 64:607–626
Petersen JH (2013) The kingdom of fungi. Princeton University Press, Princeton, p 282
Pétriacq P, López A, Luna E (2018) Fruit decay to diseases: can induced resistance and priming help? Plants 7:1–16
Piperno DR, Flannery KV (2001) The earliest archaeological maize (*Zea mays* L.) from highland Mexico: new accelerator mass spectrometry dates and their implications. Proc Natl Acad Sci U S A 98:2101–2103
Piperno DR, Moreno JE, Iriarte J, Holst I, Lachniet M, Jones JG, Ranere AJ, Castanzo R (2007) Late Pleistocene and Holocene environmental history of the Iguala Valley, Central Balsas Watershed of Mexico. Proc Natl Acad Sci U S A 104:11874–11881
Puhalla JE (1970) Genetic studies of the *b* incompatibility locus of *Ustilago maydis*. Genet Res 16:229–232
Ratcliff WC, Denison RF, Borrello M, Travisano M (2012) Experimental evolution of multicellularity. Proc Natl Acad Sci U S A 109:1595–1600
Redkar A, Hoser R, Schilling L, Zechmann B, Krzymowska M, Walbot V, Doehlemann G (2015a) A secreted effector protein of *Ustilago maydis* guides maize leaf cells to form tumors. Plant Cell 27:1332–1351
Redkar A, Villajuana-Bonequi M, Doehlemann G (2015b) Conservation of the *Ustilago maydis* effector See1 in related smuts. Plant Signal Behav 10:1–5
Rowell JB (1955a) Functional role of compatibility factors and in vitro test for sexual compatibility with haploid lines of *Ustilago zeae*. Phytopathology 45:370–374

Rowell JB (1955b) Segregation of sex factors in a diploid line of *Ustilago zeae* induced by alpha radiation. Science 121:304–306
Roy A (1972) Some microstructures in relation to Polyporaceae. Mycopathol Appl 48:11–119
Ruiz-Herrera J (1994) Polyamines, DNA methylation, and fungal differentiation. Crit Rev Microbiol 20:143–150
Ruiz-Herrera J, León CG, Guevara-Oivera G, Cárabez-Trejo A (1995) Yeast-mycelial dimorphism of haploid and diploid strains of *Ustiago maydis*. Microbiology 141:695–703
Ruiz-Herrera J, León-Ramírez CG, Martínez-Espinoza AD (2000) Morphogenesis and pathogenesis in *Ustilago maydis*. Recent Res Dev Microbiol 4:585–596
Ruiz-Herrera J, León-Ramírez C, Vera-Nuñez A, Sánchez-Arreguín A, Ruiz-Medrano R, Salgado-Lugo H, Sánchez-Segura L, Peña-Cabriales JJ (2015) A novel intracellular nitrogen-fixing symbiosis made by *Ustilago maydis* and *Bacillus* spp. New Phytol 7:769–777
Saville BJ, Donaldson ME, Doyle CE (2012) Investigating host induced meiosis in a fungal plant pathogen. In: Swan A (ed) Meiosis-molecular mechanisms and cytogenetic diversity. InTech, Croatia, pp 411–450
Schirawski J, Mannhaupt G, Münch K, Brefort T et al (2010) Pathogenicity determinants in smut fungi revealed by genome comparison. Science 330:1546–1548
Seitner D, Uhse S, Gallei M, Djamei A (2018) The core effector Cce1 is required for early infection of maize by *Ustilago maydis*. Mol Plant Pathol 19(10): 2277–2287
Steinberg G (2007) On the move: endosomes in fungal growth and pathogenicity. Nat Rev Microbiol 5:309–316
Stoll M, Piepenbring M, Begerow D, Oberwinkler F (2003) Molecular phylogeny of *Ustilago* and *Sporisorium* species (Basidiomycota, Ustilaginales) based on internal transcribed spacer (ITS) sequences. Can J Bot 81:976–984
Taylor JW, Spatafora J, O'Donnell K, Lutzoni F, Hibbet DS, Geiser D, Bruns TD, Blackwell M (2004) The fungi. In: Cracraft J, Donoghue MJ (eds) Assembling the tree of life. Oxford University Press, New York, pp 171–194
Trueheart J, Herskowitz I (1992) The *a locus* governs cytoduction in *Ustilago maydis*. J Bacteriol 174:7831–7833
Valdés-Santiago L, Ruiz-Herrera J (2015) Polyamines in fungi. Their distribution, metabolism, and role in cell differentiation and morphogenesis. CRC Press, Boca Raton, p 202
Valdés-Santiago L, Cervantes-Chávez J, Ruiz-Herrera J (2009) *Ustilago maydis* spermidine synthase is encoded by a chimeric gene, required for morphogenesis, and indispensable for survival in the host. FEMS Yeast Res 9:923–935
Valdés-Santiago L, Cervantes-Chávez JA, León-Ramírez CG, Ruiz-Herrera J (2012a) Polyamine metabolism in fungi with emphasis on phytopathogenic species. J Amino Acids 2012:1–13
Valdés-Santiago L, Cervantes-Chávez JA, Winkler R, León-Ramírez CG, Ruiz-Herrera J (2012b) Phenotypic comparison of *samdc* and *spe* mutants reveals complex relationships of polyamine metabolism in *Ustilago maydis*. Microbiology 158:674–684
Wösten HAB, Wessels JGH (2006) The emergence of fruiting bodies in Basidiomycetes. In: Fisher K (ed) The mycota I. Growth, differentiation and sexuality. Springer, Berlin/Heidelberg, pp 393–414
Zhang J et al (2015) Transcriptome analysis and its applications in identifying genes associated with fruiting body development in basidiomycete *Hypsizygus marmoreus*. PLoS One 10:1–21

Biochemical and Molecular Aspects of Dimorphism in Fungi

3

Ejaj K. Pathan, Vandana Ghormade, Redeemson Panmei, and Mukund V. Deshpande

Abstract

Most of the eukaryotic differentiation processes are unidirectional. However, fungi have the ability to grow reversibly as unicellular yeast (Y) or as filamentous hypha (H) in response to the specific strain-dependent environmental stimuli. Such a phenomenon known as "dimorphism" is not limited to a specific class of fungi. Most of the plant, human, and insect pathogenic fungi show Y-H and reversible morphogenesis, associated with their saprophytic to pathogenic change, for survival and proliferation in the host. In this chapter, we have described the factors stimulating dimorphism, the signal transduction pathways induced by these stimuli, changes in the gene/protein expression patterns due to a cascade of these signals, and, finally, translation of this genotypic effect into phenotypic change, *i.e.*, the morphological outcome. The process of fungal differentiation and formation of tumor cells follow the same regulatory series of events, involving cAMP, MAP, and RAS kinase cascades. Therefore, the molecules inhibiting Y-H transition in fungi can be explored for their anticancer potential.

Keywords

Antifungal · Dimorphic stimuli · Dimorphism · Hyphae · Signaling · Yeast cells

E. K. Pathan · R. Panmei
Biochemical Sciences Division, CSIR-National Chemical Laboratory, Pune, Maharashtra, India

V. Ghormade
Nanobioscience, Agharkar Research Institute, Pune, Maharashtra, India

M. V. Deshpande (✉)
Division of Biological Sciences, CSIR-National Chemical Laboratory, Pune, Maharashtra, India

T. Satyanarayana et al. (eds.), *Advancing Frontiers in Mycology & Mycotechnology*, https://doi.org/10.1007/978-981-13-9349-5_3

3.1 Introduction

Fungi (yeasts and filamentous forms) are the second most abundant group of species on earth with 99,000 known species, and the estimated number can go beyond 5.1 million species (Blackwell 2011). Fungi show enormous diversity in size and shape due to various processes of differentiation. Fungi not only provide a good model system to study the biochemical and molecular processes common to all forms of eukaryotic life but can yield highly relevant information to understand their biological features. Notwithstanding the diversity, many fungi have a common morphogenetic feature, i.e., dimorphism. In fungi, dimorphism is specifically referring to the ability to shift from unicellular yeast to filamentous form and vice versa. The phenomenon is reversible and dependent on environmental signals (Gow et al. 2012; Gow 1995). For instance, hyphal growth is triggered in a human pathogen *Candida albicans*, in the presence of serum (Feng et al. 1999), neutral pH (Ramon et al. 1999), and polyamines (Herrero et al. 1999).

The following sections describe the phenomenon of dimorphism and its regulation with respect to the stimuli inducing the dimorphism, the signal transduction pathways modulated by these stimuli, and the biochemical and molecular process activated by these signaling cascades, which finally leads to Y-H morphogenesis.

3.2 Dimorphism

Dimorphic fungi can grow in two alternate forms (unicellular yeast, Y, or filamentous hyphae, H) by employing two distinct cell wall deposition patterns, regulated polarized for hypha and regulated non-polarized for yeast. H-form of dimorphic fungi shows morphological transition into Y-form either by lateral or by terminal budding and sometimes by arthrospore formation. The Y-H transition initiates with the germ tube formation. However, in the case of *Mucor rouxii* and *Histoplasma capsulatum*, the H-form is comprised of true branching of hyphae, whereas *Saccharomyces cerevisiae* exhibits formation of chains of elongated Y-cells, called pseudohyphae (Gimeno et al. 1992). Hyphae or pseudohyphae may branch or develop lateral buds at the junctions between the daughter cells (sometimes called blastospores). The dimorphic fungus *Wangiella dermatitidis* displays different cell types such as Y-cells, hyphae, or pseudohyphae and thick-walled segmented sclerotic bodies (Geis and Jacobs 1985).

Dimorphic change in fungi can be triggered by different environmental conditions. Biophysical changes such as temperature, pH, and oxygen; nutritional parameters like presence of glucose, nitrogen source, and metal ions; and complex nutritional components, like serum, corn steep liquor, etc., affect the morphological changes in different fungi. Dimorphism has been reported and extensively studied in fungi of different taxonomic groups, *viz.*, *Zygomycetes* like *Benjaminiella* (Ghormade et al. 2012; Khale 1990), *Mucor* (Bartnicki-Garcia and Nickerson 1962a, b; Orlowaski 1991), and *Mycotypha* (Schulz et al. 1974), *Ascomycetes* such as *Candida* (Gow et al. 2012; Gow 1995) and *Yarrowia* (Palande et al. 2014, Zinjarde et al. 1998), and *Basidiomycetes* like *Ophiostoma* and *Ustilago* (Brunton and Gadd 1991).

Dimorphic behavior was thought to be restricted to the pathogens, which need to change their morphology to survive and proliferate in the host; however, few non-pathogenic fungi were also suggested as models to understand the biochemical basis of Y-H reversible transition and its possible applications in healthcare and other fields (Chitnis et al. 2002; Ghormade et al. 2012; Guevara-Olvera et al. 1993; Khale et al. 1992; Pathan et al. 2017).

3.3 Dimorphism-Triggering Factors

Dimorphic change in fungi can be induced by different environmental conditions, including biophysical stimuli such as pH, temperature, and oxygen and nutritional parameters such as carbon source, nitrogen source, metal ions, etc. In other words, all the organisms do not respond to different incubation conditions in a similar way (Deshpande 1998). Different physiological and nutritional factors triggering Y-H and reversible morphological changes in fungi are summarized below.

3.3.1 Temperature

Thermal dimorphism was reported in pathogenic fungi such as *Blastomyces dermatitidis*, *C. albicans*, *Coccidioides immitis*, *Histoplasma capsulatum*, *Paracoccidioides brasiliensis*, and *Sporothrix schenckii* as well as in saprophytes such as *B. poitrasii*. In general, H-form is favored at 25 °C and Y-form at 37 °C. For instance, as reported in literature, *H. capsulatum* grows in the H-form (saprophytic) at 28 °C and in the Y-form (pathogenic) at 37 °C. In *B. dermatitidis* (Cole and Sun 1985), *C. immitis* (Domer 1985), *P. brasiliensis* (San Blas and San Blas 1985), and *S. schenckii* (Travassos 1985), the increase in the temperature led to prevailing the pathogenic Y-form growth. In the case of *C. albicans*, the Y- and H-forms were induced by temperature in combination with other factors such as blood glucose and serum (Shepherd et al. 1980). *B. poitrasii* grows in the Y-form at 37 °C and as hyphae at 28 °C. However, its yeast monomorphic mutant exhibited Y-form even under hypha-favoring conditions (Khale et al. 1990).

3.3.2 Oxygen

Most of the fungi grow in a fermentative mode in a unicellular form under partial anaerobic condition (Bartnicki-Garcia 1963, Sypherd et al. 1978). For instance, under aerobic conditions, *M. rouxii* (Bartnicki-Garcia and Nickerson 1962a), *Mycotypha microspora*, and *M. africana* (Schulz et al. 1974) exhibited hyphal growth, while the Y-form was prevalent under partial anaerobiosis and/or presence of nitrogen or carbon dioxide environment (Bartnicki-Garcia and Nickerson 1962a). In *M. rouxii*, more than one factor was found to trigger the dimorphism, *viz.*, anaerobic incubation (in the presence of CO_2) always showed yeast-like development; however, depending on the nitrogen source (glycine, serine, and threonine), variable proportions of filaments were formed (Bartnicki-Garcia 1963). Unlike the above-mentioned fungi, a marine isolate of *Y. lipolytica* var. *indica* was reported to show hyphae formation under partial anaerobiosis (Palande et al. 2014; Zinjarde et al. 1998).

3.3.3 pH

In *C. albicans*, germ tube formation was prevalent at pH range 6–8, whereas acidic medium favored the Y-form (Stewart et al. 1988). It was further suggested that germ tube formation was accompanied by cytoplasmic alkalinization (Kaur et al. 1988; Stewart et al. 1988). In *B. poitrasii*, the change in pH from acidic to alkaline of the growth medium triggered Y-H transition (Ghormade and Deshpande 2000). Interestingly, in *Mycotypha*, the Y-form prevailed at the pH range of 4.5 to 7.5, while it grew in a H-form, below or above this range (Schulz et al. 1974). Interestingly, acidic pH triggered Y-H transition in *U. maydis* haploid cells (Martinez-Espinoza et al. 2004).

3.3.4 Carbon and Nitrogen Sources

Usually the presence of glucose and complex nitrogen source in the medium favors yeast morphology, while no or low glucose and inorganic nitrogen sources increase hyphae formation. Indeed, other dimorphism-triggering factors such as temperature, pH, presence of carbon dioxide or nitrogen, etc. override the above-mentioned effects on the morphological outcome, if they share common signaling pathways (Ghormade et al. 2005). In *B. poitrasii*, hyphae formation in a defined medium was induced in the presence of sugars such as lactose, mannose, sucrose, and xylose, whereas yeast growth was seen in a complex nitrogen medium with lactose and higher concentration of sucrose (Khale et al. 1990). Furthermore, low carbon/nitrogen (C:N) ratio generally favored hyphal morphology, while higher C:N ratio and the presence of organic nitrogen supported yeast morphology. In the presence of amino sugar such as *N*-acetylglucosamine (GlcNAc), the H-form growth was prominent in *C. albicans* (Mattia et al. 1982). Using organic nitrogen sources and GlcNAc, *Y. lipolytica* var. *indica* grew as H-cells, and its proportion increased under semi-anaerobic incubation conditions (Palande et al. 2014). Earlier, Ruiz-Herrera and Sentrandreu (2002) in another *Y. lipolytica* strain observed the triggering effect of increased concentration of ammonium sulfate on hyphal development, while glutamate and glutamine favored the Y-form. In the case of *U. maydis*, H-form growth was supported by NH_4NO_3, while Y-form growth was supported by KNO_3 (Ruiz-Herrera et al. 1995).

3.3.5 Metal Ions

Metal ions regulate different enzyme activities that may affect the cellular morphology. For instance, in the presence of zinc, the germ tube formation was induced in *B. poitrasii* (Doiphode 2007), *C. albicans* (Sabie and Gadd 1990), *H. capsulatum* (Pine and Peacock 1958), and *M. rouxii* (Bartnicki-Garcia and Nickerson 1962b). Interestingly, Chacko et al. (1996) observed that different commercially available peptones contained different concentrations of Mg^{++}, Zn^{++}, and inorganic phosphate. Doiphode (2007) observed that it was necessary to use specific brand to obtain maximum Y-H transition in *B. poitrasii*, which was attributed to the presence of Zn^{++}. In *Aspergillus parasiticus*, Garrison and Boyd (1974) reported that concentration of Mn^{++} in the medium was critical to obtain specific morphological outcome. For instance, for hyphal form, higher concentration of Mn^{++} was necessary. In *C. albicans*, exogenous hemin (the oxidized form of heme, containing Fe^{++}) together with a temperature shift (37 °C) was found to induce germ tube formation (Casanova et al. 1997).

3.3.6 Other Factors

The sulfhydryl-containing compounds like cysteine triggered the hyphae formation in *H. capsulatum* (Maresca and Kobayashi 1989, 2000). In *C. albicans*, serum and its derivatives were found to induce germ tube formation (Barlow et al. 1974). However, the exact component in the serum triggering Y-H transition is still unclear. In *C. tropicalis*, supplementation of ethanol in growth medium promoted hyphae formation, whereas simultaneous addition of myo-inositol prevented the morphological change (Tani et al. 1979). A similar observation was reported in the case of *B. poitrasii* and its yeast form monomorphic mutant (Khale et al. 1990).

3.3.7 Cross Talk Between Different Dimorphic Stimuli

Different environmental factors which operate through common signaling pathways lead to cross talk and possible overriding effects. In the case of Y-H change in *M. rouxii*, decreased glucose concentration favored hyphal growth in the presence of CO_2, while at higher glucose concentration, Y-form growth was prevalent even in an oxygen environment, indicating that nutritional stress overrides the effect of aerobic/anaerobic environment on the morphological outcome (Bartnicki-Garcia 1963). In *C. albicans*, acidic pH favored the Y-form; however, high concentration of serum in the medium overrides the effect of pH and induces the H-form at acidic pH (Odds 1988). Szabo and Stofanikova (2002) observed the Y-form growth of *Y. lipolytica* in an acidic environment and that neutral pH enhanced the hyphal growth in the presence of organic nitrogen sources. They further suggested that pH did not directly affect the Y-H morphogenesis in *Y. lipolytica* via pH-dependent regulation of gene expression, but rather interfered with regulation of organic nitrogen sources. *B. poitrasii* grew as Y-cells in the presence of higher glucose concentration, higher temperature, and acidic pH; however, temperature has an overriding effect over glucose as both of them share a common signaling pathway and temperature signal is upstream of glucose signal (Ghormade et al. 2012, Ghormade et al. 2005, Khale et al. 1990).

3.4 Signal Transduction Pathways Induced by Dimorphic Stimuli

Ca^{++}, cAMP, cGMP, and inositol lipids (second messengers) carry signals from various environmental stimuli such as temperature, pH, metal ions, nitrogen source, glucose, and other carbon sources, and are translated into specific intracellular responses important for growth, metabolism, and differentiation (Bahn et al. 2007, Gadd 1995, Shapiro et al. 2007). Dimorphic, pathogenic as well as saprophytic, fungi utilize signal transduction pathways to rapidly adapt and respond to changing environmental conditions. Most of these pathways are mediated by G-protein-coupled receptors (GPCRs), which have a role in modulation of the developmental and genetic networks in fungi (Bahn et al. 2007). GPCRs contain a total of nine domains including extracellular N-terminal, cytoplasmic C-terminal, and seven transmembrane helices (Ying et al. 2013). Several dozens of GPCR family have been found, and ~100 GPCR genes are reported from filamentous fungi (Li et al. 2007; Maidan et al. 2005; Xue et al. 2006); however, only a handful of GPCRs were functionally characterized from yeast (Dohlman and Thorner 2001). Despite the diversity in sequences and functions, all GPCRs act via one of the two signaling cascades, *viz.*, mitogen-activated protein kinase (MAPK) or cyclic AMP-dependent protein kinase (cAMP-PKA), both of which are downstream targets activated by receptors induced by environmental stimuli (Li et al. 2007; Ying et al. 2013). Cross talk also exists between these pathways, making them a signaling network that regulates fungal morphogenesis. Other than these, the calcium-calcineurin-mediated signal transduction pathway has also been identified to be involved in fungal growth and differentiation (Bahn et al. 2007; Chung et al. 2001; Lorenz et al. 2000; Shapiro et al. 2007). The three major signal transduction pathways involved in Y-H transition are summarized below.

3.4.1 Cyclic Adenosine Monophosphate-Dependent Protein Kinase (cAMP-PKA) Pathway

cAMP is a global second messenger controlling the cAMP-dependent protein kinase (PKA). cAMP plays a major role in a variety of fungal morphological processes, such as conidiation, dimorphism, phototropism, hyphal branching, spore germination, etc. (Bahn et al. 2007; Robson et al. 1991). For instance, in the filamentous fungus *Neurospora crassa*, cAMP and its lipophilic derivative, dibutyryl-cAMP (dbcAMP), regulate hyphal growth and conidiation (Bruno et al. 1996; Terenzi et al. 1976). cAMP via PKA also regulates cell growth in *Blastocladiella emersonii* (de Oliveira et al. 1997; Medoff et al. 1987) and sexual development in fission yeast *Schizosaccharomyces pombe* (Maeda et al. 1994).

cAMP also plays an important role in producing a yeast-like morphology in several species of *Mucor*. In other words, Y-form cells displayed high levels of endogenous cAMP, while the exogenous addition of cAMP also favored H-Y transition (Orlowski 1991). However, the H-form of *H. capsulatum* contained about five times higher level of cAMP than the Y-form (Maresca and Kobayashi 1989). The morphogenesis and virulence in several dimorphic fungi, *viz., C. albicans* (Cho et al. 1992; Niimi 1996), *H. capsulatum* (Medoff et al. 1987), *Mucor* spp. (Orlowski 1991), *O. ulmi* (Brunton and Gadd 1989), *U. maydis* (Gold et al. 1994), and *U. hordei* (Litcher and Mills 1997), were found to be regulated by cAMP via PKA signaling. cAMP also plays an important role in the appressoria formation of plant pathogen *Magnaporthe grisea* (Lee and Dean 1993), in the pseudohyphae formation of *S. cerevisiae* (Gimeno et al. 1992; Lorenz and Heitman 1997), and also in the morphogenesis of human pathogen *Cryptococcus neoformans* (Kronstad et al. 1998). In most of these fungi, cAMP signaling was directly associated with their virulence (Borges-Walmsley and Walmsley 2000).

In the industrially important filamentous fungus *Aspergillus niger*, protein phosphorylation via cAMP-PKA pathway regulates hyphal growth and conidiospore formation (Saudohar et al. 2002). In *B. poitrasii*, regulation of Y-H transition by cAMP-dependent phosphorylation of NAD- and NADP-dependent glutamate dehydrogenase was reported (Joshi et al. 2013; Khale-Kumar and Deshpande 1993). Further, it was also suggested that cAMP could exert its effect by activating membrane receptors or tubulin polymerization or regulating the activity of cAMP phosphodiesterase other than the PKA pathway.

3.4.2 Calcium Calmodulin-Calcineurin Pathway

External stimuli (like environment and nutrition) can be translated into specific intracellular effects through calcium. Ca^{++} plays an important role as a second messenger in growth and differentiation of several yeasts and filamentous fungi (Shapiro et al. 2007, Muthukumar et al. 1987). The intracellular Ca^{++} signaling is modulated by calmodulin (CaM), which regulates gene expression and also acts on several metabolic pathways, finally leading to morphological transition in fungi. The Ca-CaM pathway is regulated via calcineurin through phosphorylation and dephosphorylation of the transcription factors and enzymes including adenylate cyclase, NAD^+ kinase, phospholipase A2, phosphodiesterase, and protein kinase (Anraku et al. 1991).

The involvement of Ca-CaM-calcineurin signaling pathway in regulating the growth and morphogenesis in *A. fumigatus*, *C. albicans*, *C. neoformans*, *H. capsulatum*, *P. brasiliensis*, and others was reported (Davis et al. 2013; de Carvalho et al. 2003; Shapiro et al. 2007). The activation of this pathway in *C. albicans* was initiated by the intracellular accumulation of Ca-CaM ions (mediated by *Cch1-Mid1* channel), which led to the activation of calcineurin pathway. Further, interaction of heat shock protein Hsp90 (a molecular chaperon) with the catalytic domain of calcineurin resulted in dephosphorylation of the transcription factor *Crz1* leading to hyphae formation as well as improved antifungal tolerance in *C. albicans* (Cruz et al. 2002). In *C. neoformans* and *A. fumigatus*, homologs of this signaling pathway (identified based on sequence homology) were reported (Görlach et al. 2000; Shapiro et al. 2007; Steinbach et al. 2007). In *A. fumigatus*, a null mutant of the calcineurin catalytic subunit was reported to be defective in hyphae formation (Steinbach et al. 2006). Furthermore, inhibition of thermotolerance associated with Ca-CaM kinase 1 inhibited the yeast cell cycle in *S. schenckii* (Rodriguez-Caban et al. 2011). The Ca-CAM interactions were also reported to play an important role in the H-form growth of *C. ulmi*, and their absence led to Y-form growth (Muthukumar and Nickerson 1984). Studies with calcium calmodulin-specific inhibitors like trifluoperazine demonstrated that Ca-CaM interaction was essential for Y-H transition in human pathogen *C. albicans* (Holmes et al. 1991; Paranjpe et al. 1990; Sabie and Gadd 1990) and *S. schenckii* (Rivera-Rodriguez and Rodriguez-del Valle 1992). Similarly, calmidazolium, trifluoperazine (TFP), and W7, inhibitors of Ca-CaM signaling pathway, also inhibited the H-Y transition in *P. brasiliensis* (de Carvalho et al. 2003).

3.4.3 Mitogen-Activated Protein Kinase (MAPK) Cascade

The MAPK signaling plays an important role in a variety of fungi including *C. albicans*, *S. cerevisiae*, *U. maydis*, *C. neoformans*, *S. pombe*, and others during morphogenesis and pathogenicity (Andrews et al. 2000; Biswas et al. 2007; Cullen et al. 2004; Dean 1997; Lengeler et al. 2000; Martínez-Espinoza et al. 2004; Mayorga and Gold 1999; Wang and Heitman 1999). The MAPK cascade in dimorphic human pathogen *C. albicans* includes *Cst20* (homolog of p21-activated kinase PAK), *Hst7* (homolog of MAPK kinase [MAPKK] Ste7), *Cek1* (homologous to MAPK), and the downstream transcription factor *Cph1*, a homolog of the *S. cerevisiae* transcription factor *Ste12*.

S. cerevisiae contains at least five different MAP kinase pathways (Lengeler et al. 2000). *Ste12* is a transcription factor involved in pheromone-responsive MAPK cascade in *S. cerevisiae*. The co-operative binding of filamentation-specific factor *Tec1* with *Ste12* induced pseudohyphae (Chou et al. 2006). The *C. albicans Cph1*, a *Ste12* homolog, complemented the defect in pseudohyphae formation in diploid cells of *S. cerevisiae* (Liu et al. 1994; Malathi et al. 1994). Further, mutation in any of the genes involved in MAPK cascade or in *Cph1* resulted in defect in hyphae formation in *C. albicans* (Csank et al. 1998; Köhler and Fink 1996). In fission yeast *S. pombe*, *Mam2* (pheromone receptor) senses the P-factor, whereas M-factor was sensed by *Map3*, resulting in activation of MAPK cascade and mating (Seike et al. 2013). In *C. neoformans*, MAPK cascade was triggered by pheromone receptors

Ste3 and *Ste3a*, leading to mating and morphogenesis (Wang and Heitman 1999). The Rho-GTPase gene *Cdc42p* codes for mitogen-activated protein kinase (MAPK) in human pathogenic *P. brasiliensis*, and its regulation was found to be important for H-Y transition (Fernandes et al. 2008). By disrupting the *Ste11*, a mitogen-activating protein kinase, Cervantes-Chávez and Ruiz-Herrera (2006) demonstrated the necessity of the MAP kinase pathway for morphogenesis in ascomycete *Y. lipolytica*. The rice blast fungus *M. grisea* produces a specialized structure called appressoria to invade the host plant. The appressoria formation and invasive growth of *M. grisea* were regulated by MAP kinase-1 (*PMK1*) (Zhao et al. 2005; Xu and Hamer 1996). Zhao et al. (2005) further reported the involvement of *MST7* and *MST11* (homologs of MAP kinases *STE7* and *STE11* of yeast) in the virulence and appressoria formation of *M. grisea*. The *mst7* and *mst11* null mutants were non-pathogenic and also could not form the appressoria. The detailed understanding of these signaling pathways suggested that though different fungi respond differently to environmental stimuli, they follow the closely linked signal transduction pathways that affect the molecular and biochemical changes, which in turn regulates the Y-H morphogenesis.

3.5 Molecular Switches Regulating Y-H Dimorphism: Morphogene to Morphogenesis

In various dimorphic fungi, genes associated with different cellular functions have been extensively studied (Table 3.1). The dimorphic character of *M. circinelloides* was useful to understand the role of glucoamylase gene, important in starch degradation (Houghton-Larsen and Pedersen 2003). The extracellular enzyme secretion in the hyphal form was associated with the foraging nature for survival. The glucoamylase (*GLAM*) gene was repressed by glucose that commonly promoted the yeast form (Houghton-Larsen and Pedersen 2003). Dimorphism in *Mucor* was mainly affected by glucose that transduced its effects through cAMP-dependent protein kinase A (PKA). Binding of cAMP to the regulatory subunits (PKAR) resulted in the release of catalytic subunits (PKAC), triggering a kinase cascade affecting morphogenesis. Wolff et al. (2002) cloned and characterized *pkaR* and *pkaC* encoding regulatory and catalytic subunits of the cAMP-dependent PKA of *M. circinelloides*. The expression levels of both *pkaR* and *pkaC* were significantly higher in anaerobically grown yeast cells than the aerobically grown hyphae. However, there was a two-fold increase in expression of *pkaR* during the Y-H shift. Further, over-expression of *pkaR* resulted in multi-branched colony phenotype in *M. circinelloides* (Wolff et al. 2002). Recently, Ocampo et al. (2012) showed that protein kinase A (PKA) regulatory subunit (R) isoforms (*PKAR1*, *PKAR2*, *PKAR3*, and *PKAR4*) regulate the growth and morphogenesis in *M. circinelloides*. Further, construction of null mutants demonstrated the specific role of each isoform in growth and differentiation. They suggested that *pkaR4* is an essential gene for morphogenesis, because in knockout experiments only heterokaryons were obtained.

Table 3.1 Key genes in the regulation of cellular processes involved in yeast-hypha morphogenesis

	B. poitrasii	*C. albicans*	*H. capsulatum*	*M. racemosus*	*P. brasiliensis*	*S. cerevisiae*	*U. maydis*	*Y. lipolytica*
Cell growth and organization	*BpNADGDH BpNADPGDH I BpNADPGDH II BpODC, BpCHS1–8*	*CHS2, CHS3, CHT1–4, HEX1, ODCA, SAP1, 2, 3, 4, 5, and 6*	*CAT-A, CAT-P, MS-8, TUB-1, TUB-2*	*ADH1, CHS1, GLAM, HMGR1, HMGR2, HMGR3*	*AGN1, AGS1, BGL, CDA, CHS, GEL3P, HEX, 4-HPPD, HYD1–2, KEX2, ODC, PLB1, PLC, PLD*	*FLO8, FLO11, GCN5, MUC1, PPS1–1, SPE1–4, STA1–3*	*EGL-1, ODC, PEP4, SEP3, UBC3*	*CHS3, NAG5, ODC, XPR2, 6*
Cellular regulators		*CPH1, CZF1, EFG1, HWP-1, KRE-5, MDS-3, PHR1, PRA-1, RBF1, TUP1*	*CBP1, CDC2, RYP-1, 2, 3, YPS3*	*CNA A/B/C, CNB, PKAR1/R2/R3, PKAR4, RAS1*	*ATPS, APSK, HSP70, HSP90, PAPSR*	*ACE2P, CDK-8, ELM1–3, MFG-1, MSS10, NRG1–2, PHD1, SNF1, SKS1P, STE11–12, XBP1*	*FUZ7, KHD4, RRM4, RRM75, TUP1, UST1, UMV1–2*	*BEM1, CDC25, GPI7, HOY1, ISEC31, KEX2, MHY1,PPH21, RAC1, RAS2, SNF5, STE11, TPK-1, ZnCLp*
Transport-related genes		*RIM101*			*ISC, KTP, PCT*	*MEP2*		
Housekeeping genes	*UBC, WS-21, 18SrRNA*	*ACT*	*RPS1B, TDHI, TEF1, TEF3*	*EF1, TFC1*	*L34, TUB-b,*	*ALG9, TAF10, TFCI, UBC6*	*EF1α, GAPDH*	*EF1α*

Function of genes: *B. poitrasii* – *BpNADGDH*, NAD-dependent glutamate dehydrogenase; *BpNADPGDH I* and *II*, Y- and H-form-specific NADP-dependent glutamate dehydrogenase, respectively (Pathan 2017, Pathan et al. Unpublished data); *BpODC*, ornithine decarboxylase (Pathan et al. 2017); *CHS1–8*, chitin synthases (Chitnis et al. 2002); 18SrRNA, component of *18S* ribosome; *UBC*, ubiquitin-conjugating enzyme; *WS-21*, 40S ribosomal protein S3A (Pathan et al. 2017). ***C. albicans*** – *ACT*, actin; *CHS2, 3*, chitin synthase (Gow et al. 1993); *CHT1–4*, chitinases (McCreath et al. 1995, *Candida* genome database); *CPH*, *CZF1*, transcription factors (Biswas et al. 2007); *EFG1*, enhanced filamentous growth (Stoldt et al. 1997); *HEX1*, β-*N*-acetylglucosaminidase (Niimi et al. 1997); *HYR*, hyphal-regulated gene (Bailey et al. 1996); *HST7*, mitogen-activated protein kinase components (Gow 1995); *KRE-5*, *MDS-3*, *PRA-1*, *RBF1*, *RIM01*, transcription factors (Herrero et al. 2004; Sentandreu et al. 1998; Zacchi et al. 2010); *RBF1*, RPG box binding factor1 (Magee 1997); *SAP1–10*, aspartyl proteinase gene (Hube et al. 1994, Naglik et al. 2003); *TUP1*, transcriptional regulator (Magee 1997). ***H. capsulatum*** **–** *CAT-A*, *CAT-P*, catalases (Johnson et al. 2002); *CBP-1*, calcium binding protein (Sebghati et al. 2000); *HSP 70* and *83*, heat shock protein; *RYP2*, *3*, velvet-A family regulator (Webster and Sil 2008); *TUB1*, α-tubulin; *TUB2*, β-tubulin; *YPS 3*, yeast form-specific gene (Maresca and Kobayashi 1989); *MS8*, mold-specific gene (Tian and Shearer 2002). ***M. racemosus*** **–** *ADH1*, alcohol dehydrogenase (Rangel-Porras et al. 2005); *CNA A/B/C,CNB*, calcineurin pathway regulators (Lee et al. 2015); *CUP*, copper metallothionein gene (Cano-Canchola et al. 1992); EF1, translation elongation factor (Valle-Maldonado et al. 2015); *GLAM*, glucoamylase (Houghton-Larsen and Pedersen 2003); *HMGR1–3*, HMG-CoA-reductase (Nagy et al. 2014); *PKAR1–4*, protein kinase A regulatory subunits (Ocampo et al. 2012); *TFC1*, transcription initiation factor (Valle-Maldonado et al. 2015). ***P. brasiliensis*** **–** *AGN*, *AGS*, glucan synthase (Villalobos-Duno et al. 2013); *ATPS*, ATP sulfurylase; *APSK*, protein kinase; *PAPSR*, reductase; *BGL*, β- glucosidase; *CDA*, chitin deacetylase, *ISC*, iron-sulfur cluster-like protein; *HEX*, hexagonal peroxisome; *KTP*, cation transporter; *PCT*, P-type cation pump (Andrade et al. 2006); *PLB*, *PLC*, *PLD*, phospholipase (Soares et al. 2013); *HSP70*, heat shock protein (Da Silva et al. 1999). ***S. cerevisiae*** **–** *ACE2P*, regulator of chitinase (King and Butler 1998); *BEM*, bud emergence (Cabib et al. 1998); *CAP2*, capping protein; *CHS 2,3,5*, chitin synthase; *ELM1,2,3*, protein kinase homologs (Blacketer et al. 1994); *FLO8*, *FLO11*, floculin (Lo and Dranginis 1998); *GCN5*, histone acetyltransferase (Wang et al. 2015); *MEP2*, ammonium permease (Lorenz and Heitman 1997); *MFG-1*, transcriptional regulator (Ryan et al. 2012); *MUC1*, mucin like protein; *MSSS10*, transcriptional activator (Lambrechts et al. 1996); *MYO2*, myosin; *NRG1*, *2*, repressor protein (Kuchin et al. 2002); *PFY2*, profiling (Harold 1995); *PHD1*, DNA binding protein (Gimeno and Fink 1994); *PPS-1*, phosphoribosyl pyrophosphate synthase (Blacketer et al. 1994); *SKS1P*, serine threonine kinase (Johnson et al. 2014); *SNF1*, sucrose non-fermenting protein kinase (Kuchin et al. 2003); *STE 7,11,20*, kinases; *STE12* transcriptional factor (Liu et al. 1994); *TPM*, tropomyosin (Harold 1995); *UBC6*, ubiquitin-conjugating enzyme; *XBP1*, transcription repressor (Miles et al. 2013). ***U. maydis*** **–** *EGL-1*, filamentous growth gene (Bolker et al. 1995); *SEP3*, septin (Boyce et al. 2005); *PEP4*, vacuolar acid proteinase (Soberanes-Gutiérrez et al. 2015); *UBC3*, MAP kinase (Mayorga and Gold 1999); *UMV1,2*, velvet family regulators (Karakkat et al. 2013); *TUP1*, transcriptional repressor (Elias-Villalobos et al. 2011); *UST1*, transcription factor; *KHD4*, *RRM4*, RNA binding protein (Becht et al. 2005); *GAPDH*, glyceraldehyde-3-phosphate dehydrogenase; *EF1α*, translation elongation factor. ***Y. lipolytica*** – *NAG5*, *N*-acetylglucosamine kinase; *ODC*, ornithine decarboxylase (Campos-Góngora et al. 2018; Juan-Francisco et al. 2001); *MHY1*, *BEM1*, *RAC1*, *ISEC31*, zinc finger protein transcription factor (Hurtado and Rachubinski 2002, Hurtado and Rachubinski 1999); *KEX2*, *GPI7*, *SNF5*, *PPH21*, transcription factors (Richard et al. 2001); *XPR 2*, *6*, serine/alkaline protease (Madzak et al. 1999)

The cAMP-induced signal transduction pathway involves the RAS proteins that are assigned a putative role in cell differentiation and proliferation (Gancedo et al. 1985). Three *RAS* genes (*MRAS1*, *MRAS2*, and *MRAS3*) were cloned and characterized from *M. racemosus* and found to be similar with other ras proteins (Casale et al. 1990). Detection of high levels of *MRAS1* transcripts during Y-H morphogenesis and low levels during H-Y transition suggested its role in regulation of cAMP levels by suppression. *MRAS3* was found to be associated with the cAMP burst during spore germination, while *MRAS2* transcripts were not detected during transition or germination (Casale et al. 1990). Linz et al. (1986) cloned three distinct elongation factor lα (*EF-lα*) genes from *M. racemosus*. Since an essential role was attributed to *EF-lα* genes in eukaryotic translation, its role should be explored in Y-H morphogenesis in *M. racemosus* with expression studies.

The cell wall building blocks, chitin and chitosan, are the main determinants of morphology. In *B. poitrasii*, multiple chitin synthase genes were identified that were associated with the Y-H morphogenesis. Chitnis et al. (2002) reported eight chitin synthase genes with *BpCHS 1–4* belonging to Class I-III, *BpCHS 5–6* and *8* of Class IV, and *BpCHS7* of Class V in *B. poitrasii*. Among the eight genes (*BpCHS1–8*), *BpCHS2* and *3* were specific to the hyphal form. *C. albicans* contain four *CHS* genes, designated as *CHS1–4* (Mio et al. 1996, Munro et al. 2001). *CHS1* was primarily required for septa synthesis in Y-form cells and also for maintaining the integrity of cell wall in H-cells (Mio et al. 1996, Munro et al. 2001). The expression of *CHS2* and *CHS3* increased during Y-H transition, with *CHS2* being preferentially expressed in H-cells only (Gow et al. 1993). The *CHS3* deletion strain showed >60% reduction in chitin content, indicating that it is the major chitin synthase in *C. albicans*. However, *CHS4* did not show any involvement in Y-H transition (Sudoh et al. 1999). Three *CHS* genes were reported in *S. cerevisiae* (Bulawa and Osmond 1990, Silverman et al. 1988). *CHS1* was involved in the repair of damaged chitin; *CHS2* was required for primary septum formation, whereas *CHS3* was involved in the synthesis of chitin ring required during bud formation and also in the synthesis of spore wall. In *A. nidulans*, involvement of *CHSA* gene was suggested in conidium formation and synthesis of hyphal wall (Culp et al. 2000). In the case of *A. fumigatus*, 25% of hyphal wall chitin was synthesized by *CHS3* and also plays an important role in sporulation (Aufavre-Brown et al. 1997).

According to the Davis and Bartnicki-Garcia model proposed for the biosynthesis of cell wall polymers chitin and chitosan, nascent chitin chains synthesized by chitin synthase are simultaneously modified by chitin deacetylases (CDAs) (Orlowski 1991). The synergistic action of CS-CDA led to the formation of chitin-chitosan containing fibrils, which crystallize to form the major structural component of the cell walls. However, during autolysis, endo-chitinases first act on the cell wall to release the chitin oligosaccharides, which could be subsequently deacetylated by CDA. The first *CDA* gene was isolated, characterized, and sequenced from the zygomycete *M. rouxii* (Kafetzopoulos et al. 1993). Two *CDA* genes (termed as *CDA1* and *CDA2*) have been identified in *S. cerevisiae* by homology comparison with the *CDA* gene from the cosmid library constructed for *M. rouxii* (Christodoulidou et al. 1996; Mishra et al. 1997). *CDA* genes from *Gongronella butleri* and *Rhizopus nigricans* were also cloned and characterized (Jeraj et al. 2006; Maw et al. 2002). Bartnicki-Garcia further suggested that apical wall growth was controlled by the fine balance between cell wall synthesis and cell wall degradation. The gene coding for lytic enzymes such as chitinases (CHT) showed correlation with fungal morphogenesis. In *C. albicans*, four *CHT* genes, designated as *CHT1–4*, were reported.

The expression of *CHT1–4* genes was higher in Y-form cells and decreased during Y-H transition. The deletion of *CHT1–2* did not affect morphology in liquid media, whereas increased hyphal growth was reported during growth on solid media (McCreath et al. 1995, *Candida* genome database).

Microarray analysis of *C. albicans* was carried out under different growth conditions to investigate differential expression of genes during Y-H transition (Nantel et al. 2002). Transcriptional factors *Efg1p* and *Cph1p* were important for hyphal growth in *C. albicans* (Nantel et al. 2002). Genes coding for cellular regulators such as protein kinases (*STE12, STE20, PKC1*), transcriptional activators (*TUP1*, *EFG1*), and heat shock proteins (*HSP70*) were suggested to play an important role in the morphological transition of yeast *S. cerevisiae* (Magee 1997), *C. albicans* (Liu et al. 1994), and *H. capsulatum* (Maresca and Kobayashi 1989), respectively. The secreted aspartyl proteinases (Saps) coding multigene family of ten *SAP* genes (*SAP1–10*) showed key role in Y-H transition as well as in virulence of *C. albicans*. Out of these, *SAP4–6* were associated with hyphae formation, whereas *SAP1–3* were predominantly expressed during yeast growth (Naglik et al. 2003). In *C. albicans*, expression of genes coding for the principal structural component of the cells varied during Y-H transition (Gow 1995). For example, the *ECE1*gene (involved in cell elongation) and *CHS2* gene (coding for chitin synthase) were highly expressed in the H-form. However, the genes such as *HYR1*, a hypha-specific gene in *C. albicans* (Bailey et al. 1996), *YPS3*, a yeast-specific gene (Maresca and Kobayashi 1989), and *MS8*, a mold-specific gene (Tian and Shearer 2002), in *H. capsulatum,* which were expressed in only one of the morphological form, were reported, but the information of their products is unknown. In *M. circinelloides*, a gene coding for polyamine-metabolizing enzyme ornithine decarboxylase (*odcA*) was found to be differentially expressed during Y-H transition (Blasco et al. 2002). Further, Juan Francisco et al. (2001) demonstrated the cause-effect relationship between *odc* gene and morphological outcome in *Y. lipolytica.* The *odc* minus mutant was arrested in Y-form and produced hyphae with increasing levels of putrescine in the growth medium. Similar results were reported in *U. maydis* and *C. albicans odc* minus mutants (Guevara-Olvera et al. 1993; Herrero et al. 1999).

Recently, we have identified the set of reliable reference genes, *Ubc* (coding for ubiquitin-conjugating enzyme) and *WS-21* (coding for 40S ribosomal protein S3A), to study the gene expression during the Y-H morphogenesis in zygomycete *B. poitrasii* (Pathan et al. 2017). Further, *Ubc* and *WS-21* were used as reference genes to study the expression of ornithine decarboxylase gene (Bp*odc*) in different morphological forms of *B. poitrasii.* The expression of Bp*odc* was higher in H-cells than in Y-cells (Pathan et al. 2017). The biochemical correlation of relative proportion of NAD- and NADP-dependent glutamate dehydrogenases (GDHs; measured as NADP-/NAD-GDH ratio) with morphology was reported for the first time in *B. poitrasii* (Khale et al. 1992). Furthermore, one NAD-GDH and two form-specific NADP-GDH isoenzymes were also reported in *B. poitrasii* (Amin et al. 2004). At the molecular level, one NAD- (*BpNADGDH*) and two separate genes coding for NADP-GDH isoenzymes in *B. poitrasii* (*BpNADPGDH I* and *II*) have been identified (Pathan 2017). Under normal dimorphism-triggering conditions (glucose and temperature), *BpNADPGDH II* was not expressed in Y5 (yeast form monomorphic mutant). However, it was induced in the presence of ethanol, leading to Y-H transition in mutant Y5. The effect of ethanol was reverted by myo-inositol with subsequent repression of *BpNADPGDH II* gene. On the other hand, transformation with *BpNADPGDH II* gene also induced germ tube formation in mutant Y5. The results showed the cause-effect relationship between *BpGDH* genes and morphological outcome in *B. poitrasii* (Pathan 2017; Pathan et al. unpublished data). However, no gene was constitutively linked to the dimorphic transition.

3.6 Y-H Morphogenesis as an Antifungal Drug Target

To overcome the physiological and cellular defenses of the host, pathogenic fungi conveniently switch between the unicellular Y and filamentous H (Deshpande 1996, Georgopapadakou and Walsh 1994, Jacobsen et al. 2012, Khale et al. 1992, Ryley and Ryley 1990). In fungal pathogens, stimuli from host micro-environment trigger many biochemical events finally leading to morphogenesis. For the development of new antifungal drugs, these biochemical correlates (associated with the non−/less-pathogenic to pathogenic morphological change) can be targeted.

The cell walls of Y- and H-cells of *B. dermatitidis*, *B. poitrasii*, *C. albicans*, *H. capsulatum*, *M. rouxii*, *P. brasiliensis*, and *S. schenckii* differ significantly in their chemical composition (Khale et al. 1992; Orlowski 1991; San-Blas and San-Blas 1985). Fungal cell walls consist mainly of (1,3)-β-D-glucans that are covalently associated with (1,6)-β-D-glucans and chitin, mannans, and cell wall proteins.

During Y-H transition, the enzymes involved in cell wall synthesis and degradation, such as chitin synthase (CS), glucan synthase, chitinase, and *N*-acetylglucosaminidase, play a major role. Chitin synthase, since absent in plants and mammals, emerged as one of the promising antifungal targets. The CS inhibitors, *viz.*, polyoxin and nikkomycin, were first time isolated from the *Streptomyces* sp. culture filtrates. A number of CS inhibitors have subsequently been isolated from natural sources such as plants and microorganisms. Synthetic chitin synthase inhibitors were also designed and developed, primarily based on a diversity-oriented synthesis of UDP-GlcNAc analogs. One of the CS inhibitor, Nikkomycin Z, is under clinical trial (Chaudhary et al. 2013). The non-competitive inhibition of (1,3)-β-D-glucan synthases by echinocandins led to loss of cell wall integrity and severe cell wall stress in *C. albicans*, *C. neoformans*, *A. fumigatus*, *C. glabrata*, and *S. cerevisiae*.

Ammonia-assimilating enzymes (NAD-GDH, NADP-GDH, glutamate synthase [GOGAT], and glutamine synthetase [GS]) are present at the branch of carbon and nitrogen metabolism in chitin biosynthesis pathway. They catalyze the early steps of metabolic pathways connecting the formation of chitin. These enzymes therefore serve as important targets for antifungal drugs. Out of these, glutamate dehydrogenase levels were correlated with yeast-hypha transition in *B. poitrasii* (Joshi et al. 2013; Khale et al. 1992) and *Mucor* species. Further, the morphology of NADP-GDH null mutant of *Penicillium chrysogenum* was found to be affected (Thykaer et al. 2009). The antifungal potential of GDH inhibitors has also been evaluated (Joshi et al. 2013; Choudhury and Punekar 2007, 2009; Cunliffe et al. 1983). Most of the GDH inhibitors were L-glutamate (GDH substrate) analogs and therefore act in competitive mode. Isophthalate was the most widely reported fungal GDH inhibitor in vitro (Caughey et al. 1956; Cunliffe et al. 1983; Rogers et al. 1972; Stevens et al. 1989; Veronese et al. 1974). However, Rogers (1971) observed that L-glutamate structural analogs (e.g., glutaric acid, iminodiactic acid, oxydiglycolic acid, and thiodiglycolic acid) also inhibited the bovine GDH. Further, NADP-GDH from *A. niger* was found to be inhibited by 2-ketoglutarate analogs, *viz.*, 2-methyleneglutarate, 2-minoglutarate, 2-oxoglutarate, 2,4-pyridinedicarboxylate, and 3,5-pyrazoledicarboxylate (Noor and Punekar 2005). The chemically synthesized 1,2,3-triazole-linked β-lactam-bile acid conjugates inhibited the purified NAD-GDH of *B. poitrasii*, which also affected the Y-H transition significantly. Additionally, these compounds also affected the Y-H transition in pathogenic *C. albicans* strains and another saprophyte *Y. lipolytica* (Joshi et al. 2013).

cAMP plays a major role in various morphological processes including conidiation, dimorphism, hyphal branching, phototropism, spore germination, etc. (Bahn et al. 2007, Robson et al. 1991). Other enzymes involved in regulation of cAMP-induced cascade include adenylyl cyclase and ammonium ion permease, which can be evaluated for inhibition of Y-H morphogenesis. The cAMP-dependent protein kinases have been inhibited by compounds like phenylaminopyrimidine (PAP)-pyridines, Gleevec, BIRB796, rapamycin, triazolo-[1,5α]pyrimidine, and l-[5-isoquinolinesulfonyl]-2-methylpiperazine dihydrochloride (H7) [Brunn et al. 1996; Pillonel 2005; Richardson et al. 2006).

Another second messenger molecule Ca^{++}, which exerts its effect via Ca-CaM-calcineurin signaling, has been found to play a major role in the growth and differentiation of many yeasts and filamentous fungi (Shapiro et al. 2007; Muthukumar et al. 1987). The inhibitors of Ca-CaM pathway, *viz.*, trifluoperazine (TFP) and others, also inhibited the Y-H transition in *C. albicans* (Holmes et al. 1991; Paranjpe et al. 1990; Sabie and Gadd 1990) and *S. schenckii* (Rivera-Rodriguez and Rodriguez-del Valle 1992) and H-Y transition in *P. brasiliensis* (de Carvalho et al. 2003).

The biochemical as well as molecular correlation of polyamine metabolism with Y-H morphogenesis has been established in several dimorphic fungi. Putrescine, spermidine, and spermine are low-molecular-weight polyamines that are synthesized in the cells from their direct precursor, ornithine. Polyamines play an important role in cell growth, differentiation, and transformation (Garcia, et al. 1980; Ruiz-Herrera et al. 1983; Tabor and Tabor 1985; Calvo-Mendez et al. 1987). Ornithine decarboxylase (ODC) is the key enzyme in polyamine biosynthesis responsible for conversion of ornithine to putrescine (Ruiz-Herrera and Martinez-Espinoza 1998, Ruiz-Herrera 1994). Subsequently, spermidine synthase converts putrescine to spermidine, and spermidine is converted to spermine by spermine synthase. The addition of diaminobutanone (DAB), a competitive ODC inhibitor, affected the sporulation and germination of spores as well as Y-H morphogenesis in *Mucor* and *Phycomyces* (Ruiz-Herrera (1994). A similar effect was observed in the presence of other ODC inhibitors, viz., difluromethyl ornithine (DFMO) and dehydromonofluromethyl ornithine.

The signal transduction pathway involving cAMP, MAP, and Ras kinases leading to Y-H differentiation was also found to be involved in inducing programmed cell death (apoptosis) in fungi. In turn, differentiation and apoptosis share many biochemical events that can be exploited for development of antifungal drugs, which may act by inducing apoptosis. Currently, antifungal agents, *viz.*, amphotericin B, bleomycin, dermaseptin, farnesol, histatin, osmotin, and paclitaxel, are known to induce apoptosis in *A. fumigatus*, *C. albicans*, *Rhizoctonia solani*, and *S. cerevisiae* (Madeo et al. 2002; Phillips et al. 2003). A broad-spectrum fungicidal agent Pradimicin A (mannan synthesis inhibitor) also induced apoptosis in *S. cerevisiae* (Hiramoto et al. 2003). Further, Tupe et al. (2015) demonstrated that antifungal molecule phenazine-1-carboxamide (PC)-induced apoptosis by inhibiting the Y-H transition in *B. poitrasii* and *C. albicans*. Recently, Park et al. (2018) showed the apoptosis-mediated anticancer efficacy of azole antifungals in clinical trials and thereby opened up the possibility of exploring apoptosis-inducing antifungal molecules as new potential anticancer agents.

3.7 Epilogue

The extensive review of literature suggested the involvement of cAMP-dependent protein kinases (PKA) and/or mitogen-activated protein kinases (MAPK) in regulating the cascade, which in turn regulate the morphological outcome in fungi. The process of fungal morphogenesis and formation of tumor cells follow the same regulatory series of events, involving cAMP, MAP, and RAS kinase cascades, leading to changes from one state to another, called differentiation. Inhibition of differentiation leads to apoptosis [programmed cell death (PCD)] in normal cells, while the absence of apoptosis leads to differentiation, causing cancer in humans. Therefore, the molecules inhibiting Y-H transition in fungi can be explored for their anticancer potential. Further, the combinational therapy targeting the differentiation and inducing the apoptosis in tumor cells could be the better strategy to treat cancer. For instance, the correlates of differentiation, such as ornithine decarboxylase, glutamate dehydrogenases, etc. can be targeted to control tumor cells. On the other hand, inducer of caspases (endoproteases involved in apoptosis) can be used to enforce PCD in cancer cells. This will help in developing more specific, less toxic, and economically feasible anticancer drugs in the future.

Acknowledgments MVD is grateful to CSIR, New Delhi, for the Emeritus Scientist Scheme [21(0962)/13/EMR2] and the Department of Biotechnology (DBT-BIRAC), New Delhi, for financial support.

References

Amin A, Joshi M, Deshpande MV (2004) Morphology-associated expression of NADP-dependent glutamate dehydrogenases during yeast-mycelium transition of a dimorphic fungus *Benjaminiella poitrasii*. Ant van Leeuwenhoek 85:327–334

Andrade RV, Paes HC, Nicola AM, de Carvalho MJ, Fachin AL, Cardoso RS, Silva SS, Fernandes L, Silva SP, Donadi EA, Sakamoto-Hojo ET, Passos GA, Soares CM, Brígido MM, Felipe MS (2006) Cell organisation, sulphur metabolism and ion transport-related genes are differentially expressed in *Paracoccidioides brasiliensis* mycelium and yeast cells. BMC Genomics 7:208

Andrews DL, Egan JD, Mayorga ME, Gold SE (2000) The *Ustilago maydis ubc4* and *ubc5* genes encode members of a MAP kinase cascade required for filamentous growth. Mol Plant-Microbe Interact 13:781–786

Anraku Y, Ohya Y, Lida H (1991) Cell cycle control by calcium ad calmodulin in *Saccharomyces cerevisiae*. Biochem Biophys Acta 1093:169–173

Aufavre-Brown A, Mellado E, Gow NAR, Holden DW (1997) *Aspergillus fumigatus chsEII*: a gene related to *CHS3* of *Saccharomyces cerevisiae* and important for hyphal growth and conidiophores development but not for pathogenecity. Fungal Genet Biol 21:141–152

Bahn YS, Xue C, Idnurm A, Rutherford JC, Heitman J, Cardenas ME (2007) Sensing the environment: lessons from fungi. Nat Rev Microbiol 5:57–69

Bailey DA, Feldmann JFP, Bovey M, Gow NAR, Brown APJ (1996) The *Candida albicans HYR1* gene, which is activated in response to hyphal development, belongs to a gene family encoding yeast cell wall proteins. J Bacteriol 178:5353–5360

Barlow AJ, Aldersley T, Chattaway FW (1974) Factors present in serum and seminal plasma, which promote germ tube formation and mycelial growth of *Candida albicans*. J Gen Microbiol 82:261–272

Bartnicki-Garcia S (1963) Mold –yeast dimorphism of *Mucor*. Bacteriol Rev 27:293–304

Bartnicki-Garcia S, Nickerson WJ (1962a) Induction of yeast like development in *Mucor* by carbon dioxide. J Bacteriol 84:829–840

Bartnicki-Garcia S, Nickerson WJ (1962b) Nutrition, growth and morphogenesis of *Mucor rouxii*. J Bacteriol 84:841–858

Becht P, Vollmeister E, Feldbrügge M (2005) Role for RNA-binding proteins implicated in pathogenic development of *Ustilago maydis*. Eukaryot Cell 4:121–133

Biswas S, Van Dijck P, Datta A (2007) Environmental sensing and signal transduction pathways regulating morphopathogenic determinants of *Candida albicans*. Microbiol Mol Biol Rev 71:348–376

Blacketer MJ, Madaule P, Myers AM (1994) The *Saccharomyces cerevisiae* mutation elm4-1 facilitates pseudohyphal differentiation and interacts with a deficiency in phosphoribosyl pyrophosphate synthase activity to cause constitutive pseudohyphal growth. Mol Cell Biol 14:4671–4681

Blackwell M (2011) The fungi: 1, 2, 3... 5.1 million species? Am J Bot 98:426–438

Blasco JL, María A, García-Sánchez M, Ruiz-Herrera J, Eslavaa AP, Iturriaga EA (2002) A gene coding for ornithine decarboxylase (*odcA*) is differentially expressed during the *Mucor circinelloides* yeast-to-hypha transition. Res Microbiol 153:155–164

Bolker M, Genin S, Lehmier C, Kahmann R (1995) Genetic regulation of mating and dimorphism in *Ustilago maydis*. Can J Bot 73:320–325

Borges-Walmsley MI, Walmsley AR (2000) cAMP signalling in pathogenic fungi: control of dimorphic switching and pathogenecity. Trends Microbiol 8:133–141

Boyce KJ, Chang H, D'Souza CA, Kronstad JW (2005) An *Ustilago maydis* septin is required for filamentous growth in culture and for full symptom development on maize. Eukaryot Cell 4:2044–2056

Brunn GJ, Williams J, Sabers C, Wiederrecht G, Lawrence JC Jr, Abraham RT (1996) Direct inhibition of the signaling functions of the mammalian target of rapamycin by the phosphoinositide 3-kinase inhibitors, wortmannin and LY 294002. EMBO J 15:5256–5267

Bruno KS, Aramayo R, Minke PF, Metzenberg RL, Plamann M (1996) Loss of growth polarity and mislocalization of septa in a *Neurospora* mutant altered in the regulatory subunit of cAMP-dependent protein kinase. EMBO J 15:5772–5782

Brunton AH, Gadd GM (1989) The effect of exogenously-supplied nucleosides and nucleotides and the involvement of adenosine 3′-5′-monophosphate (cyclic AMP) in the yeast mycelium transition of *Ceratocystis* (*Ophistoma*) *ulmi*. FEMS Microbiol Lett 60:49–53

Brunton AH, Gadd GM (1991) Evidence for an inositol lipid signal pathway in the yeast-mycelium transition of *Ophiostoma ulmi*, the dutch elm disease fungus. Mycol Res 95:484–491

Bulawa CE, Osmond BC (1990) Chitin sunthase I and chitin synthase II are not required *in vivo* in *Saccharomyces cerevisiae*. Proc Natl Acad Sci U S A 87:7424–7428

Cabib E, Drgonova J, Drgon T (1998) Role of small G proteins in yeast cell polarization and wall biosynthesis. Annu Rev Biochem 67:307–333

Calvo-Mendez C, Martinez-Pacheco M, Ruiz-Herrera J (1987) Regulation of ornithine decarboxylase activity in *Mucor bacilliformis* and *Mucor rouxii*. Exp Mycol 11:270–277

Campos-G6ngora E, Palande AS, Le6n-Ramirez C, Pathan EK, Ruiz-Herrera J, Deshpande MV (2018) Determination of the effect of polyamines on an oil-degrading strain of *Yarrowia lipolytica* using an *odc* minus mutant. FEMS Yeast Res 18: foy073

Cano-Canchola C, Sosa L, Fonzi W, Sypherd P, Ruiz-Herrera J (1992) Developmental regulation of *CUP* gene expression through DNA methylation in *Mucor* spp. J Bacteriol 174:362–366

Casale WL, McConnell DG, Wang SY, Lee YJ, Linz JE (1990) Expression of a gene family in the dimorphic fungus *Mucor racemosus* which exhibits striking similarity to human *ras* genes. Mol Cell Biol 10:6654–6663

Casanova M, Cervera AM, Gozalbo D, Martinez JP (1997) Hemin induces germ tube formation in *Candida albicans*. Infect Immun 65:4360–4364

Caughey WS, Smiley JD, Hellerman L (1956) L-glutamic acid dehydrogenase: structural requirements for substrate competition: effect of thyroxine. J Biol Chem 224:591–607

Cervantes-Chávez JA, Ruiz-Herrera J (2006) STE11 disruption reveals the central role of a MAPK pathway in dimorphism and mating in *Yarrowia lipolytica*. FEMS Yeast Res 6:801–815

Chacko R, Deshpande MV, Shankar V (1996) Extracellular ribonuclease production by *Rhizopus stolonifer*: influence of metal ions. Curr Microbiol 32:246–251

Chaudhary PM, Tupe SG, Deshpande MV (2013) Chitin synthase inhibitors as antifungal agents. Min Rev Med Chem 13:222–236

Chitnis M, Munro CA, Brown AJP, Gooday GW, Gow NAR, Deshpande MV (2002) The zygomycetous fungus, *Benjaminiella poitrasii* contains large family of differentially regulated chitin synthase genes. Fungal Genet Biol 36:215–223

Cho T, Hamatake H, Kaminishi H, Hagihara Y, Watanabe K (1992) The relationship between cyclic adenosine 3′-5′-monophosphate and morphology in exponential phase *Candida albicans*. J Med Vet Mycol 30:35–42

Chou S, Lane S, Liu H (2006) Regulation of mating and filamentation genes by two distinct *Ste12* complexes in *Saccharomyces cerevisiae*. Mol Cell Biol 26:4794–4805

Choudhury R, Punekar NS (2007) Competitive inhibition of glutamate dehydrogenase reaction. FEBS Lett 581:2733–2736

Choudhury R, Punekar NS (2009) *Aspergillus terreus* NADP-glutamate dehydrogenase is kinetically distinct from the allosteric enzyme of other *Aspergilli*. Mycol Res 113:1121–1126

Christodoulidou A, Bouriotis V, Thireos G (1996) Two sporulation-specific chitin deacetylase-encoding genes are required for the ascospore wall rigidity of *Saccharomyces cerevisiae*. J Biol Chem 271:31420–31425

Chung KS, Won M, Lee SB, Jang YJ, Hoe KL, Kim DU (2001) Isolation of a novel gene from *Schizosaccharomyces pombe*: *stm1+* encoding a seven trans membrane loop protein that may couple with the heterotrimeric Galpha 2 protein, *Gpa2*. J Biol Chem 276:40190–40201

Cole GT, Sun SH (1985) Arthroconidium-spherule-endospore transformation in *Coccidiodes immitis*. In: Szaniszlo PJ (ed) Fungal dimorphism. Plenum Press, New York, pp 281–333

Cruz MC, Goldstein AL, Blankenship JR, Del Poeta M, Davis D, Cardenas ME, Perfect JR, McCusker JH, Heitman J (2002) Calcineurin is essential for survival during membrane stress in *Candida albicans*. EMBO J 21:546–559

Csank C, Schroppel K, Leberer E, Harcus D, Mohamed O, Meloche S, Thomas DY, Whiteway M (1998) Roles of the *Candida albicans* mitogen-activated protein kinase homolog, *Cek1p*, in hyphal development and systemic candidiasis. Infect Immun 66:2713–2721

Cullen PJ, Sabbagh WJ, Graham E, Irick MM, van Olden EK, Neal C, Delrow J, Bardwell L, Sprague GFJ (2004) A signaling mucin at the head of the Cdc42- and MAPK-dependent filamentous growth pathway in yeast. Genes Dev 18:1695–1708

Culp DW, Dodge CL, Miao YH, Li L, Sag-Ozkal D, Borgia PT (2000) The *chsA* gene from *Aspergillus nidulans* is necessary for maximal conidiation. Curr Microbiol 182:349–353

Cunliffe D, Leason M, Parkin D, Lea P (1983) The inhibition of glutamate dehydrogenase by derivatives of isophthalic acid. Phytochemistry 22:1357–1360

Da Silva SP, Borges-Walmsley MI, Pereira IS, Soares CMDA, Walmsley AR, Felipe MSS (1999) Differential expression of an *hsp70* gene during transition from the mycelial to the infective yeast form of the human pathogenic fungus *Paracoccidioides brasiliensis*. Mol Microbiol 31:1039–1050

Davis TR, Zucchi PC, Kumamoto CA (2013) Calmodulin binding to *Dfi1p* promotes invasiveness of *Candida albicans*. PLoS One 8:e76239

de Carvalho MJ, Amorim Jesuino RS, Daher BS, Silva-Pereira I, de Freitas SM, Soares CM, Felipe MS (2003) Functional and genetic characterization of calmodulin from the dimorphic and pathogenic fungus *Paracoccidioides brasiliensis*. Fungal Genet Biol 39:204–210

de Oliveira JCF, Borges ACC, Marques MV, Gomes SL (1997) Cloning and characterisation of the gene for the catalytic subunit of cAMP-dependent protein kinase in the aquatic fungus *Blastocladiella emersonii*. Eur J Biochem 219:555–562

Dean RA (1997) Signal pathways and appressorium morphogenesis. Annu Rev Phytopathol 35:211–234

Deshpande MV (1996) The effect of morphological changes in fungal pathogenesis. Ind J Med Microbiol 14:1–9

Deshpande MV (1998) Biochemical basis of fungal differentiation. In: Verma A (ed) Microbes: for health, wealth and sustainable environment. Malhotra Publishing House, India, pp 241–252

Dohlman HG, Thorner JW (2001) Regulation of G protein-initiated signal transduction in yeast: paradigms and principles. Annu Rev Biochem 70:703–754

Doiphode N (2007) Dimorphism in *Benjaminiella poitrasii*: role of NAD-dependent glutamate dehydrogenase in yeast-hypha transition A PhD thesis submitted to University of Pune, Pune, India

Domer JE (1985) Blastomyces dermatitidis. In: Szaniszlo PJ (ed) Fungal dimorphism. Plenum Press, New York, pp 51–67

Elias-Villalobos A, Fernández-Álvarez A, Ibeas JI (2011) The general transcriptional repressor *Tup1* is required for dimorphism and virulence in a fungal plant pathogen. PLoS Pathog 7:e1002235

Feng Q, Summers E, Guo B, Fink G (1999) Ras signaling is required for serum-induced hyphal differentiation in *Candida albicans*. J Bacteriol 181:6339–6346

Fernandes L, Paes HC, Tavares AH et al (2008) Transcriptional profile of *ras1* and *ras2* and the potential role of farnesylation in the dimorphism of the human pathogen *Paracoccidioides brasiliensis*. FEMS Yeast Res 8:300

Gadd GM (1995) Signal transduction in fungi. In: Gow NAR, Gadd GM (eds) The growing fungus. Chapman and Hall, London, pp 183–210

Gancedo JM, Mazon MJ, Eraso P (1985) Biological roles for cyclic AMP: similarities and differences between organisms. Trends Biochem Sci 10:210–212
Garcia JR, Hiatt WR, Peters J, Sypherd PS (1980) *S*-Adenosylmethionine levels and protein methylation during morphogenesis of *Mucor racemosus*. J Bacteriol 142:196–201
Garrison RG, Boyd KS (1974) Ultrastructural studies of induced morphogenesis by *Aspergillus parasiticus*. J Bacteriol 154:524–528
Geis PA, Jacobs CW (1985) Polymorphism of *Wangiella dermatitidis*. In: Szaniszlo PJ (ed) Fungal dimorphism: with emphasis on fungi pathogenic for humans. Plenum Press, New York, pp 205–233
Georgopapadakou NH, Walsh TJ (1994) Human mycoses: drugs and targets for emerging pathogens. Science 264:371–373
Ghormade V, Deshpande MV (2000) Fungal spore germination into yeast or mycelium: possible implications of dimorphism in evolution and human pathogenesis. Naturwissenschaften 87:236–240
Ghormade V, Joshi CV, Deshpande MV (2005) Regulation of polyamines: a possible model for signal transduction pathway leading to dimorphism in *Benjaminiella poitrasii*. J Mycol Pl Pathol 35:442–450
Ghormade V, Pathan E, Deshpande MV (2012) Yeast-hypha dimorphism in zygomycetous fungi. In: Ruiz-Herrera J (ed) Dimorphic fungi: their importance as models for differentiation and fungal pathogenesis. Bentham Science Publishers, USA, pp 118–139
Gimeno CJ, Fink GR (1994) Induction of pseudohyphal growth by over-expression of *PHD1*, a *Saccharomyces cerevisiae* gene related to transcriptional regulators of fungal development. Mol Cell Biol 14:2100–2112
Gimeno CJ, Ljungdah PO, Styles CA, Fink GR (1992) Unipolar cell divisions in yeast *S. cerevisiae* lead to filamentous growth regulation by starvation and RAS. Cell 68:1077–1090
Gold S, Duncan G, Barrett K, Kronstad J (1994) cAMP regulates morphogenesis in the fungal pathogen *Ustilago maydis*. Gen Dev 8:2805–2816
Görlach J, Fox DS, Cutler NS, Cox GM, Perfect JR, Heitman J (2000) Identification and characterization of a highly conserved calcineurin binding protein, CBP1/calcipressin, in *Cryptococcus neoformans*. EMBO J 19:3618–3629
Gow NAR (1995) Yeast-hyphal dimorphism. In: Gow NAR, Gadd GM (eds) The growing fungus. Chapman and Hall, London, pp 403–422
Gow NAR, Swoboda R, Bertram G, Gooday GW, Brown APJ (1993) In: Bossche VH (ed) Dimorphic fungi in biology and medicine. Plenum Press, New York, pp 61–712
Gow NAR, van de Veerdonk FL, Brown APJ, Netea MG (2012) *Candida albicans* morphogenesis and host defence: discriminating invasion from colonization. Nat Rev Microbiol 10:112–122
Guevara-Olvera L, Calvo-Mendez C, Ruiz-Herrera J (1993) The role of polyamine metabolism in dimorphism of *Yarrowia Lipolytica*. Microbiology 139:485–493
Harold FM (1995) From morphogenes to morphogenesis. Microbiology 141:2765–2778
Herrero AB, López MC, García S, Schmidt A, Spaltmann F, Ruiz-Herrera J, Dominguez A (1999) Control of filament formation in *Candida albicans* by polyamine levels. Infect Immun 67:4870–4878
Herrero AB, Magnelli P, Mansour MK, Levitz SM, Bussey H, Abeijon C (2004) *KRE5* gene null mutant strains of *Candida albicans* are avirulent and have altered cell wall composition and hypha formation properties. Eukaryot Cell 3:1423–1432
Hiramoto F, Nomura N, Furumai T, Oki T, Igarashi Y (2003) Apoptosis-like cell death of Saccharomyces cerevisiae induced by a mannose-binding antifungal antibiotic, pradimicin. J Antibiot 56:768–772
Holmes AR, Cannon RD, Shepherd MG (1991) Effect of calcium ion uptake on *Candida albicans* morphology. FEMS Microbiol Lett 77:187–194
Houghton-Larsen J, Pedersen PA (2003) Cloning and characterization of a glucoamylase gene (*GlaM*) from the dimorphic zygomycte *Mucor circinelloides*. Appl Microbiol Biotechnol 62:210–217

Hube B, Monod M, Schofield DA, Brown APJ, Gow NAR (1994) Expression of seven members of the gene family encoding secretory aspartyl proteinase in *Candida albicans*. Mol Microbiol 14:87–99

Hurtado CA, Rachubinski RA (1999) *MHY1* encodes a C2H2-type zinc finger protein that promotes dimorphic transition in the yeast *Yarrowia lipolytica*. J Bacteriol 181:3051–3057

Hurtado CA, Rachubinski RA (2002) Isolation and characterization of *YlBEM1*, a gene required for cell polarization and differentiation in the dimorphic yeast *Yarrowia lipolytica*. Eukaryot Cell 1:526–537

Jacobsen ID, Wilson D, Wächtler B, Brunke S, Naglik JR, Hube B (2012) *Candida albicans* dimorphism as a therapeutic target. Expert Rev Anti-Infect Ther 10:85–93

Jeraj N, Kunic B, Lenasi H, Breskvar K (2006) Purification and molecular characterization of chitin deacetylase from *Rhizopus nigricans*. Enz Microbial Technol 39:1294–1299

Johnson CH, Klotz MG, York JL, Kruft V, McEwen JE (2002) Redundancy, phylogeny and differential expression of *Histoplasma capsulatum* catalases. Microbiology 148:1129–1142

Johnson C, Kweon HK, Sheidy D, Shively CA, Mellacheruvu D, Nesvizhskii AI, Andrews PC, Kumar A (2014) The yeast *Sks1p* kinase signaling network regulates pseudohyphal growth and glucose response. PLoS Genet 10:e1004183

Joshi CV, Pathan EK, Punekar NS, Tupe SG, Kapadnis BP, Deshpande MV (2013) A biochemical correlate of dimorphism in a zygomycete *Benjaminiella poitrasii*: characterization of purified NAD-dependent glutamate dehydrogenase, a target for antifungal agents. Antonie Van Leeuwenhoek 104:25–36

Juan-Francisco JB, Ruiz-Herrera J, Dominguez A (2001) Disruption of gene Yl*ODC* reveals absolute requirement of polyamines for mycelial development in *Yarrowia lipolytica*. FEMS Yeast Res 1:195–204

Kafetzopoulos D, Thireos G, Vournakis JN, Bouriotis V (1993) The primary structure of a fungal deacetylase reveals the function for two bacterial products. Proc Natl Acad Sci U S A 90:8005–8008

Karakkat BB, Gold SE, Covert SF (2013) Two members of the *Ustilago maydis* velvet family influence teliospore development and virulence on maize seedlings. Fungal Genet Biol 61:111–119

Kaur S, Mishra P, Prasad R (1988) Dimorphism-associated changes in amino acid transport of *Candida albicans*. FEMS Microbiol Lett 50:97–100

Khale A (1990) Dimorphism in fungi. A PhD thesis submitted to University of Pune, Pune, India

Khale A, Srinivasan MC, Deshmukh SS, Deshpande MV (1990) Dimorphism of *Benjaminiella poitrasii*: isolation and biochemical studies of morphological mutants. Antonie Van Leeuwenhoek 57:37–41

Khale A, Srinivasan MC, Deshpande MV (1992) Significance of NADP-/NAD- glutamate dehydrogenase ratio in the dimorphic behavior of *Benjaminiella poitrasii*. J Bacteriol 174:3723–3728

Khale-Kumar A, Deshpande MV (1993) Possible involvement of cyclic adenosine 3′-5′-monophosphate in the regulation of NADP-/ NAD – glutamate dehydrogenase ratio and yeast –mycelium transition of *Benjaminiella poitrasii*. J Bacteriol 175:6052–6055

King L, Butler G (1998) Ace2p, a regulator of *CTS1* (chitinase) expression, affects pseudohyphal production in *Saccharomyces cerevisiae*. Curr Genet 34:183–191

Köhler JR, Fink GR (1996) *Candida albicans* strains heterozygous and homozygous for mutations in mitogen-activated protein kinase signaling components have defects in hyphal development. Proc Natl Acad Sci U S A 93:13223–13228

Kronstad J, de Maria A, Funnell D, Laidlaw RD, Lee N, de Sa MM, Ramesh M (1998) Signalling via cAMP in fungi: interconnections with mitogen-activated protein kinase pathway. Arch Microbiol 170:395–404

Kuchin S, Vyas VK, Carlson M (2002) *Snf1* protein kinase and the repressors *Nrg1* and *Nrg2* regulate FLO11, haploid invasive growth, and diploid pseudohyphal differentiation. Mol Cell Biol 22:3994–4000

Kuchin S, Vyas VK, Carlson M (2003) Role of the yeast *Snf1* protein kinase in invasive growth. Biochem Soc Trans 31:175–177

Lambrechts MG, Bauer FF, Marmur J, Pretorius IS (1996) Muc1, a mucin-like protein that is regulated by Mss10, is critical for pseudohyphal differentiation in yeast. Proc Natl Acad Sci U S A 93:8419–8424

Lee YH, Dean RA (1993) cAMP regulates infection structure formation in the plant pathogenic fungus. *Magnaporthe grisea*. Plant Cell 5:693–700

Lee SC, Li A, Calo S, Inoue M, Tonthat NK, Bain JM, Louw J, Shinohara ML, Erwig LP, Schumachr MA, Ko DC, Heitman J (2015) Calcineurin orchestrates dimorphic transitions, antifungal drug responses and host–pathogen interactions of the pathogenic mucoralean fungus *Mucor circinelloides*. Mol Microbiol 97:844–865

Lengeler KB, Davidson RC, D'Souza C, Harashima T, Shen WC, Wang P, Pan X, Waugh M, Heitman J (2000) Signal transduction cascades regulating fungal development and virulence. Microbiol Mol Biol Rev 64:746–785

Li L, Wright SJ, Krystofova S, Park G, Borkovich KA (2007) Heterotrimeric G protein signaling in filamentous fungi. Annu Rev Microbiol 61:423–452

Linz JE, Katayama C, Sypherd PS (1986) Three genes for the elongation factor *EF-lα* in *Mucor racemosus*. Mol Cell Biol 6:593–600

Litcher A, Mills D (1997) Fill, a G-protein alpha-subunit that acts upstream of cAMP and is essential for dimorphic switching in haploid cells of *Ustilago hordei*. Mol Gen Genet 256:426–435

Liu H, Kohler J, Fink GR (1994) Suppression of hyphal formation in *Candida albicans* by mutation of a STE12 homolog. Science 266:1723–1736

Lo WS, Dranginis AM (1998) The cell surface flocculin Flo11 is required for pseudohyphae formation and invasion by *Saccharomyces cerevisiae*. Mol Biol Cell 9:161–171

Lorenz MC, Heitman J (1997) Yeast pseudohyphal growth is regulated by GPA2, a G protein alpha homolog. EMBO J 16:7008–7018

Lorenz MC, Pan X, Harashima T, Cardenas ME, Xue Y, Hirsch JP, Heitman J (2000) The G protein coupled receptor *gpr1* is a nutrient sensor that regulates pseudohyphal differentiation in *Saccharomyces cerevisiae*. Genetics 154:609–622

Madeo F, Herker E, Maldener C, Wissing S, Lächelt S, Herlan M, Fehr M, Lauber K, Sigrist SJ, Wesselborg S, Fröhlich KU (2002) A caspase-related protease regulates apoptosis in yeast. Mol Cell 9:911–917

Madzak C, Blanchin-Roland S, Cordero-Otero RR, Gaillardin C (1999) Functional analysis of upstream regulating regions from the *Yarrowia lipolytica XPR2* promoter. Microbiology 145:75–87

Maeda T, Watanabe Y, Kunitomo H, Yamamoto M (1994) Cloning of *pka1* gene encoding the catalytic subunit of the cAMP-dependent protein kinase in *Schizosaccharomyces pombe*. J Biol Chem 269:9632–9637

Magee PT (1997) Which came first, the hypha or the yeast? Science 277:52–53

Maidan MM, Thevelein JM, Van Dijck P (2005) Carbon source induced yeast-to-hypha transition in *Candida albicans* is dependent on the presence of amino acids and on the G protein coupled receptor *Gpr1*. Biochem Soc Trans 33:291–293

Malathi K, Ganesan K, Datta A (1994) Identification of a putative transcription factor in *Candida albicans* that can complement the mating defect of *Saccharomyces cerevisiae ste12* mutants. J Biol Chem 269:22945–22951

Maresca B, Kobayashi GS (1989) Dimorphism in *Histoplasma capsulatum*: a model for the study of cell differentiation in pathogenic fungi. Microbiol Mol Biol Rev 53:186–209

Maresca B, Kobayashi GS (2000) Dimorphism in *Histoplasma capsulatum* and *Blastomyces dermatitidis*. Contrib Microbiol 5:201–216

Martinez-Espinoza AD, Ruiz-Herrera J, Leon Ramirez CG, Gold SE (2004) MAP kinase and cAMP signaling pathways modulate the pH- induced yeast to mycelium dimorphic transition in the corn smut fungus *Ustilago maydis*. Curr Microbiol 49:274–281

Mattia E, Carruba G, Angiolella L, Cassone A (1982) Induction of germ tube formation by *N*-acetyl-D-glucosamine in *Candida albicans*: uptake of inducer and germinative response. J Bacteriol 152:555–562
Maw T, Tan TK, Khor E, Wong SM (2002) Complete cDNA sequence of chitin deacetylase from *Gongronella butleri* and its phylogenetic analysis revealed clusters corresponding to taxonomic classification of fungi. J Biosci Bioeng 93:376–381
Mayorga ME, Gold SE (1999) A MAP kinase encoded by the *ubc3* gene of *Ustilago maydis* is required for filamentous growth and full virulence. Mol Microbiol 34:485–449
McCreath KJ, Specht CA, Robbins PW (1995) Molecular cloning and characterization of chitinase genes from *Candida albicans*. Proc Natl Acad Sci U S A 92:2544–2548
Medoff G, Kobayashi GS, Painter A, Travis S (1987) Morphogenesis and pathogenecity of *Histoplasma capsulatum*. Infect Immun 55:1355–1358
Miles S, Li L, Davison J, Breeden LL (2013) Xbp1 directs global repression of budding yeast transcription during the transition to quiescence and is important for the longevity and reversibility of the quiescent state. PLoS Genet 9:e1003854
Mio T, Yabe T, Sudoh M, Satoh Y, Nakajima T, Arisawa M, Yamada-Okabe H (1996) Role of three chitin synthase genes in the growth of *Candida albicans*. J Bacteriol 178:2416–2419
Mishra C, Semino C, Mckreath K, De La Vega H, Jones BJ, Specht CA, Robbins P (1997) Cloning and expression of two chitin deacetylase genes of *Saccharomyces cerevisiae*. Yeast 13:327–336
Munro CA, Winter K, Buchan A, Henry K, Becker JM, Brown AJ, Bulawa CE, Gow NAR (2001) *Chs1* of *Candida albicans* is an essential chitin synthase required for synthesis of the septum and cell integrity. Mol Microbiol 39:1414–1426
Muthukumar G, Nickerson KW (1984) Ca(II)- calmodulin regulation of fungal dimorphism in *Ceratocystis ulmi*. J Bacteriol 159:390–392
Muthukumar G, Nickerson AW, Nickerson KW (1987) Calmodulin levels in yeasts and filamentous fungi. FEMS Microbiol Lett 41:253–255
Naglik JR, Challacombe SJ, Hube B (2003) *Candida albicans* secreted aspartyl proteinases in virulence and pathogenesis. Microbiol Mol Biol Rev 67:400–428
Nagy G, Farkas A, Csernetics Á, Bencsik O, Szekeres A, Nyilasi I, Vágvölgyi C, Papp T (2014) Transcription of the three HMG-CoA reductase genes of *Mucor circinelloides*. BMC Microbiol 14:93
Nantel A, Dignard D, Bachewich C, Harcus D, Marcil A, Bouin A, Sensen CW, Hogues H, Hoog M, Gordon P, Rigby T, Benoit F, Tessier DC, Thomas DY, Whiteway M (2002) Transcription profiling of *Candida albicans* cells undergoing the yeast-to-hyphal transition. Mol Biol Cell 13:3452–3465
Niimi M (1996) Dibutyryl cyclic AMP-enhanced germ tube formation in exponentially growing *Candida albicans* cells. Fungal Genet Biol 20:79–83
Niimi M, Niimi K, Cannon RD (1997) Temperature-related expression of the vacuolar aspartic proteinase (*APR1*) gene and beta-N-acetylglucosaminidase (*HEX1*) gene during *Candida albicans* morphogenesis. FEMS Microbiol Lett 148:247–254
Noor S, Punekar NS (2005) Allosteric NADP-glutamate dehydrogenase from Aspergilli: purification, characterization and implications for metabolic regulation at the carbon-nitrogen interface. Microbiology 151:1409–1419
Ocampo J, McCormack B, Navarro E, Moreno S, Garre V, Rossi S (2012) Protein kinase a regulatory subunit isoforms regulate growth and differentiation in *Mucor circinelloides*: essential role of *PKAR4*. Eukaryot Cell 11:989–1002
Odds FC (1988) *Candida* and *candidosis*. Balliere Tindall, London
Orlowski M (1991) *Mucor* dimorphism. Microbiol Rev 55:234–258
Palande AS, Kulkarni SV, León-Ramirez C, Campos-Góngora E, Ruiz-Herrera J, Deshpande MV (2014) Dimorphism and hydrocarbon metabolism in *Yarrowia lipolytica* var. *indica*. Arch Microbiol 196:545–556
Paranjpe V, Gupta-Roy B, Datta A (1990) Involvement of calcium, calmodulin and protein phosphorylation in morphogenesis of *Candia albicans*. J Bacteriol 154:524–528

Park JH, Kim H, Kim SH (2018) Azole antifungal drugs induce cell death by suppressing mTOR through PI3K/Akt inhibition in human breast cancer. In: proceedings of the American Association for Cancer Research annual meeting. Chicago, IL

Pathan EK (2017) Molecular characterization of NAD- and NADP-dependent glutamate dehydrogenases to understand their cause-effect relationship in dimorphic zygomycete *Benjaminiella poitrasii* and its evaluation as an antifungal target. A PhD thesis submitted to Academy of Scientific and Innovative Research (AcSIR), New Delhi, India

Pathan EK, Ghormade V, Deshpande MV (2017) Selection of reference genes for quantitative real-time RT-PCR assays in different morphological forms of dimorphic zygomycetous fungus *Benjaminiella poitrasii*. PLoS One 12:e0179454

Phillips AJ, Sudbery I, Ramsdale M (2003) Apoptosis induced by environmental stresses and amphotericin B in *Candida albicans*. Proc Natl Acad Sci U S A 100:14327–14332

Pillonel C (2005) Evaluation of phenylaminopyrimidines as antifungal protein kinase inhibitors. Pest Manag Sci 61:1069–1076

Pine L, Peacock CL (1958) Studies on the growth of *Histoplasma capsulatum*. IV. Factors influencing conversion of the mycelium phase to the yeast phase. J Bacteriol 75:167–174

Ramon AM, Porta A, Fonzi WA (1999) Effect of environmental pH on morphological development of *Candida albicans* is mediated *via* the *PacC*-related transcription factor encoded by *PRR2*. J Bacteriol 181:7524–7530

Rangel-Porras RA, Meza-Carmen V, Martinez-Cadena G, Torres-Guzmán JC, González-Hernández GA, Arnau J, Gutiérrez-Corona JF (2005) Molecular analysis of an NAD-dependent alcohol dehydrogenase from the zygomycete *Mucor circinelloides*. Mol Gen Genomics 274:354–363

Richard M, Quijano RR, Bezzate S, Bordon-Pallier F, Gaillardin C (2001) Tagging morphogenetic genes by insertional mutagenesis in the yeast *Yarrowia lipolytica*. J Bacteriol 183:3098–3107

Richardson CM, Williamson DS, Parratt MJ, Borgognoni J, Cansfield AD, Dokurno P, Francis GL, Howes R, Moore JD, Murray JB, Robertson A, Surgenor AE, Torrance CJ (2006) Triazolo[1,5-a]pyrimidines as novel CDK2 inhibitors: protein structure guided design and SAR. Bioorg Medl Chem Lett 16:1353–1357

Rivera-Rodriguez N, Rodriguez-del Valle N (1992) Effects of calcium ions on the germination of *Sporothrix schenckii* conidia. J Med Vet Mycol 30:185–195

Robson GD, Weibe MG, Trinci APJ (1991) Exogenous cAMP and cGMP modulate branching in *Fusarium graminearum*. J Bacteriol 154:524–528

Rodriguez-Caban J, Gonzalez-Velazquez W, Perez-Sanchez L, Gonzalez-Mendez R, Rodriguez-del Valle N (2011) Calcium/calmodulin kinase1 and its relation to thermotolerance and HSP90 in *Sporothrix schenckii*: an RNAi and yeast two-hybrid study. BMC Microbiol 11:162

Rogers KS (1971) Molecular interaction of competitive inhibitors with bovine liver glutamate dehydrogenase. J Biol Chem 246:2004–2009

Rogers KS, Boots MR, Boots SG (1972) Molecular interaction of six aromatic competitive inhibitors with bovine liver glutamate dehydrogenase. Biochim Biophys Acta 258:343–350

Ruiz-Herrera J (1994) Polyamines, DNA methylation and fungal differentiation. Crit Rev Microbiol 20:143–150

Ruiz-Herrera J, Martinez-Espinoza AD (1998) The fungus *Ustilago maydis*, from the Aztec cuisine to the research laboratory. Int J Microbiol 1:149–158

Ruiz-Herrera J, Sentrandreu R (2002) Different effectors of dimorphism in *Yarrowia lipolytica*. Arch Microbiol 178:477–483

Ruiz-Herrera J, Ruiz A, Lopez-Romero E (1983) Isolation and biochemical analysis of *Mucor bacilliformis* monomorphic mutants. J Bacteriol 156:264–272

Ruiz-Herrera J, Leon CG, Guevera-Olvera L, Carabez-Trejo A (1995) Yeast-mycelial dimorphism of haploid and diploid strains of *Ustilago maydis* in liquid culture. Microbiol. 141:695–703

Ryan O, Shapiro RS, Kurat CF, Mayhew D, Baryshnikova A, Chin B, Lin ZY, Cox MJ, Vizeacoumar F, Cheung D, Bahr S, Tsui K, Tebbji F, Sellam A, Istel F, Schwarzmüller T, Reynolds TB, Kuchler K, Gifford DK, Whiteway M, Giaever G, Nislow C, Costanzo M, Gingras AC, Mitra RD, Andrews B, Fink GR, Cowen LE, Boone C (2012) Global gene deletion analysis exploring yeast filamentous growth. Science 337:1353–1356

Ryley JF, Ryley NG (1990) *Candida albicans*: do mycelia matter? J Med Vet Mycol 28:225–239
Sabie FT, Gadd GM (1990) Effect of zinc on the yeast mycelium transition of *Candida albicans* and examination of zinc uptake at different stages of growth. Mycol Res 94:952–958
San-Blas F, San-Blas G (1985) Paracoccidioides brasiliensis. In: Szaniszlo PJ (ed) Fungal dimorphism. Plenum Press, New York, pp 93–120
Saudohar M, Bencina M, van de Vondervoort PJ, Panneman H, Legisa M, Visser J, Ruijter GJ (2002) Cyclic AMP-dependent protein kinase is involved in morphogenesis of *Aspergillus niger*. Microbiol. 148:2635–2645
Schulz BE, Kraepelin G, Hinkel-Mann W (1974) Factors affecting dimorphism in *Mycotypha* (Mucorales): correlation with the fermentation /respiration equilibrium. J Gen Microbiol 82:1–13
Sebghati TS, Engle JT, Goldman WE (2000) Intracellular parasitism by *Histoplasma capsulatum*: fungal virulence and calcium dependence. Science 290:1368–1372
Seike T, Nakamura T, Shimoda C (2013) Distal and proximal actions of peptide pheromone M-factor control different conjugation steps in fission yeast. PLoS One 8:e69491
Sentandreu M, Elorza MV, Sentandreu R, Fonzi WA (1998) Cloning and characterization of *PRA1*, a gene encoding a novel pH-regulated antigen of *Candida albicans*. J Bacteriol 180:282–289
Shapiro RS, Robbins N, Cowen LE (2007) Regulatory circuitry governing fungal development, drug resistance, and disease. Microbiol Mol Biol Rev 75:213–267
Shepherd MG, Yin YC, Ram SP, Sullivan PA (1980) Germ tube induction in *Candida albicans*. Can J Microbiol 26:21–26
Silverman SJ, Sburlati A, Slater ML, Cabib E (1988) Chitin synthetase is essential for septum formation and cell division in *Saccaharomyces ceevisiae*. Proc Natl Acad Sci U S A 87:7424–7428
Soares DA, Oliveira MB, Evangelista AF, Venancio EJ, Andrade RV, Felipe MS, Petrofeza S (2013) Phospholipase gene expression during *Paracoccidioides brasiliensis* morphological transition and infection. Mem Inst Oswaldo Cruz 108:808–811
Soberanes-Gutiérrez CV, Juárez-Montiel M, Olguín-Rodríguez O, Hernández-Rodríguez C, Ruiz-Herrera J, Villa-Tanaca L (2015) The *pep4* gene encoding proteinase A is involved in dimorphism and pathogenesis of *Ustilago maydis*. Mol Plant Pathol 16:837–846
Steinbach WJ, Cramer RA Jr, Perfect BZ, Asfaw YG, Sauer TC, Najvar LK, Kirkpatrick WR, Patterson TF, Benjamin DK Jr, Heitman J, Perfect JR (2006) Calcineurin controls growth, morphology, and pathogenicity in *Aspergillus fumigatus*. Eukaryot Cell 5:1091–1103
Steinbach WJ, Reedy JL, Cramer RA Jr, Perfect JR, Heitman J (2007) Harnessing calcineurin as a novel anti-infective agent against invasive fungal infections. Nat Rev Microbiol 5:418–430
Stevens L, Duncan D, Robertson P (1989) Purification and characterization of NAD-glutamate dehydrogenase from *Aspergillus nidulans*. FEMS Microbiol Lett 48:173–178
Stewart E, Gow NAR, Bowen DV (1988) Cytoplasmic alkalinization during germ tube formation in *Candida albicans*. J Gen Microbiol 134:1079–1087
Stoldt VR, Sonneborn A, Leuker CE, Ernst JF (1997) *Efg1p*, an essential regulator of morphogenesis of the human pathogen *Candida albicans*, is a member of a conserved class of bHLH proteins regulating morphogenetic processes in fungi. EMBO J 16:1982–1991
Sudoh M, Tatsuno K, Ono N, Ohta A, Chibana H, YamadaOkabe H, Arisawa M (1999) The *Candida albicans CHS4* gene complements a *Saccharomyces cerevisiae skt5/chs4* mutation and is involved in chitin biosynthesis. Microbiology 145:1613–1622
Sypherd PS, Borgia PT, Paznokas JL (1978) Biochemistry of dimorphism in the fungus *Mucor*. Adv Microb Physiol 18:67–104
Szabo R, Stofanikova V (2002) Presence of organic sources of nitrogen is critical for filament formation and pH-dependent morphogenesis in *Yarrowia lipolytica*. FEMS Microbiol Lett 206:45–50
Tabor CW, Tabor H (1985) Polyamines in microorganisms. Microbiol Rev 49:81–99
Tani Y, Yamada Y, Kamihara T (1979) Morphological change in *Candida tropicalis* pK 233 caused by ethanol and its prevention by myo-inositol. Biochem Biophys Res Commun 91:351–355

Terenzi HF, Flawia MM, Tellenzinon MT, Tores HN (1976) The control of *Neurospora crassa* morphology by cyclic adenosine 3′, 5′- monophosphate and dibutyryl cyclic adenosine 3′, 5′-monophosphate. J Bacteriol 154:524–528

Thykaer J, Rueksomtawin K, Noorman H, Nielsen J (2009) Disruption of the NADPH-dependent glutamate dehydrogenase affects the morphology of two industrial strains of *Penicillium chrysogenum*. J Biotechnol 139:280–282

Tian X, Shearer G Jr (2002) The mold specific *MS8* gene is required for normal hypha formation in the dimorphic pathogenic fungus *Histoplasma capsulatum*. Eukaryot Cell 1:249–256

Travassos LR (1985) Sporothrix schenckii. In: Szaniszlo PJ (ed) Fungal dimorphism. Plenum Press, New York, pp 121–163

Tupe SG, Kulkarni RR, Shirazi F, Sant DG, Joshi SP, Deshpande MV (2015) Possible mechanism of antifungal phenazine-1-carboxamide from *Pseudomonas* sp. against dimorphic fungi *Benjaminiella poitrasii* and human pathogen *Candida albicans*. J Appl Microbiol 118:39–48

Valle-Maldonado MI, Jacome-Galarza IE, Gutierrez-Corona F, Ramirez-Diaz MI, Campos-Garcia J, Meza-Carmen V (2015) Selection of reference genes for quantitative real-time RT-PCR during dimorphism in zygomycete *Mucor circinelloides*. Mol Biol Rep 42:705–711

Veronese FM, Nyc JF, Degani Y, Brown DM, Smith EL (1974) Nicotinamide adenine dinucleotide-specific glutamate dehydrogenase of *Neurospora*: purification and molecular properties. J Biol Chem 249:7922–7928

Villalobos-Duno H, San-Blas G, Paulinkevicius M, Sánchez-Martín Y, Nino-Vega G (2013) Biochemical characterization of *Paracoccidioides brasiliensis* α-1,3-glucanase *Agn1p*, and its functionality by heterologous expression in *Schizosaccharomyces pombe*. PLoS One 8:e66853

Wang P, Heitman J (1999) Signal transduction cascades regulating mating, filamentation and virulence in *Cryptococcus neoformans*. Curr Opin Microbiol 2:358–362

Wang LC, Montalvo-Munoz F, Tsai YC, Liang CY, Chang CC, Lo WS (2015) The histone acetyltransferase *Gcn5* regulates ncRNA-ICR1 and FLO11 expression during pseudohyphal development in *Saccharomyces cerevisiae*. Biomed Res Int 2015:284692

Webster RH, Sil A (2008) Conserved factors Ryp2 and Ryp3 control cell morphology and infectious spore formation in the fungal pathogen *Histoplasma capsulatum*. Proc Natl Acad Sci U S A 105:14573–14578

Wolff A, Appel K, Petersen J, Poulsen U, Arnau J, Jacobsen M (2002) Recombinant dimorphic fungal cell. WO/2002/070721

Xu JR, Hamer JE (1996) MAP kinase and cAMP signaling regulate infection structure formation and pathogenic growth in the rice blast fungus *Magnaporthe grisea*. Genes Dev 10:2696–2706

Xue C, Bahn YS, Cox GM, Heitman J (2006) G protein-coupled receptor Gpr4 senses amino acids and activates the cAMP-PKA pathway in *Cryptococcus neoformans*. Mol Biol Cell 17:667–679

Ying S, Feng M, Keyhani NO (2013) *B. bassiana* G-protein coupled receptor. Environ Microbiol 15:2902–2921

Zacchi LF, Gomez-Raja J, Davis DA (2010) *Mds3* regulates morphogenesis in *Candida albicans* through the TOR pathway. Mol Cell Biol 30:3695–3710

Zhao X, Kim Y, Park G, Xu JR (2005) A mitogen-activated protein kinase cascade regulating infection-related morphogenesis in *Magnaporthe grisea*. Plant Cell 17:1317–1329

Zinjarde SS, Pant A, Deshpande MV (1998) Dimorphic transition in *Yarrowia lipolytica* from oil polluted seawater. Mycol Res 102:553–558

Diversity, Ecology and Utilization of Soil Fungi: Indian Scenario

4

C. Manoharachary and D. Nagaraju

Abstract

Soil is a dynamic medium and complex ecosystem harbouring millions of microbes including fungi. Soil maintains a dynamic equilibrium of fungi and other microbes in spite of the constellation of physico-chemical factors. Vegetation along with multiple and minute habitats of varied physico-chemical setup influences the soil fungal communities. Soil fungi can be isolated by conventional, non-conventional and molecular methods. Interestingly, very little is known about the patterns of soil fungal diversity and their functional roles over large geographic scales. Most dominant fungi are Aspergilli and Penicilli followed by anamorphic fungi and others. There is a need to emphasize more on the ecological grouping of soil fungi. Soil and plant health is dependent on the functional activity of soil fungi. Soil fungi are known to play an important role in the cycling of elements, biogeographical demarcation, biogeochemical transformations, recycling of stored energy, degradation of organic matter, soil fertility and nutrient mobilization. The utility of fungi in the degradation of xenobiotic compounds, waste management, in medicine, agriculture, industry, carbon sequestration, biotechnological processes and other related activities, is very well known. Mother earth has been explored all over the world for fungi and microbes, still it harbours new and rare fungi and microbes which need to be discovered and classified accurately. Many soil fungi have been cultured artificially and are made available through germ plasm conservation centres. Further, in view of the importance attached to soil fungi, soil fungal floristics need to be surveyed and observed critically on individual capacity rather than en masse.

C. Manoharachary (✉)
Mycology and Molecular Plant Pathology Laboratory, Department of Botany, Osmania University, Hyderabad, Telangana, India

D. Nagaraju
Department of Botany, Government Degree College, Warangal, Telangana, India

T. Satyanarayana et al. (eds.), *Advancing Frontiers in Mycology & Mycotechnology*,
https://doi.org/10.1007/978-981-13-9349-5_4

Keywords
Diversity · Biotechnology · Distribution · Soil · Ecology · Fungi · Identification

4.1 Introduction

Soil is a complex ecosystem and is composed of multiple and minute habitats. Soil harbours almost all major groups of fungi. Stotzky (1997) has reported that the number of genera and species existing in soil is more than in any other environment. Soil being exposed to various conditions including extreme situations encases all microorganisms present on this planet. It is a known fact that fungi are the decomposing agents, help in biogeochemical transformations and recycle the stored energy and nutrients of organic matter which has been degraded by other microbes. Therefore, fungi have been considered as major players as recyclers of biosphere. It is established that physico-chemical composition of soil influences its fertility and also plays an important role in the distribution, seasonal variations and activity of fungi inhabiting the soil. However, it is also true that the soil fertility is dependent on the qualitative and quantitative structure and function of microbes and fungi inhabiting it. But for fungi being natural scavengers the planet could have been surrounded by piles of detritus and dead plant matter. Further geo-fungi are involved in the food chain cycle through their interaction with other living biota. Fungi are also known as recyclers of waste products, chemicals, transformers and biodegraders of xenobiotics. Soil is conglomerate of abiotic compounds with diversified microscopic organisms.

4.2 Soil as an Ecosystem, Fungal Diversity and Their Distribution in Soil

Soil is a natural medium in which diversified plant groups live, multiply and die which in turn becomes a perennial source of organic matter that gets recycled by microbes and fungi for plant growth and nutrition. A variety of fungi occurs in soil which range from lower (Chytrids) to higher fungi (Agarics), saprophytes to pathogens and predaceous to mutualistic mycorrhizal fungi. Fungi belonging to all groups occur in soil as hyphae, rhizomorphs, chlamydospores, sclerotia, asexual spores such as zoospores and conidia and sexual spores such as oospores, zygospores, ascospores and basidiospores.

The fungal diversities in soil have been studied by different workers (Watanabe 2011; Manoharachary et al. 2014) to know the fungi occurring in soil both quantitatively and qualitatively, ecologically, in relation to environmental variables, colonizing habitats and also to domesticate them effectively so as to use them in manufacturing nutritional, fermentative, pharmaceutical, agriculturally important materials, cosmetic materials and others. The diversity data also provides database on the available fungi and also about the invading or imported taxa in various habitats. The fungal diversities in soil can be understood readily by going through the list of the fungi reported in various soils of the world. More accurate, reliable and critical identification of soil fungi by both traditional and modern techniques is required. However, the modern system is still in the embryonic stage; therefore, the available gene data is in poor condition. Many morpho-taxonomists have been lost, some being endemic, and the classical taxonomy has received little attention due to shifting of mycologists towards molecular taxonomy. Many of the herbaria lack type species, which has created the lacunae in the fungal taxonomy. Therefore, confusion exists due to loss of traditional morpho-taxonomists and upcoming modern molecular taxonomists who are in the embryonic stage. Presently, the fungal taxonomy is now surrounded by morpho-taxonomists and gene analysis.

Soil is an ecosystem which comprises many microhabitats and harbours diversified groups of microbes and fungi. Most of the fungi are microscopic and some of them are macroscopic (*Agaricales*), and these microbiota are the decomposers, agents of mineral cycling, recycles stored energy and components of organic matter. Soil also includes symbiotic rhizobia, actinorhiza and mycorrhizae which serve the purpose of soil fertilization and uptake of nutrients to crops and forest plants. Fungi and other microbes play an important role as recyclers of biosphere. The quantitative and qualitative nature of microbes is of paramount importance in maintaining soil fertility along with chemical composition of soils. Soil fungi are also involved in food web through their interaction with other living biotic communities. Fungi are also involved as transformers and biodegraders of xenobiotics. The microphyte diversity of terrestrial ecosystems is dependent on below ground microbial diversity. Mycota is one of the diverse groups of organisms on the earth, which are the agents that govern carbon cycling, plant nutrition and plant disease production. These are distributed in different soil ecosystems but the distribution of different fungal groups has been less documented. Several edaphic factors, namely, pH, N, P, K, Ca and others, have greater impact on fungal distribution. Many fungal taxa are cosmopolitan in distribution but their endemicity is strong in tropical conditions. They also play an important role in driving carbon cycling in forest soils, mineral nutrition of plants and alleviate carbon utilization by other soil microbes.

Soil inhabits fungi and other microorganisms such as bacteria actinobacteria, cyanobacteria, protozoa and others. There are two groups of soil fungi which include indigenous fungi being isolated only from soil and other arrivals which are frequently isolated from surrounding habitats. Fungi colonizing underground parts considered as soil-borne appear to be typical soil fungi, sometimes fungi from air get contaminated into soil and are considered as casuals. Carris et al. (1989) have isolated 63 fungal species from the cyst *Heterodera schachtii* that causes disease in soybean and often the fungi associated with roots, decomposing litter fallen seeds and other plant parts and dead animals may get associated with soil. Fungal floras of various soil types have been studied worldwide. Thousands of research publications are available on soil fungi from all over the world, and some researchers might have described the fungi in detail while others might have listed without any descriptions. Therefore, numerous fungi have been described from different parts of the world; hence, it is almost impossible to list out all such soil fungi and also to refer all publications equally. Interestingly, isolation methods, media used, incubation conditions, soil types, soil physical conditions and other factors influence the quantitative and qualitative composition of fungi and microbes. It is possible that various species of *Pythium, Saprolegnia, Achlya* and others can be detected by means of baiting methods using boiled grass blades, cucumber seeds, hemp seeds, pollen grains, insect parts and others. Some ascomycetous fungi may get isolated including basidiomycetes by Warcup soil plate method but an overgrowth of fast-growing fungi such as *Rhizopus* and *Aspergillus* may not allow other fungi to grow. Waksman (1916) dilution plate method may yield a variety of fungi but did not yield perennating fungi. Majority of soil fungi remain unidentified because of non-sporulation. Identification of soil fungi with the adaption of appropriate scientific names is important for the study of soil fungi. In view of the importance attached to soil fungi, research activity on soil fungal floras has increased. Therefore, soil fungal floristics need to be observed more on an individual capacity rather than in en masse. Mother earth though has been explored all over the world, still it harbours new fungi which need to be discovered and classified accurately.

The global fungal estimate is 0.8 million to 5.1 million (Blackwell 2011), of which only 1 lakh fungal species are described. Around 29,000 fungi are reported from India. However, majority of such fungi are from soil. The diversity and distribution of soil fungi is shaped by macro-ecological and community assembly processes. Interestingly, very little is known about patterns of soil fungal diversity and their functional roles over large geographic scales. It is also known that soil fungi may exhibit strong biogeographical patterns. Based on the available data of fungal diversity (Manoharachary et al. 2014), it is indicated that the most dominant fungi are species of *Aspergillus* followed by *Penicillium*, hyphomycetes and some members of *Ascomycota*. This indicates the available techniques are inadequate to find out fungi that are present in different soil ecosystems.

Fungi are distributed worldwide in soils and are versatile. For example, Aspergilli are more common in tropical soils, while Penicilli are abundant in temperate soils. Among zygomycetes, *Rhizopus, Mucor, Absidia* and *Cunninghamella* have often been isolated. Zoosporic fungi such as *Pythium* and *Allomyces* have been isolated more frequently, and among the ascomycetous fungi, perfect states of *Aspergillus* and *Penicillium* besides *Chaetomium* are the most common ascomycetous fungi; basidiomycetous fungi have not been isolated commonly. Anamorphic fungi are frequently isolated fungi and are richly represented by asexual states of *Aspergillus, Penicillium, Alternaria, Drechslera, Curvularia* and others. Non-sporulating fungi are usually not identified; therefore, it is difficult to classify them in spite of using molecular tools as the sterile fungi show more than one perfect stage. However, it is hoped that with increasing knowledge and techniques, unidentified fungi on agar culture may get identified more readily together with the molecular studies. The authors have isolated 104 soil fungi from wild soils, cultivated soils, forest soils, pond muds, rhizosphere soils, etc. and are listed in Table 4.1. Similarly, soil fungi that are isolated in pure culture have to be named with right names on the basis of diversified characteristic features. Fungal diversities have to be widely acknowledged on the basis of voucher specimens such as holotype for the record and

Table 4.1 List of common soil fungi

Sl. no.	Fungal species	Substance	Year and place of collection	Collected by	Accession no.
1.	*Absidia cylindrospora* Hagem	Soil	2011, Hyderabad	Kunwar	OUFH 830
2.	*Acremonium chrysogenum* (Thirum. & Sukap.) Gams	Soil	2005, Mulugu	Kunwar	OUFH 574
3.	*Allomyces arbuscula* Butler, E.J. Butler	Forest soil	1981, Maha-Boobnagar	Madhusudan Rao	OUFHS 11
4.	*Alternaria alternata* (Fr.) Keissl	Soil (*Sesamum*)	1977, Hyderabad	Manohar.	OUFHS 12
5.	*Arthrobotrys foliicola* Matsuchima	Soil	2008, Narsapur	Kunwar	OUFH 572
6.	*Aspergillus awamori* Nakaz	Soil (Tamarind)	2005, Hyderabad	Kunwar	OUFH 340
7.	*Aspergillus candidus* Link	Pond mud	1979, Nizamabad	Manohar.	OUFHS 25
8.	*Aspergillus fumigatus* Fresen.	Forest, wild, cultivated soils	1974, Anantagiri hills	Manohar.	OUFHS 35
9.	*Aspergillus nidulans* (Eidam) G. Wint.	Mud soil	1996, Khammam	Manohar.	OUFHS 40
10.	*Aspergillus niger* Tiegh	Forest soil	1974, Anantagiri hills	Manohar.	OUFHS 42
11.	*Aureobasidium pullulans* (de Bary) Arnaud	Forest soil	1989, Amrabad	Manohar.	OUFHS 58
12.	*Beltrania rhombica* Penz.	Wild soil	2002, Narsapur	Narsimha charyulu	OUFHS 60
13.	*Blakeslea trispora* Thaxt.	Forest soil	1981, Amrabad	Reddy	OUFHS 64
14.	*Chaetomella raphigera* Swift	Forest, wild, cultivated soils	1984, Anantagiri hills	Manohar	OUFHS 72
15.	*Chaetomium abuense* Lodha	Soil	2004, Hyderabad	Kunwar	OUFH 237
16.	*Chaetomium aureum* Chivers	Pond mud	1983, Hyderabad	Manohar.	OUFHS 75
17.	*Chaetomium globosum* Kunze	Paddy field soil	1979, Nanded, Nizamabad	Manohar.	OUFHS 79

(continued)

Table 4.1 (continued)

Sl. no.	Fungal species	Substance	Year and place of collection	Collected by	Accession no.
18.	*Circinella muscae* (Sorokin) Berl. De Toni	Cultivated soil (castor)	1981, Hyderabad	Ramarao	OUFHS 93
19.	*Circinella simplex* Tiegh.	Silty, clay, loamy, cultivated soil	1979, Nanded, Nizamabad	Manohar.	OUFHS 94
20.	*Cladosporium herbarum* (Pers.) Link	Rhizosphere soil (*Ocimum abscendens*)	1975, Hyderabad	Manohar.	OUFHS 96
21.	*Cladosporium macrocarpum* Preuss	Polluted pond mud	1993, Hyderabad	Narendra Babu	OUFHS 97
22.	*Cochliobolus australiensis* (Tsuda & Ueyama) Alcorn (=*Drechslera australiensis*)	Forest, wild, cultivated soils	1984, Ananthagiri hills	Manohar.	OUFHS 102
23.	*Cochliobolus geniculatus* R. Nelson (=*Curvularia geniculata*)	Soil (cashew nuts)	1980, Hyderabad	Ramarao	OUFHS 103
24.	*Cochliobolus hawaiiensis* Alcorn (=*Drechslera hawaiiensis*)	Coastal soil	1969, Chirala	Lakshmi-narsim-ham	OUFHS 104
25.	*Cochliobolus lunatus* R.R. Nelson & Haasis (=*Curvlularia lunata*)	Pond mud	1974, Hyderabad	Manohar.	OUFHS 105
26.	*Colletotrichum capsici* (Syd.) E. J. Butler & Bisby	Rhizosphere, nonrhizo. Soil (Castor)	1981, Hyderabad	Ramarao	OUFHS 109
27.	*Cunninghamella blakesleeana* Lendner	Pond mud	1976, Hyderabad	Manohar.	OUFHS 114

(continued)

Table 4.1 (continued)

Sl. no.	Fungal species	Substance	Year and place of collection	Collected by	Accession no.
28.	*Cunninghamella echinulata* (Thaxt.) Thaxt. ex Blakeslee	Cultivated soil	1976, Nanded	Manohar.	OUFHS 115
29.	*Curvularia clavata* B.L. Jain	Soil (spinach)	1995,Hyderabad	Padma	OUFHS 117
30.	*Curvularia lunata* var. *aeria* (Bat. J.A. Lima & C.T. Vasconc.) M. B. Ellis	Forest, wild, cultivated soils	1984, Ananthagiri hills	Manohar.	OUFHS 118
31.	*Emericellopsis minima* Stolk	Forest, garden soil	1977, Khammam	Manohar.	OUFHS 128
32.	*Eurotium chevalieri* Mangin	Soil (cashew nut)	1980, Hyderabad	Ramarao	OUFHS 132
33.	*Fusarium chlamydosporum* Willenw. & Reinking	Riverbank soil	1984, Gadwal	Manohar.	OUFHS 134
34.	*Fusarium oxysporum* Schlecht	Forest, wild, cultivated soils	1984, Vikarabad	Manohar.	OUFHS 137
35.	*Fusarium poae* (Peck) Wollenw.	Rhizosphere, soil (castor)	1981, Hyderabad	Ramarao	OUFHS 138
36.	*Geotrichum candidum* Link	Polluted mud	1993, Karimnagar	Narendra Babu	OUFHS 141
37.	*Gibberella fujikuroi* (Sawada) Wollenw.	Rhizosphere soil, soils (cluster bean)	1978, Hyderabad	Manohar.	OUFHS 144
38.	*Gliocladium deliquescens* Sopp	Mud	2000, Karimnagar	Shantha Devi	OUFHS 150
39.	*Graphium terricola* Manohar. Rag. Rao, Rehana & Rama Rao	Sea shore soil	1975, Bheemilipatnam	Manohar.	OUFHS 154
40.	*Humicola fuscoatra* Traaen	Soil (cashew nut, castor)	1980, Hyderabad	Ramarao	OUFHS 158
41.	*Humicola grisea* Traaen	Riverbank, cultivated soils	1977, Gadwal	Reddy	OUFHS 159
42.	*Khuskia oryzae* H.J. Huds.	Deciduous forest soil	1981, Mannanur	Reddy	OUFHS 161

(continued)

Table 4.1 (continued)

Sl. no.	Fungal species	Substance	Year and place of collection	Collected by	Accession no.
43.	*Lasiodiplodia theobromae* (Pat.) Griffiths & Maubl.	Forest soil	1977, Hyderabad	Manohar.	OUFHS 162
44.	*Macrophomina phaseolina* (Tassi) Goid.	Rhizosphere, soils (cluster bean)	1978, Hyderabad	Manohar.	OUFHS 164
45.	*Magnaporthe grisea* (T.T. Hebert) M.E. Barr	Soil	1968, Nellore	Tilak	OUFHS 165
46.	*Monodictys putredinis* (Wall.) S. Hughes	Soil	2003, Hyderabad	Kunwar	OUFH 100
47.	*Mucor hiemalis* Wehmer	Soil (cluster bean)	1978, Hyderabad	Manohar.	OUFHS 176
48.	*Mucor laysanensis* Lendn.	Forest soil	2004, Vikarabad	Manohar.	OUFHS 178
49.	*Mucor racemosus* Fresen.	Pond mud	1976,Hyderabad	Manohar.	OUFHS 180
50.	*Mucor varians* Povah	Riverbank soil	1977, Gadwal	Reddy	OUFHS 181
51.	*Myrothecium cinctum* (Corda) Sacc.	Cultivated soil	1980, Hyderabad	Manohar.	OUFHS 182
52.	*Myrothecium roridum* Tode	Forest, cultivated soils	1984, Ananthagiri	Manohar.	OUFHS 185
53.	*Myrothecium verrucaria* (Alb. & Schwein.)	Rhizosphere soil (*Datura fastuosa*)	1975, Hyderabad	Manohar.	OUFHS 186
54.	*Nectria hematococca* Berk. & Broome	Forest, wild, cultivated soils	1984, Hyderabad	Manohar.	OUFHS 188
55.	*Nectria humicola* Rama Rao	Mazie field soil	1969, Narsapur	Ramarao	OUFHS 189
56.	*Neocosmospora vasinfecta* E. F. Sm.	Paddy field soil	1977, Karimnagar	Manohar.	OUFHS 191
57.	*Paecilomyces lilacinus* (Thom.) Samson	Pond mud	1983, Hyderabad	Manohar.	OUFHS 194

(continued)

Table 4.1 (continued)

Sl. no.	Fungal species	Substance	Year and place of collection	Collected by	Accession no.
58.	*Paecilomyces variotii* Bainier	Soil (cashewnut)	1980, Hyderabad	Ramarao	OUFHS 195
59.	*Penicillium chrysogenum* Thom	Forest soil	1981, Mahaboobnagar	Madhusudan Rao	OUFHS 198
60.	*Penicillium citrinum* Thom	Cultivated soil	1979, Nanded	Manohar.	OUFHS 200
61.	*Penicillium commune* Thom	Forest, wild, cultivated soils	1984, Ananthagiri	Manohar.	OUFHS 201
62.	*Penicillium digitatum* (Pers. & Fr.) Sacc.	Mud soil	1996, Karimnagar	Alivelu- manga-mma	OUFHS 204
63.	*Penicillium italicum* Stoll	Polluted soil	1993, Hyderabad	Narendra Babu	OUFHS 212
64.	*Penicillium rubrum* Stoll	Forest, wild, cultivated soils	1984, Ananthagiri	Manohar.	OUFHS 223
65.	*Penicillium variabile* Wehmer	Rhizosphere & nonrhizo. Soils (castor)	1981, Hyderabad	Ramarao	OUFHS 229
66.	*Periconia hispidula* (Pers.) E.W. Mason & M.B. Ellis	Forest soil	2004, Bhadrachalam	Manohar.	OUFHS 233
67.	*Pestalotiopsis glandicola* (Castagne) Steyaert	Rhizosphere soil (grape)	1964, Hyderabad	Ramarao	OUFHS 235
68.	*Pestalotiopsis mangiferae* (Henn.) Steyaert	Soil	1964, Hyderabad	Ramarao	OUFHS 236
69.	*Phoma eupyrena* Sacc.	Pond mud	1977, Gadwal	Reddy	OUFHS 238
70.	*Phoma fimeti* Brunaud	Pond mud soil	1975, Vikarabad	Manohar.	OUFHS 239
71.	*Phoma herbarum* Cooke	Fresh water tank mud	1996, Karimnagar	Alivelu- manga-mma	OUFHS 241
72.	*Phoma nebulosa* (Pers.) Berk.	Seashore soil (*Casuarina*)	1996, Anakapalli	Chandra Mohan	OUFHS 244
73.	*Phytophthora palmivora* (E.J. Butler) E.J. Butler	Field soil (*Colocasia*)	1980, Hyderabad	Satya Prasad	OUFHS 246

(continued)

Table 4.1 (continued)

Sl. no.	Fungal species	Substance	Year and place of collection	Collected by	Accession no.
74.	*Pseudeurotium ovale* Stolk	Rhizosphere soil (grape)	1965, Hyderabad	Rafia Mehdi	OUFHS 251
75.	*Pythium acanthicum* Dechsler, J. Wash.	Scrub jungle soil	1975, Vikarabad	Manohar.	OUFHS 255
76.	*Pythium aphanidermatum* (Edson) Fitzp.	Pond mud	1976, Hyderabad	Manohar.	OUFHS 256
77.	*Pythium butleri* subram.	Scrub jungle soil	1975, Vikarabad	Manohar.	OUFHS 257
78.	*Pythium carolinianum* Matthews	Pond mud	1983, Hyderabad	Manohar.	OUFHS 258
79.	*Pythium elongatum* Mathews	Pond mud	1978,Hyderabad	Manohar.	OUFHS 261
80.	*Pythium spinosum* Sawada	River bank soil	1981, Gadwal	Reddy	OUFHS 265
81.	*Rhizomucor miehei* (Cooney & R. Emers.) Schipper (=*Mucor miehei*)	Soil	1984, Warangal	Veugopal Rao	OUFHS 266
82.	*Rhizopus arrhizus* A. Fish. var. *arrhizus*	Rhizosphere soil (grape)	1965, Hyderabad	Rafia Mehdi	OUFHS 268
83.	*Rhizopus microsporus* var. *chinensis* (Saito) Schipper & Stalpers	Forest soil	2004, Vikarabad	Manohar.	OUFHS 269
84.	*Rhizopus stolonifer* var. *stolonifer* (Ehrenb.) Vuill.	Pond mud	1976, Vikarabad	Manohar.	OUFHS 270
85.	*Saprolegnia monoica* Pringsheim	Maize field soil	1964, Hyderabad	Ramarao	OUFHS 274
86.	*Scolecobasidium humicola* G.L. Barron & L. V. Busch	Soil (spinach)	1995,Hyderabad	Padma	OUFHS276
87.	*Scopulariopsis brumptii* Salv.-Duval	Rhizosphere soil (grape)	1965, Hyderabad	Rafia Mehdi	OUFHS 278

(continued)

Table 4.1 (continued)

Sl. no.	Fungal species	Substance	Year and place of collection	Collected by	Accession no.
88.	*Sordaria fimicola* (Oberge ex. Desm.) Ces. & De Not.	Pond mud	1977, Vikarabad	Manohar.	OUFHS 280
89.	*Stachybotrys atra* Corda	Soil (spinach)	1995, Hyderabad	Padma	OUFHS 288
90.	*Stachybotrys bisbyi* (Sriniv.) G.L. Barron	Forest soil	1981, Mahaboobnagar	Madhusudan Rao	OUFHS 289
91.	*Stachybotrys parvispora* S. Hughes	Soil (spinach)	1995, Hyderabad	Padma	OUFHS 292
92.	*Talaromyces funiculosus* (Thom) Samson, Yilmaz, frisvad & Seifert (= *Penicillium funiculosum*)	Soil	2010, Hyderabad	Kunwar	OUFH 688
93.	*Talaromyces trachyspermus* (Samson & Abdel-Fattah) Yaguchi	Forest soil	2004, Narsapur	Manohar.	OUFHS 299
94.	*Thermoascus aurantiacus* Miehe	Soil	1984, Rampachoda-varam	Venugopal Rao	OUFHS 301
95.	*Thielavia terricola* (Gilman & Abbott) Emmons	Forest soil	1981, Amrabad	Madhusu-dan Rao	OUFHS 303
96.	*Torula caligans* (Batista & H.P. Upadhyay) M.B. Ellis	Soil	1974, Adilabad	Singh	OUFHS 304
97.	*Trichoderma atroviride* karst.	Polluted soils	2003, Hyderabad	Satyavani	OUFHS 307
98.	*Trichoderma citrinoviride* Bisset	Forest soil	2002, Vikarabad	Narasimhacharyulu	OUFHS 309
99.	*Trichoderma hamatum* (Bonord.) Bainier	Soil	2002, Vikarabad	Kunwar	OUFH 059
100.	*Trichoderma harzianum* Rifai	Forest soil	2001, Chittor	Srilakshmi	OUFHS 312

(continued)

Table 4.1 (continued)

Sl. no.	Fungal species	Substance	Year and place of collection	Collected by	Accession no.
101.	*Trichoderma virens* (Miller, Giddens & Foster) von Arx	Forest soil	2004, Vikarabad	Nagamani	OUFHS 321
102.	*Trichoderma viride* Pers. (=*Trichoderma lignorum*)	Soil (*Datura fastuosa*)	1975, Hyderabad	Manohar.	OUFHS 322
103.	*Zygorhynchus moelleri* Vuill.	Cultivated soil	1979, Nanded	Manohar.	OUFHS 332
104.	*Zygosporium masonii* Hughes	Forest soil	2005, Vikarabad	Kunwar	OUFH 282

OUFHS Osmania University Fungal Herbarium–Soil, *Manohar* Manoharachary

description of new species. Cultures are always essential for their effective use in biotechnology. The original cultures need to be single source culture.

4.3 Methodology

Soil samples need to be collected from cultivated and uncultivated soils of diversified habitats along with soil passport card. The soil fungi can be isolated by various methods, but soil dilution plate method, soil plate method by direct soil inoculation and baiting methods are the best methods for isolation of various soil fungi. By single hyphal tip from germinating reproductive unit may be placed on full pledged agar mediam and the growth of colony be observed. Czapek's agar medium, plain water agar medium, PDA medium, soil extract agar medium and others are recommended. The main drawback for dilution plate method is that it neglects slow-growing fungi. The diluted soil suspensions are poured on to appropriate selective isolation media, and single-spore cultures are established later (Davet and Rouxel 2000).

Warcup soil plate method (1950) includes direct soil inoculation on to the sterile agar medium in a sterile Petri dish. The media include Czapek's agar, tomato agar, soil extract agar and others. The drawback of this method is that it tends to neglect fungi of qualitative nature and overgrowth of the fast-growing fungi. Bating method is often used for Zoosporic fungi. Immersion slide method (Chesters 1948) has also been found useful to isolate soil fungi as this also represents the soil fungi associated with soil profile. Molecular methods such as total fungal community DNA extraction, metagenomics, phylogenetic analysis of 18S, 16S rDNA sequences, understanding of genetic diversity, PCR and several other modern techniques are available to understand and evaluate soil fungal community.

4.4 Identification of Soil Fungi

Fungal taxa thus isolated have been identified on the basis of morpho-taxonomic criteria by comparing with known species; the morphologies that are observed through a stereomicroscope, compound microscope and electron microscope along with cultural characters have to be noted. Hyphal morphology, spore morphology, ontogeny, cultural characteristics, fruiting structures, etc. have to be noted down.

Experienced mycologists may identify some fungi at a glance; however, the most suitable taxon may be assessed after repeated observations and suitable literature survey. In case of new species, the Melbourne code 2013 has to be adopted and deposition be made at gene bank.

The morphologies of fungi be maintained continuously and molecular techniques may be employed for clarifying ambiguous and vague entities.

The available monographs on geo-fungi are many (Gilman 1957; Domsch et al. 1980; Watanabe 2002); however, different monographs are available for different groups of fungi (Raper and Thom 1949; Raper and Fennel 1965; Barron 1968; Ellis 1971, 1976; Subramanian 1971). Domsch et al. (1980) has listed 450 species, and Nagamani et al. (2006) have described 332 fungal species from India. The inventorization, monitoring and biodiversity status of soil fungi have been discussed by Bills et al. (2004).

4.5 Ecological Grouping of Soil Fungi

Fungi are known to colonize cultivated crop soils, wild forest soils, soils of highest mountain peaks, deep permafrost soils, geothermal and humid soils of the volcanic horizon, mine soils and highly alkaline soils. Cold-loving fungi are restricted to Polar Regions which can tolerate 0–16 °C, and examples of this group belong to *Leptomitus, Penicillium, Cryptococcus, Chrysosporium* and others. The fungi occurring in extreme environments may be of biotechnological importance as they have the potential for the production of extremozymes, secondary metabolites, bioremediation properties and others (Nonzom and Sumbali 2015). Desert soils contain a variety of yeasts that have been documented from hot and cold deserts. It seems that man-made contaminations of Antarctica might have added *Penicillium, Aspergillus* and other common fungi from soil and air. Endolithic conidial fungi are also common (Sterflinger et al. 2012). Halophilic fungi are considered as a major source of diverse and novel metabolites. The salt-loving fungi mainly belonging to *Ascomycota* do occur in marine soils. A distinctive group of fungi exist in marine waters and marine soils. Terrestrial aquatic hyphomycetes are associated with the litter fallen on to the soil.

Thermophiles are a group of extremophiles which require relatively high temperature (41–122 °C). Successful isolation of soil thermophiles requires incubation of soil on specified media at 45 °C/50 °C. Normally, thermopiles occur in compost, hay, wood chips and also in tropical desert soils. Some of the common soil thermophiles include *Aspergillus fumigatus, Scytalidium thermophilum, Chaetomium thermophilum, Thermomyces* sp., *Humicola* sp. and others. Thermophilic fungi possess valuable enzymes such as pectinase, cellulase, xylanase and also secrete a number of secondary metabolites of biotechnological importance (Rajasekaran and Maheshwari 1993). The voluminous literature that has accumulated on soil fungi (Manoharachary et al. 2014; Taylor and Sinsabaugh 2015; Taylor et al. 2000) indicates that a number of fungal species do occur in wild and forest soils consisting of rich diversified flora and medicinal plants. This clearly indicates that mesophiles dominate the soil fungal biotic community. Cultivated soils also support rich fungal flora. Quantitatively and qualitatively, the fungi are more and diversified in species composition with reference to wild and forest soils. However, the cultivated soils no doubt support quantitatively richness of fungi but qualitatively it is represented by few fungal species. The author's experience indicates that wild and forest soils have shown quantitatively and qualitatively richness of fungi in scrub jungle forest, deciduous forests and also in grassland soils than in cultivated crop soils. However, grassland soils have shown richness of fusarial fungi (Manoharachary and Ramarao 1978; Manoharachary et al. 1989; Madhusudhan Rao and Manoharachary 1981). Submerged mud soils were rich in melanin pigment containing fungi, such as Ascomycetes, dematiaceous hyphomycetes and others. Manoharachary et al. (2014) have isolated 340 fungal species and have also indicated that anamorphic fungi formed the bulk in soil fungal biota. Further, it has also been shown that *Aspergillus* and *Penicillium* species are predominantly followed by *Chaetomium* and anamorphic fungi than other groups. Many soil-borne and root-borne pathogenic fungi were also encountered; 18 species representing the genus *Trichoderma,* a well-known biocontrol agent, were isolated. In general, it has been shown that forest and wild soils followed by rhizosphere soils and cultivated soils harbour a more number of fungi and richness of fungal species than mud soils, riverbank soils, sea shore soils, herbicide-treated soils, polluted soils and poultry farm soils (Manoharachary et al. 2014).

Temperature, pH, moisture, soil texture, soil organic matter, soil NPK and gaseous composition are known to influence the distribution and composition of soil fungi. Plant communities existing in diversified soils may have a greater impact on the quantitative and qualitative composition of soil fungi.

4.6 Soil Fungi and Soil Health

Soil health depends on soil quality and fertility. Soil health is the resultant of the interaction between different processes, properties and activities of soil microbiota including soil fungi. Soil fungi are the biological controllers, ecosystem regulators, decomposers and compound transformers. Therefore, soil fungi serve as ecosystem regulators, responsible for soil structure formation and modification of the habitats. Mycorrhizal fungi are known to stabilize the soil structure and serve as biofertilizer for plant growth. Fungi present in the soil also participate in hormone production, biological control, stress management, stabilization of soil organic matter and biodegradation of residues. Thus, soil health maintained by soil fungi is directly connected with the production of healthy food which has an impact on public and animal health (Frac et al. 2018).

Certain soils are not congenial to live for plant pathogens by limiting their survival or growth of the pathogen. The suppressive soils reduce fungal attack and are often effective against only one or two pathogens. Suppressiveness is of two types: (1) long-standing suppression which is a biological condition and appears to survive in the absence of plants and (2) inclusive suppression is initiated and sustained by crop monoculture by the addition of target pathogen. A number of soil-borne pathogens are represented by *Fusarium* sp., *Gaumannomyces* sp., *Phytophthora* sp. and *Pythium* spp. Few cause diseases in plants growing in conducive soils, and few other pathogens cause no disease in plants. Suppressiveness may be because of soil microflora and abiotic factors and may vary with the type of pathogen. Some studies have indicated that activities of antagonistic soil fungi, bacteria, actinomycetes and others are responsible for the suppression of pathogens. The formation of suppressive soils is due to the mechanisms envisaged by soil fungi and soil microbiota, namely, nutrient competition, amensalism, antagonism, parasitism and systemic-induced resistance. However, understanding of the exact mechanism in suppressive soil is still far from satisfaction and probably the application molecular assessment tools may bring more understanding of such activity (Garbeva et al. 2004).

4.7 Functions and Biotechnological Aspects of Soil Fungi

Fungi are known to play a role in organic matter production, decomposition, carbon sequestration, carbon mineralization and cycling of elements. Mycorrhizal associations in plants not only boost plant productivity but also the acquisition of water, phosphorus and nutrients. Fungal endophytes offer resistance to biotic and abiotic stress. Filamentous soil fungi promote macro-aggregate formation through soil particle binding with fungal hyphae and the fungal cell wall material as adhesive, while the most common activity of the soil fungi is the nutrient cycling. Most hydrolytic and oxidative capabilities are elaborated by soil fungi as they are principle degraders of plant cell wall material during decomposition. Production of cellulases, pectinases, laccases and others have been elaborated by many soil fungi. The potential use of chitin as a nitrogen source is widespread among fungi because the fungal cell wall includes chitin. Chitinase activity is used as an indicator of fungal biomass and metabolism. Further, the proteins get degraded by many fungi which is the largest source of nitrogen. The inorganic phosphate solubilizing fungi supply phosphorus to the plants. Mineral phosphates get degraded by *Penicillium, Aspergillus* and other soil fungi into organic soluble phosphate which then get transported to fungal hyphae and to the plants. A number of soil fungi are also known to play an important role in bioremediation. Fungi are linked to many other organisms in a complex soil food web. For example, wood decay fungi and nematodes live together. Fungal hyphae in soil secrete both extracellular and intracellular bacteria. Mycorrhizal helper bacteria help in the formation of ecto- and endo-mycorrhizas. Many insects consume fungal hyphae (Taylor and Sinsabaugh 2015).

Soil fungi are important in everyday affairs of human beings. Soil fungi are well-known degraders of raw or manufactured materials such as foodstuff, timber, textiles, leather, paint, glue, plastics, petroleum products, optical glasses and others. Some of the soil fungi such as *Aspergillus* are known to produce aflatoxins, ochratoxins and others. Soil yeasts are used in brewery industries, while mushrooms, truffles and morels are edible and some species like *Amanita, Clitocybe and Inocybe* are poisonous. Species of *Psilocybe* and *Paniolus* are hallucinogenic. Soil fungi are also used in food processing. Several antibiotics, growth hormones, organic acids, enzymes, mycoproteins and vitamins are extracted from soil fungi. Soil fungi play an important role in the biosphere by involving in the recycling of nutrients, afforestation programmes, wastewater treatments, detoxification, xenobiotics, biocontrol of diseases and in several other ways. Soil fungi are also involved in metabolic pathways and studying genetic mechanisms. However, some of the soil fungi such as *Phytophthora, Pythium, Rhizoctonia, Sclerotium, Macrophomina, Fusarium, Verticillium* and few others are also involved in causing soil-borne and root-borne diseases. Soil fungi are very successful organisms due to their great plasticity and physiological versatility. Some of the important soil-borne diseases include damping off seedlings, wilt diseases, root rot and several others.

The intense influence of soil microbes and fungi on human life and global biogeochemical cycles necessitates exploration of microbial and fungal genomes to expand our understanding of most microbial and fungal species on earth, particularly those showing low relative abundance. There is a need to understand the ecology of such rare microbes and fungal population and highlight molecular and computational methods for targeting taxonomic blind spots within the rare biosphere.

Soil fungi play an important role in pharmaceutical industries in the isolation of compounds such as penicillin, cyclosporine, lovastatin, etc. Some of the fungi such as *Trichoderma* spp. are considered as potential biocontrol agents to control parasites and predators as antagonists. Fungi such as *Arthrobotrys* have been considered as nematophagous fungi, while species of *Metarhizium* and *Beauveria* have been considered as insect pathogens. Soil fungi have gained importance in recent times and have been exploited for bioremediation of anthropogenic pollutants including pesticides, benzene, toluene, xylene, dyes, hydrocarbons and others. The unexplored soil fungi may become rich resource material for new genes and species valuable to biotechnology and medicine. Soil fungal biodiversity plays a pivotal role in sustaining growth and management of the ecosystem. In future, soil fungal diversity seems to be very challenging and advantageous to the biosphere. As on today, the number of species considered to be true soil fungi is around 15,000. It is estimated that a gram of soil may hold several thousand fungal species. The revised species in future will be additional new species which may get revealed under Genealogical Concordance Phylogenetic Species Recognition (GCPSR) programme (Taylor et al. 2000).

4.8 Rhizosphere Soil Fungi

The rhizosphere soil is the specialized ecological region, which is adjacent to the root system of the plant as influenced by the root exudates. The term rhizosphere was proposed by Hiltner (1904). The root exudates and root debris products attract many fungi and microbes. The interaction of fungi and plant root is essential for the nutrition and growth of the plant. The growth, development, productivity of many crop plants, forest plants, orchids, oilseed crops, horticultural plants, medicinal plants, cash crops and others are largely dependent on soil health, which is maintained by soil microbes, soil fungi and also rhizosphere microflora including fungi. Therefore, rhizosphere studies are of great interest to agriculturists, soil biologists, chemists, mycologists, microbiologists and molecular biologists. The rhizosphere microbes and fungi may influence the availability of nutrients, water and growth promoter and may also change the oxidation cum reduction potential. Mycorrhizae, a symbiotic association, are beneficial in the uptake of phosphorus, zinc and other minerals besides increasing the root surface area of the plant for effective ion absorption. The soil microbial and fungal interactions such as antagonism, competition, synergism occurring in soil and rhizosphere are of great importance in studying the microbial and fungal ecology of rhizosphere (Mukerji et al., 2006). The rhizosphere exudates include a wide range of organic and inorganic compounds that affect the microbial and fungal population. Therefore, the authors have studied the rhizosphere fungal flora employing soil plate, soil dilution and immersion methods. Selective media and enrichment techniques are used, which include CFU, MPN, particle counts, fluorescence microscopy, and biochemical methods such as assay for ATP, immunological and molecular techniques. The fungi isolated from rhizospheres of various plants are listed in Table 4.2.

4.9 Soil Fungi and Climate Change

Soil fungi play a critical role in the carbon cycle. Carbon which is essential for life on earth moves between air, soil and water. After the death of the plants, the carbon enters the soil, making the soil a reservoir of the carbon. The dead plant material is broken down by microbes and fungi in the soil, thus releasing the carbon into the air. The rate at which the carbon left the soil will have a major impact on the amount of atmospheric carbon, which is the key factor to drive climate change. One of the limiting factors to the growth of these decomposing fungi is the availability of nitrogen in the soil, which get solved by mycorrhizal fungi. The mycorrhizal fungi extract the nitrogen from the soil and make it available to the plants through their roots. Recently, it has been found by scientists that soils supporting ectomycorrhizal fungi contain 70 percent more carbon than the soils dominated by arbuscular mycorrhizal fungi. Thus, fungi have a greater role in the control of the global carbon cycle. Since the plants and mycorrhizae are interconnected, the future carbon cycling cannot be predicted without thinking about plants and mycorrhiza.

Table 4.2 Rhizosphere soil fungi

S. no.	Name of the fungal species
1.	*Acrophialophora fusispora* (S.B.Saksena) Samson
2.	*Alternaria alternata* (Fr.) Keissl
3.	*Aternaria humicola* Oudem
4.	*Aternaria longipes* (Elis & Everh.) E.W. Mason
5.	*Aternaria tenuissima* (Kunze) Wiltshire
6.	*Aspergillus caespitosus* Raper and Thom
7.	*Aspergillus chevalieri* Thom & Church
8.	*Aspergillus clavatus* Desm.
9.	*Aspergillus flavus* Link
10.	*Aspergillus foetidus* Thom & Raper
11.	*Aspergillus fumigates* Fresenius
12.	*Aspergillus nidulans* Eidam
13.	*Aspergillus niger* Tiegh
14.	*Aspergillus sydowii* Bainer & Sartory
15.	*Aureobasidium pullulans* (de Bary & Lowental) G. Arnand
16.	*Cephaliophora irregularis* Thaxt
17.	*Chaetomella raphigera* Swift
18.	*Chaetomium aureum* Chievers
19.	*Chaetomium globosum* Kunze
20.	*Chaetomium spirale* Zopf.
21.	*Cladosporium cladosporioides* (Fresen) G.A. de Vries
22.	*Cladosporium herbarum* (Pers.) Link
23.	*Cladosporium oxysporum* Berk. M.A. Curtis
24.	*Colletotrichum falcatum* Butler & Bisby
25.	*Cunninghamella echinulata*
26.	*Curvularia brachyspora* Boedijn
27.	*Curvularia clavata* Jain
28.	*Curvularia eragrostidis* Itenn & Mayer
29.	*Curvularia lunata* (Walker)Boedijn
30.	*Curvularia pallescens* Boedijn
31.	*Doratomyces microspores* (Sacc.) F.J. Marten & G.Sm.
32.	*Drechslera australiensis* (Bugnicourt) Subr. & Jain
33.	*Fusarium moniliforme* J.Sheld
34.	*Fusarium nivale* (Fr.) Sarauer
35.	*Fusarium oxysporum* Schltdl
36.	*Fusarium solani* Sacc.
37.	*Graphium penicillioides* Corda

(continued)

Table 4.2 (continued)

S. no.	Name of the fungal species
38.	*Humicola grisea* Traaen
39.	*Melanoma pomiformis* (Pres.ex. Fr.) Sacc
40.	*Memnoniella echinulata* (Rivolta) Galloway
41.	*Monodictys glauca* (Cooke & Harkn.) S.Hughes
42.	*Mucor racemosum* Fresen
43.	*Mucor sphaerospermum* Hagem
44.	*Mucor varians* (H.Mart.) Fr.
45.	*Myrothecium brachysporum* Nicot
46.	*Myrothecium gramineum* Lib.
47.	*Nigrospora oryzae*(Berk. & Broome) Petch.
48.	*Paecilomyces fusisporus* S.B. Saksena
49.	*Paecilomyces humicola* Onions & G.L. Barron
50.	*Penicillium chrysogenum* Thom.
51.	*Penicillium citreo-viride* Biourge
52.	*Penicillium fructigenum* Takeuchi
53.	*Penicillium funiculosum* Thom.
54.	*Penicillium oxalicum* Currie & Thom
55.	*Penicillium spinulosum* Thum
56.	*Penicillium tardum* Thom
57.	*Penicillium variabile* Wehmer
58.	*Penicillium varians* G.Sm.
59.	*Periconia atropurpurea* (Berk. M.A. Curtis) M.A. Litv.
60.	*Pestalotiopsis mangiferae* (Henn.) Steyaert
61.	*Phoma feckelli* Brackel
62.	*Phoma humicola* J.C. Gilman & E.V.Abbott
63.	*Pithomyces atro-olivaceus* (Cooke & Harkn.) M.B.Ellis
64.	*Pithomyces flavus* Berk. & Broome
65.	*Rhinocladiella basitona* (de Hoog) Arzanlou & Crous
66.	*Rhinocladiella mansonii* (Castell.) Schol-Schwarz
67.	*Rhizoctonia bataticola* (Taubenth.) E.J. Butler
68.	*Rhizopus nigricans stolonifer* (Ehrenb.) Vuill
69.	*Rhizopus nodosus arrhizus* A. Fisch.

(continued)

Table 4.2 (continued)

S. no.	Name of the fungal species
70.	*Scolecobasidium constrictum* E.V. Abbott.
71.	*Scolecobasidium humicola* G.L. Barron & L.V. Busch
72.	*Scopulariopsis brevicaulis* (Sacc.) Bainier
73.	*Scytalidium lignicola* Pesante
74.	*Spicaria elegans* (Corda) Harz.
75.	*Spicaria fumosorosea* (Wize) Vassiljevsky
76.	*Spicaria griseola* Sacc.
77.	*Spicaria helothis* Charles
78.	*Sporotrichum roseolum* Oudem. & Beij.
79.	*Stachybotrys chartarum* (Ehrenb.) S. Hughes
80.	*Stachybotrys parvispora* S.Hughes
81.	*Stachylidium bicolour* Link
82.	*Thielaviopsis paradoxa* (De Seynes) Hohn.
83.	*Torula herbarum* (Pers.) Link
84.	*Trichocladium canadense* S. Hughes
85.	*Trichoderma koningii* Oudem.
86.	*Trichoderma viride* Pers.
87.	*Trichuris spiralis* Hasselbr.
88.	*Veronaea apiculata* (J.H. Mill., Giddens &A.A. Foster) F.B. Ellis
89.	*Verticillium puniceum* Cooke & Ellis,
90.	*Wardomyces inflatus* (Marchal) Hennebert
91.	*Zygorhynchus moelleri moelleri* Vuill.

4.10 Conclusions

1. Soil is a dynamic medium for fungi and microbes and maintains balance in spite of constellation of physico-chemical factors.
2. Soil harbours diversified groups of fungi belonging to zoosporic fungi, zygomycota, *Ascomycota* and *Basidiomycota.*
3. Soil is a rich nutrient medium for the sustenance of fungi.
4. Diversified soils such as desert soils, temperate soils, tropical soils, forest soils, crop soils sand dunes, submerged soils, saline and mangrove soils, soils of high attitude and low attitude, and others support not only specific fungal taxa but also fungi common to all soils.
5. Soil physico-chemical factors, plant vegetation, altitude, meteorological conditions and other related factors influence the soil fungi both quantitatively and qualitatively.

6. There is a definite seasonal variation among soil fungi. Some fungal species are characteristically associated with one or the other soils. Further fungi like Aspergilli are distributed widely in tropics, while the Penicilli are associated with temperate soils.
7. Soil fungi are the fastest decomposing agents, help in biogeochemical transformations and recycle stored energy and nutrients. Soil fungi also help in carbon sequestration besides serving as natural scavengers, and are useful in industry, medicine, agriculture, waste management, and in biotechnology and other activities.
8. Some fungi exhibit antagonistic activity and plant growth promotion.

There is a need for in-depth studies on soil fungi, their diversity, ecology, conservation and utility for human welfare.

References

Barron GL (1968) The genera of hyphomycetes from soil. Williams and Wilkins Co, Baltimore, p 364

Bills GF, Christensen M, Power M, Thorn RG (2004) Saprobic soil fungi. In: Muellar GM, Bills GF, Foster MS (eds) Biodiversity of fungi. Inventory and monitoring methods. Elsevier Academic Press, Amsterdam, pp 271–302

Blackwell M (2011) The fungi: 1, 2, 3…..5.1 million species? Am J Bot 98:426–438

Carris LM, Glawe DA, Smyth CA, Edwards DI (1989) Fungi associated with populations of *Heterodera glycines* in two Illinois soybean fields. Mycologia 81(1):66–75

Chesters CGC (1948) A contribution to the study of fungi in the soil. Tr Br Mycol Soc 30(100):117

Davet P, Rouxel F (2000) Detection and isolation of soil fungi. Science Publishers, Enfield

Domsch KH, Gams W, Anderson TH (1980) Compendium of soil fungi, vol I and II. Academic, London

Ellis MB (1971) Dematiaceous Fungi. CMI, Kew, Surrey, p 608

Ellis MB (1976) More dematiaceous fungi. CMI, Kew, Surrey, p 507

Frac M, Hannula SE, Belka M, Jedryczka M (2018) Fungal biodiversity and their role in soil health. Front Microbiol 9:707

Garbeva P, Van Veen JA, van Elsas JD (2004) Microbial diversity in soil: selection of microbial populations by plant and soil type and implications for disease suppressiveness. Annu Rev Phytopathol 42:243–270

Gilman JC (1957) A manual of soil fungi. Iowa State University Press, Iowa

Hiltner L (1904) Über neuere Erfahrungen und Probleme auf dem Gebiete der Bodenbakteriologie unter besonderer Berücksichtigung der Gründüngung und Brache. Arb DLG 98:59–78

Madhusudhan Rao M, Manoharachary C (1981) Studies on the mycology of two pond muds and forest soils of Mannanur forest, Andhra Pradesh, India. Ind J Bot 4:137–138

Manoharachary C, Ramarao P (1978) Distribution and seasonal variation of some *Pythium* spp. in some soils, pond waters and mud soils of Hyderabad Dist. Mysore J Agric Sci 12:280–284

Manoharachary C, Reddy VR, Prasad V (1989) Population studies and phenology of fungi in scrub jungle forest soils of AP, India. Geophytology 19(2):131–139

Manoharachary C, Nagaraju D, Kunwar IK, Nagamani A (2014) Soil fungi from Andhra Pradesh, India. Indian J Bot 93(1&2):16–34

Mukerji KG, Manoharachary C, Singh J (2006) Microbial activity in the rhizosphere, vol XVI. Springer, Berlin/Heidelberg, p 349

Nagamani A, Kunwar IK, Manoharachary C (2006) Handbook of soil fungi. IK International Pvt Ltd, New Delhi, p 477

Nonzom S, Sumbali G (2015) Incidence and diversity spectrum of mitosporic fungi from the cold arid base soil of Moonland landscape (Ladakh), India. Intl J Pharma Bio Sci 6(1):252–261

Rajasekaran AK, Maheshwari R (1993) Thermophilic fungi: an assessment of their potential for growth in soil. J Biosci 18(3):345–354

Raper KB, Fennell DI (1965) The genus *Aspergillus*. Robert E Krieger Publ Co, Inc, Huntington, p 686

Raper KB, Thom C (1949) A manual of *Penicillia*. Williams and Wilkins Company, Baltimore, p 875

Sterflinger K, Tesei D, Zakharova K (2012) Fungi in hot and cold deserts with particular reference to microcolonial fungi. Fungal Ecol 5(4):453–462

Stotzky G (1997) Soil as an environment for microbial life. In: van Elsas JD, Trevors JT, Wellington EMH (eds) Modern soil microbiology. Marcel Dekker, New York, pp 1–20

Subramanian CV (1971) Hyphomycetes. ICAR Publications, New Delhi, p 930

Taylor L, Sinsabaugh RL (2015) The soil fungi: occurrence, phylogeny and ecology, chapter. In: Soil microbiology, ecology and biochemistry, pp 77–100

Taylor JW, Jacobson DJ, Kroken S, Kasuga T, Geiser DM, Hibbett DS, Fisher MC (2000) Phylogenetic species recognition and species concepts in fungi. Fungal Genet Biol 31(1):21–32

Waksman SA (1916) Do fungi live and produce mycelium in the soil? Sci NS 44:320–322

Warcup JH (1950) Soil plate method for the isolation of fungi from soil. Nature (London) 166:117–118

Watanabe T (2002) Pictorial atlas of soil and seed fungi. Morphologies of cultured fungi and key to species, 2nd edn. CRC Press, Boca Raton. 504 pp

Watanabe T (2011) Pictorial atlas of soil and seed fungi: morphologies of cultured fungi and key to species, 3rd edn. CRC Press, Boca Raton, p 426

Diversity and Bioprospecting of Yeasts from Extreme Environments

5

Shiv Mohan Singh, Nitin Adhapure, and Rohit Sharma

Abstract

The life of humans has been greatly benefited by microbes since their existence on earth. Extremophiles have unique energy transduction processes and adaptation strategies which help them to survive in an extreme environment (high or low temperature, pH, etc.). Fungi with yeast stage in their life cycle are found in various habitats including extreme environments like hot springs, alkaline lakes, hypersaline lakes, cold glaciers, deep ocean, and several others. Recently, psychrophilic yeasts have been studied widely, but contributions on thermophilic, acidophilic, and halophilic yeasts are lagging behind. *Mrakia*, *Leucosporidium*, and *Naganishia* are some examples of yeasts isolated from cold environments. In the past decades, several investigations have been undertaken on yeasts to assess their biotechnological potentials. Extremophilic yeasts produce enzymes, antifreeze proteins, heat shock proteins, PUFA, EPS, etc., which have immense applications in health, agriculture, and industry. In this chapter, diversity of extremophilic yeasts, strategies adapted for such niches, and potential applications of extremophilic yeasts in biotechnology have been discussed. Yeasts from extreme environments can be exploited in several biotechnological industries, thus helping in the bioeconomy of the country.

S. M. Singh
Banaras Hindu University (BHU), Varanasi, Uttar Pradesh, India

N. Adhapure
Department of Biotechnology and Microbiology, Vivekanand Arts, Sardar Dalip Singh Commerce and Science College, Aurangabad, Maharashtra, India

R. Sharma (✉)
National Centre for Microbial Resource (NCMR), National Centre for Cell Science (NCCS), Pune, Maharashtra, India
e-mail: rohit@nccs.res.in

T. Satyanarayana et al. (eds.), *Advancing Frontiers in Mycology & Mycotechnology*,
https://doi.org/10.1007/978-981-13-9349-5_5

Keywords
Antarctica · Biotechnology · Biopotentials · Enzymes · Extreme environments · Yeasts

5.1 Introduction

Microbes are found in almost all the habitats on earth including extreme environments (Satyanarayana et al. 2005). Many extremophilic microbes are known till date and many more yet to be discovered. The presence of microbes in extremophilic habitats has been confirmed by culturable and nonculturable (metagenomic) approaches. The cultivation of these organisms is, however, limited due to unavailability of suitable techniques. Therefore, it is desired to standardize methods for cultivating them in specific media, temperature gradients, pH, etc. for understanding the diversity of yeasts in specific habitats. Major proportion of extremophilic microbes belong to archaea and bacteria. However, there are many reports which show that blue-green algae, fungi including yeasts, and some protozoans are also present in extremophilic habitats. A yeast species is considered as extremophilic because of the following reasons: first, the yeast species should be isolated repeatedly from an extreme environment; second, the species should have physiological adaptations for environmental stresses; and third, it should show optimum growth at extreme habitat conditions. In contrast to this, extreme-tolerant yeasts grow at both normal and extreme conditions (Fig. 5.1). According to Buzzini et al. (2017, 2018), species which are occasionally observed in extreme habitats and which does not show optimum growth in extreme habitat conditions are not considered as extremophiles. Therefore, simply isolating a yeast strain from an extreme environment does not necessarily imply that it is extremophilic.

Yeasts inhabit various habitats; some are natural and some are man-made like jaggery, fruits, flower nectar, tree exudates, skin, hair, etc. (Figs. 5.2, 5.3, and 5.4). Large biodiversity of yeast also exists in the cryosphere. Several species of *Rhodotorula,* viz., *R. arctica, R. mucilaginosa, R. psychrophila, R. glacialis, and R. psychrophenolica* isolated from microspheres, are known to produce biosurfactants. Sophorolipids' biosurfactants produced by these organisms are currently used in large-scale production and commercialization. So far, the cold-adapted yeasts have not been used in biotechnology to maximum potential. Therefore, more attention is needed in this respect (Perfumo et al. 2018). *Rhodotorula mucilaginosa* is known to stimulate the oxidation of ferrous iron carried out by *Thiobacillus ferrooxidans* (Fournier et al. 1998). Application of extremophilic microorganisms depends upon the advancement in technology and the discovery of novel organisms from the extreme habitats. In polymerase chain reaction (PCR), initially, mesophilic DNA polymerase was used for every replication cycle. The discovery of *Thermus aquaticus* from extreme habitat facilitated the use of thermostable Taq DNA polymerase that made the process economical. Similarly, more discoveries on extremophiles would definitely offer good inputs where the existing technology has no solutions.

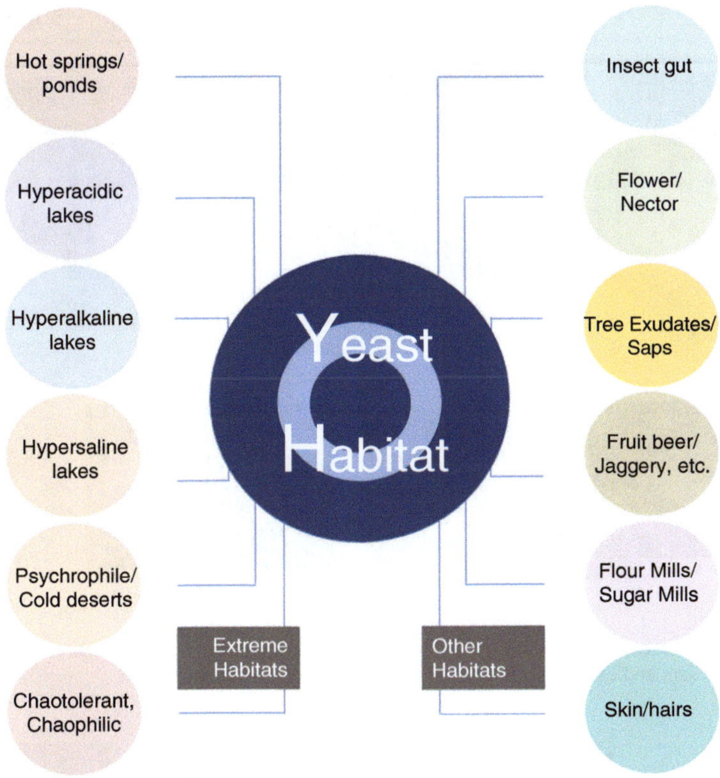

Fig. 5.1 Yeasts occupy a wide range of habitats, some are naturally extreme and some are natural/ man-made moderate

Fig. 5.2 Some examples of the moderate man-made habitats, viz., flour mill, sugar mill, jaggery sites, and fruit juice site. (Source: Compiled from Google images)

Fig. 5.3 Some examples of the moderate natural habitats, viz., banana beer, rice beer, flower nectar, tree sap, and exudates. (Source: Compiled from Google images)

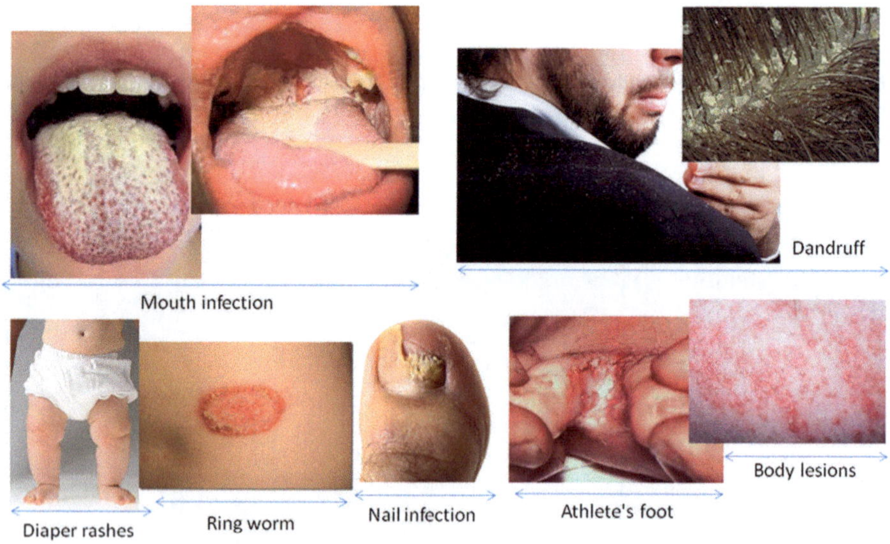

Fig. 5.4 Some examples of pathogenic yeasts inhabiting skin and hair flora. (Source: Compiled from Google images)

Major part of the biosphere has temperature below 5 °C (Feller and Gerday 2003). Polar regions cover about 14% of the earth surface (Gunot 1999). Antarctica is a part of Gondwana supercontinent currently located at the South Pole that covers about 14 million km^2. Antarctica had a warmer and green history, but currently it is cold desert. Only 2% of Antarctica is exposed to ice and snow during summer, while rest is covered by ~1.5- to 4.5-km-thick ice. Antarctica has very low temperature and one of the driest continents on earth. It supports limited and unique flora and fauna on the continent because of extreme conditions of winds, multiple freeze–thaw cycles, high UV radiation, and low levels of moisture. As per Vishniac (2006a, b) among the microbial flora, yeast species are found in abundance at Antarctica which plays an important role in nutrient recycling of the continent. There exist notable gaps in our knowledge regarding the factors that are responsible for diversity, geographical distribution, and function of yeasts in various ecosystems around the world (Spencer and Spencer 1997). In Antarctica, the factors are controlled by physical parameters such as pressure, UV radiation, salinity, temperature, and other climatic factors which in turn control the growth and metabolic activity of yeasts (Deak 2006).

The chapter discusses yeast diversity of extreme environments, their physiology, adaptation strategies, current industrial uses, and future bioprospects. It offers a general review of the knowledge added on extremophilic yeasts in the past several years. Due to climate change and other human activities, it has become essential to study these extreme ecosystems. Most of the information is obtained from the literature published and work done by us in the extreme environment.

5.2 Diversity of Yeasts in Extreme Environment

Yeasts occupy a vast number of extremophilic environments, viz., hyperalkaline, hypersaline, low and high temperature, hyperacidic, etc. (Fig. 5.5). Antarctica has the poorest biodiversity on earth, but not fully explored yet. Psychrophilic and psychrotolerant microbes inhabit in Antarctica (Russell 2006). Polar yeasts species were classified on the basis of their cultural, phenotypic, biochemical characteristics, and sequence similarity of ITS and D1/D2 domains. Several workers have shown that basidiomycetous yeasts are the dominant fungal forms in polar regions (Singh et al. 2013). *Mrakia* spp. and *Mrakiella* spp. are the most common basidiomycetous yeasts reported from various cold regions (Singh and Singh 2012; Thomas-Hall et al. 2010). In another study, Tsuji et al. (2013a, b) reported that *Mrakia* spp. constitutes ~35% of culturable fungi reported from Antarctica. Literature survey shows that considerable work has been done to characterize the bacterial diversity of Antarctic lakes (Trappen et al. 2002; Pearce 2003; Sjöling and Cowan 2003; Zhang et al. 2008). Although yeast diversity studies in Antarctica lakes began in 1960s (Stanley and Rose 1967), literature on this subject is scarce. Vaz et al. (2011) studied the yeast diversity of West Antarctic lake sediments at Port Foster bay (Deception Point) and Copacabana United States Refuge. Carrasco et al. (2012) reported 22 yeast species (*Candida, Metschnikowia, Holtermanniella,*

Fig. 5.5 Some examples of natural extremophilic habitats, viz., snow/ glaciers (Himalayan cryoconites), alkaliphilic lake (lonar lake), industrial waste (hot distillery waste), mines like coal/ bauxite (coal mine). (Source: S.M. Singh and Rohit Sharma)

Mrakia, Cryptococcus, Dioszegia, Leucosporidiella, Rhodotorula, Leuconeurospora, Wickerhamomyces, Glaciozyma, Sporidiobolus) which belonged to 12 genera isolated from King George Island, West Antarctica. D'Elia et al. (2009) isolated several yeasts along with mycelia fungi from Lake Vostok accretion ice. In East Antarctic, many genera have been isolated, viz., *Debaryomyces, Cryptococcus, Candida, Trichosporon,* and *Rhodotorula,* from soil samples of Ross Sea islands. Among the species of *Candida, C. nivalis, C. scottii, C. Gelida,* and *C. frigida* were obligate psychrophiles. Apparently 40 decades ago, Goto et al. (1969) had studied the yeast diversity of the pristine lakes in Dronning Maud Land (water- and lakeshore soil sediments). Six basidiomycetous yeasts, viz., *Candida humicola, C. famata, C. ingeniosa, C. auricularia, Rhodotorula rubra,* and *Bullera alba,* were reported from Schirmacher oasis of East Antarctica (Shivaji et al. 1994; Ray et al. 1989). In subsequent years, Tsuji et al. (2013a, b) contributed many yeast genera such as *Mrakia, Cryptococcus, Dioszegia, Rhodotorula, Leucosporidium, Glaciozyma,* and *Thelebolus* from

Skarvsnesi ice-free area, central Soya coast, Antarctica. Some cryophilic yeasts such as *Cryptococcus albidus*, *C. antarcticus*, *Mrakia blollopis, Rhodotorula microsporus*, and *Thelebolus microsporus* were also reported from Larsemann Hills, Antarctica (Singh and Nayaka 2017). Several species of *Rhodotorula* and *Cryptococcus* have been reported from the soils of South Victoria Land of Antarctica (Connell et al. 2008). Researches on yeast diversity have been carried out since long back. Ellis-Evans (1985) and Stanley and Rose (1967) isolated species of *Rhodotorula* and *Cryptococcus* from the Moss lake, Heywood lake, Deception Island lakes, and Rothera pools of West Antarctica, while Goto et al. (1969) reported their presence in Lake Vanda, East Antarctica. Species of *Rhodotorula* has been isolated from the accretion ice of Vostok lake as well lake sediments in Port Foster (D'Elia et al. 2009; Vaz et al. 2011). The genera, *Cryptococcus* and *Rhodotorula,* have been reported as dominant genera occurring in the oligotrophic lakes of Antarctica (Vishniac 2006a, b) and cold environment elsewhere (Brandão et al. 2011; de Garcia et al. 2007; Libkind et al. 2003, 2009). Cryophilic yeasts such as *Mrakia* sp., *Leucosporidium* sp., *Naganishia* sp., *Vishniacozyma* sp., etc. (Fig. 5.6) have also been recorded from glaciers of Indian Himalayas (Singh et al. Unpublished).

Besides psychrophiles, studies have also been conducted on other extreme environments. Moubasher et al. (2018) isolated 17 species belonging to 12 genera from

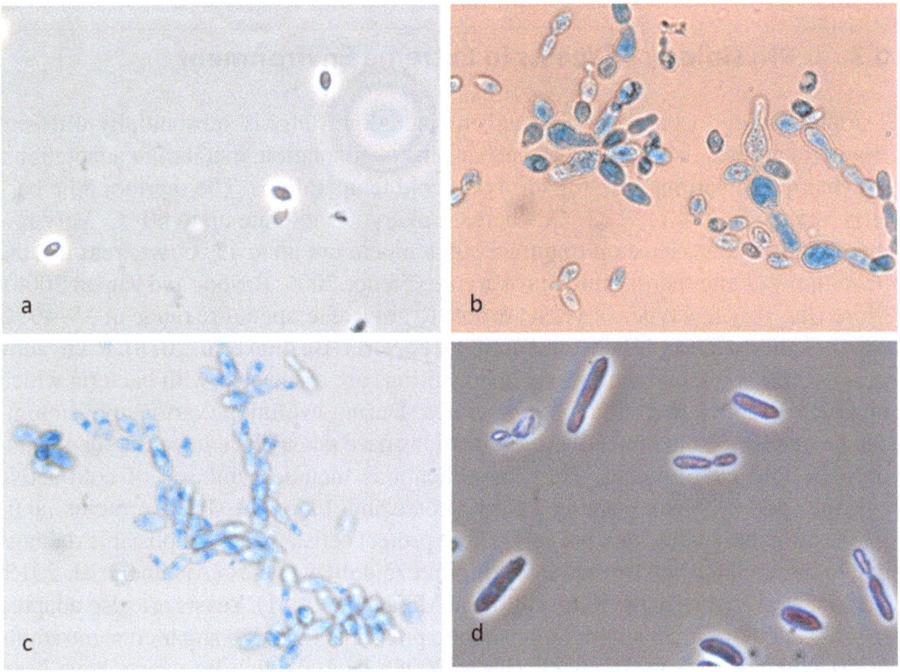

Fig. 5.6 Extremophilic yeasts isolated from cryoconites of Himalayan region, viz., (**a**) *Vishniacozyma* sp., (**b**) *Naganishia* sp., (**c**) *Leucosporidium* sp., and (**d**) *Mrakia* sp.

hypersaline alkaline lakes of Wadi El-Natrun, Egypt with *Candida* being dominant. Butinar et al. (2005) isolated yeasts from eight salterns across the world, along with the Great Salt Lake of Utah, Dead Sea of Israel, and Enriquillo Lake of Dominican Republic. They reported yeast species, viz., *Pichia guilliermondii, Debaryomyces hansenii, Yarrowia lipolytica,* and *Candida parapsilosis,* from hypersaline waters and *Rhodosporidium sphaerocarpum, R. babjevae, Rhodotorula laryngis,* and *Trichosporon mucoides* from halotolerant niches. Subsequently, Cantrell et al. (2006) isolated *Hortaea werneckii* from Cabo Rojo Solar Salterns, Puerto Rico. *Hortaea werneckii,* a melanized yeast belonging to *Capnodiales*, *Dothideomycetes*, and *Ascomycota,* causes fungal disease *tinea nigra. Wallemia ichthyophaga* (*Wallemiales, Wallemiomycetes*) is a xerophilic fungus found in a variety of habitats. Both these yeasts have been reported from hyperalkaline environment (Plemenitaš et al. 2014).

There are a few reports of thermophilic yeasts as compared to thermophilic bacteria and archaea. Kambura et al. (2016) isolated *Rhodotorula mucilaginosa,* a yeast species from the wet sediment at 81 °C. Zajc et al. (2014) introduced a new category of extreme habitat named kosmotropic with increased concentration of salts of NaCl, KCl, and $MgSO_4$ and chaotropic having high-salt concentration of NaBr, $MgCl_2$, and $CaCl_2$ making them extreme environments. *Wallemia ichthyophaga* EXF-994 and *Hortaea werneckii* EXF-225 have been isolated from such extreme environments.

5.3 Physiology of Yeasts in Extreme Environment

The physiology of eukaryotic organisms including fungi is substantially different from prokaryotes like molecular mechanisms of adaptation, metabolite adaptations, enzymatic adaptations, etc. for surviving cold temperatures. The thermophilic bacteria tolerate from 40 to 122 °C, whereas eukaryotes tolerate up to 60 °C. Very few thermophilic yeasts have optimum growth temperature up to 45 °C, whereas it is the lower limit of any thermophilic prokaryote (Stetter 2006, Raspor and Zupan 2006). Moreover, very few types of yeast are able to grow at temperature range of 35–40 °C which is almost mesophilic temperature for bacteria (Buzzini et al. 2018). Rikhvanov et al. (1999) suggested that yeasts in hot springs are associated with bacteria which provide resistance to extreme temperatures. During evolution, various physiological, biochemical, and molecular modifications have taken place in yeasts for adaptations in cold environment. These modifications include synthesis of cold-active enzymes and molecules acting as cryoprotectants like cold shock proteins, antifreeze proteins, sugars, and polyols. These protect cells against the possible damage due to intracellular ice formation during freeze – thaw cycles (Alcaino et al. 2015; Buzzini et al. 2012, 2018; Rothschild and Mancinelli 2001). Yeasts are also adapted to grow in environments with high osmotic pressures like high sugar concentrations of 50–60% (a_w up to 0.62–0.65) (Deak 2008). The alkaliphilic yeasts have been found to grow efficiently at pH 10 or above (Agno 1990; Deak 2008). Similarly, strains of *Wickerhamomyces anomalus* show growth range from pH 2 to 12.4 (Pitt and Hockin 2009). *Yarrowia lipolytica* has also a unique ability to grow rapidly in high (pH 10.0) and low (pH 3.0) pH values (Barth and Gaillardin, 1996). In addition to extremophile to a particular habitat, some yeasts also show polyextremophile, for example, *Debaryomyces hansenii* (Breuer and Harms 2006) and *Y. lipolytica* (Sekova et al. 2015). Such polyextremophilic yeasts have diverse applications in many areas (Sekova et al. 2015). Butinar et al. (2005) isolated about 45 yeasts from hypersaline lake, viz., *Rhodosporidium sphaerocarpum, Rhodosporidium babjevae, Rhodotorula laryngis, Pichia guilliermondii, Debaryomyces hansenii, Yarrowia lipolytica, Metschnikowia bicuspidata, Candida parapsilosis,* and *Trichosporon mucoides.* Although during isolations low counts of yeasts have been observed, these are present in less numbers which are true for any extreme environment. Therefore, the diversity of anti-mesophilic environment is more than any extreme environment. *Moesziomyces antarcticus* is yeast known for the production of industrially relevant molecules including extracellular enzymes (e.g., lipase) and glycolipid biosurfactants of the type mannosyl erythritol lipids (MELs) (Morita et al. 2013). In addition to the enzymes, polysaccharides from yeast would be more stable at respective extreme conditions. This property could be exploited in various ways. The foods to be preserved in cold conditions can be coated with a polysaccharide stable at cold conditions. Resistance to desiccation can be observed in *Naganishia albida* (=*Cryptococcus albidus* var. *diffluens*) which secretes a polysaccharide surrounding it and thereby protects the cell from plasmolysis (Buzzini et al. 2018).

Based on long-term transport experiments, Prista et al. (1997) have suggested different regulations of the activities of K^+ uptake and Na^+ extrusion transporters in extremophilic yeast and mesophilic yeast, viz., *D. hansenii* and *S. cerevisiae*. They observed that *D. hansenii* shows a Na^+ efflux process which has a role in uptake of potassium in high-salt environment. The main difference in both yeasts is the intracellular sodium which is not toxic to *D. hansenii.*

Several yeasts have also been reported from hyperacidic environments. According to Gadanho et al. (2006), the acidophile yeasts *Cryptococcus aciditolerans, Cryptococcus ibericus, and Cryptococcus metallitolerans* are considered the most tolerant species. Other species of yeasts which have been isolated from extreme acidic-to-moderate acidic conditions are *Rhodosporidium toruloides, Candida fluviatilis, Williopsis californica*, and three unidentified yeast species of *Rhodotorula* and *Cryptococcus*.

5.4 Adaptation Strategies of Yeasts to Extreme Environment

Due to extremely cold and oligotrophic conditions, polar microbes have different adaptation strategies such as reduced enzyme reaction rates, cold shock proteins, DNA repair mechanisms, and stable nucleic acid structures (Singh et al. 2013; Tsuji et al. 2013a, b; Thomas and Cavicchioli 2000). Polar microbes have genes for important enzymes and biosynthesis of plant growth-promoting hormones (Singh et al. 2015). Antarctic yeasts such as *Leucosporidium antarcticum* and *Rhodotorula glacialis* secrete extracellular antifreeze proteins (AFP) such as Lp AFP, Lp IRI3, and Lp IRI2 to prevent freezing of the cell when exposed to extreme cold conditions (Xiao et al. 2010; Tsuji et al. 2013a, b); similarly, Arctic yeasts also produce AFP (Pathan et al. 2010; Singh et al. 2014) in extreme environment. Robinson (2001) opined that organisms like yeasts are able to survive the polar winters due to cryoprotectant sugars, polyols, fatty acids, AFPs, and cold-active enzymes. An ascomycetous yeast, *Saccharomyces cerevisiae*, has been studied in detail for cold shock responses (Al-Fageeh and Smales 2006; Aguilera et al. 2007). A cold-tolerant fungal strain *Thelebolus microsporus* isolated from Larsemann Hills, Prydz Bay, Antarctica has been studied for pigment and fatty acid production (Singh et al. 2014). Recently, an exopolysaccharide (EPS) thelebolan was isolated, purified, and characterized from *Thelebolus* sp. which is the first ever report on bioactive EPS from yeasts (Mukhopadhyaya et al. 2014). Recently, metabolite responses of two strains of Antarctic basidiomycetous yeast *Mrakia blollopis* to cold stresses have been reported under subzero temperatures (Tsuji 2016). The unsaturated fatty acids help in membrane fluidity and considered essential for survival at extremely low temperatures. Several workers have also reported high concentrations of unsaturated fatty acids (C18:1 and C18:2) from Arctic yeasts (Pathan et al. 2010; Turk et al. 2011; Singh et al. 2013). Similar observations have also been made on *Mrakia* spp. from Antarctica (Tsuji et al. 2013a, b). The membrane fluidity in eukaryotic microorganisms at lower temperatures is also increased by relative proportion of sterols/phospholipids (Russell 2008).

Tsuji et al. (2013a, b) reported yeasts growth temperature at 1–25 °C from Skarvsnesi ice-free area, central Soya coast. Yeasts' isolates tested in Larsemann Hills grew at 4–22 °C, with best growth between 15 and 22 °C. None of the isolates survived a temperature of 30 °C, indicating that the isolates are indeed adapted to cold environments. According to Sampaio (2004), basidiomycetous yeasts are adaptable to variable nutritional conditions and therefore can survive in more extreme conditions than the ascomycetous forms. These features of basidiomycetous yeasts have made them common occupants of regions having harsh environmental conditions, such as the cold, ultra-oligotrophic habitats of Antarctica. Striking similarities have been observed in the yeast diversity of Antarctic Nella Lake sediments and other similar habitats from cold and tropical oligotrophic environments.

Yeasts are known to conquer extremophilic conditions with the help of photoprotective pigments such as carotenoids and melanins, mycosporines, glutathione, catalase or superoxide dismutase enzymes, and some vitamins. Yeast cells are protected from high level of radiation and oxidation by carotenoids in association with membrane lipids (Moliné et al. 2014; Buzzini et al. 2018). Torularhodin, torulene, β-carotene, and γ-carotene are the main carotenoids produced by basidiomycetous yeasts (*Cystobasidium*, *Cystofilobasidium*, *Dioszegia*, *Rhodotorula*, and *Sporobolomyces*) and ascomycetous yeasts (*Taphrina*). Astaxanthin is synthesized by the species *Phaffia rhodozyma* (Yurkov et al. 2008; Moliné et al. 2014). A combination of these carotenoids gives yeasts a coloration (yellow, red, orange) but colorless carotenoids like phytoene are also known (Meléndez-Martínez et al. 2015;

Buzzini et al. 2018). The cold adaptations of the yeasts have been studied in much detail as compared to the other habitats.

5.5 Biotechnological Potentials of Yeasts from Extreme Environments

Fungi are major recyclers of nutrients in colder environment and are essential to the survival of entire ecosystem (Welander 2005; Edwards et al. 2013), and have immense potential for wastewater treatment in a cold environment (Tsuji et al. 2013a, b, 2015). High esterase activity by *Cryptococcus* and *Rhodotorula* was also observed in many studies conducted in Antarctica (Brandão et al. 2011; Maharana and Singh 2018; Shivaji and Prasad 2009). Urease activity which is considered to be the diagnostic feature of the two genera was seen in all strains of *Rhodotorula* but none of the *Cryptococcus* isolated from Larsemann Hills (Fonseca et al. 2011; Sampaio 2011). *Cryptococcus* strains exhibited a good pectinase activity. Protease activity was recorded high in *Cryptococcus* strains as compared to *Rhodotorula* strains belonging to Larsemann Hills. Moreover, among all, only a single strain of *Cryptococcus albidus* Y-32 exhibited phosphatase activity. Amylase and cellulase activities in different genera showed varied response. These isolates are able to hydrolyze complex organic molecules such as protein, carbohydrate, and lipids which indicate that they are metabolically active even in extremely cold environments. And due to this they have a significant ecological role in the Antarctica ecosystems. The extracellular enzymatic activity profiles of the different yeasts are shown in Table 5.1.

As mentioned above, the acidophilic, alkaliphilic, halophilic, osmophilic, thermophilic, and psychrophilic yeasts are known till date. The applications would involve either the yeast itself or its metabolites. The industrial processes for extreme conditions such as high or low pH, temperature, or osmolarity are used; in such processes, potential extremophilic yeasts or their metabolites can be used (Table 5.2, Fig. 5.7). The metabolites from yeasts having industrial importance could be enzymes, pigments, biosurfactants, or polysaccharides.

5.6 Enzymes

There are many extremozymes currently in use but most of them belong to archaea or bacteria. Interestingly, there is a yeast, *Pseudozyma antarctica,* which is basically a psychrophile but has the ability to produce a thermostable lipase which can be stable even at 90 °C. This lipase has found to have various applications in food. In another study by Tsuji et al. (2013a, b), *Mrakia blollopis* SK-4 assimilated various carbon compounds and used various sugars for fermentation. Moreover, the lipase tolerates high temperature, a wide range of pH, and less sensitive to organic solvents and metal ions. It is also highly reactive to various chain lengths of substrates; thus, *M. blollopis* SK-4 is a promising microbe for wastewater treatment at

Table 5.1 Biotechnological potentials of Arctic yeasts

Name of yeasts	Amylase	Pectinase	Cellulase	Esterase	Protease	Phosphatase	Urease	Keratinase	References
Cryptococcus adeliensis MLB-21 (JX192658)	+	ND	+	+	++	ND	++	ND	Singh et al. (2013)
Cryptococcus saitoi MLB-22 (JX192659)	+	ND	+	++	++	ND	++	ND	Singh et al. (2013)
Cryptococcus saitoi MLB-23 (JX192660)	+	ND	+	++	++	ND	+	ND	Singh et al. (2013)
Cryptococcus saitoi MLB-26 (JX192663)	+	ND	+	++	++	ND	++	ND	Singh et al. (2013)
Cryptococcus adeliensis MLB-18 (JX192655)	+	ND	+	+	++	ND	++	ND	Singh et al. (2013)
Cryptococcus albidosimilis MLB-19 (JX192656)	+	ND	++	++	++	ND	++	ND	Singh et al. (2013)
Cryptococcus albidosimilis MLB-24 (JX192661)	+	ND	++	++	++	ND	+	ND	Singh et al. (2013)
Cryptococcus albidosimilis MLB-25 (JX192662)	+	ND	++	++	++	ND	++	ND	Singh et al. (2013)
Cryptococcus gilvescens (CCP-III-OY) Chernov & Babeva	–	–	–	+	–	–	–	ND	Singh and Singh (2012)
Rhodotorula sp. CCP-II	+	–	–	–	–	–	–	ND	Singh and Singh (2012)
Rhodotorula sp. cry-FB3	++	ND	++	+	+	ND	+	ND	Singh et al. (2016)

Rhodotorula mucilaginosa MLB-27 (JX192664)	–	ND	++	++	++	ND	++	ND	Singh et al. (2013)
Rhodosporidium lusitaniae MLB-20 (JX192657)	–	ND	++	+	++	ND	++	ND	Singh et al. (2013)
Thelebolus microspores PG278 (AB916508)	ND	ND	ND	ND	ND	ND	ND	+++	Singh et al. (2015)
Mrakia blollopis PG256 (AB916516)	ND	ND	ND	ND	ND	ND	ND	+	Singh et al. (2015)
Mrakia sp. CCP-III-WY	–	+	–	–	–	+	–	ND	Singh and Singh (2012)
Thelebolus sp. Cry-YB 240	+	ND	–	+	++	ND	+	ND	Singh et al. (2016)
Thelebolus sp. Cry-YB 241	+	ND	+	–	+	ND	+++	ND	Singh et al. (2016)

NSM Non-sporulating morphotype, Enzyme Activity (+++ = Good activity, ++ = Moderate activity, + = Low activity, – = No activity, *ND* Not detected)

Table 5.2 Potential types and applications of extremophilic yeasts

Yeast	Habitat	Metabolites/ enzymes	Possible applications	References[a]
Cryptococcus	Psychrophile	Pectinase	Food industry	Nakagawa et al. (2004)
Rhodotorulla	Psychrophile	Urease	Biosensor preparations	Fonseca et al. (2011) and Sampaio (2011)
Cryptococcus	Psychrophile	Protease	Food processing, pharmaceutical industries	Campbell et al. (2015)
Rhodotorulla	Psychrophile	Protease	Food processing, pharmaceutical industries	Campbell et al. (2015)
Cryptococcus	Psychrophile	Esterase	Degradation of industrial pollutants	Brandão et al. (2011)and Maharana and Singh (2017)
Rhodotorulla	Psychrophile	Esterase	Degradation of industrial pollutants	Brandão et al. (2011) and Maharana and Singh (2017)
Pseudozyma Antarctica	Psychrophile	Thermostable lipase	Various applications in food	Tsuji et al. (2013a, b)
M. blollopis SK-4	Psychrophile	Assimilate carbon sources, thermostable lipase	Biological agent for wastewater treatment at low temperature	Tsuji et al. (2013a, b)
Debaryomyces	Halophile	Salt-tolerant protease and lipase	Food processing, pharmaceutical industries	
Debaryomyces hansenii	Halophile	Superoxide dismutase (SOD)	Medicine, food industry: Anti-inflammation, malignant tumor regression, radiation and chemotherapy protection, premenstrual syndrome, arthritis, antiageing treatments	Garcia-Gonzalez and Ochoa (1999) and Orozco et al. (1998)
Phaffia Rhodozyma	Halophilic	Astaxanthin	Food industry	Yurkov et al. (2008) and Moliné et al. (2014)

(continued)

Table 5.2 (continued)

Yeast	Habitat	Metabolites/ enzymes	Possible applications	References[a]
Cystobasidium, Cystofilobasidium, Dioszegia, Rhodotorula, Sporobolomyces, Taphrina	Psychrophile	Torularhodin, torulene, β-carotene, and γ-carotene	Food industry	Yurkov et al. (2008) and Moliné et al. (2014)
Candida bombicola	Osmophilic (bumblebee honey)	Sophorolipid	Glycolipid surfactants Enhanced oil recovery	Elshafie et al. (2015), Ratledge and Tan (1990), and Kachholz and Schlingmann (1987)
D. hansenii	Halophile	Salt tolerance gene	For generation of salt-tolerant variety of plants	Serranno (1996)
D. hansenii	Halophile	Compatible solutes	Compatible solutes having commercial potential can be produced	Breuer and Harms (2006)
D. hansenii	Halophile	Assimilate lactate, citrate, lactose, galactose, utilize acetate as carbon source	Its abilities favor this organism as a component of starter cultures for cheese production	Welthagen and Viljoen (1998), Laubscher and Viljoen (1999), and Fatichenti et al. (1983)
D. hansenii	Halophile	Can synthesize *S*–methyl thioacetate and to a lesser extent methional	Contributes to the development of a strong Cheddar flavor	Ferreira and Viljoen (2003)
D. hansenii	Halophile	Synthesize alkali-soluble glucans	Food processing and cosmetics industries. Antitumor activity stimulates the immune system and can lower the serum cholesterol level	Dziezak (1987), Seeley (1977), Donzis (1996), Bohn and BeMiller (1995), Jamaset et al. (1996), Williams et al. (1992), and Nguyen et al. (1998)

(continued)

Table 5.2 (continued)

Yeast	Habitat	Metabolites/ enzymes	Possible applications	References[a]
D. hansenii	Halophile	Xylitol from D-xylose	Food industry for its high sweetening power, anticaries properties, and tolerance by diabetics	Cruz et al. (2000)
Rhodotorula rubra	Acidophilic	Metal extraction	Lithium extraction from lepidolite	Marcincakova et al. (2015)
Rh. mucilaginosa and *W. californica*	Acidophilic	Metal extraction	Cu and Zn removal	Lozovaia (2004)
Trichosporon	Xenobiotics	Biodegradation of xenobiotics	Bioremediation technologies	Kaszycki et al. (2006)
D. hansenii	Halophilic	Competition for nutrient	Biocontrol agent	Droby et al. (1989)

[a]All references are not listed at the end

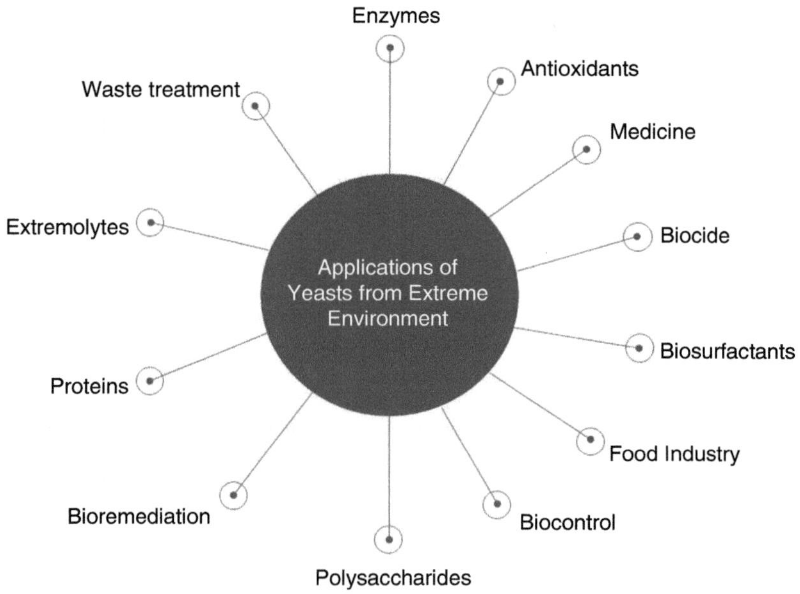

Fig. 5.7 Applications of extremophilic yeasts in various biotechnological industries

low-temperature regions (Tsuji et al. 2013a, b). The cells of *Y. lipolytica* synthesize large amounts of enzymes (lipases, proteases, esterases, and phosphatases) and organic acids (citric and α-ketoglutaric acids) and accumulate fats and proteins (Hofmeyer et al. 2014). Yeast proteases have been used to remove protein hazes from beer and wine (Fleet 1992). Moreover, other extracellular hydrolytic yeast enzymes have also been studied because of their role in wine fermentation (Charoenchai et al. 1997; Strauss et al. 2001). The genus *Debaryomyces* is able to tolerate high salt and produce proteolytic and lipolytic enzymes metabolizing milk proteins and fat. It also has capacity to grow at low temperatures and low water activities which are also thought to be an important factor for its prevalence (Besancon et al. 1992; Davenport 1980; Roostita and Fleet 1996; Fleet and Mian 1987; Tilbury 1980; Wyder and Puhan 1999). Ray et al. (1992) isolated psychrophilic yeast from Antarctica, *C. humicola* which produced acidic protease in the medium. However, the enzyme was active at temperature ranging from 0 to 45 °C and when purified has molecular mass of 36,000 Da. Some inhibitors repressing the enzyme activity were pepstatin, iodoacetamide, and sodium dodecyl sulfate.

5.7 Medicine and Antioxidants

Debaryomyces hansenii, an osmotolerant and halotolerant yeast, is a potential source of superoxide dismutase (SOD). It is a metalloenzyme which catalyzes the dismutation of superoxide radicals and is applied in medicine and food industry. Medicinal properties include anti-inflammation, malignant tumor regression, radiation and chemotherapy protection, arthritis, antiageing treatments, etc. (Garcia-Gonzalez and Ochoa 1999; Orozco et al. 1998).

5.8 Biocide and Biosurfactants

Debaryomyces hansenii shows tolerance to chlorine dioxide (ClO_2) which is a powerful biocide (Ramirez-Orozco et al. 2001). The tolerance can be used to grow yeast in non-sterile medium with ClO_2 up to 0.3 mg/l for controlling unwanted microbes. Generally, industries have shown no interests in the fatty acids produced by yeasts. Therefore, oils produced by microbe are not considered as a replacement for low-cost plant seed oils. However, yeasts produce other potentially useful lipids not found in plants like glycolipid surfactants such as a sophorose lipid produced by *Candida bombicola* and carotenoids like astaxanthin produced by *Phaffia rhodozyma* (Ratledge and Tan 1990).

5.9 Developing Salt-Resistant Plant Varieties

Research has been undertaken around the world to develop plant varieties which are resistant to salinity. This will help to increase the cultivable land area because it is difficult to grow forage and agriculturally important plants in arid and semi-arid climatic conditions. One solution can be the development of halotolerant or halophilic plants. Expression of genes with salt resistance of *D. hansenii* in plants could be an effective strategy. If such researches show some fruitful results, it will help to increase productivity in areas of high salt and reduce hunger (Serranno 1996). The ability of *D. hansenii* to grow at high rates at high-salt concentrations makes it very valuable for biotechnological applications. *D. hansenii* can be cultivated without stringent sterility measures and can use low-cost salt waste products as substrates such as salt containing glycerine/water mixtures obtained from the transesterification of rapeseed oil. In addition, it has high chemostress tolerance and thus can be highly productive. Further, potential applications stem from its ability to produce compatible solutes of commercial interest (Breuer and Harms 2006).

5.10 Food Industry

Welthagen and Viljoen (1998) and Laubscher and Viljoen (1999) suggested that *D. hansenii* has the ability to multiply in cheese, together with its capacity to assimilate lactate, citrate, lactose, and galactose, and favor this organism as a component of starter cultures for cheese production. It is capable of utilizing acetate as a sole source of carbon for assimilation only and not for fermentation of lactose. It also assimilates glucose and ferments it to a limited extent. Both these properties of this yeast species are important in cheese manufacturing (Fatichenti et al. 1983). In addition, metabolic activities of *D. hansenii* modify microenvironment in cheese to improve the effect of desired bacteria and/or *Penicillium roqueforti.* It also protects the cheese against undesired carbohydrate fermentations (Van den Tempel and Jacobsen 2000; Yamauchi et al. 1975).

D. hansenii and many cheese-ripening yeasts can synthesize *S*-methyl thioacetate and methional which are volatile sulfur compound in cheese of Cheddar and Camembert contributing to the development of a strong Cheddar flavor (Ferreira and Viljoen 2003). Flores et al. (2004) reported that *Debaryomyces* spp. can have important effects on the generation of volatile compounds during the ripening of dry-fermented sausages by inhibiting the generation of lipid oxidation products and promoting the generation of ethyl esters, processes that contribute to the development of a typical sausage aroma. *Debaryomyces* has a biotechnological advantage in its ability to use inorganic ammonium as a sole nitrogen source (Yanai et al. 1994). Moreover, there are great opportunities in this field for developing new processes and identifying new products (Breuer and Harms 2006).

5.11 Polysaccharides

Most of the ascomycetous yeasts including *Debaryomyces hansenii* are able to synthesize alkali-soluble and alkali-insoluble glucans associated with chitin. These polysaccharides have enormous applications in industries and medicines. These alkali-soluble glucans are used as thickening agents, fat substitutes, or sources of dietary fibers in food processing and the cosmetics industries (Dziezak 1987; Seeley 1977; Donzis 1996). Furthermore, they have antitumor activity, stimulate the immune system, and can lower the serum cholesterol level (Nguyen et al. 1998;

Bohn and BeMiller 1995; Jamas et al. 1996; Williams et al. 1992). Nguyen et al. (1998) investigated several yeast species for their cell wall contents. Even though *Debaryomyces* has high chitin, it synthesizes low levels of alkali-soluble glucans and the highest levels of alkali-insoluble glucans. Xylitol is used in the food industry for its high sweetening power, anticaries properties, and tolerance by diabetics. For these reasons, xylitol has been employed in the manufacture of sugar-free confections and food (Cruz et al. 2000). The ability of *Debaryomyces hansenii* to produce xylitol from commercial D-xylose and wood hydrolysates, generating high xylitol: ethanol ratios (>4) in the process, has been exploited for several decades (Girio et al. 1989; Girio et al. 1994; Parajo et al. 1995; Parajo et al. 1997; Roseiro et al. 1991).

5.12 Bioremediation and Biocontrol Agent

Biopotential of *Trichosporon* was studied by Kaszycki et al. (2006). The isolated strain can adapt to extreme levels of various xenobiotics: phenol-, formaldehyde-, methanol-, and petroleum-derived hydrocarbons. In such environments, this yeast induces metabolic potential enabling it to biodegrade at least some of these compounds. Thus, the strain seems to be a suitable microorganism for the treatment of heavy-load wastewaters originating from chemical installations. When grown as monoculture, it could be used as a biofilter to treat undiluted wastewaters since it can tolerate certain environmental contaminants at the highest concentrations generated by technological processes (Kaszycki et al. 2006). Marcincakova et al. (2015) investigated lithium extraction from lepidolite using the yeast *Rhodotorula rubra*. In another study, a yeast from acidic hot springs was resistant to high levels of heavy metals and increased the pH of acidified culture medium containing toxic metals. Some fungi (including yeasts) show a high tolerance to heavy metals and biosorption capabilities (Aksu and Donmez 2001; Podgorskii et al. 2004; Salinas et al. 2000; Abe et al. 2001). Such properties are crucial for bioremediation processes of mining industries. In a recent survey, various yeast species were able to sequester Cu and Zn. *Rhodotorula mucilaginosa* and *Williopsis californica* displayed a good capacity to accumulate these metals (Lozovaia et al. 2004). *M. blollopis* SK-4 also produces lipase which is stable at wide pH and temperature range and less sensitive to metal ions and organic solvents. Because of these reasons, it is a potential biological agent for wastewater treatment even in low-temperature regions (Tsuji et al. 2013a, b). Droby et al. (1989) demonstrated the biocontrol properties of *D. hansenii* that the addition of yeast to grapefruits results in significant inhibition of spore germination and hyphal growth of *Penicillium digitatum*.

5.13 Broad Applications of Polyextremophiles

Breuer and Harms (2006) gave a detailed review of polyextremophilic *Debaryomyces hansenii* and its applications. The species is osmo-, halo-, and xerotolerant, with diverse and versatile metabolic pathways and nonpathogenic microbe. It is an important microbe for fundamental and applied biotechnological research. The osmotolerance of *D. hansenii* is highly advantageous for some biotechnological applications because it allows quasi-non-sterile production and high product/educt concentrations. The extreme capacity of *D. hansenii* to synthesize, accumulate, and store lipids could be advantageous for the biotechnological production of both natural and artificial products. The polyextremophilic yeast *Y. lipolytica* has diverse applications in many areas (Sekova et al. 2015).

The ability of *Y. lipolytica* to adapt to various environmental conditions, including extreme ones, makes this strain very promising for biotechnological applications. The ability to grow at extreme pH values allows the creation of conditions for stable growth and the prevention of contamination upon industrial production of single-cell proteins, lipids, organic acids, lipases, bioethanol, and biodiesel (Epova et al. 2012; Sekova et al. 2015). In this way, yeasts occur in osmophilic, halophilic, acidophilic thermophilic, and psychrophilic habitats. Reports of yeast presence only from these habitats do not conclude the absence of yeast from other extreme habitats. So there remains a tremendous scope for exploration of such organisms from yet different extremophilic habitats and then recognizing the potential applications of it. As mentioned in the introduction, diverse microorganisms exist in nature and it is a human limitation that we fail to cultivate most of them. Likewise, if we fail to recognize the role of microorganisms in particular application, it will be the limitation of human beings and not of microorganisms.

5.14 Conclusions

It has increasingly been accepted that yeasts are one of the important groups of microbes which are a backbone for many industries like wine, beer, bread, and several others, thus playing an important role in sustainable development. Yeasts such as *Saccharomyces cerevisiae* and *Candida albicans* have been extensively studied due to their role in food industry and pathogenicity, respectively. With time, new species have been described from various habitats especially from the extreme environments such as high polar regions, saline lakes, high alkaline lakes, and insect gut. During our studies on psychrophiles and insect gut, we isolated many species of yeasts. This indicates that there is a need to explore the diversity of yeasts further in unexplored ecological habitats.

Isolation of yeasts from extreme environment will help biotechnological industry, thus providing economic benefits to country's bioeconomy. It is evident from the above that currently many enzymes and metabolites are industrially produced from yeasts. Processes such as waste treatment, ethanol production, and oil degradation can be improved if a thorough survey is conducted, and novel yeasts are isolated and investigated further.

Acknowledgments RS is grateful to the Department of Biotechnology, New Delhi, for financial support for the establishment of National Centre for Microbial Resource (NCMR), Pune, wide grant letter no. BT/Coord.II/01/03/2016 dated April 6, 2017. SMS acknowledges financial support from Indian Council of Agricultural Research (ICAR) [NBAIM/AMAAS/2014-17/PF/24/21] for research on the Himalayas.

References

Abe F, Miura T, Nagahama T, Inoue A, Usami R, Horikoshi K (2001) Isolation of a highly copper-tolerant yeast, *Cryptococcus* sp., from the Japan Trench and the induction of superoxide dismutase activity by Cu^{2+}. Biotechnol Lett 23:2027–2034

Agno R (1990) Taxonomic distribution of alkali-tolerant yeasts. Syst Appl Microbiol 13:394–397

Aguilera J, Randez-Gil F, Prieto JA (2007) Cold response in *Saccharomyces cerevisiae*: new functions for old mechanisms. FEMS Microbiol Rev 31:327–341

Aksu Z, Dönmez G (2001) Comparison of copper (II) biosorptive properties of live and treated *Candida* sp. J Environ Sci Health 36:367–381

Alcaíno J, Cifuentes V, Baeza M (2015) Physiological adaptations of yeasts living in cold environments and their potential applications. World J Microbiol Biotechnol 31(10):1467–1473

Al-Fageeh MB, Smales CM (2006) Control and regulation of the cellular responses to cold shock: the responses in yeast and mammalian systems. Biochem J 397:247–259

Barth G, Gaillardin C (1996) Yarrowia lipolytica. In Wolf K (ed) Handbook of nonconventional yeast in biotechnology. Springer, Berlin/Heidelberg, pp 313–388

Besancon X, Smet C, Chabalier C, Rivemale M, Reverbel JP, Ratomahenina R, Galzy P (1992) Study of surface yeast flora of Roquefort cheese. Int J Food Microbiol 17:9–18

Bohn JA, BeMiller JN (1995) (1→ 3)-ß-D-glucans as biological response modifiers: a review of structure–functional activity relationships. Carbohydr Polym 28:3–14

Brandão LR, Libkind D, Vaz ABM, Santo LCE, Moliné M, García V, Broock M, Rosa CA (2011) Yeasts from an oligotrophic lake in Patagonia (Argentina): diversity, distribution and synthesis of photoprotective compounds and extracellular enzymes. FEMS Microbiol Ecol 76:1–13

Breuer U, Harms H (2006) *Debaryomyces hansenii* – an extremophilic yeast with biotechnological potential. Yeast 23:415–437

Butinar L, Santos S, Spencer-Martins I, Oren A, Gunde-Cimerman N (2005) Yeast diversity in hypersaline habitats. FEMS Microbiol Lett 244(2):229–234

Buzzini P, Branda E, Goretti M, Turchetti B (2012) Psychrophilic yeasts from worldwide glacial habitats: diversity, adaptation strategies and biotechnological potential. FEMS Microbiol Ecol 82(2):217–241

Buzzini P, Turk M, Perini L, Turchetti B, Gunde-Cimerman N (2017) Yeasts in polar and subpolar habitats. In: Yeasts in natural ecosystems: diversity. Springer, Cham, pp 331–365

Buzzini P, Turchetti B, Yurkov A (2018) Extremophilic yeasts: the toughest yeasts around? Yeast 35(8):487–497. https://doi.org/10.1002/yea.3314

Cantrell SA, Casillas-Martinez L, Molina M (2006) Characterization of fungi from hypersaline environments of solar salterns using morphological and molecular techniques. Mycol Res 110(8):962–970

Carrasco M, Rozas JM, Barahona S, Alcaíno J, Cifuentes V, Baeza M (2012) Diversity and extracellular enzymatic activities of yeasts isolated from King George Island, the sub-Antarctic region. BMC Microbiol 12:251

Charoenchai C, Fleet GH, Henschke PA, Ben T (1997) Screening of non-*Saccharomyces* wine yeasts for the presence of extracellular hydrolytic enzymes. Austr J Grape Wine Res 3:2–8

Connell L, Redman R, Craig S, Scorzetti G, Iszard M, Rodriguez R (2008) Diversity of soil yeasts isolated from South Victoria land, Antarctica. Microb Ecol 56:448–459. https://doi.org/10.1007/s00248-008-9363-1

Cruz JM, Dominguez JM, Dominguez H, Parajo JC (2000) Dimorphic behaviour of *Debaryomyces hansenii* grown on barley bran acid hydrolysates. Biotechnol Lett 22:605–610

D'elia T, Veerapaneni R, Theraisnathan V, Rogers SO (2009) Isolation of fungi from Lake Vostok accretion ice. Mycologia 101:751–763

Davenport RR (1980) An introduction to yeasts and yeast-like organisms. Academic, London, pp 1–27

De Garcia V, Brizzio S, Libkind D, Buzzini P, Van Broock M (2007) Biodiversity of cold-adapted yeasts from glacial meltwater rivers in Patagonia, Argentina. FEMS Microbiol Ecol 59:331–341

Deak T (2006). Environmental factors influencing yeasts. In Rosa CA, Gárbor P (eds) Yeasts handbook on biodiversity and ecophysiology of yeasts. Springer, Berlin, pp 155–174

Deak T (2008) Handbook of food spoilage yeasts, 2nd edn. CRC Press, Boca Raton, pp 17–36

Donzis RA (1996). Substantially purified β-(1,3) finely ground yeast cell wall glucan composition with dermatological and nutritional uses. US Patent 5, 576,015

Droby S, Chalutz E, Wilson CL, Wisniewski M (1989) Characterization of the biocontrol activity of *Debaryomyces hansenii* in the control of *Penicillium digitatum* on grapefruit. Can J Microbiol 35:794–800

Dziezak JD (1987) Yeasts and yeast derivatives. Appl Food Technol 41:122–125

Edwards AB, Douglas AM, Anesio SM, Rassner TDL, Irvine-Fynn B, Sattler GW Griffith (2013) A distinctive fungal community inhabiting cryoconite holes on glaciers in Svalbard. Fungal Ecol 6(2): 168–176

Ellis-Evans JC (1985) Fungi from maritime Antarctic freshwater environments. Br Antarct Survey Bull 68:37–45

Epova E, Guseva M, Kovalyov L, Isakova E, Deryabina Y, Belyakova A, Zylkova M, Shevelev A (2012) Identification of proteins involved in pH adaptation in extremophile yeast *Yarrowia lipolytica*. In: Proteomic applications in biology. InTech, Rijeka

Fatichenti F, Bergere JL, Deiana P, Farris GA (1983) Antagonistic activity of *Debaryomyces hansenii* towards *Clostridium tyrobutyricum* and *C. butyricum*. J Dairy Res 50:449–457

Feller G, Gerday C (2003) Psychrophilic enzymes: hot topics in cold adaptation. Nat Rev Microbiol 1:200–208

Ferreira A, Viljoen BC (2003) Yeasts as adjunct starters in matured Cheddar cheese. Int J Food Microbiol 86:131–140

Fleet GH (1992) Spoilage yeasts. Crit Rev Biotechnol 12:1–44

Fleet GH, Mian MA (1987) The occurrence and growth of yeasts in dairy products. Int J Food Microbiol 4:145–155

Flores M, Dura MA, Marco A, Toldra F (2004) Effect of *Debaryomyces* spp. on aroma formation and sensory quality of dry-fermented sausages. Meat Sci 68:439–446

Fonseca A, Boekhout T, Fell JW (2011) *Cryptococcus* Vuillemin (1901). In: Kurtzman C, Fell JW, Boekhout T (eds) The yeasts, a taxonomic study. Elsevier, Amsterdam, pp 1661–1738

Fournier D, Lemieux R, Couillard D (1998) Essential interactions between Thiobacillus ferrooxidans and heterotrophic microorganisms during a wastewater sludge bioleaching process. Environ Pollut 101:303–309

Gadanho M, Libkind D, Sampaio JP (2006) Yeast diversity in the extreme acidic environments of the Iberian Pyrite Belt. Microb Ecol 52(3):552–563

Garcia-Gonzalez A, Ochoa JL (1999) Anti-inflammatory activity of *Debaryomyces hansenii* Cu, Zn-SOD. Arch Med Res 30:69–73

Girio FM, Peito MA, Amaral-Collaco MT (1989) Enzymatic and physiological study of D-xylose metabolism by *Candida shehatae*. Appl Microb Biotechnol 32:199–204

Girio FM, Roseiro JC, Sá-Machado P, Duarte-Reis AR, Amaral-Collaço MT (1994) Effect of oxygen transfer rate on levels of key enzymes of xylose metabolism in *Debaryomyces hansenii*. Enzym Microb Technol 16:1074–1078

Goto S, Sugiyama J, Iizuka H (1969) Taxonomic study of Antarctic yeasts. Mycologia 61:748–774

Gunot AM (1999). Microbial life in permanently cold soils. In Margesin R, Schinner F (eds) Cold-adapted organisms: ecology, physiology, enzymology and molecular biology. Springer, Berlin, pp 3–15

Hofmeyer T, Bulani SI, Grzeschik J, Krah S, Glotzbach B, Uth C, Avrutina O, Brecht M, Göringer HU, van Zyl P, Kolmar H (2014) Protein production in *Yarrowia lipolytica* via fusion to the secreted lipase Lip2p. Mol Biotechnol 56(1):79–90

Jamas S, Easson J, Davidson D, Ostroff GR (1996). Use of aqueous soluble glucan preparation to stimulate platelet production. US Patent 5, 532, 223

Kambura AK, Mwirichia RK, Kasili RW, Karanja EN, Makonde HM, Boga HI (2016) Diversity of fungi in sediments and water sampled from the hot springs of Lake Magadi and Little Magadi in Kenya. Afr J Microbiol Res 10(10):330–338

Kaszycki P, Czechowska K, Petryszak P, Międzobrodzki J, Pawlik B, Kołoczek H (2006) Methylotrophic extremophilic yeast *Trichosporon* sp.: a soil-derived isolate with potential applications in environmental biotechnology. Acta Biochimica Polonicas 53(3/2006):463–473

Laubscher PJ, Viljoen BC (1999) The occurrence, growth and survival of yeasts in matured Cheddar. University of the Free State, Bloemfontein, pp 77–95

Libkind D, Brizzio S, Ruffini A, Gadanho M, Van Broock M, Sampaio JP (2003) Molecular characterization of carotenogenic yeasts from aquatic environments in Patagonia, Argentina. Antonie Van Leeuwenhoek 84:313–322

Libkind D, Moliné M, Sampaio JP, Van Broock M (2009) Yeasts from high-altitude lakes: influence of UV radiation. FEMS Microbiol Ecol 69:353–362

Lozovaia OG, Kasatkina TP, Podgorskii VS (2004) Search of heavy metals biosorbents among yeasts of different taxonomic groups. Mikrobiol Z 66:92–101

Maharana AK, Singh SM (2018) A cold and organic solvent tolerant lipase produced by Antarctic strain *Rhodotorula* sp. Y-23. J BasicMicrobiol 58:331–342

Marcincakova R, Kadukoka J, Mrazikova A, Velgosora O, Vojtko M (2015) Lithium bioleaching from lepidolite using the yeast *Rhodoturula rubra.* J Polish Miner Eng Soc 16:1–6

Meléndez-Martínez AJ, Mapelli-Brahm P, Benítez-González A, Stinco CM (2015) A comprehensive review on the colorless carotenoids phytoene and phytofluene. Arch Biochem Biophys 572:188–200

Moliné M, Libkind D, de Garcia V, Giraudo MR (2014) Production of pigments and photo-protective compounds by cold-adapted yeasts. Cold-adapted Yeasts. Springer, Berlin/Heidelberg, pp 193–224

Morita T, Fukuoka T, Imura T, Kitamoto D (2013) Production of mannosylerythritol lipids and their application in cosmetics. Appl Microbiol Biotechnol 97(11):4691–4700

Moubasher AA, Abdel-Sater MA, Soliman ZS (2018) Diversity of yeasts and filamentous fungi in mud from hypersaline and freshwater bodies in Egypt. Czech Mycol 70(1):1–32

Mukhopadhyaya SK, Chatterjeea S, Gauria SS, Dasa SS, Mishraa A, Moumita Patraa M, Ghosha AK, Dasa AK, Singh SM, Dey S (2014) Isolation and characterization of extracellular polysaccharide Thelebolan produced by a newly isolated psychrophilic Antarctic fungus *Thelebolus.* Carbohydr Polym 104:204–212

Nguyen TH, Fleet GH, Rogers PL (1998) Composition of the cell walls of several yeast species. Appl Microbiol Biotechnol 50:206–212

Nguyen VAT, Senoo K, Mishima T, Hisamatsu M (2001) Multiple tolerance of Rhodotorula glutinis R-1 to acid, aluminum ion and manganese ion, and its unusual ability of neutralizing acidic medium. J Biosci Bioeng 92:366–371

Orozco MR, Hernandez-Saavedra NY, Valle FA, Gonzalez BA, Ochoa JL (1998) Cell yield and superoxide dismutase activity of the marine yeast *Debaryomyces hansenii* under different culture conditions. J Mar Biotechnol 6:255–259

Parajo JC, Dominguez H, Dominguez JM (1995) Production of xylitol from raw wood hydrolysates by *Debaryomyces hansenii* NRRL Y-7426. Bioproc Engin 13:125–131

Parajo JC, Dominguez H, Dominguez JM (1997) Improved xylitol production with *Debaryomyces hansenii Y-7426* from raw or detoxified wood hydrolysates. Enz Microb Technol 21:18–24

Pathan AAK, Bhadra B, Begum Z, Shivaji S (2010) Diversity of yeasts from puddles in the vicinity of midre lovenbreen glacier, arctic and bioprospecting for enzymes and fatty acids. Curr Microbiol 60(4):307–314

Pearce DA (2003) Bacterioplankton community structure in a maritime Antarctic oligotrophic lake during a period of holomixis, as determined by denaturing gradient gel electrophoresis (DGGE) and fluorescence in situ hybridization (FISH). Microb Ecol 46:92–105

Perfumo A, Banat IM, Marchant R (2018) Going green and cold: biosurfactants from low-temperature environments to biotechnology applications. Trends Biotechnol 36(3):277–289

Pitt JI, Hockin AD (2009) The ecology of fungal food spoilage. In: Fungi and food spoilage. Springer, Boston, pp 3–9

Plemenitaš A, Lenassi M, Konte T, Kejžar A, Zajc J, Gostinčar C, Gunde-Cimerman N (2014) Adaptation to high salt concentrations in halotolerant/halophilic fungi: a molecular perspective. Front Microbiol 5:199

Podgorskii VS, Kasatkina TP, Lozovaia OG (2004) Yeasts – biosorbents of heavy metals. Mikrobiol Z 66:91–103

Prista C, Almagro A, Loureiro-Dias MC, Ramos J (1997) Physiological basis for the high salt tolerance of *Debaryomyces hansenii*. Appl Environ Microbiol 63(10):4005–4009

Ramirez-Orozco M, Hernandez-Saavedra NY, Ochoa J-L (2001) *Debaryomyces hansenii* growth in non-sterile seawater ClO_2–peptone-containing medium. Can J Microbiol 47:676–679

Raspor P, Zupan J (2006) Yeasts in extreme environments. In: Biodiversity and ecophysiology of yeasts. Springer, Berlin/Heidelberg, pp 371–417

Ratledge CO, Tan KH (1990) Oils and fats: production, degradation and utilization by yeasts. In Verachtert HJ, De mot R (eds) Yeast biotechnology and biocatalysis. Marcel Dekker, New York, pp 223–253

Ray MK, Shivaji S, Rao NS, Bhargava PM (1989) Yeast strains from the Schirmacher oasis. Antarctica Polar Biol 9:305–309

Ray MK, Devi KU, Kumar GS, Shivaji S (1992) Extracellular protease from the antarctic yeast *Candida humicola*. Appl Environ Microbiol 58(6):1918–1923

Rikhvanov EG, Varakina NN, Sozinov DY, Voinikov VK (1999) Association of bacteria and yeasts in hot springs. Appl Environ Microbiol 65(9):4292–4293

Robinson CH (2001) Cold adaptation in Arctic and Antarctic fungi. New Phytol 151:341–353

Roostita R, Fleet GH (1996) The occurrence and growth of yeasts in Camembert and blue-veined cheeses. Int J Food Microbiol 28:393–404

Roseiro JC, Peito MA, Girio FM, Amaral-Collaco T (1991) The effects of the oxygen transfer coefficient and substrate concentration on the xylose fermentation by *Debaryomyces hansenii*. Arch Microbiol 156:484–490

Rothschild LJ, Mancinelli RL (2001) Life in extreme environments. Nature 409(6823):1092

Russell NJ (2006) Antarctic microorganisms: coming in from the cold. Culture 27:965–989

Russell NJ (2008) Membrane components and cold sensing. In: Margesin R, Schinner F, Marx JC, Gerday C (eds) Psychrophiles: from biodiversity to biotechnology. Springer, Berlin, pp 177–190

Salinas E, Orellano M, Rezza I, Martinez L, Marchesvky E, Tossetti M (2000) Removal of cadmium and lead from dilute aqueous solutions by *Rhodotorula rubra*. Bioresour Technol 72:107–112

Sampaio JP (2004) Diversity, phylogeny and classification of basidiomycetous yeasts. In: Agerer R, Blanz P, Piepenbring M (eds) Frontiers in Basidiomycote mycology. IHW-Verlag, Eching, pp 49–80

Sampaio JP (2011) Rhodotorula Harrison (1928). In: Kurtzman CP, Fell JW, Boekhout T (eds) The yeasts, a taxonomic study, vol 3, 5th edn. Elsevier, London, pp 1873–1927

Satyanarayana T, Raghukumar C, Shivaji S (2005) Extremophilic microbes: diversity and perspectives. Curr Sci 89(1):78–90

Seeley RD (1977) Fractionation and utilisation of baker's yeast. MBAA Tech Q 14:35–39

Sekova VY, Isakova EP, Deryabina YI (2015) Biotechnological applications of the extremophilic yeast *Yarrowia lipolytica*. Appl Biochem Microbiol 51(3):278–291

Shivaji S, Prasad GS (2009) Antarctic yeasts: biodiversity and potential applications. In: Satyanarayana T, Kunze G (eds) Yeast biotechnology: diversity and applications. Springer, Berlin, pp 3–16

Shivaji S, Chattopadhyay MK, Ray MK (1994) Bacteria and yeasts of Schirmacher Oasis, Antarctica: taxonomy, biochemistry and molecular biology. Proc NIPR Symp Polar Biol 7:173–184

Singh SM, Nayaka S (2017) Contributions to the floral diversity of Schirmacher oasis and Larsemann Hills, Antarctica. Proc Indian Natn Sci Acad 83(2):469–481

Singh P, Singh SM (2012) Characterization of yeast and filamentous fungi isolated from cryoconite holes of Svalbard, Arctic. Polar Biol 35(4):575–583

Singh SM, Singh PN, Singh SK, Sharma PK (2012) Pigment, fatty acids and extracellular enzyme analysis, of a fungal strain *Thelebolus microsporus* from Larsemann Hills Antarctica. Polar Record. https://doi.org/10.1017/S0032247412000563

Singh P, Tsuji M, Singh SM, Roy U, Hoshino T (2013) Taxonomic characterization, adaptation strategies and biotechnological potential of cryophilic yeasts from ice cores of Midre Lovénbreen glacier, Svalbard, Arctic. Cryobiology 66:167–175

Singh P, Singh SM, Tsuji M, Prasad GS, Hoshino T (2014) *Rhodotorula svalbardensis* sp. nov., a novel yeast species isolated from cryoconite holes of Ny-Ålesund, Arctic. Cryobiology 68:122–128

Singh P, Kapse N, Arora P, Singh SM, Dhakephalkar PK (2015) Draft genome of *Cryobacterium* sp. MLB-32, an obligate psychrophile from glacier cryoconite holes of high Arctic. Mar Genomics 21:25–26. https://doi.org/10.1016/j.margen.2015.01.006

Singh P, Roy U, Tsuji M (2016) Characterisation of yeast and filamentous fungi from Brøggerbreen glaciers, Svalbard. Polar Rec 52(4):442–449

Sjöling S, Cowan DA (2003) High 16S rDNA bacteria diversity in glacial meltwater lake sediment, Bratina island, Antarctica. Extremophiles 7:275–282

Spencer JFT, Spencer DM (1997) Yeasts in natural and artificial habitat. Springer, Berlin, p 381

Stanley SO, Rose AH (1967) Bacteria and yeasts from lakes on Deception Island. Proc R Soc Ser B 252:199–207

Stetter KO (2006) History of discovery of the first hyperthermophiles. Extremophiles 10(5):357–362

Strauss MLA, Jolly NP, Lambrechts MG, van Rensburg P (2001) Screening for the production of extracellular hydrolytic enzymes by non-*Saccharomyces* wine yeasts. J Appl Microbiol 91:182–190

Thomas T, Cavicchioli R (2000) Effect of temperature on stability and activity of elongation factor 2 proteins from Antarctic and thermophilic methanogens. J Bacteriol 182:1328–1332

Thomas-Hall SR, Turchetti B, Buzzini P, Branda E, Boekhout T, Threelen B, Watson K (2010) Cold-adapted yeasts from Antarctica and Italian Alps – description of three novel species: *Mrakia robertii* sp. nov., *Mrakia blollopis* sp. nov. and *Mrakiella niccombsii* sp. nov. Extremophiles 14:47–59

Tilbury RH (1980). Xerotolerant (osmophilic) yeasts. In: Skinner FA, Passmore SM, Davenport RR (eds) Biology and activities of yeasts.Academic, London, pp 153–179

Trappen SV, Mergaert J, Eygen SV, Dawyndt P, Cnockaert MC, Swings J (2002) Diversity of 746 heterotrophic bacteria isolated from microbial mats from ten Antarctic lakes. Syst Appl Microbiol 25:603–610

Tsuji M (2016) Cold-stress responses in the Antarctic basidiomycetous yeast *Mrakia blollopis*. R Soc Open Sci 3:160106

Tsuji M, Yokota Y, Shimohara K, Kudoh S, Hoshino T (2013a) An application of wastewater treatment in a cold environment and stable lipase production of Antarctic basidiomycetous yeast *Mrakia blollopis*. PLoS One 8(3):e59376

Tsuji M, Fujiu S, Xiao N, Hanada Y, Kudoh S, Kondo H, Tsuda S, Hoshino T (2013b) Cold adaptation of fungi obtained from soil and lake sediment in the Skarvsnes ice-free area, Antarctica. FEMS Microbiol Lett 346:121–130

Tsuji M, Yokota Y, Kudoh S, Hoshino T (2015) Comparative analysis of milk fat decomposition activity by *Mrakia* spp. isolated from Skarvsnes ice-free area, East Antarctica. Cryobiology 70:293–296

Turk M, Plemenita SA, Gunde-Cimerman N (2011) Extremophilic yeasts: plasmamembrane fluidity as determinant of stress tolerance. Fungal Biol 115:950–958

Van den Tempel T, Jacobsen M (2000) The technological characteristics of *Debaryomyces hansenii* and *Yarrowia lipolytica* and their potential as starter cultures for production of Danablu. Int Dairy J 10:263–270

Vaz ABM, Rosa LH, Vieira MLA, De Garcia V, Brandão LR, Teixeira LCRS, Moliné M, Libkind D, Van Broock M, Rosa CA (2011) The diversity, extracellular enzymatic activities and photoprotective compounds of yeasts isolated in Antarctica. Braz J Microbiol 42:937–947

Vishniac HS (2006a) A multivariate analysis of soil yeasts isolated from a latitudinal gradient. Microb Ecol 52:90–103

Vishniac HS (2006b) Yeast biodiversity in the Antarctic. Biodiversity and ecophysiology of yeasts. Springer, Berlin/Heidelberg, pp 419–440

Welander U (2005) Microbial degradation of organic pollutants in soil in a cold climate. Soil Sediment Contam 14:281–291

Welthagen JJ, Viljoen BC (1998) Yeast profile in Gouda cheese during processing and ripening. Int J Food Microbiol 41:185–194

Williams DL, Pretus HA, McNamee RB, Jones EL, Ensley HE, Browder IW (1992) Development of a water-soluble, sulfated (1→ 3)-ß- D-glucan biological response modifier derived from *Saccharomyces cerevisiae*. Carbohydr Res 235:247–257

Wyder M-T, Puhan Z (1999) Role of selected yeasts in cheese ripening: – an evaluation in aseptic cheese curd slurries. Int Dairy J 9:117–124

Xiao N, Inaba S, Tojo M, Degawa Y, Fujiu S, Hanada Y, Kudoh S, Hoshino T (2010) Antifreeze activities of various fungi and *Stramenophilia* isolated from Antarctica. N Am Fungi 5:215–220

Yamauchi K, Kang KH, Kaminogawa S, Komagata K (1975) Effects of yeasts isolated from cheese on the growth of lactic acid bacteria in skim milk. Jap J Zootech Sci 46:73–80

Yanai T, Tsunekawa H, Okamura K & Okamoto R (1994). Manufacture of pyruvic acid with *Debaryomyces*. JP patent 0 600 091

Yurkov AM, Vustin MM, Tyaglov BV, Maksimova IA, Sineokiy SP (2008) Pigmented basidiomycetous yeasts are a promising source of carotenoids and ubiquinone Q 10. Microbiology 77(1):1–6

Zajc J, Džeroski S, Kocev D, Oren A, Sonjak S, Tkavc R, Gunde-Cimerman N (2014) Chaophilic or chaotolerant fungi: a new category of extremophiles? Front Microbiol 5:708

Zhang H, Hosoi-Tanabe S, Nagata S, Ban S, Imura S (2008) Cultivation and characterization of microorganisms in Antarctic lakes. Conference proceedings, OCEANS Kobe: MTS/IEEE Kobe Techno-Ocean, pp 1–4

Marine Fungal Ecology in the Molecular Era

6

V. Venkateswara Sarma and Rajesh Jeewon

Abstract

The marine environment is an intriguing one and provides a range of wonderful ecological niches to explore the ecology and biodiversity of marine microorganisms. Fungi are possibly by far the most abundant "lifeforms" in the marine environments but largely unexplored. Most studies on marine fungi were from coastal habitats, and they are mainly surveys employing traditional techniques such as microscopy and/or culture-dependent methods which suggest poor diversity of marine fungi (less than 1%) predominated by *Dikarya*. In fact, open oceans were largely considered as "fungal desert" given their inaccessibility and lack of appropriate methods to recover these organisms from these harsh environments. With recent technological advances and developments in molecular techniques involving advanced DNA sequencing technologies, marine mycologists have started to unravel unseen microbial species and better understand the structural and functional diversity of environmental fungal communities. These molecular genomic tools provided insights into genetic diversity especially pertaining to recovery of uncultured fungal organisms, discovery of novel fungal lineages, as well as the metabolic diversity of these complex fungal communities. Particularly, the culture-independent techniques involving environmental cloning, next-generation sequencing are revealing a higher fungal diversity from environmental DNA samples collected from surface waters in open seas, sediments in coastal, benthic and deep sea environments, hydrothermal vents and oxygen-deficient environments. In addition to the diversity, whole genome sequencing, RNA-Seq and microarray

V. V. Sarma (✉)
Department of Biotechnology, Pondicherry University, Kalapet, Pondicherry, India

R. Jeewon
Department of Health Sciences, Faculty of Science, University of Mauritius, Reduit, Mauritius

T. Satyanarayana et al. (eds.), *Advancing Frontiers in Mycology & Mycotechnology*,
https://doi.org/10.1007/978-981-13-9349-5_6

technologies in transcriptome profiling have provided a better understanding of potentially active fungal communities. With the use of these culture-independent methods, several undescribed fungal taxa termed as "dark matter fungi" belonging mainly to zoosporic fungi such as *Blastocladiomycota*, *Chytridiomycota*, *Cryptomycota*, and *Neocallimastigomycota* and *Zygomycota* including *Entomophthoromycota*, *Kickxellomycotina*, *Mortierellomycotina*, *Mucoromycotina*, and *Zoopagomycotina* lineages have been retrieved from marine habitats. Many of these nameless and faceless taxa of the early diverging clusters are microscopic in nature with special nutritional requirements and are difficult to isolate in vitro. *Cryptomycota*, the recently described phylum, established using phylotypes based exclusively on environmental sampling, has been shown to be highly diverse, abundant and ubiquitous in distribution. The marine fungal ecology has changed paradigms in the molecular era. The diversity and ecology of marine fungi recovered from the use of molecular tools are discussed in this book chapter.

Keywords

Next-generation sequencing · Phylotypes · Environmental DNA · Deep-sea environment · Sediments · Nucleic acid primers · Environmental cloning

6.1 Introduction

Marine mycology has evolved as a specialized branch of science only in the recent past. Although marine fungi are an ecologically and physiologically defined group, they are taxonomically diverse. Kohlmeyer and Kohlmeyer (1979) have proposed a definition of marine fungi wherein those fungi that grow and sporulate exclusively in a marine or estuarine habitat were considered as obligate marine fungi and those from a freshwater or terrestrial milieu, able to grow and possibly also sporulate in the marine environment, were considered as facultative marine fungi. Marine fungi have long been studied following two techniques including (1) direct examination method where the fungi occurring on natural samples alone (e.g., driftwood, mangrove wood) were studied by observing under dissection/stereo-zoom microscopes to locate fungal propagules/fruiting structures and identify them and (2) culture techniques where soil samples beneath the mangroves or beaches are isolated into appropriate agar media (Hyde et al. 2000; Newell 1976). The former technique retrieves obligate marine fungi if the samples were inundated in marine waters or facultative marine fungi if the samples were collected from aerial parts of mangrove plants or shoreline plants, most of the fungi isolated by the latter technique, i.e., culture plates on agar media result in isolation of common terrestrial fungi, viz., species of *Aspergillus* and *Penicillium* commonly known as "marine-derived fungi." In a modified definition of marine fungi proposed by Pang et al. (2016), both the groups of fungi have been recognized as marine fungi, and their definition states that "a marine fungus is one that could be recovered repeatedly from marine habitats because (1) it is able to grow and/or sporulate (on substrata) in marine environments; (2) it forms symbiotic relationships with other marine organisms; or (3) it is shown to adapt and evolve at the genetic level or be metabolically active in marine environments."

Fungi play important roles as decomposers involved in regeneration of nutrients in detritus environments in addition to being parasites and symbionts (Webster and Weber 2007). The same is true in the case of marine fungi also where they produce organic detritus supporting large animal community (Kohlmeyer and Kohlmeyer 1979) and decompose the detrital organic matter in marine ecosystems and indulge in nutrient regeneration cycles (Fell and Master 1980; Hyde et al. 1998; Newell 1996 and Raghukumar et al. 1994, 1995) in addition to acting as breeding and nursery grounds in commercial fisheries (Jones and Alias 1997). Though fungi are thought to be major contributors of degradation of decaying plant substrata and animal remains along coastal and surface marine habitats (Kohlmeyer and Kohlmeyer 1979 and Newell 1996), only 1% (1112 species) of the known fungi are from marine environments (Jones et al. 2015), and hence questions have been raised about the importance of fungal communities in marine habitats (Richards et al. 2012) as it is suggested that fungi are nondiverse as well as low in abundance in marine habitats (Burgaud et al. 2009; Kis-Papo 2005 and Le Calvez et al. 2009). Presence of low fungal abundance in upper seawater column samples was reported by Richards et al. (2015) excepting those that occur on phytoplankton. Pang and Jones (2017), however, cautioned about interpretations of results of molecular studies as seawater is a highly dispersive and diluted medium and is not a growth substrate when compared to sediments which represent a niche for settlement of fungal propagules. They reasoned out mentioning that fungi in these substrata might not represent the actively involved marine populations since many fungal propagules could be originated from freshwater or terrestrial environments. However, relatively a higher diversity is reported from the deep-sea environments when compared to surface waters in both culture-dependent and culture-independent studies in addition to indicating their active ecological roles in deep-sea habitats (Edgcomb et al. 2011; Manohar and Raghukumar 2013; Raghukumar et al. 2004 and Singh et al. 2010, 2011, 2012). In many of these studies, it was also found that there are significant differences in the deep-sea fungal diversity between targeted environmental sequencing and conventional cultivation methods (Le Calvez et al. 2009; Singh et al. 2012 and Zhang et al. 2014). Hence a combined approach of both these methods has been suggested to get exact assessment of fungal diversity in deep-sea habitats (Xu et al. 2017).

More than 1112 marine fungal species in 472 genera are known (Jones et al. 2015). The number now stands at 1206 (Pang and Jones 2017). Jones (2011) projected that there could be more than 10,000 marine fungal species. Until the 1970s, most of the research on marine fungi concentrated on their taxonomy (Hyde et al. 2000; Jones et al. 2009). Subsequently, considerable amount of information has become available on several ecological aspects including geographical distribution, frequency of occurrence, vertical zonation, and succession (Alias et al. 1995; Booth and Kenkel 1986; Fryar 2002; Hughes 1974; Hyde and Jones 1988, 1989; Hyde

1988a, b, 1989a, b, 1990a, b; Hyde and Lee 1995; Kohlmeyer and Kohlmeyer 1979; Sarma and Vittal 2000, 2001; Sarma and Hyde 2001). However, all these studies were based on the microscopic observations and enumeration of fungal occurrences when they are in the reproductive phase. In the context of fungal succession, Fryar (2002) presented the problems in studying ecological succession studies of fungi as more often such studies remain to be "sequence of fungal sporulation" instead of "mycelial succession." The reason is at morphology and light microscopy level, where the fungi could be identified based on the fruiting structures only and not based on the mycelia or hyphae on natural samples. Furthermore, the mycelia indicate the active roles of fungi more credibly when compared to the reproductive structures/propagules. In addition to this, a single fungus may sporulate over the entire surface of the substratum and hence only presence or absence could be noted down but number of individuals cannot be quantified as such since sporulation could be from a single mycelium (Jones and Hyde 2002). One cannot deny the importance of microscopy and cultural studies in fungal diversity studies. Direct microscopic analysis and cultural studies are useful approaches for quick diversity estimates (albeit semi-quantitative) and cost-effective. These approaches, however, may not be reliable and underestimate diversity (e.g., Jeewon et al. 2018). Under cultural conditions, species that share similar physiological conditions and exhibit similar cultural characteristics may be different species (Jeewon et al. 2002, 2003a; Liu et al. 2010; Promputtha et al. 2005, 2007; Swe et al. 2009). There are also other major drawbacks such as inefficiency of growth medium used, long-time consumption, laborious and tedious laboratory procedures, lack of experts to enable proper identification, and high risk of contamination by fast-growing fungi (Jeewon and Hyde 2007, 2016). These problems could be circumvented by molecular tools which have revolutionized ecological studies of fungi in the past two decades. However, questions may come whether molecular techniques can discriminate between dormant spores, actively growing mycelium and senescing mycelium.

6.2 Molecular Methods and Marine Fungal Ecology

Molecular methods especially those based on PCR (polymerase chain reaction) to amplify taxonomically informative primer regions in environmental DNA samples clubbed with clonal library construction, DNA sequencing, and phylogenetic analyses that target fungi have usually sampled regions within the ribosomal RNA regions, particularly the small subunit rDNA (18S) and ITS regions (Richards et al. 2012). These approaches have been very useful for resolving species relationships across different taxonomic levels (Jeewon et al. 2003a; Wang et al. 2007; Hongsanan et al. 2017; Jeewon et al. 2017) and to establish novel sexual fungal species (Cai et al. 2006; Duong et al. 2004; Zhang et al. 2008), asexual fungal species (Kodsueb et al. 2007; Pinnoi et al. 2007; Tsui et al. 2006; Vijaykrishna et al. 2004), link sexual and asexual fungal species (Jeewon et al. 2003b; Karunarathna et al. 2017), or discover novel fungal species or new host records from marine environments (Devadatha et al. 2018a, b; Li et al. 2018; Swe et al. 2008a, b; Vinit et al. 2018a, b).

Though ITS regions are handy to determine species diversity and could be employed for ecosystem comparisons, when well-defined taxonomic groups are targeted, it has limited usage during interpretation of higher-level phylogenetic relationships and identifying novel groups since it provides a weak resolution among deeper branching relationships in the fungi (Richards et al. 2012; Horton and Bruns 2001). Due to this reason, some workers have concentrated on selection of SSU rRNA gene marker to study fungal diversity on higher taxonomic groups because this marker does not discriminate between closely related fungal species (Bass et al. 2007; Jebaraj et al. 2009; Richards et al. 2012). To circumvent this problem, multigene phylogenetic analyses including SSU, 5.8S, and large subunit (LSU 28S) sequences that could provide strong phylogenetic support for lower as well as higher phylogenetic levels are currently being used (Jeewon et al. 2009, 2013; Jones et al. 2011; Senanayake et al. 2018a; Wanasinghe et al. 2018; Zhao et al. 2007). In addition, to overcome bias associated with single gene phylogenies, DNA sequence-based analyses on protein-coding genes have been very common to investigate phylogenetic relationships among fungi (Luo et al. 2017; Senanayake et al. 2017, 2018b).

Studies based on morphology and culture-based studies of marine fungi have so far reported 1112 species (Jones et al. 2015). These are mainly fungi isolated from sediments and lignocellulosic substrates occurring in the coastal habitats. However, culture-dependent and culture-independent approaches from both surface waters and deep-sea environments by various workers have revealed some interesting results. For example, DNA-based sequence analyses of 49 SSU rDNA environmental clone libraries recovered totally 1077 sequences from soil, freshwater, and marine samples and also suggest that fungi, with only 124 sequences (11.5%), are comparatively nondiverse and low in abundance in upper as well as surface marine ecosystems (Richards and Bass 2005). A similar trend was observed by Massana and Pedrós-Alió (2008) when they examined 23 coastal water libraries comprising 1349 clones and 12 open ocean libraries comprising 826 clones but could recover only 16 fungal clones (0.8%) of the total marine SSU rDNA sequences processed. Usage of a different technique involving 454 sequencing of eukaryotic SSU rDNA from marine coastal waters also suggests fungal diversity to be nondiverse with less than 5% of the total OPUs (operational taxonomic units) recovered, although this technique indicated an increased diversity than clone library methods (Stoeck et al. 2010). The usage of multi-gene analyses may improve this impasse.

6.3 Marine Fungal Diversity and Abundance in Upper and Surface Open Sea Waters

Fungi have absorptive mode of nutrition (osmotrophy) that involves secretion of enzymes, breaking down of complex biopolymers, and absorption of the nutrients from the breakdown products. Such a lifestyle makes them to feed on organically rich plant and animal substrata or organic remains in soils, sediments, and detritus environments and requires attachment to respective substrata. Probably, due to these ecological characteristics, fungal diversity and abundance have been found to be low in several upper as well as surface marine water column samples studied (Kis-Papo 2005; Richards and Bass 2005). Hence, these environments are not likely to support organisms that thrive basically by attaching to larger physical substrata and adopting osmotrophy, presumably because both the enzymes secreted and their target nutrients would have been lost due to rapid diffusion into the liquid environment. Based on environmental DNA sequence analyses of marine water from upper column, Richards et al. (2015) also supported that marine fungi occur in low diversity and low abundance within this region. However, it was also clear from this study that although amplicons are detected in low numbers, unclassified OTUs can be diverse and dominated by *Chytridiomycota*, followed by *Ascomycota* and *Basidiomycota*. The other reason is the existence of the variations between photosynthesis and biomass accumulation on land as well as at sea. In terrestrial environments, the carbon is fixed into larger and composite plant tissues that are rich in energy as well as nutrients and are tough to digest. Such an aspect would have driven the evolution of osmotrophic lifestyle among the fungi with specialized plant and fungal associations leading to higher fungal diversity. However, in the open seas, the primary producers are small and unicellular lacking complex energy and nutrient-rich compounds. Accordingly, the type of fungal diversity found on land is largely missing in the open seas. Instead, as the open seas have only unicellular photosynthetic organisms and phagotrophic grazers in the surface waters, the diversity and abundance of detrital microbiota develop according to this trophic relationship which is very low (Richards et al. 2012). However, as depth increases, crossing the photozone, the particulate matter reaches the sediments thus increasing its availability for saprotrophs, including fungi, paving the way for detritus processing. In fact, the fungi have been reported to be predominantly active eukaryotic microbes in these environments (Edgcomb et al. 2011; Takishita et al. 2006).

6.4 Marine Fungal Diversity and Abundance in Deep Sea Environments

The deep-sea environmental sampling of sediments and hydrothermal vents for mycocommunities using SSU DNA clone library methods, though has shown a low diversity of fungi in the overall analysis, have indicated a predominance of ascomycete and basidiomycete forms. Further, most of these forms belong to yeast morphotypes suggesting that these taxa were easily recoverable or that they were dominant in these environments. Also, seven clusters of distinctive sequences were recovered of which six potentially represents new fungal lineages in marine environments (Bass et al. 2007). When environmental DNA (eDNA) methods using a different primer set from that in the above study was undertaken to analyze samples from deep-sea hydrothermal vents, many novel fungal lineages were recovered including three unknown phylotypes within the *Basidiomycota* and novel phylotypes close to *Chytridiomycota* (Le Calvez et al. 2009).

Culture sampling from deep-sea sediments revealed the yeast form *Cryptococcus surugensis* (Nagahama et al. 2003). On the other hand, molecular analysis of eukaryotic diversity in marine sediments by Takishita et al. (2006) revealed a different species of *Cryptococcus* (*C.curvatus*). This evidently points out that species diversity at the genus level might be higher than expected and if one relies only on cultural or only molecular data, fungal diversity estimates could be erroneous. When both DNA- and RNA-based diversity profiles were targeted to identify the eukaryotic microbial communities in deep-sea sediment cores, a high incidence of basidiomycetous yeast sequences close to existing *Cryptococcus* as well as *Malassezia* species were recovered (Edgcomb et al. 2011), with a large percentage (42%) of sequences retrieved from RNA-derived libraries cladding with *Cryptococcus* sequences. One of the advantages of conducting studies with RNA-derived libraries is that they also indicate about the metabolically active taxa in sedimentary ecosystems (Edgcomb et al. 2011).

The diversity of mycocommunities from 10 different deep-sea sediment samples was investigated by Nagano et al. (2010) using polymerase chain reaction-mediated ITS regions of rRNA gene clone analysis. The results, in addition to showing common terrestrial fungal species, indicated the presence of a group of major deep-sea phylotypes belonging to the phylum *Ascomycota* in addition to a novel phylotype cladded in deep branches within the phylum *Chytridiomycota* with *Rozella* spp. that are considered to be the closely related organisms (Table 6.1).

The fact that most of the fungi retrieved from deep-sea environmental samples clade closely to known terrestrial fungi suggesting that fungi of terrestrial or marine surface environments are adept at making transition to deep-sea habitats as evidenced in laboratory experiments which reported fungi capable of tolerating high hydrostatic pressure by altering their membrane composition (Simonato et al. 2006).

Basically, deep-sea regions are characterized by low temperatures, higher hydrostatic pressures, absence of light, and finally a very low biological diversity, and hence diversity and abundance of fungi are low when compared to coastal sediments. Further, most of the culturable fungi retrieved are common terrestrial forms (Damare et al. 2008; Singh et al. 2010). However, the number of novel phylotypes obtained from the deep-sea environments is higher when compared to other marine habitats (Manohar and Raghukumar 2013). Fungi play an active ecological role in the deep-sea environments as proven by metabolically active sequences found in RNA-based libraries (Edgcomb et al. 2011).

The ecological roles of fungi in the deep realm of Canterbury basin sediment cores, New Zealand, were studied by Rédou et al. (2014) by using 454-pyrosequencing pointed at small subunit (18S) ribosomal RNA and DNA in addition to fungal ITS1 regions. Though a total of 17,672 sequences were retrieved for five samples from 346 to 1711 mbsf (meters below seafloor) depths, only 18 operational taxonomic units (OTUs) were detected in this study but still enlightens the potentially important ecological roles of fungi in the deep-see environment. Taxa belonging to the fungal classes *Dothideomycetes*, *Exobasidiomycetes*, *Eurotiomycetes*, *Microbotryomycetes*, *Saccharomycetes*, *Sordariomycetes*, *Tremellomycetes*, and *Wallemiomycetes* were observed. The 18SrDNA analyses revealed three clusters: (I) *Cryptococcus surugaensis*, *Filobasidium globisporum*, and *Wallemia muriae* were retrieved from the deeper sediment horizon depth; (II) *Exophiala dermatitidis*,

Table 6.1 List of fungal species retrieved through culture-independent analyses from marine environments

S.No.	Name of the species	Taxonomic group	Primer	Sample nature	References
1	*Absidia glauca*	*Mucoromycetes*	18S rDNA	Near wreck of Bismarck, Brest, France in Atlantic Ocean	Bass et al. (2007)
2	*Acaulospora laevis*	*Glomeromycetes*	ITS	Sagami Bay, Japan coast	Nagano et al. (2010)
3	*Apusomonas proboscidea*	*Apusozoa*	18S rDNA	Deep-sea sediment cores, Peru Margin/ Trench	Edgcomb et al. (2011)
4	*Aspergillus flavus*	*Eurotiomycetes*	18S rDNA	Deep-sea sediment cores, Peru Margin/ Trench	Edgcomb et al. (2011)
5	*Aspergillus penicillioides*	*Eurotiomycetes*	ITS	Izu-Ogasawara Trench, Japan coast	Nagano et al. (2010)
6	*Aureobasidium pullulans*	*Dothideomycetes*	18S rDNA	deep-sea sediment cores, Peru Margin/ Trench	Edgcomb et al. (2011)
7	*Babjeviella inositovora*	*Saccharomycetes*	18S rDNA	European near shore samples	Richards et al. (2015)
8	*Basidiobolus haptosporus*	*Basidiobolomycetes*	18S rDNA	Near wreck of Bismarck, Brest, France in Atlantic Ocean	Bass et al. (2007)
9	*Basidiobolus meristosporus*	*Basidiobolomycetes*	18S rDNA	European near shore samples	Richards et al. (2015)
10	*Basidiobolus microsporus*	*Basidiobolomycetes*	18S rDNA	Near wreck of Bismarck, Brest, France in Atlantic Ocean	Bass et al. (2007)
11	*Basipetospora chlamydospora*	*Eurotiomycetes*	18S rDNA	European near shore samples	Richards et al. (2015)

(continued)

Table 6.1 (continued)

S.No.	Name of the species	Taxonomic group	Primer	Sample nature	References
12	*Batcheloromyces leucadendri*	*Dothideomycetes*	ITS 1	Deep-sea Canterbury basin sediment cores, New Zealand	Rédou et al. (2014)
13	*Bensingtonia subrosea*	*Agaricostilbomycetes*	18S rDNA	European near shore samples	Richards et al. (2015)
14	*Blastocladiella emersonii*	*Blastocladiomycetes*	18S rDNA	Near wreck of Bismarck, Brest, France in Atlantic Ocean	Bass et al. (2007)
15	*Blyttiomyces helicus*	*Chytridiomycetes*	18S rDNA	European near shore samples	Richards et al. (2015)
16	*Bullera arundinarieae*	Tremellomycetes	18SrDNA	European near shore samples	Richards et al. (2015)
17	*Candida ethanolica*	*Saccharomycetes*	18SrDNA	European near shore samples	Richards et al. (2015)
18	*Candida parapsilosis*	*Saccharomycetes*	ITS	Sagami Bay, Japan coast	Nagano et al. (2010)
19	*Candida sagamina*	*Saccharomycetes*	18S rDNA	Deep-sea sediment cores, Peru Margin/ Trench	Edgcomb et al. (2011)
20	*Candida sylvanorum*	*Saccharomycetes*	18S rDNA	European near shore samples	Richards et al. (2015)
21	*Chytridium olla*	*Chytridiomycetes*	18S rDNA	European near shore samples	Richards et al. (2015)
22	*Chytridium polysiphoniae*	*Chytridiomycetes*	18S rDNA	Deep-sea sediment cores, Peru Margin/ Trench; European near shore samples	Edgcomb et al. (2011); Richards et al. (2015)

(continued)

Table 6.1 (continued)

S.No.	Name of the species	Taxonomic group	Primer	Sample nature	References
23	*Chytriomyces angularis*	*Chytridiomycetes*	18S rDNA	Near wreck of Bismarck, Brest, France in Atlantic Ocean	Bass et al. (2007)
24	*Chytriomyces confervae*	*Chytridiomycetes*	18S rDNA	Near wreck of Bismarck, Brest, France in Atlantic Ocean	Bass et al. (2007)
25	*Clydaea vesicular*	*Lobulomycetes*	18S rDNA	European near shore samples	Richards et al. (2015)
26	*Cokeromyces recurvatus*	*Mucoromycetes*	18S rDNA	Near wreck of Bismarck, Brest, France in Atlantic Ocean	Bass et al. (2007)
27	*Colacogloea peniophorae*	*Microbotryomycetes*	18S rDNA	European near shore samples	Richards et al. (2015)
28	*Cordyceps gunnii*	*Sordariomycetes*	18S rDNA	European near shore samples	Richards et al. (2015)
29	*Cryptococcus aureus*	*Tremellomycetes*	18S rDNA	Deep-sea sediment cores, Peru Margin/ Trench	Edgcomb et al. (2011)
30	*Cryptococcus carnescens*	*Tremellomycetes*	18S rDNA	European near shore samples	Richards et al. (2015)
31	*Cryptococcus cellulolyticus*	*Tremellomycetes*	18S rDNA	European near shore samples	Richards et al. (2015)
32	*Cryptococcus curvatus*	*Tremellomycetes*	18S rDNA	Deep-sea Canterbury basin sediment cores, New Zealand	Rédou et al. (2014)
33	*Cryptococcus dimennae*	*Tremellomycetes*	18S rDNA	European near shore samples	Richards et al. (2015)
34	*Cryptococcus marinus*	*Tremellomycetes*	18S rDNA	European near shore samples	Richards et al. (2015)

(continued)

Table 6.1 (continued)

S.No.	Name of the species	Taxonomic group	Primer	Sample nature	References
35	*Cryptococcus pseudolongus*	*Tremellomycetes*	ITS 1, 18S rDNA	Deep-sea Canterbury basin sediment cores, New Zealand, deep-sea sediment cores, Peru Margin/ Trench	Rédou et al. (2014), Edgcomb et al. (2011)
36	*Cryptococcus psychrotolerans*	*Tremellomycetes*	18S rDNA	European near shore samples	Richards et al. (2015)
37	*Cryptococcus saitoi*	*Tremellomycetes*	ITS 1	Deep-sea Canterbury basin sediment cores, New Zealand	Rédou et al. (2014)
38	*Cryptococcus skinneri*	*Tremellomycetes*	ITS	Sagami Bay, Japan Coast	Nagano et al. (2010)
39	*Cryptococcus surugaensis*	*Tremellomycetes*	18SrDNA	Deep-sea Canterbury basin sediment cores, New Zealand	Rédou et al. (2014)
40	*Cyberlindnera jadinii*	*Saccharomycetes*	ITS 1	Deep-sea Canterbury basin sediment cores, New Zealand	Rédou et al. (2014)
41	*Cyberlindnera maculare*	*Saccharomycetes*	18S rDNA	European near shore samples	Richards et al. (2015)
42	*Debaryomyces hansenii*	*Saccharomycetes*	18S rDNA	European near shore samples; Arabian coast, India	Richards et al. (2015)
43	*Delfinachytridium mesopotamicum*		18S rDNA	European near shore samples	Richards et al. (2015)

(continued)

Table 6.1 (continued)

S.No.	Name of the species	Taxonomic group	Primer	Sample nature	References
44	*Diversispora spurca*	*Glomeromycetes*	18SrDNA	Near wreck of Bismarck, Brest, France in Atlantic Ocean	Bass et al. (2007)
45	*Endogone lactiflua*	*Endogonomycetes*	18SrDNA	Near wreck of Bismarck, Brest, France in Atlantic Ocean	Bass et al. (2007)
46	*Endogone pisiformis*	*Endogonomycetes*	18SrDNA	Near wreck of Bismarck, Brest, France in Atlantic Ocean	Bass et al. (2007)
47	*Elmerina caryae*	*Tremellomycetes*	ITS 1	Deep-sea Canterbury basin sediment cores, New Zealand	Rédou et al. (2014)
48	*Entophlyctis helioformis*	*Chytridiomycetes*	18S rDNA	Near wreck of Bismarck, Brest, France in Atlantic Ocean	Bass et al. (2007)
49	*Entrophospora semiglobiferus*	*Glomeromycetes*	18S rDNA	Near wreck of Bismarck, Brest, France in Atlantic Ocean	Bass et al. (2007)
50	*Exophiala dermatitidis*	*Eurotiomycetes*	18S rDNA	Deep-sea Canterbury basin sediment cores, New Zealand	Rédou et al. (2014)
51	*Exophiala spinifera*	*Eurotiomycetes*	ITS 1	Deep-sea Canterbury basin sediment cores, New Zealand	Rédou et al. (2014)

(continued)

Table 6.1 (continued)

S.No.	Name of the species	Taxonomic group	Primer	Sample nature	References
52	*Filobasidium globisporum*	*Tremellomycetes*	18S rDNA	Deep-sea Canterbury basin sediment cores, New Zealand, deep-sea sediment cores, Peru Margin/ Trench	Rédou et al. (2014), Edgcomb et al. (2011)
53	*Fimicolochytrium jonesii*	*Spizellomycetes*	18S rDNA	European near shore samples	Richards et al. (2015)
54	*Fomes fomentarius*	*Agaricomycetes*	18S rDNA	European near shore samples	Richards et al. (2015)
55	*Fusarium solani*	*Sordariomycetes*	18S rDNA	Deep-sea Canterbury basin sediment cores, New Zealand	Rédou et al. (2014)
56	*Gaertneriomyces semiglobifera*	*Spizellomycetes*	18S rDNA	Near wreck of Bismarck, Brest, France in Atlantic Ocean	Bass et al. (2007)
57	*Gaertneriomyces tenuis*	*Spizellomycetes*	18SrDNA	European near shore samples	Richards et al. (2015)
58	*Galactomyces candidum*	*Saccharomycetes*	ITS 1	Deep-sea Canterbury basin sediment cores, New Zealand	Rédou et al. (2014)
59	*Gamsiella multidivaricata*	*Mucoromycetes*	18SrDNA	European near shore samples	Richards et al. (2015)
60	*Geosiphon pyriformis*	*Zygomycetes*	18S rDNA	Near wreck of Bismarck, Brest, France in Atlantic Ocean	Bass et al. (2007)

(continued)

Table 6.1 (continued)

S.No.	Name of the species	Taxonomic group	Primer	Sample nature	References
61	*Geotrichum candidum*	*Saccharomycetes*	18S rDNA	European near shore samples	Richards et al. (2015)
62	*Geotrichum klebahnii*	*Saccharomycetes*	18S rDNA	European near shore samples	Richards et al. (2015)
63	*Glomeralla lagenaria*	*Sordariomycetes*	ITS	Izu-Ogasawara Trench, Japan coast	Nagano et al. (2010)
64	*Harpella meridianalis*	*Harpellomycetes*	18S rDNA	Near wreck of Bismarck, Brest, France in Atlantic Ocean	Bass et al. (2007)
65	*Hyaloraphidium curvatum*	*Chytridiomycetes*	18 SrDNA	European near shore samples	Richards et al. (2015)
66	*Isaria farinosa*	*Sordariomycetes*	18S rDNA	European near shore samples	Richards et al. (2015)
67	*Kondoa malvinella*	*Agaricostilbomycetes*	18S rDNA	European near shore samples	Richards et al. (2015)
68	*Kuzuhaea moniliformis*	*Zoopagomycetes*	18 SrDNA	Near wreck of Bismarck, Brest, France in Atlantic Ocean	Bass et al. (2007)
69	*Lacustromyces heimalis*	*Cladochytriomycetes*	18S rDNA	European near shore samples	Richards et al. (2015)
70	*Leptosphaerulina chartarum*	*Dothideomycetes*	ITS 1	Deep-sea Canterbury basin sediment cores, New Zealand	Rédou et al. (2014)
71	*Leucosporidiella muscorum*	*Microbotryomycetes*	ITS 1	Deep-sea Canterbury basin sediment cores, New Zealand	Rédou et al. (2014)
72	*Limacella glischra*	*Agaricomycetes*	18S rDNA	European near shore samples	Richards et al. (2015)

(continued)

Table 6.1 (continued)

S.No.	Name of the species	Taxonomic group	Primer	Sample nature	References
73	*Linderina pennispora*	*Kickxellomycetes*	18S rDNA	Near wreck of Bismarck, Brest, France in Atlantic Ocean	Bass et al. (2007)
74	*Lobulomyces angularis*	*Lobulomycetes*	18S rDNA	European near shore samples	Richards et al. (2015)
75	*Lobulomyces poculatus*	*Lobulomycetes*	18S rDNA	European near shore samples	Richards et al. (2015)
76	*Malassezia furfur*	*Exobasidiomycetes*	18S rDNA	deep-sea sediment cores, Peru Margin/ Trench	Edgcomb et al. (2011)
77	*Malassezia pachydermatis*	*Exobasidiomycetes*	18SrDNA	Deep sea Canterbury basin sediment cores, New Zealand	Rédou et al. (2014)
78	*Martensiomyces pterosporus*	*Kickxellomycetes*	18S rDNA	Near wreck of Bismarck, Brest, France in Atlantic Ocean	Bass et al. (2007)
79	*Maunachytrium keaense*	*Lobulomycetes*	18S rDNA	European near shore samples	Richards et al. (2015)
80	*Metschnikowia colocasiae*	*Saccharomycetes*	ITS	Izu-Ogasawara Trench, Japan coast	Nagano et al. (2010)
81	*Metschnikowia continentalis*	*Saccharomycetes*	ITS	Izu-Ogasawara Trench, Japan coast	Nagano et al. (2010)
82	*Metschnikowia kamakouana*	*Saccharomycetes*	ITS	Izu-Ogasawara Trench, Japan coast	Nagano et al. (2010)

(continued)

Table 6.1 (continued)

S.No.	Name of the species	Taxonomic group	Primer	Sample nature	References
83	*Meyerozyma guilliermondii*	*Saccharomycetes*	18S rDNA	Deep-sea Canterbury basin sediment cores, New Zealand; European near shore samples	Rédou et al. (2014) and Richards et al. (2015)
84	*Mortierella cystojenkini*	*Mortierellomycetes*	18 SrDNA	European near shore samples	Richards et al. (2015)
85	*Mortierella wolfii*	*Mortierellomycetes*	18S rDNA	Near wreck of Bismarck, Brest, France in Atlantic Ocean	Bass et al. (2007)
86	*Mrakia frigida*	*Tremellomycetes*	18S rDNA	Deep-sea sediment cores, Peru Margin/ Trench	Edgcomb et al. (2011)
87	*Mycogloea macrospora*	*Spiculogloeomycetes*	18S rDNA	European near shore samples	Richards et al. (2015)
88	*Neokarlingia chitinophila*	*Polychytriomycetes*	18S rDNA	European near shore samples	Richards et al. (2015)
89	*Neurospora crassa*	*Sordariomycetes*	18S rDNA	Deep-sea sediment cores, Peru Margin/ Trench	Edgcomb et al. (2011)
90	*Obelidium mucronatum*	*Chytridiomycetes*	18S rDNA	European near shore samples	Richards et al. (2015)
91	*Olpidium brassicae*	*Olpidiomycetes*	18S rDNA	European near shore samples	Richards et al. (2015)
92	*Orbilia auricolor*	*Orbiliomycetes*	18S rDNA	Arabian coast, India	Jebraj et al. (2009)
93	*Orbilia fimicola*	*Orbiliomycetes*	18S rDNA	European near shore samples	Richards et al. (2015)

(continued)

Table 6.1 (continued)

S.No.	Name of the species	Taxonomic group	Primer	Sample nature	References
94	*Orphella haysii*	*Trichomycetes*	18S rDNA	Near wreck of Bismarck, Brest, France in Atlantic Ocean	Bass et al. (2007)
95	*Paraglomus brasilianum*	*Paraglomeromycetes*	18S rDNA	Near wreck of Bismarck, Brest, France in Atlantic Ocean	Bass et al. (2007)
96	*Penicillium chrysogenum*	*Eurotiomycetes*	ITS	Izu-Ogasawara Trench, Japan coast	Nagano et al. (2010)
97	*Penicillium minioluteum*	*Eurotiomycetes*	ITS 1	Deep-sea Canterbury basin sediment cores, New Zealand	Rédou et al. (2014)
98	*Phlyctochytrium arcticum*	*Chytridiomycetes*	18S rDNA	European near shore samples	Richards et al. (2015)
99	*Physoderma maculararae*	*Physodermatomycetes*	18S rDNA	European near shore samples	Richards et al. (2015)
100	*Pichia guilliermondii*	*Saccharomycetes*	18S rDNA	Deep-sea sediment cores, Peru Margin/ Trench	Edgcomb et al. (2011)
101	*Pichia heedii*	*Saccharomycetes*	18S rDNA	European near shore samples	Richards et al. (2015)
102	*Pichia membranifaciens*	*Saccharomycetes*	18S rDNA	European near shore samples	Richards et al. (2015)
103	*Piptocephalis corymbiferae*	*Zoopagomycetes*	18SrDNA	Near wreck of Bismarck, Brest, France in Atlantic Ocean	Bass et al. (2007)

(continued)

Table 6.1 (continued)

S.No.	Name of the species	Taxonomic group	Primer	Sample nature	References
104	*Pleurostomophora richardsiae*	*Sordariomycetes*	ITS 1	Deep-sea Canterbury basin sediment cores, New Zealand	Rédou et al. (2014)
105	*Podochytrium dentatum*	*Chytridiomycetes*	18S rDNA	European near shore samples	Richards et al. (2015)
106	*Polychytrium aggregatum*	*Cladochytriomycetes*	18S rDNA	European near shore samples	Richards et al. (2015)
107	*Protomyces lactucaedebilis*	Taphrinomycetes	18S rDNA	Arabian coast, India	Jebraj et al. (2009)
108	*Puccinia poarum*	*Pucciniomycetes*	18S rDNA	European near shore samples	Richards et al. (2015)
109	*Pucciniastrum epilobii*	*Pucciniomycetes*	18S rDNA	European near shore samples	Richards et al. (2015)
110	*Radiomyces spectabilis*	*Mucoromycetes*	18S rDNA	Near wreck of Bismarck, Brest, France in Atlantic Ocean	Bass et al. (2007)
111	*Rhinocladiella similis*	*Eurotiomycetes*	ITS 1	Deep-sea Canterbury basin sediment cores, New Zealand	Rédou et al. (2014)
112	*Rhizidium endosporangiatum*	*Chytridiomycetes*	18S rDNA	Near wreck of Bismarck, Brest, France in Atlantic Ocean	Bass et al. (2007)
113	*Rhizophlyctis harderi*	*Rhizophlyctidomycetes*	18S rDNA	European near shore samples	Richards et al. (2015)
114	*Rhizophlyctis rosea*	*Rhizophlyctidomycetes*	18S rDNA	Near wreck of Bismarck, Brest, France in Atlantic Ocean	Bass et al. (2007)
115	*Rhodosporidium dacryoidum*	*Microbotryomycetes*	18S rDNA	European near shore samples	Richards et al. (2015)

(continued)

Table 6.1 (continued)

S.No.	Name of the species	Taxonomic group	Primer	Sample nature	References
116	*Rhodosporidium diobovatum*	*Microbotryomycetes*	18S rDNA	European near shore samples	Richards et al. (2015)
117	*Rhodosporidium kratochvilovae*	*Microbotryomycetes*	ITS 1	Deep-sea Canterbury basin sediment cores, New Zealand	Rédou et al. (2014)
118	*Rhodotorula bacarum*	*Microbotryomycetes*	18S rDNA	European near shore samples	Richards et al. (2015)
119	*Rhodotorula lamellibrachiae*	*Microbotryomycetes*	18S rDNA	European near shore samples	Richards et al. (2015)
120	*Rhodotorula laryngis*	*Microbotryomycetes*	ITS	Izu-Ogasawara Trench, Japan coast	Nagano et al. (2010)
121	*Rhodotorula marina*	*Microbotryomycetes*	18S rDNA	European near shore samples	Richards et al. (2015)
122	*Rhodotorula minuta*	*Microbotryomycetes*	ITS 1	Deep-sea Canterbury basin sediment cores, New Zealand; European near shore samples	Rédou et al. (2014) and Richards et al. (2015)
123	*Rhodotorula mucilaginosa*	*Microbotryomycetes*	18S rDNA	European near shore samples	Richards et al. (2015)
124	*Rhodotorula rosea*	*Microbotryomycetes*	ITS	Izu-Ogasawara Trench, Japan coast	Nagano et al. (2010)
125	*Rhodotorula yarrowii*	*Microbotryomycetes*	18S rDNA	European near shore samples	Richards et al. (2015)
126	*Rhopalomyces elegans*	*Zoopagomycetes*	18S rDNA	Near wreck of Bismarck, Brest, France in Atlantic Ocean	Bass et al. (2007)

(continued)

Table 6.1 (continued)

S.No.	Name of the species	Taxonomic group	Primer	Sample nature	References
127	*Roccella fuciformis*	*Arthoniomycetes*	18S rDNA	Arabian coast, India	Jebraj et al. (2009)
128	*Saccharomyces cerevisiae*	*Saccharomycetes*	18S rDNA	European near shore samples	Richards et al. (2015)
129	*Scutellospora calospora*	*Glomeromycetes*	18S rDNA	Near wreck of Bismarck, Brest, France in Atlantic Ocean	Bass et al. (2007)
130	*Sirobasidium brefeldianum*	*Tremellomycetes*	18SrDNA	European near shore samples	Richards et al. (2015)
131	*Spiromyces aspiralis*	*Kickxellomycetes*	18S rDNA	Near wreck of Bismarck, Brest, France in Atlantic Ocean	Bass et al. (2007)
132	*Spiromyces minutus*	*Kickxellomycetes*	18S rDNA	Near wreck of Bismarck, Brest, France in Atlantic Ocean	Bass et al. (2007)
133	*Spizellomyces acuminatus*	*Spizellomycetes*	18S rDNA	Near wreck of Bismarck, Brest, France in Atlantic Ocean	Bass et al. (2007)
134	*Spizellomyces dolichospermus*	*Spizellomycetes*	18S rDNA	European near shore samples	Richards et al. (2015)
135	*Spizellomyces plurigibbosus*	*Spizellomycetes*	18S rDNA	European near shore samples	Richards et al. (2015)
136	*Sporobolomyces hasegawianum*	*Microbotryomycetes*	18S rDNA	European near shore samples	Richards et al. (2015)
137	*Sporobolomyces inositophilus*	*Microbotryomycetes*	18S rDNA	European near shore samples	Richards et al. (2015)
138	*Syncephalis depressa*	*Zoopagomycetes*	18S rDNA	Near wreck of Bismarck, Brest, France in Atlantic Ocean	Bass et al. (2007)

(continued)

Table 6.1 (continued)

S.No.	Name of the species	Taxonomic group	Primer	Sample nature	References
139	*Torpedospora radiata*	*Sordariomycetes*	18S rDNA	European near shore samples	Richards et al. (2015)
140	*Tremella moriformis*	*Tremellomycetes*	ITS 1	Deep-sea Canterbury basin sediment cores, New Zealand,	Rédou et al. (2014),
141	*Trichosporon aquatile*	*Tremellomycetes*	18S rDNA	Deep-sea sediment cores, Peru Margin/ Trench	Edgcomb et al. (2011)
142	*Trichosporon mucoides*	*Tremellomycetes*	18S rDNA, ITS	Deep-sea Canterbury basin sediment cores, New Zealand, Izu-Ogasawara Trench, Japan coast	Rédou et al. (2014), Nagano et al. (2010)
143	*Trimorphomyces papilionaceus*	*Tremellomycetes*	18S rDNA	European near shore samples	Richards et al. (2015)
144	*Triparticalcar arcticum*	*Spizellomycetes*	18S rDNA	Near wreck of Bismarck, Brest, France in Atlantic Ocean	Bass et al. (2007)
145	*Uromyces aritriphylli*	*Pucciniomycetes*	18S rDNA	European near shore samples	Richards et al. (2015)
146	*Ustilago shiraina*	*Ustilaginomycetes*	18S rDNA	Deep-sea sediment cores, Peru Margin/ Trench	Edgcomb et al. (2011)
147	*Xerocomus chrysenteron*	*Agaricomycetes*	18S rDNA	Deep-sea sediment cores, Peru Margin/ Trench	Edgcomb et al. (2011)

(continued)

Table 6.1 (continued)

S.No.	Name of the species	Taxonomic group	Primer	Sample nature	References
148	*Wallemia muriae*	*Wallemiomycetes*	18S rDNA	Deep-sea Canterbury basin sediment cores, New Zealand	Rédou et al. (2014)
149	*Zoophagus insidians*	*Zoopagomycetes*	18S rDNA	Near wreck of Bismarck, Brest, France in Atlantic Ocean	Bass et al. (2007)
150	*Zygopolaris ephemeridarum*	*Harpellomycetes*	18S rDNA	Near wreck of Bismarck, Brest, France in Atlantic Ocean	Bass et al. (2007)
151	*Zygorhizidium planktonicum*	*Chytridiomycetes*	18S rDNA	European near shore samples	Richards et al. (2015)

Malassezia pachydermatis, *Meyerozyma guilliermondii*, *Pleurostomophora richardsiae*, and *Trichosporon mucoides* were found in sediment samples that have a low organic carbon concentration; (III) *Cryptococcus curvatus*, *Cyberlindnera jadinii*, *Fusarium solani*, *Leptosphaerulina chartarum*, and *Trichoderma* sp. were found in shallowest depths and correlated with methane concentration. Similarly, ITS1 sequence analyses also resulted in three clusters (I) *Cryptococcus saitoi*, *Leucosporidiella muscorum*, *Rhodosporidium kratochvilovae*, *Rhodotorula* sp., and *Tremella moriformis*; (II) *Batcheloromyces leucadendri*, *Chaetothyriales sp.*, *Elmerina caryae*, *Exophiala spinifera*, *P. richardsiae*, *Penicillium* sp., and *Rhinocladiella* sp.; and (III) *Cryptococcus jadinii*, *C. pseudolongus*, *Galactomyces candidum*, *L. chartarum*, and *Trichosporon* sp., and most of the OTUs were only found at a given depth (Rédou et al. 2014). The above study clearly demonstrates that the use of different primer sets yields different results. This difference in taxon recovery from the same habitats highlights that the selection of appropriate genes is crucial in determining fungal diversity and it would be far more judicious to use several gene markers as compared to only one.

Studies on culturable mycocommunities from deep-sea sediment core samples from Canterbury basin, New Zealand, resulted in isolation of more than 200 filamentous fungi (68%) and yeasts (32%) (Rédou et al. 2015). This study provides proof of deep-subsurface mycocommunities having ability to survive, adapt, grow, and interact with other microbial communities in addition to highlighting that the deep-sediment environment is one more ecological niche for fungi. Further, all the fungal taxa isolated were well-known in terrestrial habitats indicating the ability of terrestrial fungi to adapt to deep-subsurface conditions (Rédou et al. 2015). Based on 18S rDNA sequence data, it has been reported that yeasts tend to dominate fungal diversity in deep ocean floor and that huge hydrostatic pressure is not a barrier to life (Bass et al. 2007).

Employing amplicon pyrosequencing, eukaryotic 18SrRNA sequences were investigated by Orsi et al. (2013) and found a distinct set of fungi across different marine subsurface regions such as continental margins, ridge flanks, and abyssal plains near Peru Margin, Eastern Equatorial Pacific, and Mid-Atlantic Ridge. The subseafloor mycocommunities and their populations had a statistically significant correlation with various environmental parameters including nitrate, sulfide, total organic carbon (TOC), and dissolved inorganic carbon (DIC). Such correlation was also supported by terminal restriction length polymorphism (TRFLP) analyses of fungal rRNA (Orsi et al. 2013). In the North Pond (1.6 mbsf) the sequences associated with *Antrodia, Apioplagiostoma, Cordyceps, Cryptococcus, Cyberlindnera, Doassansia, Erythrobasidium, Filobasidium, Hannaella, Hyphodontia, Mycena, Powellomyces, Peyronellaea,* and *Sterium* were retrieved that represent sediment age of 0.1–2 MYA. From Hydrate Ridge (1.8 mbsf) *Cordyceps, Crinipellis, Cryptococcus, Entoloma, Leucosporidium,* and *Mycena* were retrieved, which represent sediment age of 0.1 MYA. In the Benguela Upwelling System (4.61 mbsf) *Acidomyces, Neurospora, Sordaria, Candida, Hydropus, Mycena, Steccherinium, Stereum, Filobasidium, Rhodotorula, Rhodosporidium, Diversispora,* and *Glomus* were retrieved and these represent sediment age of 0.03 MYA. From the Eastern Equatorial Pacific (45.3 mbsf), *Alternaria, Antrodia, Apioplagiostoma, Camarops, Candida, Cryptosporella, Cyberlindnera, Cryptococcus, Dioszegia, Discula, Filobasidium, Helicogloea, Mycena, Neurospora, Peyronellaea, Phruensis, Rhizoctonia, Rhodotorula, Sporobolomyces,* and *Sterigmatomyces* were retrieved that represent sediment age of 2.77 MYA. From Peru Margin (48.1 mbsf), *Cryptococcus, Glyphium, Geopyxis, Knufia, Lentinula, Mycena, Rhotorula,* and *Trichosporum* were retrieved, and these taxa represent sediment age of 2.6 MYA (Orsi et al. 2013).

The deep sea environment, in particular, the sediments, must be inhabited by many undiscovered species, but our knowledge is still limited and fragmentary because our methods available do not allow us to fully recover all species from those habitats. High-throughput approaches have shed a new perception of deep-sea fungal community ecology. One of the latest studies dealing with deep-sea sediment samples is that of Nagano et al. (2018) in Brazil. The authors reported similar fungal diversity as other previous work with *Aspergillus*, *Penicillium*, psychrotrophic fungi, and red-pigmented basidiomycetous yeasts as dominant fungi. Interestingly, ubiquitous taxa such as *Aspergillus*, *Pestalotiopsis*, and *Trichoderma* have also been recovered, and these have been reported to be able to degrade petroleum hydrocarbons. This points out that many of these fungi could be significant bioremediators that can potentially be exploited to combat pollution problems.

6.5 Deep-Sea Hydrothermal Vents

The fungal diversity of deep-sea hydrothermal samples from East Pacific Rise at the Elsa site, Pacific Ocean, and Mid-Atlantic Ridge at Menez Gwen site, employing small-subunit rRNA (18SrDNA) gene sequences amplified by culture-independent PCR using DNA extracts, was investigated by Le Calvez et al. (2009). The results showed that *Chytridium polysiphoniae*, *Rhizophlyctis rosea*, *Spizellomycete* sp., *Powellomyces* sp., *Rhizophydium* sp., *Kappamyces laurelensis* belonging to *Chytridiomycota*; *Cryptococcus* sp., *Filobasidium* sp., *Bullera nakhonratchasimensis*, *Kockovaella kimperatae*, *Feliomyces ogasawarensis*, *Exida uvapsassa*, *Auricularia polytricha*, and *Fibulorhizoctonia* sp. belonging to *Basidiomycota* were present. In addition to this culture-dependent studies employing molecular techniques revealed the presence of *Penicillium tardum*, *Wangiella dermatidis*, *Phaeococcomyces exophialae*, *Capronia acutiseta*, *Graphium* sp., *Exophiala jeanselmei*, *Phialophora* sp., *Rhinocladiella atrovirens*, *Exophiala* sp., *Helicoon fuscosporium*, *Spliocaea oleagina*, *Tricodelitschia munkii*, *Phaeotrichum benjamini*, *Hortaea werneckii*, *Aureobasidium pullulans*, *Diaporthe eres*, *Acremonium* sp., *Tritirachium* sp., *Engyodontium album*, *Nadsionella nigra*, *Pochonia chlamydosporia*, and *Tolypocladium cylindrosporum* (Le Calvez et al. 2009). Further, many of the species recorded in this study through culture-independent methods were not known even at higher taxonomic levels in the *Chytridiomycota*, *Ascomycota*, and *Basidiomycota*. A combined culture-dependent and culture-independent sequence-based study on fungal distribution and diversity at a deep-sea hydrothermal vent site at the Mid-Atlantic Ridge of the South Atlantic Ocean undertaken by Xu et al. (2017) revealed that the mycocommunity was dominated by members belonging to *Ascomycota* and *Basidiomycota*. In addition, several novel phylotypes, i.e., 28 of 65 total fungal OTUs from clone library construction and 2 out of 19 cultural fungal phylotypes, were recovered indicating the presence of unrevealed diversity of fungi in this habitat. Also, they found that the mycocommunities in the chimney samples were different from those found in three sulfide samples. Further, their qPCR studies have shown that fungal LSU rRNA gene copy numbers ranged from 5.88×10^5 to 6.77×10^6 copies/gram rock of wet weight. Also, their results showed that *Ascomycota* was 2–3 times more abundant than the *Basidiomycota* (Xu et al., 2017). Within the *Ascomycota* they found *Sordariomycetes* to be the most dominant group followed by *Dothideomycetes*, *Saccharomycetes*, *Eurotiomycetes*, and *Leotiomycetes*.

6.6 Anoxic Marine Environments

Fungi are often found as saprotrophs in terrestrial environments that are low in oxygen and have cellular and genomic adaptations for survival in anoxic environments (Embley et al. 2006; Gojkovic et al. 2004). Vast areas of marine environments are anoxic. Fungal sequences closely related to *Fusarium* species in deep-sea anoxic environments, using SSU DNA clone libraries, have been reported by Jebraj et al. (2009). Further, Jebaraj and Raghukumar (2009) have demonstrated that *Fusarium* strains isolated from anoxic deep-sea samples could utilize nitrate for respiration and accumulate nitrite thus indicating a role in anaerobic denitrification in marine environments. When multiple fungal-specific SSU rDNA primer sets and a universal eukaryotic-specific marker were used to amplify SSU sequences from samples of oxygen-depleted regions of the Arabian sea, a greater diversity of fungi was found in the fungal-specific marker genes than the clone libraries constructed using universal eukaryotic primers (Jebaraj et al. 2009). Further, this study also indicated the necessity of using different primers to regulate PCR biases, and chances of missing fungal diversity when universal primer sets alone were used. In the same study, 56% of the total 48 phylotypes branched within the ascomycete radiation and 41% within basidiomycetes with only 2% belonged to lower fungi (Jebraj et al. 2009). While culture-dependent studies suggest the presence of a higher percentage of filamentous forms in deep-sea environments, the molecular analyses of sediment and water samples from marine environments using SSU rDNA marker genes by Richards et al. (2012) indicate that *Dikarya* (Ascomycetes and Basidiomycetes) are the most likely recoverable marine fungal lineages with a high percentage of sequences recovered within *Dikarya* belonging to well-known yeast groups. Earlier Alexander et al. (2009) also reported predominance of fungi from sediment samples from the hypersaline anoxic deep-sea basin of L'Atalante using SSU rDNA sequence analyses. The detection of fungal hyphae in anoxic sediments in mangrove habitat using Calcoflour staining and epifluorescene microscopy and using 454 pyrosequencing of nuclear ribosomal ITS regions with the latter technique revealing dominance of *Agaricomycetes* in the *Basidiomycota* has been reported (Arfi et al. 2012). Clades close to the basidiomycetous yeast *Malassezia*; *Rozella* and allied clusters belonging to *Cryptomycota*; well-known fungal taxa belonging to *Penicillium*, *Eupenicillium,* and *Aspergillus*; and basal clones close to *Chytridiomycota*, in addition to several novel sequences have been reported from different anoxic environments including coastal anoxic sediments of Bolinas tidal flat (Dawson and Pace 2002), L'Atlantic basin at a depth of 3500 m (Alexander et al. 2009), Arabian sea from depths of 25–200 m (Jebaraj et al. 2010; Manohar and Raghukumar 2013), Methane cold seeps, South China Sea at depths ranging from 350 to 3000 m (Lai et al. 2007), Methane cold seep, Sagami Bay at a depth of 1080 m (Nagahama et al. 2011; Takishita et al. 2007), Kagoshima bay at a depth of 204 m (Takishita et al. 2005), Cariaco basin at a depth of 340 m (Stoeck et al. 2006), and Gotland deep, Baltic Sea at depths of 200–240 m (Stock et al. 2009). In addition to the diversity of fungi, an active role for fungi also has been reported in marine anoxic habitats including denitrification (Manohar and Raghukumar 2013; Stief et al. 2014), tolerating hypersalinity in the deep hypersaline anoxic basins (DHABs) [Alexander et al. 2009].

6.7 Fungal Pathogens in Marine Environment

Fungal pathogens closely related to terrestrial fungi are specifically present in marine mammals. These include *Aspergillus* (aspergillosis), *Blastomyces* (blastomycosis), *Candida* (candidiasis), *Coccidioides* (coccidioidomycosis), *Cryptococcus* (cryptococcosis), *Fusarium* (fusariomycosis), *Malassezia* (dermatitis), and several disease-causing zygomycetes (Higgins 2000). *Aspergillus sydowii*, which is common in terrestrial environments, is found in coral reeves as a pathogen of sea fan corals (Alker et al. 2001). An unidentified fungal pathogen occurring in shallow reef habitat that spreads dense black fungal bands was found causing the death of coralline algae (Raghukumar and Ravindran 2012). Fungi parasitize virtually all groups of marine animals in the marine environment and attack both wild and cultivated marine animals (Hatai 2012; Marano et al. 2012; Porter 1986; Ramaiah 2006; Shields and Overstreet 2007). Among the several groups, Oomycetan taxa cause diseases on a wide range of hosts but prominently attack invertebrates during seed production of marine crustaceans including shrimps and crabs (Hatai 2012; Marano et al. 2012; Beakes et al. 2014). The prominent members are *Halipthoros*, *Halioticida*, and *Halocrusticida* belonging to Haliphthorales, which are exclusively marine (Beakes et al. 2014). Other genera of the Oomycetes that cause diseases among marine animals are *Atkinsiella*, *Lagenidium*, and *Sirolpidium* (Beakes et al. 2014). A *Fusarium* sp. has been identified as a disease-causing agent of American lobsters (Cawthorn 2011). Interestingly this genus has been represented in various environmental DNA analyses (Richards et al. 2012). Several ascomycete fungi are known to be pathogens of marine algae (Kohlmeyer and Kohlmeyer 1979). Pathogenic fungal signatures belonging to *Exophiala dermatitidis* and *Trichosporon dermatis* (Gadanho and Sampaio 2005) from hydrothermal vents in Mid-Atlantic Rifts southwest of the Azores archipelago employing metagenomics were reported. Clones related to the yeast *Malassezia furfur* have been reported from hydrothermal vents of the Lost city (Lopez-Garcia et al. López-Garcia et al. 2007), Rainbow (Bass et al. 2007) in the Mid-Atlantic Ridge.

6.8 Marine Fungal Diversity in Coastal Regions

Coastal regions have a large availability of organic matter to consumers as detritus due to terrestrial run-off and high primary production often leading to eutrophication (Danovaro and Pusceddu 2007). Mangroves, coral reefs, salt marshes, shore line plants, and different halophytic plants are different habitats having different niches in the coastal habitats that offer substrata for colonization of marine fungi. Driftwood was considered to support a higher number of fungi in the past (Kohlmeyer and Kohlmeyer 1979). However, this was overtaken by mangroves which support relatively a higher number of marine fungi (Hyde et al. 2000). Fungal diversity in mangroves has been explored by several workers (see Jones and Alias 1997; Hyde and Lee 1995; Sarma and Hyde 2001). Around 656 fungi were reported from mangroves of the world up to 2003 (Schmidt and Shearer 2003). In mangroves, the taxa belonging to the group *Dothideomycetes* are rich and diverse when compared to *Sordariomycetes* (Suetrong et al. 2009). Earlier only microscopic observations were used for identification of marine fungi from mangrove substrata. In recent times, the culturable fungi are also supplemented with DNA sequence data, and several new genera and new species are published with support from phylogeny (Abdel-Wahab et al. 2010; Devadatha et al. 2017, 2018a, b, c). Molecular diversity analysis using pyrosequencing method revealed a high fungal diversity with *Agaricomycetes* predominating in coastal mangrove regions (Arfi et al. 2012). Spatiotemporal variations play a role in determining the fungal community compositions in the ocean. Water temperature and salinity drive community compositions of wood-inhabiting marine fungi (Booth and Kenkel 1986) in addition to log attachment (fixed vs. free-floating) and location (Rama et al. 2014).

Various ecological aspects of marine fungi in mangroves have been investigated. These include (1) frequency of occurrence (Alias et al. 1995; Borse 1988; Hyde 1988a, 1989a, b, 1992; Hyde and Jones 1989; Hyde et al. 1990; Jones and Alias 1997; Sarma and Hyde 2001; Sarma et al. 2001; Sridhar and Maria 2006; Volkmann-Kohlmeyer and Kohlmeyer 1993), (2) host and substrate specificity (Alias and Jones 2000; Hyde 1990a; Hyde and Alias 2000; Hyde and Jones 1988; Hyde and Lee 1995; Poonyth et al. 1999; Sarma and Vittal 2000, 2001), (3) succession (Hyde 1991; Leong et al. 1991, and Kohlmeyer et al. 1995), (4) spatiotemporal variations (Aleem 1980; Hyde 1989a; Alias et al. 1995; Sarma and Vittal 1998–1999), (5) vertical distribution (Hyde 1988b, 1989b,c 1990b; Hyde and Jones 1988; Jones and Tan 1987; Kohlmeyer et al. 1995; Sarma and Vittal 2002), (6) salinity and horizontal distribution (Aleem 1980; Hyde 1992; Hyde and Sarma 2006; Kohlmeyer and Kohlmeyer 1979), and (7) geographical distribution (Hyde and Lee 1995; Jones and Alias 1997; Kohlmeyer 1983, 1987; Schmidt and Shearer 2004). However, all these studies are based on morphology and microscopy. It is well known that fungi could be identified at a microscopic level based on their fruiting structures/reproductive propagules. Many mycologists consider that fruiting structures by and large do not indicate the active roles of fungi unlike the mycelia. To this extent, the location and identity of mycelia on natural samples are difficult. Several novel techniques are explored to study the mycelial fungi including particle filtration technique (Bills and Polishook 1994; Polishook et al. 1996). A large number of fungi isolated through the particle filtration technique remain nonsporulating and hence require molecular analyses to identify them. Several of these fungi seem to be novel but due to lack of sporulation they remain uncharacterized. Nevertheless, these morphological investigations provide qualitative assessments of diversity and ecological data, though the quantitative data is often questioned due to the reason that "only mycelial state is considered as functionally involved in various metabolic activities and not the sporulation stage." Future studies should include attempts of direct isolation of fungal DNA from lignocellulosic substrata and check whether "yet to be cultured fungi" also could be found. The same would also expand the scope of diversity and ecological investigations of fungi colonizing lignocellulosic substrata from mangroves and other such habitats. It is surprising that recent research review papers omitted a discussion on the ecological data of morphological investigations (e.g.,

Richards et al. 2012). There is a need to integrate the diversity and ecological information from both morphological and molecular analyses.

Few studies employing molecular tools to understand the ecology of marine fungi in sediments and water samples have been carried out near mangroves and other coastal habitats. Molecular studies on planktonic diversity of fungi from some coastal locations of Brazil (Cury et al. 2011), coral reefs of Hawaii (Gao et al. 2008; Gao et al. 2010), and mangrove regions (Arfi et al. 2012) have resulted in the identification of fungal organisms belonging to *Ascomycota*, *Basidiomycota*, and *Chytridiomycota* in addition to few cladding into novel environmental clusters (Manohar and Raghukumar 2013). Though fungi were largely considered to be pathogens in the coral reef ecosystems (Kim et al. 2006; Vega Thurber et al. 2009; Yarden et al. 2007), the metagenomics and functional diversity analyses of microbiota in this ecosystem have revealed that fungi are dominant community and are involved even in the nitrogen cycling (Wegley et al. 2007). Many fungi isolated from sponges and other organisms in corals have shown varied secondary metabolites that have wide applications (Zhuo et al. 2011).

Most of the fungal OTU clusters identified from European near shore water column samples predominantly belonged to chytrid-like and yeast *Dikarya* phylotypes (Richards et al. 2015). Filamentous fungal forms such as Pezizomycotina seem to be less suited for marine water column environments than soils and sediments which have solid substrates rich in organic matter (Richards et al. 2012, 2015). The investigations involving ion semiconductor sequencing (Ion Torrent) of the ribosomal large subunit (LSU/28S) to explore fungal diversity from water as well as sediment samples taken from four habitats in North Carolina revealed dominance of taxa belonging to *Ascomycota* and *Chytridiomycota* (Picard 2017). Further, this study also revealed that sand flats and wetland sediments have the highest diversity, although benthic sediments could harbor a higher proportion of novel sequences. Another interesting study focusing on metatranscriptomics analysis of mangrove habitats around Mauritius have also led to an interesting study where the potential roles of different microorganisms could be elucidated (Rampadaruth et al. 2018).

6.9 Coral-Associated Marine Fungi

The facultative marine fungi seem to play ecologically important roles in coral environments of Lakshadweep Islands, India, and Great Barrier Reef, Australia (Ravindran et al. 2001; Morrison-Gardiner 2002; Yarden 2014; Yarden et al. 2007). Species belonging to *Alternaria*, *Aspergillus*, *Aureobasidium*, *Cladosporium*, *Fusarium*, *Penicillium*, *Hormonema*, *Phialophora*, and *Phoma* have been reported from different healthy scleractinian corals (Ravindran et al. 2001). The obligate marine fungi reported from corals include *Koralionastes* and *Corallicola* of Koralionastaceae family (Kohlmeyer and Volkmann-Kohlmeyer 1992). A study on coral-associated fungal ribosomal DNA amplicons of *Acropora hyacinthus* coral colonies from neighboring natural pools with varying water temperatures suggested a high diversity of *Basidiomycetes* and *Ascomycetes* such as novel lineages. Colonies from a warmer pool comprised phylogenetically more diverse mycocommunities than from colder pool (Amend et al. 2012). Four taxa were retrieved in all coral colonies sampled in this study and they seemed to represent obligate associates. Further, the DNA and RNA (mRNA sequences) analyses indicated a metabolically active and diverse marine fungal communities in the corals (Amend et al. 2012). Ecological roles of endolithic fungi such as denitrification have been reported from corals (Wegley et al. 2007).

The mycocommunities, along with *Symbiodinium* and bacteria, related to the Caribbean coral *Siderastrea siderea*, taken from two depths viz., 17 and 27 m on Conch Reef in the Florida Keys, were studied by Bonthond et al. (2018) employing the high-throughput amplicon sequencing targeting ITS rRNA gene. In this study, even though they used fungal-specific primers (Nikolcheva and Bärlocher 2004), only 22.2% of the 790,398 quality-filtered ITS2 reads were fungal while the remaining belonged to other eukaryotes. Although 184 fungal OTUs were assumed to belong to fungal kingdom using the UNITE reference database in the Mothur pipeline, their additional comparison of the complete fungal OUT sequence set along the GenBank database revealed 145 OTUs have high similarity to nonfungal sequences. Hence they could relate only 39 fungal OTUs having similarity hits to fungi or Mesomycetozoa. Among these, a large number (34) of OTUs belonged to *Ascomycota*, while only 2 OTUs belonged to *Basidiomycota*, 2 to Mesomycetozoa, and 1 OUT to Entomorphthoromycota. It was also found by these workers that ITS2 similarity with sequences from GenBank was notably low, i.e., 91%, which is indicative of the presence of a higher novel diversity connected with *S. siderea*. The most abundant fungus was found most similar (76%) to sequences belonging to Lulworthiales in the class *Sordariomycetes* (Bonthond et al. 2018). While endolithic fungi (fungi thriving within the skeleton) may invade coral tissues at times of stress (Yarden 2014), the fungi also have been shown to parasitize on endolithic algae (Le Campion-Alsumard et al. 1995). Samples of *S. siderea* collected from the same reef but with differences in the depth did not show any difference in mycocommunities (Bonthond et al. 2018). These authors recommended that more conserved markers such as the SSU or LSU rRNA should be employed to characterize fungal diversity colonizing corals.

6.10 Conclusions and Future Perspectives

A list of marine fungi retrieved through culture-independent studies from different marine habitats is shown in Table 1. It could be seen from the table that (1) many fungi are known terrestrial fungi, (2) several new phylotypes are encountered from the marine habitats and they await formal recognition, and (3) *Dikarya* dominate the list of fungi of which *Ascomycota* are rich in diversity. Though the table is still not exhaustive, it could be deduced that the number and diversity of marine fungi retrieved through culture-independent studies is also low. The molecular tools are helping us to identify the ecosystem functioning and the role of fungi in marine environments in addition to the diversity.

Employing novel molecular tools to discover fungi and their ecological significance is exciting but the laborious nature, time, and cost of sophisticated machinery deter further research. Next-generation sequencing has revolutionized microbial ecology, but we are still recovering some common fungal organisms similar to terrestrial organisms. Are those studies still reflecting only a fraction of those cryptic taxa? Are these "DNA-based taxa" especially those novel phylotypes which are genetically different from known ones more ecologically and functionally important than common ubiquitous fungi? There is a dire need to assess their physiological roles in vitro, but the problem is that most of the taxa cannot be cultured. Hence novel methods must be brought forward so that these organisms can be grown under laboratory conditions to permit further experimentations and use approaches discussed by Reich and Labies (Reich and Labes 2017). The overwhelming majority of marine fungal taxa, especially from deep sea and harsh environments still remain to be described. How to translate the loner sequences (OTUs) into official taxonomic names and make them nomenclaturally valid to facilitate taxonomy is an aspect which is still in a transitional stage. The other problem is how to properly identify these novel phylotypes. At some point, one would be very tempted to label them as new species but based on what DNA sequence similarities or differences? These aspects need to be worked out. Whenever any taxonomic novelties and potential phylogenetic relationships are proposed based on environmental DNA, one should do so with extreme precautions as it could lead to a number of future problems (Hongsanan et al. 2018). So far, most DNA-based studies have used only one gene region which represents only a short fragment of the whole genome. It is common practice nowadays to use multigene phylogeny, especially those from protein-coding genes as they are more informative. How far can we go in designing primers for these regions and target a wider diversity of marine fungi? From a phylogenetic perspective, most studies have revealed that these marine organisms possibly have complex evolutionary scenarios and these need to be dealt with. In addition, given that those fungi adapt to extreme environments, the possibility that they can be exploited for medicinal, pharmaceutical, bioprospecting, and bioremediation potential should not be neglected.

Acknowledgments VVS would like to dedicate this chapter to Late Prof. B.P.R. Vittal, C.A.S. in Botany, University of Madras, Guindy Campus, Chennai, India, for introducing him to marine mycology and for being a great mentor and to Dr. Seshagiri Raghukumar and Dr. Chandralatha Raghukumar, formerly with National Institute of Oceanography, Goa, India, for their encouragement and inspiration.

References

Abdel-Wahab MA, Pang K-L, Nagahama T, Abdel-Aziz FA, Jones EBG (2010) Phylogenetic evaluation of anamorphic species of *Cirrenalia* and *Cumulospora* with the description of eight new genera and four new species. Mycolog Prog 9:537–558

Aleem AA (1980) Distribution and ecology of marine fungi in Sierra Leone (Tropical WestAfrica). Bot Mar 23:679–688

Alexander E, Stock A, Breiner H-W, Behnke A, Bunge J, Yakimov MM, Stoeck T (2009) Microbial eukaryotes in the hypersaline anoxic L'Atalante deep-sea basin. Environ Microbiol 11:360–381

Alias SA, Jones EBG (2000) Colonization of mangrove wood by marine fungi at Kuala Selangor mangrove stand. Malaysia Fungal Divers 5:9–21

Alias SA, Kuthubutheen AJ, Jones EBG (1995) Frequencyofoccurrenceoffungi onwood in Malaysian mangroves. Hydrobiologia 295:97–106

Alker AP, Smith GW, Kim K (2001) Characterization of *Aspergillus sydowii* (Thom et Church), a fungal pathogen of Caribbean Sea fan corals. Hydrobiologia 460:105–111

Amend AS, Barshis DJ, Oliver TA (2012) Coral associated marine fungi form novel lineages and heterogeneous assemblates. ISME J 6:291–301

Arfi Y, Marchand C, Wartel M, Record E (2012) Fungal diversity in anoxic-sulfidic sediments in a mangrove soil. Fungal Ecol 5:282–285

Bass D, Howe A, Brown N, Barton H, Demidova M, Michelle H, Li L, Sanders H, WatkinsonSCC WS, Richards TAA (2007) Yeast forms dominate fungal diversity in the deep occans. Proc R Soc B 274:3069–3077

Beakes GW, Honda D, Thines M (2014) Systematics of the Straminipila: Labyrinthulomycota, Hyphochytriomycota, and Oomycota. In: McLaughlin DJ, Spatafora JW (eds) The Mycota VII. Part A. Systematics and evolution, 2nd edn. Springer, Berlin/Heidelberg, pp 39–96

Bills GF, Polishook JD (1994) Abundance and diversity of microfungi in leaf litter of a lowland rain forest in Costa Rica. Mycologia 86:187–198

Bonthond G, Merselis DG, Dougan KE, Graff T, Todd W, Fourqurean JW, Rodriguez-Lanetty M (2018) Inter-domain microbial diversity within the coral holobiont *Siderastrea siderea* from two depth habitats. Peer J 6:e4323. https://doi.org/10.7717/peerj.4323

Booth T & Kenkel N (1986) Ecological studies of lignicolous marine fungi: a distribution model based on coordination and classification. In The Biology of Marine Fungi (ed.S.T. Moss), Cambridge University Press, Cambridge, pp. 297-310

Borse BD (1988) Frequency of occurrence of marine fungi from Maharashtra coast, India. Ind J Mar Sci 17:165–167

Burgaud G, Le Calvez T, Arzur D, Vandenkoornhuyse P, Barbier G (2009) Diversity of culturable marine filamentous fungi from deep-sea hydrothermal vents. Environ Microbiol 11:1588–1600

Cai L, Jeewon R, Hyde KD (2006) Molecular systematics of *Zopfiella* and allied genera: evidence from multigene sequence analyses. Fungal Biology 110:359–368

Cawthorn RJ (2011) Diseases of American lobsters (*Homarus americanus*): a review. J Invertebr Pathol 106:71–78

Cury JC, Araujo FV, Coelho-Souza SA, Peixoto RS, OliveiraJAL SHF, Davila AMR, Rosado AS (2011) Microbial diversity of a Brazilian coastal region influencedby an upwelling system and anthropogenic activity. PLoS One 6:e16553

Damare SR, Nagarajan M, Raghukumar C (2008) Spore germination of fungi belonging to *Aspergillus* species under deep-sea conditions. Deep-Sea Res 55:670–678

Danovaro R, Pusceddu A (2007) Biodiversity and ecosystem functioning in coastal lagoons: does microbial diversity play any role? Estuar Coast Shelf Sci 75:4–12

Dawson SC, Pace NR (2002) Novel kingdom-level eukaryotic diversity in anoxic environments. Proc Natl Acad Sci USA 99:8324–8329

Devadatha B, Sarma VV, Wasnasinghe DN, Hyde KD & Jones EBG (2017) Introducing the new Indian mangrove species, *Vaginatispora microarmatispora* (Lophiostomataceae) based on morphology and multigene phylogenetic analysis. Phytotaxa. 329: 139-149. https://doi.org/10.11646/phytotaxa.329.2.4

Devadatha B, Sarma VV, Ariyawansa HA, Jones EBG (2018a) *Deniquelata vittalii* sp. nov., a novel Indian saprobic marine fungus on *Suaeda monoica* and two new records of marine fungi from Muthupet mangroves, East coast of India. Mycosphere 9:565–582. https://doi.org/10.5943/mycosphere/9/3/8

Devadatha B, Sarma VV, Jeewon R, Jones EBG (2018b) *Morosphaeria muthupetensis* sp.nov. (Morosphaeriaceae) from India: Morphological characterisation and multigene phylogenetic inference. Bot Mar 61(4):395–405. https://doi.org/10.1515/bot-2017-0124

Devadatha B, Sarma VV, Jeewon R, Wanasinghe DN, Hyde KD, EBG J (2018c) *Thyridariella*, a novel marine fungal genus from India: morphological characterization and phylogeny inferred

from multigene DNA sequence analyses. Mycolog Progr 17:791–804. https://doi.org/10.1007/s11557-018-1387-4
Duong LM, Lumyong S, Hyde KD, Jeewon R (2004) *Emarcea castanopsicola* gen. et sp. nov. from Thailand, a new xylariaceous taxon based on morphology and DNA sequences. Stud Mycol 50:253–260
Edgcomb VP, Beaudoin D, Gast R, Biddle JF, Teske A (2011) Marine subsurface eukaryotes: the fungal majority. Environ Microbiol 13:172–183
Fell JW, Master IM (1980) The association and potential role of fungi in mangrove detrital systems. Bot Mar 23:257–263
Fryar SC (2002) Fungal succession or sequence of fruit bodies? In: Hyde KD, Jones EBG (eds) Fungal succession. Fungal Divers 10:5–10
Gadanho M, Sampaio JP (2005) Occurrence and diversity of yeasts in the mid-Atlantic ridge hydrothermal fields near the Azores Archipelago. Microb Ecol 50:408–417
Gao Z, Li BL, Zheng CC, Wang G (2008) Moleculardetection of fungal communities in the Hawaiian marine sponges *Suberites zeteki* and *Mycale armata*. Appl Environ Microbiol 74:6091–6101
Gao Z, Johnson ZI, Wang G (2010) Molecular characterizationof the spatial diversity and novel lineages of mycoplankton in Hawaiian coastal waters. ISME J 4:111–120
Gojkovic Z, Knecht W, Zameitat E, Warneboldt J, Coutelis JB, Pynyaha Y, Neuveglise C, Moller K, Loffler M, Piskur J (2004) Horizontal gene transfer promoted evolution of the ability to propagate under anaerobic conditions in yeasts. Mol Genet Genom 271:387–393
Hatai K (2012) Diseases of fish and shell fish caused by marine fungi. In: Raghukumar C (ed) Biology of marine fungi. Springer, Berlin, pp 15–52
Hongsanan S, Maharachchikumbura S, Hyde KD, Samarakoon M, Jeewon R, Zhao Q, Bahkali A (2017) An updated phylogeny of Sordariomycetes based on phylogenetic and molecular clock evidence. Fungal Divers 84:84. https://doi.org/10.1007/s13225-017-0384-2
Hongsanan S, Jeewon R, Purahong W, Xie N, Liu J-K, Jayawardane RS, Ekanayake RS, Dissanayake A, Raspe O, Hyde KD, Stadler M, Persoh D (2018) Can we use environmental DNA as holotypes? Fungal Divers 92:1–30
Hughes GC (1974) Geographical distribution of the higher marine fungi. Veroeffentlichender Institute ftir Meeres forschung, Bremerhaven Supplement 5:419–441
Hyde KD (1988a) Studies on the tropical marine fungi of Brunei. Bot J Linn Soc 98:135–151
Hyde KD (1988b) Observations on the vertical distribution of marine fungi on *Rhizophora* spp. at Kampong Danau mangrove, Brunei. Asian Mar Biol 5:77–81
Hyde KD (1989a) Ecology of tropical marine fungi. Hydrobiologia 178:199–208
Hyde KD (1989b) Intertidal mangrove fungi from North Sumatra. Can J Bot 67:3078–3082
Hyde KD (1989c) Vertical zonation of intertidal mangrove fungi. In: Hattori T, lshida Y, Maruyama Y, Moria R, Uchida A (eds) Recent advances in microbial ecology. Japan Scientific Societies Press, Tokyo, pp 302–306
Hyde KD (1990a) A comparison of the intertidal mycota of five mangrove tree species. Asian Mar Biol 7:93–107
Hyde KD (1990b) A study of the vertical zonation of intertidal fungi on *Rhizophora apiculata* at Kampong Kapok mangrove, Brunei. Aquat Bot 36:255–262
Hyde KD (1991) Fungal colonization of *Rhizophora apiculata* and *Xylocarpus granatum*poles in Kampong Kapok mangrove Brunei. Sydowia 43:31–38
Hyde KD (1992) Intertidal mangrove fungi from the West Coast of Mexico including onenew genus and two new species. Mycol Res 96:25–30
Hyde KD, Alias SA (2000) Biodiversity and distribution of fungi associated with decomposing *Nypafruticans*. Biodivers Conserv 9:393–402
Hyde KD, Jones EBG(1988) Marine mangrove fungi. Marine Ecol (P.S.Z.N.I.) 9:15–33
Hyde KD, Jones EBG (1989) Ecological observations on marine fungi from the Seychelles. Bot J Linn Soc 100:237–254
Hyde KD, Lee SY (1995) Ecology of mangrove fungi and their role in nutrient cycling. What gaps occur in our knowledge? Hydrobiologia 295:107–118

Hyde KD, Sarma VV (2006) Biodiversity and ecological observations on filamentous fungi of *Nypa fruticans* along the Tutong River, Brunei. Ind J Mar Sci 35(4):297–307

Hyde KD, Chalermongse A, Boonthavikoon T (1990) Ecology of intertidal fungi at Ranong mangrove, Thailand. Trans Mycol Soc Japn 31:17–28

Hyde KD, Jones EBG, Leano E, Pointing SB, Poonyth AD, Vrijmoed LLP (1998) Role of marine fungi in marine ecosystems. Biodivers Conserv 7:1147–1161

Hyde KD, Sarma VV, Jones EBG (2000) Morphology and taxonomy of higher marine fungi. In: Hyde KD, Pointing SB (eds) Marine mycology a practical approach. Fungal Diversity Press, Hong Kong, pp 172–204

Jebaraj CS, Raghukumar C (2009) Anaerobic denitrification in fungi from the coastal marine sediments off Goa, India. Mycol Res 113:100–109

Jebaraj CS, Raghukumar C, Behnke A, Stoeck T(2009) Fungal diversity in oxygen-depleted regions of the Arabian Sea revealed by targeted environmental sequencing combined with cultivation. FEMS Microbiol Ecol 71:399–412

Jebaraj CS, Raghukumar C, Behnke A, Stoeck T (2010) Fungal diversity in oxygen-depleted regions of the Arabian Sea revealed by targeted environmental sequencing combined with cultivation. FEMS Microbiol Ecol 71:399–412

Jeewon R, Hyde KD (2007). Detection and diversity of fungi from environmental samples: traditional versus molecular approaches. In: Varma A, Oelmüller R (eds) Advanced techniques in soil microbiology. Soil biology, vol 11. Springer, Berlin/Heidelberg, pp 427. https://doi.org/10.1007/978-3-540-70865-0_1

Jeewon R, Hyde KD (2016) Establishing species boundaries and new taxa among fungi: recommendations to resolve taxonomic ambiguities. Mycosphere 7(11):1669–1677

Jeewon R, Liew ECY, Hyde KD (2002) Phylogenetic relationships of *Pestalotiopsis* and allied genera inferred from ribosomal DNA sequences and morphological characters. Mol Phylogenet Evol 25:378–392

Jeewon R, Liew ECY, Simpson JA, Hodgkiss IJ, Hyde KD (2003a) Phylogenetic significance of morphological characters in the taxonomy of *Pestalotiopsis* species. Mol Phylogenet Evol 27:372–383

Jeewon R, Cai L, Zhang K, Hyde KD (2003b) *Dyrithiopsis lakefuxianensis* gen et sp. nov. from Fuxian Lake, Yunnan, China and notes on the taxonomic confusion surrounding *Dyrithium*. Mycologia 95:911–920

Jeewon R, Yeung QSY, Hyde KD (2009) Novel phylogenetic group within *Thozetella* (Chaetosphaeriaceae): a new taxon based on morphology and DNA sequence analyses. Can J Microbiol 55:680–687

Jeewon R, Ittoo J, Mahadeb D, Jaufeerally-Fakim Y, Kai WH, Liu A-R (2013) DNA based identification and phylogenetic characterisation of endophytic and saprobic fungi from *Antidesma madagascariense*, a medicinal plant in Mauritius. J Mycol 2013:1. https://doi.org/10.1155/2013/781914

Jeewon R, Wanasinghe DN, Rampadaruth S, Puchooa D, Zhou L-G, Liu A-R, Wang H-K (2017) Nomenclatural and identification pitfalls of endophytic mycota based on DNA sequence analyses of ribosomal and protein genes phylogenetic markers: A taxonomic dead end? Mycosphere 8(10):1802–1817

Jeewon R et al (2018) Hidden mycota of pine needles: Molecular signatures from PCR-DGGE and Ribosomal DNA phylogenetic characterization of novel phylotypes. Sci Rep 8:18053. https://doi.org/10.1038/s41598-018-36573-z

Jones EBG, Alias SA (1997) Biodiversity of mangrove fungi. In: Hyde KD (ed) Biodiversity of tropical microfungi. Hong Kong University Press, Hong Kong, pp 71–92

Jones EBG, Hyde KD (2002) Succession: where do we go from here? In Hyde KD, Jones EBG (eds) Fungal succession. Fungal Divers 10:241–253

Jones EBG, Tan TK (1987) Observations on manglicolous fungi from Malaysia. Tran Br Mycol Soc 89:390–392

Jones EBG, Sakayaroj J, Suetrong S, Somrithipol S, Pang KL (2009) Classification of marine Ascomycota, anamorphic taxa and Basidiomycota. Fungal Divers 35:1–187

Jones MDM, Forn I, Gadelha C, Egan MJ, Bass D, Massana R, Richards TA (2011) Discovery of novel intermediate forms redefines the fungal tree of life. Nature 474:200–203

Jones EBG, Suetrong S, Sakayaroj J, Bahkali AH, Abdel-Wahab MA, Boekhout T, Pang K-L (2015) Classification of marine Ascomycota, Basidiomycota, Blastocladiomycota and Chytridiomycota. Fungal Diver 73:1–72

Karunarathna A, Papizadeh M, Senanayake IC, Jeewon R, Phookamsak R, Goonasekara ID, Wanasinghe DN, Wijayawardene NN, Amoozegar MA, Shahzadeh Fazeli SA, Camporesi E, Hyde KD, Weerahewa HLD, Lumyong S, McKenzie EHC (2017) Novel fungal species of Phaeosphaeriaceae with an asexual/sexual morph connection. Mycosphere 8(10):1818–1834

Kim K, Alker AP, Shuster K, Quirolo C, Harvell CD (2006) Longitudinal study ofaspergillosis in sea fan corals. DisAquat Organ 69:95–99

Kis-Papo T (2005) Marine fungal communities. In: Dighton J, White JF, Oudemans P (eds) The fungal community: its organisation and role in the ecosystem. Boca Raton, Taylor & Francis, pp 61–92

Kodsueb R, McKenzie EHC, Lumyong S, Hyde KD, Jeewon R (2007) Molecular phylogeny of *Aquaticheirospora broccolii*; a new synnematous hyphomycete taxon from Thailand and its teleomorphic affinities to Massarinaceae. Bot J LinnSoc 155:283–296

Kohlmeyer J (1983) Geography of marine fungi. Austr J BotSuppl Ser 10:67–76

Kohlmeyer J (1987) Marine fungi from Aldabra, the Galapagos and other tropical islands. Can J Bot 65:571–582

Kohlmeyer J, Kohlmeyer E (1979) Marine mycology: The higher fungi. Academic, London/New York/San Francisco, p 690

Kohlmeyer J, Volkmann-Kohlmeyer B (1992) Two Ascomycotina from coral reefs in the Caribbeanand Australia. Cryptogam Bot 2:367–374

Kohlmeyer J, Bebout B, Volkmann-Kohlmeyer B (1995) Decomposition of mangrove wood by marine fungi and teredinids in Belize. Mar Ecol (P.S.Z.N.I.) 16:27–39

Lai X, Cao L, Tan H, Fang S, Huang Y, Zhou S (2007) Fungal communities from methane hydrate bearing deep-sea marine sediments in South China Sea. ISME J 1:756–762

Le Calvez T, Burgaud G, Mahé S, Barbier G, Vandenkoornhuyse P (2009) Fungal diversity in deep-seahydrothermal ecosystems. Appl Environ Microbiol 75:6415–6421

Le Campion-Alsumard T, Golubic S, Priess K (1995) Fungi in corals: symbiosis or disease?Interaction between polyps and fungi causes pearl-like skeleton biomineralization. Mar Ecol Progr Ser 117:137–147. https://doi.org/10.3354/meps117137

Leong WF, Tan TX, EBG J (1991) Fungal colonization of submerged *Bruguiera cylindrica* and *Rhizophora apiculata* wood. Bot Mar 34:69–76

Li JF, Jeewon R, Phookamsak R, Bhat DJ, Mapook A, Chukeatirote E, Hyde KD, Lumyong S, McKenzie EHC (2018) *Marinophialophora garethjonesii* gen. et sp. nov.: a new hyphomycete associated with Halocyphina from marine habitats in Thailand. Phytotaxa 345(1):1–12

Liu AR, Chen SC, Wu SY, Xu T, Guo LD, Jeewon R, Wei JG (2010) Cultural studies coupled with DNA based sequence analyses and its implication on pigmentation as a phylogenetic marker in *Pestalotiopsis* taxonomy. Mol Phylogenet Evol 57:528–535

López-Garcia P, Vereshchaka A, Mozeira D (2007) Eukaryotic diversity associated with carbonatesand fluid-seawater interface in Lost City hydrothermal field. Environ Microbiol 9:546–554

Luo ZL, Bhat DJ, Jeewon R, Boonmee S, Bao D-F, Zhao Y-C, Chai H-M, Su H-Y, Su X-J, Hyde KD (2017) Molecular phylogeny and morphological characterization of asexual fungi (Tubeufiaceae) from freshwater habitats in Yunnan, China. Cryptogamie Mycol 38:1–28

Manohar CS, Raghukumar C (2013) Fungal diversity from various marine habitats deduced through culture independent studies. FEMS Microbiol Lett 341:69–78

Marano AV, Pires-Zottarelli CLA, de Souza JI, Glockling SL, Leano EM, Gachon CMM,Strittmatter M, Gleason FH (2012) Hyphochytriomycota, Oomycota and Perkinsozoa (SupergroupChromalveolata). In Jones EBG, Pang K-L (eds) Marine fungi andfungal-like organisms. De Gruyter, Berlin/Boston, pp 167–213

Massana R, Pedrós-Alió C (2008) Unveiling new microbial eukaryotes in the surface ocean. Curr Opin Microbiol 11:213–218

Morrison-Gardiner S (2002) Dominant fungi from Australian coral reefs. Fungal Divers 9:105–121
Nagahama T, Hamamoto M, Nakase T, Horikoshi K (2003) *Rhodotorula benthica* sp nov and *Rhodotorula calyptogenae* sp nov, novel yeast species from animals collected from the deepsea floor, and *Rhodotorula lysiniphila* sp nov, which is related phylogenetically. Int J Syst Evol Microbiol 53:897–903
Nagahama T, Takahashi E, Nagano Y, Abdel-Wahab MA, Miyazaki M (2011) Molecular evidencethat deep-branching fungi are major fungal components in deep-sea methane cold-seep sediments. Environ Microbiol 13:2359–2370
Nagano Y, Nagahama T, Hatada Y, Nunoura T, Takami H, Miyazaki J, Takai K, Horikoshi K (2010) Fungal diversity in deep-sea sediments – the presence of novel fungal groups. Fungal Ecol 3:316–325
Nagano Y, Miura T, Nishia S, Lima AO, Nakayama C, Pellizari VH, Fujikura K (2018) Fungal diversity in deep-sea sediments associated with asphalt seeps at the Sao Paulo Plateau. Deep-Sea Res Pt II 146:59–67
Newell SY (1976) Mangrove fungi: the succession in the mycoflora of red mangrove (*Rhizophora mangle* L.) seedlings. In: Jones EBG (ed) Recent advances in aquatic mycology. Wiley, New York, pp 51–91
Newell SY (1996) Established and potential impacts of eukaryotic mycelial decomposers in marine/terrestrial ecotones. J Exp Mar Biol Ecol 200:187–206
Nikolcheva LG, Bärlocher F (2004) Taxon-specific fungal primers reveal unexpectedly high diversity during leaf decomposition in a stream. Mycol Progr 3:41–49. https://doi.org/10.1007/s11557-006-0075-y
Orsi W, Biddle JF, Edgcomb V (2013) Deep sequencing of sub-seafloor eukaryotic rRNA reveals active fungi across marine subsurface provinces. PLoS One 8(2):e56335. https://doi.org/10.1371/journal.pone.0056335
Pang K-L, Jones EBG (2017) Recent advances in marine mycology. Bot Mar 60:361–362
Pang KL, Overy DP, Jones EBG, Calado MDL, Burgaud G, Walker AK, Johnson JA, Kerr RG, Cha HJ, Bills GF (2016) 'Marine fungi' and 'marine-derived fungi' in natural product chemistry research: toward a new consensual definition. Fungal Biol Rev 30:163–175
Picard KT (2017) Coastal marine habitats harbor novel early-diverging fungal diversity. Fungal Ecol 25:1–13
Pinnoi A, Jeewon R, Sakayaroj J, Hyde KD, Jones EBG (2007) *Berkleasmium crunisia* sp. nov. and its teleomorphic affinities to the Pleosporales based on 18S, 28S and ITS-5.8S rDNA sequence analyses. Mycologia 99:378–384
Polishook JD, Bills GF, Lodge DJ (1996) Microfungi from decaying leaves of two rain forest trees in Puerto Rico. J Ind Microbiol Biotechnol 17:284–294
Poonyth AD, Hyde KD, Peerally A (1999) Intertidal fungi in Mauritian mangroves. Bot Mar 42:243–252
Porter D (1986) Mycoses of marine organisms: an overview of pathogenic fungi. In: Moss ST (ed) The biology of marine fungi. Cambridge University Press, Cambridge, pp 141–153
Promputtha I, Jeewon R, Lumyong S, McKenzie EHC, Hyde KD (2005) Ribosomal DNA fingerprinting in the identification of non-sporulating endophytes from *Magnolia liliifera* (Magnoliaceae). Fungal Divers 20:167–186
Promputtha I, Lumyong S, Vijaykrishna D, McKenzie EHC, Hyde KD, Jeewon R (2007) A phylogenetic evaluation of whether endophytes become saprotrophs at host senescence. Microbial Ecol 53:579–590
Raghukumar C, Ravindran J (2012) Fungi and their role in corals and coral reef ecosystems. In: Raghukumar C (ed) Biology of marine fungi. Springer, Berlin, pp 89–114
Raghukumar S, Sharma S, Raghukumar C, Sathe-Pathak V (1994) Thraustochytrid and fungal component of marine detritus. IV. Laboratory studies on decomposition of leaves of themangrove *Rhizophora apiculata* Blume. J Exp Mar Biol Ecol 183:113–131
Raghukumar S, Sathe-Pathak V, Sharma S, Raghukumar C (1995) Thraustochytrid and fungalcomponent of marine detritus. III. Field studies on decomposition of leaves of the mangrove *Rhizophora apiculata* Blume. Aquat Microb Ecol 9:117–125

Raghukumar C, Raghukumar S, Sheelu G, Gupta SM, Nath BN, Rao BR (2004) Buried in time: culturable fungi in a deep-sea sediment core from the Chagos Trench. Indian Ocean. Deep-Sea Res Pt I 51:1759–1768
Rama T, Norden J, Davey ML, Mathiassen GH (2014) Fungiahoy! Diversity on marine wooden substrata in the high North. Fungal Ecol 8:46–58
Ramaiah N (2006) A review on fungal diseases of algae, marine fishes, shrimps and corals. Ind J Mar Sci 35:380–387
Rampadaruth S,Bandhoa K, Puchooa D, Jeewon R, Bal S (2018) Metatranscriptomics analysis of mangroves habitats around Mauritius. World J Microbiol Biotechnol 34:59. https://doi.org/10.1007/s11274-018-2442-7
Ravindran J, Raghukumar C, Raghukumar S (2001) Fungi in *Porites lutea*: association withhealthy and diseased corals. Dis Aquat Org 47:219–228
Rédou V, Ciobanu MC, Pachiadaki MG, Edgcomb V, Alain K, Barbier G, Burgaud G (2014) In-depth analyses of deep subsurface sediments using 454 pyrosequencing reveals a reservoir of buried fungal communities at record-breaking depths. FEMS Microbiol 90:908–921
Rédou V, Navarri M, Meslet-Cladiere L, Barbier G, Bergaud G (2015) Species richness and adaptation of marine fungi from deep-sub seafloor sediments. Appl Environ Microbiol 81:3571–3583
Reich M, Labes A (2017) How to boost marine fungal research: A first step towards a multidisciplinary approach by combining molecular fungal ecology and natural products chemistry. Mar Genomics 36:57–75
Richards TA, Jones MDM, Leonard G, Bass D (2012) Marine fungi: their ecology and molecular diversity. Annu Rev Mar Sci 4:495–522
Richards TA, Leonard G, Mahé F, del Campo J, Romac S, Jones MDM, Maguire F, Dunthorn M, De Vargas C, Massana R, Chambouvet A (2015) Molecular diversity and distribution of marine fungi across 130 European environmental samples. Proc R Soc Lond [Biol] 282:20152243. https://doi.org/10.1098/rspb.2015.2243
Sarma VV, Hyde KD (2001) A review on frequently occurring fungi in mangroves. Fungal Divers 8:1–34
Sarma VV, Vittal BPR (1998–1999) Ecological studies on mangrove fungi from east coast of India. Observations on seasonal occurrence. Kavaka 26&27:105–120
Sarma VV, Vittal BPR (2000) Biodiversity of mangrove fungi on different substrata of *Rhizophora apiculata* and *Avicennia* spp. from Godavari and Krishna deltas, east coast of India. In Hyde KD, Ho WH, Pointing SB (eds) Aquatic mycology across the millennium. Fungal Divers 5:23–41
Sarma VV, Vittal BPR (2001) Biodiversity of fungi on selected mangrove plants in the Godavari and Krishna deltas, east coast of India. Fungal Divers 6:113–129
Sarma VV, Vittal BPR (2002) A preliminary study on vertical distribution of manglicolous fungi on prop roots of *Rhizophora apiculata* Blume at Krishna delta, east coast of India. Kavaka 30:21–29
Sarma VV, Hyde KD, Vittal BPR (2001) Frequency of occurrence of mangrove fungi from the East coast of India. Hydrobiologia 455:41–53
Schmidt JP, Shearer CA (2003) A checklist of mangrove–associated fungi, their geography and known host plants. Mycotaxon 80:423–477
Schmidt JP, Shearer CA (2004) Geographical and host distribution of lignicolous mangrove microfungi. Bot Mar 47:496–500
Senanayake IC, Crous PW, Groenewald JZ, SSN M, Jeewon R, AJL P, Bhat JD, Perera RH, Li QR, Li WJ, Tangthirasunun N, Norphanphoun C, Karunarathna SC, Camporesi E, Manawasighe I, Al-Sadi AM, Hyde KD (2017) Families of Diaporthales based on morphological and phylogenetic evidence. Stud Mycol 86:217–296
Senanayake IC, Jeewon R, Chomnunti P, Wanasinghe D, Norphanphoun C, Karunarathna A, Pem D, Perera RH, Camporesi E, McKenzie EHC, Hyde KD, Karunarathna SD (2018a) Taxonomic circumscription of Diaporthales based on multigene phylogeny and morphology. Fungal Divers 93:241. https://doi.org/10.1007/s13225-018-0410-z

Senanayake IC, Jeewon R, Camporesi E, Hyde KD, Zeng Y-J, Tian S-L, Xie N (2018b) *Sulcispora supratumida* sp. nov. (Phaeosphaeriaceae, Pleosporales) on *Anthoxanthum odoratum* from Italy. Myco Keys 38:35–46

Shields JD, Overstreet RM (2007) Diseases, parasites, and other symbionts. Chapter 8. In Kennedy, VS, Cronin LE (eds) The blue crab, *Callinectes sapidus*. Maryland Sea Grant, College Park, pp 299–417

Simonato F, Campanaro S, Lauro FM, Vezzi A, D'Angelo M, Vitulo N, Valle G, Barlett DH (2006) Piezophilic adaptation: a genomic point of view. J Biotechnol 126:11–25

Singh P, Raghukumar C, Verma P, Shouche Y (2010) Phylogenetic diversity of culturablefungi from the deep-sea sediments of the Central Indian Basin and theirgrowth characteristics. Fungal Divers 40:89–102

Singh P, Raghukumar C, Verma P, Shouche Y (2011) Fungal community analysis inthc dccp-sea sediments of the Central Indian Basin by culture-independent approach. Microb Ecol 61:507–517

Singh P, Raghukumar C, Meena RM, Verma P, Shouche Y (2012) Fungal diversityin deep-sea sediments revealed by culture-dependent and culture-independent approaches. Fungal Ecol 5:543–553

Sridhar KR, Maria GL (2006) Fungal diversity on mangrove woody litter *Rhizophora mucronata* (Rhizophoraceae). Ind J Mar Sci 35:318–325

Stief P, Fuchs-Ocklenburg S, Kamp A, Manohar CS, Houbraken J, Boekhout T, de Beer D, Stoeck T (2014) Dissimilatory nitrate reduction by *Aspergillus terreus* isolated from the seasonal oxygen minimum zone in the Arabian Sea. BMC Microbiol 14:35

Stoeck T, Hayward B, Taylor GT, Varela R, Epstein SS (2006) A multiple PCR-primer approachto access the microeukaryotic diversity in environmental samples. Protist 157:31–43

Stoeck T, Bass D, Nebel M, Christen R, Jones MDM, Briner H-W, Richards TA (2010) Multiple marker parallel tag environmental DNA sequencing reveals a highly complex eukaryotic community in marine anoxic water. Mol Ecol 19(1):21–31. https://doi.org/10.1111/j.1365-294X.2009.04480

Suetrong S, Schoch CL, Spatafora JW, Kohlmeyer J, Volkmann-Kohlmeyer B, Sakayaroj J, Phongpaichit S, Tanaka K, Hirayama K, Jones EBG (2009) Molecular systematics of the marine Dothideomycetes. Stud Mycol 64:155–173

Swe A, Jeewon R, Hyde KD (2008a) Nematode-trapping fungi from mangrove habitats. Cryptogam Mycol 29:333–354

Swe A, Jeewon R, Pointing SB, Hyde KD (2008b) Taxonomy and phylogeny of *Arthrobotrys mangrovispora,* a new marine nematode-trapping fungal species. Bot Mar 51:331–338

Swe A, Jeewon R, Pointing SB, Hyde KD (2009) Diversity and abundance of nematode-trapping fungi from decaying litter in terrestrial, freshwater and mangrove habitats. Biodivers Conserv 18:1695–1714

Takishita K, Miyake H, Kawato M, Maruyama T (2005) Genetic diversity of microbial eukaryotes in anoxic sediment around fumaroles on a submarine caldera floor based on the small-subunit rDNA phylogeny. Extremophiles 9:185–196

Takishita K, Tsuchiya M, Reimer JD, Maruyama T (2006) Molecular evidence demonstrating the basidiomycetousfungus *Cryptococcus curvatus* is the dominant microbial eukaryote in sediment at the Kuroshima Knollmethane seep. Extremophiles 10:165–169

Takishita K, Yubuki N, Kakizoe N, Inagaki Y, Maruyama T (2007) Diversity of microbial eukaryotes in sediment at a deep-sea methane cold seep: surveys of ribosomal DNA libraries from raw sediment samples and two enrichment cultures. Extremophiles 11:563–576

Tsui CKM, Berbee M, Jeewon R, Hyde KD (2006) Molecular phylogeny of *Dictyosporium* and allied genera inferred from ribosomal DNA. Fungal Divers 21:157–166

Vega Thurber R, Willner-Hall D, Rodriguez-Mueller B, Desnues C, Edwards RA, Angly F, Dinsdale E, Kelly L, Rohwer F (2009) Metagenomic analysis of stressed coralholobionts. Environ Microbiol 8:2148–2163

Vijaykrishna D, Mostert L, JeewonR HKD, Crous PW (2004) *Pleurostomosphora,* an anamorph of *Pleurostoma* (Calosphaeriales), a new anamorph genus morphologically similar to *Phialophora*. Stud Mycol 50:387–398

Vinit K, Cheewangkoon R, Thambugala KM, Jones EBG, Jeewon R, Doilom M, Brahmanage RS, Hyde KD (2018a) *Rhytidhysteron mangrovei,* (Hysteriaceae) a new species from mangroves in Phetchaburi Province, Thailand. Phytotaxa. (in press)

Vinit K, Doilom M, Wanasinghe DN, Bhat DJ, Brahmanage RS, Jeewon R, Xiao Y, Hyde KD (2018b) Phylogenetic placement of *Akanthomyces muscarius*, a new endophyte record from *Nypa fruticans* in Thailand. Curr Res Environ Appl Mycol J 8(3):404–417

Volkmann-Kohlmeyer B, Kohlmeyer J (1993) Biogeographic observations on Pacificmarine fungi. Mycologia 85:337–346

Wanasinghe DN, Phukhamsakda C, Hyde KD, Jeewon R, Lee HB, Jones EBG, Tibpromma S, Tennakoon DS, Dissanayake AJ, Jayasiri SC, Gafforov Y, Camporesi E, Bulgakov TS, Ekanayake AH, Perera RH, Samarakoon MC, Goonasekara ID, Mapook A, Li WJ, Senanayake IC, Li JF, Norphanphoun C, Doilom M, Bahkali AH, Xu JC, Mortimer PE, Tibell L, Savic ST, Karunarathna SC (2018) Fungal diversity notes 709–839: taxonomic and phylogenetic contributions to fungal taxa with an emphasis on fungi on Rosaceae. Fungal Divers 89:1–236

Webster J, Weber WS (2007) Introduction to fungi. Cambridge University Press, Cambridge, p 841

Wegley L, Edwards R, Rodriguez-Brito B, Liu H, Rohwer F (2007) Metagenomic analysis of the microbial community associated with the coral *Porites astreoides*. Environ Microbiol 9:2707–2719

Xu D, Jiao N, Ren R and Warren A (2017). Distribution and diversity of microbial eukaryotes in bathypelagic waters of the South China Sea. J. Eukaryot. Microbiol. 64: 370–382. https://doi.org/10.1111/jeu.12372

Yarden O (2014) Fungal association with sessile marine invertebrates. Front Microbiol 5:228. https://doi.org/10.3389/fmicb.2014.00228

Yarden O, Ainsworth TD, Roff G, Leggat W, Fine M, Hoegh-Guldberg O (2007) Increased prevalence ofubiquitous ascomycetes in an acropoid coral (*Acropora formosa*) exhibiting symptoms of brown band syndromeand skeletal eroding band disease. Appl Environ Microbiol 73:2755–2757

Zhang Y, Jeewon R, Fournier J, Hyde KD (2008) Multi-gene phylogeny and morphotaxonomy of *Amniculicola lignicola*: novel freshwater fungus from France and its relationships to the Pleosporales. Fungal Biol 112:1186–1194

Zhang XY, Tang GL, Xu XY, Nong XH, Qi SH (2014) Insights into deep-seasediment fungal communities from the East Indian Ocean using targetedenvironmental sequencing combined with traditional cultivation. PLoS One 9(10):e109118. https://doi.org/10.1371/journal.pone.0109118

Zhao RL, JeewonR GE, Desjardin D, Soytong K, Hyde KD (2007) Ribosomal DNA phylogenies of *Cyathus*: Is current morphotaxonomic classification robust? Mycologia 99:385–395

Zhuo R, Ma L, Fan F, Gong Y, Wan X, Jiang M, Zhang X, Yang Y (2011) Decolorization of different dyes by a newlyisolated white-rot fungi strain *Ganoderma* sp.En3 and cloning and functional analysis of its laccase gene. J Hazard Mater 192:855–873

Saccharomyces cerevisiae: Oscillatory Orchestration of Growth

7

David Lloyd

Abstract

It is important to understand the structure and function of Baker's yeast as it serves as an excellent model, not only for many other fungi but also for almost all eukaryotic cell systems, including those of humans and hence biomedicine. In 1996, *Saccharomyces cerevisiae* became the first genetically defined eukaryotic organism to be sequenced: 6604 genes (> 400 orthologous with, and replaceable by human ones). Although its evolution has diverged for approximately 1.5 billion years from that of human cells, yeast research provides basic clues and insights into the molecular deficiencies and disorders of many human conditions: mitochondrial dysfunction, cellular division, apoptosis, diabetes, obesity, the accompaniments of old age, cancers, and "dynamic diseases" (neuropsychiatric conditions, e.g., many sleep disorders, depression). Dynamic maintenance of redox status and balanced energy supply and demand is crucial for optimum function and survival of yeast and of all cells. Recent advances in optoelectronics enable fast and continuous dynamic interrogation of processes in vivo. In this chapter, I outline the exquisite time order of molecular, metabolic, and biosynthetic events and processes that are organized with a precision that is commensurate with both spatial and temporal organization and that finds a coherence and resonance universally in life processes.

Keywords

Time structure · Oscillations · Rhythms · Ultradian · Mitochondria · Metabolism · Energetics · Oxygen

D. Lloyd (✉)
School of Biosciences, Cardiff University, Cardiff, Wales, UK
e-mail: lloydd@cardiff.ac.uk

T. Satyanarayana et al. (eds.), *Advancing Frontiers in Mycology & Mycotechnology*,
https://doi.org/10.1007/978-981-13-9349-5_7

7.1 Introduction

Research on yeast has played a long and central position in the elucidation of cellular metabolism, structure, and function (Barnett 1998–2010; Fig. 7.1). However, our understanding of the biochemistry of fermentation and the biogenesis and bioenergetics of membranes and organelles as presented and taught in textbooks still largely lacks an appreciation of recent rapid advances in the possibilities of access to dynamic information. In the past four decades, rather than from work with separated organelles, membranes and "non-sedimentable fractions" from cell-free extracts, more and more can be achieved by working with whole organisms, cultured cells, "organoids," or whole organisms (Aon et al. 2014). Minimally invasive,

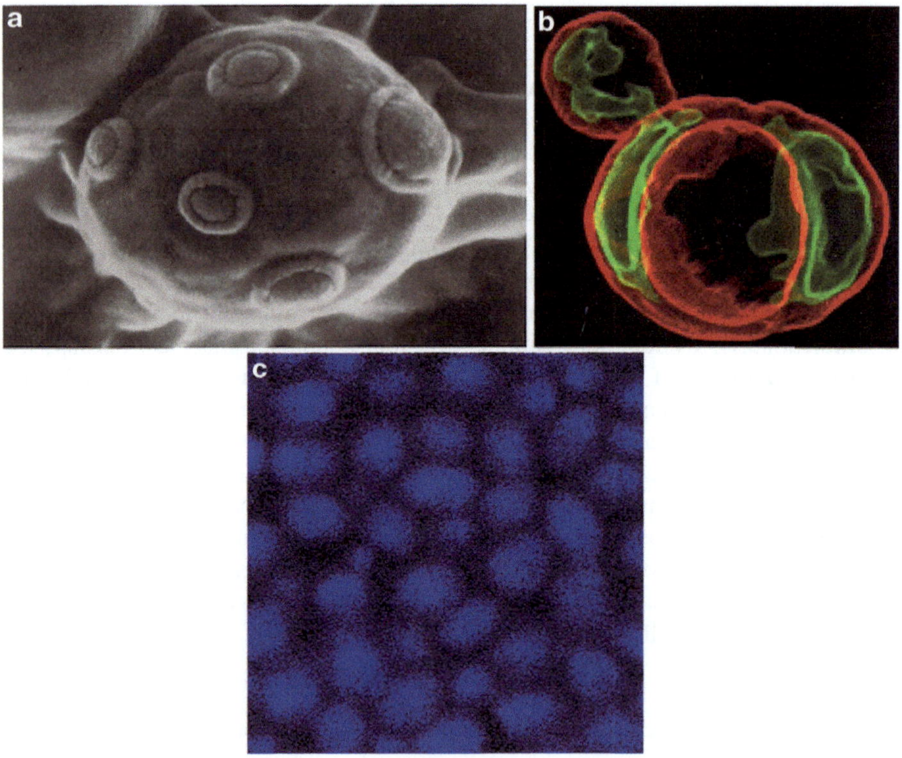

Fig. 7.1 *Saccharomyces cerevisiae* strain IFO 0233 from the Institute of Fermentation, Osaka. (**a**) Scanning electron micrograph: surface of a mature mother organism bearing many bud scars. (**b**) A large mitochondrial branching into the bud in a cider yeast: $DiOC_6$ (3) fluorescence intensity, color coded (green), vacuolar membrane (orange), and plasma membrane (red); provided courtesy of Dr. AJ Hayes (**c**) (NAD(P)H autofluorescence ("the bellwether of intracellular redox states"). In this cluster of yeast cells (attached in a single layer to poly-lysine-coated well of a perfusion chamber with aerated PBS, 5 mM glucose), their closely apposed cell walls are not apparent. Two-photon fluorescence excitation at 740 nm, emission at <490 nm. (Please see also Aon et al. (2007) for on-line videos of rotation of the image of the yeast in (**b**), and the synchronous oscillation of the yeast cluster in (**c**), at doi: 10.1016/j.febslet.2006.11.068)

rapidly responding and continuous techniques of optical, spectroscopic (Cartledge et al. 1972; Poole et al. 1974), and magnetic methods (Ohnishi et al. 1972) have revolutionized our appreciation of cellular biophysics, biochemistry, and physiology (Li 2011; Xu et al. 2016). The gap between molecular biology and a functional understanding of the organization of cells has thereby become much narrower in great leaps and accelerating ways forward.

7.1.1 Time in the Living World

Truly steady or stationary states are rare in biological systems. Homeostasis is a term that should be replaced by homeodynamic, as the apparent constancy of many aspects of living systems is a consequence of the operation of the high efficiency of their dynamic regulatory systems (Yates 1982, 1992, 1993; Yates and Yates 2008; Aon and Cortassa 1997; Lloyd 2016; Cortassa et al. 2003, 2012). Otherwise, apparently steady states are often only an artifact of time-averaging within a sampled population of quite different cells: this obscures the fine-grained time structure of each individual.

Dynamic interrogation of the time course of changes in reactant concentrations, events, and processes in living systems has revealed at least four distinct types of dynamic behavior: oscillatory, rhythmic, clocklike timekeeping, and synchronizing. Nonlinear dynamics detected as being chaotic and fractal is predicted on cell division cycle, circadian, and ultradian time scales (Lloyd and Lloyd 1992, 1995; Lloyd et al. 1992) and experimentally demonstrated in yeast (Roussel and Lloyd 2007; Aon et al. 2008a, b; Lloyd 2009).

7.1.2 Oscillations

Oscillatory operation of enzyme catalyzed reactions, metabolic, biosynthetic, and signaling pathways are commonplace in yeasts and other fungi (Fricker et al. 2007). Oscillations in concentrations (pool sizes) of metabolites or coenzymes are typical of triggered responses to perturbation. Respiratory oscillations with periods longer than 1h have been repeatedly described in yeast cultures in the earlier literature (Cortassa et al. 2012). They are usually not autonomous or self-sustained and can be inconvenient consequences of external perturbation by temperature, pH or medium flow instabilities. Their periods are almost always highly temperature sensitive, and their amplitudes decrease with each cycle, i.e., they are highly damped, fading away after a few cycles. Their waveforms are distinctive, but can be modified. Thus, they can be reset, i.e., phase advanced or delayed by effectors (positive or negative, respectively); they can also be amplitude modulated.

7.1.3 Rhythms

Rhythms are self-sustained oscillations (emergent and autonomous) and persistent even if their characteristics become modified (as just described for oscillations), but their "free-running" periods are robust and amplitudes fully recoverable.

7.1.4 Biological Clocklike Timekeepers

Biological clocklike timekeepers, like physical clocks, are temperature compensated.

The term "biological clock", in popular use, usually refers to the ~ 24 h circadian clock that has evolved in almost all animals, plants, and many fungi and cyanobacteria. It matches many hundreds of physiological and behavioral functions to the rotation of the earth.

7.1.5 Ultradian (Circahoralian) Clocks

Clocks or timekeepers that cycle many times in a day are called ultradian (a general term), or more specifically circahoralian (~ hourly ones), and it is inevitable that they are also faster than hourly timekeepers. In yeast, a 40 min rhythm is a basic timer mechanism. Longer period oscillations have been repeatedly described in earlier literature. Some 12 h and 8h rhythmic transcripts seen *in vivo* in mammalian systems appear as harmonics of circadian rhythms, but these disappear in cultured cells (Hughes et al. 2009).

7.1.6 The Cell-Division Cycle

The cell-division cycle is not a clock but an emergent biological process (Goodwin 1963; Hess and Bioteux 1971; Lloyd et al. 1982a, b, 1992; Rensing et al. 2001; Klevecz et al. 2008). It is often regarded as depending on a series of stepwise progressions, i.e., like a set of dominoes that must fall in a specific sequence, controlled at specific "checkpoints" (Murray and Kirschner 1989). Eukaryotes have evolved prior to the cyclin/cyclin kinase mechanism, and respiratory oscillations can trigger and progress the cell division cycle (Papagiannakis et al. 2017). Many others have always regarded the cell cycle (and it is now evident) to be a process dependent on an interwoven ensemble of oscillators, rhythms, and timekeepers acting on different time scales but integrated as a heterarchy with no central control.

7.2 Glycolytic Oscillations

By far the most studied periodic biochemical behavior, other than the cell division cycle, and currently the most thoroughly understood as a mechanism, is oscillatory glycolysis. This was first observed in intact anaerobic yeast cells in the presence of glucose as rapidly fluctuating levels of NADH fluorescence (Duysens and Amesz 1957; Ghosh and Chance 1964). Between then and 1971, experiments in Chance's group established the phenomena of oscillating concentrations of the intermediates of glucose entry and utilization (Maitra 1966). Concentrations of the metabolic intermediates (Maitra and Estabrook 1964) and of adenine and pyridine nucleotides (Estabrook and Maitra 1962) showed distinct phase relationships. Furthermore, the oscillations (period ~ 1 min, Olsen et al. 2009; Fig. 7.2) were usually highly damped, but less so in single cells than in a population, they were maintained in extracts and highly temperature dependent (temperature coefficient of 2.2 per 10 °C, between 20 and 35 °C, and phase shifted by ADP and acetaldehyde) (Chance et al. 1973). Details of early work have been reviewed (Lloyd and Stupfel 1991). Glycolytic mutants in yeast, first described by Maitra (1971), have proved invaluable in research on metabolic pathways and thus on oscillatory mechanisms (Lobo and Maitra 1983). Subsequent research described in at least 300 papers, has provided essential information on metabolic control mechanisms. In yeast as a model, but not solely there, and also more generally throughout biology, feedback and feed-forward loops are essential in the metabolism of sugars for the physiology of organ function, e.g., in the heart (Higgins et al. 1973; Achs et al. 1979).

Glycolytic and mitochondrial membrane oscillations are linked (Andersen et al. 2007; Poulsen et al. 2008; Olsen et al. 2009), and many hypotheses for possible functions of oscillating glycolysis in yeast have been proposed, but there is no definite evidence for any of these suggestions. They may simply be a reflection of "sloppy" control (Degn 1972). But, based on a theory proposed by Heimburg (2017), Thoke et al. (2018) have asked whether oscillating metabolism and the maintenance of low entropy in living systems are inevitable bedfellows. This group (Thoke et al. 2015, 2017) has uncovered an interesting new aspect, that strong coupling of the intracellular dynamics of water and glycolytic oscillations (including ATP) occurs in yeast cell populations and in single yeast cells in a scale invariant manner. Both characteristics are similarly affected by D_2O.

7.3 Other Short-Term Oscillations

Electrically detected sustained oscillations (periods ranging from 3 to 9 min) are described in growing yeast cultures (Male et al. 1999). Periodic changes in the levels of intermediate metabolites in glycolysis as well as changes in pH of the media have been measured, which demonstrate this phenomenon. Here we observe that the conductivity of the media also changes periodically when yeast cells are cultured under similar conditions. As conductivity is easily measured, this provides a simple, more quantitative method to study these changes than those currently used. An

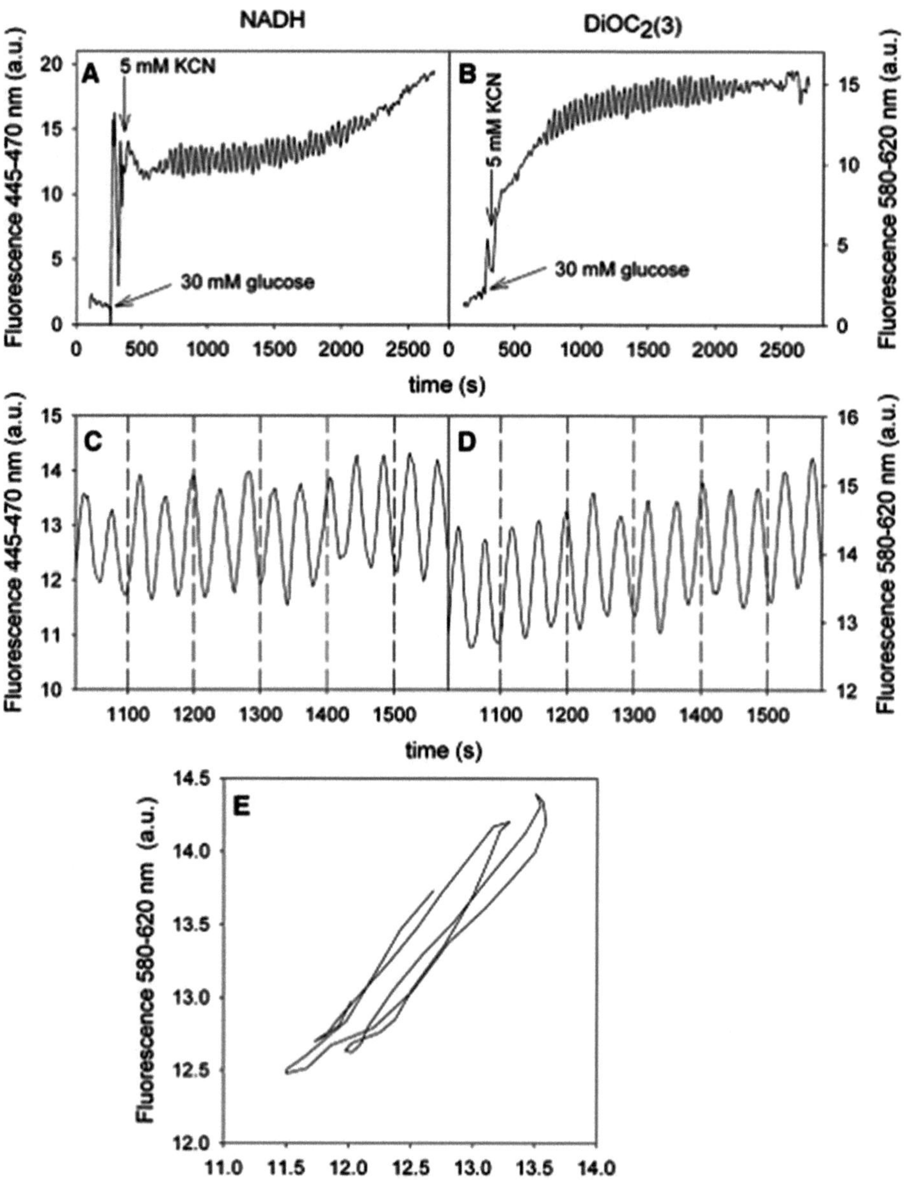

Fig. 7.2 Glycolytic oscillations: Time series of NADH fluorescence (**a**, **c**), $DiOC_2(3)$ fluorescence (**b**, **d**), and phase plot of $DiOC_2(3)$ fluorescence vs. NADH fluorescence (**e**). Yeast cells were suspended in 100 mM phosphate buffer, pH 6.8, to a cell density of 10% wet weight. At time 0, 4 μM $DiOC_2(3)$ were added to the suspension. At times 180 s and 240 s, 30 mM glucose and 5 mM KCN were added to the suspension, respectively. The temperature was 25 °C. (Olsen et al. 2009)

electrical biosensor referred to as ECIS (electrical cell surface impedance sensing) was used to study the small conductivity changes (in the order of 0.1%). No significant differences in the observed periods were found in the two yeast strains or the commercially purchased yeast extract studied. Presumably, changes in ac impedance are related to plasma membrane function extensively used to follow cell numbers and measure vitality of organisms during large-scale fermentation processes. Monitoring a cell culture process with online RF spectroscopy creates new opportunities to understand cellular changes throughout the length of the process. Oscillations in growth rates in permittistatically controlled continuous cultures (Davey et al. 1996) are interesting as chaotic behavior was confirmed by a variety of time-series analyses.

Monitoring the culture directly reduces the errors in secondary assays of samples to the medium as lag the production mechanism or only be indirectly related to the formation of the desired product (Carvell 2011).

Metabolic oscillations involving pulsatile NH4 release have been measured during colony growth of yeast (Palková et al. 2002); thereby they alternately acidify and alkalinize the medium. Metabolic changes in mitochondrial and peroxisomal activities are likened to this signaling process (Palková and Váchová 2016). These data imply that mitochondria can enter different states, which can divergently affect subsequent cellular development and have possible implications for use of yeast as a model for retrograde signaling in tumor growth.

Interacting oscillatory processes in yeast use calcium as a secondary messenger, generating calcium pulses and fluxes in response to a variety of signals such as hyperosmolarity, hypoosmolarity, cell stress, glucose, and alpha factor. After entering via plasma membrane channels, it activates a signal transduction cascade, including calmodulin, calcineurin, and a transcription factor, Crz1p. Sequestration within intracellular membranes and organelles bind Ca^{2+}, so as to maintain low cytosolic concentrations (Espeso 2016).

Pulsatile behavior of transcription factors is pervasive, heterogeneous, and unsynchronized in clonal yeast populations (Dalal et al. 2014).

7.4 Ultradian Rhythms

7.4.1 Respiratory Oscillations

The discovery of oscillatory O_2 consumption in synchronized cultures of *S. cerevisiae* was first described by Nosoh and Takamiya (1962), but not followed up until similar observations of Wiemken et al. (1970) and Dharmalingham and Jayaraman (1973). Mochan and Pye (1973) described high amplitude respiratory oscillations in yeast cultures adapting from glucose to ethanol utilization with a period of approximately 30 min and a twofold change in respiration rate as O_2 became depleted (but was still above 20 μM). Changes in cellular redox levels of NADH and cytochromes accompanied these events, thereby confirming the participation of mitochondria and emphasizing that these are very different from glycolytic oscillations.

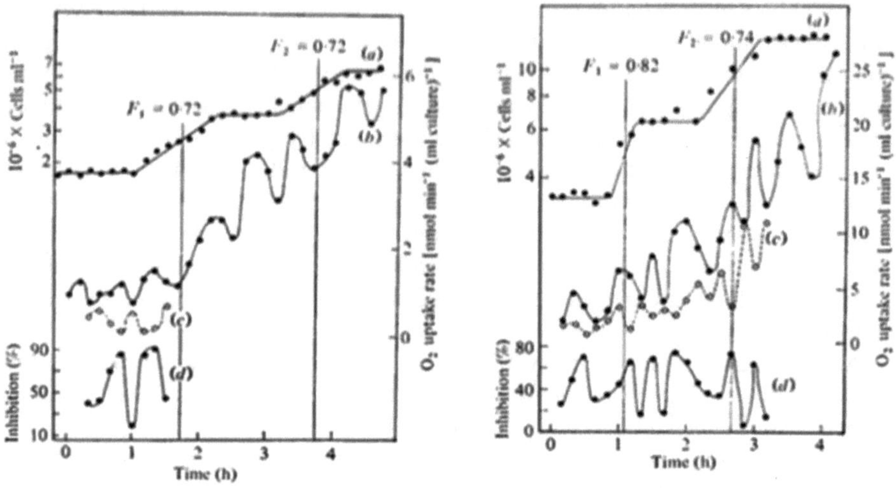

Fig. 7.3 Respiratory oscillations in *Candida utilis*. Oxygen consumption in samples taken from synchronous cultures of the yeast growing in chemically defined media in the presence of (**a**) acetate (3.88 g/L) or (**b**) glycerol (12.2 g/L): the effects of 30 μM DCCD or 50 μM KCN. F1 and F2 denote the synchrony indices. (a) cell counts, (b) controls with no inhibitors added, (c) sample respiratory rates, and (d) inhibition (%) in the presence of either inhibitor. Both the mitochondrial electron transport and ATP synthase inhibitor inhibit respiration maximally at the peaks of O_2 consumption as expected of respiratory control. (Kader and Lloyd 1979)

Poole et al. (1973), using a sedimentation velocity centrifugation size selection method for synchrony in *Schizosaccharomyces pombe* growing in a chemically defined medium containing 1% glucose showed that O_2 consumption doubled per cell division cycle but via two distinct maxima. The mitochondrial basis for this behavior was suggested by stimulation by 16 μM-CCCP an uncoupler of oxidative phosphorylation. A double-step per cell cycle when glycerol replaces glucose as respiratory substrate indicates that the oscillatory pattern previously observed depended on interaction between the glycolytic pathway and mitochondria. Results with *Candida utilis* (Fig. 7.3, Kader and Lloyd 1979) verified this, while those with glycerol or acetate in this highly aerophilic yeast showed clearly the purely non-fermentative responses, with a straightforward control of respiration by adenine nucleotide balances (respiratory control) (Chance and Williams 1956). Interaction between glycolytic and mitochondrial ATP generation in *S. cerevisiae* has been confirmed (Olsen et al. 2009): cytoplasmic membrane interactions also occur (Andersen et al. 2007; Poulsen et al. 2008) indicating the complexity of its cellular energetic network.

In cell-free samples extracts of the sample, through the cell division cycle of glucose-grown *Schiz. pombe*, cytochrome oxidase activity was shown as only a single maximum (Poole and Lloyd 1973). Spectrophotometrically measured mitochondrial cytochromes $a{+}a_3$ and b_{563} were expressed twice per cell cycle whereas c_{548} only once. These latter results were obtained by cell cycle fractionation using size separation in a sucrose gradient in a zonal rotor (Poole et al. 1974; Lloyd et al.

1982a, b). Oligomycin-sensitive ATPase activity in both glucose-repressed and de-repressed synchronous cultures oscillated with an amplitude of between 13% and 80% of its total four times per cell cycle (i.e., in 3 h), and inhibitor sensitivity over a four-fold range (Edwards and Lloyd 1977). Confirmation of timings was noted after rate zonal cell cycle fractionation. Inhibitor sensitivities to six different inhibitors all showed interesting differences in binding sites, but that to efrapeptin, a compound known to bind to the catalytic center (F_1), showed parallel behavior to ATPase activity (Lloyd and Edwards 1977).

7.4.2 Respiratory Timekeeping

Lloyd et al. (1982a, b) showed temperature compensation of the respiratory oscillation period in the small soil amoeba, *Acanthamoeba castellanii*, and in *Schizosaccharomyces pombe* (Kippert and Lloyd 1995). Quantal increments in cell cycle times as growth temperature was increased were also indicated in both these studies (Lloyd and Kippert 1993), as had previously been shown to be the case in mammalian cells (Klevecz 1976). Adenine nucleotide measurements in *Crithidia, Acanthamoeba*, and *Tetrahymena* also showed high amplitude oscillations with periods almost identical with those of their characteristic mitochondrial respiratory oscillations. Phase relationships revealed *in vivo* mitochondrial respiratory control (Edwards and Lloyd 1978). In *Acanthamoeba*, total protein and RNA (Edwards and Lloyd 1980, 1982), as well as mitochondrial F_1-ATPase, and peroxisomal catalase, like respiration, showed 69 min ultradian rhythms (Edwards et al. 1981, 1982). Experiments with *Candida utilis* (Kader and Lloyd 1979 (Fig. 7.3), Lloyd et al. 1975, 1977) showed the highest frequency of oscillations (period, 33 min), as might be expected of the fastest growing eukaryotic species of those tested (cell division time 90 min). Both respiration and total protein of *Candida* confirmed that the widespread rhythms in the amoeba were not curious anomalies but occur more generally and are important and necessary physiological characteristics (Lloyd et al. 1981). The current dogma at that time was that "balanced" microbial growth must involve continuous increases of all cellular constituents, with rates dependent on available nutrients and O_2. Populations of asynchronously dividing (or budding) individuals conveyed this false impression as differences between cells are smoothed out (time-averaged) in bulk assays. That extensive turnover is not confined to non-growing or differentiating cells but was also shown to be so in proliferating cells in a dedicated lifetime of research by the groups of Brodsky (1975, 1992, 2014) and Luzikov (1984, 2009) in Moscow. The novelty of these observations and the conclusion that extensive degradation of newly synthesized proteins is considerable met with widespread doubt and disbelief as the received wisdom prevalent in those days was being contradicted. Publication thus proved difficult. Although work on size selected *Candida utilis* (Lloyd et al. 1977) was successful, that on the asymmetrical cell-budding cycle in *S. cerevisiae* rendered size selection synchrony imprecise.

Kuriyama (Fig. 7.4a, b) initiated a new era by discovering a large-scale continuous self-synchronization procedure (Satroutdinov et al. 1992; Keulers et al. 1996a, b; Keulers and Kuriyama 1998; Murray 2004 (Fig. 7.4c). The yeast used in all the

Fig. 7.4 The Kuriyama continuous self-synchronized culture system of *Saccharomyces cerevisiae* (**a**) Dr. Hiroshi Kuriyama. (**b, c**) The continuous culture system. The impellors were driven by a 24 V motor (K; green line), and controlled by a power supply (A). Temperature was measured using a Pt 100 probe (yellow line), and controlled (B) to a user set-point by a 333 W heater (pink line). A water jacket (at 10 °C) consisting of rubber tubing wrapped around the glass vessel (G) provides cooling (not shown). pH was measured using a steam sterilizable gel electrode connected (cyan line) to a pH meter (C). pH was automatically controlled (D) by the addition of 2.5 M NaOH (dark blue line) or 2 N phosphoric acid (red line). An oxygen meter (E) was connected to a polarographic oxygen electrode (magenta line). Sterile dry air was fed (black line) to the fermentor via a sparge arm, its flow rate regulated by an Aalborg mas flow controller (F). Waste air was passed through a condenser (10 °C) and exited via two sterile filters (J).

studies described below was *S. cerevisiae*, strain IFO 0233, from the Institute of Fermentation, Osaka, Japan, grown under standard conditions. Other growth media, aeration rates, and strains of *S. cerevisae* are diverse in their rhythmic dynamics, but central redox metabolic progression and timings of biogenesis of membranes and organelle structure are comparable (Machné and Murray 2012).

Temperature compensated periodicity is a necessary characteristic of all time-keepers, be they physical, chemical, or biological. Clock control of the self-synchronized yeast populations was demonstrated over a range of temperatures suitable for growth (25–35 °C) with a Q_{10} of 1.07 (Murray et al. 2001, Fig. 7.5). Temperature dependence of the oscillation was also revelatory: robust persistence of rhythmic performance (temperature compensation) was observed between 26 and 35 °C, but above and below that temperature, a longer waveform (4 h at 35 °C, 8 h at 26 °C) became superimposed on the stable rhythm (Murray and Lloyd 2006).

A mutated yeast with an oscillatory period of 18 min has also a deficient temperature compensation mechanism (Q_{10}) = 1.6 as opposed to 1.07 for the wild type (Adams et al. 2003). The Gts1 protein product has homology with the circadian gene product PER in higher organisms, but is highly pleiotropic, having many diverse effects. Xu and Tsurugi (2007a, b) suggest that one effect of deletion of GTS1 and TPS1 in influencing the respiratory rhythm involves inhibition of the metabolism of storage carbohydrates (glycogen and trehalose) and consequences for synchronization by ethanol. In batch cultures on trehalose, both short period and long period autonomous oscillations can coexist (Jules et al. 2005). Correlation of the latter with budding indices indicates a strict relationship to the cell budding cycle and also provides fundamental new insights into the mechanisms of trehalose mobilization (Jules et al. 2007).

Another way of altering cellular phenotypes of redox cycles is to change the glucose concentration in the growth medium (Chin et al. 2012). Lowering of glucose by 40% (from 1.75% to 1%) reversibly lengthens the oscillation period from 2 to 4 h, and at 0.8% glucose, the culture cell count decreased and oscillation ceased,

Fig. 7.4 (continued) Sterile medium was stored in a reservoir (10 l; L) and pumped into the fermentor by low flow peristaltic pump (I; Microperpex, LKB). To avoid back contamination, the medium was pumped through an anti-growback device prior to entry to the vessel (G). Waste was collected from the fermentor by a glass arm weir. A peristaltic pump (H; 101U Watson- Marlow) then pumped the waste into a collection vessel (white line). In addition, there was a double-valved sample assembly and three auxiliary ports with rubber septa that were used for extra apparatus or pulse addition of perturbating agents. The data was acquired by an in-house software that controlled the data acquisition board (DAS 16, computer boards) and was programmed by the user to collect and store data (up to 1 s intervals). In some experiments (not shown), a membrane-inlet mass spectrometric probe was used (for measuring gases and low molecular mass volatiles directly), either by immersion in the culture, or exposed to the headspace gas, and a rotating disc dual-channel fluorimeter (for measuring redox levels of NAD(P)H and/or flavins; Johnson Foundation, University of Pennsylvania) was linked to the vessel using light guides Standard conditions employed were: working vol. 800 ml; dilution rate, 0.085 h^{-1}; stirrer rate, 800 rpm; airflow rate, 180 ml min^{-1}; temperature 30 °C (±0.2) and pH controlled at 3.4. Pulse injections were carried out through a sterile filter (0.2 μm porosity) (Lloyd et al. 2008)

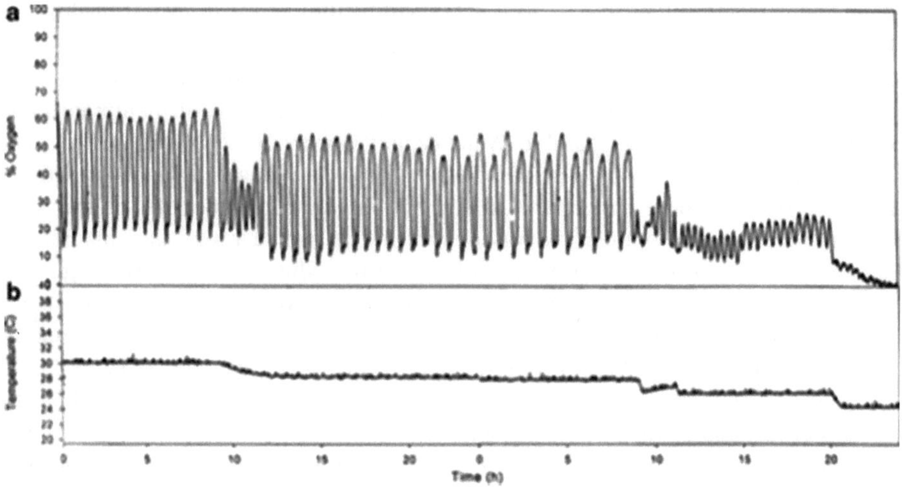

Fig. 7.5 Effects of step temperature changes on the respiratory oscillations in a continuous culture of *S. cerevisiae* grown in ethanol-containing medium. (**a**) Dissolved O_2 and (**b**) temperature were measured continuously on-line. (Murray et al. 2001)

presumably due to attenuation of intercellular signaling and loss of population synchrony.

7.4.3 Self-Sustained (Autonomous) Synchrony and Its Perturbation

Maintenance of synchrony depends on intercellular communication by volatile easily diffusible small molecules. That acetaldehyde (Fig. 7.6) was a prime candidate, previously suggested for populations of yeasts showing oscillatory glycolysis (Chance et al. 1973), was also demonstrated for ultradian rhythms (Keulers et al. 1996a, b; Keulers and Kuriyama 1998). A key criterion for effectors of oscillatory behavior is the construction of phase response curves, i.e., plots of altered phase as a function of the ~40 min ultradian cycle stage at which an effector is injected. In Kuriyama's laboratory, H_2S was found to be a suitable synchronizing substance (Sohn et al. 2000). Phase-response measurements (Murray et al. 2003, Fig. 7.7) indicated that concentration-dependent resetting of the yeast ultradian clock was produced weakly by acetaldehyde at 1.0 μM and increasingly so by 3 μM. Injections of $(NH_4)_2S$ additions also required 3 μM to produce an effect (Murray et al. 2003). Sulfite additions also gave phase resetting, and the investigators concluded that acetaldehyde and sulfite fine-tune the oscillatory states and sulfide is the prime synchrony mediator.

A biological redox cycle was proposed by Rapkine (1931) in which thiol-disulfide reactions are key mediators. A similar mechanism accounts for the core

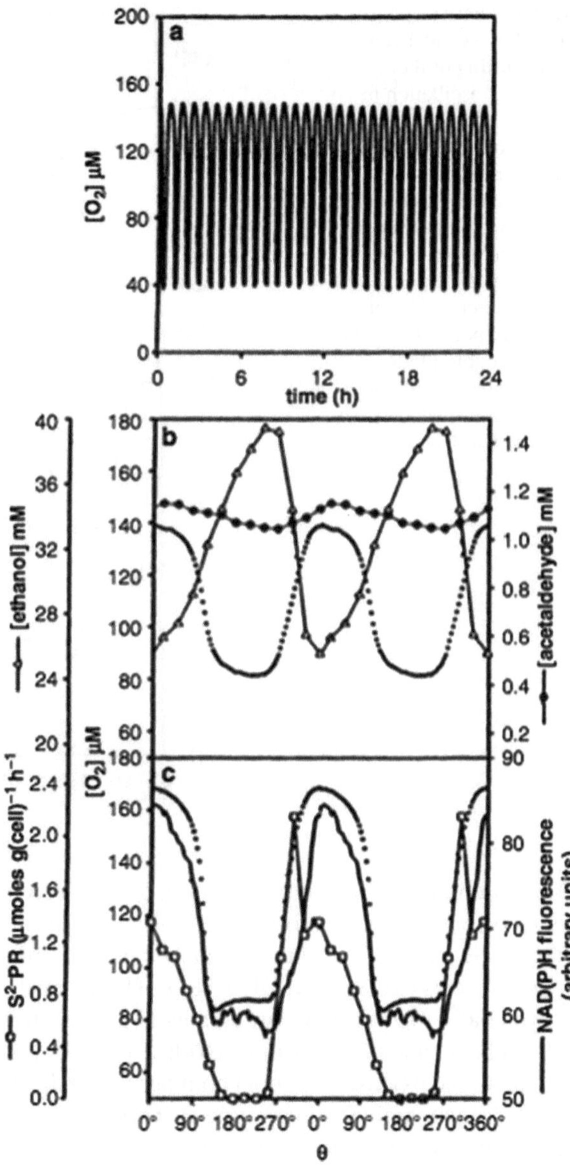

Fig. 7.6 Continuously monitored oscillatory variables (dissolved O_2 and NAD(P)H and metabolites sampled from the aerobic continuous cultures of yeast. (**a**) The respiratory oscillation. (**b, c**) Phase angle plots with respect to dissolved O_2 (…., on both panels), of acetaldehyde (▲) and ethanol (●); (**c**) S^{2-}PR (sulfide production rate, □; and NAD(P)H fluorescence (-). The data are double-plotted (side-by-side) in order to clarify phase-related events. (Murray et al. 2003)

reactions of respiratory rhythms (Murray and Lloyd 2005, 2006; Lloyd and Murray 2007).

Perturbation of the 40 min timekeeping respiratory rhythm in *S. cerevisiae* by agents that disturb the redox cycle implicated (Lloyd and Murray 2000) participation of the predominant yeast mitochondrial non-protein thiol, glutathione (GSH, Murray et al. 1999c). Thus, the actions of a GSH reductase inhibitor (5-nitrofuraldehyde), an inhibitor of the synthesis of the tripeptide GSH,

Fig. 7.7 Phase response curves for acetaldehyde phase-shifting of the respiratory oscillation in yeast. The dissolved O (···) plot on both panels provides a guide to the phase of the injection. The straight line on the upper panel (3 mM) is the linear regression of the points obtained after and before the "break-point" (0°). The data are double-plotted (side by side) in order to clarify phase-related events (Murray et al. 2003)

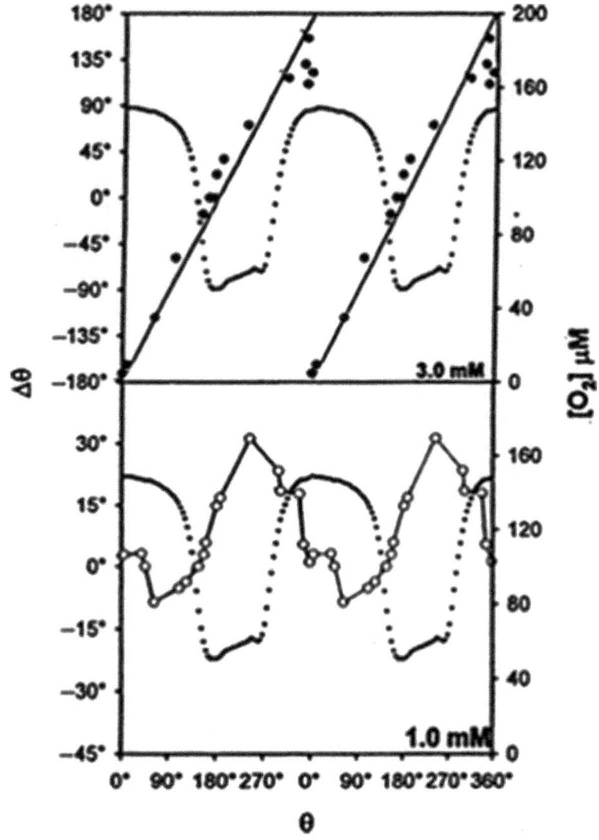

(D,L-butathionine-S,R)-sulphoximine), or thiol-reacting agents (NO^+, Murray et al. 1999a, b), are such as to perturb the oscillation. The re-oxidative balance of this core redox cycle depends on intracellular O_2 and its partial reduction products: reactive oxygen species (ROS: $O^{2-\cdot}$, OH–, H_2O_2) (Kwak et al. 2003) and possibly singlet O_2 (Lloyd et al. 1979) are key participants in this reaction network.

Li^+, phenelzine (a monoamine oxidase type-A+B inhibitor) and ipronazide (type A monoamine oxidase inhibitor) both perturbed the 40 min cycle by prolonging the period. A monoamine oxidase type-B inhibitor, pargyline, was ineffective, as were melatonin, serotonin, dopamine, or tyramine. Glutamate altered the amplitude of the rhythm but not its period (Salgado et al. 2002). In *Schizosaccharomyces pombe*, inositol monophospha-tase inhibition is a likely target for the period lengthening action of Li^+ on the cell division cycle as measured by the frequency of septum formation (Kippert 1997). A similar mechanism of action on signaling pathways for this anti-depressant agent was proposed for both the human and plant circadian cycles (Engelmann et al. 1976). Thus, the phosphoinositol pathway is an ancient highly conserved evolutionary mechanism.

Mitochondrial inner membranes and cristae show conformational changes between condensed (energized) and orthodox (resting) states due to alternating swelling of the inter-membrane space during the respiratory rhythm (Lloyd et al. 2002a, b).

Uncouplers of oxidative phosphorylation injected into the self-synchronized culture stimulated respiration (decreased residual dissolved O_2) within seconds as a result of depolarizing the mitochondrial inner membranes and releasing an increased flux of electron transport along the respiratory chain (Lloyd 2003). Energetic deficiency that results leads to a drastic attenuation of oscillatory amplitude of the ultradian clock for about 5 h. As the dilution of uncoupling agent proceeds in the continuous flow system, partial recovery of respiration is followed by completion of the delayed ongoing cell division cycle evident between 12 and 15 h later seen as the enveloping slow wave.

7.4.4 Transcriptomics

Analyses of the sequence of transcription in synchronous yeast cultures reveal the molecular details of a remarkably organized orchestration (Klevecz and Murray 2001; Klevecz et al. 2004, Fig. 7.8). Thus, wavelet analysis of time series data from expression arrays gave promise of detection of "dynamic architecture of phenotype." This was realized as a genome-wide oscillation in transcription with maxima at three almost equal intervals in the respiratory cycle of self-synchronized cultures. Two groups of temporally clustered genes (comprised of 4679 out of a total of 5329 expressed genes) are maximally recorded during low oxygen consumption (LOC, i.e., high residual dissolved O_2 in the culture), "the reductive phase" of the 40 min cycle. A third cluster (650 genes) is maximally expressed during the phase of high oxygen consumption (HOC). Synchronous bursts of DNA replication were measured using flow cytometry: these individuals accounted for the fraction of the total cell population ready for cell cycle progression to budding were gated by the transcriptional 40 min cycle. It was suggested that the process of DNA replication is thereby confined to a protective reduced intracellular environment to minimize damaging oxidative reactions (e.g., by reactive oxygen species: superoxide and hydroxyl radicals, H_2O_2, and singlet oxygen).

Continued research by Machné and Murray (2012) on transcriptome time series, from separate cultures growing under conditions that elicit either 40 min or 5 h period respiratory oscillations, both showed seven co-expressed gene clusters representing two superclusters. These encode for expression during phases of high and low O_2 consumption, respectively. Their promoters each have different nucleotide contents, nucleosome promoter configurations, and ATP-dependent remodeling complexes. ATP/ADP ratios oscillate as metabolic activities of the two complexes alternate, suggesting a possible mechanism for delayed negative feedback of phenotypic expression and adaptation to environmental change.

Machné et al. (2017) have devised a method for segmentation of time series given estimates of typical segment sizes and noise levels. The published example

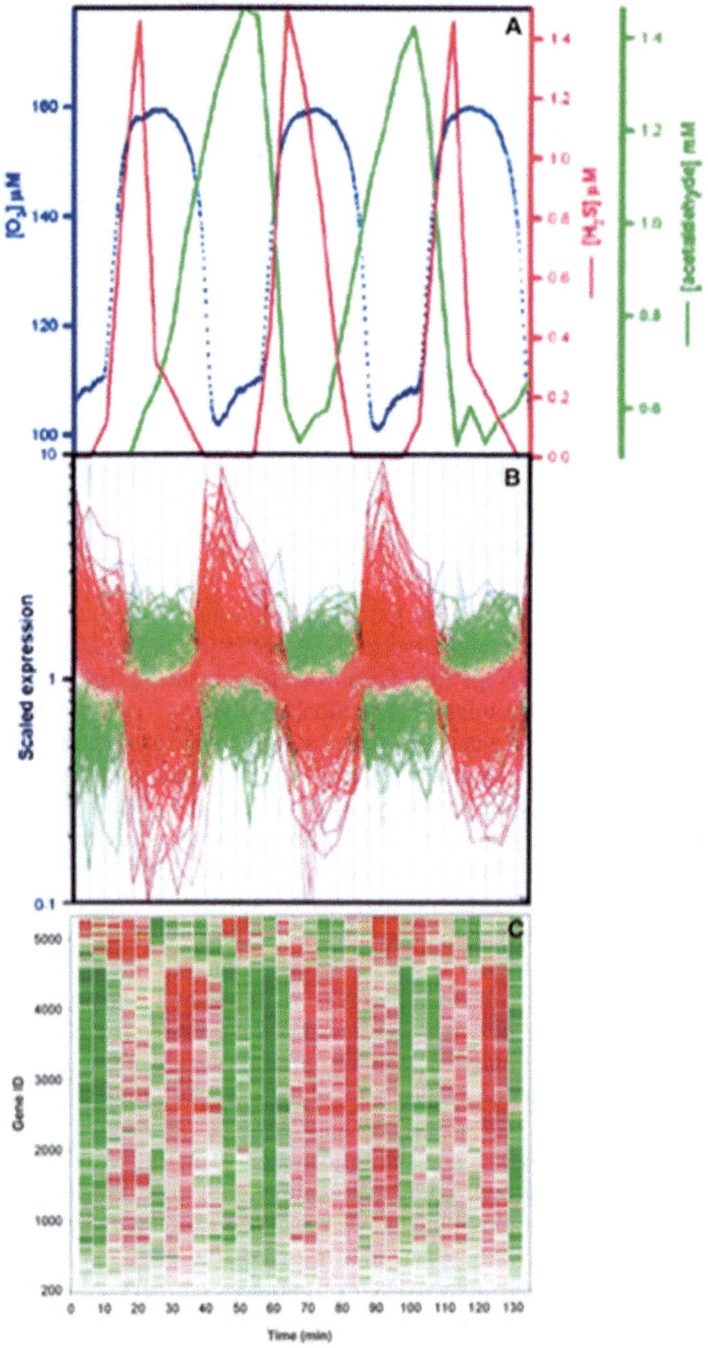

Fig. 7.8 The transcriptome of *Saccharomyces cerevisiae* in continuous culture: activity with respect to three cycles of the respiratory oscillations. (**a**) Dissolved O_2 in the culture, acetaldehyde and H_2S. (**b**) Transcripts in phase with the oxidative phase of the culture (red), and those in phase with the reductive phase (green). (**c**) The scale for the transcripts was obtained by dividing the intensity of expression of that gene at that point by the median intensity for each experiment. (Data derived from Klevecz et al. 2004)

required minimal notion of similarity of transcriptome sequencing data from budding yeast; in high temporal resolution, over ca. 2.5 cycles of the 40 min respiratory oscillation illustrate the power of this approach. The algorithm was used with a similarity measure focusing on periodic expression profiles across the metabolic cycle rather than coverage per time point.

7.4.5 Metabolomics

Sampling of synchronous culture every 12 min and careful handling, disruption, and extraction of metabolic intermediates and coenzymes by minimally destructive procedures enabled elucidation of a sequence of altered states closely corresponding to in vivo transformations. Rapid sampling and quenching in methanol at −80 °C was quickly flash centrifuged at −9 °C for 10 min. Mixing 1:1 with chloroform was then followed by centrifugal filtration through a 5kDa filter. Supernatants accumulated through three successive 40 min cycles were analyzed using capillary electrophoresis-mass spectrometry, using anionic, cationic, and nucleotide methodologies. In correspondence with the defined order of metabolite production, the results indicated that 70% of molecular species showed maximal pool sizes occurred with increasing NAD(P)H and biosynthetic and respiratory rates (Murray et al. 2007, Fig. 7.9). Oscillatory molecular concentrations predominate to such an extent that this periodic behavior may be considered as being metabolome-wide. Sasidharan et al. (2012a, b, c) and Amariei et al. (2013) modified the metabolite extraction procedure by using a dry-ice ethanol bath (−80 °C) with methanol containing N-ethylmaleimide (to stabilize thiols) and bead-beating (that circumvents the centrifugal step) in the extraction procedure. A total of 1650 oscillating chemical species can now be presented on a phase diagram (Sasidharan et al. 2012c).

7.4.6 Heterarchical Cellular Coherence

Amariei et al. (2014b) have developed a Fourier-based method that provides a suitable method for resolution of periodic behavior in noisy data sets. Relationships between chromatin occupancy, transcriptional activity, and energetic states of synchronous cultures revealed a global transcriptional slowing down, indicating a "reset point" in the transcriptional cycle. This begins as the catabolic phase and stress-response transcriptional stage end and finishes at the start of the anabolic and cell-growth transcriptional program. Histones from both catabolic and anabolic superclusters were deacetylated at this point.

The microarray data was combined with other transcriptomics data and was statistically analyzed. Thus, together with transcription factor activities, biosynthetic events were seen to possess a definite order, and a large transcription complex formed part of the timing circuit for integrated biosynthetic, reductive, and cell division cycle programs. It had previously been shown that 90% of the mRNA species concentrations peaked in the reductive phase (Klevecz et al. 2004). However,

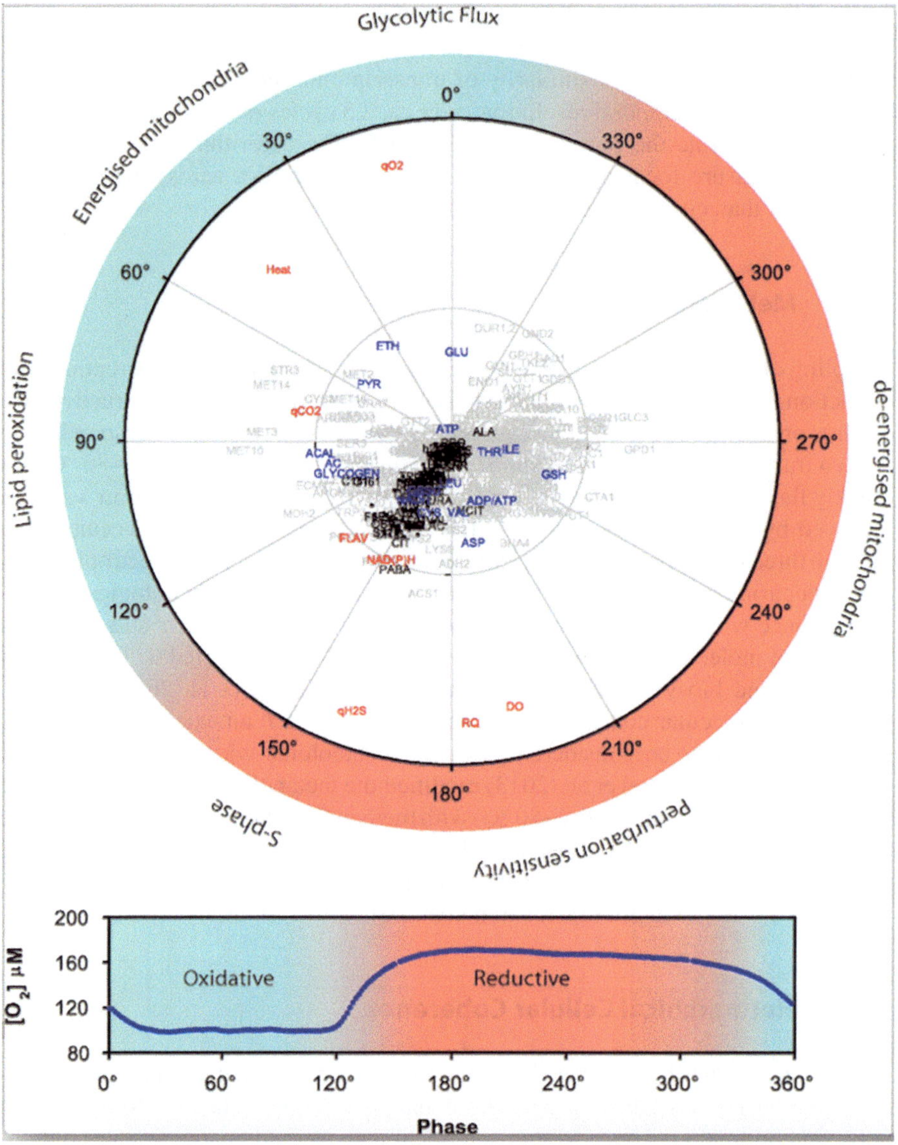

Fig. 7.9 A map of the ultradian clock (τ ~ 40 min) in *Saccharomyces cerevisiae*. Phase reconstruction of the signal-to-noise ratio (S/N) was measured for variables during the respiratory oscillation in continuously growing yeast. The polar plot was constructed from the fast Fourier transform analyses on numerous data sets. The red text represents the online parameters where the S/N had to be divided by 10. The blue text represents metabolites measured in low throughput enzymatic or HPLC methods. The black text shows data from high throughput GC-MS measurements. The grey text represents transcripts involved in metabolism measured using microarray studies. The timing of transcription constructed from phase angles and oscillation strength values. The blue region indicates the oxidative phase and the red region represents reductive phase. Physiological markers are indicated in text centered on the peak of the phenotype. For example, perturbation sensitivity refers to the sensitivity of this region to redox altering compounds such as ROS and glutathione, as well as other chemical agents. The lower plot represents a phase-normalized cycle of dissolved O_2 and is meant to guide the reader. The transcript names are common names found in the yeast genome database and the metabolite names are from a model of yeast metabolism. (Data from Murray et al. 2007)

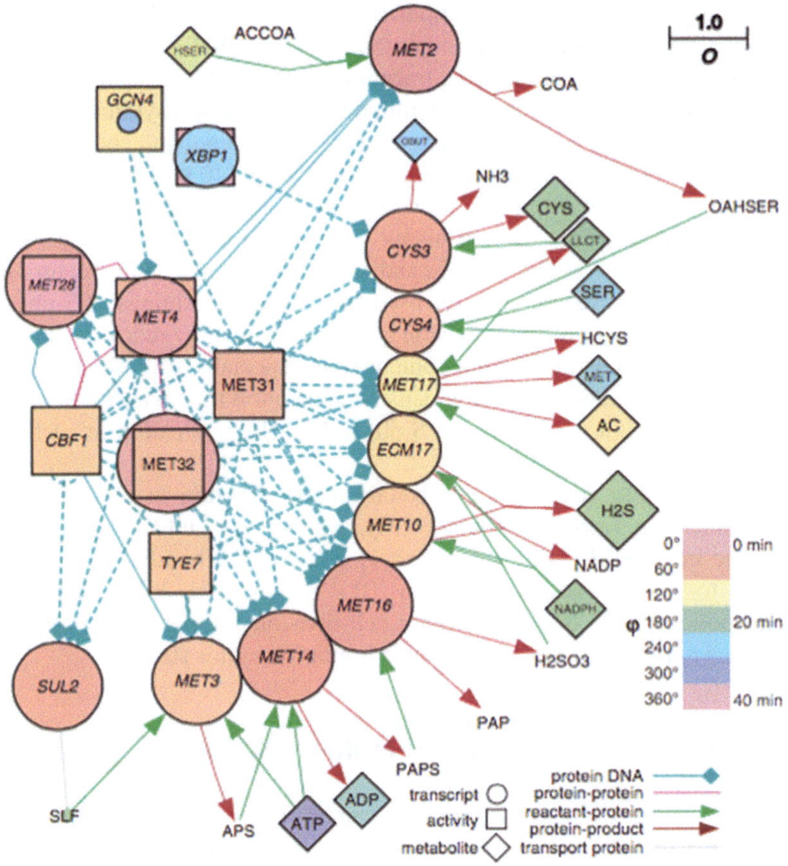

Fig. 7.10 A small central portion of the yeast interactome. Network derived for sulfur assimilation from top oscillating transcripts and metabolites. The figure key provides a guide to the network. *SLF* sulfate, *LLCT* cystathione, *HSER* homoserine, *OASER* o-acetylhomoserine, *OBUT* 2 oxobutanoate, *AC* acetate, *ACCOA* acetyl- coenzyme A, *COA* coenzyme A (color-coding shows phase angles of maxima in the ~40 min ultradian respiratory cycle; for further details, see Murray et al. 2007)

translational, proteomic, and metabolic controls also contribute (Fig. 7.10), so that the network is extremely complex but cannot be considered in hierarchical terms: thus cellular physiology and biochemistry is a heterarchical network (Yates 1982, 1992, 1993; Yates and Yates 2008; Cortassa et al. 2012; Aon et al. 2014). There is no central master oscillator: the operation of the interactome emerges as a consequence of the interactions of many sub-graphs (Fig. 7.10), each with the potential to oscillate, thereby providing adaptability and robust defenses against environmental disturbances (Lloyd and Kippert 1993).

Population asynchrony time-averages bulk observations in stirred fermenters unless the conditions routinely employed to maintain rhythmic behavior are observed. Lowering glucose levels by 40% or changing the dilution rate switches the respiratory rhythm to longer periods (Chin et al. 2012). Within a 24 h experiment, lowering media components other than glucose: $(NH_4)_2 SO_4$, $CaCl_2$, $MgSO_4$, or dilution of the solution containing trace ions (Fe^{3+}, Zn^{2+}, Cu^{2+} and Mn^{2+}) by 40% in cultures displaying ~ 2 or ~ 4 h ultradian respiratory rhythms produced no changes. Transcripts involved in processes were still ordered and coupled in a similar sequence to those described for cultures expressing ~40 min rhythms. However, peak-to-trough ratios of amplitudes were increased and the proportion of un-budded cells was increased. Thus the organisms spend more time to traverse the G_0/G_1 phase of the cell cycle at low glucose and spend longer time in respiratory energy generation. Flow cytometric analyses used to evaluate the timings of S phase suggest that the duration of the cell division cycle is regulated by redox oscillations. In the standard conditions of medium composition, especially glucose concentration, aeration rates (kLa), and population density, generation time was 8.2 h so that 8% of the cell population undergoes the S phase of DNA replication in each 40 min ultradian cycle.

In a prototrophic yeast strain, a batch culture growing in phosphate-limited ethanol-containing medium spontaneously synchronizes into multiple ultradian cycles. The fraction of organisms in G0/G1 phase of the budding cycle increased from 90% to 99% (Slavov and Botstein 2011; Slavov et al. 2011), demonstrating that metabolic cycling does not require cell-division cycling and metabolic synchrony does not necessitate carbon-source limitation in this strain.

7.4.7 Nonlinear Behavior of the Multi-oscillator

The robust capacity of the cellular system can be challenged by progressive lowering of the pH of the continuous culture in 0.1 pH unit steps (Murray and Lloyd 2006). The dynamic reorganization capabilities are observed as an alteration of waveform into increasing complexity of respiratory activity and cell division cycle controls. The usual 40–43 min period and amplitude variations (~5%) routinely observed as the "stable" oscillatory state at pH 3.4 was progressively modified, and next-amplitude plots indicated a chaotic component in the period-generating mechanisms and not in amplitude attenuation. Below pH 2.9, outputs became unstable with a 390 min period at pH 2.7. When 0.2 increments were employed from pH 3.4 to pH 6.0, the usual stable oscillatory behavior was maintained.

Observation on longer and shorter time scales enables the uncovering of nonlinear performance. Membrane-inlet mass spectrometry (Bohátka et al. 1983; Bohátka 1985; Lloyd et al. 1983a, b, 1985, 1996) proves ideal for highly precise rapid and/or long-term monitoring of the self-synchronized continuous cultures with stable ~40 minute periods. As points are registered every 12 s for m/z values at 32, 34, 40, and 44, lability or stability of the output enables minute by minute, hour by hour, day after day measurement to be followed for several months (Roussel and Lloyd 2007; Fig. 7.11a). After a disturbance, a multi-oscillatory mode in which dissolved O_2 and CO_2 were recorded showed 13 h as well as 36 and 4 min modes, not continuously seen, but returning at regular intervals during the 13 h cycle. This behavior

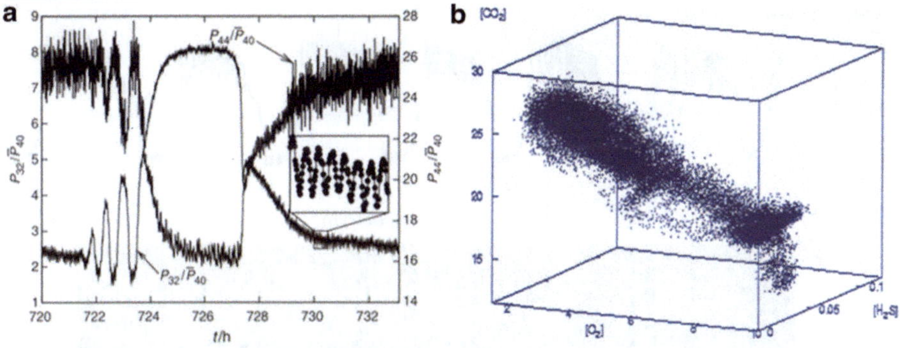

Fig 7.11 (**a**) The partially resolved yeast ultradian multi-oscillator using the MIMS probe. Phase relationships between the continuously monitored parameters and dissolved O_2 in the self-synchronous yeast continuous culture. (**b**) Plots of dissolved O_2 and CO_2 at m/z values 32, 44 using Ar at m/z 40 as an inert reference gas (Roussel and Lloyd 2007). The inset shows an amplified section of the O_2 trace. (**b**) A chaotic attractor revealed in the self-synchronous yeast continuous culture. This plot was obtained by monitoring of dissolved O_2, CO_2, and H_2S at m/z values 32, 44, and 34, respectively, using Ar at m/z 40 as an inert reference gas. Recording of the output of the MIMS probe was set to a cycle time of 12 s. Sampling was over a period of 3 months resulting in 36 000 points as shown (Roussel and Lloyd 2007)

was modeled as a metabolic attractor (Fig. 7.11b). Extended collection of data on dissolved O_2, CO_2 and H_2S, using dissolved argon as inert reference provided data every 15 s for 3 months (37, 374 data points). The chaotic behavior (leading Lyapunov exponent $0.752 \pm 0.004\ h^{-1}$) indicates sensitivity to initial conditions, a hallmark of deterministic chaos (Lloyd and Lloyd 1992, 1995; Kumar 1996; Aon et al. 2011). This might be a consequence of a loan chaotic oscillator, or a result of more than one or many interacting oscillators, the latter case seeming quite likely from what we know of the predominance of multi-oscillatory performance comprising the interactome.

Some indications of rhythms with periods shorter than 4 min seen in data analyzed using wavelet methods (Kurz et al. 2017) would repay further studies of more rapidly sampled culture, i.e., at intervals below the Nyquist limit.

7.4.8 Fluorescence Imaging: Mitochondrial Rhythmicity

The ~4 min respiratory oscillation observed in using MIMS is likely to arise from mitochondria as oscillations of a similar frequency in NADH and mitochondrial inner membrane electrochemical potential ($\Delta\Psi$m) have been imaged in tethered yeasts incubated with 5 mM glucose under aerobic conditions after pre-loading with 5 nM tetramethylrhodamine ethyl ester (TMRE) for 30 min (Fig. 7.12).

The intense auto-fluorescence in *Schiz. pombe* of both mitochondrial NAD(P)H and flavin emission was first measured using two-laser scanning (Bashford et al. 1980).

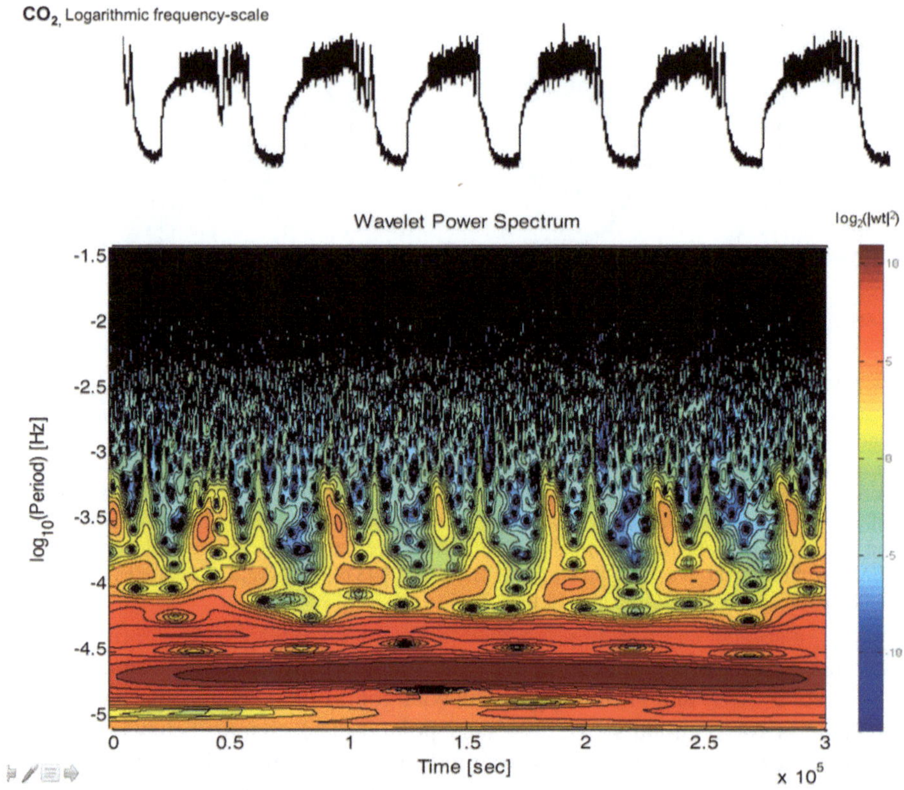

Fig. 7.12 Wavelet analysis of the CO_2 trace from the self-synchronous continuous culture data explained in Fig. 7.11. Note the fast oscillatory modes of periods< 4 min. (Unpublished analysis of F. Kutz , courtesy of MA Aon)

Scanning 2-photon excitation indicated that NAD(P)H and ($\Delta\Psi$m) oscillate in phase and discrimination between cytosolic and mitochondrial emission is possible in a single *S. cerevisiae* organism (Aon et al. 2007). This imaging technique has also been invaluable for tracing oscillations of glutathione (after reaction with monochlorobimane and reactive oxygen species (mitosox for O_2^- , 5-(-6)-chloromethyl-2′,7′-dichlorohydrofluorescine diacetate for H_2O_2 and hydroxyl radicals)) in both *S. cerevisiae* and *Candida albicans* (Aon et al. 2007; Lemar 2003; Lemar et al. 2005, 2007; Lloyd 2006; Lloyd et al. 2012) (Fig. 7.13). The importance of mitochondrial oscillations in mammalian tissues (Boiteux and Chance 1970) is recognized now as of pivotal importance in cardiac rhythmicity (Aon et al. 2008a, b; Lloyd et al. 2012; Kurz et al. 2017).

Individual mitochondria isolated from human (HeLa) cells trapped in nanofluidic channels using an inner membrane-specific lipophilic anionic fluorophore (TMRM) show typical polarized potentials and responses to substrates, or Ca^{2+} (Zand et al. 2013). A similar dye (JC-1) indicates fast fluctuations in $\Delta\Psi$m

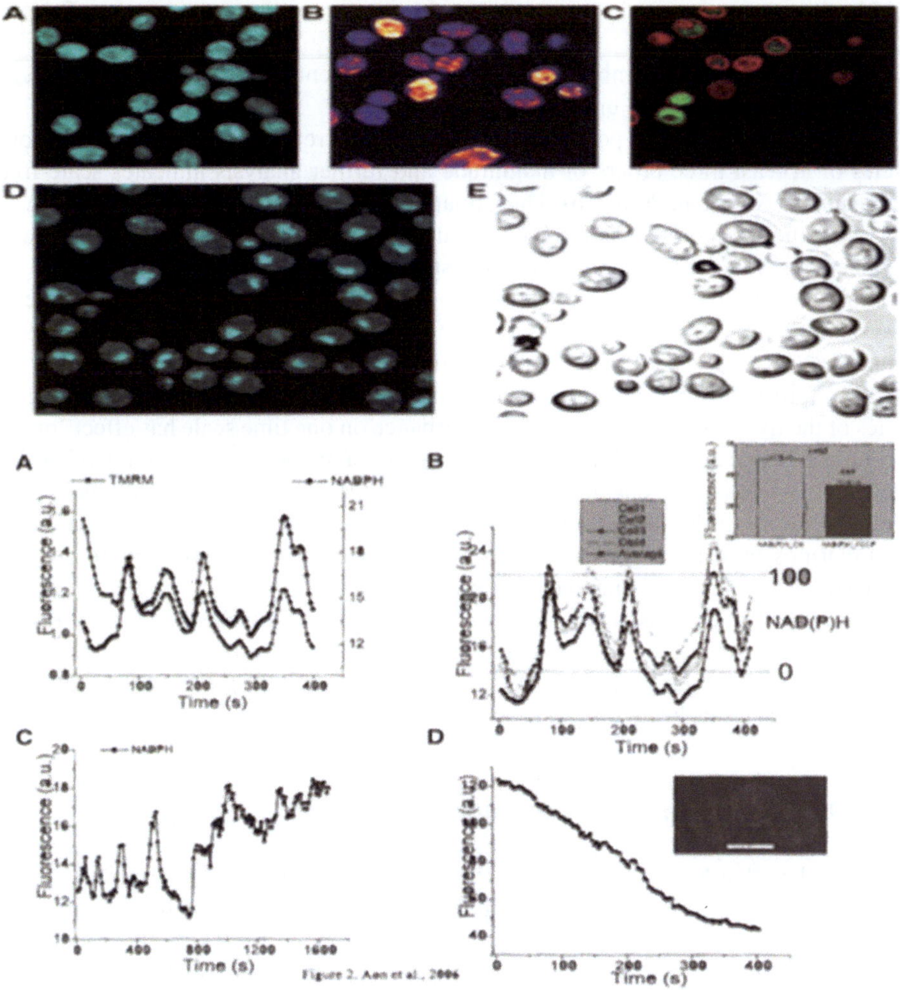

Fig. 7.13 Spontaneous synchronized oscillations in a synchronized yeast. Images of organisms (E), monitored by 2-photon scanning laser microscopy: mitochondrial membrane potential imaged for (**a**) NAD(P)H (auto-fluorescence), (**b**) ΔΨm (TMRM), (**c**) reactive oxygen species (5-(-6)-chloromethyl-2′,7′-dichlorohydrofluorescine diacetate, green), and glutathione (monochlorobimane). Please see also Lemar et al. (2007) and Aon et al. (2007) for on-line video of the synchronous oscillation of this yeast cluster at doi: 10.1016/j.febslet.2006.11.068

Spontaneous, synchronized oscillations in a contiguous synchronized layer of about 30 *S. cerevisiae* cells incubated with (**a**, **b**) or without (**c**) aeration of the perfusion buffer, monitored by 2-photon scanning laser microscopy. (**a**) Aerobic, whole cell ΔΨm and NAD(P)H oscillations exhibited by a microscopic field of <30 yeasts perfused with aerated PBS, pH 7.4, in the presence of 5 mM glucose. (**b**) Synchronous NAD(P)H oscillations of individual yeast cells as well as the average fluorescence from the whole microscopic field. The inset shows a histogram of whole cell NAD(P)H fluorescence intensity in the presence of 5 mM cyanide (CN) or 2 μM FCCP, a ΔΨm uncoupler, for $n = 60$ yeasts, $^{***}P < 0.001$. (**c**) Whole yeast cell NAD(P)H oscillations under perfusion conditions similar as in panel A but in the absence of aeration. (**d**) Fluorescent microspheres of approximately the same diameter of a yeast cell were imaged in the same microscopic field of yeast and their fluorescence followed simultaneously during the oscillatory response. A monotonic decrease in the microspheres' fluorescence intensity can be noticed and were essentially due to photobleaching. Three fluorescent microspheres are shown in the inset: the bar corresponds to 5 μm (Aon et al. 2007, see also supplementary information)

(mitochondrial inner membrane electrochemical potential), previously described as "flickering," but not well understood.

The multi-oscillatory performance of yeast cells represents a range of frequencies of at least three orders of magnitude and further analysis indicates scale-free dynamics (Aon et al. 2008a, b). Thus, relative dispersional and power spectral analyses of the time series of dissolved O_2 and CO_2 signals obtained by MIMS gave values of similarity in the periods demonstrated in the inverse relationship between amplitude and frequency (i.e., the origin of the inverse power law governing fractal systems) of $D_f = 1.0$ ($r = 0.86$) and $\beta = -1.95$ ($r = 0.85$) for O_2 and $D_f = 1.0$ ($r = 0.98$) and $\beta = -1.40$ ($r = 0.72$) for CO_2, respectively. This validation of self-similar, i.e., fractal, properties provides functional understanding for the observed characteristics of the dynamics. It indicates that disturbance on one time scale has effects on all time scales that operate in the system and predicts a memory-like retention of earlier events. The detection, characteristics, and plausible functions of deterministic chaotic systems in biology, and their relationships with observed temporal self-similarities have been extensively discussed, since earlier descriptions of chaotic physical, chemical, and enzyme reactions.

Single yeast and clusters of yeast organisms examined attached to the poly-lysine coated glass surface of the perfusion well (Aon et al. 2007), or in microfluidic chambers (Silverman et al. 2010), show oscillatory "metabolic" redox rhythms, indicating that this behavior is a property of individual cells.

7.4.9 Respiratory Timekeeping: A Summary

It is evident that Kuriyama's self-synchronous continuous culture methodology has enabled highly significant advances in the resolution of time structure of yeast, an invaluable experimental model system, and at present the most fully investigated of any organism. A global system of transcriptional and transcriptional regulator oscillators provides a map of a global dynamic system. This mechanism is sensitive to initial conditions (e.g., perturbation by phenazine, Salgado et al. 2002; Li and Klevecz 2006). More than 1600 oscillating metabolites (Murray et al. 2007; Sasidharan et al. 2012a, b, c; Murray et al. 2014; Amariei et al. 2014a) reveal coordination with metabolic functions, organelle elaboration and function, and the cell division cycle. Redox control is central in the cellular network and implicit to rhythmicity (Lloyd and Murray 2005, 2007; Murray and Lloyd 2006; Murray et al. 2011; Murray et al. 2014).

This makes a rational progression through the 40 min ultradian redox rhythm, respiratory (high oxygen consumption), early reductive, and late reductive phases (low oxygen consumption), summarized as follows:

In HOC: ribosome biogenesis, RNA metabolic process, gene expression, RNA biosynthetic process, and S-amino acid metabolism

In early LOC: DNA-dependent DNA replication and nitrogen compound metabolic process

In late LOC: proteolysis, stress response, and proteolysis involved in cellular protein catabolism

The Kuriyama aerobic fermentation system thus presents us with a procedure providing deep understanding of whole-system physiology as well as detailed dissection of dynamic biochemistry. Some other groups (e.g., Tu et al. 2005, Tu and McKnight 2006) have not paid adequate attention to the strictly defined requirements of the details of the culture technique and implementation of its control systems: misinterpretations have necessitated corrections (Lloyd 2006; Murray 2006; Chin et al. 2012).

The temporal controls revealed by Kuriyama's procedures cover almost all aspects of cellular function. It establishes a definite order to metabolic and biosynthetic processes and also organelle functions and development. It also provides confirmation of earlier ideas on the *Saccharomyces cerevisiae* cell division cycle as well as those in other fungi and lower eukaryotes (protozoa and algae) as collections of interacting oscillators (Lloyd et al. 1982a, b), with deterministic chaotic attractors (Klevecz 1976; Klevecz and Li 2007; Klevecz 1992; Klevecz et al. 2008). The origins of muti-cellularity and the emergence of cellular differentiation may have been organized by synchrony via mitochondrial oscillations (Fig. 7.13).

The fractal multi-oscillator behavior of mitochondrial respiration in yeast and the mammalian heart muscle indicates highly conserved physiology and biochemistry over 1.5 billion years of evolution (Aon et al. 2011; Lloyd et al. 2012). This comparison further extends the concept of the value of yeast research into a widening biomedical arena especially in the relevance of mitochondrial energetics and molecular genetics to human disease (Wallace 2012, 2015, Picard et al. 2015, 2016) (Fig. 7.14).

7.5 Circadian (Diurnal) Oscillations

In an organism capable of budding every 90 min, it is not unexpected that a special mechanism for generating ~24 h rhythms has not been discovered. Of course, under environmental conditions, even with plentiful nutrients, but at low temperature or at high light intensity, such a rhythm of slow cellular proliferation might prove an advantage.

Yeast respiration is inhibited by light (Ninnemann et al. 1970), and Wille showed in 1974 that *Candida albicans* can be photo-entrained into a circadian growth mode. Work on *S. cerevisiae* mutants lacking mitochondrial and nuclear encoded cytochrome components also lacks this capacity (Woodward et al. 1978, Ułaszewski et al. 1979). Photochemical action spectra for the light-inhibition of growth gave a maximum at 408 nm in the Soret region of the cytochrome spectrum. Resistance to light inhibition was greatest in yeasts lacking all three types of cytochromes (*b*, *c*, and *a*).

Photic input for synchronization to entrainment requires a b-type cytochrome as shown in a mutant of *Schizosaccharomyces pombe* (Kippert et al. 1990). *Candida*

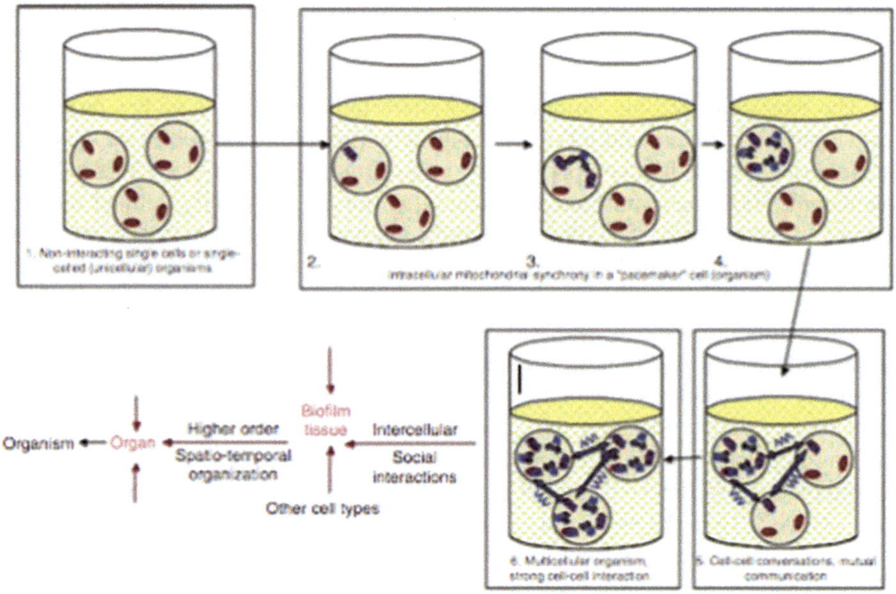

Fig. 7.14 The evolution of multicellular organisms has utilized ultradian rhythms for both intracellular coherence and intercellular signaling; this has led to population synchronization, social interactions, and the development of tissues and organs leading to higher levels of complexity in metazoa and plants. (Illustration by Dr. Victoria Gray)

albicans uses porphyrins and flavins as photoreceptors (405 nm or 445 nm, respectively), and these wavelengths are used for photoinactivation (Plavskii et al. 2018). Photolyases have been extensively characterized and contain two chromophores, a reduced flavin (FADH) and non-covalently linked fluorescent conjugated pterine (5, 10-methenyl tetrahydrofolate) absorbing maximally at 385 nm (Ninnemann 1995).

Wild-type strains of *S. cerevisiae* growing slowly (with14 h generation times) can be entrained to periodic light/dark 10 h/14 h regimes. These cultures give subsequent free-running cycles with a period of 26 h and persist for 7–8 days in complete darkness (Edmunds 1980). However, this yeast lacks most of the "canonical" genetic determinants. Thus, PER1, 2, and 3, TIM, and the total of 15 genes present in mouse, rat, and human genomes are absent, as are the 25 characterized transcription factors. Casein kinases (numbering 5), the CREB signaling factors (23), and the two light-sensing proteins, melatonin receptors (Mtn 1a and 1b) are also not detected. Those genes present in *Neurospora* (the most studied fungal circadian system, Dunlap and Loros, 2017) and many other fungi. FRQ, WC-1, and WC-2 are also absent from *S. cerevisiae*, whereas FRH and FWD1 are present but serve non-circadian functions. The plant circadian genes ELF3 and 4, LUX, PRFS7 and 9, and also VIVID are also absent from baker's yeast, and no opsonins, phytochromes, or cryptochromes have been characterized. Next-generation sequence analyses (Li et al. 2015; Hughes et al. 2017) may be expected to reveal further details of genetic controls of circadian clocks in fungi and of prominent differences in yeasts from those of other fungi and higher organisms. Circadian rhythms of intracellular Mg^{++} are widely implicated in enzyme activities and membrane ion channels: the whole cell is a circadian oscillator (Dunlap and Loros 2017).

Circadian "cryptic clock"-like functions revealed in *S. cerevisiae* by using entrainment to temperature cycles rather than by growing organisms under constant conditions in the laboratory (Eelderink-Chen et al. 2010, 2015; Merrow and Raven, 2010) find interest in consideration of the behavior of yeasts under natural environmental conditions. In *Schizosaccharomyces pombe*, growing on slowly metabolized sugars (e.g., mannose) or low temperatures, quantized cell division times have been observed (Kippert 2001).

Acknowledgments DL would like to thank all the students and coworkers for permission to summarize their contribution to work on temporal organization of the growth of yeasts.

References

Achs MJ, Kohn MC, Garfinkel (1979) Computer simulation of metabolism in pyruvate-perfused rat heart. IV. Model behavior. Am J Phys 237:R174–R180

Adams CA, Kuriyama H, Lloyd D, Murray (2003) The Gts1 protein stabilizes the autonomous oscillator in yeast. Yeast 20:463–470

Amariei C, Machné R, Stoic V, Soga T, Tomita M, Lloyd D, Murray DB (2013) Time resolved DNA occupancy dynamics during the respiratory oscillation uncover a global reset point in the yeast growth program. Microb Cell 1(9):279–288

Amariei C, Machné R, Sasidharan K, Gottstein W, Tomita M, Lloyd D, Murray DB (2014a) The dynamics of cellular energetics during continuous yeast culture. Conf Proc IEEE Eng Med Biol Soc 2013:2708–2711

Amariei C, Tomita M, Murray DB (2014b) Quantifying periodicity in omics data. Front Cell Dev Biol 2:40. https://doi.org/10.3389/fcell.2014.00040

Andersen AZ, Poulsen AK, Brasen JC, Olsen LF (2007) On-line measurements of oscillating mitochondrial membrane potential in glucose-fermenting *Saccharomyces cerevisiae*. Yeast 24:731–739

Aon MA, Cortassa S (1997) Dynamic biological organization: fundamentals as applied to cellular systems. Springer, New York

Aon MA, Cortassa S, Lemar KM, Hayes A, Lloyd D (2007) Single and cell population respiratory oscillations in yeast: a 2-photon study. FEBS Lett 58:6–14

Aon MA, Roussel MR, Cortassa S, O'Rourke B, Murray DB, Beckmann M, Lloyd D (2008a) The scale-free dynamics of eukaryotic cells. PLoS One 3(11):e3624

Aon MA, Cortassa S, O'Rourke B (2008b) Mitochondrial oscillations in physiology and pathophysiology. Adv Exp Med Biol 641:98–117

Aon MA, Cortassa S, Lloyd D (2011) Chaos in biochemistry and physiology. In: Meyers RA (ed) Encyclopedia of biochemistry and molecular cell biology and molecular medicine: systems biology, 2nd edn. Wiley-VCH Verlag, Weinheim, pp 239–276

Aon MA, Lloyd D, Saks V (2014) From physiology, genomes, systems, self-organization to systems biology: the historical roots of a twenty-first century approach to complexity. In: Aon MA, Saks V, Schlattner U (eds) Systems biology of metabolic and signaling networks: energy, mass and information transfer. Springer, Berlin/Heidelberg, pp 3–17

Barnett JA (1998–2010). A history of research on yeasts: parts 1–14. Yeast 14:1439–1451 (1998) to 27: 875–904 (2010)

Bashford CL, Poole RK, Lloyd D, Chance B (1980) Oscillations of redox state in synchronized dividing cultures of *Acanthamoeba castellanii* and *Schizosaccharomyces pombe*. Biophys J 29:1–11

Bohátka S (1985) Quadrupole mass spectrometric measurement of dissolved and free gases. In: Degn H, Cox RP, Toflund H (eds) Gas enzymology. D. Reidel, Dordrecht, pp 1–16

Bohátka S, Langer G, Szilágyi J, Berecz I (1983) Gas concentration determination in fermenters with quadrupole mass spectrometer. Int J Mass Spec Ion Phys 47:277–274

Boiteux A, Chance B (1970) Eighth Int Con Biochem, Luzern Abs

Brodsky VY (1975) Protein synthesis rhythm. J Theor Biol 55:167–200

Brodsky VY (1992) Rhythms of protein synthesis and other cirahoralian oscilations: the possible involvement of fractals. In: Lloyd D, Rossi EL (eds) Ultradian rhythms in life processes. An inquiry into fundamental principles of chronobiology and psychology. Springer, London, pp 23–40

Brodsky VY (2014) Circahoralian (ultradian) rhythms. Biochemistry (Mosc) 79(6):483–495

Cartledge TG, Lloyd D, Erecińska M, Chance B (1972) The development of the respiratory chain of *Saccharomyces carlbergensis* during respiratory adaptation. Biochem J 130:739–747

Carvell J (2011) Creating new opportunities in process control through radio frequency impedence spectroscopy. BMC Proc 5(Suppl 8):P57

Chance B, Williams GR (1956) The respiratory chain and oxidative phosphorylation. Adv Enzymol 17:65–134

Chance B, Williamson G, Lee IY, Mela L, DeVault D, Ghosh AK, Pye EK (1973) Synchronization phenomena in oscillations in yeast cells and isolated mitochondria. In: Chance B, Pye EK, Ghosh AK, Hess B (eds) Biological and biochemical oscillators. A Colloquium of the Johnson Foundation. Academic, New York/London, pp 285–302

Chin SL, Marcus IM, Klevecz RR, Li CM (2012) Dynamics of oscillatory phenotypes in *Saccharomyces cerevisiae* reveal a network of genome-wide transcriptional oscillators. FEBS J 279:1119–1130

Cortassa S, Aon MA, Marbán E, Winslow RL, O'Rourke (2003) An integrated model of cardiac mitochondrial metabolism. Biophys J 84:2734–2755

Cortassa S, Aon MA, Iglesias AA, Aon JC, Lloyd D (2012) An introduction to metabolic and cellular engineering, 2nd edn. World Scientific, Singapore, pp xiv+ 428

Dalal CK, Cai L, Lin Y, Rahbar K, Elowitz MB (2014) Pulsatile dynamics in the yeast proteome. Curr Biol 24(18):2189–2194

Davey HM, Davey CL, Woodward AM, Edmonds AN, Lee AW, Kell DB (1996) Oscillatory, stochastic and chaotic growth rate fluctuations in permittistatically controlled yeast cultures. Biosystems 39:43–61

Degn H (1972) Oscillating chemical reactions in homogeneous phase. J Chem Educ 49:302–307

Dharmalingham, Jayaraman (1973) Mitochondriogenesis in synchronous cultures of yeast. I. Oscillatory pattern of respiration. Arch Biochem Biophys 157:197–202

Dunlap JC, Loros J (2017) Yes, circadian rhythms actually affect everything. Cell Res 26:759–760

Duysens LNM, Amesz J (1957) Fluorescence spectrophotometry of reduced phosphopyridine nucleotide in intact cells in the near ultraviolet and visible region. Biochem Biophys Acta 24:19–26

Edmunds LN Jr (1980) Blue light photoreception in the inhibition and synchronization of growth and transport the yeast *Saccharomyces*. In: Senger H (ed) The blue light syndrome. Springer, Berlin, pp 584–596

Edwards SW, Lloyd D (1977) Mitochondrial adenosine triphosphate of the fission yeast *Schizosaccharomyces pombe* 972h. Biochem J 162:39–46

Edwards SW, Lloyd D (1978) Oscillations of respiration and adenine nucleotides in synchronous cultures of *Acanthamoeba castellanii*: mitochondrial respiratory control *in vivo*. J Gen Microbiol 108:197–204

Edwards SW, Lloyd D (1980) Oscillations of protein and RNA content during synchronous growth of *Acanthamoeba castellanii*. FEBS Lett 109:21–26

Edwards SW, Lloyd D (1982) The mitochondrial adenosine triphosphatase of *Acanthamoeba castellanii*: oscillatory accumulation of enzyme activity, enzyme protein and F1-inhibitor during the cell cycle. Biochem J 202:453–458

Edwards SW, Evans JB, Lloyd D (1981) Oscillatory accumulation of catalase during the cell cycle of *Acanthamoeba castellanii*. J Gen Microbiol 125:459–462

Edwards SW, Evans JB, Williams JL, Lloyd D (1982) The mitochondrial adenosine triphosphatase of *Acanthamoeba castellanii*: oscillatory accumulation of enzyme activity, enzyme protein and F1-inhibitor during the cell cycle. Biochem J 202:453–458

Eelderink-Chen Z, Mazzotta G, Sturre M, Bosman J, Roenneberg T, Merrow M (2010) A circadian clock in *Saccharomyces cerevisiae*. Proc Natl Acad Sci U S A 107:2043–2047

Eelderink-Chen Z, Olmedo M, Bosman J, Merrow M (2015) Chapter 4 – Using circadian entrainment to find cryptic clocks. Methods Enzymol 551:73–93

Engelmann W, Bollig I, Hartmann R (1976) The effects of lithium ions on circadian rhythms. Arznemitteforschung 26(6):1085–1086

Espeso EA (2016) The CRaZy calcium cycle. Adv Exp Med Biol 892:169–186

Estabrook RW, Maitra PK (1962) A fluorimetric method for the quantitative analysis of adenine and pyridine nucleotides. Anal Biochem 3:369–382

Fricker MD, Tlalka M, Bebber D, Tagaki S, Watkinson SC, Darrah PR (2007) Fourier-based spatial mapping of oscillatory phenomena in fungi. Fungal Genet Biol 44(11):1077–1084

Ghosh AK, Chance B (1964) Oscillation of glycolytic intermediates in yeast cells. Biochem Biophys Res Commun 16:174–181

Goodwin BC (1963) Analytical physiology of cells and developing organisms: a dynamic theory of cellular control processes. Academic, London/New York, pp x+249

Heimburg T (2017) Linear nonequilibrium thermodynamics of reversible periodic processes and chemical oscillations. Phys Chem Chem Phys 19:17331–19341

Hess B, Bioteux A (1971) Oscillatory phenomena in biochemistry. Annu Rev Biochem 53:237–258

Higgins J, Frenkel R, Hulme E, Lucas A, Rangazas (1973) The control theoretic approach to the analysis of glycolytic oscillators. In: Chance B, Pye EK, Ghosh AK, Hess B (eds) Biological and biochemical oscillators. A Colloquium of the Johnson Foundation. Academic, New York/London, pp 127–175

Hughes ME, DiTacchio, Hayes KR et al (2009) Harmonics of circadian gene transcripts. PLoS Genet 5(4):e1000442

Hughes ME, Arbruzzi KC, Allada R et al (2017) Guidelines for genome-scale analysis of biological rhythms. J Biol Rhythm 32:380–393

Jules M, François J, Parrou JL (2005) Autonomous oscillations in *Saccharomyces cerevisiae* during cultures on trehalose. FEBS J 272:1490–1500

Jules M, Beltran G, François J, Parrou JL (2007) New insights into trehalose metabolism by NTH2 encodes a functional cytosolic trehalase, and deletion of TPS1 reveals Ath1p-dependent trehalose mobilization. Appl Environ Microbiol 74(3):605–613

Kader J, Lloyd D (1979) Respiratory oscillations and heat evolution in synchronous cultures of *Candida utilis*. J Gen Microbiol 114:455–461

Keulers M, Kuriyama H (1998) Extracellular signaling in an yeast culture. In: Holcombe WLM, Paton R, Holcombe M (eds) Information processing in cells and tissues. Plenum, New York, pp 85–94

Keulers M, Satroutdinov AD, Kuriyama H (1996a) Oscillations in ethanol grown *Saccharomyces cerevisiae*. FEBS Microbiol Lett 142:253–258

Keulers M, Satroutdinov AD, Suzuki T, Kuriyama H (1996b) Synchronization affector of autonomous short-period-sustained oscillation of *Saccharomyces cerevisiae*. Yeast 12:673–682

Kippert F (1997) Lithium affects the ultradian clock of *Schizosaccharomyces pombe* by inhibition of inositol monophosphatase. Biochem Soc Trans 25(4):S602

Kippert F (2001) Cellular signaling and the complexity of biological timing: insights from the ultradian clock of *Schizosaccharomyces pombe*. Phil Trans Roy Soc Lond B Biol Sci 356:1725–1733

Kippert F, Lloyd D (1995) A temperature-compensated clock ticks in *Schizosaccharomyces pombe*. Microbiology 141:883–890
Kippert F, Engelmann W, Ninnemann H (1990) Blue light synchronizes and inhibits the growth of *Schizosaccharomyces pombe*. Prog Clin Biol Res 341B:271–280
Klevecz RR (1976) Quantized generation times in mammalian cells as an expression of the cellular clock. Proc Natl Acad Sci U S A 73:4012–4016
Klevecz RR (1992) A precise circadian clock from chaotic cell cycle oscillations. In: Lloyd D, Rossi EL (eds) Ultradian rhythms in life processes. An inquiry into fundamental principles of chronobiology and psychology. Springer, London, pp 41–69
Klevecz RR, Li CM (2007) Evolution of the clock from yeast to man by period-doubling folds. Cold Spring Harb Symp Quant Biol 72:421–429
Klevecz RR, Murray DB (2001) Genome wide oscillations: wavelet analysis of time series analysis in yeast expression arrays uncovers the dynamic architecture of phenotype. Mol Biol Rep 28:73–82
Klevecz RR, Bolen J, Forest G, Murray DB (2004) A genome-wide oscillation in transcription gates DNA replication and the cell cycle. Proc Natl Acad Sci U S A 101:1200–1205
Klevecz RR, Li CM, Marcus I, Frankel PH (2008) Collective behavior in gene regulation: the cell is an oscillator, the cell cycle a developmental process. FEBS J 275:2372–2384
Kumar N (1996) Deterministic chaos: complex chance out of simple necessity. University Press (India), Jawaharlal Nehru Centre for Advanced Research, Hyderabad, pp xi+96
Kurz FT, Kembro JM, Flesia AG, Armoundas AA, Cortassa S, Aon MA, Lloyd D (2017) Network dynamics: quantitative analysis of complex behavior in metabolism, organelles, and cells, from experiments to models and back. Wiley Interdiscip Rev Syst Biol Med 9(1). https://doi.org/10.1002/wsbm.1352
Kwak WJ, Kwong GS, Jin I, Kuriyama H, Sohn HY (2003) Involvement of oxidative stress in the regulation of H_2S production during ultradian metabolic oscillation of *Saccharomyces cerevisiae*. FEBS Microbiol Lett 219:99–104
Lemar KM (2003) Cell death mechanisms in the human opportunistic pathogen, *Candida albicans*. J Eukaryot Microbiol 50(Suppl):685–686
Lemar KM, Passa O, Aon MA, Cortassa S, Muller CT, Plummer S, O'Rourke B, Lloyd D (2005) Allyl alcohol and garlic (*Allium sativum*) produce oxidative stress in *Candida albicans*. Microbiology 151:3257–3265
Lemar KM, Aon MA, Cortassa S, O'Rourke B, Muller CT, Lloyd D (2007) Diallyl disulphide depletes glutathione in *Candida albicans* : oxidative-stress mediated cell death studied by two-photon microscopy. Yeast 24:695–706
Li LZ (2011) Special section in memory of Professor Britton Chance: celebrating the life and legasy of Britton Chance. J Innov Optical Health Sci 4(2):v–vii
Li CM, Klevecz RR (2006) A rapid genome-scale response to the transcriptional oscillator to perturbation reveals a period-doubling path to phenotypic change. Proc Natl Acad Sci U S A 103:16254–16259
Li J, Grant GR, Hogenesch JB, Hughes ME (2015) Chapter sixteen-considerations for RNA-seq analysis of circadian rhythms. Methods Enzymol 551:349–367
Lloyd D (2003) Effects of uncoupling of mitochondrial energy conservation on thevultradian clock-driven oscillations in *Saccharomyces cerevisiae* continuous culture. Mitochondrion 3:139–146
Lloyd D (2006) The ultradian clock: not to be confused with the cell cycle. Nat Rev Mol Cell Biol 7. https://doi.org/10.1038/hrm1980-cl
Lloyd D (2009) Oscillations, synchrony and deterministic chaos. In: Lüttge U, Beyschlag W, Büdel B, Francis D (eds) Progress in botany, vol 70. Springer, Berlin, pp 69–92
Lloyd D (2016) Biological timekeeping: the business of a blind watchmaker. Sci Prog 99(2):113–132
Lloyd D, Edwards SW (1977) Mitochondrial adenosine triphosphate of the fission yeast *Schizosaccharomyces pombe* 972h. Biochem J 162:581–590
Lloyd D, Kippert F (1993) Intracellular coordination by the ultradian clock. Cell Biol Int 17:1047–1052

Lloyd AL, Lloyd D (1992) Hypothesis: the central oscillator of the circadian clock is a controlled chaotic attractor. Biosystems 29:77–85

Lloyd AL, Lloyd D (1995) Chaos: its detection and significance in biology. Biol Rhythm Res 26:233–252

Lloyd D, Murray DB (2000) Redox cycling of intracellular thiols: state variables for ultradian, cell division and circadian cycles. In: Van den Driessche T, Guisset JL, DeVries GP (eds) Redox behaviour of Circadian systems. Kluwer, Amsterdam, pp 85–94

Lloyd D, Murray DB (2005) Ultradian metronome: timekeeper for orchestration of cellular coherence. Trends Biochem Sci 30(7): 373–377

Lloyd D, Murray DB (2007) Redox rhythmicity: clocks at the core of temporal coherence. Bioessays 29:465–473

Lloyd D, Stupfel M (1991) The occurrence and functions of ultradian rhythms. Biol Rev 66:275–299

Lloyd D, John L, Edwards C, Chagla AH (1975) Establishment of large scale synchronous cultures of microorganisms by continuous flow size selection. J Gen Microbiol 88:153–158

Lloyd D, John L, Hamill M, Phillips C, Kader J, Edwards SW (1977) Continuous-flow cell cycle fractionation of eukaryotic microorganisms. J Gen Microbiol 99:223–227

Lloyd D, Boveris A, Reiter R, Filipkowski M, Chance B (1979) Chemiluminescence of *Acanthamoeba castellanii*. Biochem J 184:149–156

Lloyd D, Edwards SW, Williams JL (1981) Oscillatory accumulation of total cellular protein in *Candida utilis*. FEMS Microbiol Lett 12:295–298

Lloyd D, Poole RK, Edwards SW (1982a) The cell division cycle: temporal organization and control of cellular growth and reproduction. Academic, London, pp xii+513

Lloyd D, Edwards SW, Fry JC (1982b) Temperature-compensated oscillations in respiration and cellular protein in synchronous cultures of *Acanthamoeba castellanii*. Proc Natl Acad Sci U S A 79:3785–3788

Lloyd D, Kristensen B, Degn H (1983a) Glycolysis and respiration in yeasts: the Pasteur effect studied by mass spectrometry. Biochem J 212:749–754

Lloyd D, Edwards SW, Williams JL, Evans JB (1983b) Mitochondrial cytochromes of *Acanthamoeba castellanii*: oscillatory accumulation haemoproteins, immunological determinants and activity during the cell cycle. FEMS Microbiol Lett 16:307–312

Lloyd D, Bohátka S, Szilágyi J (1985) Quadrupole mass spectrometry in the monitoring and control of fermentations. Biosensors 1:179–212

Lloyd D, Lloyd AL, Olsen LF (1992) The cell division cycle: a physiologically plausible dynamic model can exhibit chaotic solutions. Biosystems 27:17–24

Lloyd D, Thomas K, Price D, O'Neill W, Oliver K, Williams TN (1996) A MIMS miniprobe for the direct simultaneous measurement of multiple gas species. J Microbiol Methods 25:145–151

Lloyd D, Salgado LE, Turner MP, Murray DB (2002a) Respiratory oscillations in yeast: clock-driven mitochondrial cycles of energization. FEBS Lett 519:41–44

Lloyd D, Salgado LEJ, Turner MP, Suller MTE, Murray DB (2002b) Cycles of mitochondrial energization driven by the ultradian clock in a culture of *Saccharomyces cerevisiae*. Microbiology 148:3715–3724

Lloyd D, Murray DB, Klevecz RR, Wolf J, Kuriyama H (2008) The ultradian clock (~ 40 min) in yeast. In: Lloyd D, Rossi EL (eds) Ultradian rhythms from molecules to mind: a new vision of life. London, Springer, pp 11–42

Lloyd D, Cortassa S, O'Rourke B, Aon MA (2012) What yeast and cardiomyocytes share: ultradian oscillatory redox mechanisms of cellular coherence and survival. Integr Biol (Camb) 4:65–74

Lobo Z, Maitra PK (1983) Genetics of yeast glucokinase. Genetics 105:501–515

Luzikov VN (1984). Mitochondrial biogenesis and breakdown (trans. Galkin AV, Roodyn DB). Consultants Bureau, New York/London

Luzikov VN (2009) Principles of control over formation of structures responsible for respiratory functions of mitochondria. Biochem Mosc 74:1443–1456

Machné R, Murray DB (2012) The yin and yang of yeast transcription : elements of a global feedback system between metabolism and chromatin. PLoS One 7(6):e37906

Machné R, Murray DB, Stadler PF (2017) Similarity-based segmentation of multi-dimensional signals. Sci Rep 7(1):12355

Maitra PK (1966) Pulsating glucose flux in yeast. Biochem Biophys Res Commun 25:462–467

Maitra PK (1971) Glucose and fructose metabolism in a phosphoglucoisomeraseless mutant of *Saccharomyces cerevisiae*. J Bacteriol 107:759–769

Maitra PK, Estabrook RW (1964) A fluorimetric method for the enzymic determination of glycolytic intermediates. Anal Biochem 7:472–484

Male T, Feder J, Giaever GN, Giaever I (1999) Oscillations in yeast observed electrically. Biol Rhythm Res 30(4):361–370

Merrow M, Raven M (2010) Finding time: a daily clock in yeast. Cell Cycle 9:1671–1672

Mochan E, Pye K (1973) Respiratory oscillations in adapting yeast cultures. Nat New Biol 242:177–179

Murray DB (2004) On the temporal organization of *Saccharomyces cerevisiae*. Curr Genom 5:665–671

Murray DB (2006) The respiratory oscillation in yeast: phase definitions and periodicity. Nature Rev Mol Cell Biol 7. https://doi.org/10.1038/hrm1980-cl

Murray AW, Kirschner MW (1989) Dominoes and clocks: the union of two views of the cell cycle. Science 246:914–621

Murray DB, Lloyd D (2005) Ultradian metronome: timekeeper for orchestration of cellular coherence. Trends Biochem Sci 30:373–377

Murray DB, Lloyd D (2006) A tunable attractor underlies yeast respiratory dynamics. Biosystems 90:287–294

Murray DB, Engelen FA, Keulers M, Kuriyama H, Lloyd D (1999a) NO+ but not NO inhibits respiratory oscillations in ethanol-grown chemostat cultures of *Saccharomyces cerevisiae*. FEBS Lett 431:297–299

Murray DB, Engelen FA, Keulers M, Kuriyama H, Lloyd D (1999b) NO+ but not NO inhibits respiratory oscillations in ethanol-grown chemostat cultures of *Saccharomyces cerevisiae*. Biochem Soc Trans 26(4):S339

Murray DB, Keuler M, Englen F, Lloyd D, Kuriyama H (1999c) Involvement of glutathione in the regulation of respiratory oscillation during a continuous culture of *Saccharomyces cerevisiae*. Microbiology 145:2739–2745

Murray DB, Roller S, Kuriyama H, Lloyd D (2001) Clock control of respiratory oscillations found during yeast continuous culture. J Bacteriol 183:7253–7259

Murray DB, Klevecz RR, Lloyd D (2003) Generation and maintenance of synchrony in *Saccharomyces cerevisiae* continuous culture. Exp Cell Res 287:10–15

Murray DB, Beckmann M, Kitano H (2007) Regulation of yeast oscillatory dynamics. Proc Natl Acad Sci U S A 104:2241–2246

Murray DB, Haynes K, Tomita T (2011) Redox regulation in respiring *Saccharomyces cerevisiae*. Biochim Biophys Acta 1819:845–958

Murray DB, Amariei C, Sasidharan K, Machne R, Aon MA, Lloyd D (2014) Temporal partitioning of the yeast cellular network. In: Aon MA, Saks V, Schlattner U (eds) Systems biology of metabolic and signaling networks: energy, mass and information transfer, Springer series in biophysics, 16. Springer, Berlin/Heidelberg, pp 323–349

Ninnemann H (1995) Some aspects of blue light research. Photochem Photobiol 61:22–31

Ninnemann H, Butler WL, Epel BL (1970) Inhibition of respiration in yeast by light. Biochem Biophys Acta 205:499–506

Nosoh Y, Takamiya A (1962) Synchronisation of budding cycle in yeast cells and effect of carbon monoxide and nitrogen deficiency on the synchrony. Plant Cell Physiol 3:59–66

Ohnishi T, Cartledge TG, Lloyd D (1972) The development of mitochondrial iron-sulphur protens during the respiratory adaptation of *Saccharomyces carlsbergensis*. FEBS Lett 52:90–94

Olsen LF, Andersen AZ, Lunding A, Brasen JC, Poulsen AK (2009) Regulation of glycolytic oscillations by mitochondrial and plasma membrane H^+-ATPases. Biophys J 96:3850–3861

Palková Z, Váchová L (2016) Mitochondria in aging cell differentiation. Aging 8(7):1287–1288

Palková Z, Devaux F, Icicová M, Mináriková L, Le Crom S, Jacq C (2002) Ammonia pulses and metabolic oscillations guide yeast colony development. Mol Biol Cell 13:3901–3914
Papagiannakis A, Niebel B, Wit EC, Heinemann M (2017) Autonomous metabolic oscillations robustly gate the early and late cell cycle. Mol Cell 65:285–295
Picard M, McManus MJ, Gray JD, Nasca C, Moffat C, Kopinski PK, Seifert EL, McEwen BS, Wallace DC (2015) Mitochondrial functions modulate neuroendocrine, metabolic, inflammatory and transcriptional responses to acute psychological stress. Proc Natl Acad Sci U S A 112(48):E6614–E6623
Picard M, Wallace DC, Burelle Y (2016) The rise of mitochondria in medicine. Mitochondrion 30:105–116
Plavskii VY, Mikulich AV, Tretyakova AI et al (2018) Porphyrins and flavins are endogenous acceptors of optical radiation determining photoinactivation of microbial cells. J Photochem Photobiol 183:172–183
Poole RK, Lloyd D (1973) Changes in enzyme activities in synchronously dividing cultures of *Schizosaccharomyces pombe* h. Biochem J 136:195–207
Poole RK, Lloyd D, Kemp RB (1973) Respiratory oscillations and heat evolution in synchronous cultures of fission yeast, *Schizosaccharomyces pombe* 972 h. J Gen Microbiol 77:209–290
Poole RK, Lloyd D, Chance B (1974) The development of the cell division cycle of a glucose-repressed fission yeast, *Schizosaccharomyces pombe* 972 h. Biochem J 138:201–210
Poulsen AK, Andersen AZ, Brasen JC, Scharff-Poulsen AM, Olsen LF (2008) Probing glycolytic and membrane potential oscillations in *Saccharomyces cerevisiae*. Biochemistry 47:7477–7484
Rapkine L (1931) Sur les processes chimiques au course de la division cellulaire. Ann Physiol Physicochem 7:382–418
Rensing L, Meyer-Grahl U, Ruoff P (2001) Biological timing and the clock metaphor: oscillatory and hour-glass mechanisms. Chronobiol Int 18:329–369
Roussel MR, Lloyd D (2007) Observation of a chaotic multi-oscillatory metabolic attractor by real-time monitoring of a yeast continuous culture. FEBS J 274:1011–1018
Salgado EL, Murray DB, Lloyd D (2002) Some antidepressants (Li^+, monoamine oxidase type-A inhibitors) perturb the ultradian clock in *Saccharomyces cerevisiae*. Biol Rhythm Res 33:351–361
Sasidharan K, Soga T, Tomita M, Murray DB (2012a) A yeast metabolite extraction protocol optimized for time-series analyses. PLoS One 7:e44283
Sasidharan K, Amariei C, Tomita M, Murray DB (2012b) Rapid DNA, RNA and protein extraction protocols optimized for slow continuously growing yeast cultures. Yeast 29:311–322
Sasidharan K, Tomita M, Aon MA, Lloyd D, Murray DB (2012c) The time-structure of yeast metabolism *in vivo*. In: Goryanin I, Goryachev A (eds) Advances in systems biology. Springer, New York
Satroutdinov AD, Kuriyama H, Kobayashi H (1992) Oscillatory metabolism of *Saccharomyces cerevisiae* in continuous culture. FEMS Microbiol Lett 98:261–268
Silverman SJ, Petti AA, Slavov N, Parsond L, Briehof R, Thiberge S, Zenklusen D, Gandhi SJ, Larson DR, Singer RH, Botstein DP (2010) Metabolic cycling in single yeast cells from unsynchronized steady-state populations limited on glucose or phosphate. Proc Natl Acad Sci U S A 107:6946–6951
Slavov N, Botstein D (2011) Coupling among groth rate response, metabolic cycle and cell division cycle in yeast. Mol Biol Cell 22:1997–2009
Slavov N, Macinskas J, Caudy A, Botstein D (2011) Metabolic cycling without cell division in respiring yeast. Proc Natl Acad Sci U S A 108:19090–19095
Sohn HY, Murray DB, Kuriyama H (2000) Ultradian oscillation of *Saccharomyces cerevisiae* during aerobic continuous culture: hydrogen sulfide mediates population synchrony. Yeast 16:1185–1190
Thoke HS, Tobiesen A, Brewer J, Hansen PL, Stock RP, Olsen LF, Bagatolli LA (2015) Tight coupling of metabolic oscillations and intracellular water dynamics. PLoS One 10:e0117308
Thoke HS, Thorsteinsson S, Stock RP, Bagatolli LA, Olsen LF (2017) The dynamics of intracellular water constrains glycolytic oscillations in *Saccharomyces cerevisiae*. Sci Rep 7:16250

Thoke HS, Olsen LF, Duelund L, Stock RP, Heimburg T, Bagatolli LA (2018) Is a constant low-entropy process at the root of glycolytic oscillations? J Biol Phys 44:419–431

Tu BP, McKnight SL (2006) Metabolic cycles as an underlying basis of biological oscillations. Nature Rev Mol Cell 7:696–701

Tu BP, Kudlicki A, Rowika M, McKnight SL (2005) Logic of the cell metabolic cycle: temporal compartmentation of cellular processes. Science 310:1152–1158

Ułaszewski S, Mamouneas T, Shen W-K, Rosenthal OJ, Woodward JR, Cirillo VP, Edmunds LN Jr (1979) Light effects in yeast: evidence for participation of cytochromes in photoinhibition of *Saccharomyces cerevisiae* cultured at low temperatures. J Bacteriol 138:523–529

Wallace DC (2012) Mitochondria and cancer. Nat Rev Cancer 12:685–698

Wallace DC (2015) Mitochondrial DNA in human variation and disease. Cell 163:33–38

Wiemken A, von Meyenburg HK, Matile P (1970) Properties of the vacuole in baker's yeast synchronized with a new method. Acta Fac Med Uni Brunensis 37:47–52

Wille JJ Jr (1974) Light entrained circadian oscillations of growth rate in the yeast *Candida utilis*. In: Scheving LE, Halberg F, Pauly JE (eds) Chronobiology. Igaku Shoin, Tokyo, pp 72–77

Woodward JR, Cirillo VP, Edmunds LN Jr (1978) Light effects in yeast: inhibition by visible light of growth and transport in *Saccharomyces cerevisiae* grown at low temperatures. J Bacteriol 133:692–698

Xu Z, Tsurugi K (2007a) Role of Gts1 in regulation of energy-metabolism oscillation in continuous cultures of the yeast *Saccharomyces cerevisiae*. Yeast 24:161–170

Xu Z, Tsurugi K (2007b) De-stabilisation of energy-metabolism oscillation in the absence of trehalose synthesis in the chemostat culture of yeast. Arch Biochem Biophys 464(2):350–358

Xu HN, Tchou J, Feng M, Zhao H, Li LZ (2016) Optical redox imaging indices discriminate human breast cancer from normal tissues. J Biomed Opt 21(11):114003

Yates FE (1982) Outline of a physical theory of physiological systems. Can J Physiol Pharmacol **60**:217–248

Yates FE (1992) Fractal applications in biology: scaling time in biochemical networks. In: Brand L, Johnson ML (eds) Numerical methods, methods enzymol, vol 210. Academic, New York, pp 636–676

Yates FE (1993) Self-organizing systems. In: Boyd CAR, Noble D (eds) The logic of life: the challenge of integrative physiology. Oxford University Press, Oxford, pp 189–218

Yates FE, Yates LB (2008) Ultradian rhythms as the signatures of life. In: Lloyd D, Rossi EL (eds) Ultradian rhythms from molecules to mind, a new vision of life. Springer, Dordrecht, pp 249–260

Zand K, Pham T, Davila A Jr, Wallace DC, Burke PJ (2013) Nanofluidic platform for single mitochondria analysis using fluorescence microscopy. Anal Chem 85(12):6018–6025

Neurospora Genetic Backgrounds Differ in Meiotic Silencing by Unpaired DNA (MSUD) Strength: Implications for *Dp*-Mediated Suppression of Repeat-Induced Point Mutation (RIP)

8

Durgadas P. Kasbekar

Abstract

Neurospora strains bearing chromosome segment duplications (*Dp* strains) can be generated in the laboratory and are also recoverable from natural populations. Two "genome defense" processes, RIP (repeat-induced point mutation) and MSUD (meiotic silencing by unpaired DNA), impart some unusual phenotypes to *Dp*-heterozygous crosses. MSUD is an RNAi-based gene silencing of any gene not properly paired with a homologous sequence during meiosis. Efficient MSUD, as in the highly inbred *N. crassa* Oak Ridge (OR) background used for most genetic studies, can make *Dp*-heterozygous crosses barren and very non-productive. Our recent studies showed, however, that strains with inefficient MSUD are more widespread in nature than OR-type strains, and in them, *Dp*-heterozygous crosses can be non-barren. RIP occurs in the haploid nuclei of the pre-meiotic dikaryon and alters, via multiple G:C to A:T transition mutations, duplicated DNA sequences present in an otherwise haploid genome, and thus destroys duplicated genes. RIP-induced mutants can be recovered among the progeny of *Dp*-heterozygous crosses. Additionally, the presence of large *Dp*s (> 250 kb) suppresses RIP in smaller gene-sized duplications (< 5 kb) in the same cross, presumably by titration of the RIP machinery. Thus, in an inefficient MSUD background a large Dp would enable a small duplication to escape RIP, whereas efficient MSUD induces barrenness and thereby suppresses the production of escapees. Our findings enable us to explain why only one *Neurospora* strain was found to contain an active transposable element, whereas all the other strains examined contained only its RIP-inactivated copies.

Dedicated to Kevin McCluskey for his excellent contributions to the *Neurospora* community as Curator of the Fungal Genetics Stock Center and to mark 27 years of our friendship, especially on the squash courts of UofA, Tucson.

D. P. Kasbekar (✉)
Centre for DNA Fingerprinting and Diagnostics, Hyderabad, Telangana, India
e-mail: kas@cdfd.org.in

T. Satyanarayana et al. (eds.), *Advancing Frontiers in Mycology & Mycotechnology*,
https://doi.org/10.1007/978-981-13-9349-5_8

Keywords
Neurospora · *Neurospora crassa* · Repeat-induced point mutation (RIP) · Meiotic silencing · Meiotic silencing by unpaired DNA (MSUD) · Transposable element

8.1 Introduction

Uniquely among fungi, the *Neurospora* genome has practically no repeated DNA (Galagan et al. 2003). Wrapped inside this mystery lurked a riddle: one *Neurospora* strain had been found which contained multiple copies of a transposon (Kinsey and Helber 1989). The mystery was solved by the discovery of "genome defense" by repeat-induced point mutation (RIP; Selker 1990) and the riddle by the demonstration that the transposon-bearing strain also has a dominant suppressor of RIP phenotype (Noubissi et al. 2001). RIP alters duplicated DNA sequences present in an otherwise haploid genome and thus destroys duplicated genes. Dominant RIP suppression is effected by large chromosome segment duplications (*Dp*s) in a fundamentally stoichiometric manner (Bhat and Kasbekar 2001; Singh and Kasbekar 2008). However, up until now it was generally accepted that crosses with *Dp* strains are barren and produce very few progeny (Perkins 1997; Perkins et al. 1997). Therefore, one could argue that *Dp*-mediated RIP suppression was only a laboratory artifact and might not be relevant in nature. Barrenness is caused by MSUD (meiotic silencing by unpaired DNA), an RNAi-based gene silencing process (Shiu et al. 2001). Our recent work suggests that that "*Dp* crosses" are, in fact, often not barren (Giri et al., 2019). The blanket misattribution of the barren phenotype to them happened because the highly inbred *N. crassa* Oak Ridge (OR) background was used for most genetic studies in *Neurospora* (Perkins 2004), and in which MSUD is unusually efficient (Ramakrishnan et al. 2011). I will briefly describe MSUD, RIP, and *Dp*s and then summarize the findings that led to the revision of our views vis-à-vis *Dp*s and barrenness. Hammond (2017) gives an excellent review on MSUD and Gladyshev (2017) an equally contemporary one on RIP.

8.2 Meiotic Silencing by Unpaired DNA (MSUD)

In MSUD, an unknown mechanism detects homologous DNA sequences that are misaligned ("unpaired") in meiosis and is presumed to transcribe them into "aberrant RNA" (aRNA). An RNA-dependent RNA polymerase (RdRP) encoded by the *sad-1* (*suppressor of ascus dominance-1*) gene is then believed to make the aRNA double-stranded. The double-stranded RNA is then processed into single-stranded MSUD-associated small interfering RNA (masiRNA) that guides a perinuclear silencing complex to degrade complementary mRNA as it exits the nucleus. This effects silencing of the misaligned genes as well as any of their homologues, regardless of whether the homologues were themselves paired. A subset of mutant *sad-1* alleles was found to semi-dominantly suppress MSUD (Shiu et al. 2001). It included the UV-induced *Sad-1*^{UV} allele in which the 3' half of the gene is deleted, the *Sad-1*$^{\Delta}$ allele which is deleted for the whole gene, and the *Sad-1*^{RIP141} allele that contained 828 GC to AT transition mutations induced by the RIP mutational process (see below) and hence shared only 87% identity with *sad-1*$^{+}$. These alleles are presumed to cause the wild-type *sad-1*$^{+}$ allele to become unpaired in meiosis, which triggers its autogenous silencing, and the consequent reduction in RdRP level reduces the accumulation of masiRNAs against other unpaired genes. In other words, the suppressor alleles "silence the silencer." In contrast, the *sad-1*^{RIP73} allele that contained only 194 RIP-induced GC to AT changes did not suppress MSUD, presumably because its sequence was not sufficiently severely altered to unpair the *sad-1*$^{+}$ allele and trigger its autogenous silencing. Crosses homozygous mutant for *sad-1* were arrested in ascus development in the meiotic prophase, and the asci aborted and degenerated.

Several more MSUD genes were subsequently identified based on the "silencing the silencer" phenotype of their deletion alleles. The screens typically used one or more of the MSUD tester strains *::act*, *::asm-1*, *::Bml*^{r}, *::mei-3*, and *::r*$^{+}$ in which an additional copy of the *act* (*actin*), *asm-1*$^{+}$ (*ascospore maturation-1*), *Bml* (*β-tubulin*), *mei-3*, or *r*$^{+}$ (*round ascospores*) gene was inserted at an ectopic location (usually the *his-3* locus on chromosome 1) and its unpairing in *tester*-heterozygous crosses instigated masiRNA production that reduced actin, ASM-1, β-tubulin, MEI-3, or R protein levels and resulted in striking ascus or ascospore phenotypes. In contrast, homozygous *tester mat-A* × *tester mat-a* crosses did not show MSUD. While heterozygosity for semi-dominant mutant alleles of the MSUD genes *sad-1*, *sad-2*, *sad-3*, *sms-2* (*suppressor of meiotic silencing-2*), *dcl-1* (*dicer-like-1*), and *qip* (*QDE-2 interacting protein*) suppressed MSUD via the "silencing the silencer" route, their homozygous crosses failed to make ascospores. Thus, the proteins encoded by these genes are required for MSUD and also appear to have additional roles in sexual reproduction. In contrast, crosses homozygous for the MSUD mutations *sad-4*, *sad-5*, and *sad-6* did not show MSUD but could produce ascospores, which suggested that their gene products were only required for MSUD but were otherwise dispensable for sexual reproduction. Hammond (2017) presented a model outlining how the different MSUD proteins might be involved in masiRNA generation. Briefly, the hypothesized aRNAs made from the unpaired DNA segments are converted into dsRNAs by the SAD-1 RdRP and the dicer protein DCL-1 then processes the dsRNAs into masiRNAs; QIP removes the passenger strand from the masiRNAs, which are then used by the Argonaut protein SMS-2 in the silencing complex to identify complementary mRNAs for silencing. The SAD-3 helicase might help in dsRNA formation or in another process that might require nucleic acid unwinding. The SAD-1, DCL-1, SMS-2, SAD-3, and QIP proteins assemble on a perinuclear scaffold made by the SAD-2 protein. Since masiRNAs are absent from SAD-4-deficient crosses, the SAD-4 protein also plays a role in their production. The SAD-5 and SAD-6 proteins are located inside the nucleus and their roles are still obscure. Additionally, the *cpc-20* and *cpc-80* genes that encode the CPC-20 and CPC-80 proteins of the cap-binding complex (CBC) are required for efficient MSUD (Decker et al. 2017). The CBC associates with the 5′ cap of mRNA transcripts, and its absence might make it harder for the silencing complex to recognize and capture target mRNAs exiting the nucleus, allowing some to reach the translational machinery and thus lowering MSUD efficiency (Decker et al. 2017).

8.3 Repeat-Induced Point Mutation (RIP)

The RIP (repeat-induced point mutation) mutational process occurs during a sexual cross, in the haploid nuclei of the pre-meiotic dikaryon that forms following fertilization and lasts until karyogamy, and targets DNA sequences duplicated in an otherwise haploid nucleus with multiple G:C to A:T transition mutations and methylates many of the remaining cytosine residues (Selker 1990; Irelan and Selker 1996). Crosses homozygous mutant for the *rid* (*RIP-defective*) gene are deficient for RIP (Freitag et al. 2002).

Our serendipitous discovery that colonies produced from *erg-3* (*ergosterol-3*) mutant ascospores on Vogel's–sorbose agar medium have a distinct growth morphology that makes them unambiguously distinguishable from sibling wild-type colonies by mere inspection under a dissection microscope enabled us to conveniently assay for RIP (Noubissi et al. 2001). The transgene *Dp(erg-3)* (also called *Dp* 1.3^{ec}*hph*) contains an *hph*-tagged duplication of a 1.3 kb segment of the *erg-3* gene. When *Dp(erg-3)* strains were crossed with OR strains of the opposite mating type, 8–13% of the progeny were RIP-induced *erg-3* mutants, and DNA sequencing revealed that they contained numerous G:C to A:T transition mutations, many of which introduced in-frame stop codons. When we crossed the *Dp(erg-3)* strains with > 440 wild-isolated strains and scored the frequency of RIP-induced *erg-3* mutants in the progeny, we found that crosses with seven wild strains yielded far fewer than 0.5% RIP-induced *erg-3* mutant progeny, suggesting that they dominantly suppressed RIP. Interestingly, the seven dominant RIP suppressor wild strains included the Adiopodoumé strain (FGSC #430) isolated in Ivory Coast, West Africa. It was the only *Neurospora* strain found bearing an active transposable element, called *Tad* (Kinsey and Helber 1989), while all the other hundreds of strains examined carried only RIP-inactivated relics of *Tad* (Kinsey 1989). The Adiopodoumé strain's dominant RIP suppressor phenotype explained the survival of the *Tad* transposon.

8.4 *Dp* Strains and MSUD-Induced Barrenness of *Dp* × *N* Crosses

A large number of chromosome rearrangements have been identified and studied in *N. crassa*, including several insertional translocations (*IT*), quasiterminal translocations (*QT*), and reciprocal translocations (*RT*) (Perkins 1997). *IT*s transfer a segment of a donor chromosome into a recipient chromosome without any reciprocal exchange, *QT*s move a distal segment of a donor chromosome to the tip region of the recipient chromosome distal to any essential gene and presumably cap the donor

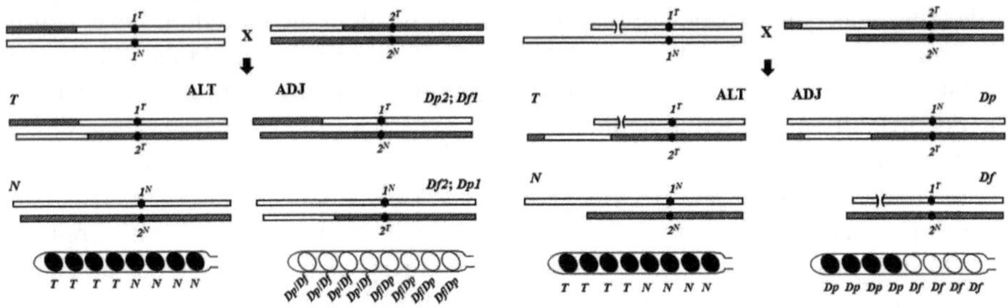

Fig. 8.1 Alternate versus adjacent-1 segregation in $T \times N$ crosses. In alternate segregation the centromeres 1^T and 2^T, derived from the translocation parent, go to one spindle pole, and centromeres 1^N and 2^N, derived from the normal sequence parent, go to the other pole; in adjacent-1 segregation, 1^T and 2^N go to one pole, and 2^T and 1^N go to the other pole. Alternate segregation in $IT \times N$ (right panels) produces 4 *IT* plus 4 *N* parental-type viable ascospores, and the asci have the 8B:0W phenotype; adjacent-1 segregation produces asci with 4 *Dp* plus 4 *Df* ascospores, and since the *Df* ascospores fail to turn black, the ascus phenotype is 4B:4W (B = black ascospores, W = white ascospores). Alternate and adjacent-1 segregation are equally likely; therefore $IT \times N$ crosses produce equal numbers of the two ascus types, i.e., 8B:0W = 4B:4W. The left panels show the $RT \times N$ cross. Again, alternate segregation produces asci with 4 *RT* plus 4 *N* parental-type viable ascospores, and the ascus phenotype is 8B:0W, whereas adjacent-1 segregation produces ascospores containing either deficiency *Df1* or *Df2*, together with, respectively, the complementary duplication *Dp2* or *Dp1*. The *Df*s cause inviability; therefore all the ascospores remain white, and the ascus phenotype is 0B:8W. Consequently, $RT \times N$ crosses produce 8B:0W = 0B:8W

chromosome's breakpoint with the recipient chromosome's tip, and *RT*s reciprocally interchange the terminal segments of chromosomes. Thus, a *QT* is formally like an *RT* but behaves genetically like an *IT*. Significantly, one-third of the viable progeny from $IT \times N$ and $QT \times N$ crosses (*N* is a normal sequence strain) contain a duplication of the translocated segment, that is, they are *Dp* type (Fig. 8.1).

It was known for a long time that in *N. crassa*, *Dp*-heterozygous crosses (i.e., $Dp \times N$ or $Dp \times T$) have a characteristic barren phenotype, that is, they produce normal-looking perithecia but generate exceptionally few ascospores (Perkins 1997). Perkins et al. (1997) had speculated that the barrenness might be caused by the inactivation of *Dp*-borne essential genes by RIP. However, this idea was questioned when we found *Dp* × Adiopodoumé crosses to be as unproductive as their *Dp* × OR counterparts (Noubissi et al. 2001, cited by Shiu et al. 2001). Recall that the Adiopodoumé strain has a dominant RIP suppressor phenotype. Oddly, residual support for RIP's involvement in barrenness persisted in my laboratory. One student, not involved in the Adiopodoumé experiments, was diehard enough to publish a dissenting note (Prakash 2001), which was inopportune, because shortly thereafter Shiu et al. (2001) reported that *Dp* × *Sad-1* crosses were 50-fold to 8000-fold more productive than the corresponding *Dp* × *sad*$^+$ crosses, and a little later we confirmed that the barrenness was not alleviated by homozygosity for the RIP suppressor mutation *rid* (Bhat and Kasbekar 2004; Vyas et al. 2006). Presumably, in *Dp*-heterozygous crosses the *Dp*-borne genes, including those required for meiosis or ascus development, are unable to pair properly with a homologue, and their silencing makes the cross barren. [Happily, the recalcitrant student did excellently in his own research (Prakash et al. 1999; Prakash and Kasbekar 2002a, 2002b) and went on to build a successful research career. Clearly, my laboratory harbored a lively diversity of academic opinion.]

Somewhat surprisingly, *Dp*-homozygous crosses also were barren, and *Sad-1* failed to increase their productivity (Shiu et al. 2001; Bhat and Kasbekar 2004). We have proposed a model to explain why *Sad-1* might not be as effective in suppressing MSUD in a *Dp*-homozygous cross (Bhat and Kasbekar 2004; Singh et al. 2009). However, given that the *sad-5* mutant is now available, whose homozygous cross is fertile but does not show MSUD, it can be used to test the homozygous *Dp*; *sad-5* × *Dp*; *sad-5* cross and conclusively settle this issue.

8.5 *Dps* Suppress RIP, Presumably by Titration of the RIP Machinery

Perkins et al. (1997) were the first to show that RIP can occur in crosses involving strains bearing sizeable *Dps*. They examined 11 different *Dp* × *N* crosses, which because of their barren phenotype produced very few progenies (<500, in fact, six crosses produced <100 progenies each); nevertheless, they obtained RIP-induced mutant alleles in 17 different genes. Of these, 13 mapped to the segment duplicated in the corresponding *Dp* parent, one was outside the duplication but closely linked, and three were unlinked to the duplication. Two mutant alleles were examined for DNA sequence alterations and found to contain G:C to A:T changes as expected from RIP. However, the density of mutant sites was lower than what was seen when RIP occurred in gene-sized duplications (such as *Dp(erg-3)*). Indeed, it was this observation that led Perkins et al. (1997) to (incorrectly) speculate that the few ascospores produced in the barren crosses might be the ones in which RIP either had not occurred or was very inefficient, and thereby account for the low frequency of RIP-induced alterations observed in the survivors. Their results motivated us to ask whether the production of RIP-induced *erg-3* mutant progeny is affected in *Dp* × *Dp(erg-3)*.

Our first *Dp* × *Dp(erg-3)* crosses were made prior to the revelation that *Sad-1* can increase the productivity of *Dp*-heterozygous crosses, and we were constrained to use "brute force" (i.e., set up multiple replicates for each cross) to obtain sufficient progeny numbers. The frequency of RIP-induced *erg-3* mutants was reduced if the *Dps AR17*, *OY329*, or *S1229* were present in either the same or the other parental nucleus of the cross, which suggested that the larger *Dps* (all > 100 kb) act as sinks to titrate out the RIP machinery from the smaller (1.3 kb) *Dp(erg-3)* duplication (Bhat and Kasbekar 2001). The barrenness of the *Dp(IBj5)* × OR cross was so severe that it defied "brute force," and we resorted to screening for wild strains whose crosses with *Dp(IBj5)* might yield progeny ascospores in sufficient numbers to do the test (Fehmer et al. 2001). Crosses with four wild strains (Lahore-1 (FGSC #1824), Dagguluru-1 (FGSC #3360), Okeechobee (FGSC # 3968), and Tiassale (FGSC # 4825)) with the *Dps AR17*, *OY329*, and *S1229* indeed yielded more progeny ascospores, in contrast, crosses with two, Golikro (FGSC# 4830) and Costa Rica (FGSC# 852), produced approximately 100x fewer ascospores. Moreover, the wild strains did not interfere with the *Dps'* ability to suppress RIP in *Dp(erg-3)*. We performed crosses of *Dp(IBj5)* with the four "productive" wild strains and obtained sufficient progeny numbers to enable us to conclude that *Dp(IBj5)* also is a RIP suppressor (Fehmer et al. 2001).

Subsequently, the *Sad-1* mutant made it convenient to survey many more *Dp*s (Singh et al. 2008). All told, of the 39 *Dp*s we examined, 33 suppressed RIP, 5 (EB4, B362i, UK14-1, NM169d, and R2394) did not, and 1 (CJS1) gave inconclusive results. We could estimate the size of 27 suppressor and 3 non-suppressor *Dp*s. The former were all greater than 270 kb, whereas the latter were all less than 200 kb (Singh et al. 2008, 2010). It was also possible to achieve RIP suppression by combining multiple non-suppressor *Dp*s in a cross, to thereby increase the total duplication size to greater than 270 kb. For example, the *Dp*s EB4, B362i, and R2394 were individually non-suppressing, but their double and triple heterozygous crosses could suppress RIP (Singh and Kasbekar 2008). Together our results support the idea that *Dp*-mediated RIP suppression might occur by titration of the RIP machinery. Perkins et al. (2007) independently confirmed our contention that *Dp*s can suppress RIP.

However, the use of the *Sad-1* mutant, or particular wild strains, or "brute force" to obtain sufficient ascospore numbers invites the criticism that *Dp*-mediated RIP suppression might be only a laboratory-bound curiosity, and it might not be significant in natural populations. Our recent results now enable us to rebut this criticism (see below).

Two of the seven wild-isolated dominant RIP suppressor strains (see Sect. 8.3), Sugartown (P0854) and Golur-1 (P0334), were barren in crosses with normal sequence strains, and the RIP suppressor and barren phenotypes of Sugartown were tightly linked, suggesting that a naturally occurring duplication might underlie both phenotypes (Bhat et al. 2003). Additionally, crosses with three other wild strains, Georgetown-6 (P2622), Batu Ferringi-1 (P2681), and Brabadougou (P4296), also were barren although without an associated RIP suppressor phenotype. These three strains might contain duplications smaller than 270 kb hence making them unable to suppress RIP. The *Sad-1* mutation suppressed the barren phenotype of Sugartown, Golur-1, and Georgetown-6 (but not of Batu Ferringi-1 and Brabadougou), providing additional support to the idea that the Sugartown, Golur-1, and Georgetown-6 strains contain naturally occurring duplications. It should now be possible to verify this by whole genome sequencing.

8.6 *Neurospora* Genetic Backgrounds Harbor Wide Genetic Variation in MSUD Strength

All the studies described thus far had used *N. crassa* strains of the standard OR genetic background, except, of course, the screens for the dominant RIP suppressor wild strains, and the wild strains that increased the productivity of the *Dp(IBj5)*-heterozygous crosses. The tacit assumption being that what was true for OR would be generally true for *Neurospora*. Unexpectedly, MSUD was often not as robust when the OR-derived MSUD tester strains (*tester*^{OR}, specifically, *::Bml*^{r} and *::mei-3*) were crossed with diverse wild strains (Kasbekar et al. 2011; Ramakrishnan et al. 2011). Of 80 wild strains examined, crosses with only 8, designated as the "OR" type, showed a silencing phenotype comparable to that in the *tester*^{OR} × OR crosses; crosses with 4 wild strains, designated as the "Sad" type, failed to silence *bml* and *mei-3*$^{+}$; and the remaining 68 showed an intermediate phenotype, silencing *bml* but not *mei-3*$^{+}$, and were designated the "Esm" type (for "early suppression in meiosis"). Further, while crosses of *Dp(EB4)* and *Dp(IBj5)* strains with the OR-type wild strains were barren, their crosses with the Sad-type wild strains were fertile, and with the Esm-type wild strains, respectively, fertile and barren. The two *Dp*s duplicate genome segments of, respectively, 35 and 115 genes.

One hypothesis (model 1) to explain the OR/Sad/Esm difference was that sequence heterozygosity between the *tester*^{OR} and wild genomes might cause a natural asynapsis and consequent self-silencing of one or more MSUD gene. This model generalizes from the observation that deletion alleles of several MSUD genes can act as semi-dominant suppressors of MSUD (e.g., *Sad-1*, *Sad-2*) and also from results showing that the r^{+} allele exhibits nearly complete silencing in a cross between an r^{+} strain and one carrying the r^{RIP93} allele, which was only 6% polymorphic over 4.5 kb (Pratt et al. 2004). An alternative hypothesis (model 2) is that natural populations harbor wide genetic variation in MSUD strength, and the OR strains represent the MSUD-conducive extreme.

To distinguish between the models, we used the Sad-type wild strains Bichpuri-1 *a* (B) and Spurger *A* (S) to construct the novel "B/S" background in which the *mat A* and *mat a* strains were near-isogenic. Next, we constructed new testers in this background (*tester*$^{B/S}$), specifically, *::r1*$^{B/S1}$ and *::r3*$^{B/S1}$, that were exactly analogous to the *::r1*^{OR} and *::r3*^{OR} testers made in the OR background by Samarajeewa et al. (2014). Thus, the *::r*^{OR} × OR and the *::r*$^{B/S1}$ × BS1 crosses were fully comparable, and if model 1 was correct we would expect them to show similar MSUD efficiency, and if model 2 is correct, then the *::r*$^{B/S1}$ × BS1 crosses should not show the high MSUD efficiency characteristic of *::r*^{OR} × OR. The *::r*^{OR} × OR crosses produced more than 90% round ascospores, whereas the *::r*$^{B/S}$ × B/S crosses produced fewer than 50% round ascospores, which showed that MSUD was more efficient in OR than in B/S. Reassuringly, the *::r3*$^{B/S1}$ × *::r3*^{OR} crosses produced fewer than 5% round ascospores, which confirmed that the *::r*$^{B/S}$ and *::r*^{OR} transgenes were detected as being allelic. These results allowed us to reject model 1 and support model 2 and to conclude that different *Neurospora* genetic backgrounds can differ widely in MSUD efficiency, and that the OR background represents an MSUD-conducive extreme (Giri et al. 2019).

8.7 Weak MSUD Genetic Backgrounds Show Non-barren *Dp*-Heterozygous Crosses

Strain 85 is the standard genetic background of the related species *N. tetrasperma*, and it shows weak MSUD (Jacobson et al. 2008; Ramakrishnan et al. 2011; Giri et al. 2019). We had introgressed four *N. crassa* translocations (EB4, IBj5, UK14-1, and B362i) into *N. tetrasperma* strain 85, and from the *T* × *N* crosses we were able to obtain the corresponding *N. tetrasperma Dp* strains (Giri et al. 2015, 2016). This made it possible to compare *T* × *N* and *Dp* × *N* crosses in weak and strong MSUD backgrounds, namely, *N. tetrasperma* strain 85 and *N. crassa* OR. In the strain 85 background the *T*- and *Dp*-heterozygous crosses showed comparable productivity (Giri et al. 2019). Recall that the barrenness of *Dp(IBj5)*-heterozygous crosses was especially severe in the OR background (see Sect. 8.5). In contrast, the *Dp(IBj5)* × 85 and *T(IBj5)* × 85 crosses produced, respectively, 1.3×10^5 and 7.1×10^5 ascospores, demonstrating that *Dp*-heterozygous crosses can be non-barren in a weak MSUD background. These results, together with those in Sect. 8.6, enable us to rebut the criticism that *Dp*-mediated RIP suppression is merely a laboratory artifact and rather support the assertion that *Dp*-mediated RIP suppression is likely active in natural populations.

Recall that two of the seven wild-isolated RIP suppressor strains were barren in crosses and the barrenness was suppressed by the *Sad-1* mutation, which suggested that a naturally occurring duplication underlay their RIP suppressor phenotype (see Sects. 8.3 and 8.5). It is also possible that duplications confer the RIP suppressor phenotype of some of the other strains, but do so in a weak-MSUD background, thus causing their crosses to be non-barren. It should now be possible to test this via whole genome sequencing.

8.8 A-to-I mRNA Editing: More Grist for the Sexual Variation Mill?

A-to-I mRNA editing, known for some time in animals, was discovered only recently in fungi, following RNA-Seq studies, and predicted to affect both MSUD and RIP (Liu et al. 2016, 2017; see Kasbekar 2016 for a Commentary; Bian et al. 2018 for a review). The editing converts specific adenosine residues (A) in mRNA to inosine (I) via hydrolytic deamination. Since I is recognized as guanosine (G), therefore it has a similar effect as an A-to-G substitution. In animals, the editing is catalyzed by enzymes called ADARs (adenosine deaminases acting on RNA). Fungi lack ADAR orthologs and hence were assumed to lack A-to-I mRNA editing. Also, mRNA editing is not seen in *Saccharomyces cerevisiae* and *Schizosaccharomyces pombe*, the two most intensively studied fungi. Fungal A-to-I mRNA editing was first reported in *Fusarium graminearum* and shortly thereafter in *N. crassa* and *N. tetrasperma*, and in *Sordaria macrospora* and *Pyronema confluens* (Liu et al. 2016, 2017; Teichert et al. 2017; Cao et al. 2017). In all these species, the editing occurs in the sexual stage and preferentially targets A's in hairpin loops, which is similar to the anticodon loop of tRNA targeted by ADATs (adenosine deaminases acting on tRNA). ADATs catalyze A-to-I editing in tRNAs, and metazoan ADARs are believed to have evolved from an ADAT that acquired dsRNA binding domains. It is speculated that the fungal editors might be the ADAT2 and ADAT3 enzymes acting along with sexual-stage-specific co-factors. Unlike in animals, most A-to-I editing sites in fungi occur in coding regions and are non-synonymous (result in amino acid changes). Editing allows both edited and unedited versions to co-exist in a cell, and the presence of many editing sites can generate many protein variants. Interestingly, genes that are under strong negative selection appear to be selected for editing to increase their proteomic diversity. Additionally, many editing sites are conserved among different fungi, and several appeared to be favored by positive selection. The frequency of DNA base replacement at non-synonymous editing sites was lower than at synonymous editing sites compared with unedited A sites. This suggests it is advantageous to the fungus to maintain proteome diversity during the sexual stage.

Editing of a premature stop codon (PSC) generates a functional protein from the *F. graminearum rid* ortholog, and although *N. crassa rid* does not have a PSC, it contains two nonsynonymous editing sites. Therefore, editing is essential for RIP in *Fusarium* and also possibly in *Neurospora*. Additionally, multiple non-synonymous editing sites were found in the *sad-1*, *sad-2*, *sad-4*, *sad-5*, and *sad-6*, *sms-2*, *dcl-1/sms-3*, and *qip* genes, although *sad-3* had only one. Therefore, it is very likely that the efficiency of both MSUD and RIP is affected by mRNA editing.

8.9 Concluding Remarks

Early work (pre-2011) in my laboratory showed that *Dp* strains can suppress RIP in *Neurospora*, while more recent work (post-2011) showed that *Neurospora* populations harbor wide genetic variation in MSUD strength and that *Dp*-heterozygous crosses are not barren in "weak MSUD" genetic backgrounds. Hence *Dp*-mediated RIP suppression is likely to be more significant than was previously appreciated. *Dp*-mediated RIP suppression can explain the otherwise mysterious persistence of the *Tad* retrotransposon in only the *N. crassa* Adiopodoumé strain, whereas all the other *Neurospora* strains examined contained only RIP-altered relics of *Tad* (Kinsey 1989). Conceivably, *Tad* was sheltered from RIP by an ancestral *Dp*, with possibly the *Dp*-generating rearrangement itself being produced by crossover between non-allelic *Tad* copies. As the copy number of *Tad* increased, *Dp*-mediated titration of the RIP machinery became superfluous, as titration could now be achieved by the increased transposon copy-number. In parallel, the original *Dp* could breakdown and possibly disappear from the Adiopodoumé strain.

The discovery of mRNA editing in *Neurospora* and other filamentous fungi and the evidence that it alters transcripts from genes important for MSUD and RIP suggest that MSUD and RIP efficiency will depend on factors that affect editing. We have initiated mapping studies of the genetic difference underlying the strong and weak MSUD phenotypes of the OR and B/S1 backgrounds (Giri et al. 2019). Future studies will further illuminate the role of *Dp*-mediated RIP suppression in genome organization, maintenance, and evolution in *Neurospora* and other fungi.

Acknowledgements I thank Dev Ashish Giri for preparing Fig. 8.1. I receive support as an Indian National Science Academy (INSA) Senior Scientist, in the Centre for DNA Fingerprinting and Diagnostics (CDFD), and I am an Honorary Visiting Scientist at the Centre for Cellular and Molecular Biology (CCMB). *Neurospora* strains were obtained from the Fungal Genetics Stock Center, Manhattan, Kansas, USA (McCluskey et al. 2010).

References

Bhat A, Kasbekar DP (2001) Escape from repeat-induced point mutation of a gene-sized duplication in *Neurospora crassa* crosses that are heterozygous for a larger chromosome segment duplication. Genetics 157:1581–1590

Bhat A, Kasbekar DP (2004) Why are *Neurospora crassa* crosses that are homozygous for a large duplication barren? Fungal Genet Newslett 51:15–16

Bhat A, Noubissi FK, Vyas M, Kasbekar DP (2003) Genetic analysis of wild-isolated *Neurospora crassa* strains identified as dominant suppressors of repeat-induced point mutation (RIP). Genetics 164:947–961

Bian Z, Ni Y, Xu JR, Liu H (2018) A-to-I mRNA editing in fungi: occurrence, function, and evolution. Cell Mol Life Sci. https://doi.org/10.1007/s00018-018-2936-3

Cao S, He Y, Hao C, Xu Y, Zhang H, Wang C, Liu H, Xu JR (2017) RNA editing of the AMD1 gene is important for ascus maturation and ascospore discharge in *Fusarium graminearum*. Sci Rep 7:4617

Decker LM, Xiao H, Boone EC, Vierling MM, Shanker BS, Kingston SL, Boone SF, Haynes JB, Shiu PKT (2017) The nuclear cap-binding complex mediates meiotic silencing by unpaired DNA. G3 7:1149–1155

Fehmer M, Bhat A, Noubissi FK, Kasbekar DP (2001) Wild-isolated *Neurospora crassa* strains that increase fertility of crosses with segmental aneuploids used to establish that a large duplication suppresses RIP in a smaller duplication. Fungal Genet Newslett 48:13–14

Freitag M, Williams RL, Kothe GO, Selker EU (2002) A cytosine methyltransferase homologue is essential for repeat-induced point mutation in *Neurospora crassa*. Proc Natl Acad Sci U S A 99:8802–8807

Galagan, J. E., S. E. Colvo, K. A. Borkovich, E. U. Selker, [...] B. Birren 2003 The genome sequence of the filamentous fungus *Neurospora crassa*. Nature 422: 859–868.

Giri DA, Rekha S, Kasbekar DP (2015) Neurospora heterokaryons with complementary duplications and deficiencies in their constituent nuclei provide an approach to identify nucleus-limited genes. G3 5:1263–1272

Giri DA, Rekha S, Kasbekar DP (2016) Crosses heterozygous for hybrid Neurospora translocation strains show transmission ratio distortion disfavoring homokaryotic ascospores made following alternate segregation. G3 6:2593–2600

Giri DA, Pankajam AV, Nishant KT, Kasbekar DP (2019) The *Neurospora crassa* standard oak ridge background exhibits an atypically efficient meiotic silencing by unpaired DNA. G3 9:1487–1496

Gladyshev E (2017) Repeat-induced point mutation and other genome defense mechanisms in fungi. Microbiol Spectr 5. https://doi.org/10.1128/microbiolspec.FUNK-0042-2017

Hammond TM (2017) Sixteen years of meiotic silencing by unpaired DNA. Adv Genet 97:1–42

Irelan JT, Selker EU (1996) Gene silencing in filamentous fungi: RIP, MIP and quelling. J Genet 75:313–324

Jacobson DJ, Raju NB, Freitag M (2008) Evidence for the absence of meiotic silencing by unpaired DNA in *Neurospora tetrasperma*. Fungal Genet Biol 45:351–362

Kasbekar DP (2016) RNA-Seq, and ye shall find: sexual-stage-specific A-to-I RNA editing in fungi. J Biosci 41:171–172

Kasbekar DP, Singh PK, Ramakrishnan M, Kranthi Raj B (2011) Carrefour Mme. Gras: a wild-isolated *Neurospora crassa* strain that suppresses meiotic silencing by unpaired DNA and uncovers a novel ascospore stability defect. Fungal Genet Biol 48:612–620

Kinsey JA (1989) Restricted distribution of the Tad transposon in strains of *Neurospora*. Curr Genet 15:271–275

Kinsey JA, Helber J (1989) Isolation of a transposable element from *Neurospora crassa*. Proc Natl Acad Sci U S A 86:1929–1933

Liu H, Wang Q, He Y, Chen L, Hao C, Jiang C, Li Y, Dai Y, Kang Z, Xu JR (2016) Genome-wide A-to-I RNA editing in fungi independent of ADAR enzymes. Genome Res 26:499–509

Liu H, Li Y, Chen D, Qi Z, Wang Q, Wang J, Jiang C, Xu JR (2017) A-to-I RNA editing is developmentally regulated and generally adaptive for sexual reproduction in *Neurospora crassa*. Proc Natl Acad Sci U S A 114:E7756–E7765

McCluskey K, Wiest A, Plamann M (2010) The Fungal Genetics Stock Center: a repository for 50 years of fungal genetics research. J Biosci 35:119–126

Noubissi FK, McCluskey K, Kasbekar DP (2001) Repeat-induced point mutation (RIP) in crosses with wild-isolated strains of *Neurospora crassa*: evidence for dominant reduction of RIP. Fungal Genet Biol 31:91–97

Perkins DD (1997) Chromosome rearrangements in Neurospora and other filamentous fungi. Adv Genet 36:239–398

Perkins DD (2004) Wild type *Neurospora crassa* strains preferred for use as standards. Fungal Genet Newslett 51:7–8

Perkins DD, Margolin BS, Selker EU, Haedo SD (1997) Occurrence of repeat induced point mutation in long segmental duplications of *Neurospora*. Genetics 147:125–136

Perkins DD, Freitag M, Pollard VC, Bailey-Shrode LA, Selker EU, Ebbole DJ (2007) Recurrent locus-specific mutation resulting from a cryptic ectopic insertion in Neurospora. Genetics 175:527–544

Prakash A (2001) An alternative explanation for the reduced repeat-induced point mutation (RIP) efficiency in a gene-sized duplication in *Neurospora crassa* crosses heterozygous for a larger chromosome segment duplication. Fungal Genet Newslett 48:35

Prakash A, Kasbekar DP (2002a) Genes encoding chimeras of *Neurospora crassa erg-3* and human TM7SF2 proteins fail to complement *Neurospora* and yeast sterol C-14 reductase mutants. J Biosci 27:105–112

Prakash A, Kasbekar DP (2002b) The sterol C-14 reductase encoded by the *Neurospora crassa erg-3* gene: essential charged and polar residue identified by site specific mutagenesis. Mol Gen Genomics 266:787–795

Prakash A, Sengupta S, Aparna K, Kasbekar DP (1999) The *erg-3* (sterol $\Delta^{14,15}$-reductase) gene of *Neurospora crassa*: generation of null mutants by repeat-induced point mutation and complementation by proteins chimeric for human lamin B receptor sequences. Microbiology 145:1443–1451

Pratt RJ, Lee DW, Aramayo R (2004) DNA methylation affects meiotic trans-sensing, not meiotic silencing, in Neurospora. Genetics 168:1925–1935

Ramakrishnan M, Sowjanya TN, Raj KB, Kasbekar DP (2011) Meiotic silencing by unpaired DNA is expressed more strongly in the early than the late perithecia of crosses involving most wild-isolated *Neurospora crassa* strains and in self-crosses of *N. tetrasperma*. Fungal Genet Biol 48:1146–1152

Samarajeewa DA, Sauls PA, Sharp KJ, Smith ZJ, Xiao H, Groskreutz KM, Malone TL, Boone EC, Edwards KA, Shiu PKT, Larson ED, Hammond TM (2014) Efficient detection of unpaired DNA requires a homolog from the Rad54-like family of homologous recombination proteins. Genetics 198:895–904

Selker EU (1990) Premeiotic instability of repeated sequences in Neurospora crassa. Annu Rev Genet 24:597–613

Shiu PKT, Raju NB, Zickler D, Metzenberg RL (2001) Meiotic silencing by unpaired DNA. Cell 107:905–916

Singh PK, Kasbekar DP (2008) Titration of repeat-induced point mutation (RIP) by chromosome segment duplications in *Neurospora crassa*. Genetica 134:267–275

Singh PK, Iyer SV, Ramakrishnan M, Kasbekar DP (2009) Chromosome segment duplications in *Neurospora crassa*: barren crosses beget fertile science. BioEssays 31:209–219

Singh PK, Iyer SV, Nagasowjanya T, Kranthi Raj B, Kasbekar DP (2010) Translocations used to generate chromosome segment duplications in Neurospora can disrupt genes and create novel open reading frames. J Biosci 35:539–546

Teichert I, Dahlmann TA, Kuck U, Nowrousian M (2017) RNA editing during sexual development occurs in distantly related filamentous ascomycetes. Genome Biol Evol 9:855–868

Vyas M, Ravindran C, Kasbekar DP (2006) Chromosome segment duplications in *Neurospora crassa* and their effects on repeat-induced point mutation (RIP) and meiotic silencing by unpaired DNA. Genetics 172:1511–1519

Part II

Environmental Sustainability

Fungal Bioremediation: A Step Towards Cleaner Environment

9

Darshan M. Rudakiya, Archana Tripathi, Shilpa Gupte, and Akshaya Gupte

Abstract

Mycoremediation is a technique wherein various fungi are used in the remediation of hazardous contaminants like polycyclic aromatic hydrocarbons, dyes, hazardous phenolics, heavy metals, and several others. Fungi are the ubiquitous and diverse group of organisms that produce a wide array of enzymes and metabolites (organic acids, exopolysaccharides, etc.). By synergic action of these metabolites, fungi can efficiently degrade or transform the hazardous contaminants. Additionally, live or dead fungal biomass has the higher sorption capability of contaminants as compared to the other microorganisms. In this chapter, application of fungi in degrading dyes and polycyclic aromatic hydrocarbons, remediation of heavy metals, and other contaminants is discussed, wherein fungi degrade the organics and aid in remediation of metals using a variety of strategies.

Keywords

Bioremediation · Dyes · Pesticides · Polycyclic aromatic hydrocarbons · White rot fungi

---

D. M. Rudakiya · A. Tripathi · A. Gupte (✉)
Department of Microbiology, N V Patel College of Pure & Applied Sciences, Anand, Gujarat, India

S. Gupte
Department of Microbiology, Ashok & Rita Patel Institute of Integrated Study & Research in Biotechnology and Allied Sciences, Anand, Gujarat, India

T. Satyanarayana et al. (eds.), *Advancing Frontiers in Mycology & Mycotechnology*,
https://doi.org/10.1007/978-981-13-9349-5_9

9.1 Introduction

Overall quality of environment is directly connected with the quality of life on the globe. Earlier, people believed that they have infinite resources of land and water; however, these resources are not finite, and still, they are being used extensively all over the globe (Bouwer 2017). Contaminated lands and water bodies are outcomes of industrial activities with the disposal of hazardous substances. Certain hazardous pollutants such as polycyclic aromatic hydrocarbons (PAHs), textile dyes, toluene, ethylbenzene, trinitrotoluene (TNT), pentachlorophenol (PCP), polychlorinated biphenyls (PCB), pesticides, benzene, and xylene are insistent in surroundings. Additionally, they possess the carcinogenic and/or mutagenic effects. Unwanted substances are generally released into the environment by man, which damage either health or resources (Reddy and Mathew 2001). The problem has increased worldwide, and the estimated numbers of contaminated sites are significantly increasing. A rapid, cost-effective, ecologically responsible method of cleanup of the pollutants is urgently needed due to the magnitude of this issue and the lack of a reasonable solution (Cheng et al. 2018; Rudakiya et al. 2018).

Bioremediation, degradation of toxic compounds using biological agents, provides an effective solution and convert contaminants into non-toxic or less toxic form. Bioremediation is defined as the productive use of microbes to remove or detoxify pollutants, usually contaminants in soil, water, or sediments that otherwise threaten public health (Tsekova et al. 2010). Bioremediation is a process that removes xenobiotic compounds from the biosphere. Bioremediation process employs various microorganisms or plants to eliminate hazardous contaminants by metabolizing them and by converting them into CO_2 and water. The main aim of the bioremediation is to eliminate contaminants to undetectable concentration limit, which is established by regulatory agencies (Lun et al. 2018).

Various organisms have been employed to remove the variety of contaminants, which include bacteria, fungi, plants, algae, etc. Among all organisms, fungi are a diverse group of organisms, which are ubiquitous in the environment. Their major contribution ranges from various industrial applications to remediation purposes (de Lorenzo 2018). Fungi can easily survive in most of the habitats and play dynamic role in ecosystem. They regulate the flow of energy and nutrients through their mycelial networks. Additionally, their effect to the environment is seen microscopically, even though impact is unseen from outer biosphere. In the soil, mycelial networks are generally covered from several meters to several hectares of forest, which decay and convert the organics and other metabolites into another form. Thus, these fungi are considered the natural and true ecosystem engineers (Kumar et al. 2018). Fungi microscopic and macroscopic eukaryotic organisms, can easily grow on different substrates and are capable of continuing their function almost indefinitely. Fungi include molds, yeasts, ascomycetes, and basidiomycetes and are unique microorganisms and macroorganisms, which can be employed in the remediation of wastes and wastewaters. Some molds, yeasts, and other fungi are highly tolerant to extreme conditions such as higher temperature, acidic or alkaline pH, higher concentration of metals, etc. Fungi are highly plastic bodies, and most of the fungal cells are totipotent, so that the entire organism can regenerate from spores and hyphal fragments. Mycoremediation is the process wherein fungi degrade or cause to deteriorate the variety of materials and compounds (Bharath et al. 2019). The degradation efficiency of different hazardous contaminants using fungi has been shown in Table 9.1, wherein degradative compounds are dyes, PAHs, biomass, heavy metals, etc.

9.2 Degradation of Organic Contaminants

(a) *Dyes*

A dye or dyestuff is colorant or colored organic compound or mixture, which is generally utilized to impart color to a substrate like cloth, paper, plastic, or leather in a practically permanent manner. Due to the different chromophore groups, property and color of the compound have been changed. Benzene structures containing chromophore groups are called chromogens. These compounds are colored, but they not the dye as they do not hold the affinity to unite with fibers and tissues. The color may be easily removed by mechanical methods. Some important chromophores are -N=N, (CH=CH), -C=N, and -N=O. Dye must contain not only a chromophore group but also another group that imparts to the compound the property of electrolytic dissociation. Such auxiliary groups are known as auxochromes. Some are -OH, -NH, -NHR, -COOH, -SOH, and -NR. A chromogen without auxochrome can never act as a dye (Robinson et al. 2001).

Most of the dyes are organics and complex in nature. Synthesis of these dyes is initiated with azulene synthesis. As a result, dyes can be classified as natural and synthetic according to their sources of origin (Van der Zee and Villaverde 2005).

One of the ways of classifying dyes is based on their chemical constitution, particularly considering the chromophoric system present in dye molecule (Table 9.2). The main classes of dyes according to their chemical constitutions are given in the Fig. 9.1.

The worldwide production of synthetic dyes increases annually. Dyes are extensively utilized for textile, paper printing, color photography, cosmetic, pharmaceutical, and leather industries. Inefficiencies in the dyeing process and some technical problems can result into the discharge of dyestuff in wastewater during textile processing, which ultimately finds way into the environment. Over the globe, it is assessed that 5–10% dyes are lost in the effluent during the dyeing process. In India, wastewater generated from textile processing is 450,000 m^3/day, while dyeing facility alone generates 32,000 m^3/day wastewater (Liu et al. 2005).

Dyes are more difficult to degrade because of their synthetic origin and complex aromatic structure. Various methods have been developed to remove synthetic dyes from water and wastewater to decrease their impact on the environment and are divided into four major categories, which are shown in Fig. 9.2.

The physico-chemical treatments are found to be effective, but their application is limited due to excess usage of chemicals, sludge production, subsequent disposal

Table 9.1 List of chemical classes efficiently degraded by fungi

Chemical class	Fungus	Compounds	References
Polycyclic aromatic hydrocarbons	*Cunninghamella elegans*	Acenaphthene	Pothuluri et al. (1992)
	Bjerkandera sp., *Cunninghamella elegans, Naematoloma frowardii, Phanerochaete chrysosporium, Phanerochaete laevis, Pleurotus ostreatus, Pleurotus sajor-caju, Ramaria* sp., *Rhizoctonia solani, Trametes versicolor*	Anthracene	Kadri et al. (2017), Johanne and Majcherczyk (2000), Kotterman et al. (1998), and Sack and Günther (1993)
	Aspergillus niger, Agrocybe aegerita, Candida parapsilosis, Crinipellis maxima, Pleurotus ostreatus AGHP-1	Pyrene	Hammel et al. (1986), Manilla-Pérez et al. (2011), and Patel et al. (2010)
	Laetiporus sulphureus, Penicillium sp., *Pleurotus ostreatus* AGHP-1	Fluoranthene	Sack and Günther (1993) and Patel et al. (2009)
Dyes	*Ganoderma cupreum* AG1	Reactive Violet 1	Gahlout et al. (2017)
	Trametes hirsuta	Acid Blue 225, Basic Red 9, Direct Blue 71, Crystal Violet, Indigo, Poly R-478, Reactive Blue 19, Reactive Blue 221	Nyanhongo et al. (2002), Abadulla et al. (2000), Campos et al. (2001), and Maceiras et al. (2001)
	Pseudolagarobasidium acaciicola AGST3	Violet P3P, Green ME4BL, Blue 3R, Direct black 22, Green HE4G, Reactive red M5B	Thakur and Gupte (2015)
	Aspergillus niger	Congo Red	Fu and Viraraghavan (2002)
	Aspergillus foetidus	Drimarene Black HFGR1, Drimarene Navy BF Blue F2G1, Drimarene Red BR F3B1	Sumathi and Manju (2000)
Biomass degradation	*Pseudolagarobasidium acaciicola* AGST3	*Pithecellobium dulce* and *Tamarindus indica*	Rudakiya and Gupte (2017, 2019)
	Tricholoma giganteum AGDR1		
	Pleurotus ostreatus and Trametes versicolor	Wheat straw	Shrivastava et al. (2011)

(continued)

Table 9.1 (continued)

Chemical class	Fungus	Compounds	References
Pesticides	*Aspergillus niger*	Endosulfan	Bhalerao and Puranik (2007)
	Coriolus versicolor	Aldicarb, atrazine, alachlor	Hai et al. (2013)
	Pleurotus sajor-caju, *Trametes pubescens*	Chlorophenols	Denizli et al. (2005)
Metals	*Phanerochaete chrysosporium*	As, Cr, Cd	Shah et al. (2018) and Rudakiya et al. (2018)
	Schizophyllum commune	Cr, Cu, Zn, Ni	Javaid et al. (2010)
	Beauveria bassiana	Zn, Cu, Cd, Cr, Ni	Gola et al. (2016)
Polychlorinated dioxins and dibenzofurans	*Stropharia rugosoannulata*, *Phanerochaete velutina*	Polychlorinated dibenzo-p-dioxins, polychlorinated dibenzofurans	Anasonye et al. (2014)
Endocrine disrupting chemicals	*Fusarium incarnatum* UC-14, *Irpex lacteus*, *Trametes polyzona*, *Pleurotus ostreatus*	Bisphenol A	Chhaya and Gupte (2013), Shin et al. (2007), Chairin et al. (2013), and Hirano et al. 2000
Pharmaceutical active compounds	*Trametes versicolor*	Analgesics and anti-inflammatories (ibuprofen, salicylic acid, acetaminophen, diclofenac, naproxen, phenazone); psychiatric drugs (carbamazepine, venlafaxine, citalopram, diazepam); β-blockers (atenolol, sotalol, propranolol); antibiotics (ciprofloxacin, ofloxacin, sulfamethoxazole, clarithromycin, trimethoprim); antihypertensives (valsartan, irbesartan, losartan, amlodipine); diuretics (furosemide, hydrochlorothiazide, torasemide); H2 antagonist (ranitidine)	Badia-Fabregat et al. (2015)

Table 9.2 Classification of dyes and their possible applications

Dye class	Description	Method	Fibers typically applied to	Typical fixation (%)	Typical pollutants associated with various dyes
Acid	Water-soluble anionic compounds	Exhaust/beck/continuous (carpet)	Wool, nylon	80–93	Color; organic acids; unfixed dyes
Basic	Water soluble, applied in weakly acidic dye baths; very bright dyes	Exhaust/beck	Acrylic, some polyesters	97–98	N/A
Direct	Water-soluble anionic compounds; can be applied directly to cellulosic without mordant (or metal-like chromium and copper)	Exhaust/beck/continuous	Cotton, rayon, other cellulosic	70–95	Color; salt; unfixed dye; cationic fixing agents; surfactant; defoamer; leveling and retarding agents; finish; diluents
Disperse	Not water soluble	High temperature exhaust, continuous	Polyester, acetate, other synthetics	80–92	Color; organic acids; carriers; leveling agents; phosphates; defoamers; lubricants; dispersants; delustrants; diluents
Reactive	Water-soluble anionic compounds; largest dye class	Exhaust/beck cold pad batch/continuous	Cotton, other cellulosic, wool	60–90	Color; salt; alkali; unfixed dye; surfactants; defoamer; diluents; finish
Sulfur	Organic compounds containing sulfur or sodium sulfide	Continuous	Cotton, other cellulosic	60–70	Color; alkali; oxidizing agent; reducing agent; unfixed dye
Vat	Oldest dyes; more chemically complex; water insoluble	Exhaust/package/continuous	Cotton, other cellulosic	80–95	Color; alkali; oxidizing agents; reducing agents

problems, high installation, as well as operating costs (Sarioglu et al. 2007). Therefore, as an efficient alternative, biological processes have received increasing attention due to their cost-effectiveness, ability to produce less sludge, and maintain environmental safety (Banat et al. 1996).

Several fungi are well documented for aerobic degradation of organic contaminants including azo dyes. Among all fungi, white rot fungi produce lignocellulolytic enzymes including laccase, Mn peroxidase, and lignin peroxidase, which are mainly

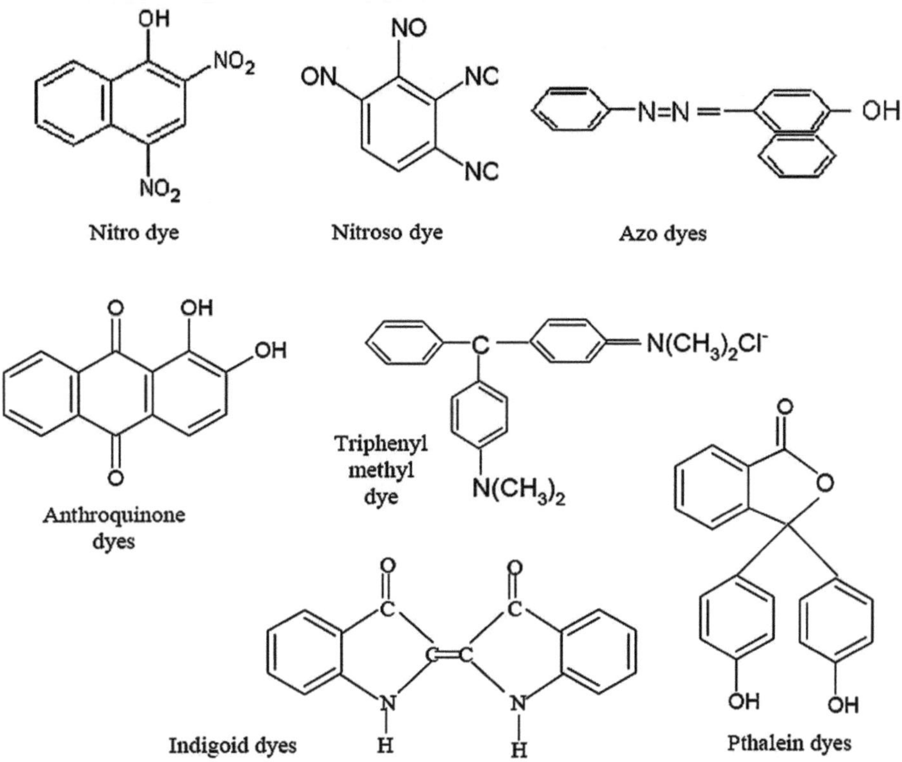

Fig. 9.1 Classification of dyes based on the chemical composition

involved in degradation of lignin that are present in lignocellulosic substrates. Ligninolytic system is directly involved in the degradation of dyes. The ability of white rot fungi is to degrade dye can be correlated with its ability to degrade lignin as the dye molecules have the structural similarity of phenolic ring with lignin. Thus, utilization of white rot fungi is the most inimitable technology of bioremediation due to their ability to degrade structurally diverse xenobiotic and textile dyes (Christian et al. 2005). *P. chrysosporium* is able to decolorize several industrial dyes and polymeric dyes. *P. chrysosporium* cultures, extracellular fluid, and purified lignin peroxidase were able to degrade crystal violet and six other triphenylmethane dyes by sequential N-demethylation. Azo dyes Orange II, Tropaeolin O, and Congo red and the heterocyclic dye Azure B were decolorized by cultures of *P. chrysosporium*. Most studies on the reactive dye biodegradation by fungi have been focused on lignin degrading white rot fungi. Swamy and Ramsay (1999) reported the role of different white rot spp. like *Bjerkandera* sp., *P. chrysosporium*, and *T. versicolor* in decolorization of reactive dyes.

Various fungal species such as *T. versicolor* (Erkurt et al. 2007), *Pleurotus ostreatus* (Faraco et al. 2009), *T. trogii*, and *T. villosa* (Levin et al. 2010) are known to degrade and decolorize various synthetic dyes. In 2006, Vijaykumar et al. isolated

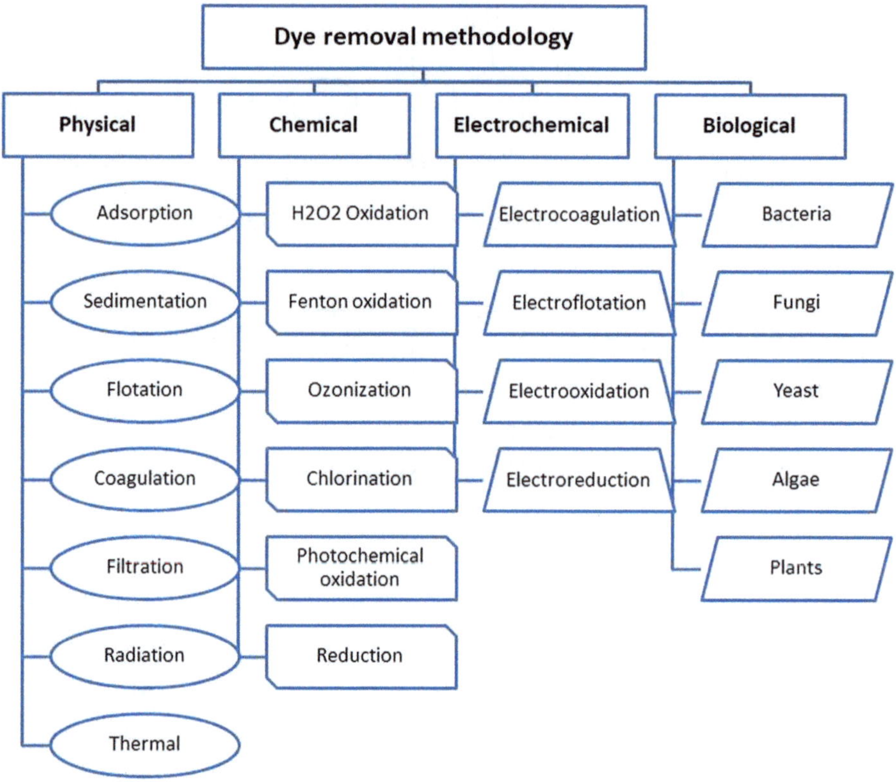

Fig. 9.2 Removal of different dyes by physical, chemical, electrochemical, and biological methodologies

Cladosporium cladosporioides from coal sample, which is utilized for degradation and decolorization of five different azo and triphenylmethane dyes such as acid black 210, acid blue 193, reactive black B(S), crystal violet, and reactive black BL/LPR. Capabilities of *P. chrysosporium* and *P. ostreatus* of free and immobilized laccase mixtures from *P. ostreatus* on industrial dye wastewaters have been demonstrated by Faraco et al. (2009). Wastewater model containing various dyes with complex trisazo, polyazo, and tilbene structures was degraded by *P. chrysosporium* and, acid wastewater model containing acid dyes was decolorized by *P. ostreatus*. Fungal-based cleaning systems have been suffering from drawback of adsorption; thus, in order to overcome this limitation, the cell-free enzyme extracts obtained from fungal cultures have been also used to treat the textile dyes by Chander and Arora (2007).

(b) *Polycyclic Aromatic Hydrocarbons (PAHs)*

In this era of industrialization, the problem faced is the contamination of soil, air, and water by chemicals having disastrous effect on human health and environment.

With the wide use of pesticides in agriculture, the soil and ground water are being polluted with organic compounds. Use of fungi to transform variety of harmful chemicals to non-toxic forms has been a topic of interest nowadays, known as bioremediation (Riva 2006). Laccase secreted from various white rot fungi has a wide range of application in degradation of this organic contaminant which has advantage over the various physico-chemical methods because of its cost, efficiency, and environmental benignity. It has also been reported that immobilized laccase is also useful in removing phenolic and chlorinated phenolic pollutants. Laccase from *Trametes versicolor* and *Pleurotus ostreatus* was used for degradation of PCBs and phenol (Keum and Li 2004). It was observed that the chlorination increased and so the degradation rate decreased. The 3-hydroxy biphenyl was much sturdier to laccase than 2- or 4-hydroxy. When glucose and fructose were added as co-substrates, the degradation rate increased (Desai and Nityanad 2011). In total, 71% of p-hydroxy benzoic acid and 56% of protocatechuic acid were degraded. *Trametes versicolor* is able to oxidize 3–5 ring PAHs with use of mediators like ABTS and HBT. Mixture of xenobiotic compounds like PCP, 2-chlorophenol, 2,4-dichlorophenol, and 2,4,6-trichlorphenol was degraded using laccase from *Trametes pubescens* (Gaitan et al. 2011). Zhao et al. (2010) showed the degradation of dichlorodiphenyltrichloroethane (DDT) in soil.

PAHs are an important environment pollutant, which are universally distributed, having fused aromatic rings (Ukiwe et al. 2013). PAHs are formed due to partial combustion of organic matter containing carbon and hydrogen. PAHs are toxic, carcinogenic, and mutagenic, and so their presence in atmosphere is of great concern and has harmful effect on human health. So, the US Environmental Protection Agency has listed 16 PAHs as "consent decree" priority pollutants. According to US EPA, these priority pollutants need to be monitored regularly to avoid human exposure to contaminated sites. Bioremediation is technology that uses the microbes (typically, heterotrophic, bacteria, and fungi) to reduce or transform hazardous contaminants to materials such as CO_2, water, inorganic salts, microbial biomass, and other by-products that may be less hazardous than the parent materials. Biodegradation of polycyclic aromatic hydrocarbon (PAHs) has been studied in bacteria (Hamamura et al. 2013), fungi (Cerniglia and Sutherland 2010), or algae (Muñoz et al. 2003).

Microbes have the capability to degrade basic compound and substituted PAHs in various environments under aerobic/anaerobic condition (Yang et al. 1978). Three different mechanisms are studied in the aerobic metabolism of PAHs by bacteria and fungi (Ligninolytic & non-Ligninolytic). Basidiomycetes are a group of organisms that broadly mineralize the recalcitrant PAHs because of their ability to produce ligninolytic enzymes (laccase, peroxidase). Anaerobic metabolism of PAHs occurs via hydrogenation of the aromatic ring. White rot fungi have the ability to withstand toxic levels as compared to bacteria. Two types of fungal metabolism of PAHs are studied, which is mediated by ligninolytic (white rot fungi) and non-ligninolytic fungi.

Non-ligninolytic Fungi Mechanism of PAHs metabolism by non-ligninolytic fungi involves the oxidation of aromatic ring using Cytochrome P_{450} monooxygenase enzyme to yield arene oxide. Similarly, PAHs are metabolised by mammals by the oxidation of aromatic ring using dioxygenase enzymes to form cis-dihydrodiols, wherein the monooxygenase incorporates only one oxygen atom onto the ring to form an arene oxide (Sutherland et al. 1995). The reaction is mediated by epoxide hydrolase to form a trans-dihydrodiol. Most of the non-ligninolytic fungi do not have the capability to mineralize PAHs completely. These PAH conjugates are less toxic and more soluble than their parent compounds. The non-ligninolytic fungi such as *Chrysosporium pannorum*, *Cunninghamella elegans*, and *Aspergillus niger* use Cytochrome P_{450} monooxygenase for oxidative degradation of PAH. An example of oxidation of pyrene by *Aspergillus niger* is shown in (Fig. 9.3). Fungal metabolites produced after degradation are often more water soluble and chemically reactive than parent PAH, thus increasing their potential for mineralization by indigenous soil bacteria.

Ligninolytic Fungi Ligninolytic fungi are commonly known as white rot fungi as they cause white rot of woods. White rot fungi have been known for PAH degrading abilities when grown under ligninolytic and non-ligninolytic culture condition. There are two types of ligninolytic enzymes, peroxides and laccase. White rot fungi produce ligninolytic enzymes which are involved in oxidation of lignin present in wood and various other organic matters. The genera of *Phanerochaete*, *Trametes*, *Bjerkandera*, and *Pleurotus* have shown significant potential for the degradation of PAH. The three main LMEs are lignin peroxidase, Mn-dependent peroxidase, and laccase. The enzymes laccase, MnP, and LiP are secreted extracellularly and oxidize the organic matters via non-specific radical-based reaction. Ligninolytic fungi have application in biodegradation of PAH to CO_2 and O_2. *Phanerochaete chrysosporium* has been generally used as model organism to study potential for degradation of PAH. Initial attack of *Phanerochaete chrysosporium* under culture condition that favors ligninolysis is catalyzed by LiP and MnP. Some white rot fungi metabolize PAH to their quinones. White rot fungi have the ability to produce metabolites with higher water solubility and chemical reactivity than the parent PAH. Mono- or dihydroxylated PAH metabolites could be used as substrates for bacteria, and further their degradation in soil is caused by indigenous microflora. Manganese peroxidase is thought to work through a similar mechanism of one-electron oxidation with diffusible oxidizing agent Mn^{+3}- chelate, which is a weak oxidant (Böhmer et al. 1998). It was incapable of oxidizing PAHs with ionization potentials equal to or greater than chrysene (5–7.8 eV). The system mediated by the peroxidation of unsaturated lipids by MnP and Mn^{+3} was studied to catalyze the oxidative cleavage of phenanthrene to diphenic acid. Laccase has the ability to catalyze one-electron oxidation of PAHs such as anthracene and benzo(a) pyrene both having ionization potentials 57.55 eV. Collins et al. (1996) reported degradation of anthracene and benzo(a) pyrene using *Trametes versicolor* by crude as well as two purified isoenzymes. Level of oxidation of anthracene was higher by two purified laccases with ABTS. It

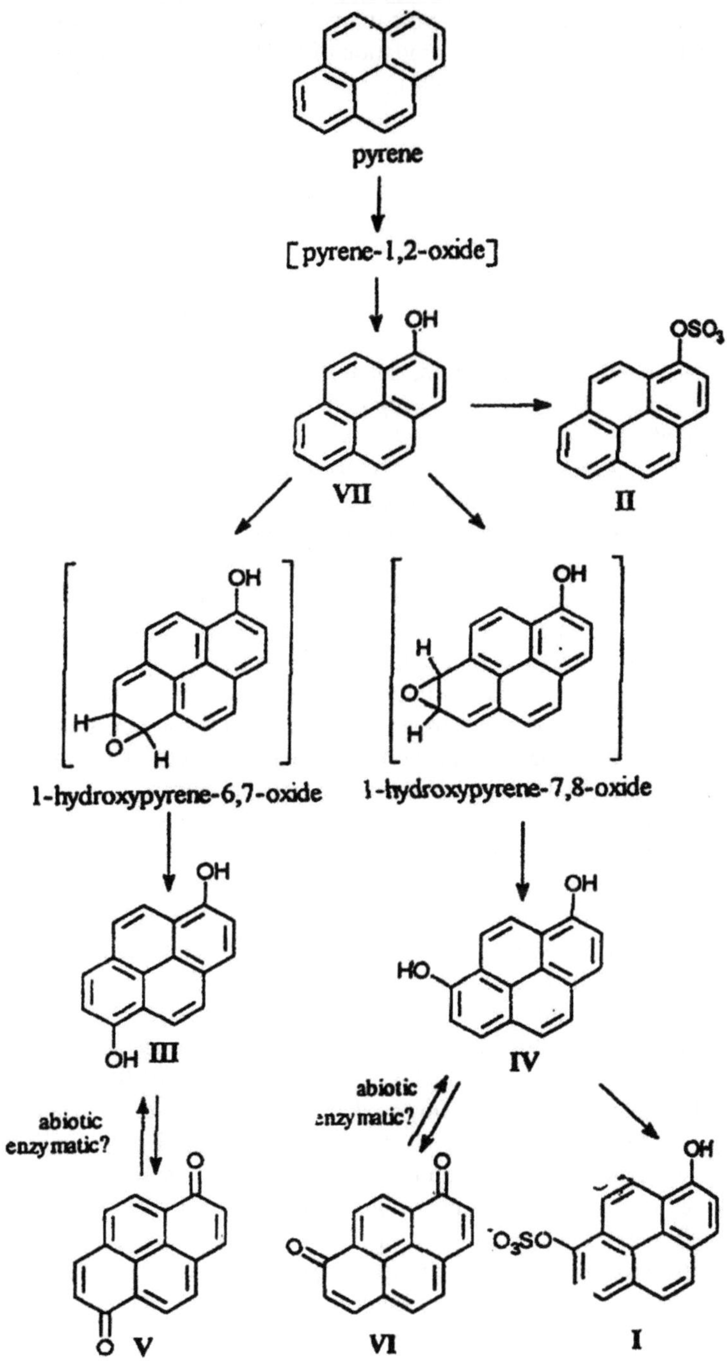

Fig. 9.3 Pathway for the degradation of pyrene by *Aspergillus niger* SK 9317 (Wunder et al. 1994)

has been verified that such laccase/mediating substrate couples oxidize PAHs and that the IP threshold value for the oxidation of PAHs by laccase appears to be similar to that of LiP (Böhmer et al. 1998).

(c) *Metals*

The natural aquatic reservoirs contaminated with industrial effluent of heavy metals such as Pb, Cr, Hg, U, Se, Zn, As, Cd, Ni, etc. pose a serious threat to human health and ecological systems (Ahalya et al. 2005; Xuejiang et al. 2006; Rudakiya and Pawar 2013; Sharaf and Alharbi 2013). Conventional physico-chemical methods such as chemical precipitation, solvent precipitation, reverse osmosis, lime coagulation, oxidation-reduction, evaporation, incineration, and electrodialysis have been applied earlier to remediate heavy metals from the wastewater (Alhuwalia and Goyal 2007; Zahoor and Rehman 2009). All of these methodologies have several drawbacks like generation of toxic sludge, recurring expenses, incomplete removal, high reagent, and energy requirements, and disposal of waste after treatment is the major problem (Bishnoi et al. 2004; Pandey et al. 2013). Therefore, attempts have been made for bioremediation of heavy metals using microbial systems as an attractive method because such techniques are cost-effective and environmentally compatible (Yin et al. 2008).

Microorganisms such as bacteria, fungi, algae, and yeast are considered as potential alternative to the traditional treatment technologies for remediation of heavy metals (Gadd 2010). Usually, sequestering of the heavy metals from the solution by living or dead microbial biomasses through a passive mode (metabolism independent) is described as biosorption, while the removal of metals via an active transport mode (metabolically mediated system) is described as bioaccumulation. Fungi, in particular under submerged conditions, have been widely used for the removal of heavy metals. They show versatility for growth towards the different pH, temperature, nutrient availability, and high concentrations of heavy metals (Gadd 1987). The applicability of the fungi for removal of heavy metals has some advantages due to high percentage of cell wall materials (Lacina et al. 2003).

Fungal cell wall contains a large quantity of polysaccharides and proteins, which contain many functional groups such as carboxyl, amine, hydroxyl, and phosphate groups, which are known to be involved in the metal chelation (Baldrian 2003; Couto et al. 2004). Moreover, most of the microorganisms produce extracellular polymeric substances (EPS) in natural environment for attachment, tolerance, and reduction of heavy metals. Anionic property of EPS imparts the electrostatic interactions with heavy metals (Guibaud et al. 2005). The acidic property of EPS is mainly contributed by several functional groups such as hydroxyl, carboxyl, uronic acids, amino acids, phenolic, and sulfhydryl groups, which interact with positively charged metal ions (Gutnick and Bach 2000).

A. niger has been found to be capable of removing heavy metals like Pb, Cd, and Cu (Kapoor et al. 1999). *Agaricus macrosporus* efficiently extract Cd, Hg, and Cu from contaminated wastes. Huang et al. (1989) also studied removal of Cd using

Table 9.3 Metal removal efficiency by different fungi and their mechanisms

Metal	Fungi	Removal	Mechanism	References
Cu, Cr, Cd, Zn, Ni	*Beauveria bassiana*	74.10% (Cu), 61% (Cr), 63.4% (Cd), 67.8% (Zn), 75% (Ni)	Sorption and accumulation process	Gola et al. (2016)
As, Cd, Cr	*Phanerochaete chrysosporium*	98% (As), 29% (Cd), 56% (Cr)	Biosorption	Shah et al. (2018)
Cu, Zn, Cd, Pb, Cr	*Penicillium simplicissimum*	64% (Cu), 31% (Zn), 32% (Cd), 71% (Pb), 92% (Cr)	Biosorption	Chen et al. (2019)
Cr	*Aspergillus* sp.	98.96%	Sorption and accumulation	Chakraborty et al. (2018)
Au	*Aspergillus niger*	95%	Sorption	Kapoor and Viraraghavan (1995)
Hg	*Penicillium chrysogenum*	43%	Sorption	Nemec et al. (1977)
U	*Geotrichum* sp. dwc-1	96%	Sorption	Zhao et al. 2016

nine different species of fungi both in batch and continuous reactors. *P. chrysosporium* was also utilized for completive sorption of Cu, Cd, and Pb with an initial concentration of 100 mg/l and pH 6.0 (Say et al. 2001). Uptake efficiency of *Saccharomyces cerevisiae* and *Penicillium chrysogenum* in ternary system is 50% as compared to single metal sorption, which is due to the presence of different co-cations (Bakkaloglu et al. 1998). Table 9.3 shows the cation metal removal efficiency by different fungi with their mechanisms. Majority of research is carried out on the removal of metal cations; however, very few studies show the removal of anion metals. Co-culture of *Fusarium solani* and *Trichoderma polysporum* efficiently removed tetracyanonickelate [$K_2Ni(CN)_4$] at pH 4.0. Likewise, *Fusarium oxysporum*, *Scytalidium thermophilum*, and *Penicillium miczynski* could remove 50% of hexacyanoferrate [$K_4Fe(CN)_6$] at pH 4.0. To remove this metal, pH, Fe(III)–cyanide complex anions, and sorbent concentration play a crucial role in binding capacity to *Rhizopus arrhizus* (Aksu and Balibek 2007).

(d)*Pesticides*

Pesticide pollution is one of the major concerns regarding contamination of the environment. Pesticides are a diverse group of chemicals that have inorganic as well as organic groups. They mainly contain insecticides, nematicides, herbicides, fungicides, and soil fumigants (Verma et al. 2014). Pesticides are classified into two major groups: degradable pesticides and non-degradable pesticides. Degradable pesticides are herbicides (Treflan, Dalapon, Dacthal, Paraquat, etc.), insecticides (Sevin, Malathion, Methoxychlor, etc.), and fungicides (Zineb, Captan, Mancozeb, Benlate, etc.), while non-degradable pesticides are insecticides (chlordane, DDT, aldrin, dieldrin, etc.), herbicides (turbacil, simazine, tordon, etc.), and fungicides (PMAS, Calo-Clor, Cd compounds, etc.) (Vargas 1975).

Frequent and intense use of the pesticides leads to the accumulation of pesticides into the soil and possibly transfers into different water bodies. For instance, metribuzin is regularly used as it is an easily water-soluble compound (Hernandez et al. 2004) and linuron type of pesticides is non-degradable in soil (Guzzella et al. 2006). These pesticides inhibit electron transport at photosystem II receptor site, which can regulate the weeds in several crops. Higher usage of these pesticides leads to the accumulation of these chemicals in different water bodies, which makes groundwater contaminated (Hernandez et al. 2004; Guzzella et al. 2006). According to the World Health Organization (WHO), the toxicity of pesticides is divided into four major categories based on the LD_{50} values: (a) extremely hazardous (Type IA); (b) highly hazardous (Type IB); (c) moderately hazardous (Type II); and (d) slightly hazardous (Type III). In India, 4 pesticides belong to Class IA, 15 pesticides belong to class IB, and 76 pesticides belong to class II, which constituted 40% of the total registered pesticides (Verma et al. 2014).

Due to the highest toxicity, several bacterial strains and their co-culture were extensively used to degrade the different types of pesticides. However, very less data are available on the degradation of pesticides as compared to the bacteria. In case of fungi, pH, soil type, fungal biomass, moisture, organic matter, and aeration are chief factors, which can affect the degradation of pesticides. Generally, pesticide degradation occurs through the Krebs cycle. In fungi, the transformation of pesticides takes place through cometabolism. Biochemical reactions in the fungal degradation of pesticides are alkylation, dealkylation, amide or ester hydrolysis, dehalogenation, dehydrogenation, hydroxylation, ether cleavage, ring cleavage, oxidation, reduction, condensation, and conjugate formation (Bollag 1974).

Mucor alternans (Anderson and Lichtenstein 1971), *Fusarium oxysporum* (Engst and Kujawa 1968), and *Trichoderma viride* (Matsumura and Boush 1968) are known to degrade DDT. Similarly, various soil fungi are known for the degradation of endosulfan. Endosulfan sulfate is produced as a major metabolite due to endosulfan degradation by *P. chrysosporium* BU-1 (Kullman and Matsumura 1996) (Fig. 9.4). Initially, endosulfan is either oxidized to endosulfan sulfate, a terminal end product, or hydrolyzed to non-sulfur-containing endosulfan metabolites. Initial hydrolysis of endosulfan produces endosulfan diol, which is further oxidized to endosulfan hydroxyether followed by the formation of either endosulfan lactone or the putative endosulfan dialdehyde (Martens 1976). Fungi also degrade the organophosphorus pesticides, wherein *Aspergillus oryzae* transformed malathion to β-monoacid and dicarboxylic acid (Lewis et al. 1975).

(e) *Dioxins*

Dioxins and furans are one of the most hazardous compounds, which have higher toxicity (Fig. 9.5). Dioxin compounds are the major by-products of various industrial processes such as pesticide preparation, paper and pulp bleaching, waste incineration, etc. In addition, these compounds are also obtained from the impurities of

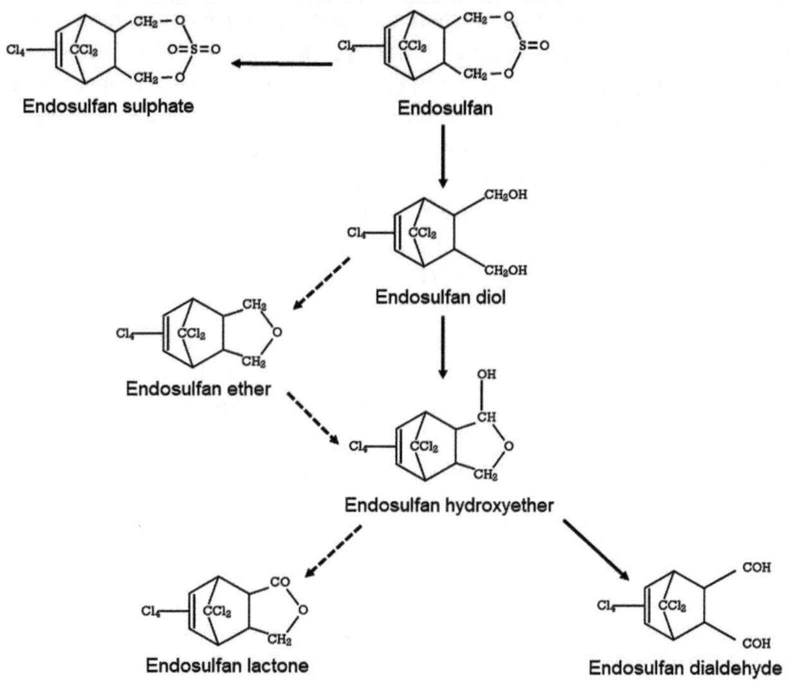

Fig. 9.4 Pathways for the degradation of endosulfan by *Phanerochaete chrysosporium* BU-1. Solid arrows show the major metabolic pathways; dashed arrows depict the minor metabolic pathways (Adapted from Kullman and Matsumura 1996)

Fig. 9.5 Chemical structures of dioxins and furans

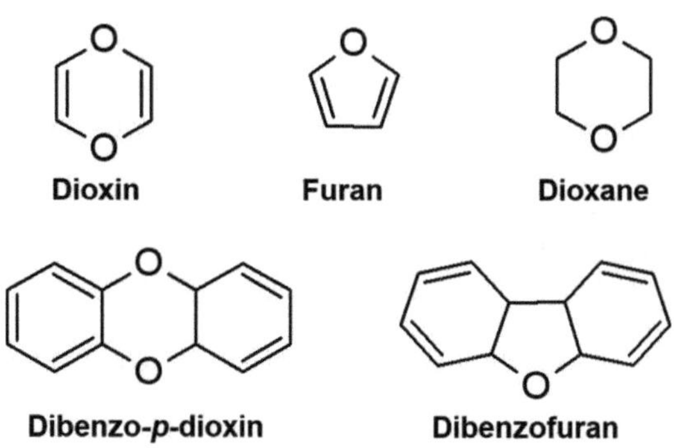

chlorophenols, e.g., PCP, tetrachlorophenol, and trichlorophenol, which are generally utilized as a wood preservative (Anasonye et al. 2014). Dioxins and furans are highly stable compounds, which are lipophilic in nature. Nevertheless, fungi have great potential to degrade dioxins as compared to other microorganisms. In filamentous fungi, dibenzofurans have been hydrolyzed using the monooxygenase system (Hammer and Schauer 1997). Eight fungal strains can be used for the degradation of dibenzofuran, which is to be converted into some intermediates with hydrophilic characteristics. Among them, *Trichosporon mucoides* converts six major metabolites by transforming dibenzofuran (Hammer et al. 1998). *Cordyceps sinensis* is a cyclic ether-degrading fungus that has been shown a new degradation pathway for dioxins (Nakamiya et al. 2005). Dibenzo-p-dioxin degradation results into some catechol products, which are further metabolized to cis-muconates. Transformation of polychlorinated dibenzo-p-dioxins, such as 2,7-dichloro-, 2,3,7-trichloro-, and 1,2,6,7- and 1,2,8,9-tetrachlorodibenzo-p-dioxins, to hydroxylated and methoxylated compounds has been carried out by *Phlebia lindtneri* (Kamei and Kondo 2005).

9.3 Conclusion

The persistence of various hazardous contaminants such as dyes, PAHs, heavy metals, and other pollutants in the environment directs to an expedition for novel methodologies to remove them. Fungi possess different types of strategies to remove the contaminants. In this chapter, bioremediation of dyes, PAHs, heavy metals, pesticides, and dioxin by fungi has been highlighted, and thus they have been successfully utilized in the last few decades. Recently, data is available regarding the degradation of contaminants using different fungi and their strategies to remove them. The large-scale development of these processes is available; however, it is still in early stages. To explore the single or multiple compound degradation, some combinations should be used such as fungal inoculum, environmental condition, and choice of pollutants. Additionally, dead fungal biomass should be used to establish an alternative for ion-exchange resins. The role of fungi in biosorption of pollutants can lead to the progress of new technology for the treatment of wastewater.

References

Abadulla E, Tzanov T, Costa S, Robra KH, Cavaco-Paulo A, Gubitz GM (2000) Decolorization and detoxification of textile dyes with a laccase from *Trametes hirsuta*. Appl Environ Microbiol 66:3357–3362

Ahalya N, Kanamadi RD, Ramachandra TV (2005) Biosorption of chromium (VI) from aqueous solutions by the husk of Bengal gram (*Cicer arientinum*). Electron J Biotechnol 8:258–264

Aksu Z, Balibek E (2007) Chromium (VI) biosorption by dried *Rhizopus arrhizus*: effect of salt (NaCl) concentration on equilibrium and kinetic parameters. J Hazard Mater 145:210–220

Alhuwalia SS, Goyal D (2007) Microbial and plant derived biomass for removal of heavy metals from wastewater. Bioresour Technol 98:2243–2257

Anasonye F, Winquist E, Kluczek-Turpeinen B, Räsänen M, Salonen K, Steffen KT, Tuomela M (2014) Fungal enzyme production and biodegradation of polychlorinated dibenzo-p-dioxins and dibenzofurans in contaminated sawmill soil. Chemosphere 110:85–90

Anderson JPE, Lichtenstein EP (1971) Effect of nutritional factors on DDT-degradation by. Can J Microbiol 17(10):1291–1298

Badia-Fabregat M, Lucas D, Gros M, Rodríguez-Mozaz S, Barceló D, Caminal G, Vicent T (2015) Identification of some factors affecting pharmaceutical active compounds (PhACs) removal in real wastewater.Case study of fungal treatment of reverse osmosis concentrate. J Hazard Mater 283:663–671

Bakkaloglu I, Butter TJ, Evison LM, Holland FS, Hancock IC (1998) Screening of various types biomass for removal and recovery of heavy metals (Zn, Cu, Ni) by biosorption, sedimentation and desorption. Water Sci Technol 38:269–277

Baldrian P (2003) Interactions of heavy metals with white-rot fungi. Enzym Microb Technol 32:78–91

Banat IM, Nigam P, Singh D, Marchant R (1996) Microbial decolorization of textile-dyecontaining effluents: a review. Bioresour Technol 58(3):217–227

Bhalerao TS, Puranik PR (2007) Biodegradation of organochlorine pesticide, endosulfan, by a fungal soil isolate, *Aspergillus niger*. Int Biodeterior Biodegradation *59*(4):315–321

Bharath Y, Singh SN, Keerthiga G, Prabhakar R (2019) Mycoremediation of contaminated soil in MSW sites. In: Waste management and resource efficiency. Springer, Singapore, pp 321–329

Bishnoi NR, Bajaj M, Sharma N, Gupta A (2004) Adsorption of Cr(VI) on activated rice husk carbon and activated alumina. Bioresour Technol 91:305–307

Böhmer S, Messner K, Srebotnik E (1998) Oxidation of Phenanthrene by a fungal laccase in the presence of 1-Hydroxybenzotriazole and unsaturated lipids. Biochem Biophys Res Commun 244(1):233–238

Bollag JM (1974) Microbial transformation of pesticides. Adv Appl Microbiol 18:75–130

Bouwer EJ (2017) Bioremediation of chlorinated solvents using alternate electron acceptors. In: Handbook of bioremediation (1993). CRC Press, Boca Raton, pp 149–176

Campos R, Kandelbauer A, Robra KH, Cavaco-Paulo A, Gubitz GM (2001) Indigo degradation with purified laccases from Trametes hirsuta and *Sclerotium rolfsii*. J Biotechnol 89:131–139

Cerniglia CE, Sutherland JB (2010) Degradation of polycyclic aromatic hydrocarbons by fungi. In: Handbook of hydrocarbon and lipid microbiology, pp 2079–2110

Chairin T, Nitheranont T, Watanabe A, Asada Y, Khanongnuch C, Lumyong S (2013) Biodegradation of bisphenol A and decolorization of synthetic dyes by laccase from white-rot fungus, Trametes polyzona. Appl Biochem Biotechnol 169:539–545

Chakraborty V, Sengupta S, Chaudhuri P, Das P (2018) Assessment on removal efficiency of chromium by the isolated manglicolous fungi from Indian Sundarban mangrove forest: Removal and optimization using response surface methodology. Environ Technol Innov 10:335–344

Chander M, Arora DS (2007) Evaluation of some white-rot fungi for their potential to decolourise industrial dyes. Dyes Pigments 72(2):192–198

Chen SH, Cheow YL, Ng SL, Ting ASY (2019) Mechanisms for metal removal established via electron microscopy and spectroscopy: a case study on metal tolerant fungi *Penicillium simplicissimum*. J Hazard Mater 362:394–402

Cheng M, Zeng G, Huang D, Yang C, Lai C, Zhang C, Liu Y (2018) Tween 80 surfactant-enhanced bioremediation: toward a solution to the soil contamination by hydrophobic organic compounds. Crit Rev Biotechnol 38(1):17–30

Chhaya U, Gupte A (2013) Possible role of laccase from *Fusarium incarnatum* UC-14 in bioremediation of Bisphenol A using reverse micelles system. J Hazard Mater 254:149–156

Christian V, Shrivastava R, Shukla D, Modi HA, Vyas BRM (2005) Degradation of xenobiotic compounds by lignin-degrading white-rot fungi: enzymology and mechanisms involved. Indian J Exp Biol 45:301–312

Collins PJ, Kotterman M, Field JA, Dobson A (1996) Oxidation of anthracene and benzo [a] pyrene by laccases from Trametes versicolor. Appl Environ Microbiol 62(12):4563–4567

Couto SR, Sanroman MA, Hoefer D, Gubitz GM (2004) Stainless steel: A novel career for the immobilization of the white rot fungus *Trametes hirsute* for decolorization of textile dyes. Bioresour Technol 95:67–72

de Lorenzo V (2018) Biodegradation and bioremediation: an introduction: consequences of microbial interactions with hydrocarbons, oils, and lipids. Biodegr Bioremed:1–21

Denizli A, Cihangir N, Tüzmen N, Alsancak G (2005) Removal of chlorophenols from aquatic systems using the dried and dead fungus *Pleurotus sajorcaju*. Bioresour Technol 96:59–62

Desai SS, Nityanand C (2011) Microbial laccases and their applications: a review. Asian J Biotechnol 3(2):98–124

Engst R, Kujawa M (1968) Enzymatischer Abbau des DDT durch Schimmelpilze. 3. Mitt. Darstellung des 2,2-Bis(p-chlorphenyl)acetaldehyds (DDHO) und seine Bedeutung im Abbaucyclus. Food Nahrung 12(8):783–785

Erkurt EA, Ünyayar A, Kumbur H (2007) Decolorization of synthetic dyes by white rot fungi, involving laccase enzyme in the process. Process Biochem 42(10):1429–1435

Faraco V, Pezzella C, Miele A, Giardina P, Sannia G (2009) Bio-remediation of colored industrial wastewaters by the white-rot fungi Phanerochaete chrysosporium and Pleurotus ostreatus and their enzymes. Biodegradation 20(2):209–220

Fu Y, Viraraghavan T (2002) Removal of Congo red from an aqueous solution by fungus Aspergillus Niger. Adv Environ Res 7(1):239–247

Gadd GM (1987) Fungal response towards heavy metals. In: Hebert RA, Codd GA (eds) Microbes in extreme environments. Academic, London, pp 84–109

Gadd GM (2010) Metals, minerals and microbes: Geomicrobiology and bioremediation. Microbiology 156:609–643

Gahlout M, Rudakiya DM, Gupte S, Gupte A (2017) Laccase-conjugated amino-functionalized nanosilica for efficient degradation of Reactive Violet 1 dye. Int Nano Lett 7(3):195–208

Gaitan IJ, Medina SC, González JC, Rodríguez A, Espejo ÁJ, Osma JF, Sarria V, Alméciga-Díaz CJ, Sánchez OF (2011) Evaluation of toxicity and degradation of a chlorophenol mixture by the laccase produced by Trametes pubescens. Bioresour Technol 102(3):3632–3635

Gola D, Dey P, Bhattacharya A, Mishra A, Malik A, Namburath M, Ahammad SZ (2016) Multiple heavy metal removal using an entomopathogenic fungi Beauveriabassiana. Bioresour Technol 218:388–396

Guibaud G, Comte S, Bordas F, Dupuy S, Baudu M (2005) Comparison of the complexation potential of extracellular polymeric substances (EPS), extracted from activated sludges and produced by pure bacteria strains, for cadmium, lead and nickel. Chemosphere 59:629–638

Gutnick DL, Bach H (2000) Engineering bacterial biopolymers for the biosorption of heavy metals; new products and novel formulation. Appl Microbiol Biotechnol 54:451–460

Guzzella L, Pozzoni F, Giuliano G (2006) Herbicide contamination of surficial groundwater in Northern Italy. Environ Pollut 142(2):344–353

Hai FI, Yamamoto K, Nakajima F, Fukushi K, Nghiem LD, Price WE, Jin B (2013) Degradation of azo dye acid orange 7 in a membrane bioreactor by pellets and attached growth of Coriolusversicolour. Bioresour Technol 141:29–34

Hammel KE, Kalyanaraman B, Kirk TK (1986) Oxidation of polycyclic aromatic hydrocarbons and dibenzo[p]-dioxins by *Phanerochaete chrysosporium* ligninase. J Biol Chem 261:16948–16952

Hammer E, Schauer F (1997) Fungal hydroxylation of dibenzofuran. Mycol Res 101:433–436

Hammer E, Krowas D, Schafer A, Specht M, Francke W, Schauer F (1998) Isolation and characterization of dibenzofuran-degrading yeast: identification of oxidation and ring cleavage products. Appl Environ Microbiol 64:2215–2219

Hamamura N, Ward DM, Inskeep WP (2013) Effects of petroleum mixture types on soil bacterial population dynamics associated with the biodegradation of hydrocarbons in soil environments. FEMS Microbiol Ecol 85(1):168–178

Hernández-Romero AH, Tovilla-Hernández C, Malo EA, Bello-Mendoza R (2004) Water quality and presence of pesticides in a tropical coastal wetland in southern Mexico. Marine Pollut Bull 48:1130–1141

Hirano T, Honda Y, Watanabe T, Kuwahara M (2000) Degradation of bisphenol A by the lignin-degrading enzyme, manganese peroxidase, produced by the white-rot basidiomycete, Pleurotus ostreatus. Biosci Biotechnol Biochem 64:1958–1962

Huang C, Huang CP, Morehart AL, Westman DC (1989) Removal of toxic heavy metals from contaminated groundwater by a fungal adsorption process, Final Technical Report. Contract 14-08-0001-G1292, U.S. Geological Survey, Washington, DC. 121 pp

Javaid A, Bajwa R, Javaid A (2010) Biosorption of heavy metals using a dead macro fungus *Schizophyllum commune* fries: evaluation of equilibrium and kinetic models. Pak J Bot *42*(3):2105–2118

Johanne C, Majcherczyk A (2000) Natural mediators in the oxidation of polycyclic aromatic hydrocarbons by laccase mediator systems. Appl Environ Microbiol 66:524–528

Kadri T, Rouissi T, Brar SK, Cledon M, Sarma S, Verma M (2017) Biodegradation of polycyclic aromatic hydrocarbons (PAHs) by fungal enzymes: a review. J Environ Sci 51:52–74

Kamei I, Kondo R (2005) Biotransformation of dichloro-, trichloro-, and tetrachlorodibenzo-*p*-dioxin by the white-rot fungus *Phlebia lindtneri*. Appl Microbiol Biotechnol 68:560–566

Kapoor A, Viraraghavan T (1995) Fungal biosorption — an alternative treatment option for heavy metal bearing wastewaters: a review. Bioresour Technol 53(3):195–206

Kapoor A, Viraraghavan T, Cullimore DR (1999) Removal of heavy metals using the fungus Aspergillus niger. Bioresour Technol 70:95–104

Keum YS, Li QX (2004) Fungal laccase-catalyzed degradation of hydroxy polychlorinated biphenyls. Chemosphere 56(1):23–30

Kotterman MJ, Vis EH, Field JA (1998) Successive mineralization and detoxification of benzo[a] pyrene by the white rot fungus Bjerkandera sp. strain BOS55 and indigenousmicroflora. Appl Environ Microbiol 64:2853–2858

Kullman SW, Matsumura F (1996) Metabolic pathways utilized by *Phanerochaete chrysosporium* for degradation of the cyclodiene pesticide endosulfan. Appl Environ Microbiol 62:593–600

Kumar MN, Ravikumar R, Sankar MK, Thenmozhi S (2018) New insight into the effect of fungal mycelia present in the bio-pretreated paddy straw on their enzymatic saccharification and optimization of process parameters. Bioresour Technol 267:291–302

Lacina C, Germain G, Spiros AN (2003) Utilization of fungi for biotreatment of raw wastewater. Afr J Biotechnol 2:620–630

Levin L, Melignani E, Ramos AM (2010) Effect of nitrogen sources and vitamins on ligninolytic enzyme production by some white-rot fungi. Dye decolorization by selected culture filtrates. Bioresour Technol 101(12):4554–4563

Lewis DL, Paris DF, Baughman GL (1975) Transformation of malathion by fungus, *Aspergillus oryzae* isolated from a freshwater pond. Bull Environ Contam Toxicol 13:596–601

Liu Y, Chen X, Li J, Burda C (2005) Photocatalytic degradation of azo dyes by nitrogen-doped TiO2 nanocatalysts. Chemosphere 61(1):11–18

Lun JH, Hewitt J, Sitabkhan A, Eden JS, Tuipulotu DE, Netzler NE et al (2018) Emerging recombinant noroviruses identified by clinical and waste water screening. Emerg Microbes Infect 7:50

Maceiras R, Rodriguez-Couto S, Sanroman A (2001) Influence of several activators on the extracellular laccase activity and in vivo decolourization of poly R-478 by semi-solid-state cultures of Trametes versicolor. Acta Biotechnol 21:255–264

Manilla-Pérez E, Lange AB, Luftmann H, Robenek H, Steinbüchel A (2011) Neutral lipid production in *Alcanivorax borkumensis* SK2 and other marine hydrocarbonoclastic bacteria. Eur J Lipid Sci Technol 113:8–17

Martens R (1976) Degradation of [8,9-14C]endosulfan by soil microorganisms. Appl Environ Microbiol 31:853–858

Matsumura F, Boush GM (1968) Degradation of insecticides by a soil fungus *Trichoderma viride*. J Econ Entomol 61:610–612

Muñoz R, Guieysse B, Mattiasson B (2003) Phenanthrene biodegradation by an algal-bacterial consortium in two-phase partitioning bioreactors. Appl Microbiol Biotechnol 61(3):261–267

Nakamiya K, Hashimoto S, Ito H, Edmonds JS, Yasuhara A, Morita M (2005) Degradation of dioxins by cyclic ether degrading fungus, *Cordyceps sinensis*. FEMS Microbiol Lett 248:17–22

Nemec P, Prochazka H, Stamberg K, Katzer J, Stamberg J, Jilek R, Hulak P (1977) U.S. Patent No. 4,021,368. Washington, DC: U.S. Patent and Trademark Office

Nyanhongo GS, Gomes J, Gubitz GM, Zvauya R, Read J, Steiner W (2002) Decolorization of textile dyes by laccases from a newly isolated strain of *Trametes modesta*. Water Res 6:1449–1456

Pandey AK, Jamalludin A, Pandey AK (2013) Biosorption characteristics of *Aspergillus flavus*in removal of Nickel from an aqueous solution. Int J Green Herb Chem 2:47–51

Patel H, Gupte A, Gupte S (2009) Biodegradation of Fluoranthene by Basidiomycetes fungal isolate Pleurotus Ostreatus HP-1. Appl Biochem Biotechnol 157(3):367–376

Patel H, Gupte A, Gupte S (2010) Biodegradation of pyrene by Pleurotus ostreatus HP-1 and its correlation with extracellular ligninolytic enzyme production. Int J Agric Environ Biotechnol 3:313–319

Pothuluri JV, Freeman JP, Evans FE, Cerniglia CE (1992) Fungal metabolism of acenaphthene by *Cunninghamella elegans*. Appl Environ Microbiol 58:3654–3659

Reddy CA, Mathew Z (2001) Bioremediation potential of white rot fungi. Br Mycol Soc Symp Ser 23:52–78

Riva S (2006) Laccases: blue enzymes for green chemistry. Trends Biotechnol 24(5):219–226

Robinson T, McMullan G, Marchant R, Nigam P (2001) Remediation of dyes in textile effluent: a critical review on current treatment technologies with a proposed alternative. Bioresour Technol 77(3):247–255

Rudakiya DM, Gupte A (2017) Degradation of hardwoods by treatment of white rot fungi and its pyrolysis kinetics studies. Int Biodeterior Biodegradation 120:21–35

Rudakiya DM, Gupte A (2019) Assessment of white rot fungus mediated hardwood degradation by FTIR spectroscopy and multivariate analysis. J Microbiol Methods 157:123–130

Rudakiya DM, Pawar K (2013) Evaluation of remediation in heavy metal tolerance and removal by Comamonas acidovorans MTCC 3364. IOSR J Environ Sci Toxicol Food Technol 5(5):26–32

Rudakiya DM, Iyer V, Shah D, Gupte A, Nath K (2018) Biosorption potential of phanerochaete chrysosporium for arsenic, cadmium, and chromium removal from aqueous solutions. Global Chall. https://doi.org/10.1002/gch2.201800064

Sack U, Günther T (1993) Metabolism of PAH by fungi and correlation with extracellular enzymatic activities. J Basic Microbiol 33:269–277

Sarioglu M, Bali U, Bisgin T (2007) The removal of C.I. basic red 46 in a mixed methanogenic anaerobic culture. Dyes Pigments 74(1):223–229

Say R, Denizli A, Arica MY (2001) Biosorption of cadmium(II), lead(II) and copper(II) with the filamentous fungus Phanerochaete chrysosporium. Bioresour Technol 76:67–70

Shah D, Rudakiya DM, Iyer V, Gupte A (2018) Simultaneous removal of hazardous contaminants using polyvinyl alcohol coated *Phanerochaete chrysosporium*. Int J Agric Environ Biotechnol 11:235–241

Sharaf EF, Alharbi E (2013) Removal of heavy metals from waste water of tanning leather industry by fungal species isolated from polluted soil. Afr J Biotechnol 12:4351–4355

Shin E, Choi HT, Song H (2007) Biodegradation of endocrine-disrupting bisphenol A by white rot fungus *Irpex lacteus*. J Microbiol Biotechnol 17:1147

Shrivastava B, Thakur S, Khasa YP, Gupte A, Puniya AK, Kuhad RC (2011) White-rot fungal conversion of wheat straw to energy rich cattle feed. Biodegradation 22:823–831

Sumathi S, Manju BS (2000) Uptake of reactive textile dyes by Aspergillus foetidus. Enzym Microb Technol 27(6):347–355

Sutherland JB, Rafii F, Khan AA, Cerniglia CE (1995) Mechanisms of polycyclic aromatic hydrocarbon degradation. In: Microbial transformation and degradation of toxic organic chemicals. Wiley-Liss, New York, pp 269–306

Swamy J, Ramsay JA (1999) The evaluation of white rot fungi in the decoloration of textile dyes. Enzym Microb Technol 24(3–4):130–137

Thakur S, Gupte A (2015) Optimization and hyper production of laccase from novel agaricomycete *Pseudolagarobasidium acaciicola* AGST3 and its application in in vitro decolorization of dyes. Ann Microbiol 65:185–196

Tsekova K, Todorova D, Ganeva S (2010) Removal of heavy metals from industrial wastewater by free and immobilized cells of *Aspergillus niger*. Int Biodeterior Biodegradation 64:447–451

Ukiwe LN, Egereonu UU, Njoku PC, Nwoko CIA, Allinor JI (2013) Polycyclic aromatic hydrocarbons degradation techniques: a review. Int J Chem 5(4)

van der Zee FP, Villaverde S (2005) Combined anaerobic–aerobic treatment of azo dyes—a short review of bioreactor studies. Water Res 39(8):1425–1440

Vargas JM (1975) Pesticide degradation. J Arboricult 1:232–233

Verma JP, Jaiswal DK, Sagar R (2014) Pesticide relevance and their microbial degradation: a-state-of-art. Rev Environ Sci Biotechnol 13:429–466

Wunder T, Kremer S, Sterner O, Anke H (1994) Metabolism of the polycyclic aromatic hydrocarbon pyrene by Aspergillus Niger SK 9317. Appl Microbiol Biotechnol 42(4):636–641

Xuejiang W, Ling C, Siqing Z, Jianfu JM, Chovelon R, N.J. (2006) Biosorption of Cu (II) and Pb (II) from aqueous solutions by dried activated sludge. Min Eng 19:968–971

Yang SK, Deutsch JOSEPH, Gelboin HV (1978) Benzo (a) pyrene metabolism: activation and detoxification. In: Polycyclic Hydrocarbons and Cancer, vol 1, pp 205–231

Yin H, He B, Peng H, Ye J, Yang F, Zhang N (2008) Removal of Cr(VI) and Ni(II) from aqueous solution by fused yeast: study of cations release and biosorption mechanism. J Hazard Mater 158:568–576

Zahoor A, Rehman A (2009) Isolation of Cr(VI) reducing bacteria from industrial effluents and their potential use in bioremediation of chromium containing wastewater. J Environ Sci 21:814–820

Zhao C, Li X, Ding C, Liao J, Du L, Yang J, Yang Y, Zhang D, Tang J, Liu N, Sun Q (2016) Characterization of uranium bioaccumulation on a fungal isolate Geotrichum sp. dwc-1 as investigated by FTIR, TEM and XPS. J Radioanal Nucl Chem 310(1):165–175

Zhao YC, Yi XY, Zhang M, Liu L, Ma WJ (2010) Fundamental study of degradation of dichlorodiphenyltrichloroethane in soil by laccase from white rot fungi. Int J Environ Sci Technol 7(2):359–366

Exploring Fungi-Associated Lignocellulose Degradation: Secretomic and Proteomic Approaches

10

Akshay Shankar, Shruti Ahlawat, and Krishna Kant Sharma

Abstract

Filamentous fungi of phyla *Basidiomycota* and *Ascomycota* are the group of microorganisms that are capable of secreting a variety of proteins and other secondary metabolites depending on the environment and culture conditions. The protein constitutes the hydrolytic enzymes which cause the deconstruction of the plant cell wall and has applicability in several biotechnological processes including second-generation ethanol production. Secretomic and proteomic analysis of the fungi is an excellent tool to find out the biological mechanisms of lignocellulose degradation. Furthermore, it is also an important tool to search for novel enzymes or metabolites of the biotechnology field. Extracellular secretion of the protein from different fungal species has been studied using different high-throughput techniques such as 2-D PAGE, MALDI-ToF/ToF, LC-MS/MS, iTRAQ technique using LC-MS/MS, and Nano-LC-MS/MS protein mass spectrometry. Bioinformatics tools have equal importance in the prediction and profiling of the expressed proteins, according to the current database. For this reason, publications documenting the fungal secretome and proteome have increased significantly in the past few years. Herein, we have updated the development and evolution of the proteome/secretome technology and its application in the protein profiling and functional genomics of the economically important filamentous fungi.

Keywords

Fungi · Proteome · Secretome · Lignocelluloses · Mass spectrometry · Liquid chromatography

A. Shankar · S. Ahlawat · K. K. Sharma (✉)
Department of Microbiology, Maharshi Dayanand University, Rohtak, Haryana, India
e-mail: kksharma.microbiology@mdurohtak.ac.in

T. Satyanarayana et al. (eds.), *Advancing Frontiers in Mycology & Mycotechnology*,
https://doi.org/10.1007/978-981-13-9349-5_10

10.1 Introduction

Fungi constitute a crucial and distinct component of the soil microbial communities' vis-à-vis soil health. They perform essential functions of an ecosystem like nutrient cycling, organic matter decomposition, and transfer of nutrients from the mycorrhizal species to plants. Earlier reports demonstrate an important role played by both *Ascomycetes* and *Basidiomycetes* in various biogeochemical cycles (Kellner and Vandenbol 2010; Sharma et al. 2013). *Basidiomycetes* are an ecologically most important group of fungi in the terrestrial environments, which are involved in litter components breakdown and in effective degradation of various recalcitrant organic compounds in humic layer and soil litter (Dix and Webster 1995; Steffen et al. 2007). Few fungi are known to cause disease in plants and humans, while others have an evident potential in controlling the insect pests.

Most importantly, fungi are an abundant source of the therapeutic metabolites and industrially essential enzymes. Till date, the proteomic studies performed for *Ascomycetes* exceed those of *Basidiomycetes* (Fig. 10.1c) because of the prevalence

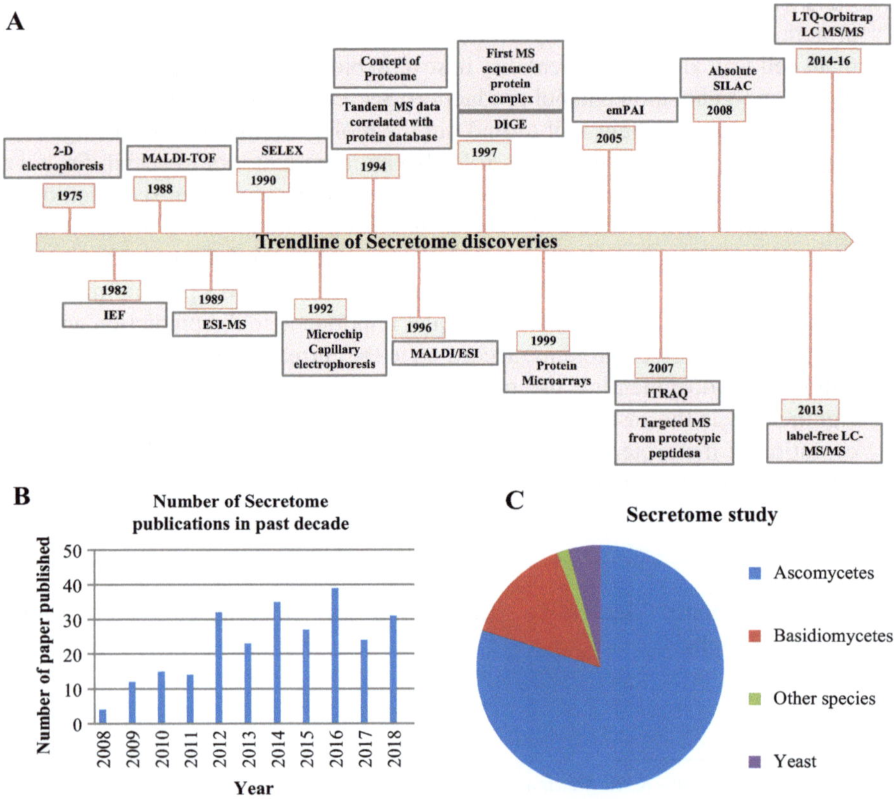

Fig. 10.1 (**a**) Trendline representing secretome discoveries. (**b**) Number of secretome publications in past decade. (**c**) Pie diagram representation of secretome studies conducted on various fungal classes

of the former in causing diseases and since many of them are known as "cell factories" of the enzyme secretion, e.g., *Aspergillus niger* (Adav et al. 2010), *A. flavus* MDU-5, *A. oryzae* MDU-4, *Trichoderma citrinoviride* MDU-1, *T. harzianum* MDU-2, and *T. longibrachiatum* MDU-6 (Chutani and Sharma 2015; Chutani and Sharma 2016; Miura and Ueda 2018).

Various technologies such as the protein mass spectrometry, genome sequencing, transcriptome sequencing (Jain et al. 2019), and bioinformatics advances have now enabled the detailed analysis of the fungal biochemistry (Doyle 2011; de Oliveira et al. 2011; Sharma 2016; Kumar et al. 2017). With time upgraded technologies have also increased which are listed in Fig. 10.1a. Till now, hundreds of fungal genomes are publicly accessible (http://genome.jgi.doe.gov/programs/fungi/index.jsf; http://jgi.doe.gov/our-science/science-programs/fungal-genomics/recent-fungal-genome-eleases/) by 1000 Fungal Genome (1KFG) project (Sharma, 2016). The fungal proteome analysis speed was enhanced due to sequenced genomes availability and the progress in the instrumental resolution of the mass spectrometry. Currently, the major challenges faced in the modern fungal biology are the comprehension of expression, regulation, and function of fungal genomes encoded protein sets. The proteomics is becoming an essential method of functional genomics, because it gives the information at the protein level, which not indeed corresponds to the transcriptomics information (Bianco and Perrotta 2016). In this chapter, the main focus is on the evolution of proteomic techniques, extracellular proteome (secretome) of the filamentous fungi and its involvement in bio-based processes including deconstruction of the lignocellulosic biomass and xenobiotic degradation.

10.2 Fungal Lignocellulolytic Enzyme Secretion

The lignocellulose-degrading enzymes and their biotechnological applications have been discussed previously in several articles (Sharma et al. 2013). *Basidiomycetes* have tremendous potential in the deconstruction of the lignocellulosic materials and biodegradation of the industrial pollutants. Extended range of the biodegradable pollutants can be degraded by the extracellular, intracellular, and other nonspecific enzyme systems of the white rot fungi, which includes versatile peroxidase (VP), lignin peroxidase (LiP), manganese peroxidase (MnP), laccases, and other accessory enzymes (Table 10.1). Protein secretion is necessary for the fungal growth and survival as the fungal species survive by absorbing different available environmental nutrients acted by various secretory hydrolyzing enzymes. Industry exploits these high secretions of the enzyme for various valuable products.

The secretome study can be explained as the collection of proteins released by an organism or a cell under different environmental or nutritional stress. The fungal secretome has two features:

i) the extracellular secretion with intracellular secretion or integral membrane protein. In extracellular secretome, the proteins are secreted into the medium. However, in the intracellular secretome, the secreted proteins remain attached to

Table 10.1 Lignocellulolytic enzymes secreted by different fungal species

Enzyme	Enzyme classification no.	Reaction catalyzed	Applications	References
Laccase (benzenediol: oxygen oxidoreductase)	1.10.3.2	Phenol oxidation	Pathogenesis; formation of rhizomorph and fruiting body; spore resistance; pigment synthesis; degradation of lignin	Mayer and Staples (2002), Claus (2004), Minussi et al. (2007), and Maciel and Ribeiro (2010)
Lignin peroxidase (LiP)	1.11.1.14	Phenol polymerization	Lignin biodegradation; fungal defense against pathogens	Trejo-Hernandez et al. (2001), Piontek et al. (2001), Erden et al. (2009), and Maciel and Ribeiro (2010)
Manganese peroxidase (MnP)	1.11.1.13	Phenol oxidation; oxidize Mn^{2+} to Mn^{3+}	Lignin degradation; interspecific fungal interactions	Trejo-Hernandez et al. (2001), Hofrichter (2002), and Maciel and Ribeiro (2010)
Aryl alcohol oxidase	1.1.3.7	H_2O_2 production	–	Martinez et al. (2009)
Glyoxal oxidase (GLOX)	1.2.3.5	H_2O_2 production	–	Martinez et al. (2009)
Manganese-independent peroxidase	1.11.1.7	Oxidize aromatic substrates	–	Ruiz-Dueñas and Martínez (2009)
Versatile peroxidase (VP)	1.11.1.16	Oxidizes Mn^{2+} and high redox potential aromatic compounds	Oxidize peroxidases substrates as phenolic compounds and dyes	Ruiz-Dueñas et al. (2009)
Cellobiose dehydrogenase	1.1.99.18	Unite hydrolytic and oxidative systems; dispose Mn II for MnP; degradation of lignin	–	Henriksson et al. (2000a, b), Kersten and Cullen (2007), and Carvalho et al. (2009)
Glycoxal hydrolase	3.2.1.21	Breakdown of lignin by synergic increase in enzyme	Lignocellulosic breakdown	de Gouvea et al. (2018)

the cell membrane or cell wall rather than releasing outside. Such proteins are called integral membrane proteins.

ii) the fungal secreted proteins vary with the fungal species as their growth pattern changes with different environmental and growth substrate conditions.

Lignocellulose is abundant in the plant cell wall. It is the largest source of fixed global carbon and is increasing as a potential source for biofuels and new biomaterial products. Readily accessible and economical carbohydrates can be formed with the lignin-forming phenyl propanoid units obtained by decomposing the lignocelluloses. Currently, the biological toolbox of the microorganisms was employed to degrade the lignocelluloses for carbon and energy source. *Ascomycetes* and *Basidiomycetes* are the most economical degraders of the lignocelluloses. White rot fungi degrade lignin and hemicellulose, while the brown rot fungi degrade celluloses and some part of hemicelluloses. The common white rots (*Phanerochaete chrysosporium*, *Trametes versicolor*, *Ganoderma lucidum*, *Lentinula edodes*) (Bouws et al. 2008) and brown rot fungi (*Postia placenta*, *Serpula lacrymans*) (Wymelenberg et al. 2010; Eastwood et al. 2011) have the ability to uptake different carbon and nitrogen sources to cover the diverse ecological places such as wood, soil, and organic waste materials. These wood-colonizing white rot basidiomycetes can degrade wood lignocellulose by producing different ligninolytic enzymes (laccases and peroxidases) along with several accessory enzymes including heme peroxidases such as cytochrome c peroxidases, chloroperoxidases, glyoxal oxidases (GLOX), methanol oxidases, pyranose dehydrogenases, and aryl alcohol oxidases (Sharma and Kuhad 2008; Alfaro et al. 2014). Filamentous fungi produce significant quantities of the carbohydrate-degrading enzymes which release fermentable sugars from the plant waste that is the demand of the industry to produce cellulases in the large quantities (Chutani and Sharma 2016; Mathieu et al. 2016).

10.3 Methodology and Advancement

Secretome gives the idea about the whole set of predicted and unpredicted genes which codes for the regulatory proteins present in a genome. It predicts the detailed knowledge of the protein sets released under various culture conditions with diverse environmental factors. In recent years, the numbers of fungal genome or proteome sequences are rising because of the advancement in the research technology on the fungal secretome. There are different methods as well as workflows for the analysis of the fungal secretome, majorly based on two ways: separation and identification of the produced proteins. Separation methods depend on the electrophoresis or liquid chromatography (LC) techniques while the identification methods involve the mass spectrometry (MS). In updated research technology, both the separation and identification methods can be done using liquid chromatography-mass spectrometry (LC-MS).

10.3.1 Techniques for Secreted Protein Secretion Analysis

10.3.1.1 Proteome Analysis Approach

Extracellular secretions of the proteins from different *Ascomycetes*, *Basidiomycetes*, and other fungal or bacterial cultures are now studied in detail by the different high-throughput methods such as protein mass spectrometry, transcriptome, and proteome. In proteomic methods, the researcher gets the information about all upregulated or downregulated proteins with different other functional and nonfunctional proteins of that organism. In the proteomic analysis, the peptide and protein identification is based mainly on MS (Han et al. 2008; Graham et al. 2011). It gives the accurate data on the protein or a peptide molecular mass from the given sample by matching or searching online where the mass of the complete set of protein or peptide and genome sequences are known.

For the fungal proteome analysis, two different strategies from the protein extraction to peptide identification, i.e., bottom-up MS and top-down MS, are very important (Fig. 10.2). In top-down MS, the mass spectrometer provides the analysis result from the intact protein giving information regarding its molecular mass, sequence of amino acids, position, and type of the posttranslational modifications. Although, its utilization to the large-scale proteomics remains a challenge due to the requirement of very expensive mass spectrometers with exceptionally high resolution

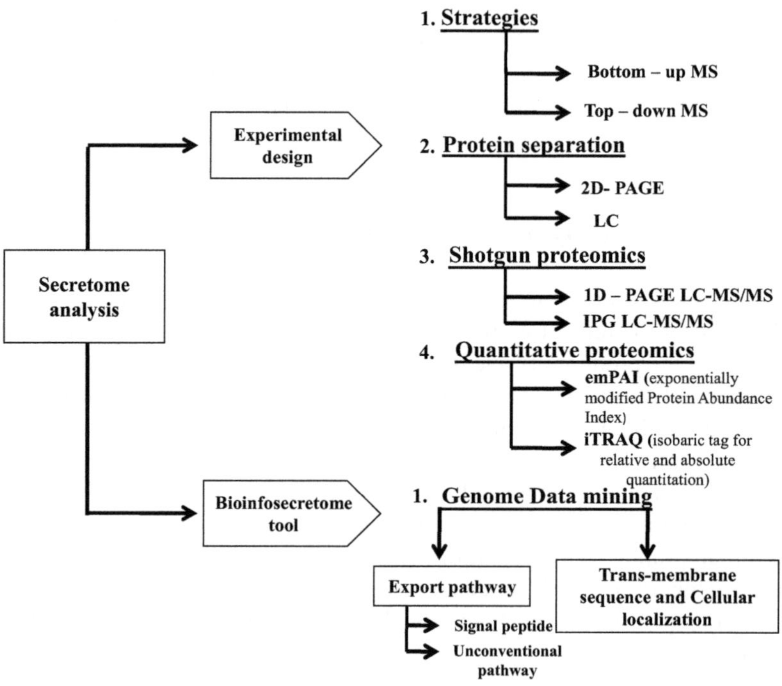

Fig. 10.2 Different tools and techniques used for the proteome and secretome analysis

(Waanders et al. 2007; Collier et al. 2008). Bottom-up MS approach is applied for the peptide analysis such that the proteins are extracted, proteolytic digested and then passed through the mass spectrometer. It can be done by two methods: either the digestion is done after the protein separation (sort-then-break method) or is performed directly in the protein sample followed by proteolytic peptides separation (break-then-sort method). In the former strategy, the peptide mixture is analyzed directly to generate a peptide mass fingerprint (Henzel et al. 1993) or be further sorted by liquid chromatography integrated with a tandem MS. While in the latter approach the digestion mixture peptides are first separated and then examined via MS. It is a high-throughput technique as discussed in the shotgun proteomics.

10.3.2 2-D Gel Electrophoresis

The earlier systematic approach for the protein quantification in protein complexes uses high-resolution two-dimensional electrophoresis (2-DE), a principal technique explaining the secretome from the last 10 years. By this method, the secretary proteins from a sample are separated in two dimensions where the protein is generally carried out in a cross-linked polymer, i.e., polyacrylamide for separation based on their charge-to-mass ratio. In 2-D PAGE, at first, they are separated depending on their isoelectric focusing (IEF) and thereafter according to their molecular mass (SDS-PAGE). By using IEF in 2-D PAGE, Mattow and coworkers (Mattow et al. 2003) separated 300 protein spots. The right combination of two-dimensional separation procedures allows new protein identification and evaluation of their relative abundance in a sample. The 2-D PAGE technique also allows the identification of the posttranslational modifications of the sample protein because they are separated before trypsin digestion (Jungblut et al. 2008). After obtaining the high-quality gel, the secretome analysis further involves the protein spots identification, isolation, and sequencing, such that the protein spot visualization needs staining techniques with high sensitivity (e.g., silver staining) which can be performed with MS analysis of the sample (Jungblut et al. 2008). Shevchenko and other researchers designed the procedure through which the 2-D PAGE identified protein spots could either be excised individually from the gel or be trypsin digested within the gel. The obtained tryptic peptides can be analyzed using mass spectrometry (Shevchenko et al. 1996).

2-D PAGE protein analysis with MS has some limitations such as peptides transfer from the gel to the mass spectrometer, resolving proteins which are too basic or acidic or too large or small, and detecting low amounts of the proteins (low sensitivity, 10^3–10^7 copies per cell) (Lamer and Jungblut 2001; Minden 2007; Jungblut et al. 2008). Because of these limitations, the use of this technology for comparing various samples became difficult. Difference gel electrophoresis (DIGE) approach helped to overcome this problem, where the two different protein samples covalently labeled with separate fluorescent dyes are electrophoresed together on the same 2-D gel (Minden 2007).

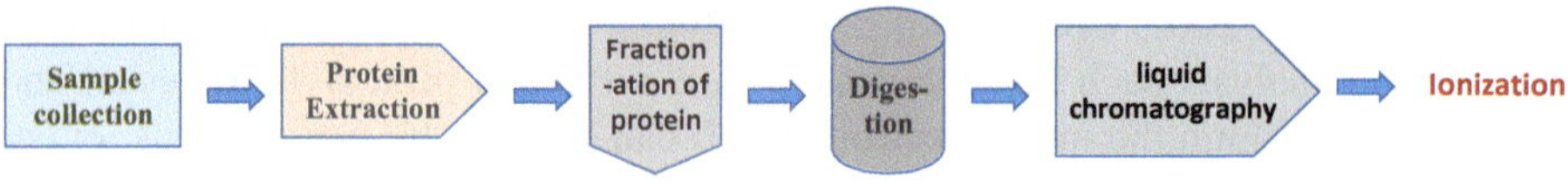

Fig. 10.3 Steps for liquid chromatography sample preparation of protein

10.3.3 Liquid Chromatography

The chromatographic techniques are mainly used to separate the proteins, nucleic acids, or small molecules from a complex mixture. Liquid chromatography (LC), a powerful approach for proteomic study, involves the separation of proteins or peptides. It is a peptide-centric approach where the peptide mixture components obtained by tryptic digestion of proteins are separated via liquid chromatography before its analysis using MS. These LC-based proteomics and secretomics techniques are comparatively more valuable for the identification of unambiguous proteins where samples consist of thousands of proteins and hundreds or thousands of peptides obtained from their tryptic digestion.

Proteomic study based on LC is completed via various steps of sample preparation which includes cell lysis for breaking cells apart and fractionation to eliminate the high-abundance proteins or other unnecessary proteins (Fig. 10.3). The used proteins are then trypsin digested to split the intact proteins into peptide components. These peptides are commonly introduced to a LC column for separation. Upon eluting from the LC column, the peptides are ionized. Mass spectrometers connected LC columns are used for analyzing the resolved peptides. Powerful proteomics analysis requires high sensitivity and high resolving power to reach a minimum dynamic peak range of 10^5. Two-dimensional nano-liquid chromatographic (2-D nLC) methods are used to obtain these resolution and sensitivity levels (Nagele et al. 2003), which involves sample enrichment, strong cation exchange, reverse-phase chromatography, and nanospray ion trap MS with data-dependent tandem MS spectra acquisition (Alfaro et al. 2014).

10.4 Shotgun Proteomics

When bottom-up approaches are performed on the protein mixtures, it is named as shotgun proteomics. In a shotgun proteomics approach, protein analysis involves the protein characterization by analyzing the protein released peptides via proteolysis (Zhang et al. 2013). All proteolytic peptides collected as the mass spectra are compared against the whole group of peptide masses from an organism(s) genome sequence obtained from different online tools (like Uniport) and analyzed using bioinformatics tools like MASCOT, X!Tandem, SEQUEST, and proteome discoverer (Craig and Beavis 2004; Ravalason et al. 2008; Efstathiou et al. 2017). In shotgun proteomics, different strategies applied for the peptide separation are immobilized pH gradient isoelectric focusing (IPG-LC-MS/MS), pre-fractionation using one-dimensional electrophoresis (1-D PAGE-LC-MS/MS), and the gel-free LC-MS shotgun method (LC-LC-MS/MS).

10.4.1 One-Dimensional Electrophoresis

Gel electrophoresis and column chromatography are the most commonly used methods for protein fractionation. This is the most used technology for the secretome identification in various cell types of fungi, plants, and animals (Mukherjee and Mani 2013). It involves the protein pre-fractionation via 1-D gel electrophoresis followed by its in-gel digestion and fixation using LC-MS/MS. Secreted proteins from the lignocellulose-degrading fungi are fractionated via 1-D SDS-PAGE, where most of the proteins are separated according to the molecular mass and the separated bands are cut from the gel lanes. Then, the gel slice is trypsin digested followed by peptide separation using reverse-phase liquid chromatography (RP-HPLC) and then MS analyzed.

10.4.2 Immobilized pH Gradient Isoelectric Focusing

Under this approach, IPG strip pH range is a major drawback in the protein identification. In this technique, OFFGEL fractionators permit liquid-phase peptide IEF (Horth et al. 2006). The separation is performed using a two-phase system with a compartment divided carrier ampholyte and buffer-free solution upper liquid phase and an IPG strip lower phase. After loading the sample and applying the voltage gradient, the peptides move via IPG strip till they reach their pI in a given compartment. After isoelectric focusing, for further analysis, the peptides can be recovered easily in a solution. The IEF strip is cut into pieces and peptides in a piece are eluted and analyzed using LC/MS. This approach gives better peptides separation than the classic 2-D PAGE (Poland et al. 2003; Abdallah et al. 2012).

10.4.3 Gel-Free LC-MS Shotgun Approach

Gel-free methods of the shotgun approach can be considered as a direct consequence of the numerous innovative developments in the past two or three decades (Zargar et al. 2016). These methods exclude few of the major experimental errors observed in the case of gel based approaches. Although earlier 2-D electrophoresis is a mature and well-established technique, it faces few limitations on quantitative reproducibility and difficulty in studying certain classes of proteins (Abdallah et al. 2012). Therefore, researchers have to take out the idea for alternative approaches of gel-free proteomics. In this approach, the extracted proteins are first digested in a solution to obtain a peptide mixture which are then separated via multidimensional-LC that are strong cation exchange and reversed-phase LC columns for separating the peptides into fractions, prior to tandem mass spectrometry analysis (Link et al. 1999; Washburn et al. 2001). This technique has the capacity to detect the greatest secreted proteins count in the fungal secretomes. In *Postia placenta*, both the bottom-up and top-down MS strategies were used to detect the proteins where 19 proteins were detected using 1-D PAGE-LC-MS/MS and 63 proteins by LC-LC-MS/

MS (Wymelenberg et al. 2010). Similarly, in *Phanerochaete chrysosporium* 30 proteins were identified via 1-D PAGE LC-MS/MS and 73 by LC-LC-MS/MS.

On comparing various separation techniques like 1-D PAGE-LC-MS/MS, 2-D PAGE, and IPG-IEF shotgun, IPG-IEF-LC-MS/MS method was found to identify the largest number of proteins (Abdallah et al. 2012). However, there is a need to merge many analytical approaches to detect various protein samples because some small secreted proteins like glycosylphosphatidyl inositol (GPI) were identified via 1-D LC-MS/MS and IPG but not via 2-D PAGE due to difficulty in separating wide molecular weight range proteome (Alfaro et al. 2014).

10.5 Quantitative Proteomics

Accurate quantitation of proteins is currently the most difficult and dynamic areas of proteomics. In quantitative proteomics, major methods depend on the multiple factors such as samples source, samples number, number of treatments studied, time requirement, equipment type, and cost on sample analysis. The quantitation methods involving isotope labeling approaches such as Isotope-Coded Protein Labeling (ICPL), Isotope-Coded Affinity Tag (ICAT), $^{15}N/^{14}N$ metabolic labeling, Tandem Mass Tags (TMT), Stable Isotope Labeling by Amino Acids in Cell Culture (SILAC), $^{18}O/^{16}O$ enzymatic labeling, and Isobaric Tags for Relative and Absolute Quantitation (iTRAQ) have been adopted well in the quantitative proteomics [Bantscheff et al. 2007; Xie et al. 2011]. Proteins labeled with specific reagents having a protein reactive group, a linker with light or heavy isotopes, and a biotin purification tag recover the proteolytic digested peptides. In quantitative approach, majority of MS involves protein digestion using trypsin and peptide level measurements to estimate a summarized value for a corresponding protein. These methods can be categorized into two: one using stable isobaric or isotope tags and other is label-free quantitation.

10.5.1 Isotope-Coded Affinity Tags (ICAT)

ICAT approach is used for the differential isotope labeling which involves three functional elements: a light or heavy isotopes coded linker, a cysteinyl residues sulfhydryl groups binding specific chemical reactive group, and a biotin tag for affinity purification (Gygi et al. 1999). Proteins having cysteine residues are tagged either by the light or heavy isotopes, where heavy isotopes have eight ^{13}C atoms. Thereafter, the isotope-labeled samples are combined and cleaved proteolytically. Before MS analysis, purification of the tagged cysteine-containing peptides using biotin-avidin affinity columns was performed to reduce the sample complexity. Specificity of the ICAT labeling to cysteinyl groups is exploited to study the redox status of the proteins. Thus, permit the quantitative analysis of the redox proteome and the ozone stress in the plants (Stroher and Dietz 2006; Hagglund et al. 2008). Some limitations for ICAT technique are: cysteine residues requirement restricts the label count per protein, not applicable to cysteine-free proteins and the results are not robust enough (Wiese et al. 2007).

10.5.2 Isobaric Tags for Relative and Absolute Quantitation (iTRAQ)

Many drawbacks of the quantitative techniques like ICPL and ICAT were overcome in iTRAQ technique. iTRAQ is a strong enzymatic toolbox for the deconstruction of lignocellulose biomass. In this, the peptides obtained from separate samples are tagged with separate compounds devised to obtain a similar total mass followed by their cleavage to generate the separate fragments. Peptide N-terminal and ε-amino group of lysine side chains are labeled with the iTRAQ reagents where it allows the comparison of up to eight samples in one experiment (Alfaro et al. 2014). In classical approach, two different isobaric labels are used for comparing the peptides N-terminal and lysine residues of the two samples. Such that after mixing the two samples, distinctly labeled peptides develop as one peak in MS analysis. In the next step, peptides fragment into smaller peptides followed by the differentiation of the reporter molecules such that the relative intensities of the corresponding reporters were used to deduce the relative proportion of each peptide (Ross et al. 2004).

This high performance technique has gained huge popularity in proteomics attributed to increased MS sensitivity when compared with ICAT and other quantitative techniques because of the contribution of all samples to the precursor ion signal. This can be used for both intact proteins and proteolytic peptides from a protein sample (Wiese et al. 2007). It has a great propensity in determining the most abundant proteins with a limitation for MS/MS, limiting the gathered information to the most abundant peptides. *P. chrysosporium* proteomes produced through various carbon sources or natural biomass has been compared successfully using iTRAQ approach (Manavalan et al. 2011). iTRAQ is also used to study the lignocellulolytic fungi secreted proteins in single cultures and in consortia (Adav et al. 2012a, b) and to investigate the change in protein expression in reaction to *Aspergillus fumigatus* obtained drug, caspofungin.

10.6 Secretome

Fungi consume the lignocellulosic substrates by secreting extracellular hydrolytic enzymes, known as secretome. Fungal secretome is a highly advanced proteomics technology that can reveal highly specific known and unknown potential lignocellulose-degrading enzymes, providing a better understanding level in the mechanism of lignocellulose hydrolysis. The fungi released enzymes have been investigated from different viewpoints which have ample of biotechnological applications. In fungal secretome analysis method, different *Basidiomycetes* and other related fungi (listed in Table 10.2) produce different cellulases, peroxidases, lignin peroxidases (LiP), MnP, and other enzymes that can degrade complex sugar

Table 10.2 Details of sequencing methods and protein sequences reported from different fungal species

S. No.	White rot *Basidiomycetes*	Sequencing method	Separation column	No. of proteins	References
1.	*Phanerochaete chrysosporium*	Nanoscale CapLC/MS/MS	C18 silica column	16	Abbas et al. (2005)
		1-D LC-MS/MS using 1100 series LC/MSD Trap SL spectrometer	Reversed phase C18 column	40	Wymelenberg et al. (2006)
		SDS-PAGE/LC-MS/MS	–	16	Sato et al. (2007)
		SDS-PAGE MS for solid, 2-D MS for liquid	–	18	Alfaro et al. (2014)
		2-D MS	–	14	Ravalason et al. (2008)
		SDS-PAGE-MS	–	190	Wymelenberg et al. (2009)
		LC-LC-MS/MS	Fused silica tubing, C_{18} particles packed column	73	Wymelenberg et al. (2010)
		SDS-PAGE/LC-MS/MS	Fused silica tubing, C_{18} particles packed column	30	Alfaro et al. (2014)
		Nano-LC-LC-MS/MS	–	356	Vandem Wymelenberg et al. (2011)
		iTRAQ after separation on SDS-PAGE	PolySULFOETHYL A SCX column	224	Manavalan et al. (2011)
		2-D MS	reversed-phase capillary HPLC column	47	Hori et al. (2011)
		iTRAQ technique using LC-MS/MS	Nanobored C18 column	329	Adav et al. (2011)
		LTQ-FT Ultra LC-MS/MS	capillary column packed with C18 AQ	289	Adav et al. (2014)
		Nano-LC-MS/MS	ZB-5HT GC capillary column	112	Zhu et al. (2016)

(continued)

Table 10.2 (continued)

2.	*Ganoderma lucidum*	SDS-PAGE-LTQ-FT	Capillary column with C18 AQ	71	Manavalan et al. (2012)
	Ganoderma sp.	Nano-LC-LTQ-Orbitrap MS	Agilent flow system 1100 micro	105	Hori et al. (2013)
	G. trabeum	Nano-LC-MS/MS	Trap column and analytical column with fused silica capillary	109	Zhu et al. (2016)
	G. lucidum	Nano-LC-MS/MS LTQ-OrbitrapVelos	PepMap 100 C18-nano viper	142 (induced)	Jain et al. (2019)
3.	*Ceriporiopsis subvermispora*	LC-LC-MS/MS	Agilent flow system 1100 micro	121	Fernandez-Fueyo et al. (2012)
		Nano-LC-MS/LTQ-Orbitrap	–	248	Hori et al. (2014)
4.	*Trametes* sp.	Western blot and LC-MS/MS with LTQ-Orbitrap MS	Nano column Pep map C18	2	Lebrun et al. (2011)
		Nano-LC-MS/MS with LTQ-Orbitrap MS	Agilent flow system 1100 micro flow system	218	Floudas et al. (2012)
5.	*Trametes trogii*	SDS followed by Micro TOF-Q	HPLC Column ZORBAX SB-C18	64	Ji et al. (2012)
6.	*Bjerkandera adusta*	Nano-LC-LTQ-Orbitrap MS	Agilent flow system 1100 micro flow system	187	Hori et al. (2013)
7.	*Phlebia brevispora*		–	178	Alfaro et al. (2014)
8	*Schizophyllum commune*	Nano-LC-MS/MS	Trap column and analytical column with fused silica capillary	229	Zhu et al. (2016)
9	*Ceriporiopsis subvermispora*		Trap column and analytical column with fused silica capillary	95	Zhu et al. (2016)
10	*Lentinula edodes*	Nano-LC coupled with tandem MS	Nano Easy HPLC	230	Cai et al. (2017)

(continued)

Table 10.2 (continued)

S. No.	Brown rot *Basidiomycetes*	Sequencing method		No. of proteins	Reference
11	*Serpula lacrymans*	IPG LC-MS/MS	–	39	Eastwood et al. (2011)
12	*Postia placenta*	LC-LC-MS/MS connected with LTQ-Orbitrap MS	HPLC using fabricated column with fused silica tubing	63	Alfaro et al. (2014)
		1-D LC-MS-MS	–	19	Wymelenberg et al. (2010)
		Nano-LC tandem MS with LTQ-Orbitrap MS	–	413	Alfaro et al. (2014)
			C18 resin reversed-phase column	231	San Ryu et al. (2011)
Symbiotic fungi					
13	*Laccaria bicolor*	SDS-PAGE gel digestion with MS	–	77	Vincent et al. (2011)
		IPG-IEF shotgun/MS	C18 micro elution column	142	
		Nanospray LTQ IT MS	C18 PepMap column	116	
Other fungi					
14	*Coprinopsis cinerea*	2-D PAGE/MS	–	76	Hoegger et al. (2007)
Correlation of white and brown rot fungi secretome					
15	*Coniophora puteana*	Nano-LC-MS/MS using Agilent flow system 1100 micro flow system	HPLC using Fabricated column with fused silica tubing	269	Alfaro et al. (2014)
16	*Fomitopsis pinicola*			253	Floudas et al. (2012)
17	*Trametes*			218	Floudas et al. (2012)
18	*Stereum hirsutum*			208	Floudas et al. (2012)
19	*Dichomitus squalens*			180	Alfaro et al. (2014)
20	*Gloeophyllum*			174	Floudas et al. (2012)
21	*Wolfiporia cocos*			171	Floudas et al. (2012)
22	*Auricularia delicata*			151	Alfaro et al. (2014)
23	*Dacryopinax* sp.			145	Floudas et al. (2012)
24	*Punctularia strigosozonata*			135	Floudas et al. (2012)
25	*Fomitiporia mediterranea*			85	Alfaro et al. (2014)

(cellulose, hemicelluloses, and lignin) into simple sugar (Sripuan et al. 2003; Juhasz et al. 2005; Adsul et al. 2007; Adav et al. 2012a, b; Manavalan et al. 2012; Alfaro et al. 2014). After secretome study, it was revealed that *G. lucidum* has four new glycoside hydrolases (such as two endo 1, 4-β mannanases, glycoside hydrolase family 35 and 74 family) which cleaves the glycosidic bond between a carbohydrate and non-carbohydrate moiety (Manavalan et al. 2012).

10.6.1 White Rot Basidiomycetes

Wood-decaying *Basidiomycetes*, usually classified as brown or white rot fungi, are inhabitants of forest litter, where they have a major role in carbon cycling.

1. *Phanerochaete chrysosporium:* White rot basidiomycete, such as *P. chrysosporium*, secretes extracellular hydrolases that attack cellulose and hemicellulose and depolymerize the lignin by oxidative mechanisms. It is a model white rot fungi having manganese (MnP) and lignin peroxidases (LiP). *P. chrysosporium* genome has been sequenced and its secretome has been investigated under various conditions (Martinez et al. 2004) such that their automatic annotated genome estimated 10,048 gene models (http://genome.jgi-psf.org/Phchr1/Phchr1.home.html). Microarray analysis for whole-genome expression and LC-tandem MS of extracellular proteins of this fungus revealed a total of 545 genes showing the ligninolytic or cellulolytic degradation depending on the remarkable changes in accumulation of transcript and/or peptide sequences of secreted proteins (Wymelenberg et al. 2009). Among these proteins, most upregulated groups were of lignin and manganese peroxidase, carbohydrate-active enzymes, six members of glycoside hydrolase family 61, many carbohydrate esterases, and polysaccharide lyases. Proteins of unknown function are predicted to encode by above 190 upregulated genes. Different scientist studied and reported the secretome work differently by switching in/or out different artificial medium. They found that the different proteins are upregulated or down regulated and some are constitutively expressed under different conditions.

P. chrysosporium secretome studied via 1-D PAGE-LC-LC-MS/MS in shaking submerged cultures of synthetic media under controlled C or N sources recognized 40 secreted proteins containing 13 peptidases, 8 GHs, 6 esterase-lipases, 5 LiPs, and 3 MnPs (Wymelenberg et al. 2006). Further, secretome studied by iTRAQ-based approach in shaking submerged cultures added with synthetic cellulose, lignin or a mixture of both, secreted proteins included 66 GHs, 16 proteases, and 16 lignin-degrading enzymes. In the experiment, MnPs and LiPs were absent, declaring non-efficiency of the synthetic lignin as an inducer for these two enzymes (Manavalan et al. 2011). The effect of xylan or starch supplementation in shaking submerged cultures of *P. chrysosporium* having cellulose as a source of carbon was studied, identifying 47 secretome proteins such that the addition of xylan raised the synthesis of various GHs (Hori et al. 2011). In another experiment, various complex lignocellulosic substrates such as hay, corn stove, wheat bran, sugarcane bagasse, sawdust, and wood chips were used for the quantitative study of protein secretion in shaking submerged cultures and identified 52% GHs, 10% lignin-depolymerizing enzymes, and 20% proteases secretion (Adav et al. 2011). In the presence of glucose, cellulose, and lignin substrate for the growth of *P. chrysosporium*, 236 ± 30, 279 ± 74, and 289 ± 66 secretory proteins were found. It showed 48 glycosylated lignocellulolytic secretory proteins accompanied by glucose-(20 proteins) and cellulose-containing media (34 proteins) (Adav et al. 2014).

2. *Ganoderma lucidum*: *G. lucidum* is a lignocellulose-degrading white rot basidiomycete of immense medical importance and great wood degradation ability which helps in the production of bioenergy. The genome and proteome study presented 16,113 annotated gene models containing a collection of cytochrome P_{450} encoding genes with high-value wood-degrading enzymes (Chen et al. 2012). Another research group performed secretome study for 40 days, grown fruiting body on solid substrate of sugarcane bagasse. The analysis by 1-D PAGE-LC-MS/MS described 71 proteins classified into GHs (24% cellulases, 5% hemicellulases, and 10% others), lignin-hydrolyzing enzymes (24% of one MnP and five laccases), 7% phosphatases, 2% proteases, 10% transport proteins, and 10% hypothetical proteins (Manavalan et al. 2012). Yu and coworkers analyzed three different stages of protein sample (16 days mycelia, 60 and 90 days fruiting body) by LC-MS/MS (Yu et al. 2015). Previous database revealed 803 identified proteins. Among those, 61 lignocellulose-degrading proteins were identified, and majority of them (49 proteins) were found in fruiting bodies of 90 days. 470 proteins out of 803 proteins studied had GO annotations and were grouped into 36 GO terms, with "binding activity," "catalytic activity," and "hydrolase activity" having high percentage value. The basic proteomic and sequence alignment analysis of a new potent immunomodulatory protein, GL18769, was found to have strong immunomodulatory activity in *G. lucidum* (Yu et al. 2015).
3. *Pleurotus sapidus*: It is a member of the family of oyster mushrooms, having pleasant flavor and high nutritional value. Secreted proteins were analyzed in a shaking submerged culture with limited nitrogen medium containing glass wool or peanut shells as carrier material and glucose as source of carbon (Zorn et al. 2005). Secreted enzymes by *P. sapidus* include various hydrolases and lignolytic enzymes of the MnP/VP family. Peptidases dominated in the peanut shells grown cultures, whereas peroxidase type enzymes were secreted by cultures grown on glass wool. Using nano-LC coupled to tandem MS, secreted whole protein of *P. ostreatus* grown on wheat straw or poplar wood as a principal C/N source was identified over 500 proteins. Out of these, 34%, 15%, and 6% were unique for wheat straw, glucose, and poplar wood cultures, respectively. 20% were produced under different conditions and additional 19% were shared between two lignocellulose cultures (Fernandez-Fueyo et al. 2016).
4. *Ceriporiopsis subvermispora*: It is a member of *Polyporales* having lignin- as well as cellulose-degrading ability. Fernandez-Fueyo and coworkers performed

LTQ-Orbitrap LC-LC-MS/MS and analyzed the culture filtrate of 5-day-old culture for comparing the protein abundance profiles of *P. chrysosporium* and *C. subvermispora* (Fernandez-Fueyo et al. 2012). 60 and 121 proteins were identified in filtrate, out of which 18 and 3 were GHs with 3 MnP identified in the *C. subvermispora* filtrate. Extracellular secreted protein of *C. subvermispora* grown in ball-milled aspen containing liquid medium was analyzed for various days sample by nano-LC-tandem MS. Enzymes related to lignin degradation such as manganese peroxidases and aryl alcohol oxidase were detected in early 3 days sample. Similarly, various cellulose- and xylan-degrading enzymes like GH5 endoglucanase, GH7 cellobiohydrolase, and GH10 endoxylanase were found after 5 days and GH12, GH45, cellobiose dehydrogenase, and lytic polysaccharide monooxygenase after 7 days (Hori et al. 2014).

5. *Trametes* sp.: *Trametes* is a white rot Basidiomycete, usually found on tree stumps. It is responsible for the degradation of less-degradable lignin rich unused plant debris. Different species of *Trametes* (*T. pubescens*, *T. multicolor*, *T. hirsute*, *T. trogii*, and *T. versicolor*) have the ability to grow on agricultural waste and degrade lignin, cellulose, and hemicellulose (Lebrun et al. 2011; Ji et al. 2012; Knezević et al. 2013; Vasina et al. 2016). Genome size of *T. versicolor* is 44.8 Mbp which presents the annotated gene models of 14,296 (http://genome.jgi.doe.gov/Trave1/Trave1.info.html) (Floudas et al. 2012). The secretome study identified 218 proteins with 65 GHs, 27 peptides, 8 lignin-degrading oxidases, 5 lipases, 4 proline oxidases (POX), 3 MnPs, and 44 unrecognized proteins. Secretome study of different *Trametes* sp. in various medium and agro-waste had identified proteins such as laccase and peroxidase from *T. hirsute* st. 072 (Tomsovsky and Homolka 2003; Vasina et al. 2013). 20% carbohydrate metabolism, 11% fatty acid metabolism, and 11% protein metabolism proteins with 29% cell wall- and lignin-degrading enzymes and 29% of other proteins were identified from *T. trogii* (Ji et al. 2012). Two laccase isoenzymes, LacA and LacB, were reported to be secreted by *T. versicolor* in the presence of metal salt (Lebrun et al. 2011).
6. *Postia placenta*: It is a wood-decaying basidiomycete, closely related to *P. chrysosporium.* It is the common inhabitants of forest ecosystems, largely causing destructive decay of cellulose in wood specially conifer wood (Hibbett and Donoghue 2001) with an important role in carbon cycling. It is of considerable economic importance due to being the principal agent for decay of wooden structures. In a study, its genome was sequenced and 17,173 gene models had been annotated. Interestingly, its genome had two POX genes while no LiP, VP, and MnP genes (Martinez et al. 2009). Transcripts and secreted protein of *P. placenta* were analyzed by LC-tandem MS of growth medium with cellulose as the main carbon source and identified the gene of hemicelluloses, β-1,4-endoglucanase, and upregulated protein of putative iron reductases, quinone reductase, and structurally divergent oxidases. In comparison, secretome study of the *P. chrysosporium* and *P. placenta* identified total of 73 and 67 secreted proteins, respectively. *P. placenta* secreted various hemicellulases, while *P. chrysosporium* secreted a variety of extracellular GHs for attacking the cellu-

loses and hemicelluloses together (Wymelenberg et al. 2010). In presence of either pine or aspen containing media, a total 143 secreted proteins were detected from *P. placenta* (Wymelenberg et al. 2011). The LC-LC-MS/MS analysis of *P. placenta* secreted proteins in aspen chips in solid culture found one endo-1,4-β-xylanase and three endo-1,4-β-D-glucanases that participates in cellulose degradation (San Ryu et al. 2011).

10.6.2 Ascomycetes

Ascomycetes have a potential role in lignocelluloses biomass recycling and are highly exploitable in enzyme industry. Similar to *Basidiomycetes*, *Ascomycota* were used to develop cost-effective technology for converting plant cell wall waste (i.e., lignocellulose) to the useful form of bioenergy and biochemicals, which are used for the production of organic acids and pharmaceutical and industrial important enzymes.

Secretome study of *Ascomycota* group of fungi analyzed the enzyme cocktail of different strains (*Aspergillus wentii*, *A. niger*, *Neocosmospora haematococca*, *Penicillium variabile*) and identified 46–159 number of proteins, which revealed a large number of carbohydrate-active enzymes (CAZymes) (Couturier et al. 2016). Many other *Ascomycetes* fungi and their secretome suggest them to be a good producers of lignocellulose degrading enzymes, for example, *T. reesei* contains 201 glycoside hydrolase genes, 99 glycosyl transferase genes, 22 carbohydrate esterase genes, and 5 polysaccharide lyase genes (Martinez et al. 2008; Hakkinen et al. 2012) and *Neurospora crassa* produces various extracellular and intracellular beta-glucosidase (Znameroski et al. 2012). Out of 33 quantified proteins, 4 were enzymes including 2 cellobiohydrolases (CBH-1 and GH6-2), a β-glucosidase (GH3-4), and an endoglucanase (GH5-1) (Phillips et al. 2011), revealing highly overlapped groups of 430 genes with a large proportion of induced CAZy genes and 252 biomass regulon (BR) genes (Galagan et al. 2003; Wang et al. 2015). *Neocosmospora haematococca* (as *Nectria haematococca*) had 15,707 predicted genes (Coleman et al. 2009). Among several aspergilli, sequence of *Aspergillus nidulans* was become available by the end of 2005 (Galagan et al. 2005). *A. niger* analyzed through shotgun proteomics approach identified a total of 102 secreted proteins and 1,126 of microsomal proteins (de Souza et al. 2011). *A. oryzae* under submerged culture conditions produced intermediates of α-glucosidase A, glucoamylase A, cellulase B, and xylanase G2 whereas in solid-state culture produced glucoamylase B, alanyl dipeptidyl peptidase, and α-mannosidase (Machida et al. 2005). *A. fumigatus* produced 32.9% cellulases and glycoside hydrolases; 16.2% hemicellulases; 11.7% proteases and peptidases; 8.1% lignin-degrading enzymes; 7.6% chitinases, phosphatases, and lipases; 0.9% amylases; and 22.5% proteins with unknown function (Adav et al. 2015).

10.7 Comparative Proteomics

In a comparative study of fungal proteomics, there are many techniques to find out the secreted proteins from the extracellular or intracellular of the fungi. A pipeline can be developed for identifying and comparing the fungal secretome of 49 fungi including 41 *Basidiomycetes*, 6 *Ascomycetes*, 1 *Zygomycota*, and 1 *Chytridiomycota*. Total genes analyzed in fungal genomes range from 4000 to 25,000. The relative amount of secreted proteins ranges from 3% to 10% of the total proteome (Pellegrin et al. 2015). Fungi contain a diverse range of secreted proteins such as pectinases, cellulases, hemicellulases, and lignin-degrading accessory enzymes which converts lignocellulosic agriculture polymer to monomers.

In comprehensive experimental secretome study based on protein harvesting and LC/MS/MS analysis involving iTRAQ studies, more filamentous fungi were predicted to have more or less than 350 proteins revealed in the secretome of filamentous *Ascomycetes* (Lu et al. 2010; Saykhedkar et al. 2012; Nitsche et al. 2012; Zeiner et al. 2016) and *Basidiomycetes* (Wymelenberg et al. 2010; Hori et al. 2014). In the comparative study with 4 ascomycota fungi, approximately 900 proteins for *Pezizomycotina* were identified. Other saprotrophs and some plant pathogens secreted approximately 600 and 1200 proteins, respectively (Krijger et al. 2014). For comparing *Ascomycetes* with *Basidiomycetes*, secretome data was classified in a similar manner that found 11 (Nitsche et al. 2012), 20 (Phalip et al. 2005), or 39 (Saykhedkar et al. 2012) GH families in *Ascomycetes* and 31 (Hori et al. 2014) and 35 (Rohr et al. 2013) GH families in white rot *Basidiomycetes*. *A. fumigatus* Z5 produces more cellulases and hemicellulases while growing on cellulosic substrates and few of them identified oxidoreductases when the glucose was used as the carbon source (Liu et al. 2013). Similarly, white rot basidiomycete *C. subvermispora* secreted large ratio of GHs to peptidases while grown on aspen wood (Hori et al. 2014). A large number of LiPs and MnPs were found in the culture of *P. chrysosporium* of white rot fungi, but in *Ascomycetes* fungi, these enzymes were less or not identified. Further, a large number of oxygen to water reducing laccases, phenol oxidases, peroxidases, and H_2O_2-generating enzymes were revealed in *Basidiomycetes* culture supernatants (Glenn et al. 1983; Hoegger et al. 2006; Bouws et al. 2008). Recently, two white rot basidiomycetes such as *Ganoderma* sp. kk-02 and *G. lucidum* MDU-7 were used to compare laccase isozymes, which were produced extracellularly in the presence of copper. Further, six different types of laccase isoenzymes (Galc H1, Glac L1, Glac L2, Glac L3, Glac L4, and Glac L5) were identified using MALDI-TOF peptide fingerprinting technique (Kumar et al. 2015, 2017).

10.8 Future Prospectives

Secretome and proteome analyses were started earlier with the use of methods based on 2-D gel electrophoresis (O'Farrell 1975). In the recent time of scientific era of advances in software and in silico tools, experimental methods have now changed from 2-D gel electrophoresis to very updated tools like LC-MS/MS, iRTAQ, and next-generation sequencing to describe fungal secretomes, analyze expression profile, and identify new enzymes (Kim et al. 2007; Sato et al. 2007; Jain et al. 2019). The use of secretomic and proteomic techniques is the best method for obtaining the broad range of complex and highly dynamic protein consortia secreted from *Ascomycetes* and *Basidiomycetes* to degrade the lignocellulosic biomass.

Recent proteomic approach is used to identify putative secreted peptides that belong into several categories, like mycotoxins, hormones, and effector molecules, that facilitate colonization during host-microbe interactions (Le Marquer et al. 2019). In the future, this broad study will open new avenues in the area of small-secretory fungal peptides that regulates their cellular biology and modulates the fungal interaction with the environment. Bioinformatics tools also have equal importance in this study which helps in the prediction of significant functions according to the past data. The main motive behind *Ascomycetes* and *Basidiomycetes* secretome analysis is to critically study the valuable data on the importance of extracellular enzymes from the cellular part of the microorganisms that can be further utilized in the deconstruction of the waste lignocellulosic material. Also, the quantitative proteomics supports valuable wet lab data on the importance of every enzyme or protein.

Acknowledgment The authors acknowledge Maharshi Dayanand University for the infrastructure facilities. FIST-DST grant to the Department of Microbiology is sincerely acknowledged.

References

Abbas A, Koc H, Liu F, Tien M (2005) Fungal degradation of wood: initial proteomic analysis of extracellular proteins of *Phanerochaete chrysosporium* grown on oak substrate. Curr Genet 47:49–56

Abdallah C, Dumas-Gaudot E, Renaut J, Sergeant K (2012) Gel-based and gel-free quantitative proteomics approaches at a glance. Int J Plant Genom 2012:494572

Adav SS, Li AA, Manavalan A, Punt P, Sze SK (2010) Quantitative iTRAQ secretome analysis of *Aspergillus niger* reveals novel hydrolytic enzymes. J Proteome Res 9(8):3932–3940

Adav SS, Ravindran A, Chao LT, Tan L, Singh S, Sze SK (2011) Proteomic analysis of pH and strains dependent protein secretion of *Trichoderma reesei*. J Proteome Res 10:4579–4596

Adav SS, Ravindran A, Cheow ES, Sze SK (2012a) Quantitative proteomic analysis of secretome of microbial consortium during saw dust utilization. J Proteome 75(18):5590–5603

Adav SS, Ravindran A, Sze SK (2012b) Quantitative proteomic analysis of lignocellulolytic enzymes by *Phanerochaete chrysosporium* on different lignocellulosic biomass. J Proteome 75(5):1493–1504

Adav SS, Ravindran A, Sze SK (2014) Study of *Phanerochaete chrysosporium* secretome revealed protein glycosylation as a substrate-dependent post-translational modification. J Proteome Res 13(10):4272–4280

Adav SS, Ravindran A, Sze SK (2015) Quantitative proteomic study of *Aspergillus fumigatus* secretome revealed deamidation of secretory enzymes. J Proteome 119:154–168

Adsul MG, Bastawde KB, Varma AJ, Gokhale DV (2007) Strain improvement of *Penicillium janthinellum* NCIM 1171 for increased cellulase production. Bioresour Technol 98(7):1467–1473

Alfaro M, Oguiza JA, Ramirez L, Pisabarro AG (2014) Comparative analysis of secretomes in basidiomycete fungi. J Proteome 102:28–43

Bantscheff M, Schirle M, Sweetman G, Rick J, Kuster B (2007) Quantitative mass spectrometry in proteomics: a critical review. Anal Bioanal Chem 389(4):1017–1031

Bianco L, Perrotta G (2016) Fruit development and ripening: proteomic as an approach to study Olea Europaea and other non-model organisms. Agric Proteomics 1:53–65

Bouws H, Wattenberg A, Zorn H (2008) Fungal secretomes-nature's toolbox for white biotechnology. Appl Microbiol Biotechnol 80(3):381

Cai Y, Gong Y, Liu W, Hu Y, Chen L, Yan L, Zhou Y, Bian Y (2017) Comparative secretomic analysis of lignocelluloses degradation by *Lentinula edodes* grown on microcrystalline cellulose, lignosulfonate and glucose. J Proteome 163:92–101

Carvalho W, Canilha L, Ferraz A, Milagres AM (2009) An overview of wood structure, composition and biodegradation. Nova Chem 32(8):1–5

Chen S, Xu J, Liu C, Zhu Y, Nelson DR, Zhou S, Li C, Wang L, Guo X, Sun Y, Luo H (2012) Genome sequence of the model medicinal mushroom *Ganoderma lucidum*. Nat Commun 3:913

Chutani P, Sharma KK (2015) Biochemical evaluation of xylanases from various filamentous fungi and their application for the deinking of ozone treated newspaper pulp. Carbohydr Polym 127:54–63

Chutani P, Sharma KK (2016) Concomitant production of xylanases and cellulases from *Trichoderma longibrachiatum* MDU-6 selected for the deinking of paper waste. Bioprocess Biosyst Eng 39(5):747–758

Claus H (2004) Laccases: structure, reactions, distribution. Micron 35(1-2):93–96

Coleman JJ, Rounsley SD, Rodriguez-Carres M, Kuo A, Wasmann CC, Grimwood J, Schmutz J, Taga M, White GJ, Zhou S, Schwartz DC (2009) The genome of *Nectria haematococca*: contribution of supernumerary chromosomes to gene expansion. PLoS Genet 5(8):e1000618

Collier TS, Hawkridge AM, Georgianna DR, Payne GA, Muddiman DC (2008) Top-down identification and quantification of stable isotope labeled proteins from *Aspergillus flavus* using online nano-flow reversed-phase liquid chromatography coupled to a LTQ-FTICR mass spectrometer. Anal Chem 80(13):4994–5001

Couturier M, Navarro D, Favel A, Haon M, Lechat C, Lesage-Meessen L, Chevret D, Lombard V, Henrissat B, Berrin JG (2016) Fungal secretomics of ascomycete fungi for biotechnological applications. Mycosphere J 7(10):1546–1553

Craig R, Beavis RC (2004) TANDEM: matching proteins with tandem mass spectra. Bioinformatics 20(9):1466–1467

de Gouvea PF, Bernardi AV, Gerolamo LE, de Souza SE, Riano-Pachon DM, Uyemura SA, Dinamarco TM (2018) Transcriptome and secretome analysis of *Aspergillus fumigatus* in the presence of sugarcane bagasse. BMC Genomics 19(1):232

de Oliveira JM, van Passel MW, Schaap PJ, de Graaff LH (2011) Proteomic analysis of the secretory response of *Aspergillus niger* to D-maltose and D-xylose. PLoS One 6(6):e20865

de Souza WR, de Gouvea PF, Savoldi M, Malavazi I, de Souza Bernardes LA, Goldman MHS, Goldman GH (2011) Transcriptome analysis of Aspergillus niger grown on sugarcane bagasse. Biotechnol Biofuels 4(1):40

Dix NJ, Webster J (1995) Colonization and decay of wood. In: Fungal ecology. Springer, Dordrecht, pp 145–171

Doyle S (2011) Fungal proteomics: from identification to function. FEMS Microbiol Lett 321(1):1–9

Eastwood DC, Floudas D, Binder M, Majcherczyk A, Schneider P, Aerts A, Asiegbu FO, Baker SE, Barry K, Bendiksby M, Blumentritt M (2011) The plant cell wall–decomposing machinery underlies the functional diversity of forest fungi. Science 333(6043):762–765

Efstathiou G, Antonakis AN, Pavlopoulos GA, Theodosiou T, Divanach P, Trudgian DC, Thomas B, Papanikolaou N, Aivaliotis M, Acuto O, Iliopoulos I (2017) ProteoSign: an end-user online differential proteomics statistical analysis platform. Nucleic Acids Res 45(W1):W300–W306

Erden E, Ucar MC, Gezer T, Pazarlioglu NK (2009) Screening for ligninolytic enzymes from autochthonous fungi and applications for decolorization of Remazole Marine Blue. Braz J Microbiol 40(2):346–353

Fernandez-Fueyo E, Ruiz-Duenas FJ, Ferreira P, Floudas D, Hibbett DS, Canessa P (2012) Comparative genomics of *Ceriporiopsis subvermispora* and *Phanerochaete chrysosporium* provide insight into selective ligninolysis. Proc Natl Acad Sci U S A 109:5458–5463

Fernandez-Fueyo E, Ruiz-Duenas FJ, Lopez-Lucendo MF, Perez-Boada M, Rencoret J, Gutierrez A, Pisabarro AG, Ramirez L, Martinez AT (2016) A secretomic view of woody and nonwoody lignocellulose degradation by *Pleurotus ostreatus*. Biotechnol Biofuels 9(1):49

Floudas D, Binder M, Riley R, Barry K, Blanchette RA, Henrissat B, Martínez AT, Otillar R, Spatafora JW, Yadav JS, Aerts A (2012) The Paleozoic origin of enzymatic lignin decomposition reconstructed from 31 fungal genomes. Science 336(6089):1715–1719

Galagan JE, Calvo SE, Borkovich KA, Selker EU, Read ND, Jaffe D, FitzHugh W, Ma LJ, Smirnov S, Purcell S, Rehman B, Elkins T, Engels R, Wang S, Nielsen CB, Butler J, Endrizzi M, Qui D, Ianakiev P, Bell-Pedersen D, Nelson MA, Werner-Washburne M, Selitrennikoff CP, Kinsey JA, Braun EL, Zelter A, Chulte U, Kothe GO, Jedd G, Mewes W, Staben C, Marcotte E, Greenberg D, Roy A, Foley K, Naylor J, Stange-Thomann N, Barrett R, Gnerre S, Kamal M, Kamvysselis M, Mauceli E, Bielke C, Rudd S, Frishman D, Krystofova S, Rasmussen C, Metzenberg RL, Perkins DD, Kroken S, Cogoni C, Macino G, Catcheside D, Li W, Pratt RJ, Osmani SA, DeSouza CP, Glass L, Orbach MJ, Berglund JA, Voelker R, Yarden O, Plamann M, Seiler S, Dunlap J, Radford A, Aramayo R, Natvig DO, Alex LA, Mannhaupt G, Ebbole DJ, Freitag M, Paulsen I, Sachs MS, Lander ES, Nusbaum C, Birren B (2003) The genome sequence of the filamentous fungus *Neurospora crassa*. Nature 422(6934):859–868

Galagan JE, Calvo SE, Cuomo C, Ma LJ, Wortman JR, Batzoglou S, Lee SI, Başturkmen M, Spevak CC, Clutterbuck J, Kapitonov V, Jurka J, Scazzocchio C, Farman M, Butler J, Purcell S, Harris S, Braus GH, Draht O, Busch S, D'Enfert C, Bouchier C, Goldman GH, Bell-Pedersen D, Griffiths-Jones S, Doonan JH, Yu J, Vienken K, Pain A, Freitag M, Selker EU, Archer DB, Penalva MA, Oakley BR, Momany M, Tanaka T, Kumagai T, Asai K, Machida M, Nierman WC, Denning DW, Caddick M, Hynes M, Paoletti M, Fischer R, Miller B, Dyer P, Sachs MS, Osmani SA, Birren BW (2005) Sequencing of *Aspergillus nidulans* and comparative analysis with *A. fumigatus* and *A. oryzae*. Nature 438(7071):1105–1115

Glenn JK, Morgan MA, Mayfield MB, Kuwahara M, Gold MH (1983) An extracellular H_2O_2-requiring enzyme preparation involved in lignin biodegradation by the white rot basidiomycete *Phanerochaete chrysosporium*. Biochem Biophys Res Commun 114(3):1077–1083

Graham C, McMullan G, Graham RL (2011) Proteomics in the microbial sciences. Bioeng Bugs 2(1):17–30

Gygi SP, Rist B, Gerber SA, Turecek F, Gelb MH, Aebersold R (1999) Quantitative analysis of complex protein mixtures using isotope-coded affinity tags. Nat Biotechnol 10:994

Hagglund P, Bunkenborg J, Maeda K, Svensson B (2008) Identification of thioredoxin disulfide targets using a quantitative proteomics approach based on isotope-coded affinity tags. J Proteome Res 7(12):5270–5276

Hakkinen M, Arvas M, Oja M, Aro N, Penttilä M, Saloheimo M, Pakula TM (2012) Re-annotation of the CAZy genes of *Trichoderma reesei* and transcription in the presence of lignocellulosic substrates. Microb Cell Factories 11(1):134

Han X, Aslanian A, Yates JR III (2008) Mass spectrometry for proteomics. Curr Opin Chem Biol 12(5):483–490

Henriksson G, Johansson G, Pettersson G (2000a) A critical review of cellobiose dehydrogenases. J Biotechnol 78(2):93–113

Henriksson G, Zhang L, Li J, Ljungquist P, Reitberger T, Pettersson G, Johansson G (2000b) Is cellobiose dehydrogenase from *Phanerochaete chrysosporium* a lignin degrading enzyme? Biochim Biophys Acta Protein Struct Mol Enzymol 1480(1–2):83–91

Henzel WJ, Billeci TM, Stults JT, Wong SC, Grimley C, Watanabe C (1993) Identifying proteins from two-dimensional gels by molecular mass searching of peptide fragments in protein sequence databases. Proc Natl Acad Sci 90(11):5011–5015

Hibbett DS, Donoghue MJ (2001) Analysis of character correlations among wood decay mechanisms, mating systems, and substrate ranges in homobasidiomycetes. Syst Biol 50(2):215–242

Hoegger PJ, Kilaru S, James TY, Thacker JR, Kües U (2006) Phylogenetic comparison and classification of laccase and related multicopper oxidase protein sequences. FEBS J 273(10):2308–2326
Hoegger P, Majcherczyk A, Dwivedi R, Svobodova K, Kilaru S, Kües U (2007) Enzymes in wood degradation. In: Kües U (ed) Wood prod. Wood technol. Biotechnol. Impacts. Universitätsverlag Göttingen, Göttingen, pp 383–432
Hofrichter M (2002) Lignin conversion by manganese peroxidase (MnP). Enzym Microb Technol 30(4):454–466
Hori C, Igarashi K, Katayama A, Samejima M (2011) Effects of xylan and starch on secretome of the basidiomycete *Phanerochaete chrysosporium* grown on cellulose. FEMS Microbiol Lett 321(1):14–23
Hori C, Gaskell J, Igarashi K, Samejima M, Hibbett D, Henrissat B, Cullen D (2013) Genome wide analysis of polysaccharides degrading enzymes in 11 white-and brown-rot Polyporales provides insight into mechanisms of wood decay. Mycologia 105(6):1412–1427
Hori C, Gaskell J, Igarashi K, Kersten P, Mozuch M, Samejima M, Cullen D (2014) Temporal alterations in secretome of selective ligninolytic fungi *Ceriporiopsis subvermispora* during growth on aspen wood reveal its strategy of degrading lignocellulose. Appl Environ Microbiol 17:AEM-03652
Horth P, Miller CA, Preckel T, Wenz C (2006) Efficient fractionation and improved protein identification by peptide OFFGEL electrophoresis. Mol Cell Proteomics 5(10):1968–1974
Jain KK, Kumar A, Shankar A, Pandey D, Chaudhary B, Sharma KK (2019) De novo transcriptome assembly and protein profiling of copper-induced lignocellulolytic fungus *Ganoderma lucidum* MDU-7 reveals genes involved in lignocellulose degradation and terpenoid biosynthetic pathways. Genomics. https://doi.org/10.1016/j.ygeno.2019.01.012
Ji XL, Zhang WT, Gai YP, Lu BY, Yuan CZ, Liu QX, Mu ZM (2012) Patterns of lignocellulose degradation and secretome analysis of *Trametes trogii* MT. Int Biodeterior Biodegradation 75:55–62
Juhasz T, Szengyel Z, Reczey K, Siika-Aho M, Viikari L (2005) Characterization of cellulases and hemicellulases produced by *Trichoderma reesei* on various carbon sources. Process Biochem 40(11):3519–3525
Jungblut PR, Holzhütter HG, Apweiler R, Schlüter H (2008) The speciation of the proteome. Chem Cent J 2(1):16
Kellner H, Vandenbol M (2010) Fungi unearthed: transcripts encoding lignocellulolytic and chitinolytic enzymes in forest soil. PLoS One 5(6):e10971
Kersten P, Cullen D (2007) Extracellular oxidative systems of the lignin-degrading Basidiomycete *Phanerochaete chrysosporium*. Fungal Genet Biol 44(2):77–87
Kim Y, Nandakumar MP, Marten MR (2007) Proteomics of filamentous fungi. Trends Biotechnol 25(9):395–400
Knezević A, Milovanović I, Stajić M, Vukojević J (2013) Potential of *Trametes* species to degrade lignin. Int Biodeterior Biodegradation 85:52–56
Krijger JJ, Thon MR, Deising HB, Wirsel SG (2014) Compositions of fungal secretomes indicate a greater impact of phylogenetic history than lifestyle adaptation. BMC Genomics 15(1):722
Kumar A, Sharma KK, Kumar P, Ramchiary N (2015) Laccase isozymes from Ganoderma lucidum MDU-7: isolation, characterization, catalytic properties and differential role during oxidative stress. J Mol Catal B Enzym 113:68–75
Kumar A, Singh D, Sharma KK, Arora S, Singh AK, Gill SS, Singhal B (2017) Gel-based purification and biochemical study of laccase isozymes from *Ganoderma* sp. and its role in enhanced cotton callogenesis. Front Microbiol 8:674
Lamer S, Jungblut PR (2001) Matrix-assisted laser desorption-ionization mass spectrometry peptide mass fingerprinting for proteome analysis: identification efficiency after on-blot or in-gel digestion with and without desalting procedures. J Chromatogr B Biomed Sci Appl 752(2):311–322
Le Marquer M, Clemente HS, Roux C, Savelli B, dit Frey NF (2019) Identification of new signalling peptides through a genome-wide survey of 250 fungal secretomes. BMC Genomics 20(1):64. https://doi.org/10.1186/s12864-018-5414-2

Lebrun JD, Demont-Caulet N, Cheviron N, Laval K, Trinsoutrot-Gattin I, Mougin C (2011) Secretion profiles of fungi as potential tools for metal ecotoxicity assessment: a study of enzymatic system in *Trametes versicolor*. Chemosphere 82(3):340–345

Link AJ, Eng J, Schieltz DM, Carmack E, Mize GJ, Morris DR, Garvik BM, Yates JR (1999) Direct analysis of protein complexes using mass spectrometry. Nat Biotechnol 17(7):676

Liu D, Li J, Zhao S, Zhang R, Wang M, Miao Y, Shen Y, Shen Q (2013) Secretome diversity and quantitative analysis of cellulolytic *Aspergillus fumigatus* Z5 in the presence of different carbon sources. Biotechnology for Biofuels 6(1):149

Lu X, Sun J, Nimtz M, Wissing J, Zeng AP, Rinas U (2010) The intra-and extracellular proteome of *Aspergillus niger* growing on defined medium with xylose or maltose as carbon substrate. Microb Cell Factories 9(1):23

Machida M, Asai K, Sano M, Tanaka T, Kumagai T, Terai G, Kusumoto KI, Arima T, Akita O, Kashiwagi Y, Abe K (2005) Genome sequencing and analysis of *Aspergillus oryzae*. Nature 438(7071):1157

Maciel MJ, Ribeiro HC (2010) Industrial and biotechnological applications of ligninolytic enzymes of the basidiomycota: a review. Electron J Biotechnol 13(6):14–15

Manavalan A, Adav SS, Sze SK (2011) iTRAQ-based quantitative secretome analysis of *Phanerochaete chrysosporium*. J Proteome 75(2):642–654

Manavalan T, Manavalan A, Thangavelu KP, Heese K (2012) Secretome analysis of *Ganoderma lucidum* cultivated in sugarcane bagasse. J Proteom 77:298–309

Martinez D, Larrondo LF, Putnam N, Gelpke MD, Huang K, Chapman J, Helfenbein KG, Ramaiya P, Detter JC, Larimer F, Coutinho PM (2004) Genome sequence of the lignocellulose degrading fungus *Phanerochaete chrysosporium* strain RP78. Nat Biotechnol 22(6):695

Martinez D, Berka RM, Henrissat B, Saloheimo M, Arvas M, Baker SE, Danchin EG (2008) Genome sequencing and analysis of the biomass-degrading fungus *Trichoderma reesei* (syn. *Hypocrea jecorina*). Nature Biotechnol 26(5):553

Martínez D, Challacombe J, Morgenstern I, Hibbett DS, Schmoll M, Kubicek CP, Ferreira P, Ruiz-Duenas FJ, Martinez AT, Kersten P, Hammel KE, Vanden-Wymelenberg A, Gaskell J, Lindquist E, Sabat G, Bondurant SS, Larrondo LF, Canessa P, Vicuna R, Yadav J, Doddapaneni H, Subramanian V, Pisabarro AG, Lavin JL, Oguiza JA, Master E, Henrissat B, Coutinho PM, Harris P, Magnuson JK, Baker SE, Bruno K, Kenealy W, Hoegger PJ, Kues U, Ramaiya P, Lucas S, Salamov A, Shapiro H, Tu H, Chee CL, Misra M, Xie G, Teter S, Yaver D, James T, Mokrejs M, Pospisek M, Grigoriev IV, Brettin T, Rokhsar D, Berka R, Cullen D (2009) Genome, transcriptome, and secretome analysis of wood decay fungus *Postia placenta* supports unique mechanisms of lignocellulose conversion. Proc Natl Acad Sci U S A 106:1954–1959

Mathieu Y, Piumi F, Valli R, Aramburu JC, Ferreira P, Faulds C, Record E (2016) Secreted aryl alcohol quinone oxidoreductases from *Pycnoporus cinnabarinus* provides new activities and insights into fungal degradation of plant biomass. Appl Environ Microbiol 12:AEM-03761

Mattow J, Schaible UE, Schmidt F, Hagens K, Siejak F, Brestrich G, Haeselbarth G, Müller EC, Jungblut PR, Kaufmann SH (2003) Comparative proteome analysis of culture supernatant proteins from virulent Mycobacterium tuberculosis H37Rv and attenuated M. bovis BCG Copenhagen. Electrophoresis 24(19-20):3405–3420

Mayer AM, Staples RC (2002) Laccase: new functions for an old enzyme. Phytochemistry 60(6):551–565

Minden J (2007) Comparative proteomics and difference gel electrophoresis. Biotechniques 43(6):739–745

Minussi RC, Miranda MA, Silva JA, Ferreira CV, Aoyama H, Marangoni S, Rotilio D, Pastore GM, Durán N (2007) Purification, characterization and application of laccase from *Trametes versicolor* for colour and phenolic removal of olive mill wastewater in the presence of 1-hydroxybenzotriazole. Afr J Biotechnol 6(2):1–10

Miura N, Ueda M (2018) Evaluation of unconventional protein secretion by *Saccharomyces cerevisiae* and other fungi. Cell 7(9):128

Mukherjee P, Mani S (2013) Methodologies to decipher the cell secretome. Biochim Biophys Acta Protein Proteomics 1834(11):2226–2232

Nagele E, Vollmer M, Hörth P (2003) Two-dimensional nano-liquid chromatography-mass spectrometry system for applications in proteomics. J Chromatogr A 1009(1-2):197–205

Nitsche BM, Jørgensen TR, Akeroyd M, Meyer V, Ram AF (2012) The carbon starvation response of *Aspergillus niger* during submerged cultivation: insights from the transcriptome and secretome. BMC Genomics 13(1):380

O'Farrell PH (1975) High resolution two-dimensional electrophoresis of proteins. J Biol Chem 250(10):4007–4021

Pellegrin C, Morin E, Martin FM, Veneault-Fourrey C (2015) Comparative analysis of secretomes from ectomycorrhizal fungi with an emphasis on small-secreted proteins. Front Microbiol 18(6):1278

Phalip V, Delalande F, Carapito C, Goubet F, Hatsch D, Leize-Wagner E, Dupree P, Van Dorsselaer A, Jeltsch JM (2005) Diversity of the exoproteome of *Fusarium graminearum* grown on plant cell wall. Curr Genet 48(6):366–379

Phillips CM, Iavarone AT, Marletta MA (2011) Quantitative proteomic approach for cellulose degradation by *Neurospora crassa*. J Proteome Res 10(9):4177–4185

Piontek K, Smith AT, Blodig W (2001) Lignin peroxidase structure and function. Biochem Soc Trans 29:111–116

Poland J, Cahill MA, Sinha P (2003) Isoelectric focusing in long immobilized pH gradient gels to improve protein separation in proteomic analysis. Electrophoresis 24(7–8):1271–1275

Ravalason H, Jan G, Molle D, Pasco M, Coutinho PM, Lapierre C, Pollet B, Bertaud F, Petit-Conil M, Grisel S, Sigoillot JC (2008) Secretome analysis of *Phanerochaete chrysosporium* strain CIRM-BRFM41 grown on softwood. Appl Microbiol Biotechnol 80(4):719–733

Rohr CO, Levin LN, Mentaberry AN, Wirth SA (2013) A first insight into *Pycnoporus sanguineus* BAFC 2126 transcriptome. PLoS One 8(12):e81033

Ross PL, Huang YN, Marchese JN, Williamson B, Parker K, Hattan S, Khainovski N, Pillai S, Dey S, Daniels S, Purkayastha S (2004) Multiplexed protein quantitation in *Saccharomyces cerevisiae* using amine-reactive isobaric tagging reagents. Mol Cell Proteomics 3(12):1154–1169

Ruiz-Dueñas FJ, Martínez AT (2009) Microbial degradation of lignin: how a bulky recalcitrant polymer is efficiently recycled in nature and how we can take advantage of this. Microb Biotechnol 2(2):164–177

Ruiz-Dueñas FJ, Pogni R, Morales M, Giansanti S, Mate MJ, Romero A, Martínez AT (2009) Protein radicals in fungal versatile peroxidase catalytic tryptophan radical in both compound I and compound II and studies on W164Y, W164H, and W164S variants. J Biol Chem 284(12):7986–7994

San Ryu J, Shary S, Houtman CJ, Panisko EA, Korripally P, John FJ, Crooks C, Siika-aho M, Magnuson JK, Hammel KE (2011) Proteomic and functional analysis of the cellulase system expressed by *Postia placenta* during brown rot of solid wood. Appl Environ Microbiol 77:7933–7941

Sato S, Liu F, Koc H, Tien M (2007) Expression analysis of extracellular proteins from *Phanerochaete chrysosporium* grown on different liquid and solid substrates. Microbiology 153(9):3023–3033

Saykhedkar S, Ray A, Ayoubi-Canaan P, Hartson SD, Prade R, Mort AJ (2012) A time course analysis of the extracellular proteome of *Aspergillus nidulans* growing on sorghum stover. Biotechnology for biofuels. 5(1):52

Sharma KK (2016) Fungal genome sequencing: basic biology to biotechnology. Crit Rev Biotechnol 36(4):743–759

Sharma KK, Kuhad RC (2008) Laccase: enzyme revisited function redefined. Indian J Microbiol 48:309–316

Sharma KK, Singh D, Singh B, Kuhad RC (2013) Ligninolytic enzymes in environmental management. In: Biotechnology for environmental management and resource recovery. Springer, India, pp 219–238

Shevchenko A, Wilm M, Vorm O, Mann M (1996) Mass spectrometric sequencing of proteins from silver-stained polyacrylamide gels. Anal Chem 68(5):850–858

Sripuan T, Aoki K, Yamamoto K, Tongkao D, Kumagai H (2003) Purification and characterization of thermostable α-galactosidase from *Ganoderma lucidum*. Biosci Biotechnol Biochem 67(7):1485–1491
Steffen W, Crutzen PJ, McNeill JR (2007) The Anthropocene: are humans now overwhelming the great forces of nature. AMBIO 36(8):614–621
Stroher E, Dietz KJ (2006) Concepts and approaches towards understanding the cellular redox proteome. Plant Biol 8(4):407–418
Tomsovsky M, Homolka L (2003) Laccase and other ligninolytic enzyme activities of selected strains of *Trametes* spp. from different localities and substrates. Folia Microbiol 48(3):413
Trejo-Hernandez MR, Lopez-Munguia A, Ramirez RQ (2001) Residual compost of *Agaricus bisporus* as a source of crude laccase for enzymic oxidation of phenolic compounds. Process Biochem 36(7):635–639
Vasina DV, Pavlov AR, Koroleva OV (2016) Extracellular proteins of *Trametes hirsuta* s t. 072 induced by copper ions and a lignocellulose substrate. BMC Microbiol 16(1):106
Vincent D, Kohler A, Claverol S, Solier E, Joets J, Gibon J, Lebrun MH, Plomion C, Martin F (2011) Secretome of the free-living mycelium from the ectomycorrhizal basidiomycete *Laccaria bicolor*. J Proteome Res 11(1):157–171
Vasina DV, Loginov DS, Koroleva OV (2013) Comparative proteomic study of the basidiomycete *Trametes hirsuta* grown on different substrates. Biochem Mosc 78(5):477–484
Waanders LF, Hanke S, Mann M (2007) Top-down quantitation and characterization of SILAC-labeled proteins. J Am Soc Mass Spectrom 18(11):2058–2064
Wang B, Cai P, Sun W, Li J, Tian C, Ma Y (2015) A transcriptomic analysis of *Neurospora crassa* using five major crop residues and the novel role of the sporulation regulator rca-1 in lignocellulase production. Biotechnology for Biofuels 8(1):21
Washburn MP, Wolters D, Yates JR III (2001) Large-scale analysis of the yeast proteome by multidimensional protein identification technology. Nat Biotechnol 19(3):242
Wiese S, Reidegeld KA, Meyer HE, Warscheid B (2007) Protein labeling by iTRAQ: a new tool for quantitative mass spectrometry in proteome research. Proteomics 7(3):340–350
Wymelenberg AV, Minges P, Sabat G, Martinez D, Aerts A, Salamov A, Grigoriev I, Shapiro H, Putnam N, Belinky P, Dosoretz C (2006) Computational analysis of the *Phanerochaete chrysosporium* v2. 0 genome database and mass spectrometry identification of peptides in ligninolytic cultures reveal complex mixtures of secreted proteins. Fungal Genet Biol 43(5):343–356
Wymelenberg AV, Gaskell J, Mozuch M, Kersten P, Sabat G, Martinez D, Cullen D (2009) Transcriptome and secretome analyses of *Phanerochaete chrysosporium* reveal complex patterns of gene expression. Appl Environ Microbiol 75(12):4058–4068
Wymelenberg AV, Gaskell J, Mozuch M, Sabat G, Ralph J, Skyba O, Mansfield SD, Blanchette RA, Martinez D, Grigoriev I, Kersten PJ (2010) Comparative transcriptome and secretome analysis of wood decay fungi *Postia placenta* and *Phanerochaete chrysosporium*. Appl Environ Microbiol 76(11):3599–3610
Wymelenberg AV, Gaskell J, Mozuch M, BonDurant SS, Sabat G, Ralph J, Skyba O, Mansfield SD, Blanchette RA, Grigoriev IV, Kersten PJ (2011) Significant alteration of gene expression in wood decay fungi *Postia placenta* and *Phanerochaete chrysosporium* by plant species. Appl Environ Microbiol 77(13):4499–4507
Xie F, Liu T, Qian WJ, Petyuk VA, Smith RD (2011) Liquid chromatography-mass spectrometry-based quantitative proteomics. J Biol Chem 286:25443–25449
Yu GJ, Yin YL, Yu WH, Liu W, Jin YX, Shrestha A, Yang Q, Ye XD, Sun H (2015) Proteome exploration to provide a resource for the investigation of *Ganoderma lucidum*. PLoS One 10(3):e0119439
Zargar SM, Gupta N, Mir RA, Rai V (2016) Shift from gel based to gel free proteomics to unlock unknown regulatory network in plants: a comprehensive review. J Adv Res Biotechnol 1(1):19
Zeiner CA, Purvine SO, Zink EM, Pasa-Tolic L, Chaput DL, Haridas S, Wu S, LaButti K, Grigoriev IV, Henrissat B, Santelli CM (2016) Comparative analysis of secretome profiles of manganese (II)-oxidizing Ascomycete fungi. PLoS One 11(7):e0157844

Zhang Y, Fonslow BR, Shan B, Baek MC, Yates JR III (2013) Protein analysis by shotgun/bottom-up proteomics. Chem Rev 113(4):2343–2394

Zhu N, Liu J, Yang J, Lin Y, Yang Y, Ji L, Li M, Yuan H (2016) Comparative analysis of the secretomes of *Schizophyllum commune* and other wood-decay basidiomycetes during solid-state fermentation reveals its unique lignocellulose-degrading enzyme system. Biotechnol Biofuels 9(1):42

Znameroski EA, Coradetti ST, Roche CM, Tsai JC, Iavarone AT, Cate JH, Glass NL (2012) Induction of lignocellulose-degrading enzymes in *Neurospora crassa* by cellodextrins. Proc Natl Acad Sci 109(16):6012–6017

Zorn H, Bouws H, Takenberg M, Nimtz M, Getzlaff R, Breithaupt DE, Berger RG (2005) An extracellular carboxylesterase from the basidiomycete *Pleurotus sapidus* hydrolyses xanthophyll esters. Biol Chem 386(5):435–440

Fungi the Crucial Contributors for Nanotechnology: A Green Chemistry Perspective

11

Aliesha Moudgil and Bhushan P. Chaudhari

Abstract

Myconanotechnology has gained an exceptional momentum in today's world. It encompasses a wide range of applications like in agriculture, biomedical, electronics, textiles, cell and molecular biology, nanodevices, and many more. This fast-growing research field has a list of many nanoparticles that have been synthesized from different kinds of fungi. It is an eco-friendly approach that has sidelined the conventional methods of chemical and physical synthesis that requirè high energy and are toxic and expensive. The fungal-based synthesis gains functionality due to its discrete and diversified advantages that are being explored. Various parameters like temperature, pH, time and concentrations of fungal biomass or extract as well as of the precursor play a vital role in the protocols. These factors have to be optimized for the laboratory as well as industrial scale production. A greater importance to the use of cell-free extracts for nanosynthesis has been given. They contain the enzymes that are reported to be involved in the synthesis mechanisms as reducing agents and capping agents. The role of fungi in nanosynthesis and nanotechnology per se has many patents under its name. Although there are voids in the existing synthesis methods as well as lacunae in the application part that need to be addressed and filled. With an extensive research and work, myconanoparticles can be exploited to the maximum and lead its advancement in new areas.

Keywords

Mycosynthesis · Nanoparticles · Reductase · Synthesis · Applications

A. Moudgil · B. P. Chaudhari (✉)
Biochemical Sciences Division, CSIR-National Chemical Laboratory, Pune, Maharashtra, India
e-mail: bp.chaudhari@ncl.res.in

T. Satyanarayana et al. (eds.), *Advancing Frontiers in Mycology & Mycotechnology*,
https://doi.org/10.1007/978-981-13-9349-5_11

11.1 Introduction

Nanotechnology is an interdisciplinary area involving the convergence of fields like biology, chemistry, and material sciences. It owes its emergence to the esteemed and distinguished physicist Dr. Richard Feynman, who in his talk, "There's plenty of room at the bottom," had put forth this concept. At the nanoscale level, a change in the physical, electrical, mechanical, optical, and other fundamental properties of the nanoparticles is observed that can be exploited to distinct and diverse applications. This active research area utilizes the structures at 1 nm to 100 nm for the development and production of novel materials and devices. Nanotechnology employs chemical and physical synthesis of the nanoparticles that are explored to the maximum. The chemical mode is exercised most recurrently by dispersing the preformed polymers, polymerization of monomers, and ionic gelation of materials (Ingale 2013). Chemical reduction, microemulsion technique, UV-initiated photo reduction, photo-induced reduction, electrochemical synthetic method, irradiation method, and microwave-assisted synthesis are various other practices that are listed under chemical synthesis (Iravani et al. 2014), while the physical synthesis is generally carried out by evaporation-condensation and laser ablation (Iravani et al. 2014). Reports on ball milling techniques are available for the synthesis of ZnO nanoparticles (Salah et al. 2011) and Fe-Cu nanoparticles (Todaka et al. 2002). These methods not only pose toxicity as a by-product, impacting the human health, but are also cost-intensive with a surplus energy requirement (Fatima et al. 2015). Parameters like the stability and size of the nanoparticles are compromised along with aggregation of the particles leading to high polydispersity (Chen et al. 2003). This led to a mandatory shift of protocols to an approach that could supersede the shortcomings of the ongoing practices. Thus researchers steered to an eco-friendly line using plant extracts, microbes, and various biological particles like proteins, peptides, enzymes, and viruses for the nanoparticle synthesis (Ingale 2013). Microorganisms (bacteria, actinomycetes, fungi, yeasts, and algae) are used as biogenerators for this purpose. Talking about the biogenic production, there has been a growing trend of producing vaccines from cell lines. The huge cost of the vaccine production can be cut down and the scalable industrial production can be carried out (Hoeksema et al. 2018). This alternative approach is gaining tremendous impetus due to its cost-effectiveness, easy availability of the microbes, rapid and effortless synthesis, less energy requirement, and a less toxic mechanism (Saxena et al. 2014). The stability issues are alleviated in the bioinspired nanoparticles that show exceptional stability and solubility because of capping proteins and reducing agents present on their surface (Aziz et al. 2016). Thus preventing aggregation and excluding the use of capping agents that can cause toxicity like in chemical synthesis procedures (Syed 2012). The biogenic systems not only advance on physiological conditions of pH, temperature, and ionic strength (Alghuthaymi et al. 2015) but also have an impact and control over the size and shape of the particles (Balaji et al. 2009; Saxena et al. 2014). Bioremediation and bioleaching are known processes and the role of microorganisms in the same is well established. Researchers have exploited this property of the microbes in the area of nanotechnology for the nanosynthesis protocols. Extraction and accumulation of metals which in turn reduces the environmental pollution (Chen et al. 2003) and their ability of stress-induced reduction of toxic metals (Bhambure et al. 2009) by the high redox potential of the microbial enzymes is well harnessed (Fatima et al. 2015).

The relevance of fungi for this role has attained an eminent position due to its varied advantages over other microbes and plants. This has led to the advancement of mycosynthesis or myconanotechnology, where fungi are used as nanofactories for nanoparticle synthesis (Alghuthaymi et al. 2015). This position of fungi is because of the hefty amount of proteins secreted that catalyze heavy metals and lead to a comparable greater yield of nanoparticles at a quicker rate (Das et al. 2012; Saxena et al. 2014). Easy isolation and subculturing techniques at basic nutritional demands augment the use of fungi. Being totipotent, they can be grown by hyphae or spores (Saxena et al. 2014). The nanoparticles synthesized are highly monodisperse in comparison to other microbes derived particles (Hamedi et al. 2014). The increased surface area provided by the fungal mycelial mesh and the fact that mycelia can combat the flow pressure and agitation in the bioreactors or other chambers more effectively than the other biogenic factories, make fungi a good choice (Saha et al. 2010; Alghuthaymi et al. 2015). Furthermore, they are remarkably tolerant to metals and show a high wall binding tendency (Chen et al. 2003).

11.2 Fungi in Nanosynthesis

Intracellular and extracellular syntheses of nanoparticles are the two known mechanisms. In the intracellular scheme, electrostatic interactions form between the heavy metals and proteins or enzymes at the fungal cell wall. Further, the enzymes at the cell wall reduce the metal ions resulting in their aggregation and ultimately in the formation of nanoparticles (Kashyap et al. 2013). A study on silver nanoparticles synthesized from the fungus *Verticillium* revealed that the Ag^{+} ions interact with the carboxylate groups of the enzymes of the mycelia. These on reduction form silver nuclei, which eventually grows due to the accumulation of the reduced metal ions (Mukherjee et al. 2001). While during the extracellular synthesis a communication of the enzyme reductase that is released by the cells in the medium with the metal ions is pressed upon (Khan et al. 2018). This approach is more fitted for the purpose since it is prone to produce an immense variety of nanoparticles (Basavaraja et al. 2008). The intracellular synthesis is believed to have a strong domination over the shape and size of the nanoparticles. But lysis of the cells is an obligatory step to release the formed nanoparticles and require further purification that adds to its cost and increases inconvenience. On the contrary, the extracellularly precipitated nanoparticles are free from the unneeded cellular components culminating in their direct application (Aziz et al. 2016) (Fig. 11.1).

Amidst these two methods, fungi exhibits high extracellular synthesis making the downstream processing and biomass handling easier than bacteria and viruses, thus rendering it more economical (Das et al. 2012; Saxena et al. 2014). Mycogenic production can be practiced by either top-down or bottom-up approaches. A

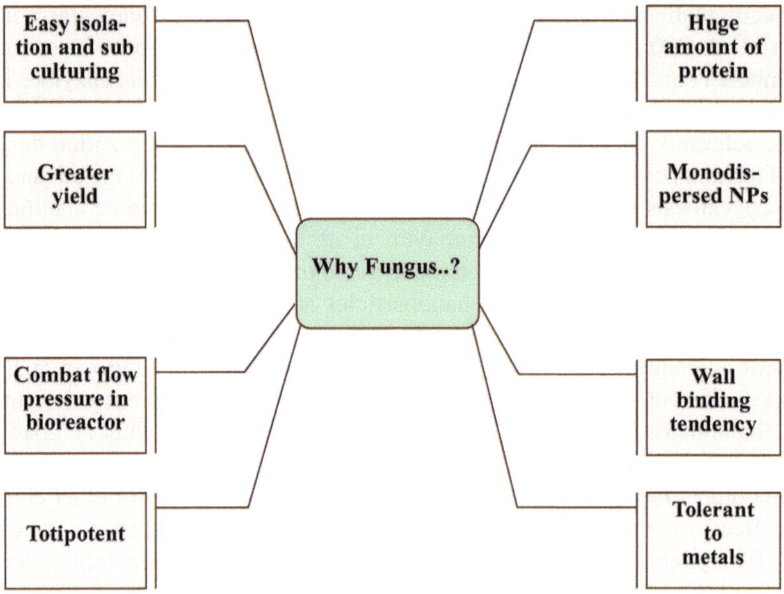

Fig. 11.1 Advantages of fungi in nanosynthesis

step-by-step construction or the self-assembly of atoms, molecules, or clusters is regarded as a bottom-up approach. The initially formed nanoparticles eventually assemble together as per their respective protocols either by chemical or biological methods. On the other hand, the top-down approach advances with the reduction of the specific material in size to form the final product. The bottom-up approach provides uniformity in the product with lesser defects as compared to the alternative approach that comes with surface imperfections of the product. Thus the former has an upper hand in the production of metal nanoparticles (Thakkar et al. 2010).

11.3 Mycosynthesis of Nanoparticles

Investigations have been carried out for the determination of the potential mechanisms of the fungi-mediated nanoparticle synthesis. All the probable ways convene to the enzymatic metal reduction by fungi. Based on the researches done experimenters have put forth the following possibilities. Extracellular synthesis for silver nanoparticles is reported in *Fusarium oxysporum* by the action of α-NADPH-dependent reductase, nitrate-dependent reductase, and extracellular quinone shuttle (Alghuthaymi et al. 2015). The same is corroborated in a study that was carried out where under anaerobic conditions nitrate reductase was incubated at 25 °C along with freshly prepared $AgNO_3$, phytochelatin, 4-hydroxyquinoline, and α-NADPH. The confirmation of the formed silver nanoparticles was made by the color change from colorless to brown at the end of the reaction. Spectrophotometric analysis revealed an absorption band at 260–270 nm of proteins, hydroxyquinoline, and α-NADPH. At 413 nm, a time-dependent increase in the intensity of surface plasmon resonance was observed. No surface plasmon resonance at 413 nm in the absence of the enzymes, α-NADPH, or hydroxyquinoline or phytochelatin in the reaction mixture confirmed the role of the enzymes in the reduction of silver from nitrate to nitrite and reduction of α-NADPH to α-$NADP^+$. It is proposed that hydroxyquinoline acts as an electron shuttle like quinones, transferring electrons to Ag^{2+} after nitrate reduction to form Ag (Anil Kumar et al. 2007). NADH-dependent reductase is considered to reduce the Ag^+ ions to form silver nanoparticles. Specificity of this reductase to *Fusarium oxysporum* was stated. A reaction with *Fusarium moniliforme* showed neither intracellular nor extracellular silver nanoparticle synthesis (Ahmad et al. 2003). Studies on *Fusarium semitectum* deny the specificity of this reductase to just *Fusarium oxysporum*, where reduction of Ag^+ to Ag^0 is observed under the same experimental conditions (Basavaraja et al. 2008). Another report dealing with *Fusarium oxysporum*-based synthesis suggests the action of reductase enzyme on Cd^{2+} and SO_4^{2-} to form CdS nanoparticles (Ahmad et al. 2002). *Fusarium oxysporum*, as per another investigation, is believed to form silica (SiO_2) and titania (TiO_2) by extracellular protein-mediated hydrolysis of SiF_6 and TiF_6. The possibility of hydrolysis by any non-proteinaceous component of the fungi like gums and gels was negated by adding only the non-proteinaceous parts in the reaction mixture, which resulted in no particle formation (Bansal et al. 2005). The hydrolyzing property of the extracellular enzymes secreted by the plant pathogen *Fusarium oxysporum* has been believed to be responsible for the synthesis of nanoscale ternary oxides particles like barium titanate (BT). A reaction of *F. oxysporum* with barium acetate and potassium hexaflurotitanate at room temperature led to the formation of nanocrystalline BT. These ferroelectric particles display ferroelectric-paraelectric transition at room temperature (Bansal et al. 2006). A rapidly growing oil fungus, *Mucor hiemalis*, is known to produce varied proteins and lipids. Its intrinsic property to tolerate extensive environmental conditions was taken advantage for the production of silver nanoparticles. The extracellular enzymes so produced extract Ag^+ ions from $AgNO_3$ environment and further reduce them to Ag^0 (elemental state) to form nanoscale silver (Aziz et al. 2016).

A strain of marine yeast *Yarrowia lipolytica* is known to produce reductase that has shown to reduce nitro groups of organic solvents. This property was explored to produce gold nanoparticles. Possibilities of the role of protease in the mechanism is under consideration as per earlier reports that reveal the action of proteases on gold precursors (Agnihotri et al. 2009). The fungi *Fusarium oxysporum* and *Verticillium* sp. were imposed upon with $K_3[Fe(CN)_6]$ and $K_4[Fe(CN)_6]$ salts for the production of nanoscale magnatites. Other iron salts were also used to curb the effect of toxicity caused by cyanide salts. Cationic proteins were reported for the extracellular hydrolysis of the iron salts for the synthesis of magnetite particles at room temperature (Bharde et al. 2006). Based on all the above mentioned reports a heavy weightage is levied upon the significant role of enzymes or proteins in the mycosynthesized nanoparticles protocols. This finding paved the way for the use of cell-free extracts for the same.

The use of cell-free extracts in the protocols instead of the microbial cells in the reaction mixture is gaining much momentum. In a study on gold nanoparticles formed from *Rhizopus oryzae*, AuNPs were reported to be formed by the enzyme glutathione reductase with NADH acting as a cofactor. The role of proteins in the respective mechanism was confirmed by using heat treated protein extract with the gold precursor ($HAuCl_4$), which resulted in no particle formation. It was believed that heating caused the proteins to denature (Das et al. 2012). This confirmed not only the eminent role of proteins in the mechanism but also the efficiency of the cell-free extracts. A recent leap from the cell-based reaction mixtures to the cell-free extracts is seen. The cell-mediated protocols are levied upon with the lysing of the microbial cells and then further isolation and purification of the nanoparticles, while the nanoparticles from the enzymatic extracts reduce the downstream processing step of purification, making the product suitable for industrial applications. As per a report enzyme nitrate reductase involved in the mechanism of nanosynthesis can be immobilized and can be recycled for the process, cutting down the cost and making it economical (Gholami-Shabani et al. 2014). Spectrophotometric, FTIR, and DLS analysis of the silver nanoparticles from the cell-free filtrate of *Neurospora intermedia* provided the confirmation of three steps in the nucleation mechanism of the formed monodisperse AgNPs. Initially the reducing agents in the reaction mixture reduce silver ions to their silver atoms (Ag^0). These form nucleation centers that form clusters by aggregation of atoms and also catalyze the further reduction process. This reduction process occurs frequently so as to attain larger particles. Lastly, the capping agents present stabilize the formed particles and prevent the further aggregation (Hamedi et al. 2014).

Marine organisms are known to produce exceptional nanoparticles in the size range of 1 nm to 100 nm that comprises of the nanofabrics like pearls. Marine bionanotechnology undoubtedly has a wide scope although the research in this area has not much undertaking. The property of the marine microorganisms present at the sea bottom since ages of reducing the elements can be explored to the extreme in the area of nanotechnology (Asmathunisha and Kathiresan 2013). In a study, gold nanoparticles were synthesized from *Aspergillus sydowii* which is a marine fungus. A relation of the size of the nanoparticles with the concentration of the gold precursor was shown. *Aspergillus sydowii* has been shown to effectively and efficiently synthesize nanoparticles. At lower precursor concentration there was synthesis of larger sized particles while at lower precursor concentrations the particles were of smaller size (Vala 2015). Another report suggests the synthesis of silver nanoparticles from a fungus isolated from mangrove root-soil, namely, *Penicillium fellutanum*. The extracellularly synthesized nanoparticles thus formed were said to be synthesized at a very rapid rate as compared to other microorganisms when controlled with basic parameters of pH, silver ion concentration, temperature, and time (Kathiresan et al. 2009).

Almost no signs of aggregation in a time-dependent manner show the high stability of the biologically synthesized nanoparticles. Extracellularly synthesized gold nanoparticles from *Aspergillus niger* have a reported zeta potential of −32.10 mV, which comments on the high stability of the AuNPs. No signs of aggregation are

believed to be the action of the proteins that act as capping agents. Based on the FTIR analysis, bands for amide linkages are observed in the infrared region. This confirms the presence of proteins in the product solution (Bhambure et al. 2009). This finding can be corroborated with another study of synthesis of silver nanoparticles from *Trichoderma viride*. An intense peak of the surface plasmon resonance at 421 nm specific to the silver nanoparticles confirms its synthesis, while an additional absorbance band at 270 nm is a characteristic peak of aromatic amino acids (tryptophan and tyrosine) that suggest the presence of proteins in the product. The FTIR measurements for the same show a band at 1650 cm^{-1}, specific to primary amine NH band. Secondary amine NH band at 1540 cm^{-1} and a primary amine CN stretch vibration of protein at 1040 cm^{-1} are visible in the spectra. Another band at 1425 cm^{-1} confirms the methyl scissoring vibration from protein. These results show close resemblance to the native proteins and specify no change in the secondary structure after their interaction with the silver nanoparticles (Mohammed Fayaz et al. 2009b). The FTIR analysis of the gold nanoparticles from *Humicola* sp. shows a spectrum with bands 1654 cm^{-1} and 1541 cm^{-1} that are specific for amide I and amide II of proteins, respectively. A possibility of the stability of the particles by the proteins that bind to the surface either by a free amino group or by cysteine residues is stated (Syed 2012). Results from another study of gold nanoparticles from *Penicillium* sp. concur with others. Absorption peaks at 1053.89, 1412.95, and 1626.37 cm^{-1} are observed for *P. citrinum* while 1061.02, 1407.11, and 1637.27 cm^{-1} are observed for *P. waksmanii*. Peaks near 3401 and 2919 cm^{-1} are for OH stretching. The 1626 cm^{-1} band is for the amide I that occur due to the carbonyl stretch in proteins, while 1041 cm^{-1} is the C-N stretching vibration of amine. Thus the presence of amide groups linkages and –COO- which is believed to be from the protein is confirmed. Thus through all the studies done the action of proteins as capping agents is well established and their role in the stabilization is also well acknowledged.

11.4 Factors Affecting Nanoparticle Synthesis

A number of known parameters can affect the fungus-mediated synthesis of nanoparticles. The effect of time on the synthesis of the silver nanoparticles by *Chrysosporium tropicum* and *Fusarium oxysporum* shows a production enhancement in a time-dependent manner based on micro-scan spectrum (Heinrich et al. 2011). A different study on silver nanoparticles from *Fusarium oxysporum* produced the similar kind of results. The fluorescence emission spectrum showed an increase in the intensity with time, suggesting increase in the production of particles (Ahmad et al. 2003).

A time reliant increase in the synthesis of nanoparticles with the increasing concentration of the *C. tropicum* and *F. oxysporum* also establishes that the concentration of the fungus or the fungal extract is an important parameter in the synthesis protocols (Heinrich et al. 2011).

The pH of solution is again important that plays a vital role. A study on gold nanoparticles synthesized from *Rhizopus oryzae* revealed an interesting change in the absorbance of the SPR band, size, and zeta potential of the particles with variation in the pH. They reported maximum stability of the protein capped gold nanoparticles at neutral, basic and slightly acidic conditions. The SPR absorbance at 538 nm observed a dip on decreasing the pH below 5.0, while on further lowering the pH below 3.0 an additional SPR band was observed at 670 nm. With this change in the pH conditions not only did the size increase from 26 nm at pH 7.0 to 250 nm at pH 2.0 but also switched the values of zeta potential from −25.3 ± 0.7 mV at pH 7.0 to +12.5 ± 2.5 mV at pH 2.0. The reason for this aggregation is believed to be due to lowering of the electrostatic repulsion by the protonation of the protein carboxylate groups of pKa~5. The intermolecular H bond thus formed between the protonated groups of the capped protein is also thought to be a cause for the aggregation (Das et al. 2012).

Temperature is another consideration during the fungal-based nanosynthesis. Silver nanoparticles from *Trichoderma viride* show SPR bands at different wavelengths at different temperatures. At 40 °C a narrow and sharp peak is recorded at 405 nm indicating the formation of the AgNPs. But the peak at 27 °C is observed at 420 nm and that of 10 °C at 451 nm. This pinpoints an increase in the size of the particles. Thus the study shows an inverse relationship between the reaction temperature and the size of the particles (Mohammed Fayaz, Balaji, Kalaichelvan, et al. 2009). Based on another study of gold nanoparticles from *Rhizopus oryzae*, no change in either the size or the charge of the particles was observed over a range of 30–100 °C. Particle aggregation was detected during prolonged heating at 100 °C. This can be implied to the denaturation of the proteins on the surface of the particles that imparts them stability (Das et al. 2012).

The ionic strength also plays an eminent role in determining the size of the particles. On increasing the concentration of NaCl in the suspension from 0 to 1.5 M, a shift in the SPR peak is observed which directly comments on the change in the size of the gold nanoparticles produced from *R. oryzae.* The increase in size is due to the aggregation of the particles at a higher ionic strength. This change is attributed to the neutralization of the charge present on the surface, which is confirmed by the zeta potential analysis. Thus the masking of the electrostatic repulsion between the negative charges of the proteins of the nanoparticles lead to the salt mediated effect (Das et al. 2012).

In a study of the synthesis of the gold nanoparticles from *V. luteoalbum*, an effect of the age of the cells in the culture is related to the size of the particles although there was no change noted in their shape. Cells in the later stage of the growth cycle yielded a comparatively lower number of particles than the cells earlier in the growth cycle. This can be supported by the possibility that cells produce a higher concentration of enzymes and proteins in their exponential phase leading to a much greater output (Thakkar et al. 2010). The precursor concentration in the reaction mixture also should be considered, as at higher concentrations the morphologies might get changed and there is a possibility of aggregation as well (Thakkar et al. 2010).

Thus the enlisted parameters need to be optimized according to the fungus used and also depending on the application of the product so as to achieve the desired results. Optimization of the parameters affecting the synthesis is mandatory during the industrial scale up as well. This leads to the synthesis of nanoparticles of specific shapes and sizes. Table 11.1 has a list of various nanoparticles of a specific shape and size synthesized from different fungal species.

11.5 Characterization of Nanoparticles

After the synthesis of nanoparticles from the respective methods, there is a need to characterize them so as to ensure the confirmatory production of the desired particle as well as its shape and size. The preliminary size determination is carried out by dynamic light scattering. DLS can also be used for the determination of the polydispersity index of the nanoparticles. Further, the size confirmation can be done using transmission electron microscopy (TEM), scanning electron microscopy (SEM), and atomic force microscopy (AFM). These techniques help to elucidate the size, shape, and morphology of the formed nanoparticles. By energy-dispersive X-ray (EDX) analysis, the components of the specific nanoparticles can be determined. In a few studies, there is a need of high-resolution microscopy for which HRTEM, i.e., high-resolution TEM, is performed. The surface charge can be analyzed by measuring the zeta potential. The crystal structure of the particle along with its composition can be deduced by X-ray diffraction (XRD). The preliminary step to verify the synthesis can be done by UV-Vis spectroscopy which works on the surface plasmon resonance that is characteristic to particular nanoparticles. Thus the optical properties and the concentration can be effectively elucidated by this technique. Fourier transform infrared (FTIR) spectroscopy can be carried out to determine the functional groups. The presence of proteins on the surface of the nanoparticles can be proved and determined with this (Ingale 2013; Rai and Posten 2013; Mourdikoudis et al. 2018; Zhao et al. 2018).

11.6 Applications of Myconanoparticles

The optical, electrical, photochemical, and electronic characteristics of the fungal-derived nanoparticles have paved their path to varied applications. These fungal-based nanoparticles have been put into use industrially as well as in small scale. Scribbled below are the various applications of the fungal-mediated nanoparticles:

- The extracellular biosynthesis of gold nanoparticles since free from the toxic elements can be used in a huge scale production, which is not possible in intracellular synthesis as the nanoparticles are attached to the cells. The particles so produced have an upper hand in the homogeneous catalysis and nonlinear optics (Bhambure et al. 2009).

Table 11.1 Synthesis of nanoparticles from fungi

Nanoparticle	Fungi	Shape	Size	Synthesis mode	References
Au	*Humicola* spp.	Spherical	18–24 nm	Extracellular	Syed (2012)
Au	*Bipolaris tetramera*	Spherical	58.4 nm	Extracellular	Fatima et al. (2015)
		Triangular	110.13 nm		
		Hexagonal	261.73 nm		
Au	*Trichothecium* sp.	Spherical, triangular, and hexagonal	5–200 nm	Extracellular	Ahmad et al. (2005)
		Spherical	10–25 nm	Intracellular	
Au	*Neurospora crassa*	Triangular, hexagonal, and pentagonal	10–200 nm	Neurospora crassa extract	Quester et al. (2013)
		Quasi-spherical	3–12 nm		
Au	*Penicillium* sp.	Spherical	45 nm	Extracellular	Du et al. (2011)
			50 nm	Intracellular	
Au	*Fusarium semitectum*	Spherical	18–80 nm	Extracellular	Dasaratrao Sawle et al. (2008)
Au-Ag			10–35 nm		
Au	*Lentinula edodes*	Spherical	5–50 nm	Intracellular	Vetchinkina et al. (2018)
Au	*Aspergillus Niger*	Spherical, elliptical	12.79 ± 5.61 nm	Extracellular	Bhambure et al. (2009)
Au	*Rhizopus oryzae*	Spherical	5–65 nm	*Rhizopus oryzae* extract	Das et al. (2012)
Ag	*Rhizopus stolonifera*	Spherical	5–50 nm	Extracellular	Afreen and Ranganath (2011)
Ag	*Cladosporium cladosporioides*	Spherical	10–100 nm	Extracellular	Balaji et al. (2009)
Ag	*Fusarium semitectum*	Spherical	10–60 nm	Extracellular	Basavaraja et al. (2008)
Ag	*Trichoderma viride*	Spherical	2–4 nm	*Trichoderma viride* filtrate	Mohammed Fayaz et al. (2009b)
		Spherical and rod shaped	10–40 nm		
		Nanoplates	80–100 nm		
Ag	*Fusarium oxysporum*	Spherical	50 nm	Purified enzymes	Gholami-Shabani et al. (2014)

Ag	*Mucor hiemalis*	Spherical	5–15 nm	Extracellular	Aziz et al. (2016)
Ag	*Bipolaris nodulosa*	Spherical, hexahedral, triangular, and semi-pentagonal	10–60 nm	Extracellular	Saha et al. (2010)
Ag	*Bipolaris tetramera*	Spherical	109.4 nm	Extracellular	Fatima et al. (2015)
Ag	*Trichoderma longibrachiatum*	Spherical	24.43 nm	Extracellular	Elamawi et al. (2018)
Ag	*Fusarium oxysporum*	Spherical	20–50 nm	Extracellular	Durán et al. (2005)
Ag	*Aspergillus niger* and *Aspergillus terreus*	Spherical	26.5–100 nm	Extracellular	Baskar et al. (2015)
Ag	*Beauveria bassiana*	Spherical	36.88–60.93 nm	Extracellular	Banu and Balasubramanian (2014)
Ag	*Aspergillus fumigatus*	Spherical and triangular	5–25 nm	Extracellular	Bhainsa and D'Souza (2006)
Ag	*Alternaria alternata*	Spherical	20–60 nm	Extracellular	Gajbhiye et al. (2009)
Ag	*Aspergillus Niger*	Spherical	20 nm	Extracellular	Gade et al. (2008)
Gd_2O_3 (gadolinium oxide)	*Humicola* sp.	Quasi-spherical	3–8 nm	Extracellular	Khan et al. (2014)
$BaTiO_3$ (barium titanate)	*Fusarium oxysporum*	Quasi-spherical	4–5 nm	Extracellular	Bansal et al. (2006)
CdS	*Fusarium oxysporum*	Hexagonal	5–20 nm	Extracellular	Ahmad et al. (2002)
SiO_2 (silica)	*Fusarium oxysporum*	Quasi-spherical	5–15 nm	Extracellular	Bansal et al. (2005)
TiO_2 (titania)	*Fusarium oxysporum*	Spherical	6–13 nm	Extracellular	Bansal et al. (2005)
Magnetite	*Fusarium oxysporum*	Quasi-spherical	20–50 nm	Extracellular	Bharde et al. (2006)
PbS (lead sulfide)	*Rhodosporidium diobovatum*	Spherical	2–5 nm	Intracellular	Seshadri et al. (2011)

- *Trichoderma viride* and *Hypocrea lixii* synthesized gold nanoparticles have been reported to show exceptional biocatalyst characteristic as they reduce 4-nitrophenol to 4-aminophenol in the presence of $NaBH_4$. These particles have also been reported to be used as antimicrobial agents against pathogenic bacteria (Mishra et al. 2014).
- An experimental study of biosynthesized gold nanoparticles from an entomopathogenic fungus *Beauveria bassiana* shows their biolarvicide action against dengue vector *Aedes aegypti* (Banu and Balasubramanian 2014). Similarly, silver nanoparticles synthesized from *Cochliobolus lunatus* show the same application to the same effect against *Aedes aegypti* and *Anopheles stephensi* (Salunkhe et al. 2011; Banu and Balasubramanian 2014). There have been a report mentioning the antimosquito activity as well (Soni and Prakash 2012). Thus the nanoparticles can be utilized for vector control.
- The use of silver nanoparticles as an antimicrobial agent is quite evident and proven. But the use of these biosynthesized nanoparticles have been reported as wound healing agents that are believed to regulate the cytokines in the mechanism (Sundaramoorthi et al. 2009).
- A proclaimed use of the silver nanoparticles from *Fusarium oxysporum* in materials such as cloth has been mentioned. The sterility attained by the incorporation of the silver nanoparticles is exploited by their use in hospitals against pathogenic bacteria like *S. aureus* (Durán et al. 2007).
- The role of nanoparticles in the preservation of fruits and vegetables has been stated. Based on a study silver nanoscale particles from *Trichoderma viride* when added with sodium alginate film shows antibacterial activity. This film when layered on carrots and pear increase their shelf life (Mohammed Fayaz et al. 2009a).
- In addition to the role of myconanoparticles in the preservation of fruits and vegetable, their role in other areas of agriculture is well reported. This can be achieved by nanoclays and nanofilms. These fungal nanoparticles have an established suited role as a delivery agent. They have a function in the delivery of fertilizers, fungicides, and varied chemicals in the form of nanocapsules or nanoemulsions. Their antifungal and antibacterial property is explored and exploited in the field of agriculture. They are used for antimicrobial sterilization as well. Myconanoparticles mediated improvement in the plant resistance against diseases is noted. The successful genetic transformation of the plants by these nanoparticles is confirmed by the gene expression studies. Thus loopholes in the area of plant-pathogen interactions as well as disease management can be addressed by myconanotechnology (Alghuthaymi et al. 2015).
- An innovative approach to PCR using a conjugate of the gold and silver nanoparticles from *F. oxysporum* with the master mix and *Candida* sp. DNA has been put forth. This shows an effectively high specificity and sensitivity for the identification of *Candida* species (Bansod et al. 2013).
- Myconanoparticles have invaded the ferroelectric applications as well. BaTiO3 (BT), an electroceramic dielectric material is used in thin-film capacitors, pyroelectric detectors, infrared imaging systems, nonvolatile memories and

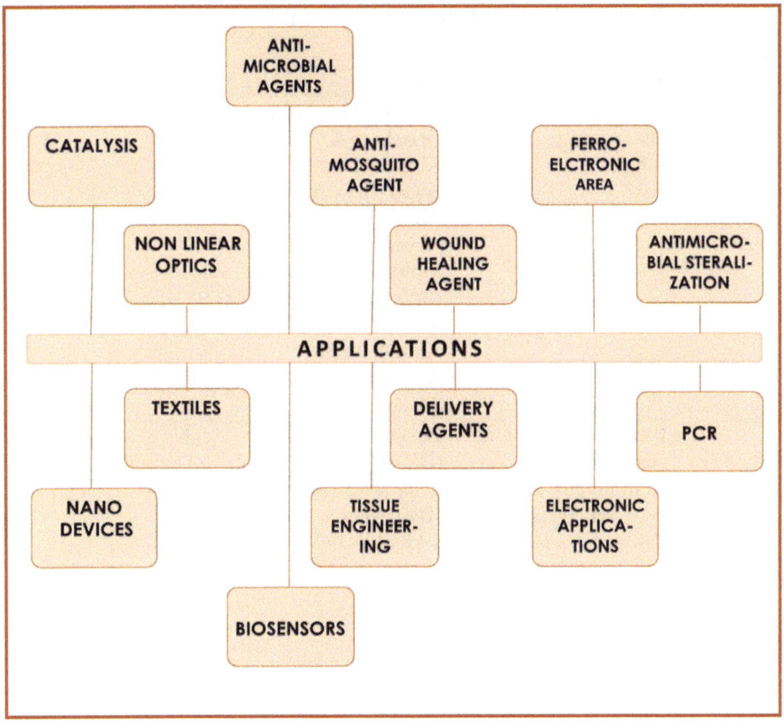

Fig. 11.2 Applications of myconanoparticles

microelectronic industry. An eco-friendly approach for the synthesis of BT has been adopted to curb the need of extreme temperature, pressure, and pH conditions. Biological synthesis aims at proceeding with an energy saving strategy thus making it more economical (Bansal et al. 2006).

- The significant characteristic of fungus as nanofactories is its control over the shape and size of the nanoparticles. This is taken advantage in using mycosynthesized nanoparticles in nanodevices like single electron transistors, light-emitting diodes, and photoelectrochemical cells (Seshadri et al. 2011).
- Fungal-based nanowires have also been developed that have their role in electronic devices at a nanoscale. Myconanoparticles have also had their way in biosensors, tissue engineering, and drug delivery as nanofibrous mats (Kashyap et al. 2013) (Fig. 11.2).

11.7 Patents

Patents pertaining to the role of fungi in the field of nanotechnology are in Table 11.2. A hike in the production efficiencies of sulfur-free spherical particles was achieved by yeast grown in a medium where the respective element for the product

Table 11.2 Patents on the role of fungi in nanotechnology

Publication number	Title	Inventors	Publication date	References
US 009701552 B1	Synthesis of silver nanoparticles using fungi.	Khalid Mustafa Osman Ortashi (SA), Manal Ahmed Gasmelseed Awad (SA), Awatif Ahmed Hendi (SA), Abeer Ramadan Mohamed Abdelaziz (SA), Ahmed Sameer Ahmed Hendi (SA), Abdulhakeem Abdulmuhseen Alahmed (SA)	11 July 2017	Osman Ortashi et al. (2016)
US 20140363871 A1	Process for obtaining copper nanoparticles from fungus selected between *Hypocrea lixii* and *Trichoderma koningiopsis* in bioremediation of wastewater and production of copper nanoparticles	Benedito CORREA (BR), Claudio Augusto Oller NASCIMENTO (BR), Marcia Rengina SALVADORI (BR)	11 December 2014	Corrêa et al. (2014)
US 008986975 B2	Production of sulfur-free nanoparticles by yeast	Zoltan Mester (CA), Laurent Ouerdane (FR)	24 March 2015	Ouerdane and Laurent (2009)
US 008394421 B2	Synthesis of nanoparticles by fungi	G. Ali Mansoori (US)	12 March 2013	Mansoori (2008)
US 20120070376 A1, WO 2012024229 A1	Yeast cell wall particles for receptor-targeted nanoparticle delivery	Gray R. OSTROFF (US), Ernesto SOTO (US)	22 March 2012	Ostroff and Soto (2019)

nanoparticle is reduced to a lower oxidation state (US 8986975B2) [Ouerdane and Laurent 2009]. Metal nanoparticles were prepared by fungi from the reaction mixture containing the fungal biomass and metal salt over a temperature range of 35–60 °C (US 9701552B1) [Osman Ortashi et al. 2016]. Silver nanoparticles were extracellularly synthesized from Trichoderma reesei. The mechanism occurs through catalytic method where enzymes or other metabolites act as reducing agents (US 8394421B2) (li Mansoori 2008). A cost-effective and an eco-friendly approach has been developed for the production of copper nanoparticles from *Hypocrea lixii* or *Trichoderma koningiopsis*. An effective scale up of these particles along with the bioremediation of wastewater has been carried out by the dead biomass of either of the two microorganisms (US 2014/0363871A1) [Corrêa et al. 2014].

11.8 Challenges

Although luring and charming the whole mechanism seems there are still challenges to be tackled in fungal nanotechnology, which need attention. Various reactants can impose toxicity to the microorganisms that can lead to low production or no production at all. Even the products formed can be toxic. If the sterility of the environment is compromised it can lead to undesirable outcomes. Maintaining a sterile and contamination-free setup again adds up to the cost of production. The incurring media cost and the cost of the huge reaction setups for a scalable process can cause a hindrance (Juibari et al. 2011). Silver nanoparticles have been observed to take a longer time when synthesized from fungi as compared to other microorganisms (Zhao et al. 2018). During the intracellular synthesis of nanoparticles from fungi, the particles since formed inside the cells need to be recovered. After the effective recovery, it takes an additional step for purification during the downstream processing. Due to these added steps not only the cost of production hits a hike but also the whole process becomes very tiresome and requires additional instruments for the process (Saxena et al. 2014). Further while scaling up the whole process for an industrial application, the use of the fungal cells for the synthesis can become quite inconvenient (Gholami-Shabani et al. 2014). Although myconanotechnology has advanced in the field of agriculture, there are many shortcoming and drawbacks that need to be intensely taken up. Nanoscale aluminum oxide inhibits root elongation in plants like cucumber, soybean, etc. Zinc oxide nanoparticles have been shown to be toxic inhibiting root elongation in some plants maize, carrots, etc. There are reports on the inhibition of germination and seedling lengths. These myconanoparticles have been seen to hamper the growth of beneficial bacteria in the wastewater treatment. With the increased use of the nanoparticles in varied fields, their degradation is necessary so as to prevent the accumulation in the environment (Kashyap et al. 2013).

11.9 Conclusions and Future Prospects

The chapter makes the role and benefits of myconanoparticles quite evident. A switch from the conventional techniques to a greener approach that too a fungal-mediated synthesis has not only yielded nanoparticles of the desired shape and size but also has levied a control over the process. The cost-effectiveness and its scalable property has bought it in a list of varied applications and yet more to be explored. The high stability incurring due to the surface bound proteins provides an added advantage. All the stated and proved facts put fungi at a preferable position to be used as nanofactories. The advent of these fungal produced particles in numerous fields makes this approach very attractive and highly functional. But this area is also tagged along with shortcomings for more research can be carried out to curb the limitations. The use of extremophilic microorganisms that can thrive in extreme and adverse conditions can be explored more that can acknowledge a few enlisted shortcomings. The use of marine fungus in nanotechnology is not much explored. Various marine fungi can be used to synthesize varieties of desired nanoparticles and can be put into different applications. There has been noted loophole in the area of agriculture where myconanoparticles have been used. There is a need to address these lacunae so as to ensure efficient growth and protection of the plants. Various other fields can be ventured into where the application of these particles can be explored to an utmost level. For the industrial scaling-up strategies, all the limitations addressed above should be taken up so as to ensure a cost-effective production. There is a great scope in this area to make it an industrial buzz and highly economical.

References

Afreen VR, Ranganath E (2011) Synthesis of monodispersed silver nanoparticles by Rhizopus Stolonifer and its antibacterial activity against MDR strains of pseudomonas aeruginosa from burnt patients. Int J Environ Sci 1:1830–1840. Integrated Publishing Association

Agnihotri M, Joshi S, Kumar AR, Zinjarde S, Kulkarni S (2009) Biosynthesis of gold nanoparticles by the tropical marine yeast *Yarrowia lipolytica* NCIM 3589. Mater Lett 63:1231–1234. https://doi.org/10.1016/J.MATLET.2009.02.042.. North-Holland

Ahmad A, Mukherjee P, Mandal D, Senapati S, Islam Khan M, Rajiv K, Murali S (2002) Enzyme mediated extracellular synthesis of CdS nanoparticles by the fungus, fusarium oxysporum. Am Chem Soc 124:12108. https://doi.org/10.1021/JA027296O

Ahmad A, Mukherjee P, Senapati S, Mandal D, Islam Khan M, Kumar R, Sastry M (2003) Extracellular biosynthesis of silver nanoparticles using the fungus fusarium oxysporum. Colloids Surf B: Biointerfaces 28:313–318. https://doi.org/10.1016/S0927-7765(02)00174-1. Elsevier

Ahmad A, Senapati S, Khan MI, Kumar R, Sastry M (2005) Extra-/intracellular biosynthesis of gold nanoparticles by an alkalotolerant fungus, Trichothecium sp. J Biomed Nanotechnol 1:47–53. https://doi.org/10.1166/jbn.2005.012

Alghuthaymi MA, Almoammar H, Rai M, Said-Galiev E, Abd-Elsalam KA (2015) Myconanoparticles: synthesis and their role in phytopathogens management. Biotechnol Biotechnol Equip 29:221–236. https://doi.org/10.1080/13102818.2015.1008194

Anil Kumar S, Abyaneh MK, Gosavi SW, Kulkarni SK, Pasricha R, Ahmad A, Khan MI (2007) Nitrate reductase-mediated synthesis of silver nanoparticles from AgNO3. Biotechnol Lett 29:439–445. https://doi.org/10.1007/s10529-006-9256-7

Asmathunisha N, Kathiresan K (2013) A review on biosynthesis of nanoparticles by marine organisms. Colloids Surf B: Biointerfaces 103:283–287. https://doi.org/10.1016/J.COLSURFB.2012.10.030. Elsevier

Aziz N, Pandey R, Barman I, Prasad R (2016) Leveraging the attributes of Mucor hiemalis-derived silver nanoparticles for a synergistic broad-Spectrum antimicrobial platform. Front Microbiol 7. https://doi.org/10.3389/fmicb.2016.01984

Balaji DS, Basavaraja S, Deshpande R, Bedre Mahesh D, Prabhakar BK, Venkataraman A (2009) Extracellular biosynthesis of functionalized silver nanoparticles by strains of Cladosporium cladosporioides fungus. Colloids Surf B: Biointerfaces 68:88–92. https://doi.org/10.1016/J.COLSURFB.2008.09.022. Elsevier

Bansal V, Rautaray D, Bharde A, Ahire K, Sanyal A, Ahmad A, Sastry M (2005) Fungus-mediated biosynthesis of silica and titania particles. J Mater Chem 15:2583. https://doi.org/10.1039/b503008k. The Royal Society of Chemistry

Bansal V, Poddar P, Ahmad A, Sastry M (2006) Room-temperature biosynthesis of ferroelectric barium titanate nanoparticles. Am Chem Soc 128:11958. https://doi.org/10.1021/JA063011M

Bansod S, Bonde S, Tiwari V, Bawaskar M, Deshmukh S, Gaikwad S, Gade A, Rai M (2013) Bioconjugation of gold and silver nanoparticles synthesized by Fusarium oxysporum and their use in rapid identification of Candida species by using bioconjugate-nano-polymerase chain reaction. J Biomed Nanotechnol 9:1962–1971

Banu AN, Balasubramanian C (2014) Myco-synthesis of silver nanoparticles using Beauveria bassiana against dengue vector, Aedes aegypti (Diptera: Culicidae). Parasitol Res 113:2869–2877. https://doi.org/10.1007/s00436-014-3948-z

Basavaraja S, Balaji SD, Lagashetty A, Rajasab AH, Venkataraman A (2008) Extracellular biosynthesis of silver nanoparticles using the fungus *Fusarium semitectum*. Mater Res Bull 43:1164–1170. https://doi.org/10.1016/J.MATERRESBULL.2007.06.020. Pergamon

Baskar G, Chandhuru J, Fahad KS, Praveen AS, Bharathi R, Fyna S (2015) Mycological synthesis and characterization of silver nanoparticles by *Aspergillus* species. J Chem Pharm Res 7:300–306

Bhainsa KC, D'Souza SF (2006) Extracellular biosynthesis of silver nanoparticles using the fungus *Aspergillus fumigatus*. Colloids Surf B: Biointerfaces 47:160–164. https://doi.org/10.1016/j.colsurfb.2005.11.026

Bhambure R, Bule M, Shaligram N, Kamat M, Singhal R (2009) Extracellular biosynthesis of gold nanoparticles using aspergillus Niger - its characterization and stability. Chem Eng Technol 32:1036–1041. https://doi.org/10.1002/ceat.200800647. Wiley-Blackwell

Bharde A, Rautaray D, Bansal V, Ahmad A, Sarkar I, Yusuf SM, Sanyal M, Sastry M (2006) Extracellular biosynthesis of magnetite using fungi. Small 2:135–141. https://doi.org/10.1002/smll.200500180

Chen JC, Lin ZH, Ma XX (2003) Evidence of the production of silver nanoparticles via pretreatment of *Phoma* sp.3.2883 with silver nitrate. Lett Appl Microbiol 37:105–108. https://doi.org/10.1046/j.1472-765X.2003.01348.x

Corrêa B, Nascimento CAO, Salvadori MR (2014) Process for obtaining copper nanoparticles from a fungus selected between *Hypocrea lixii* and trichoderma koningiopsis and use of fungi selected between *Hypocrea lixii* and *Trichoderma koningiopsis* in bioremediation of wastewater and production of copper na. United States

Das SK, Dickinson C, Lafir F, Brougham DF, Marsili E (2012) Synthesis, characterization and catalytic activity of gold nanoparticles biosynthesized with Rhizopus oryzae protein extract. Green Chem 14:1322. https://doi.org/10.1039/c2gc16676c. Royal Society of Chemistry

Dasaratrao Sawle B, Salimath B, Deshpande R, Bedre MD, Prabhakar BK, Venkataraman A (2008) Biosynthesis and stabilization of Au and Au-Ag alloy nanoparticles by fungus, Fusarium semitectum. Sci Technol Adv Mater 9:035012. https://doi.org/10.1088/1468-6996/9/3/035012. Taylor & Francis

Du L, Xian L, Feng J-X (2011) Rapid extra−/intracellular biosynthesis of gold nanoparticles by the fungus *Penicillium* sp. J Nanopart Res 13:921–930. https://doi.org/10.1007/s11051-010-0165-2

Durán N, Marcato PD, Alves OL, De Souza GIH, Esposito E (2005) Mechanistic aspects of biosynthesis of silver nanoparticles by several *Fusarium oxysporum* strains. J Nanobiotechnol 3:8. https://doi.org/10.1186/1477-3155-3-8. BioMed Central

Durán N, Marcato PD, De Souza GIH, Alves OL, Esposito E (2007) Antibacterial effect of silver nanoparticles produced by fungal process on textile fabrics and their effluent treatment. J Biomed Nanotechnol 3:203–208. https://doi.org/10.1166/jbn.2007.022

Elamawi RM, Al-Harbi RE, Hendi AA (2018) Biosynthesis and characterization of silver nanoparticles using *Trichoderma longibrachiatum* and their effect on phytopathogenic fungi. Egypt J Biol Pest Control 28:28. https://doi.org/10.1186/s41938-018-0028-1. Springer Berlin Heidelberg

Fatima F, Bajpai P, Pathak N, Singh S, Priya S, Verma S (2015) Antimicrobial and immunomodulatory efficacy of extracellularly synthesized silver and gold nanoparticles by a novel phosphate solubilizing fungus Bipolaris tetramera. BMC Microbiol 15:52. https://doi.org/10.1186/s12866-015-0391-y

Gade AK, Bonde P, Ingle AP, Marcato PD, Durán N, Rai MK (2008) Exploitation of *Aspergillus niger* for synthesis of silver nanoparticles. J Biobaased Mater Bioenergy 2:243–247. https://doi.org/10.1166/jbmb.2008.401

Gajbhiye M, Kesharwani J, Ingle A, Gade A, Rai M (2009) Fungus-mediated synthesis of silver nanoparticles and their activity against pathogenic fungi in combination with fluconazole. Nanomedicine 5:382–386. https://doi.org/10.1016/j.nano.2009.06.005

Gholami-Shabani M, Akbarzadeh A, Norouzian D, Amini A, Gholami-Shabani Z, Imani A, Chiani M, Riazi G, Shams-Ghahfarokhi M, Razzaghi-Abyaneh M (2014) Antimicrobial activity and physical characterization of silver nanoparticles green synthesized using nitrate reductase from *Fusarium oxysporum*. Appl Biochem Biotechnol 172:4084–4098. https://doi.org/10.1007/s12010-014-0809-2

Hamedi S, Shojaosadati SA, Shokrollahzadeh S, Hashemi-Najafabadi S (2014) Extracellular biosynthesis of silver nanoparticles using a novel and non-pathogenic fungus, Neurospora intermedia: controlled synthesis and antibacterial activity. World J Microbiol Biotechnol 30:693–704. https://doi.org/10.1007/s11274-013-1417-y

Heinrich N, Prakash S, Soni N, Prakash S (2011) Factors affecting the geometry of silver nanoparticles synthesis in *Chrysosporium Tropicum* and fusarium Oxysporum. Am J Nanotechnol 2:112–121. https://doi.org/10.3844/ajnsp.2011.112.121.. Science Publications

Hoeksema F, Karpilow J, Luitjens A, Lagerwerf F, Havenga M, Groothuizen M, Gillissen G et al (2018) Enhancing viral vaccine production using engineered knockout vero cell lines – a second look. Vaccine 36:2093–2103. https://doi.org/10.1016/j.vaccine.2018.03.010

Ingale AG (2013) Biogenic synthesis of nanoparticles and potential applications: an eco- friendly approach. J Nanomed Nanotechnol 04:1–7. https://doi.org/10.4172/2157-7439.1000165

Iravani S, Korbekandi H, Mirmohammadi SV, Zolfaghari B (2014) Synthesis of silver nanoparticles: chemical, physical and biological methods. Res Pharmaceut Sci 9:385–406. Wolters Kluwer -- Medknow Publications

Juibari M, Motamedi S, Abbasalizadeh GSJ, Noruzi M (2011) Intensified biosynthesis of silver nanoparticles using a native extremophilic *Ureibacillus thermosphaericus* strain. Mater Lett 65:1014–1017. https://doi.org/10.1016/J.MATLET.2010.12.056.. North-Holland

Kashyap PL, Kumar S, Srivastava AK, Sharma AK (2013) Myconanotechnology in agriculture: a perspective. World J Microbiol Biotechnol 29:191–207. https://doi.org/10.1007/s11274-012-1171-6. Springer

Kathiresan K, Manivannan S, Nabeel MA, Dhivya B (2009) Studies on silver nanoparticles synthesized by a marine fungus, *Penicillium fellutanum* isolated from coastal mangrove sediment. Colloids Surf B: Biointerfaces 71:133–137. https://doi.org/10.1016/j.colsurfb.2009.01.016

Khan SA, Gambhir S, Ahmad A (2014) Extracellular biosynthesis of gadolinium oxide (Gd2 O3) nanoparticles, their biodistribution and bioconjugation with the chemically modified anticancer drug taxol. Beilstein J Nanotechnol 5:249–257. https://doi.org/10.3762/bjnano.5.27. Beilstein-Institut

Khan AU, Malik N, Khan M, Cho MH, Khan MM (2018) Fungi-assisted silver nanoparticle synthesis and their applications. Bioprocess Biosyst Eng 41:1–20. https://doi.org/10.1007/s00449-017-1846-3

Mansoori GA (2008) Synthesis of nanoparticles by fungi. United States

Mishra A, Kumari M, Pandey S, Chaudhry V, Gupta KC, Nautiyal CS (2014) Biocatalytic and antimicrobial activities of gold nanoparticles synthesized by *Trichoderma* sp. Bioresour Technol 166:235–242. https://doi.org/10.1016/j.biortech.2014.04.085

Mohammed Fayaz A, Balaji K, Girilal M, Kalaichelvan PT, Venkatesan R (2009a) Mycobased synthesis of silver nanoparticles and their incorporation into sodium alginate films for vegetable and fruit preservation. J Agric Food Chem 57:6246–6252. https://doi.org/10.1021/jf900337h.. American Chemical Society

Mohammed Fayaz A, Balaji K, Kalaichelvan PT, Venkatesan R (2009b) Fungal based synthesis of silver nanoparticles—an effect of temperature on the size of particles. Colloids Surf B: Biointerfaces 74:123–126. https://doi.org/10.1016/J.COLSURFB.2009.07.002. Elsevier

Mourdikoudis S, Pallares RM, Thanh NTK (2018) Characterization techniques for nanoparticles: comparison and complementarity upon studying nanoparticle properties. Nanoscale 10:12871–12934. https://doi.org/10.1039/C8NR02278J. Royal Society of Chemistry

Mukherjee P, Ahmad A, Mandal D, Satyajyoti Senapati SR, Sainkar M, Renu Parishcha IK et al (2001) Fungus-mediated synthesis of silver nanoparticles and their immobilization in the mycelial matrix: a novel biological approach to nanoparticle synthesis. Nano Lett 1:515–519. https://doi.org/10.1021/nl0155274.. American Chemical Society

Osman Ortashi KM, Awad MAG, HENDI AA, Abdelaziz ARM, Hendi ASA, Alahmed AA (2016) Synthesis of silver nanoparticles using fungi. United States: King Saud University

Ostroff, Gary R, and Ernesto Soto. 2019. Yeast cell wall particles for receptor-targeted nanoparticle delivery. http://www.patentbuddy.com/Patent/20120070376

Ouerdane Z, Laurent M (2009) Production of sulfur-free nanoparticles by yeast. United States.

Quester K, Avalos-Borja M, Vilchis-Nestor AR, Camacho-López MA, Castro-Longoria E (2013) SERS properties of different sized and shaped gold nanoparticles biosynthesized under different environmental conditions by *Neurospora crassa* extract. Edited by Richard G. Haverkamp. PLoS One 8:e77486. https://doi.org/10.1371/journal.pone.0077486. Public Library of Science

Rai M, Posten C (2013) Green biosynthesis of nanoparticles: mechanisms and applications. CABI, Wallingford

Saha S, Sarkar J, Chattopadhyay D (2010) Production of silver nanoparticles by a phytopathogenic fungus Bipolaris nodulosa and its antimicrobial activity. Dig J Nanomater 4:887–895

Salah N, Habib SS, Khan ZH, Memic A, Azam A, Alarfaj E, Zahed N, Al-Hamedi S (2011) High-energy ball milling technique for ZnO nanoparticles as antibacterial material. Int J Nanomed 6:863–869. https://doi.org/10.2147/IJN.S18267. Dove Press

Salunkhe RB, Patil SV, Patil CD, Salunke BK (2011) Larvicidal potential of silver nanoparticles synthesized using fungus *Cochliobolus lunatus* against Aedes aegypti (Linnaeus, 1762) and Anopheles stephensi Liston (Diptera; Culicidae). Parasitol Res 109:823–831. https://doi.org/10.1007/s00436-011-2328-1

Saxena, J, MM Sharma, S Gupta, A Singh - World J Pharm Sci, and undefined 2014. 2014. Emerging role of fungi in nanoparticle synthesis and their applications. World J Pharm Sci 3: 2278–4357.

Seshadri S, Saranya K, Kowshik M (2011) Green synthesis of lead sulfide nanoparticles by the lead resistant marine yeast, *Rhodosporidium diobovatum*. Biotechnol Prog 27:1464–1469. https://doi.org/10.1002/btpr.651

Soni N, Prakash S (2012) Fungal-mediated nano silver: an effective adulticide against mosquito. Parasitol Res 111:2091–2098. https://doi.org/10.1007/s00436-012-3056-x

Sundaramoorthi C, Kalaivani M, Mathews DM, Palanisamy S, Kalaiselvan V, Rajasekaran A (2009) Biosynthesis of silver nanoparticles from *Aspergillus niger* and evaluation of its wound healing activity in experimental rat model. Int J Pharm Tech Res 1:1523–1529. Sphinx Knowledge House

Syed A (2012) Extracellular biosynthesis of monodispersed gold nanoparticles, their characterization, cytotoxicity assay, biodistribution and conjugation with the anticancer drug doxorubicin. J Nanomed Nanotechnol 04:1–7. https://doi.org/10.4172/2157-7439.1000156. OMICS International

Thakkar KN, Mhatre SS, Parikh RY (2010) Biological synthesis of metallic nanoparticles. Nanomedicine 6:257–262. https://doi.org/10.1016/J.NANO.2009.07.002. Elsevier

Todaka Y, McCormick PG, Tsuchiya K, Umemoto M (2002) Synthesis of Fe-Cu nanoparticles by mechanochemical processing using a ball mill. Mater Trans 43:667–673. https://doi.org/10.2320/matertrans.43.667. The Japan Institute of Metals and Materials

Vala AK (2015) Exploration on green synthesis of gold nanoparticles by a marine-derived fungus *Aspergillus sydowii*. Environ Progress Sustain Energy 34:194–197. https://doi.org/10.1002/ep.11949. Wiley

Vetchinkina EP, Burov AM, Ageeva MV, Dykman LA, Nikitina VE (2018) Biological synthesis of gold nanoparticles by the xylotrophic basidiomycete *Lentinula edodes*. Prikladnaia biokhimiia i mikrobiologiia 49:402–408. http://www.ncbi.nlm.nih.gov/pubmed/24455867. Accessed 15 Dec

Zhao X, Zhou L, Rajoka MSR, Yan L, Jiang C, Shao D, Zhu J et al (2018) Fungal silver nanoparticles: synthesis, application and challenges. Crit Rev Biotechnol 38:817–835. https://doi.org/10.1080/07388551.2017.1414141

Part III

Interactions with Plants, Animals and Humans

Recent Developments in Ectomycorrhizal Research

12

Tanveer Kaur and M. Sudhakara Reddy

Abstract

Symbiotic associations are known to be established by rhizospheric fungi with the root systems of host plants and trees. Ectomycorrhizae, the symbiotic fungi, provide growth-limiting micronutrients to host plants and enable plants and trees to colonize temperate and boreal regions. The symbiotic associates reciprocally exchange nutrients at the interface of ectomycorrhizae known as Hartig net. The ectomycorrhizal association leads to various root fabrication modifications such as development of plentiful short and inflated lateral roots ensheathed by ectomycorrhizal fungal mantle. Ectomycorrhizae have been known to be evolved from 60 independent saprophytic lineages. The last few years saw the development of latest techniques to gain insight into evolution, diversity, and reciprocal trades occurring between symbiotic partners and molecular mechanisms playing role beneath these phenomena. The whole genomes of several ectomycorrhizal fungi have been sequenced leading to improved perception of behaviour of these fungi and their genes in ectomycorrhizal associations. In this chapter, latest developments in biodiversity of ectomycorrhizal fungi, novel genes involved in symbiosis, molecular mechanisms behind survival of ectomycorrhizal fungi in metal-contaminated regions and extreme environments such as Arctic and Antarctic regions have been summarized.

Keywords

Ectomycorrhizae · Hartig net · Symbiosis · Metal tolerance · Genes · Extreme environment

---

T. Kaur · M. S. Reddy (✉)
Department of Biotechnology, Thapar Institute of Engineering & Technology, Patiala, Punjab, India
e-mail: msreddy@thapar.edu

T. Satyanarayana et al. (eds.), *Advancing Frontiers in Mycology & Mycotechnology*,
https://doi.org/10.1007/978-981-13-9349-5_12

12.1 Introduction

Plant growth and development relies on the availability of nutrients in the rhizosphere, which is often a limiting aspect. To address this issue, land plants have unfolded various strategies to optimize nutrient acquisition like beneficial interactions with soil-inhabiting microorganisms (Garcia et al. 2016). Ectomycorrhizal (ECM) fungi take up macronutrients from the soil and swap these macronutrients with the host plant for photosynthetically derived carbohydrates (Smith and Read 2008). These mutual exchanges take place at the symbiotic collocation of distinct fungal structures termed ectomycorrhizae, which develop between root cortical cells forming a structure called Hartig net (Becquer et al. 2018a) (Fig. 12.1). The plants colonized with ECM have the benefit, over the non-colonized ones, of getting nutrients from outside the zones of nutrient depletion created by their roots (Smith and Read 2008).

The functioning of this guild can be described by the knowledge of structure and types of mycorrhizal symbiosis (Allen 1991). Earlier two types of mycorrhizae based on whether the fungus penetrated the root cortical cell walls or not were reported by Frank (1887, 1891). Those in which fungal hyphae penetrate the cell wall are known as endomycorrhizae (e.g. Arbuscular mycorrhizae), and those in which the fungus remains outside the plant cell are known as ectomycorrhizae (Allen 1991). The pictorial description of two broad symbiotic associations can be seen in Fig. 12.2. The ectomycorrhizal symbiosis involves in most of the temperate, tropical, and boreal forest trees (Wang and Qiu 2006). Ectomycorrhizae are

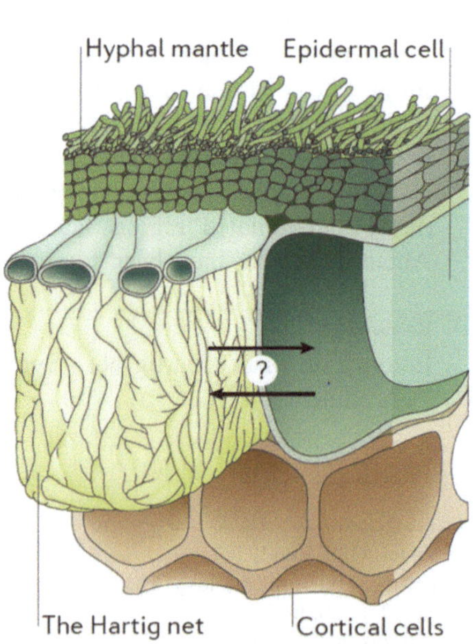

Fig. 12.1 Ectomycorrhizal interaction showing the hyphal mantle formed by layers of aggregated hyphae that cover the root surface and the Hartig net. The proliferation of membranes favours the bi-directional movement of signals and nutrients between symbiont partners. (Adapted by permission from Springer Nature: Nature Reviews Microbiology (Unearthing the roots of ectomycorrhizal symbioses) [Martin et al. (2016)])

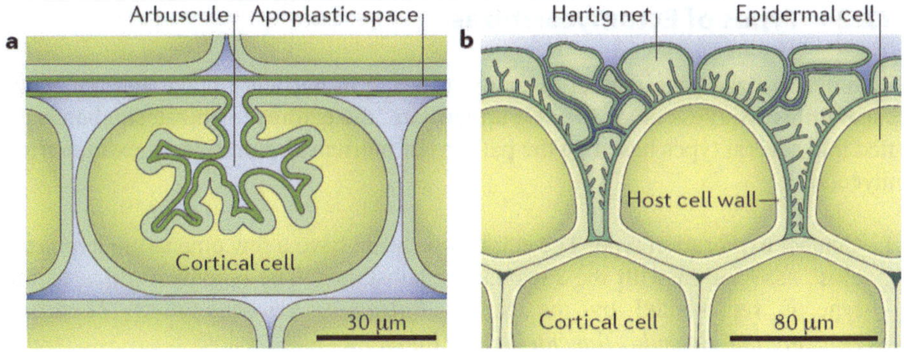

Fig. 12.2 (**a**) In arbuscular mycorrhiza, hyphae emanate from germinated spores grow in the apoplastic space between plant root cells and then penetrate cortical cells, forming tree-like arbuscules, (**b**) In ectomycorrhiza, hyphae from basidiomycetes and ascomycetes from a soil propagule, attach onto epidermal cells of emerging lateral roots, proliferate and differentiate into a series of hyphal layers, known as the sheathing mantle. The network of hyphae formed is known as the Hartig net. (Adapted by permission from Springer Nature: Nature Reviews Microbiology (Unearthing the roots of ectomycorrhizal symbioses) [Martin et al. (2016)])

commonly formed by *Basidiomycota* and to a lesser extent by *Ascomycota*. Their association forms grid-like communities in which an ECM fungus forms a cohort with several trees (Selosse et al. 2006; Beiler et al. 2010), from different species, and sometimes a single tree collaborates with different ECM fungal species (Bahram et al. 2011). These rampant mycorrhizal networks promote ecosystem stability (Simard et al. 2012).

Ectomycorrhizal community diversification and architecture transit over time impromptu (Henry et al. 2014) as they are susceptible to environmental turmoil, which can lead to their partial or total destruction (Lazaruk et al. 2005). Some ECM have a tendency to tolerate these disturbances whether they are metal contamination or below zero temperatures and emerge as they possess unique genes responsible for this behaviour.

ECM fungi adequately accumulate heavy metals from soils (Gadd et al. 2012). They promote plant dynamism in metal disturbed environments because the heavy metal tolerant ECM function as an impediment for the passage of metals into plant tissues (Colpaert et al. 2011; Khullar and Reddy 2018). They also stimulate the plant response to abiotic stress (Urban 2011). While the capacity of ECM fungi to form symbiotic associations with land plants is known for decades, and their role in the accumulation of heavy metals in large amounts has been reported since the 1970s, decoding the molecular mechanism behind this paradox is being understood only recently (Smith and Read 2008). In this review, a number of recent researches, review articles, and book chapters have been thoroughly reviewed for unique genes involved in symbiosis and heavy metal tolerance of ECM fungi. The recent reporting of ECM communities harbouring Arctic and Antarctic environments has also been reviewed.

12.2 Types of Ectomycorrhizae

Ectomycorrhizae have a great diversity differing in nutrient uptake and relocation capabilities (Burgess et al. 1993). Furthermore, the ectomycorrhizae can be classified into several types based on the patterns of nutrient exploration by extrametrical mycelia:

1. Contact exploration type: The ectomycorrhizae have a few emanating hyphae in close association with the surrounding substrates. The ectomycorrhiza of this type is represented by the genera *Balsamia*, *Chroogomphus*, *Lactarius*, *Leucangium*, *Russula*, *Tomentella* and *Tuber.*
2. Short distance exploration type: These ectomycorrhizae have an ample number of emanating hyphae, but without any rhizomorphs. The ectomycorrhizal members of this type are *Cenococcum*, *Elaphomyces*, *Genea*, *Humaria*, *Sphaerozona*, *Tricharina*, *Hebeloma*, *Descolea*, *Inocybe*, *Rozites*, *Tylospora* and some *Tuber* species with long cystidia.
3. Medium distance exploration type: These ectomycorrhizae form rhizomorphs and can be further divided into three subtypes, based on different features of rhizomorphs:
   (i) Fringe subtype: The ectomycorrhiza form hairy rhizomorphs, which split and interconnect repeatedly. The emanating hyphae have extended contact to the soil. The genera *Amphinema*, *Dermocybe*, *Laccaria*, *Lyophyllum and Piloderma.*
   (ii) Mat subtype: The rhizomorphs of this subtype are cognate and have a limited range of nutrient exploration, even after mat forming capabilities. The genera *Bankera*, *Boletopsis*, *Gomphus*, *Phellodon*, *Piloderma and Ramaria* come under this subtype.
   (iii) Smooth subtype: The rhizomorphs of this subtype are either undifferentiated or marginally differentiated from a central thick hypha. The ectomycorrhizal mantles appear smooth due to lack of emanating hyphae. Some species of *Amanita*, *Tricholoma*, *Lactarius*, *Tomentella* and *Albatrellus* form this subtype.
4. Long distance exploration type: this type of ectomycorrhizae can be marked by a few, but highly demarcated and lengthy rhizomorphs. They have the tendency to spread in the soil up to several decimetres (Raidl 1997). Ectomycorrhiza belonging to the genera *Suillus*, *Pisolithus*, *Rhizopogon*, *Scleroderma*, *Boletinus* and *Leccinum* are included under this exploration type.

12.3 Ectomycorrhiza in Extreme Environments

12.3.1 Ectomycorrhizae in Arctic and Antarctic Continents

Due to the global industrialization, the fossil fuels such as coal and petroleum are being used at an alarming rate. These fossil fuels release huge quantities of greenhouse gases, comprising majorly of carbon monoxide (CO). These gases are responsible for global warming, which has resulted in the increase of surrounding temperature. This change in temperature has also been witnessed by the polar continents. On an average, 0.34 °C rise in the surrounding temperature has been observed in the maritime Antarctic since the last decade (Newsham et al. 2016).

Deep snow insulates the soil beneath and forbids it from becoming awfully cold, which has good consequences for vegetation growth and soil processes in winters and consecutive summers (Schimel et al. 2004). To study the diversification in the arctic plant communities and associated mycorrhizal and root-associated fungi, 18-year-long summer warming experiment (+2 °C per year, c.a. 2 months per year) was performed by Morgado et al. (2015) in Alaska. Consequently, the soil fungal communities were reorganized in the moist tundra region. The ectomycorrhizal basidiomycetes decline in richness due to increase in temperature that affected fungal metabolism and fungus-plant interactions. Ectomycorrhizal genera *Inocybe* and *Sebacina* growing in Arctic tundra under summer warming showed a strong decline in comparison to those growing in control sites (Morgado et al. 2015).

Glaciers represent exposed substrates, that comprise exclusive habitat for primary succession (Alfredsen and Høiland 2001). These habitats indicate disturbance, in which primary production is low due to constrained plant growth in Polar Regions. For efficient plant growth, nutrients, water and optimum temperature are the vital factors, which are lacking in these regions (Billings 1987). Thus there is a slow nutrient turnover (Haselwandter and Read 1980). The glaciers contain low levels of organic matter and nitrogen. Fungi play a significant role in the initial colonization of these habitats through spore dispersal approaches and, hence, in aiding the early establishment of mycorrhizal host plants (Cazares and Trappe 1994). Fungi inhabiting these regions have adapted to extreme life conditions by various physiological and biochemical transformations such as alteration in the distribution of viable cytoplasm within hyphae to survive in such freezing temperatures (Addy et al. 1994). Decreased size of fruiting bodies and reduced number of gills are other adaptive features displayed by alpine ectomycorrhiza (Maggi et al. 2013). Ectomycorrhizae are cardinal for the durability of plant species in such nutrient limited polar environments, as they provide the latter with ample nitrogen and phosphorous (Dahl 1956; Tedersoo et al. 2010).

The alpine environments appertaining to the presence of glaciers in the Apennines and Alpes in Lombardi (Italy) have been studied with the motive of the conservation of fungal biodiversity in cold terrains, threatened from global warming. A total of 27 ectomycorrhizal basidiomycota have been identified. *Lactarius salicis-herbaceae* was isolated from two alpine grassland plots with snow bed vegetation and for the first time on calcareous soil (Maggi et al. 2013). It has also been reported earlier from acidic soils by Jamoni (2008). This finding widens the ecological preferences of this fungus. The dwarf form of the fungus, *Inocybe microfastigiata* has also been reported in Lombardy. The basidiomes of this fungus growing in Lombardy were found to be much smaller, truly representing the case of fungal alpine dwarfism (Maggi et al. 2013). Another ectomycorrhizal fungus found was *Laccaria* in Montana as reported previously in eastern and western Italian Alps (Jamoni 2008).

In another study conducted in alpine grassland situated in southern piedmont, *Helianthemum nummularium* subsp. *grandiflorum* had been reported. A number of ectomycorrhizal fungi associated with these plants have been recorded (*A. pantherina*, *A. vaginata*, *Russula pascua*, *Boletus luridus*, *Cortinarius anomalus* and a new species *Amanita helianthemicola*). Earlier ectomycorrhizal ascomycetes associated with *Heliamthemum* were found in Mediterranean range (Turgeman et al. 2011). Ectomycorrhizal fungi associated with *Helianthemum* have hardly been described in alpine grasslands (Barden 2007).

Ectomycorrhizal fungi are the prominent fungi in the Arctic region as stated earlier; the soils of the Arctic region are low in nutrients, so the plants growing there rely on symbiosis with mycorrhizal fungi for their endurance (Hobbie et al. 2009; Bjorbækmo et al. 2010). ECM fungi provide 61–86% N to arctic tundra plants. Recent below ground studies in Arctic regions have indicated that *Tomentella*, *Cortinarius*, *Inocybe*, *Hebeloma* and *Russula* are the most common ECM genera found in the tundra region (Geml et al. 2012; Timling and Taylor 2012; Timling et al. 2012). One hypothesis states that species with extrametrical mycelium of medium distance and long distance exploration types may possibly explore the nutrients by producing extracellular enzymes. Morgado et al. (2015) reported *Tomentella*, *Inocybe*, *Russula and Cortinarius* as the dominant ECM genera in a study undertaken in moist and dry tundra.

12.3.2 Ectomycorrhizal Genes in Metal-Contaminated Sites

ECM can be employed in phytoextraction and phytostabilization of sites disturbed with heavy metals as ECM curtail the noxious consequences of heavy metals on host plants by diluting heavy metal concentration, enhancing nutritional status of plants, thereby boosting plant growth (Tang et al. 2019). ECM thrives in such metal-contaminated sites possibly by the mode of metal chelating ligands called metallothioneins and glutathione.

12.3.2.1 Metallothioneins

Metallothioneins (MTs) are a group of cytosolic peptides which bind heavy metals through cysteinyl-thiolate bonds and help some eukaryotes and prokaryotes in tolerance to various heavy metals (Vašák and Meloni 2011; Leszczyszyn et al. 2013). Silver in the form of Ag^+ is one of the most pernicious metals to various organisms. Ag^+ turns the cellular components such as enzymes inactive or malfunctioning by binding to them. When studying the accumulation of Ag^+ in ECM thriving in silver-contaminated environments, highest Ag^+ concentration of 304–692 mg/kg was found in *Amanita* species. Hyperaccumulation of Ag in the ectomycorrhizal fungus *A. submembranacea* is due to its binding to metallothioneins. Cellular components extracted from the tissue of *A. submembranacea* revealed no H^+ exchangeable Ag^+ illustrating that accumulated Ag^+ is not stashed extracellularly via ion exchange bonds. Size exclusion chromatography (SEC) on cell-free extracts revealed that intracellular complex formation is involved in sequestration of Ag^+ in

A. submembranacea. The Ag^+ containing fractions from SEC were pooled and processed to spectrofluoremetric analysis and denaturing PAGE to support the fact that Ag^+ sequestration in *A. submembranacea* involved metallothioneins (Borovička et al. 2010). Apart from contaminated environments, high levels of accumulated Ag^+ (100–1200 mg/kg) have been reported in two ECM basidiomycetes, *Amanita strobiliformis* and *Amanita solitaria*, from pristine soils too (Gryndler et al. 2012). Gryndler et al. (2012) hypothesized that mellowed fruiting bodies of Ag hyperaccumulating basidiomycetes decompose and release the metal in the surrounding environment resulting in alteration of composition of soil microbial flora. Through in vitro experiments, they had reported that decomposing fruiting bodies rich in high concentration in silver favoured an embellished populace of soil fungi and pampered bacterial community. When sublethal forms of mineral Ag were applied, fungal communities were replaced by bacterial communities. Overall implying that fungi are more tolerant to Ag in comparison to bacteria (Gryndler et al. 2012).

Osobová et al. (2011) investigated three isoforms of metallothionein genes, AsMT1 isolated from fruiting bodies of Ag hyperaccumulating *Amanita strobiliformis.* qRT-PCR analysis of three MT encoding genes, AsMT1a, AsMT1b and AsMT1c, revealed that AsMT1a mRNA mirrored 75% of all AsMT1 transcripts, thus concluding AsMT1a to be the major ligand responsible for the binding of Ag in fruiting bodies of *A. strobiliformis*. Yeast complementation assay revealed that cDNAs of all three AsMT1 induced by Ag complemented the CUP1 and ycf-1 null genotypes rendering cadmium and copper tolerance to mutant yeast cells. There is no confirmation on their function in Ag detoxification (Mehra et al. 1988).

In a recent study, four copper transporter (CTR) proteins, AsCTR1 and AsCTR4 showing similarity to vacuolar CTRs and AsCTR2 and AsCTR3 to plasma membrane have been predicted in wild grown *A. strobiliformis* sporocarps with high amounts of accumulated Ag and Cu. Yeast complementation assay involving AsCTR2 and AsCTR3 genes led to increased Ag and Cu uptake by yeast mutants (Beneš et al. 2016). Ramesh et al. (2009) identified two MT genes, HcMT1 and HcMT2, from the ECM fungus *Hebeloma cylindrosporum.* HcMT1 and HcMT2 gene expression in *H. cylindrosporum* under metal stress plight was analyzed by competitive RT-PCR analysis. Functional complementation assays in yeast mutants revealed that HcMT1 and HcMT2 genes lend increased tolerance against Cd and Cu, respectively. Metalothionein genes, HmMT1, HmMT2 and HmMT3 have been characterized from sporocarps of *Hebeloma mesophaeum* growing on a metal disturbed site (Lead smelter site). Heterologous complementation assays in yeast mutants expressed tolerance of yeast mutants to Zinc and it was upregulated by only HmMT1. Tolerance of yeast mutants to Cd and Cu was upregulated highly by HmMT1 followed by HmMT2 and HmMT3. Increased tolerance to Ag was conferred by HmMT2 and HmMT3 but not HmMT1. HmMT3 gene was strongly induced in the presence of Ag. HmMT1 was strongly promoted in the presence of Zn and Cd. HmMT2 gene was not detected in *H. mesophaeum* thriving on a control unpolluted site (Sácký et al. 2014).

Reddy et al. (2014) characterized two metallothionein genes, LbMT1 and LbMT2 from the ECM fungus *Laccaria bicolor* under the plight of metal stress.

```
ShMT1   --------------------MSTATEVPV-SNNNCGSSSCSCGTSCQCKPGECKC   34
ShMT2   --------------------MSTPTEVPV-SNNNCGSSSCSCVTSCQCKPGQCKC   34
PaMT1   --------------------MQSVNAVLVNNNGNCGSAACACGSNCACKPGECKC   35
RaMT1   --------------------MSPVIQNPV-NEHHCGNSSCTCGDSCQCKPGECKC   34
AsMT1   --------------------MQSESQSLV-SFANCGSNSCNCGASCACKPGDCKC   34
PiMT1   --------------------MNTITSVPV-NFNNCGSNSCGCGSSCACKPGECKC   34
GlMT1   --------------------MYSTTDVVK--NAACGSSSCNCGATCACKPGECKC   33
LbMT1   --------------------MISTINVPV--SQTCGSSSCNCGESCACKPGECKC   33
SiMT1   --------------------MISETIVPV--NQNCGNSSCSCGDSCQCKPGECKC   33
PmMT1   --------------------MQSVNAVLVNNNDKCGSAACTCGSSCACKPGECKC   35
```

Fig. 12.3 Multiple sequence alignment of Metallothionein genes, ShMT1 and ShMT2 genes with various homologous sequences retrieved by BLASTp analysis. Both *Suillus himalayensis* MT genes (ShMT1) KY775394 and (ShMT2) KY775395 showed homology to metallothionein genes from various basidiomycetes *Pisolithus albus* (PaMT1) AJO67962, *Russula atropurpurea* (RaMT1) AHA31882, *Amanita strobiliformis* (AsMT1) AGO04615, *Paxillus involutus* (PiMT1) AAS19463, *Ganoderma lucidum* (GlMT1) ABP02008, *Laccaria bicolor* (LbMT1) AHI43933, *Serendipita indica* (SiMT1) ACT83730, and *Pisolithus microcarpus* (PmMT1) EST N25. Multiple sequence alignment showed conserved C-X-C motifs (Highlighted in blue). (Reproduced by permission from Microbiology Society from (Metal induction of two metallothionein genes in the ectomycorrhizal fungus *Suillus himalayensis* and their role in metal tolerance) [Kalsotra et al. (2018)])

Multiple sequence alignment of LbMT1 gene with various homologous sequences by data retrieved using BLASTp is shown in Fig. 12.3. The levels of expression of both LbMT1 and LbMT2 genes increased as a function of increased concentration of Cu; only LbMT1 was expressed in response to Cd, while Zn did not influence the transcription of either gene. Both LbMT1 and LbMT2 conferred high resilience to both Cu and Cd as observed by heterologous yeast complementation assays. Another two MT genes, *ShMT1* and *ShMT2* were characterized by Kalsotra et al. (2018) from the ECM fungus *Suillus himalayensis*. Multiple sequence alignment of ShMT1 and ShMT2 genes with various homologous sequences is shown in Fig. 12.3. The antiphon of these MT genes to the external Cu and Cd concentrations was studied by qRT-PCR analysis. ShMT1 and ShMT2 genes were reported to be more inducible by copper than cadmium. Functional complementation in yeast mutant strains sensitive to Cu, Cd and Zn, *cup1*$^{\Delta}$, *ycf1*$^{\Delta}$and *zrc1*$^{\Delta,}$ respectively revealed that two genes conferred copper, cadmium and zinc tolerance and could perform an important act in their detoxification.

Russula and *Hebeloma* species engage different approaches for intracellular accumulation of high Zn contents in their sporocarps (Pikalova et al. 2011). Two genes of cation diffusion facilitator family (CDF), RaCDF1 and RaCDF2 have been identified and characterized in Zn accumulating ECM fungus *Russula vatropurpurea*. Fusion of translational product of RaCDF1 with Green fluorescent protein (GFP) indicated that vacuolar sequestration of the metals lead to this gene mediated Zn and Co tolerance of yeast mutants. Zn- efflux action of RaCDF2 was propounded due to downregulation of Zn and upregulation of Cd. Transcriptional product fusion protein fluorescence was localized on plasma membrane. It also increased the sensitivity of yeast mutants to Co and Cd (Pikalova et al. 2011).

In *H. cylindrosporum*, a constitutively expressed gene belonging to a cation diffusion facilitator family, HcZnT1 has been identified which confers tolerance to Zn in otherwise sensitive mutant yeast cells by functional complementation assay (Blaudez and Chalot 2011). Kubrová et al. (2014) reported that ECM basidiomycete thriving on Uranium-polluted soil does not accumulate U higher than 3 mg/kg, i.e. very limited as compared or suggested by earlier studies (Gadd and Fomina 2011).

Mycorrhizal fungi may also outweigh the adverse brunts of aluminium (Al) in the territory of roots. Response of Al on the growth and nutrition of *Populus deltoides* in the presence of the ECM fungus was studied. When mycorrhizal (*Paxillus involutus*) and non-mycorrhizal *Populus deltoides* plants were subjected to different concentrations of Al, the biomass of mycorrhizal plants upsurged as compared to non-mycorrhizal plants. The levels of phosphorus, magnesium and calcium were also increased in mycorrhizal plants in presence of Al. Decreased Al content in shoots and induction of organic acid production in mycorrhizal plants compared to non-mycorrhizal plants suggested that ectomycorrhizal colonization bestows Al tolerance to *P. deltoides* (Khosla et al. 2009).

Accumulation of various heavy metals by ECM fungi associated with Norway spruce growing on smelter polluted area was studied by Cejpková et al. (2016) and compared with non-mycorrhizal fine roots. ECM of *Imleria badia* showed higher concentration of Cl, Cd, Zn and Ag (1510–37,100 mg/kg of dry biomass). Vanadium was accumulated by ECM fungus *Amanita muscaria* only.

12.3.2.2 Glutathione

Glutathione is a tripeptide (L-γ-glutamyl-L-cysteinylglycine) (307 Da) consisting of 3 amino acids, glutamate, cysteine and glycine (Townsend 2007). Both prokaryotes and eukaryotes produce glutathione. Glutathione is a key component in metal scavenging, due to the high affinity of metals for its thiol (-SH) group. During metal-induced oxidative stress, it also acts as an antioxidant (Khullar and Reddy 2018). Glutathione sequesters the toxic metal(loid)s by forming non-toxic glutathione-metal(loid) conjugates. These conjugates are further compartmentalized into the vacuoles through ABC (ATP-binding cassette) transporters (Schlunk et al. 2015). If we look at biosynthesis of glutathione, L-glutamate and L-cysteine form γ-glutamylcysteine in the presence of γ-glutamylcysteine synthetase and ATP. At the C-terminal of γ-glutamylcysteine, glycine is added by another enzyme, glutathione synthetase in the presence of ATP, finally forming glutathione. The biosynthesis of glutathione is illustrated in Fig. 12.4.

The γ-glutamatecysteine synthetase is a rate-limiting enzyme in glutathione synthesis. Cd^{2+} ions enhance the γ-glutamatecysteine synthetase activity inside the cell. Other heavy metals such as As, Hg and Cr also trigger the expression of genes in glutathione biosynthesis pathway (Vido et al.2001; Thorsen et al. 2007). Location wise, glutathione synthetase is localized in cytosol, but γ-glutamatecysteine synthetase is restricted to plastids (Wachteret al. 2005). Sulphur availability and γ-glutamatecysteine synthetase activity are the two rate-limiting steps of this

Fig. 12.4 A two-step reaction catalysed by two enzymes, γ-glutamatecysteine synthetase and glutathione synthetase, showing biosynthesis of glutathione

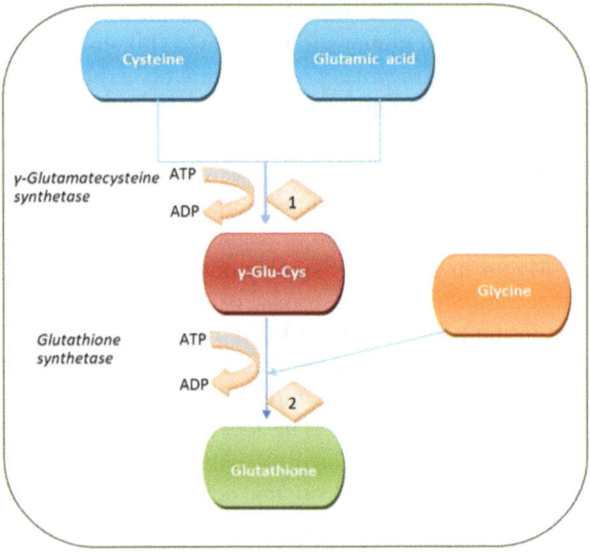

pathway. During metal toxicity, both sulphur availability and γ-glutamatecysteine synthetase activity upsurge fulfilling the appeal for higher glutathione synthesis must be there for heavy metal detoxification and endurance (Jozefczaket al. 2012). The glutathione synthesis in cell is controlled by the feedback inhibition. When there is metal stress, glutathione is synthesized inside the cell, which is oxidized and utilized in phytochelatin synthesis, thus depleting the cellular glutathione followed by the release of feedback inhibition and increase in glutathione synthesis (Hibi et al. 2004).

Courbot et al. (2004) reported augmentation in glutathione synthesis in response to cadmium stress in *Paxillus involutus*. Gallie et al. (1993) reported similar response in *Laccaria laccata*. Ilyas and Rehman (2015) reported that glutathione production in ECM is mainly activated by cadmium and arsenic stress followed by chromium, lead and copper. Furthermore, genes encoding glutathione synthetase and γ-glutamylcysteine synthetase have been detected in the genomes of *Hebeloma cylindrosporum*, *Laccaria bicolor* and *Paxillus involutus* (Bellion et al. 2006). Khullar and Reddy (2019a) reported a boost in glutathione levels inside *Laccaria bicolor* cells in response to external cadmium and arsenic stress. qPCR analysis revealed that the expression of glutathione production genes, γ-glutamylcysteine synthetase (Lbγ-GCS) and glutathione synthetase (LbGS) increased when exposed to cadmium and arsenic. The role of both the genes was studied in yeast mutants sensitive to arsenic and cadmium using functional complementation assays. Both the genes provided cadmium and arsenic tolerance to otherwise sensitive yeast mutants. Recently, Khullar and Reddy (2019b) reported γ-glutamylcysteine synthetase mediated bioaccumulation of glutathione in another ECM fungus, *Hebeloma cylindrosporum*, in response to cadmium stress. The activity of γ-glutamylcysteine synthetase increased as the concentration of cadmium increased that resulted in increase in glutathione production, clearly depicting its role in protection from oxidative stress. Heavy metal tolerance imparted by genes was validated by complementation assays in sensitive yeast mutants. In this manner, γ-glutamylcysteine synthetase and glutathione synthetase genes play a role in heavy metal stress tolerance and detoxification by ECM fungi.

12.4 Novel Ectomycorrhizal Genes in Symbiosis

ECM fungi have incurred the loss of function of their saprotropic ancestors of decaying the lignocelluloses in organic matter present in soil for deriving carbon (Treseder et al. 2006). Therefore, they have to depend on their plant partners for carbon and other components. All ECM have at least one gene coding for an oxidoreductase, lytic polysaccharide monooxygenase, which is essential for decomposition of chitin and cellulose (Lundell et al. 2014). The ECM symbiosis needs some morphological modifications in the two symbiotic associates necessary for the formation of symbiotic organization. This is done through the mode of regulation of some genes (Martin et al. 2007; Kohler et al. 2015). The fungal part of ECM root is formed by two fungal tissues, the mantle which establishes outside the root and the Hartig net which is present in the apoplastic region of root. It has been reported that the mantle is responsible for the uptake of minerals and water from the soil, while in the Hartig net, there is a raised expression of several transporter genes (Hacquard et al. 2013; Balestrini and Kottke 2016). Jumpponen and Egerton-Warburton (2005) discussed certain determinants of plant-ECM establishment which have 'compatibility' between host plants and fungal species, the 'environment' and the 'biotic factors', which facilitate communication between ECM and host plant species.

12.4.1 Vast Array of Novel Genes in *Laccaria bicolor*

The *L. bicolor* genome contains about 10,000 genes including the genes involved in symbiosis development with host plants (Martin and Nehls 2009). *Laccaria bicolor-Populus* species is widely used as an *in vitro* model system for the exploration of development of mutualistic associations in ectomycorrhizal symbioses as shown in Fig.12.5a.

Recently some novel genes involved in symbiosis establishment of *L. bicolor* with host trees/plants have been deciphered, which are outlined below.

12.4.1.1 Mycorrhizal-Induced Small Secreted Proteins *MiSSps*

The MiSSPs secreted by ECM fungi interact with plant receptors of various hormones such as auxins, gibberellins and salicylic acid to modify root development (Martin et al. 2016). Mycorrhizal induced small secreted protein 7 (MiSSP7) has been characterized in *L. bicolor* (Huang et al. 2006). Upon receiving diffusible signals from plant roots, the fungus secretes MiSSP7, which is actively endocytosed into plant cells nucleus and alters transcriptome of the host plant/trees. It amends

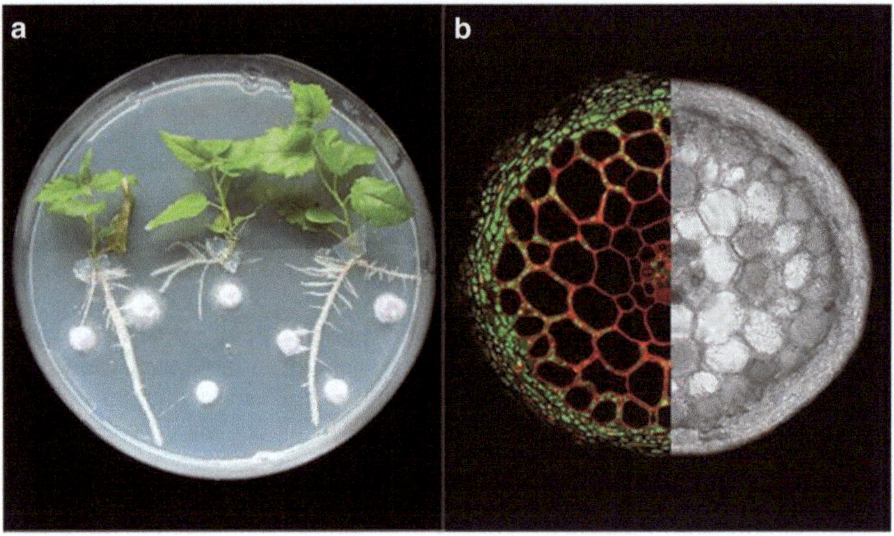

Fig. 12.5 (**a**) The *in vitro* growth of an ectomycorrhizal symbiosis between *P. tremula x alba* plantlets and *L. bicolor* hyphae can be seen emanating from agar plugs. (**b**) A transverse section of a rootlet of *Populus vtrichocarpa* showing colonization by *L. bicolor* (in green colour). Intrusions of *L. bicolor* hyphae can be seen between the cell walls (red) of host plant collectively forming a fungal network in plant root cells known as the Hartig net. (Adapted by permission from Springer Nature: Nature Reviews Microbiology (Unearthing the roots of ectomycorrhizal symbioses) [Martin et al. 2016])

plant hormone signals to facilitate fungal hyphae penetration into plant cells and formation of Hartig net. After entering the host nucleus, MiSSP7 collaborates with the regulator of jasmonate signalling pathway (Plett et al. 2011; Kazan and Manners 2012), i.e. jasmonate ZIM domain protein 6(JAZ6) (Plett et al. 2014). When fungal hyphae penetrate the host apoplast, jasmonate is released due to generation of wounding signals. Accumulation of jasmonate, JAZ6 could get activated, but the interaction between JAZ6 and MiSSP7 stabilizes the former, repressing the defence mechanism related to JAZ6, thus allowing the colonization of plant root apoplast by fungal hyphae, which otherwise could be prevented due to JAZ6 involved host defence mechanisms (Plett et al. 2014). The role of MiSSP7 in helping in *L. bicolor*-*Populus* symbiosis is confirmed by observing that accumulation of MiSSP7 leads to inhibition of jasmonate signalling pathway in poplar and others (Plett et al. 2011; Kazan and Manners 2012).

12.4.1.2 Endoglucanase LbGH5-CBM1 (Domain Gene)

Earlier it was hypothesized that colonization of host plant apoplast by the symbiotic fungal hyphae depends on the mechanical force resulting from the growth of hyphal tip (Peterson and Massicotte 2004). Host apoplast penetration was also said to be supported by the secretion of fungal plant cell wall degrading enzymes like polygalacturonases and endoglucanases (Martin et al. 2008; Veneault-Fourrey et al. 2014).

Recently, an endo-1,4-β mannase encoded by LbGH5-CBM1 from the genome of *L. bicolor* has been characterized. The increased expression of this gene during establishment of symbiosis has been confirmed by RNAi- knockdown of expression of this gene. It has been suggested that mutant lines with decreased level of expression of this enzyme enact a reduced number of mycorrhizal roots in host plant. The CBM1 domain of Lb GH5-CBM1 cleaves internal cellulose bonds in the plant cell wall contributing to fungal hyphae penetration of root cells and Hartig net formation (Zhang et al. 2018).

12.4.1.3 Secreted Transcriptional Activator Proteins STAPs

A new class of effector proteins STAPs, secreted by *L. bicolor* has been reported by Daguerre et al. (2017). STAPs play a role in controlling host plant gene expression during ECM development.

12.4.1.4 Carbohydrate-Active Enzyme (CAZyme)

Carbohydrate-active enzymes (CAZyme) produced by the fungi participate in ECM formation. The regulation of CAZyme of *L. bicolor* was studied using genome-wide transcriptome profiling during symbiosis development. As the fungal hyphae surround the root cells and start penetrating the apoplast, i.e. the first step of colonization of root cells, the CAZymes actively present on fungal cell wall are upregulated promoting symbiosis development. AA9 or GH12 CAZymes are involved in slackening of cellulose microfibrils, whereas GH8 or GH12 CAZymes play a role in relenting middle lamella of host plant tissue (Veneault-Fourrey et al. 2014).

12.4.1.5 Indole-3 Acetic Acid (IAA)

Metabolic profiling shows that during ECM formation, the mycelium of *L. bicolor* produces high concentrations of the indole-3-acetic acid. The analysis of auxin response gene expression in ECM roots suggested that symbiosis-dependent auxin signalling is activated during the colonization by *L. bicolor* (Vayssières et al. 2015). The ectomycorrhizal roots of a *P. tremula* x *alba–L. bicolor* S238N symbiosis is shown in Fig. 12.6.

12.4.2 Novel Genes in *Tricholoma vaccinum*

12.4.2.1 Aldehyde Dehydrogenases

Aldehyde dehydrogenases play an important role in stress response of the symbiotic association. In ectomycorrhizal basidiomycete, *Tricholoma vaccinum*, eight aldehyde dehydrogenases, Ald1 to Ald7 and TyrA, have been marked. From the functional analysis, it was revealed that Ald1 overexpressing ECM strains increased the width of the apoplast of host tree spruce *Picea abies*, facilitating the accommodation of the Hartig net of the *T. vaccinum* transformants (Henke et al. 2016).

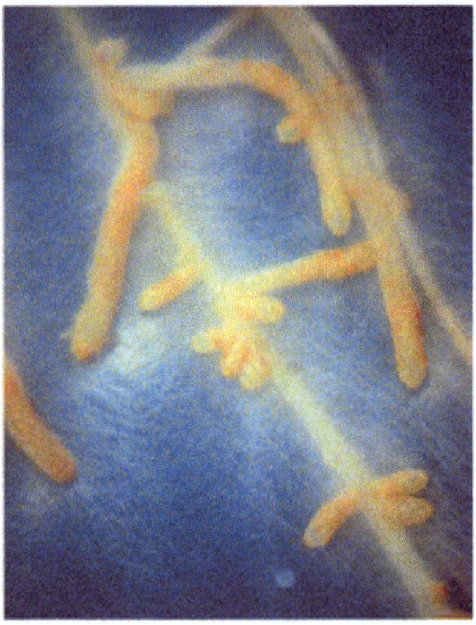

Fig. 12.6 The extensive clustering of ensheathed ectomycorrhizal rootlets is seen resulting from substantial changes in auxin metabolism. Adapted by permission from Springer Nature: Nature Reviews Microbiology (Unearthing the roots of ectomycorrhizal symbioses) [Martin et al. 2016]

12.4.2.2 Hydrophobins

Hydrophobins are minuscule cysteine-rich protein molecules which foment interactions of fungal hyphae with hydrophobic surfaces and play role in the development of aerial hyphae (Sammer et al. 2016). They are amphiphilic small secreted proteins often present as certain gene transcripts in most fungal genomes. It has been predicted that in 41% of the cases, hydrophobins unfold from duplication actions (Rineau et al. 2017). Phylogenetic analyses of 9 *T. vaccinum* hydrophobin coding genes, *hyd1 to hyd9*, were performed to derive their roles in different stages of ECM formation. In aerial mycelium, *hyd8* was upregulated. While *hyd5* was induced after metal stress. In the early stage of mycorrhization, the root exudates are produced due to injury; this is followed by *hyd4* and *hyd5* induction. While *hyd5* was upregulated during Hartig net and mantle formation accompanied by suppression of *hyd8* and *hyd9*. Hydrophobins, *hyd3* and *hyd8*, were activated during fruit body formation (Rineau et al. 2017).

12.4.2.3 Indole-3 Acetic Acid (IAA)

Another gene expression and bioinformatic analysis revealed that indole-3-acetic acid (IAA) secreted from fungal hyphae acts as a diffusible signal that induces branching in fungal cultures and also enhances Hartig net formation. The gene ald1 coding for aldehyde dehydrogenase is upregulated by indole-3-acetaldehyde and distinctly expressed in ectomycorrhiza. The multidrug and toxic extrusion (MATE) transporter, Mte1 found in *T. vaccinum* supports the export of IAA from mycorrhizal fungal hyphae to host cells (Krause et al. 2015).

12.4.3 Novel Genes in *Hebeloma cylindrosporum*

12.4.3.1 Tandem-Pore Outward-Rectifying K^+ (TOK) Channels

The ectomycorrhizal fungus *H. cylindrosporum* boosts the K^+ nutrition of *Pinus pinaster* under mineral shortage situations. Earlier the molecular members responsible for the transfer of Pi towards the host plant were not known. The genome analysis of *H. cylindrosporum* revealed the tandem-pore outward-rectifying K^+ (TOK) channels potentially responsible for K^+ transport in host plant, *P. pinaster*. The expression pattern of these channels as analyzed by yeast complementation assays revealed that HcTOK is embroiled in the symbiotic transfer of K^+ from the fungus enroute the plant (Guerrero-Galán et al. 2018).

12.4.3.2 Pi Transporter (HcPT2)

Transcriptomic approach, microscopy, whole plant physiology, X-ray fluorescence mapping, and ^{32}P labelling have unravelled the performance of a fungal Pi transporter, HcPT2, during the *H. cylindrosporum-Pinus pinaster* ectomycorrhizal association. HcPT2 was confined to the extra-radical hyphae and the Hartig net. It was established that host plant induced HcPT2 expression played a crucial role in symbiosis by influencing both Pi influx into the mycelium and efflux towards roots under the control of host plant *P. pinaster* (Becquer et al. 2018b).

12.4.4 Novel Genes in Russulaceae Symbiosis

Russulaceae is one of the families of the order *Russulales*, a group exhibiting great diversity in the mode of nutrition and morphology of basidiocarps (Miller et al. 2006). Russulaceae contains about 2000 species. They are late-stage homesteaders of forests, signifying their influence in sustaining nutrient networks in mature forests (Twieg et al. 2007). Ectomycorrhizal members of Russulaceae produce lignin peroxidases (Chen et al. 2001; Bödeker et al. 2009) and laccases in high abundance (Chen et al. 2003), even higher than those produced by their saprotrophic ancestoral fungi (Luis et al. 2005; Liao et al. 2014). *Lactarius quietus* secreted laccases have been associated with the bud break of *Quercus* trees, delivering carbon to the sapling when young and translate the regulation to oxidative reactions as the plants develop (Courty et al. 2007).

12.5 Recent ECM Fungal Species Discovered

Kumari et al. (2011) reported *Cantharellus pseudoformosus* as a novel species associated with *Cedrus deodara* from *Chamba* district of Himachal Pradesh, India.

The basidiome of this ECM is shown in Fig. 12.7. Reddy and Verma (2014) and Verma and Reddy (2015) presented first-ever report of two new *Suillus* species,

Fig. 12.7 Basidiome of *Cantharellus pseudoformosus*. Scale bar a: 2 cm. (Adapted by permission of Elsevier: Mycoscience (*Cantharellus pseudoformosus*, a new species associated with *Cedrus deodara* from India), [Kumari et al. (2011)])

Fig. 12.8 *Suillus indicus* basidiomes: (**a**) Young basidiome showing umbo and very few appressed fibrillose squamules on the pileus. (**b**) Basidiome showing white partial veil and absence of glandular dots on the surface of stipes. (**c**) Mature basidiome with appressed fibrillose squamules and a low obtuse umbo on pileal surface. (**d**) Stipe with annulus and no glandular dots/smears. Scale bars: (**a**–**b**) = 1 cm, (**c**–**d**) = 2 cm. (Reproduced by permission of Taylor and Francis from [*Suillu sindicus* sp. nov. (Boletales, Basidiomycota), a new boletoid fungus from Northwestern Himalayas, India) [Verma and Reddy 2015])

Suillus triacicularis sp. nov. growing as ECM with *Pinus roxburghii* and *P. wallichiana* and *Suillus indicus* growing in association with *Cedrus deodara* in Northwestern Himalayan region of India (Fig.12.8). Morphological features of fruiting bodies and phylogenetic analysis of internal transcribed spacer (ITS) region revealed these species to be different from closely related *Suillus* species. Two new species of *Russsula*, *Russula chiui* and *Russsula pseudopectinatoides* have been identified from South Western China based on the morphology of fruiting bodies and the phylogenetic analysis of ITS region (Liu and Hall 2004). *Russula* species differ from *Lactarius* by the absence of lactifers (Buycket al. 2018). Twelve rare species of *Russula* (*R. periglypta*, *R. atropurpurea*, *R.congoana*, *R. hygrophytica*, *R. adusta*, *R. delicula*, *R. luteotacta*, *R. martinica*, *R. cinerella*, *R. mariae*, *R. michiganensis*, *R. aciculocystis*, and 1 species of *Lactarius*, latex emanating *Lactarius nebulosus*) are the new reported ECM fungi from different forest ecosystems of Western Ghats. All the *Russula* species have been found to exhibit ectomycorrhizal association with tree species like *Vateria indica*, *Hopea ponga*, *H. parviflora*, *Calophyllum apetalaum* and *Myristica malabarica* (Mohanan 2014).

12.6 Future Perspectives and Conclusions

Ectomycorrhizal (ECM) fungi secrete several enzymes which improve soil organic matter degradation and nitrogen uptake. *Piloderma olivaceum*, a common ECM fungus produces extracellular proteases which improve the ability of host plant *Pinus sylvestris* to utilize N from extracellular proteins. Due to the ability to access proteinaceous N, ECM can play a potential role in forest C and N cycling even after death of fungal hyphae due to natural recycling of N (Heinonsalo et al. 2015).

A co-cultivation system in axenic tube culture has been developed for investigating the synergy between an ectomycorrhizal plant and its symbiotic fungal partner. This system could also be employed as a pre-treatment for promoting robust mycorrhiza along with optimal plant-fungal symbiosis preceding planting in environmental sites (Arduini et al. 1994). Unquestionably, the composition of ECM community is affected due to environmental disturbance in terms of operational taxonomic units (OTUs), but even then few OTUs are detected both in natural and degraded environments (Henry et al. 2016). Similar finding was recorded by Karpati et al. (2011), when ECM associated with *Quercus rubra* thriving on natural forest sites were compared with those in disturbed urban sites by morphological and PCR-RFLP analysis. Such OTUs can be employed for ecological reclamation of highly disturbed mining sites by implanting ectomycorrhizal seedlings or plants involving these OTUs (Henry et al. 2016).

New data on ECM fungal communities has been made available due to preface of high-throughput sequencing techniques and metagenomics approach. Attenuated total reflectance-Fourier transform infrared spectroscopy (ATR-FTIR) followed by multivariate cluster analysis has been recently utilized to study the diversity of ECM communities in environmental samples (Pena et al. 2014). Barcodes of fungal ITS region have been developed for studying ECM fungal diversity in various ecosystems by the various researchers working in this area. Next-generation sequencing systems such as the Illumina platform and Ion torrent PGM and FUNguild have been recently developed to study fungal OTUs (Mello and Balestrini 2018). Several pipelines for processing the fungal OTUs have also been developed. In spite of development of novel bioinformatics tools for high-throughput sequencing, the fidelity in assessing ECM fungal communities is still a great challenge. Data from Tedersoo et al. (2014) revealed that high-throughput sequencing technique followed by DNA metabarcoding identified major phyla and classes of fungi in all ecosystems along with several lineages of ECM which have not been described or sequenced previously.

Genome sequencing is only the early stride to gather information on how an organism connects with the surrounding environment and with other organisms. The combination of functional, structural, cellular and bioinformatics is aimed at furnishing knowledge on the function of genes and proteins in actual processes. A lot more research in this area is necessary for exercising the effective role of plant and fungal genes in ECM establishment and development (Garcia et al. 2015).

References

Addy HD, Schaffer GF, Miller MH, Peterson RL (1994) Survival of the external mycelium of a VAM fungus in frozen soil over winter. Mycorrhiza 5:1–5

Alfredsen G, Høiland K (2001) Succession of terrestrial macrofungi along a deglaciation gradient at Glacier Blåisen, South Norway. Nord J Bot 21:19–37

Allen MF (1991) The ecology of mycorrhizae. Cambridge University Press, Cambridge, pp 20–25

Arduini I, Godbold DL, Onnis A (1994) Heavy metal uptake and distribution in tree seedlings. G Bot Ital 128:219–219

Bahram M, Põlme S, Kõljalg U, Tedersoo L (2011) A single European aspen (Populus tremula) tree individual may potentially harbour dozens of Cenococcum geophilum ITS genotypes and hundreds of species of ectomycorrhizal fungi. FEMS Microbiol Ecol 75:313–320

Balestrini R, Kottke I (2016) Structure and development of ectomycorrhizal roots. In: Molecular mycorrhizal symbiosis, pp 47–62

Barden N (2007) *Helianthemum* grasslands of the Peak District and their possible mycorrhizal associates. Field Mycol 8:119–126

Becquer A, Garcia K, Amenc L, Rivard C, Doré J, Trives-Segura C, Szponarski W, Russet S, Baeza Y, Lassalle-Kaiser B, Gay G, Zimmermann SD, Plassard C (2018a) The Hebeloma cylindrosporum HcPT2 Pi transporter plays a key role in ectomycorrhizal symbiosis. New Phytol 220:1185–1199

Becquer A, Garcia K, Amenc L, Rivard C, Doré J, Trives-Segura C, Szponarski W, Russet S, Baeza Y, Lassalle-Kaiser B, Gay G, Zimmermann SD, Plassard C (2018b) The *Hebeloma cylindrosporum* HcPT2 Pi transporter plays a key role in ectomycorrhizal symbiosis. New Phytol 220:1185–1199

Beiler KJ, Durall DM, Simard SW, Maxwell SA, Kretzer AM (2010) Architecture of the wood-wide web: Rhizopogon spp. genets link multiple Douglas-fir cohorts. New Phytol 185:543–553

Bellion M, Courbot M, Jacob C, Blaudez D, Chalot M (2006) Extracellular and cellular mechanisms sustaining metal tolerance in ectomycorrhizal fungi. FEMS Microbiol Lett 254:173–181

Beneš V, Hložková K, Matěnová M, Borovička J, Kotrba P (2016) Accumulation of Ag and Cu in Amanita strobiliformis and characterization of its Cu and Ag uptake transporter genes AsCTR2 and AsCTR3. Biometals 29:249–264

Billings W (1987) Constraints to plant growth, reproduction, and establishment in arctic environments. Arct Alp Res 19:357–365

Bjorbækmo M, Carlsen T, Brysting A, Vrålstad T, Høiland K, Ugland K, Geml J, Schumacher T, Kauserud H (2010) High diversity of root associated fungi in both alpine and arctic Dryas octopetala. BMC Plant Biol 10:244

Blaudez D, Chalot M (2011) Characterization of the ER-located zinc transporter ZnT1 and identification of a vesicular zinc storage compartment in Hebeloma cylindrosporum. Fungal Genet Biol 48:496–503

Bödeker ITM, Nygren CMR, Taylor AFS, Olson Å, Lindahl BD (2009) ClassII peroxidase-encoding genes are present in a phylogenetically wide range of ectomycorrhizal fungi. ISME J 3:1387–1395

Borovička J, Kotrba P, Gryndler M, Mihaljevič M, Řanda Z, Rohovec J, Cajthaml T, Stijve T, Dunn CE (2010) Bioaccumulation of silver in ectomycorrhizal and saprobic macrofungi from pristine and polluted areas. Sci Total Environ 408:2733–2744

Burgess TI, Malajczuk N, Grove TS (1993) The ability of 16 ectomycorrhizal fungi to increase growth and phosphorus uptake of Eucalyptus globulus Labill. And E. diversicolor F. Muell. Plant Soil 153:155–164

Cazares E, Trappe JM (1994) Spore dispersal of ectomycorrhizal fungi on a glacier forefront by mammal mycophagy. Mycologia 86:507

Cejpková J, Gryndler M, Hršelová H, Kotrba P, Řanda Z, Synková I, Borovička J (2016) Bioaccumulation of heavy metals, metalloids, and chlorine in ectomycorrhizae from smelter-polluted area. Environ Pollut 218:176–185

Chen DM, Bastias BA, Taylor AFS, Cairney JWG (2003) Identification of laccase-like genes in ectomycorrhizal basidiomycetes and transcriptional regulation by nitrogen in Piloderma byssinum. New Phytol 157:547–554

Chen DM, Taylor AFS, Burke RM, Cairney JWG (2001) Identification of genes for lignin peroxidases and manganese peroxidases in ectomycorrhizal fungi. New Phytol 152:151–158

Colpaert JV, Wevers JHL, Krznaric E, Adriaensen K (2011) How metal-tolerant ecotypes of ectomycorrhizal fungi protect plants from heavy metal pollution. Ann For Sci 68:17–24

Courbot M, Diez L, Ruotolo R, Chalot M, Leroy P (2004) Cadmium responsive thiols in the ectomycorrhizal fungus Paxillus involutus. Appl Environ Microbiol 70:7413–7417

Courty P-E, Bréda N, Garbaye J (2007) Relation between oak tree phenology and the secretion of organic matter degrading enzymes by Lactarius quietus ectomycorrhizas before and during bud break. Soil Biol Biochem 39:1655–1663

Daguerre Y, Levati E, Ruytinx J, Tisserant E, Morin E, Kohler A, Montanini B, Ottonello S, Brun A, Veneault-Fourrey C, Martin F (2017) Regulatory networks underlying mycorrhizal development delineated by genome-wide expression profiling and functional analysis of the transcription factor repertoire of the plant symbiotic fungus Laccaria bicolor. BMC Genomics 18:737

Dahl E (1956) Rondane Mountain vegetation in South Norway and its relation to the environment. Aschehoug, Oslo. In: Rodwell JS (ed) British plant communities. Cambridge University Press, Cambridge, pp 615–616

Frank B (1887) Sind die Wurzelanschwellungen der Erlen und Eläagnaceen Pilzgallen. Berichte der Dtsch Bot Gesellschaft, Bd V 50

Frank B (1891) Ueber die auf Verdauung von Pilzen abzielende Symbiose der mit endotrophen Mykorhizen begabten Pflanzen, sowie der Leguminosen und Erlen. Ber Dtsch Bot Ges 9:244–253

Gadd GM, Fomina M (2011) Uranium and Fungi. Geomicrobiol J 28:471–482

Gadd GM, Rhee YJ, Stephenson K, Wei Z (2012) Geomycology: metals, actinides and biominerals. Environ Microbiol Rep 4:270–296

Gallie U, Meire M, Brunold C (1993) Effect of cadmium on nonmycorrhizal and mycorrhizal Norway spruce seedlings *Picea abies* (L) Karst and its ectomycorrhizal fungi Laccaria laccata (Scop ex Fr) Bk and Br- sulfate reduction, thiols and distribution of the heavy-metals. New Phytol 125:837–843

Garcia K, Delaux P-M, Cope KR, Ané J-M (2015) Molecular signals required for the establishment and maintenance of ectomycorrhizal symbioses. New Phytol 208:79–87

Garcia K, Doidy J, Zimmermann SD, Wipf D, Courty P-E (2016) Take a trip through the plant and fungal transportome of mycorrhiza. Trends Plant Sci 21:937–950

Geml J, Timling I, Robinson CH, Lennon N, Nusbaum HC, Brochmann C, Noordeloos ME, Taylor DL (2012) An arctic community of symbiotic fungi assembled by long-distance dispersers:

phylogenetic diversity of ectomycorrhizal basidiomycetes in Svalbard based on soil and sporocarp DNA. J Biogeogr 39:74–88

Gryndler M, Hršelová H, Soukupová L, Borovička J (2012) Silver release from decomposed hyperaccumulating Amanita solitaria fruit-body biomass strongly affects soil microbial community. Biometals 25:987–993

Guerrero-Galán C, Delteil A, Garcia K, Houdinet G, Conéjéro G, Gaillard I, Sentenac H, Zimmermann SD (2018) Plant potassium nutrition in ectomycorrhizal symbiosis: properties and roles of the three fungal TOK potassium channels in Hebeloma cylindrosporum. Environ Microbiol 20:1873–1887

Hacquard S, Tisserant E, Brun A, Legué V, Martin F, Kohler A (2013) Laser microdissection and microarray analysis of Tuber melanosporum ectomycorrhizas reveal functional heterogeneity between mantle and Hartig net compartments. Environ Microbiol 15:1853–1869

Haselwandter K, Read DJ (1980) Fungal associations of roots of dominant and sub-dominant plants in high-alpine vegetation systems with special reference to mycorrhiza. Oecologia 45:57–62

Heinonsalo J, Sun H, Santalahti M, Bäcklund K, Hari P, Pumpanen J (2015) Evidences on the ability of mycorrhizal genus Piloderma to use organic nitrogen and deliver it to Scots pine. PLoS One 10(e0131561):1–17

Henke C, Jung E-M, Voit A, Kothe E, Krause K (2016) Dehydrogenase genes in the ectomycorrhizal fungus Tricholoma vaccinum: a role for Ald1 in mycorrhizal symbiosis. J Basic Microbiol 56:162–174

Henry C, Raivoarisoa J-F, Razafimamonjy A, Ramanankierana H, Andrianaivomahefa P, Ducousso M, Selosse M-A (2016) Characterization of ectomycorrhizal communities of *Asteropeia mcphersonii* seedlings spontaneously growing in natural forest and in open disturbed areas. Bot Lett 163:273–279

Henry C, Sellosse M-A, Richard F, Ramanankierana H, Ducousso M (2014) Comprendre la dynamique des communautés mycorhiziennes lors des successions végétales. Première partie : méthodes d'étude, caractérisations et fonctionnement (revue bibliographique). Rev For Française Fr.], ISSN 0035

Hibi T, Nii H, Nakatsu T, Kimura A, Kato H, Hiratake J, Oda J (2004) Crystal structure of gammaglutamylcysteine synthetase: insights into the mechanism of catalysis by a key enzyme for glutathione homeostasis. Proc Natl Acad Sci U S A 101:15052–15057

Hobbie JE, Hobbie EA, Drossman H, Conte M, Weber JC, Shamhart J, Weinrobe M (2009) Mycorrhizal fungi supply nitrogen to host plants in Arctic tundra and boreal forests: 15 N is the key signal. Can J Microbiol 55:84–94

Huang G, Dong R, Allen R, Davis EL, Baum TJ, Hussey RS (2006) A root-knot nematode secretory peptide functions as a ligand for a plant transcription factor. Mol Plant-Microbe Interact 19:463–470

Ilyas S, Rehman A (2015) Oxidative stress, glutathione level and antioxidant response to heavy metals in multi-resistant pathogen, *Candida tropicalis*. Environ Monit Assess 187:1–7

Jamoni P (2008) Funghi alpini delle zone alpine superiori e inferiori. Associazione Micologica Bresadola; Fondazione Centro Studi Micologici dell' A.M.B., Trento Italy; Vicenza Italy

Jumpponen A, Egerton-Warburton L (2005) Mycorrhizal Fungi in successional environments: a community assembly model incorporating host plant, environmental, and biotic filters. In: Dighton J, White JF (eds) The fungal community, its organization and role in the ecosystem, 3rd edn. CRC Press, Boca Raton, pp 159–188

Kalsotra T, Khullar S, Agnihotri R, Reddy MS (2018) Metal induction of two metallothionein genes in the ectomycorrhizal fungus Suillus himalayensis and their role in metal tolerance. Microbiology 164:868–876

Karpati AS, Handel SN, Dighton J, Horton TR (2011) Quercus rubra-associated ectomycorrhizal fungal communities of disturbed urban sites and mature forests. Mycorrhiza 21:537–547

Kazan K, Manners JM (2012) JAZ repressors and the orchestration of phytohormone crosstalk. Trends Plant Sci 17:22–31

Khosla B, Kaur H, Reddy MS (2009) Influence of ectomycorrhizal colonization on the growth and mineral nutrition of *Populus deltoides* under aluminum toxicity. J Plant Interact 4:93–99

Khullar S, Reddy MS (2018) Ectomycorrhizal fungi and its role in metal homeostasis through metallothionein and glutathione mechanisms. Curr Biotechnol 7:231–241

Khullar S, Reddy MS (2019a) Cadmium and arsenic responses in the ectomycorrhizal fungus Laccaria bicolor: glutathione metabolism and its role in metal(loid) homeostasis. Environ Microbiol Rep 11:53–61

Khullar S, Reddy MS (2019b) Cadmium induced glutathione bioaccumulation mediated by γ-glutamylcysteine synthetase in ectomycorrhizal fungus Hebeloma cylindrosporum. Biometals 32:101–110

Kohler A, Kuo A, Nagy LG et al (2015) Convergent losses of decay mechanisms and rapid turnover of symbiosis genes in mycorrhizal mutualists. Nat Genet 47:410–415

Krause K, Henke C, Asiimwe T, Ulbricht A, Klemmer S, Schachtschabel D, Boland W, Kothe E (2015) Biosynthesis and secretion of indole-3-acetic acid and its morphological effects on tricholoma vaccinum-spruce ectomycorrhiza. Appl Environ Microbiol 81:7003–7011

Kubrová J, Žigová A, Řanda Z, Rohovec J, Gryndler M, Krausová I, Dunn CE, Kotrba P, Borovička J (2014) On the possible role of macrofungi in the biogeochemical fate of uranium in polluted forest soils. J Hazard Mater 280:79–88

Kumari D, Reddy MS, Upadhyay RC (2011) Cantharellus pseudoformosus, a new species associated with Cedrus deodara from India. Mycoscience 52:147–151

Lazaruk LW, Kernaghan G, Macdonald SE, Khasa D (2005) Effects of partial cutting on the ectomycorrhizae of Picea glauca forests in northwestern Alberta. Can J For Res 35:1442–1454

Leszczyszyn OI, Imam HT, Blindauer CA (2013) Diversity and distribution of plant metallothioneins: a review of structure, properties and functions. Metallomics 5:1146

Liao H-L, Chen Y, Bruns TD, Peay KG, Taylor JW, Branco S, Talbot JM, Vilgalys R (2014) Metatranscriptomic analysis of ectomycorrhizal roots reveals genes associated with Piloderma-Pinus symbiosis: improved methodologies for assessing gene expression in situ. Environ Microbiol 16:3730–3742

Liu YJ, Hall BD (2004) Body plan evolution of ascomycetes, as inferred from an RNA polymerase II phylogeny. Proc Natl Acad Sci U S A 101:507–512

Luis P, Kellner H, Zimdars B, Langer U, Martin F, Buscot F (2005) Patchiness and spatial distribution of laccase genes of ectomycorrhizal, saprotrophic, and unknown basidiomycetes in the upper horizons of a mixed forest cambisol. Microb Ecol 50:570–579

Lundell TK, Mäkelä MR, de Vries RP, Hildén KS (2014) Genomics, lifestyles and future prospects of wood-decay and litter-decomposing basidiomycota. Adv Bot Res 70:329–370

Maggi O, Tosi S, Angelova M, Lagostina E, Fabbri AA, Pecoraro L, Altobelli E, Picco AM, Savino E, Branda E, Turchetti B, Zotti M, Vizzini A, Buzzini P (2013) Adaptation of fungi, including yeasts, to cold environments. Plant Biosyst – An Int J Deal All Asp Plant Biol 147:247–258

Martin F, Aerts A, Ahrén D et al (2008) The genome of Laccaria bicolor provides insights into mycorrhizal symbiosis. Nature 452:88–92

Martin F, Kohler A, Duplessis S (2007) Living in harmony in the wood underground: ectomycorrhizal genomics. Curr Opin Plant Biol 10:204–210

Martin F, Kohler A, Murat C, Veneault-Fourrey C, Hibbett DS (2016) Unearthing the roots of ectomycorrhizal symbioses. Nat Rev Microbiol 14:760–773

Martin F, Nehls U (2009) Harnessing ectomycorrhizal genomics for ecological insights. Curr Opin Plant Biol 12:508–515

Mehra RK, Tarbet EB, Gray WR, Winge DR (1988) Metal-specific synthesis of two metallothioneins and gamma-glutamyl peptides in Candida glabrata. Proc Natl Acad Sci U S A 85:8815–8819

Mello A, Balestrini R (2018) Recent insights on biological and ecological aspects of ectomycorrhizal fungi and their interactions. Front Microbiol 9:216

Miller SL, Larsson E, Larsson K-H, Verbeken A, Nuytinck J (2006) Perspectives in the new Russulales. Mycologia 98:960–970

Mohanan C (2014) Macrofungal diversity in the Western Ghats, Kerala, India: members of Russulaceae. J Threat Taxa 6:5636–5648

Morgado LN, Semenova TA, Welker JM, Walker MD, Smets E, Geml J (2015) Summer temperature increase has distinct effects on the ectomycorrhizal fungal communities of moist tussock and dry tundra in Arctic Alaska. Glob Chang Biol 21:959–972

Newsham KK, Hopkins DW, Carvalhais LC, Fretwell PT, Rushton SP, O'Donnell AG, Dennis PG (2016) Relationship between soil fungal diversity and temperature in the maritime Antarctic. Nat Clim Chang 6:182–186

Osobová M, Urban V, Jedelský PL, Borovička J, Gryndler M, Ruml T, Kotrba P (2011) Three metallothionein isoforms and sequestration of intracellular silver in the hyperaccumulator Amanita strobiliformis. New Phytol 190:916–926

Pena R, Lang C, Naumann A, Polle A (2014) Ectomycorrhizal identification in environmental samples of tree roots by Fourier-transform infrared (FTIR) spectroscopy. Front Plant Sci 5:229

Peterson RL, Massicotte HB (2004) Exploring structural definitions of mycorrhizas, with emphasis on nutrient-exchange interfaces. Can J Bot 82:1074–1088

Pikalova T, Sacky J, Briksi A, Borovicka J, Kotrba P (2011) Zinc accumulation and different ways to sequestration of intracellular zinc in fruit-bodies of ectomycorrhizal fungi Russula spp. and hebeloma spp. In: Proceedings of the 7th international conference on mushroom products

Plett JM, Daguerre Y, Wittulsky S, Vayssières A, Deveau A, Melton SJ, Kohler A, Morrell-Falvey JL, Brun A, Veneault-Fourrey C, Martin F (2014) Effector MiSSP7 of the mutualistic fungus Laccaria bicolor stabilizes the Populus JAZ6 protein and represses jasmonic acid (JA) responsive genes. Proc Natl Acad Sci U S A 111:8299–8304

Plett JM, Kemppainen M, Kale SD, Kohler A, Legué V, Brun A, Tyler BM, Pardo AG, Martin F (2011) A secreted effector protein of Laccaria bicolor is required for symbiosis development. Curr Biol 21:1197–1203

Raidl S (1997) Studien zur Ontogenie an Rhizomorphen von Ektomykorrhizen. J Cramer

Ramesh G, Podila GK, Gay G, Marmeisse R, Reddy MS (2009) Different patterns of regulation for the copper and cadmium metallothioneins of the ectomycorrhizal fungus Hebeloma cylindrosporum. Appl Environ Microbiol 75:2266–2274

Reddy MS, Prasanna L, Marmeisse R, Fraissinet-Tachet L (2014) Differential expression of metallothioneins in response to heavy metals and their involvement in metal tolerance in the symbiotic basidiomycete Laccaria bicolor. Microbiology 160:2235–2242

Reddy MS, Verma B (2014) Suillus triacicularis sp. nov., a new species associated with Pinus roxburghii from northwestern Himalayas, India. Phytotaxa 162:157

Rineau F, Lmalem H, Ahren D, Shah F, Johansson T, Coninx L, Ruytinx J, Nguyen H, Grigoriev I, Kuo A, Kohler A, Morin E, Vangronsveld J, Martin F, Colpaert JV (2017) Comparative genomics and expression levels of hydrophobins from eight mycorrhizal genomes. Mycorrhiza 27:383–396

Sácký J, Leonhardt T, Borovička J, Gryndler M, Briksí A, Kotrba P (2014) Intracellular sequestration of zinc, cadmium and silver in Hebeloma mesophaeum and characterization of its metallothionein genes. Fungal Genet Biol 67:3–14

Sammer D, Krause K, Gube M, Wagner K, Kothe E (2016) Hydrophobins in the life cycle of the ectomycorrhizal basidiomycete Tricholoma vaccinum. PLoS One 11:e0167773

Schimel JP, Bilbrough C, Welker JM (2004) Increased snow depth affects microbial activity and nitrogen mineralization in two Arctic tundra communities. Soil Biol Biochem 36:217–227

Schlunk I, Krause K, Wirth S, Kothe E (2015) A transporter for abiotic stress and plant metabolite resistance in the ectomycorrhizal fungus Tricholoma vaccinum. Environ Sci Pollut Res 22:19384–19393

Selosse M-A, Richard F, He X, Simard SW (2006) Mycorrhizal networks: des liaisons dangereuses? Trends Ecol Evol 21:621–628

Simard SW, Beiler KJ, Bingham MA, Deslippe JR, Philip LJ, Teste FP (2012) Mycorrhizal networks: mechanisms, ecology and modelling. Fungal Biol Rev 26:39–60

Smith SE, Read DJ (2008) Mycorrhizal symbiosis. Elsevier, New York, p 120
Tang Y, Shi L, Zhong K, Shen Z, Chen Y (2019) Ectomycorrhizal fungi may not act as a barrier inhibiting host plant absorption of heavy metals. Chemosphere 215:115–123
Tedersoo L, Bahram M, Polme S et al (2014) Global diversity and geography of soil fungi. Science 346:1256688–1256688
Tedersoo L, May TW, Smith ME (2010) Ectomycorrhizal lifestyle in fungi: global diversity, distribution, and evolution of phylogenetic lineages. Mycorrhiza 20:217–263
Thorsen M, Lagniel G, Kristiansson E, Junot C, Nerman O, Labarre J, Tamás MJ (2007) Quantitaive transcriptome, proteome, and sulfur metabolite profiling of the *Saccharomyces cerevisiae* response to arsenite. Physiol Genomics 30:35–43
Timling I, Dahlberg A, Walker DA, Gardes M, Charcosset JY, Welker JM, Taylor DL (2012) Distribution and drivers of ectomycorrhizal fungal communities across the North American Arctic. Ecosphere 3:art111
Timling I, Taylor DL (2012) Peeking through a frosty window: molecular insights into the ecology of Arctic soil fungi. Fungal Ecol 5:419–429
Townsend DM (2007) S-Gluathionylation: indicator of cell stress and regulator of the unfolded protein response. Mol Interv 7:313–324
Treseder KK, Torn MS, Masiello CA (2006) An ecosystem-scale radiocarbon tracer to test use of litter carbon by ectomycorrhizal fungi. Soil Biol Biochem 38:1077–1082
Turgeman T, Ben Asher J, Roth-Bejerano N, Kagan-Zur V, Kapulnik Y, Sitrit Y (2011) Mycorrhizal association between the desert truffle Terfezia boudieri and Helianthemum sessiliflorum alters plant physiology and fitness to arid conditions. Mycorrhiza 21:623–630
Twieg BD, Durall DM, Simard SW (2007) Ectomycorrhizal fungal succession in mixed temperate forests. New Phytol 176:437–447
Urban A (2011) Metal elements and the diversity and function of ectomycorrhizal communities, pp 231–254
Vašák M, Meloni G (2011) Chemistry and biology of mammalian metallothioneins. JBIC J Biol Inorg Chem 16:1067–1078
Vayssières A, Pěnčík A, Felten J, Kohler A, Ljung K, Martin F, Legué V (2015) Development of the Poplar-Laccaria bicolor Ectomycorrhiza modifies root auxin metabolism, signaling, and response. Plant Physiol 169:890–902
Veneault-Fourrey C, Commun C, Kohler A, Morin E, Balestrini R, Plett J, Danchin E, Coutinho P, Wiebenga A, de Vries RP, Henrissat B, Martin F (2014) Genomic and transcriptomic analysis of Laccaria bicolor CAZome reveals insights into polysaccharides remodelling during symbiosis establishment. Fungal Genet Biol 72:168–181
Verma B, Reddy MS (2015) *Suillus indicus* sp. nov. (Boletales, Basidiomycota), a new boletoid fungus from northwestern Himalayas, India. Mycology 6:35–41
Wang B, Qiu Y-L (2006) Phylogenetic distribution and evolution of mycorrhizas in land plants. Mycorrhiza 16:299–363
Zhang F, Anasontzis GE, Labourel A, Champion C, Haon M, Kemppainen M, Commun C, Deveau A, Pardo A, Veneault-Fourrey C, Kohler A, Rosso M-N, Henrissat B, Berrin J-G, Martin F (2018) The ectomycorrhizal basidiomycete Laccaria bicolor releases a secreted β-1,4 endoglucanase that plays a key role in symbiosis development. New Phytol 220:1309–1321

Rumen Microbiome and Plant Secondary Metabolites (PSM): Inhibition of Methanogenesis and Improving Nutrient Utilization

13

D. N. Kamra and B. Singh

Abstract

Plants contain a large number of secondary compounds which are not required for the primary activities of plants but act as a defense against pathogenic microbes and dust particles. These plant secondary metabolites (PSM) include saponins, tannins, essential oils, alkaloids, terpene compounds, etc. These PSM have strong anti-methanogenic activity, and a few of them have also fiber degradation stimulating activity, but many of these have no effect on feed degradation or have an adverse effect on nutrient release. A proper combination of these PSM might have a balanced activity against methane inhibition and improve fiber degradation, making the process of livestock production economic and eco-friendly.

Keywords

Rumen microbes · Diversity · Methanogenesis · Improving fiber degradation · Saponins · Tannins · Essential oils

13.1 Introduction

There are several thousands (about 200,000) of secondary metabolites are present in plants, which protect them from the invasion of foreign particles and pathogenic microbes. Primary metabolites of plants are directly involved in the growth and development of a plant, while secondary metabolites are compounds produced in other metabolic pathways and are not essential to the functioning of the plants.

D. N. Kamra (✉)
Animal Nutrition Division, ICAR-Indian Veterinary Research Institute, Izatnagar, Uttar Pradesh, India

B. Singh
ICAR-Indian Veterinary Research Institute, Palampur, Himachal Pradesh, India

T. Satyanarayana et al. (eds.), *Advancing Frontiers in Mycology & Mycotechnology*,
https://doi.org/10.1007/978-981-13-9349-5_13

These compounds are synthesized during secondary metabolism. Therefore, these are classified as plant secondary metabolites. The major plant secondary metabolites are saponins, tannins, essential oils, and alkaloids.

13.2 Plant Secondary Metabolites (PSM)

13.2.1 Saponins

The naturally occurring chemicals are found in plants, which have soap-like qualities and produce lather with water just like detergent, and are present in more than 100 families of plants. In the digestive tract, saponins produce an emulsification of fat-soluble molecules. Specifically, saponins bind to bile acids and help eliminate them from the body, preventing cholesterol from being reabsorbed. In addition to this, saponins can help boost the immune system, exhibit an antioxidant effect, and may even support bone strength. Saponins have several qualities that act against cancer cells. In particular, some saponins have an antioxidant effect and may be directly toxic to cancer cells. Other beneficial characteristics of saponins include encouraging normal detoxification. Saponins found in oats and spinach support digestion by accelerating the body's ability to absorb calcium and silicon. They modify ruminal fermentation by suppressing ruminal protozoa and selectively inhibiting some bacteria. The symbiosis of protozoa with methanogenic bacteria in the rumen is well established, and the selective suppression of protozoa has been suggested to be a promising approach to reduce methane production.

Using *Yucca* as a source of saponins, Sliwiński et al. (2002) did not record methane reduction, while a decrease was observed in other studies. On the other hand, Holtshausen et al. (2009) at 1% of *Yucca* extract in the diet (saponin content 0.06% in diet) did not observe a reduction in methane production. The saponin content in the *Yucca* extract used in the study is not given. Although *Yucca* extracts used in the studies of Santoso et al. (2004) and Holtshausen et al. (2009) were obtained from the same commercial company, the products used could be different. However, it may be noted that Holtshausen et al. (2009) used 100-fold higher amount of the extract than that used by Santoso et al. (2004), and also in the former study, saponins used were fivefold higher than the amount of the extract, but no methane reduction was observed. Although these results are difficult to explain, the difference in effects could be due to different diets used; effects might be higher for silage-based diet used by Santoso et al. (2004). Hess et al. (2004) used dried fruits of *Sapindus saponaria* and recorded a decrease in methane production. The level of saponins in the diet in this study was 0.75%, which is much higher than the levels of *Yucca* extracts/saponins that elicited methane reduction.

Saponins are reported in plants like soapnut/soapwort and a few others. The saponins are soluble in water and form foam. These contain glycosides, steroid, and terpenoid glycosides as one of the major constituents. It can act as a detergent and a fire extinguisher. These consist of a polycyclic aglycones, which are either choline steroids or triterpenoids which are attached with C3 and ether bonds to sugar side chains (Güçlü-Ustündağ and Mazza 2007). Some of the saponins reduce feed intake and growth rate of nonruminant animals, while others are not very harmful.

Saponins are not harmful in tropical forage legumes, but they are common in several temperate forage legumes. Alfalfa contains several saponins (medicagenic acid, soyasapogenol A, soyasapogenol B, lucernic acid), and the seeds and foliage of chickpeas (*Cicer arietinum*), soybeans, and common beans contain saponins. *Yucca* contains sarsaponins and are occasionally grazed by cattle. In monogastric animals like poultry and swine, poisoning is reported due to saponins: irritated mucous membranes of the mouth and digestive tract, reduced feed intake, decreased performance, anorexia, weight loss, rough hair coat, gastroenteritis, diarrhea, and possibly abortion. On consuming fresh alfalfa, saponins cause bloat in ruminants. Bloat only occurs in animals grazing temperate legumes that contain saponins but not in livestock grazing tropical legumes or temperate legumes like bird's-foot trefoil that does not contain saponins. However, low-saponin cultivars of alfalfa can cause bloating.

13.2.2 Tannins

Tannins are defined as phenolic compounds of high molecular weight ranging from 500 Da to more than 3000 Da which are found in plant leaves, bark, fruit, wood, and roots. These are used in making leather, and gallic acid is employed in the production of inks. Tannins in tea give the beverage its astringency. Teas with high levels of tannins have a bitter taste especially in green and black tea. Tannins also remove harmful microbes from the body and fight against harmful bacteria, viruses, and fungi. On regular use, tannic acid can cause side effects such as stomach irritation, nausea, vomiting, and liver damage.

Tannins are active for plant defense mechanisms against mammalian herbivores, birds, and insects (Hassanpour et al., 2011). Tannins are soluble in water (20–35°C), leaving aside high molecular weight tannins. These are water-soluble polyphenolic substances and have the ability to bind with proteins that form insoluble or soluble tannin-protein complexes. Tannins are able to make complexes with polysaccharides (cellulose, hemicelluloses, and pectin) and nucleic acids, steroids, alkaloids, and saponins (Chaichi Semsari et al., 2011). Now it has also been observed that tannins have beneficial effects on animals in having antimicrobial and anthelmintic effects in ruminants (Hassanpour et al. 2011).

As per chemical structure and properties, tannins are divided into two main groups: hydrolysable (HT) and condensed tannins (CT) (Chaichi Semsari et al. 2011). The chemical structures of HTs (gallotannins and ellagitannins) are molecules which contain carbohydrate, generally D-glucose, as a central core (Min and Hart 2003). The hydrolysable groups of these carbohydrates are esterified with phenolic groups, such as ellagic acid or gallic acid (Haslem 1989).

In the dry season of tropics, the quality of lignocellulosic/tanniniferous feeds is poorer, and livestock production is limited due to poor availability of nutrients, i.e., 2.5–7.0% crude protein (dry matter basis) and low dry matter digestibility of 40–50%. The ruminants lose weight and milk production drops (Patra and Saxena 2010). Improved animal performance has been frequently reported in response to the use of high-quality tanniniferous forages as supplements for ruminants fed with low-quality roughage diets. In addition to that, moderate amounts of CTs have been reported to exert beneficial effects on protein metabolism in ruminants, decreasing rumen degradation of dietary protein and increasing absorption of amino acids in the small intestine (Hervás et al. 2003). The CT may enable dietary protein bypass from the rumen for digestion in the lower digestive tract (Hassanpour et al. 2011). An increase in flow of metabolizable protein or essential amino acids to the small intestine has been observed in animals grazing forages of high-CT content compared to those grazing a low-CT diet (Waghorn 2008). The plant secondary metabolites have beneficial effects on protein metabolism in ruminants, decreasing rumen degradation of dietary protein and increasing absorption of amino acids in the small intestine.

13.2.3 Essential Oils

A concentrated hydrophobic liquid extracted from plants containing aromatic compounds is known as essential oil, which is also known as volatile oils, ethereal oils, aetherolea, or simply as the oils of the plants. These oils are extracted by steam distillation and are used in perfumes, soaps, cosmetics, household cleaning products, and flavoring of foods and drinks. The essential oils are used in aroma therapy for skin treatment. Improper use of essential oils may cause harms including allergic reactions and skin irritation, and children may be particularly susceptible to the toxic effects of improper use. The essential oils are also named after the plants from which these are extracted like garlic oil, mentha oil, sweet orange oil, lemon oil, cedarwood oil, clove oil, eucalyptus oil, jasmine oil, rose oil, etc., which are used for various reasons.

Patra and Yu (2012) studied five EOs with different chemical structures, e.g., clove oil (CLO) containing eugenol (phenylpropanoid), eucalyptus oil (EUO) cineole (bicyclic monoterpenoid), garlic oil (GAO) alliin and allicin (organosulfur compounds), origanum oil (ORO) thymol (monoterpenoid monocyclic phenol), and peppermint oil (PEO) menthol (monoterpenoid monocyclic nonphenol) on different fermentation characteristics and rumen microbiome in the rumen. This study demonstrated that different EOs vary in their potencies in modulating rumen microbial populations and fermentation.

All the EOs exhibited an adverse effect on the three rumen cellulolytic bacteria, i.e., *Fibrobacter succinogenes*, *Ruminococcus flavefaciens*, and *R. albus*, which were significantly reduced by all EOs. The *F. succinogenes* population suffered from more inhibition than the populations of *R. flavefaciens* and *R. albus* for all EOs. The results showed that EOs could significantly decrease methane production, ammonia production, and the abundance and diversity of archaea with increasing doses. Calsamiglia et al. (2007) compiled a review on the effect of essential oils to

Table 13.1 Essential oils with antimicrobial activity, their active components, and susceptible microorganisms

Essential oil	Name	Active components	Susceptible microorganisms	References
Allium sativum	Garlic	Allicin, diallyl sulfite	Enteropathogenic bacteria	Ross et al. (2001)
Cinnamomum cassia	Cassia	Cinnamaldehyde	*Escherichia coli, Staphylococcus aureus*	–
Origanum vulgare	Oregano	Carvacrol, thymol	Gram-positive and gram-negative bacteria	Sivropoulou et al. (1996) and Dorman and Deans (2000)
Syzygium aromaticum	Clove	Eugenol	*E. coli, Staph. Aureus, L. monocytogenes, S. enteritidis, C. jejuni*	Smith-Palmer et al. (1998)
Zingiber officinale, ginger zingiberene	Ginger	Zingerone	Gram-positive and gram-negative bacteria	–

[a]*Based on* Calsamiglia et al. (2007)

study their effect on methane production and microbial ecosystem, and results are summarized in Table 13.1.

13.2.4 Rumen Microbiome

Herbivorous animals retain within their gastrointestinal tract microbiome that specialize in hydrolysis and fermentation of lignocellulosic plant biomass. The gut microflora is exceedingly diverse and contains representatives of all three domains – Eukarya, Archaea, and Bacteria.

Next-generation sequencing (NGS) and DNA microarrays have revealed that diversity and population density and range of microbial communities present in any ecosystem are much diverse and complex than explained based on customary culture-based methods and conventional molecular biological methods.

13.2.5 Conventional Techniques

The rumen microbial ecosystem consists of a vast majority of different microbes like bacteria, protozoa, fungi, bacteriophages, archaea, and mycoplasma, but a large number of them (more than 90%) are not culturable. Therefore, until now we have been playing only with less than 10% of the total microbes present in the ecosystem, while all these microbes had been active biologically in the system. Some of the genera present in the ecosystem are listed below (based on Kamra 2005).

13.2.6 Rumen Microbes

Bacteria (adapted from Kamra 2005)

Cellulose degraders: *Fibrobacter succinogenes* (*Bacteroides succinogenes*), *Ruminococcus flavefaciens, Ruminococcus albus, Clostridium longisporum, Clostridium lochheadii, Eubacterium cellulosolvens* (*Cillobacterium cellulosolvens*)

Hemicellulose degraders: *Butyrivibrio fibrisolvens, Prevotella ruminicola (Bacteroides ruminicola), Eubacterium xylanophilum, E. uniformis*

Starch degraders: *Streptococcus bovis, Ruminobacter amylophilus (Bacteroides amylophilus)*, (*Bacteroides ruminicola*)

Sugar/dextrin fermenters: *Succinivibrio dextrinosolvens, Succinivibrio amylolytica, Selenomonas ruminantium, Lactobacillus acidophilus, L. casei, L. fermentum, L. plantarum, L. brevis, L. helveticus, Bifidobacterium globosum, B. longum, B. thermophilum, B. ruminantium*

Pectin degraders: *Treponema saccharophilum, Lachnospira multiparus*

Protein degraders: *Prevotella ruminicola, Ruminobacter amylophilus, Clostridium bifermentans*

Urea hydrolysers: *Megasphaera elsdenii, Micrococcus*

Acid utilizers: *Megasphaera elsdenii* (*Peptostreptococcus elsdenii*), *Wolinella succinogenes* (*Vibrio succinogenes*), *Veillonella gazogenes* (*Veillonella alcalescens, Micrococcus lactolytica*)

Oxalic acid degraders: *Oxalobacter formigenes*

Sulfate-reducing bacteria: *Desulfovibrio desulfuricans, Desulfotomaculum ruminis*

Succinic acid utilizers: *Succiniclasticum ruminis*

Lipolytic bacteria: *Anaerovibrio lipolytica*

Acetogenic bacteria: *Eubacterium limosum, Acetitomaculum ruminis*

Tannin degraders: *Streptococcus caprinus, Eubacterium oxidoreducens*

Mimosine degraders: *Synergistes jonesii*

Archaea

Archaea: *Methanobrevibacter ruminantium, Methanobacterium formicicum, Methanosarcina barkeri, Methanomicrobium mobile*

Mycoplasma

Anaeroplasma bactoclasticum, Anaeroplasma abactoclasticum

Protozoa

Rumen anaerobic protozoa (ciliates): *Isotricha prostoma, I. intestinalis, Dasytricha ruminantium, Oligoisotricha bubali*

Entodiniomorphid protozoa: *Entodinium bovis, E. bubalum, E. caudatum, E. longinucleatum, Diplodinium dendatum, D. indicum, Eremoplastron asiaticus, E. bubalus, Eudiplodinium maggii, Ostracodinium trivesiculatum, Polyplastron multivesiculatum, Metadinium medium, Epidinium caudatum, Ophryoscolex caudatus, Caloscolex camelicus*

Rumen Fungi

Rumen anaerobic fungi: *Neocallimastix frontalis, N. patriciarum, N. hurleyensis, Sphaeromonas communis (Caecomyces communis), Caecomyces equi, Orpinomyces bovis, Anaeromyces mucronatus (Ruminomyces mucronatus), Ruminomyces elegans, Piromyces communis*

The ruminants consume lignocellulosic feeds like cereal straws and stovers, sugarcane-based agricultural by-products, and mature green fodders in India and other tropical countries. The ruminants are not able to digest these feeds by themselves, but microbes (bacteria, ciliate protozoa, fungi, and archaea present in the fermentation sacs of the gastrointestinal tract like rumen, pseudo rumen, and caecum) help in digesting these feeds and convert them into volatile fatty acids (the major source of energy for the ruminants) and microbial protein which serve as nitrogen source in the ruminant animals (Kamra 2005).

13.3 Fiber Degradation and Microbial Diversity

The rumen microbiome can be studied in more details by rRNA sequencing and consists of several thousand microbes belonging to three different domains like Bacteria, Archaea, and Eukarya (fungi and protozoa). Bacteria are the most diverse and represent about 95% of total microbes (Flint et al. 2008). *Prevotella* was the predominant bacterium representing about 30% of the total rumen bacteria for cellulose degradation. The already known and the most important key fibrolytic bacteria, viz., *R. flavefaciens*, *R. albus*, *F. succinogenes*, *Butyrivibrio*, *Clostridium, and Eubacterium,* represented only ~2% of ruminal bacterial 16S rRNA.

Like any other microenvironment, the rumen microbial ecosystem too has more than 90% of microbes which are unculturable; therefore, till date it is not possible to understand the mechanism of its functioning, complexity, and interaction among the microbes. By advancement in molecular biology tools, like the construction of 16S rDNA or 16S rRNA libraries, metagenomics, metatranscriptomics, etc., the researchers have tried to explore the unculturable microbes of rumen. Looking at the richness of rumen microbiome in terms of diversity and microbial enzymes, the metagenomic studies on rumen microbiome are an emerging research area. The metagenomic analysis of rumen microbiome of Surti buffalo fed on four different feeding schedules using pyrosequencing has been reported by Singh et al. (2011). The sequences were analyzed using Metagenome Rapid Annotation using Subsystems Technology (MG-RAST). The distribution of phylotypes and environmental gene tags (EGTs) detected within each rumen sample was dominated by Bacteroidetes/Chlorobi, Firmicutes, and Proteobacteria in all the samples irrespective of the type of diet, and most of the genes belonged to Bacteroidetes/Chlorobi group. A report from the author's laboratory from the rumen of Murrah buffalo also showed dominance of these three phylotypes followed by Actinobacteria representing bacteria-specific EGTs (~about 80% of total EGTs) with the abundance of supporting EGTs represented by *Bacteroides* and *Ruminococcus* with very few *Fibrobacter* which is considered to be the major fiber-degrading bacteria in the rumen (unpublished data). The Surti buffalo rumen microbiome was dominated by carbohydrate metabolism (20%) in high-fiber diet (100% roughage) and lowest (13%) in high-concentrate diet (75% concentrate) (Singh et al. 2011).

It has been observed in the literature that F/B (Firmicutes/Bacteroidetes) ratio was higher in high-fiber diet (low TDN diet). This higher F/B ratio indicated higher fiber utilization which was supposed to be due to an increase in the population density of *Ruminococcus flavefaciens*. Similarly, in our study, a numerically higher F/B ratio and higher *Ruminococcus* count (by real-time PCR) were observed with a high-fiber diet (Kala et al. 2017). The majority of bacterial genera reported were able to degrade lignocellulosic feeds. *Prevotella* was the predominant bacterium representing about 30% of the total rumen bacteria. *Ruminococcus* and *Fibrobacter* were only 2–3% of rumen bacteria community of buffaloes irrespective of the diet. Predominance of *Prevotella* and very little representation of *Ruminococcus* and *Fibrobacter* were also found in the goat rumen microbiome as reported in the literature. The already known three most important key fibrolytic bacteria, viz., *R. flavefaciens*, *R. albus*, and *F. succinogenes*, represented only ~2% of the ruminal bacterial 16S rRNA. Very little representation of these fibrolytic bacteria might be the reason for no impact of diet variation on these microbes in most of the studies.

13.3.1 Fungi

Among the rumen microbes, anaerobic fungi are the most efficient fiber degraders, in spite of the fact that fungi contribute only a small fraction of total biomass. It might be because all the fungi reported are fibrolytic, and the enzymes required for fiber degradation have high specific activities. In addition to that, the fungi make the substrate ready for degradation of feed by bacteria and protozoa. The fiber-degrading enzymes secreted by the rumen fungi are more active as compared to rumen bacteria. Only six genera of rumen anaerobic fungi have been identified so far, namely, *Neocallimastix*, *Piromyces* (previously known as *Piromonas*), *Caecomyces* (previously known as *Sphaeromonas*), *Orpinomyces*, *Anaeromyces* (previously known as *Ruminomyces*), and *Cyllamyces* (Akin and Rigsby 1987).

13.3.2 Methanogens

The archaea appear to be the most important microorganisms in the rumen as these have the capacity of converting carbon dioxide and hydrogen into methane. Only eight species of ruminal methanogens have been isolated in pure cultures: *Methanobacterium formicicum*, *Methanobacterium bryantii*, *Methanobrevibacter ruminantium*, *Methanobrevibacter millerae*, *Methanobrevibacter olleyae*, *Methanomicrobium mobile*, *Methanoculleus olentangyi*, and *Methanosarcina barkeri* (Janssen and Kirs 2008).

The herbivore gut, especially the rumen, is essentially obligatory anaerobic fermentation chamber hosting a highly dynamic state of microorganisms (archaea, bacteria, fungi, and protozoa) that helps in the digestion of the ingested plant biomass. The rumen microbial population is very dense, comprising of around 10^{10} bacteria/ml, 10^6 protozoa/ml, and 10^3 fungi/ml. Henderson et al. (2015) looked into the microbial diversity of the rumen from 35 countries to study the effect of diet, host species, and geography in which the animals are reared. The archaeal diversity of these animals is highly conserved which make it possible to mitigate methane emission by controlling only the dominant species of methanogens. The microbial community composition is more affected by diet (Kala et al. 2017), but the hosts are less influential in the diversity of microbes in the rumen. But by the latest technique of meta-transgenomics, 23 genera of methanogens/hydrogen utilizers have been reported, i.e., *Methanobrevibacter*, *Methanothermobacter*, *Methanoplanus*, *Sulfolobus*, *Methanosarcina*, *Methanospirillum*, *Pyrococcus*, *Methanoculleus*, *Aciduliprofundum*, *Methanoregula*, *Methanosphaera*, *Methanosphaerula*, *Methanococcoides*, *Methanocaldococcus*, *Methanocorpusculum*, *Thermoplasma*, *Methanococcus*, *Methanobacterium*, etc. In archaea, Euryarchaeota was the most predominant phylum comprising major methane-producing archaea; *Methanobrevibacter* was the most abundant genus in the rumen of buffaloes. Also, the community structure of methanogenic archaea was not influenced by TDN level in the diet (Kala et al. 2017).

There appears no correlation of the number of methanogens with methanogenesis in the rumen; it was hypothesized that more H_2 production contributes to the higher methane emissions in cattle as compared to yak (Mi et al. 2017).

13.4 Global Warming and Methanogenesis

Methane, a greenhouse gas (GHG), is emitted from different sources including natural sources, viz., wetlands, oceans and freshwater ecosystem, wild fires, and the digestive system of wild herbivores and herbivorous insects. Methane-generating agriculture activities include wild or captive animals (St-Pierre and Wright 2012), domesticated ruminants (Kelly et al. 2016; Yatoo et al. 2018), animal manure (Ozbayram et al. 2018), burning of crop residues, and paddy fields. Methanogenesis is an anaerobic respiration process that produces methane as an end product. In general aerobic respiration, the glucose is converted to CO_2, O_2, and H_2O. In contrast, during hydrogenotrophic methanogenesis, a multistep metabolic cycle, H_2 is oxidized to H^+, and CO_2 is reduced to CH_4. Notably, the methanogenic archaea live in endosymbiosis with certain ciliates in the gut ecosystem. This is because many anaerobic ciliate protozoa possess hydrogenosomes that generate molecular hydrogen that is consumed by methanogenic archaea (Lewis et al. 2018). Methanogens also thrive in the cytoplasm of anaerobic unicellular eukaryotes and in the gastrointestinal tracts of animals and humans. The protozoa-associated methanogen community is a less explored area that should be viewed as a strategy to mitigate methane emission. The methanogenesis is a low energy-yielding process catalyzed by obligate methane producers or methanogens. The energy yield is ≤ 1 ATP per methane generated (Lyu et al. 2018). Methane production through enteric fermentation is of concern due to its contribution to the accumulation of greenhouse gases (GHG) in the atmosphere. Methane emissions from various major sources is as wetland (217 Tg), fossil fuel (96 Tg), geological and fresh water (94 Tg), gut fermentation of domestic ruminants (89 Tg), livestock manure and sewage (75 Tg), rice cultivation and plant biomass burning (71 Tg) per year.

Rumen methanogen population in mid-lactating cows maintained on a total mixed ration (forage 35%, corn silage 30%, and concentrate 35%) consisted of unassigned *Methanobrevibacter* sp. (36.6%), *Methanosphaera* sp. (26.8%), *Methanobrevibacter ruminantium* (21.9%), and *Methanosarcinales* (14.6%) (Whittford et al. 2001). The numbers of methanogens per gram of rumen contents estimated by using denaturing gradient gel electrophoresis (DGGE) and 16sRNA clone gene library analysis were found to be higher than that estimated by conventional culturing methods (10^6cells per ml) and the reports of Orpin (1988) who reported methanogen population ranging from 10^6 to 10^7 cells/ml of rumen contents. 16S rRNA gene libraries constructed from pooled PCR products obtained from rumen contents of Holstein and Jersey rumen, fed analogous diets, showed that sequences representing *Methanobrevibacter millerae* were more prominent in Jersey cattle, while *Methanosphaera* sp. and novel uncultured methanogens dominated in Holstein species. *Methanobrevibacter ruminantium* was detected in both the species (King et al. 2011). Using NGS, a core methanogenic archaea community comprising of *Methanobrevibacter* (*Mbr.*) *smithii*, *Mbr. thaueri*, *Mbr. ruminantium*, and *Mbr. Millerae* was identified in primiparous Holstein, Jersey, and Holstein-Jersey crossbred cattle (Cersosimo et al. 2016). The study shows that a core methanogenic community is present among dairy cattle breeds, which is influenced by factors such as the breed of animal and lactation stage (Cersosimo et al. 2016).

Non-dairy or beef cattle might have a difference in their methanogenic archaea. Methanogen populations were studies in Hereford-cross beef cattle fed with different sources of starch. *Methanobrevibacter*-related methanogens were found to be the predominant genera, of which *M. ruminantium* was the most abundant being 51.2%, Thermoplasmatales-related methanogens were around 37.8%, while other unassigned genera such as Methanosarcinales were around 9.4% of total methanogens (Wright et al. 2007). Two novel species of formate-utilizing archaea, viz., *Methanobrevibacter millerae* sp. nov., and *Methanobrevibacter olleyae* sp. nov., have been isolated from sheep and cow rumen (Rea et al. 2007). *Methanobrevibacter* spp. and *Methanosphaera stadtmanae* phylotypes were reported in crossbred Karan Fries cattle (Sirohi et al. 2013). Archaeal sequences corresponding to *M. Ruminantium* and *M. millerae* have been identified from rumen of yak (*Bos grunniens*) (An et al. 2005).

Water and swamp buffaloes are two important species used for milk and meat (Singh et al. 2009) production. *Methanomicrobium mobile* was found as dominating methanogen in Murrah (Chaudhary and Sirohi 2009) and Surti (Singh et al. 2012) buffaloes. *Methanobrevibacter*-related phylotypes were more common in Mediterranean buffaloes (Franzolin et al. 2012). Like other ruminants, the methanogens are reported from sheep, whose diversity varies depending on the breed and feeding habits in sheep. 16S rRNA gene sequence analysis of mixed rumen microbiome of Marino sheep from Australia harbored *Methanobrevibacter*-related archaea, especially the *M. millerae* as predominant species (Wright et al. 2004). In another study, *Methanobrevibacter* sp., especially *Methanobrevibacter gottschalkii*, and *Methanobacterium* sp. were found in sheep (Wright et al. 2008).

13.5 Camelids and Other Non-ruminants/Pseudo-Ruminants

Camelids perform better on the vegetation of poor quality and arid tropical or arid temperate zones. In view of their characteristic dietary habits and substantial production performance, interest is in unravelling their gut microbiota. 16S rRNA analysis of fecal samples of Bactrian camels (*Camelus bactrianus*) kept in zoos revealed the presence of *Methanobrevibacter* sp. as dominating methanogens (Turnbull et al. 2012). Fecal samples of Sumatran orangutans, given frugivorous diets at the Perth Zoo, revealed the presence of 37 different methanogen-specific 16S rRNA sequences. Major methanogenic genera detected included *Methanosphaera stadtmanae*, *Methanobrevibacter smithii*, and *Methanobacterium beijingense* (Facey et al. 2012). Real-time PCR analysis showed the foregut of hoatzin (*Opisthocomus hoazin*) was found to have methanogenic *Methanobrevibacter ruminantium* (Wright et al. 2009) (Table 13.2).

Table 13.2 Diversity of methanogenic archaea in the rumen of different animals

Methanogens	Characteristics	References
Total methanogens	*Methanosphaera stadtmanae* from human feces, methanogens, *Methanobrevibacter gottschalkii*, *Methanobrevibacter thaueri*, *Methanobrevibacter woesei*, and *Methanobrevibacter wolinii* have been cultured from the feces of horse, cow, goose, and sheep, respectively	Dridi et al. (2009)
General	Domain – Archaea, phylum – Euryarchaeota, pseudomurein present in *Brevibacterium*, and *Methanobacterium* and protein in *Methanomicrobium*, heteropolysaccharide in *Methanosarcina*, and protein in *Methanomicrobium*	Balch et al. (1979)
Methanobacterium, *Methanobrevibacter ruminantium*	Rod shaped, forms methane from hydrogen and carbon dioxide, and formate as substrates	Balch et al. (1979)
Methanobacterium formicicum	Rod shaped, non-motile	Balch et al. (1979)
Methanomicrobium mobile	Rod shaped, forms methane from hydrogen and carbon dioxide, and formate as substrates	Balch et al. (1979)
Methanosarcina barkeri	*Methanosarcina barkeri* and *M. mazei* produce methane from hydrogen and carbon dioxide, acetate, methylamines, and methanol	Rouviere et al. (1983)
Methanosarcina mazei	Same as above except for hydrogen and carbon dioxide	Balch et al. (1979)

The complex microbiome of the gut ecosystem converts indigestible plant biomass to microbial proteins, short-chain organic acids, and gases such as CO_2, H_2, and CH_4. Methanogenic archaea are of particular interest as they eliminate surplus metabolic hydrogen generated during plant fiber digestion. The control of methanogenesis is because of the current need to minimize methane emission from ruminant livestock.

Methanogens are prokaryotic microorganisms of domain Archaea, which fall within kingdom Euryarchaeota (Woese et al. 1990), that produce methane as the major end product through complex biochemical pathways. The methanogens are strictly anaerobic archaea, which occupy a wide range of anoxic habitats and harsh conditions and play an important role in biogeochemical cycles with potential biotechnological applications (Sharma et al. 2018). Another concern is that methane is produced through the normal process of fiber digestion; hence, it is the wastage of energy generated from dietary forage. Methanogens reduce hydrogen levels via the production of methane, thereby stimulating food fermentation by saccharolytic bacteria. On the other hand, colonization by archaea is suggested to promote gastrointestinal and metabolic diseases such as colorectal cancer, inflammatory bowel disease, and obesity. Compared to other forms of respiration, the methanogens operate at a very low reducing potential.

Draft genome sequences of methanogens, namely, *Methanobacterium bryantii*, *Methanosarcina spelaei*, *Methanosphaera cuniculi*, and *Methanocorpusculum parvum*, represent a diverse set of isolates capable of methylotrophic, acetoclastic, and hydrogenotrophic methanogenesis. The genome analysis of these methanogenic archaea displays a shift toward energy conservation. In addition, the analysis of their membrane proteins and transporters distinguished various energy conservation modes. The analysis of the predicted membrane proteins and transporters distinguished differing energy conservation methods utilized during methanogenesis, such as chemiosmotic coupling in *Msar. spelaei* and electron bifurcation linked to chemiosmotic coupling in *Mbac. bryantii* and *Msph. cuniculi* (Gilmore et al. 2017).

The interest in methane-producing microorganisms has resulted from the fact that methane has a connection with global warming and that around 6% of the ingested fiber is wasted in the form of methane (Johnson and Johnson 1995). Geographical locations, species, and diet of animals have an impact on methanogenic populations. In addition, methanogenesis is of great concern in evolving strategies to control and manipulate its production. Rumen methanogens differ depending on diet and geographical location of the host, as does the methanogenesis. Further, it is relatively difficult to measure methane emissions from the rumen; it can be estimated from the stoichiometry of the fermentation of volatile fatty acids. The methane production of the forage-fed cattle is 1.5-fold greater than the concentrate-fed cattle.

13.6 Effect of PSM on Rumen Fungi

13.6.1 Application of Metagenomics in the Rumen

There were many questions unanswered before the advent of metagenomics like the diversity of rumen/gut of animals, their role in fiber degradation, the exact enzyme profile needed for degradation of lignocellulosic feed, generation of methane in the rumen, etc. It does not mean that all these questions have been answered, but the majority of pathways are now easy to understand.

- It is now confirmed that the earlier known fiber-degrading bacteria, like *Ruminococcus*, *Fibrobacter*, *Butyrivibrio*, etc., constituting only 3–4% of the total microbial biomass, are not the only microbes responsible for the degradation of fibrous feeds. There are several other groups like *Prevotella* and Firmicutes, contributing a significant part of total biomass, which play a vital role in fiber degradation.
- Esterases are another group of enzymes which break the bond between lignin and carbohydrates and release the latter free for degradation by cellulolytic microbes and release of energy for the utilization by the animals.
- The archaeal diversity of the ruminants are highly conserved which makes it possible to mitigate methane emission by controlling only the dominant species of methanogens.
- In the majority of the cases, there is no correlation between methanogens and methane synthesis. By the use of metagenomic techniques, it has now been confirmed that hydrogen-producing microbes determine the quantity of methane synthesis. If the hydrogen emitters are inhibited (as in defaunation), methane synthesis is reduced considerably.

13.6.2 Methanogenesis in Rumen

As discussed above, the chemical composition of plants reveals that in addition to normal constituents like cellulose, hemicellulose, soluble sugars, proteins, fats, etc., there are also some unique molecules present like saponins, tannins, essential oils, alkaloids, etc. As these molecules are not synthesized as a result of primary metabolism of plants, such compounds are classified as plant secondary metabolites, which are usually meant for providing protection to the plants against predators, pathogens, invaders, etc. Several thousands of such metabolites have been identified. Majority of these compounds fall in the category of lignins, tannins, saponins, terpenoids/volatile essential oils, alkaloids, etc. These plant secondary metabolites have antimicrobial activity, but their mechanism of action and inhibition of microbial growth are very specific, and therefore these are active against a specific group of microbes. This specificity of these plant secondary metabolites against microbial groups can be used for selective manipulation of rumen fermentation. The methanogens which are classified as archaea have a distinctly different chemical composition of the cell walls from that of the other true bacteria present in the rumen. Therefore, there is a possibility that any one of the plant secondary compounds might act as a selective inhibitor of methanogens and can be used as a feed additive for the manipulation of rumen fermentation. The role of tannins, saponins, and essential oils has been in the inhibition of methanogens or the process of methanogenesis in the rumen.

In the "RUMEN-UP" project in Europe, 450 plants (mainly foliage) have been screened in vitro for their potential to inhibit methanogenesis by the rumen microbes at a concentration of 50 mg/500 mg of substrate incubated in 120 ml serum bottles. The selection of plants was restricted to those which are either growing or could be grown in European countries, therefore excluding most of the plants growing in the tropical regions of the world. Out of 450 plants examined in this project, 35 plants inhibited methane more than 15% and only six (*Carduus pycnocephalus*, *Populus tremula*, *Prunus avium*, *Quercus robur*, *Rheum nobile* Hook. F. and Thoms., and *Salix caprea*) more than 25% in comparison to their controls.

Another major project has been funded by FAO/IAEA Joint Division of the United Nations, which ran for 6 years (2003–2009) at eight international locations on using rumen molecular biology techniques to search for methods to inhibit methane emission by the livestock. The Indian participant of the project (IVRI, Izatnagar) worked on plant secondary metabolites to determine their potential as a rumen modifier to reduce methane emission. Various combinations of plants containing secondary metabolites have been tested in in vivo using buffalo as an experimental animal, and it has been observed that in vivo methane emission by the animal can be inhibited to the extent of 20–30% without affecting feed conversion efficiency and health of animals (Table 13.3).

13.6.3 In Vivo Feeding Trials

Most of the studies on the effect of plant secondary metabolites have been conducted in in vitro conditions. There are only a few experiments conducted to report methane inhibition in vivo. The results indicate that there are many plants which contain secondary metabolites and are active against rumen methanogenesis. Many times methane inhibition by secondary metabolites in in vitro conditions might not be translated into similar effects in in vivo conditions. This might happen due to improper selection of the dose of these metabolites in the ration of animals.

An in vivo experiment in sheep with tea saponins revealed that saponins inhibited protozoa, methane emission, and improved rumen fermentation, where the reduction of methane emission was mediated through an inhibitory effect on protozoa (Zhou et al. 2010). Some other experiments indicate positive results of including plant/plant extracts on inhibition of methanogenesis. In one experiment, *Terminalia chebula*, *Allium sativum*, and the mixture of these two plants were fed to sheep at the rate of 1% of DMI resulted in decreased ($p = 0.09$) methane production

Table 13.3 Effect of plant extracts on inhibition of in vitro methanogenesis*

Plant	Common name	Plant part	In vitro inhibition of methanogenesis (%)		
			Ethanol extract	Methanol extract	Water extract
Acacia concinna	Shikakai	Seed pulp	5.32	17.60	19.61
Allium cepa	Onion	Bulb	8.76	16.38	32.64
Allium sativum	Garlic	Bulb	61.31	69.73	19.88
Azadirachta indica	Neem	Seed cake	34.59	21.89	−14.80
Cannabis indica	Bhang	Leaves	34.42	30.67	3.33
Citrus limonum	Lemon	Peel extract	12.90	8.67	11.84
Emblica officinalis	Amla	Seed pulp	19.51	27.68	−26.69
Eugenia jambolana	Jamun	Leaves	5.61	24.27	5.66
Foeniculum vulgare	Fennel	Seed	39.42	70.72	−14.54
Mangifera indica	Mango	Leaves	23.17	35.67	9.15
Populus deltoides	Poplar	Leaves	8.49	85.86	−7.72
Psidium guajava	Guava	Leaves	81.79	9.29	9.44
Quercus incana	Oak	Leaves	−1.37	29.53	4.18
Sapindus mukorossi	Soapnut	Seed pulp	95.80	20.18	39.40
Syzygium aromaticum	Clove	Flower bud	46.96	85.61	2.37
Terminalia belerica	Baheda	Seed pulp	5.54	28.11	13.18
Terminalia chebula	Harad	Seed pulp	58.54	99.79	6.43
Trachyspermum ammi	Ajwain	Seed	42.28	−2.68	−11.35

Adapted from Patra et al. (2006)*, Agarwal et al. (2006), and *Kamra et al. 2008.

by 24, 11, and 23.5% in *T. chebula*, *A. sativum*, and the mixture of the two, respectively, when expressed as L/kg digestible DM intake (Patra and Saxena 2010). *T. chebula* is a rich source of tannin (4.89% of DM), whereas *A. sativum* is rich in essential oils. The data indicated that *T. chebula* was more effective as compared to garlic. The reason for low *A. sativum* activity might be due to the instability of allicin, the main secondary metabolite responsible for antimicrobial activity.

In another experiment, a mixture of three plants (Mix 3) fed to buffalo calves at the rate of 1, 2, and 3% of DMI resulted in a dose-dependent inhibition in methane emission (l/kg DDM) since percent inhibition increased with an increase in dose of the feed additive (Chaudhary and Sirohi 2009) without affecting dry matter digestibility at any of the levels of feed additives tested. The VFA and fiber-degrading enzyme activities were not affected, whereas there were a few changes in the rumen microbial profile as estimated by real-time PCR, but these were not responsible for any significant change in rumen fermentation. Many other experiments with growing animals have been conducted, out of which a few of these experiments have been listed in Table 13.4.

Table 13.4 Effect of plant extracts on inhibition of in vivo methane production (l/kg DDMI)

Plant	Methane inhibition (%)	Body weight gain (%)	Animal	References
Anti-methane	32	–	Buffalo	Kamra et al. (2010)[a]
Methane-suppressor	23	16	Buffalo	Kamra et al. (2012)[a]
BEO (blend of essential oils)	14.6	7.4	Buffalo	Yatoo et al. (2018)
Terminalia chebula	24.6	–	Sheep	Patra et al. (2011)
Ajwain oil and lemongrass oil in 1:1 ratio	16.7	No effect	Buffalo	Samal et al. (2016)
Garlic and soapnut in 2:1 ratio	12.9	No effect	Buffalo	
Garlic, soapnut, Harad and ajwain in 2:1:1:1 ratio	8.4	No effect		
Ficus benghalensis leaves	21.8		Sheep	Malik et al. (2017)
Artocarpus heterophyllus leaves	20.6		Sheep	
Azadirachta indica leaves	24.07		Sheep	

[a]Patents submitted

The results of experiments conducted so far with plant secondary metabolites indicate that there are some plants, which appear to have a good potential for use as a feed supplement to inhibit methane emission in the ruminants. There might be many more such plants which have not yet been tested. Therefore, screening of plants should be a continuous process to search for more useful plants, which can be used for rumen manipulation. In the secondary screening process, only selected plants should be tested in in vivo experiments to examine their potential for practical application.

13.7 Relation between Hydrogen Producers and Methanogens

Yak is a lower methane producer than cattle, in spite of the fact that both animals are fed similar diets and there are only small variations between the microbiomes of both animals. The methane and hydrogen yields in yak vs cattle are 0.26 vs 0.33 mmol methane/g dry matter intake and 0.28 vs 0.86 mmol/d hydrogen generation. Hydrogen recovery from cattle was significantly higher than that from yak (Mi et al. 2017). The relative abundance of methanogens was not different between the two animal species. It was hypothesized that more H_2 production is the reason for the higher methane emission in cattle as compared to yak. Kittlemann et al. (2013) were of the view that abundance of fibrolytic bacteria (major hydrogen producers) is related to the methanogen communities and consequently with methane production. Therefore, the abundance of methanogens does not have a direct correlation with methane production, but the partial pressure of hydrogen is more important. Minimizing metabolic H_2 production in the rumen might reduce the availability of H_2 to methanogens. Suppression in ruminal H_2 producers is usually accompanied with a concurrent decrease in feed fermentation. This can be achieved by the intensification of propiogenesis and rumen biohydrogenation or promoting reductive acetogenesis in the rumen. Targeting H_2 utilizing protozoa or other microbes accountable for interspecies H_2 transfer to the methanogen can be a fruitful strategy to reduce methane emissions.

An Australian animal, the tammar wallaby (*Macropus eugenii*) harbors unique gut bacteria and produces 20% of the amount of methane produced by ruminants per unit of digestible energy intake. Pope et al. (2011) isolated a dominant bacterial species (WG-1) from wallaby which was affiliated to the family *Succinivibrionaceae* and implicated in lower methane emissions from starch-containing diets. Pure-culture studies confirm that the bacterium is capnophilic and produces succinate, further explaining a microbiological basis for lower methane emissions from macropodids. The abundance of WG-1 is variable in samples collected from animals in winter and spring; their results show that these bacteria will be numerically dominant when the plane of nutrition is rich in starch and soluble sugars.

As discussed above, hydrogen produced during fermentation of feed is responsible for methane production in the rumen, and the minimized metabolic H_2 production during enteric fermentation and diversion of H_2 away from the methanogenesis might result into useful energy-rich metabolites.

The most important process of methane reduction is killing of unproductive animals; this approach is neither ethical nor possible in India where the slaughter of cattle is not permissible in some of the states of country. Minimizing metabolic H_2 production in the rumen might reduce the availability of H_2 to methanogens. Suppression in ruminal H_2 producers is usually accompanied with concurrent decrease in feed fermentation and diversion of metabolic H_2 away from methanogenesis. This can be achieved by the intensification of propiogenesis and rumen biohydrogenation or promoting reductive acetogenesis in the rumen. Targeting H_2-utilizing protozoa or other microbes accounts for interspecies H_2 transfer to the methanogen which could be a good alternate for eradicating enteric methane emission. However, a significant reduction in rumen ciliates might lead to reduced fiber degradation. Another important way to tackle the emission of methane is to directly target rumen archaea through various approaches. By doing so, the enteric methane emission will decrease, and additional H_2 will also be adequate to stimulate alternate hydrogenotrophic pathways, i.e., reductive acetogenesis.

13.8 Future Perspectives

Although a large number of plants have been tested for their potential to reduce methanogenesis and/or increase degradability of feed under in vitro conditions, a few of them either alone or in combination have been evaluated in in vivo conditions and have been found that these plants containing secondary metabolites can cause

20–30% decrease in methane emission and result in improved growth rate of animals by 8–10% as compared to control animals. Therefore, further studies are required for screening more plants and feeding them in large number of animals for a longer duration for practical application to achieve eco-friendly and economic livestock production.

References

Agarwal N, Kamra DN, Chaudhary LC, Patra AK (2006) Effect of *Sapindus mukorossi* extracts on *in vitro* methanogenesis and fermentation characteristics in buffalo rumen liquor. J Appl Anim Res 30:1–4

Akin DE, Rigsby LL (1987 Sep) Mixed fungal populations and lignocellulosic tissue degradation in the bovine rumen. Appl Environ Microbiol 53(9):1987–1995

An D, Dong X, Dong Z (2005) Prokaryote diversity in the rumen of yak (Bos grunniens) and Jinnan cattle (Bos taurus) estimated by 16S rDNA homology analyses. Anaerobe 11(4):207–215. Epub 2005 Mar 27

Balch WE, Fox GE, Magrum LJ (1979) Methanogens: reevaluation of a unique biological group, *Microbiol.* Rev 43(2):260–296

Calsamiglia S, Busquet M, Cardozo PW, Castillejos L, Ferret A (2007) Invited review: essential oils as modifiers of rumen microbial fermentation. J Dairy Sci 90:2580–2595

Cersosimo LM, Bainbridge ML, Kraft J, Wright AD (2016) Influence of periparturient and postpartum diets on rumen methanogen communities in three breeds of primiparous dairy cows. BMC Microbiol 16:78. https://doi.org/10.1186/s12866-016-0694-7

Chaichi Semsari M, MaheriSis N, Sadaghian M, Eshratkhah B, Hassanpour S (2011) Effects of administration of industrial tannins on nutrient excretion parameters during naturally acquired mixed nematode infections in Moghani sheep. J Amer Sci 7(6):245–248

Chaudhary PP, Sirohi SK (2009). Dominance of *Methanomicrobium* phylotype in methanogen population present in Murrah buffaloes (Bubalus bubalis). Lett Appl Microbiol;49(2):274–277. doi: 10.1111/j.1472-765X.2009.02654.x.Epub 2009 May 27

Dorman HJD, Deans SG (2000) Antimicrobial agents from plants: antibacterial activity of plant volatile oils. J Appl Microbiol 88:308–316. [PubMed].s

Dridi B, Henry M, El Kh'echine A, Raoult D, Drancourt M (2009) High prevalence of *Methanobrevibacter smithii* and *Methanosphaera stadtmanae* detected in the human gut

Facey HV, Northwood KS, Wright AD (2012) Molecular diversity of methanogens in fecal samples from captive Sumatran orangutans (*Pongo abelii*). Am J Primatol 74(5):408–413. https://doi.org/10.1002/ajp.21992

Flint HJ, Bayer EA, Rincon MT, Lamed R, White BA (2008) (2008). Polysaccharide utilization by gut bacteria: potential for new insights from genomic analysis. Nat Rev Microbiol 6(2):121–131. https://doi.org/10.1038/nrmicro1817

Franzolin R, St-Pierre B, Northwood K, Wright AD (2012) Analysis of rumen methanogen diversity in water buffaloes (*Bubalus bubalis*) under three different diets. Microb Ecol 64(1):131–139. https://doi.org/10.1007/s00248-012-0007-0

Gilmore SP, Henske JK, Sexton JA, Solomon KV, Seppälä S, Yoo JI, Huyett LM, Pressman A, Cogan JZ, Kivenson V, Peng X, Tan Y, Valentine DL, O'Malley MA (2017) Genomic analysis of methanogenic archaea reveals a shift towards energy conservation. BMC Genomics 18(1):639. https://doi.org/10.1186/s12864-017-4036-4

Güçlü-Ustündağ O, Mazza G (2007) Saponins: properties, applications and processing

Haslam E (1989) Plant polyphenols- vegetable tannins revisited. Cambridge University Press, Cambridge

Hassanpour S, Sadaghian M, MaheriSis N, Eshratkhah B, Chaichi SM (2011) Effect of condensed tannin on controlling faecal protein excretion in nematode-infected sheep: in vivo study. J Amer Sci 7(5):896–900

Henderson G, Cox F, Ganesh S, Jonker A, Young W (2015 Oct 9) Global rumen census collaborators, Janssen PH. Rumen microbial community composition varies with diet and host, but a core microbiome is found across a wide geographical range. Sci Rep 5:14567. https://doi.org/10.1038/srep14567. Erratum in: Sci Rep. 2016;6:19175. Yanez-Ruiz, David R [corrected to Yáñez-Ruiz, David R]; Pinares-Patino, Cesar [corrected to Pinares-Patiño, Cesar]; Munoz, Camila [corrected to Muñoz, Camila].

Hess HD, Beuret RA, Lötscher M, Hindrichsen IK, Machmüller A, Carulla JE, Lascano CE, Kreuzer M (2004) Ruminal fermentation, methanogenesis and nitrogen utilization of sheep receiving tropical grass hay-concentrate diets offered with *Sapindus saponaria* fruits and Cratylia argentea foliage. Anim Sci 79:177–189

Hervás G, Frutos P, Giráldez FJ, Mantecón ÁR, Del Pino MC (2003) Effect of different doses of quebracho tannins extract on rumen fermentation in ewes. Anim Feed Sci Technol 109:65–78

Holtshausen L, Chaves AV, Beauchemin KA, McGinn SM, McAllister TA, Odongo NE, Cheeke PR, Benchaar C (2009 Jun) Feeding saponin-containing Yucca schidigera and *Quillaja saponaria* to decrease enteric methane production in dairy cows. J Dairy Sci 92(6):2809–2821. https://doi.org/10.3168/jds.2008-1843. Erratum in: J Dairy Sci. 2009 Jul;92(7):3543. Odongo, N E [added]

Janssen PH, Kirs M (2008) Structure of the archaeal community of the rumen. Appl Environ Microbiol 74(12):3619–3625. https://doi.org/10.1128/AEM.02812-07. Epub 2008 Apr 18

Johnson KA, Johnson DE (1995) Methane emissions from cattle. J Anim Sci 73(8):2483–2492

Kala A, Kamra DN, Kumar A, Agarwal N, Chaudhary LC, Joshi CG (2017) Impact of levels of total digestible nutrients on microbiome, enzyme profile and degradation of feeds in buffalo rumen. PLoS One. https://doi.org/10.1371/Journal.pone.0172051

Kamra DN, Patra AK, Chatterjee PN, Kumar R, Agarwal N, Chaudhary LC (2008) Effect of plant extracts on methanogenesis and microbial profile of the rumen of buffalo: a brief overview. Aust J Exp Agric Res Aust J Exp Agric 48:175–178

Kamra DN (2005) Rumen microbial ecosystem. Curr Sci 89:124–135

Kamra DN, Zadbuke S, Agarwal N, Choudhary LC, Bhar R (2010) Anti methane. Patent submitted

Kamra DN, PawarM, Agarwal N, Choudhary LC, Chaturvedi VB (2012) Methane Suppressor, Patent submitted

Kelly WJ, Pacheco DM, Li D, Attwood GT, Altermann E, Leahy SC (2016) The complete genome sequence of the rumen methanogen *Methanobrevibacter millerae* SM9. Stand Genomic Sci 11:49. https://doi.org/10.1186/s40793-016-0171-9. eCollection 2016

King EE, Smith RP, St-Pierre B, Wright AD (2011) Differences in the rumen methanogen populations of lactating Jersey and Holstein dairy cows under the same diet regimen. Appl Environ Microbiol 77(16):5682–5687. https://doi.org/10.1128/AEM.05130-11. Epub 2011 Jun 24

Kittelmann S, Seedorf H, Walters WA, Clemente JC, Knight R, Gordon JI, Janssen PH (2013) Simultaneous amplicon sequencing to explore co-occurrence patterns of bacterial, archaeal and eukaryotic microorganisms in rumen microbial communities. PLoS One 8(2):e47879. https://doi.org/10.1371/journal.pone.0047879

Lewis WH, Sendra KM, Embley TM, Esteban GF (2018) Morphology and phylogeny of a new species of anaerobic ciliate, *Trimyemafinlayi* n. sp., with endosymbiotic methanogens. Front Microbiol 9:140. https://doi.org/10.3389/fmicb.2018.00140. eCollection 2018

Lyu Z, Shao N, Akinyemi T, Whitman WB (2018) Methanogenesis. Curr Biol 28(13):R727–R732. https://doi.org/10.1016/j.cub.2018.05.021

Malik PK, Kolte AP, Bakshi B, Baruah L, Bhatta R (2017). Enteric methane mitigation in sheep through selected tanniniferous tropical tree leaves. Livestock Science. 2016

Mi J, Zhou J, Huang X, Long R (2017) Lower Methane Emissions from Yak Compared with Cattle in Rusitec Fermenters. PLoS One 12(1):e0170044. https://doi.org/10.1371/journal.pone.0170044

Min BR, Hart SP (2003) Tannins for suppression of internal parasites. J Anim Sci 81:102–109

Orpin CG (1988) Nutrition and biochemistry of anaerobic Chytridiomycetes. Biosystems 21(3–4):365–370

Ozbayram EG, Ince O, Ince B, Harms H, Kleinsteuber S (2018). Comparison of Rumen and Manure microbiomes and implications for the inoculation of anaerobic digesters. Microorganisms 6(1). pii: E15. https://doi.org/10.3390/microorganisms6010015

Patra AK, Saxena J (2010) Exploitation of dietary tannins to improve rumen metabolism and ruminant nutrition. J Sci Food Agric 91:24–37

Patra AK, Yu Z (2012) Effects of essential oils on methane production and fermentation by, and abundance and diversity of, rumen microbial populations. Appl Environ Microbiol 78(12):4271–4280. https://doi.org/10.1128/AEM.00309-12

Patra AK, Kamra DN, Agarwal N (2006) Effect of spices on rumen fermentation, methanogenesis and protozoa counts in *in vitro* gas production test. Int Congr Ser 1293:176–179

Patra AK, Kamra DN, Bhar R, Kumar R, Agarwal N (2011) Effect of *Terminalia chebula* and *Allium sativum* on *in vivo* methane emission by sheep. J Anim Physiol Anim Nutr 95:187–191

Pope PB, Smith W, Denman SE, Tringe SG, Barry K, Hugenholtz P, McSweeney CS, McHardy A, Morrison M (2011) Isolation of Succinivibrionaceae implicated in low methane emissions from tammar wallabies. Science 333:646–648

Rea S, Bowman JP, Popovski S, Pimm C, Wright AD (2007) *Methanobrevibacter millerae* sp. nov. and *Methanobrevibacter olleyae* sp. nov., methanogens from the ovine and bovine rumen that can utilize formate for growth. Int J Syst Evol Microbiol 57(Pt 3):450–456

Ross ZM, O'GaraEA HDJ, SleightholmeHV MDJ (2001) Antimicrobial properties of garlic oil against human enteric bacteria: evaluation of methodologies and comparisons with garlic oil sulfides and garlic powder. Appl Environ Microbiol 67:475–480

Rouviere W, Fiebig K, Hippe H (1983) Distribution of cytochromes in methanogenic bacteria. FEMS Microbiol Lett 20(3):407–410

Samal L, Chaudary LC, Agarwal N, Kamra DN (2016) Impact of phytogenic feed additives on growth performance, nutrient digestion and methanogenesis in growing buffaloes. Anim Prod Sci 55:1056–1063

Santoso B, Mwenya B, Sar C, Gamo Y, Kobayashi T, Morikawa R, Kimura K, Mizukoshi H, Takahashi J (2004) Effects of supplementing galacto-oligosaccharides, *Yucca schidigera* or nisin on rumen methanogenesis, nitrogen and energy metabolism in sheep. Livest Prod Sci 91:209–217

Sharma S, Ding Y, Jarrell KF, Brockhausen I (2018) Identification and characterization of the 4-epimerase AglW from the archaeon Methanococcus maripaludis. Glycoconj J 35(6):525–535. https://doi.org/10.1007/s10719-018-9845-4

Singh B, Chauhan MS, Singla SK, Gautam SK, Verma V, Manik RS, Singh AK, Sodhi M, Mukesh M (2009) Reproductive biotechniques in buffaloes (Bubalus bubalis): status, prospects and challenges. Reprod Fertil Dev 21(4):499–510. https://doi.org/10.1071/RD08172

Singh KM, Ahir VB, Tripathi AK, Ramani UV, Sajnani M, Koringa PG, Jakhesara S, Pandya, PR, Rank DN, Murty DS, Kothari RK, Joshi C.G. (2011). Metagenomic analysis of Surti buffalo (*Bubalus bubalis*) rumen: a preliminary study. Mol Biol Rep 39(4): 4841–4848 https://doi.org/10.1007/s11033-011-1278-0. Epub 2011 Sep 27

Singh KM, Ahir VB, Tripathi AK, Ramani UV, Sajnani M, Koringa PG, Jakhesara S, Pandya PR, Rank DN, Murty DS, Kothari RK, Joshi CG (2012 Apr) Metagenomic analysis of Surti buffalo (*Bubalus bubalis*) rumen: a preliminary study. Mol Biol Rep 39(4):4841–4848. https://doi.org/10.1007/s11033-011-1278-0. Epub 2011 Sep 27

Sirohi SK, Chaudhary PP, Singh N, Singh D, Puniya AK (2013) The 16S rRNA and mcrA gene based comparative diversity of methanogens in cattle fed on high fibre based diet. Gene 523(2):161–166. https://doi.org/10.1016/j.gene.2013.04.002. Epub 2013 Apr 18

Sliwiński BJ, Kreuzer M, Wettstein HR, Machmüller A (2002 Dec) Rumen fermentation and nitrogen balance of lambs fed diets containing plant extracts rich in tannins and saponins, and associated emissions of nitrogen and methane. Arch Tierernahr 56(6):379–392

Sivropoulou A, Papanikolaou E, Nikolaou C, Kokkini S, Lanaras T, Arsenakis M (1996) Antimicrobial and cytotoxic activities of *Origanum* essential oils. J Agric Food Chem 44:1202–1205

Smith-Palmer A, Stewart J, Fyfe L (1998 Feb) Antimicrobial properties of plant essential oils and essences against five important food-borne pathogens. Lett Appl Microbiol 26(2):118–122

St-Pierre B, Wright AD (2012 Jan 5) Molecular analysis of methanogenic archaea in the forestomach of the alpaca (*Vicugna pacos*). BMC Microbiol 12(1). https://doi.org/10.1186/1471-2180-12-1

Turnbull KL, Smith RP, St-Pierre B, Wright AD (2012) Molecular diversity of methanogens in fecal samples from Bactrian camels (*Camelus bactrianus*) at two zoos. Res Vet Sci 93(1):246–249. https://doi.org/10.1016/j.rvsc.2011.08.013. Epub 2011 Sep 13

Waghorn G (2008) Beneficial and detrimental effects of dietary condensed tannins for sustainable sheep and goat production- Progress and challenges. Anim Feed Sci Technol 147:116–139

Whitford V, Ennos AR, Handley JF (2001). 'City form and natural process' – indicators for the ecological performance of urban areas and their application to Merseyside, UK. Landsc Urban Plan 57(2):91–103

Woese CR, Kandler O, Wheelis ML (1990 Jun) Towards a natural system of organisms: proposal for the domains archaea, Bacteria, and Eucarya. Proc Natl Acad Sci U S A 87(12):4576–4579

Wright AD, Williams AJ, Winder B, Christophersen CT, Rodgers SL, Smith KD (2004) Molecular diversity of rumen methanogens from sheep in Western Australia. Appl Environ Microbiol 70(3):1263–1270

Wright AD, Auckland CH, Lynn DH (2007) Molecular diversity of methanogens in feedlot cattle from Ontario and Prince Edward Island, Canada. Appl Environ Microbiol 73(13):4206–4210. Epub 2007 May 4

Wright AD, Ma X, Obispo NE (2008 Aug) Methanobrevibacter phylotypes are the dominant methanogens in sheep from Venezuela. Microb Ecol 56(2):390–394. https://doi.org/10.1007/s00248-007-9351-x. Epub 2007 Dec 29

Wright AD, Northwood KS, Obispo NE (2009) Rumen-like methanogens identified from the crop of the folivorous. South American bird, the hoatzin (Opisthocomus hoazin) 3(10):1120–1126. https://doi.org/10.1038/ismej.2009.41. Epub 2009 Apr 23

Yatoo MA, Chaudhary LC, Agarwal N, Chaturvedi VB, Kamra DN (2018) Effect of feeding of blend of essential oils on methane production, growth, and nutrient utilization in growing buffaloes. Asian-Australas J Anim Sci 31(5):672–676. https://doi.org/10.5713/ajas.16.0508. Epub 2017 Feb 23

Zhou Y-Y, Mao H-L, Jiang F et al (2010). Tea saponins inhibit ruminal methane emission through the inhibitory effect on protozoa in Hu sheep. In: Proceedings of Fourth Greenhouse Gases and Animal Agric. Conference

Class B-Trichothecene Profiles of *Fusarium* Species as Causal Agents of Head Blight

14

Emre Yörük and Tapani Yli-Mattila

Abstract

Fusarium head blight is a destructive disease of all small grain cereals worldwide. More than ten *Fusarium* species cause the diseases; *F. graminearum* and *F. culmorum* are the major causal agents in European, Asian and American continents. The disease leads to losses in crop quality and quantity including contamination with mycotoxins in particular class B-trichothecenes. Class B-trichothecenes include deoxynivalenol (DON), nivalenol (NIV) and their acetylated derivatives (15-acetylated deoxynivalenol (15-ADON), 3-acetylated deoxynivalenol (3-ADON) and 4-acetylated nivalenol (4-ANIV). Distributions of these toxin profiles have been detected in many regions around the world. 15-ADON, *F. graminearum* and wheat seem to have been the major chemotype, causal agent, and host plant, respectively. Moreover, more than five host plants including barley, rice, and maize, which are economically important, have been associated with *Fusarium* spp. 3-ADON chemotype is predominating Northern Europe, while NIV chemotype has been reported as locally predominating profile in certain geographic regions. Current modern techniques including PCR and chromatographic analysis present reliable, fast, and informative output data worldwide. Since the toxicity of mycotoxins, aggressiveness of species, and scab resistance of host plants vary, chemotyping studies could efficiently contribute to disease management strategies.

E. Yörük
Department of Molecular Biology and Genetics, Faculty of Arts and Sciences, Istanbul Yeni Yuzyil University, Istanbul, Turkey

T. Yli-Mattila (✉)
Molecular Plant Biology, Department of Biochemistry, University of Turku, Turku, Finland
e-mail: tymat@utu.fi

T. Satyanarayana et al. (eds.), *Advancing Frontiers in Mycology & Mycotechnology*, https://doi.org/10.1007/978-981-13-9349-5_14

Keywords
Chemotype · Detection methods · *Fusarium culmorum* · *Fusarium graminearum* · Mycotoxin · profiling, Trichothecenes.

14.1 Introduction

Plants are interacting with other organisms (bacteria, fungi, nematodes and several others) and environmental factors (humidity, temperature, light, etc.) throughout their life. Under particular/specific conditions, the incidence of diseases could occur as a result of these interactions. Phytopathogenic organisms could lead to devastating diseases on economically important plants when favourable temperature and moisture conditions and susceptible host are present (Dyakov et al. 2007).

Infection of plants by phytopathogenic fungi causes yield losses and reduction in crop quality and quantity. By this way, economic loss occurs, and the losses could reach millions of dollar even in local areas (Salas et al. 1999; Windels 2000). Besides, mycotoxins, produced by fungal pathogens that accumulate in crops and contaminated food and live feed, are hazardous for human and animals. Some of these toxic molecules have a stable structure even when they are exposed to elevated high temperatures (Gutleb et al. 2002; Lauren and Smith 2001). Thus, the contamination possibility of crops by these stable molecules and disease epidemics could reach high levels. Characterization of fungal diseases and mycotoxins is significant in terms of treatment strategies for diseases and the development of specific diagnostic approaches. Here we describe the class B-trichothecene profiles in *Fusarium* spp. associated with head blight disease.

14.2 *Fusarium* Head Blight

Fusarium head blight (FHB), also known as scab of wheat or ear blight of maize, is one of the most serious fungal diseases worldwide. The disease is at first reported in England in the year 1884, and it is associated with all small grain cereals in particular wheat, barley and maize (Goswami and Kistler 2004; Parry et al. 1995). In Russian Far East, FHB was reported already in 1882 (Yli-Mattila and Gagkaeva 2016), and even long before this year, poisoning of local animals and people has occurred. The disease has devastating effects on wheat and barley especially in humid and semi-humid areas throughout the world (Bai and Shaner 2004). FHB leads to yield losses and reduction in quality and quantity of cereals. The disease becomes widespread because of changes in climatic conditions, crop rotations, and other agronomic practices (Bottalico and Perrone 2002; Miedaner et al. 2008; Parry et al. 1995; Saharan et al. 2004). As a result of epidemics, billion dollars of losses have been recorded in a great number of countries (Bai and Shaner 2004; Lori et al. 2009; Windels 2000). Especially in the 1990s, losses arising from FHB of wheat were estimated over $3 billion in more than 10 states in America and Canada. An example is that from the year 1993 to 1998 where economic losses was reached $300 million in Manitoba. Similar economic losses have also been recorded in European and Asian countries (Bai and Shaner 2004; Windels 2000).

During the last years, FHB epidemics on maize, wheat, barley and oats were reported by several researchers in Asian (Ji et al. 2007; Li et al. 2009; Yörük and Albayrak 2012), European (Gagkaeva and Yli-Mattila 2004; Jennings et al. 2004a, b; Yli-Mattila and Gagkaeva 2010, 2016; Yli-Mattila 2012) and American (Scoz et al. 2009; Walker et al. 2000) continents which harbour more than half of human community. Besides, causal agents of head blight were isolated from scabby kernels planted in Australia and North Africa (Kammoun et al. 2010; Obanor et al. 2010). The disease has become a leading threat to agronomy worldwide, and efforts have been currently tried to develop novel strategies against the disease.

All small grain cereals are potent host for FHB causal agents. If favourable temperature, moisture and host plant are available, FHB pathogens could infect cereals at different stages in the host plant's life. After the flowering stage, the anthesis is the most common time to be infected by pathogen microorganisms (Bai and Shaner 2004; Parry et al. 1995). Disease symptoms are similar in different cereals. Shrivelled kernels and accumulation of mycotoxins, in particular, deoxynivalenol (DON), nivalenol (NIV), and their acetylated derivatives, are two characteristics of the disease (Saharan et al. 2004). At the same time, DNA levels of *F. graminearum* and other *Fusarium* species are increased (Yli-Mattila and Gagkaeva 2010; Yli-Mattila and Gagkaeva, 2016). First, symptoms are seen as the brown small lesions that occurred along rachis and glumes. This is followed salmon pink to red fungal growth, which could be easily seen along the spike. Discoloration spreads along spikelet. At the later stage of the disease, kernels become shrunken (Fig. 14.1), and even spikes could have no grain (Parry et al. 1995).

The infection causes field losses. The level of yield losses could depend on several factors. Genotypic characteristics of fungal pathogens and natural resistance level of host plants are the most important factors (Bai and Shaner 2004; Parry et al. 1995; Saharan et al. 2004). Quellet and Seifert (1993) initiated genotyping studies by using random-amplified polymorphic DNA (RAPD) and polymerase chain reaction (PCR) markers in *Fusarium graminearum* isolates associated with head blight in Canada. More than two decades ago, *Fusarium* species were isolated from scabby

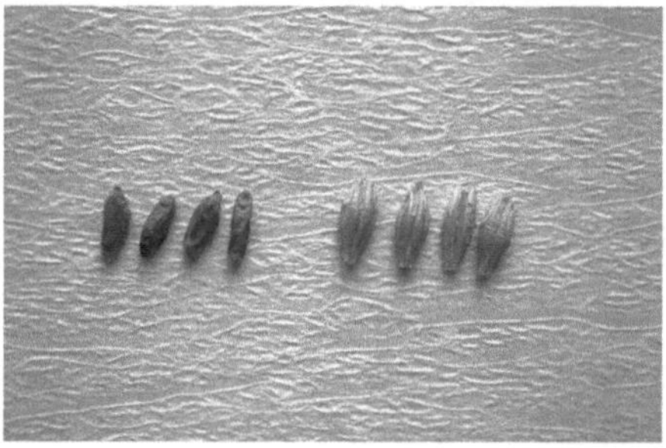

Fig. 14.1 Shrivelled kernels obtained from scabby barley spikes in Çorum region of Turkey (four grains left side) and healthy kernels of *Hordeum vulgare* L. cv. Tokak157/37

spikes, and genetic variation in local and global isolate collections was investigated by many authors (Chehri et al. 2011; Chung et al. 2008; Gürel et al. 2010; Laday et al. 2004; Mishra et al. 2006; Yli-Mattila 2010; Yörük and Albayrak 2013). Genetic characterization studies are indirectly important in terms of development of the novel disease control strategies, and the number of genotyping studies is being increased day by day. The second factor is the resistance level of the host plant. The number of highly FHB-resistant cultivars is limited, and *Triticum aestivum* L. cv. Sumai-3 is the most widely used and well-known resistant host (Anand et al. 2003; Bernardo et al. 2007; Wisniewska and Kowalczyk 2005). Five (I–V) types of resistance to FHB are present in wheat. Type I resistance is associated with the percentage of spikelets with symptoms. Type II resistance is the number of infected spikelets below the inoculation initiation point. The remaining resistance levels are not directly quantified. However, damaged kernels, mycotoxin tolerance and accumulation levels are the markers for these remaining three resistance levels (Bai and Shaner 2004; Foroud and Eudes 2009). Since the development of plant varieties resistant to head blight with favourable agronomic traits is hard and time-consuming, approaches including genetic characterization of FHB pathogens, mycotoxin measurement, DNA measurements by qPCR and chemotyping and genetic manipulations in pathogen in particular quelling of mycotoxigenic genes seem to be more useful and beneficial.

14.3 *Fusarium* Species Associated with FHB

Fusarium genus has been linked to head blight and ear blight of all small grain cereals.

Fusarium genus was first described by Link (1809: Mag.Ges. Naturf. Freunde, Berlin 3:10), and sanctioned under Fries1821.*Fusarium* genus is a member of phylum Ascomycota. The genus whose natural habitat is soil could survive in depth up to 1 metre of soil (Goswami and Kistler 2004; Parry et al. 1995). Optimum pH and temperature ranges are 5–7 and 24–32 °C, respectively, and species can adapt to low temperature ranges and could carry on producing mycotoxins (Doohan et al. 1999; Yörük and Albayrak 2014). Also, infected host species shows great variation among the genera. While several species have potential to infect a wide range of plants which are not closely related including ornamental to forest plants without distinguishing organs, some others have limited host species including cereals (Leslie and Summerell 2006). The genus has more than 50 species. However, *Fusarium* head blight disease is associated with 17 *Fusarium* species, *Fusarium graminearum* species complex and *Microdochium nivale* (Bottalico and Perrone 2002; Davari et al. 2013; Larone 2011; Saharan et al. 2004). *Fusarium graminearum* and *F. culmorum* are reported as major causal agents of FHB and ear blight diseases worldwide (Bai and Shaner 2004; Miedaner et al. 2008; Parry et al. 1995).

Fusarium graminearum (anamorph) is the major causal agent of FHB in many regions worldwide. The teleomorph of this fungus is *Gibberella zeae*. In the future, only one name should be used for each fungal species (Hawksworth 2011).

According to Geiser et al. (2013), the anamorph name *Fusarium* should be used as the only genus name for all *Fusarium* species instead of different teleomorph names, including *Gibberella*.

Fusarium graminearum is particularly reported as primary FHB pathogen in many countries located in Asia including China (mainly *F. asiaticum*), Iran and Turkey (Haratian et al. 2008; Ji et al. 2007; Tunali et al. 2006), America including the USA and Canada (Goswami and Kistler 2004; Mishra et al. 2004; Salas et al. 1999) and most countries in Europe including Germany, Russia, Hungary, etc. (Miedaner et al. 2001, 2008; Yli-Mattila 2010; Yli-Mattila and Gagkaeva 2016). As it is in many fungal species, the haploid phase is predominant in the life cycle of *F. graminearum*. The genome of homothallic species possesses two mating type alleles, *MAT-1* and *MAT-2* (Kerényi et al. 2004). This hemi-biotrophic fungal species is predicted as a model organism, and life cycle could be completed in 2 weeks in laboratory conditions (Bowden and Leslie 1999; Trail et al. 2002). Genome project of *F. graminearum* was released in the year 2006. Two strains, PH-1 and GZ3639, were used in plasmid, fosmid and bacterial artificial chromosome vectors derived from whole genome sequencing process, and genome size was noted as 36.1 Mb comprising of 13,939 genes, 37,575 exon regions and 48.33% GC content. In regard to genome profiles of closely related fungi whose genome projects have been completed including *Neurospora crassa*, *Magnaporthe grisea*, *Aspergillus nidulans*, *F. oxysporum* and *F. verticillioides*, *F. graminearum* genome possesses genes associated with transcription factors, hydrolytic enzymes and transmembrane proteins with high number. Single nucleotide polymorphism (SNP) frequency between two strains is detected as 0–17.5 SNP/kb according to normalization analysis, and telomere regions include SNPs with high frequency. Moreover, a limited number of microsatellite and transposon regions have been linked to repeat-induced point mutation (RIP). Microarray and bioinformatics analysis showed that the species include 382 species-specific genes and 1368 functionally characterized genes, and 10,000 EST records were deposited in the National Center for Biotechnology Information database (Cuomo et al. 2007; Trail 2009).

Morphological and molecular investigations resulted in interesting and useful output about the taxonomy of *Fusarium graminearum* species. Aoki and O'Donnell (1999) described a new *Fusarium* species, *F. pseudograminearum*, which is formerly known as *F. graminearum* population 1. This novel species recognition was supported by data associated with differences in colony growth rate, conidia width, absence of homothallic perithecia and also *β-tubulin* gene sequence data. Moreover, detailed researches on the primary causal agent of FHB, *F. graminearum*, lead to discovering of lineage differentiation and/or recognition of species complex for this fungus. *F. graminearum* species complex is comprised of phylogenetically different species with undistinguishable morphological characteristics, distinct chemotypes, and agro-ecological origins (Wang et al. 2008, 2011). O'Donnell et al. (2000) recognized seven different clades of *F. graminearum* species with distinct genetic structure and geographical distribution by characterization of nucleotide sequence belonging to six nuclear genes. Lineages 2, 3, and 5, lineages 1 and 4 and lineages 6 and 7 were frequently observed in African, American and Asian continents, respectively.

RAPD-derived SCAR and restriction fragment length polymorphism markers were adopted in lineage differentiation (Carter et al. 2002). Later these lineages were called phylogenetic species. At the moment 15 phylogenetically distinct species have been recognized among *F. graminearum* species complex via multilocus genotyping and molecular marker technologies (O'Donnell et al. 2000; Wang et al. 2008; Yli-Mattila et al. 2009). *F. graminearum* sensu stricto (*F. graminearum s.s.*) is distributed worldwide, whereas the remaining species are spread among restricted agro-ecological regions. *F. asiaticum*, *F. vorosii*, *F. nepalense* and *F. ussurianum* seem to be endemic to Asia, while *F. austroamericanum*, *F. meridionale*, *F. cortaderiae*, *F. brasilicum*, *F. boothii*, *F. mesoamericanum*, *F. louisianense* and *F. gerlachii* are specific to America, and *F. acacia-mearnsii* and *F. aethiopicum* are specific to Africa and Oceania. Prevalence of these species could vary dramatically in association with the geographic region, chemotype and host species. However, *F. asiaticum* has been detected in North America, South America and Europe (Przemieniecki et al. 2014).

In *F. graminearum* and *F. asiaticum*, the chemotype profile includes all three chemotypes, and in five species the chemotype profile includes 3-ADON and NIV, while only 3-ADON, 15-ADON or NIV chemotype has been detected in the rest of the species (Astolfi et al. 2011; Boutigny et al. 2011; Fernández-Ortuño et al. 2013; Gale et al. 2011; Karugia et al. 2009; Nielsen et al. 2012; O'Donnell et al. 2008; Pasquali and Migheli 2014; Sampietro et al. 2011; Wang et al. 2008, 2011; Zhang et al. 2012). Distribution of chemotypes among *F. graminearum* species complex is variable according to each species. To our knowledge, NIV chemotype is not represented by each species of *F. graminearum* species complex (Przemieniecki et al. 2014). Isolates of *F. graminearum s.s.* and *F. asiaticum* belonging to three class B-trichothecene chemotypes have been recorded, whereas 3-ADON and 15-ADON distribution shows great variation for the remaining species (Table 14.1). Similarly, wheat, barley and maize have shown to be hosts of *F. graminearum s.s.* and *F. asiaticum*. 3-ADON and 15-ADON chemotypes of *F. graminearum s.s.* have also been found in oats (Yli-Mattila et al. 2009; Yli-Mattila and Gagkaeva 2010). Some of the remaining 13 species have not associated with maize or barley. However, this could depend on the restricted and limited number of isolates studied and characterized on these 13 species (Desjardins and Proctor 2011; Fernández-Ortuño et al. 2013; Pasquali and Migheli 2014; Wang et al. 2011; Yli-Mattila and Gagkaeva 2010).

Fusarium culmorum is the secondary predominant pathogen of *Fusarium* head blight worldwide, and it is phylogenetically closely related to *F. graminearum* species complex. Necrotrophic *Fusarium culmorum* has a wide range of host species compared to *F. graminearum*, and it also causes several diseases in addition to FHB and root rot (Hornok et al. 2007; Scherm et al. 2011). *Fusarium culmorum* isolates were obtained from scabby kernels planted in many geographically different regions (Bakan et al. 2002; Chung et al. 2008; Jennings et al. 2004b; Kammoun et al. 2010; Miedaner et al. 2001; Scherm et al. 2011; Scoz et al. 2009; Tunali et al. 2006). Especially, *F. culmorum* has been prevalent in several European countries, and the species has been predominating especially in regions with cooler climatic conditions compared to *F. graminearum* which has been predominant in

Table 14.1 Generalized chemotype profile and suggested origin for the species of *F. graminearum* species complex

Species	Origin	Host			Chemotype		
		Wheat	Maize	Barley	3-ADON	15-ADON	NIV
F. graminearum s.s.	Cosmopolitan	+	+	+	+	+	+
F. asiaticum	Asia	+	+	+	+	+	+
F. vorosii	Asia	+	nr[a]	+	nr	+	nr
F. ussurianum	Asia	+	nr	+	+	nr	nr
F. nepalense	Asia	+	+	nr	nr	+	nr
F. austroamericanum	South America	+	+	+	+	nr	+
F. meridionale	South America	+	+	+	nr	nr	+
F. cortaderiae	South America	+	+	+	+	nr	+
F. brasilicum	South America	+	nr	+	+	nr	+
F. boothii	Central America	+	+	+	nr	+	nr
F. mesoamericanum	Central America	nr	nr	nr	+	nr	+
F. gerlachii	North America	+	nr	nr	nr	nr	+
F. louisianense	Central America	nr	nr	nr	nr	nr	+
F. acacia-mearnsii	Africa/Oceania	+	nr	nr	+	nr	+
F. aethiopicum	Africa	+	nr	nr	nr	+	nr

[a]nr means there is no present report about specific host or chemotype

warmer and more humid regions. However, generalization about *F. graminearum/F. culmorum* with warmer/cooler regions could not be always accurate because *F. graminearum* has been replacing *F. culmorum* in cereals in Northern Europe (Bai and Shaner 2004; Bottalico and Perrone 2002; Parry et al. 1995; Saharan et al. 2004; Yli-Mattila 2010; Yli-Mattila et al. 2013). No sexual stage has been reported in *F. culmorum*, but presence of parasexual stage has been suggested by Miedaner et al. (2001, 2008). Only 3-ADON and NIV chemotypes of *F. culmorum* have been found (e.g. Yli-Mattila and Gagkaeva 2010). Even if there is no released genome project of *F. culmorum*, output related to genome information of *F. culmorum* FcUK99 strain which is 3-acetyldeoxynivalenol (3-ADON) producer has been currently accumulated (Baldwin et al. 2010; Scherm et al. 2011). Moreover, nucleotide sequences of specific genes, microsatellite markers, RNA polymerase enzyme genes, genes associated with mycotoxin production, and pathogenicity-related genes have been released on NCBI database.

In addition to two global head blight causal agents, *F. graminearum* and *F. culmorum*, *F. sporotrichioides*, *F. poae*, *F. pseudograminearum* and *F. cerealis* are reported to be primary trichothecene-producing pathogens associated with FHB worldwide. These species are generally co-isolated from scabby plants with *F. graminearum* and *F. culmorum*. Predominancy of these species could be correlated with alterations in climatic conditions in local agro-economic regions and changes in crop rotations. Studies on these four species are still limited, and present studies have been focused on genotyping and chemotyping (Chandler et al. 2003; Chung et al. 2008; Dinolfo et al. 2010; Isebaert et al. 2009; Mishra et al. 2006; Osborne and Stein 2007; Tomczak et al. 2002). Moreover, no genome project or genetic manipulation studies revealing functions of specific genes as reported in *F. graminearum* or *F. culmorum* (Brown et al. 2002; Chandler et al. 2003; Kim et al. 2008; Lee et al. 2001) have been published in these four species or in the remaining FHB-related *Fusarium* species (except for *F. avenaceum* and *F. langsethiae*, Lysøe et al. 2016a, b). However, even if these species have been represented by low infection frequencies in cereals, dramatic and negative effects on economy, mycotoxin contamination profiles and yield loss ranges have been almost similar to that in *F. graminearum* and *F. culmorum* (Bai and Shaner 2004; Bottalico and Perrone 2002; Saharan et al. 2004; Windels 2000). *F. sambucinum* and *F. equiseti* are also class B-trichothecene producer phytopathogens. Studies on their chemotype distribution determination have been currently limited. However, they have been used in general molecular genetic studies including mycoprotein production, species-specific identification, trichothecene production analysis, mating type analysis, etc. (Adejumo et al. 2007; Desjardins and Beremand 1987; Jurado et al. 2005; Kristensen et al. 2007; Miller and Mackenzie 2000).

14.4 Mycotoxins Produced by Causal Agents of FHB

Mycotoxins are secondary metabolites that are produced by fungi. These metabolites have toxigenic effects on both prokaryotic and eukaryotic organisms. *Fusarium* genus is one of the five main mycotoxin producer fungal genera (Gutleb et al. 2002; Niessen 2007). *Fusaria* produces several types of mycotoxins including trichothecenes, zearalenone, fumonisin, fusarins, butenolid, enniatins and others. Seven mycotoxin biosynthetic pathways have been reported in *Fusaria*, including common genes associated with secondary metabolite production such as polyketide synthase, terpene cyclase, cyclic peptide synthetase and cytochrome P450 (Desjardins and Proctor 2007). However, it should be mentioned that intraspecific variation in terms of mycotoxins produced and chemotype profiles occurs in different *Fusarium* species.

For example, *F. avenaceum* isolates have a potential to produce enniatins and moniliformin, whereas *F. graminearum* produces class B-trichothecenes and zearalenone. The interspecific variation among *Fusarium* species surely results in different levels of agronomic losses since mycotoxins produced show diversity in toxicity (Kimura et al. 2007; Ryu et al. 1988). Another important note is that *Fusarium* species co-produce mycotoxins meaning that more than one mycotoxin type accumulate on small grain cereals. Despite that, produced mycotoxin profiles for species are generally known and specific. *Fusarium graminearum*, *F. culmorum* and most other *Fusarium* sp. associated with FHB mainly produce class B-trichothences, zearalenone, fusarin C and butenolide, and the most commonly accumulated on cereal mycotoxin are deoxynivalenol, nivalenol, and their acetylated derivatives. Class A-trichothecenes include T-2 and HT-2 mycotoxins with more toxigenic effects according to class B-trichothecenes (Desjardins and Proctor 2007; Foroud and Eudes 2009). The most important class A-trichothecene producers in cereals are *F. sporotrichioides*, *F. langsethiae*, which have been mainly found in Europe, and *F. sibiricum*, which has been mainly found in Asia (Yli-Mattila et al. 2011, 2013, 2015).

14.5 Class B-Trichothecenes and *tri5* Gene Cluster

Trichothecenes are sesquiterpenoid secondary metabolites consisting of incorporated cyclohexane/tetrahydrofuran circles. These molecules are classified into four groups: class A-, B-, C- and D-trichothecenes. Class A- and B-trichothecenes include more than 100 variants (Foroud and Eudes 2009; Sudakin 2003). Most frequently isolated types are class A- and B-trichothecenes, and class A molecules differ from class B-trichothecenes by possessing an additional carbonyl group at C-8 position. Trichothecenes carry five functionally different groups, and generally hydroxyl and acetyl groups are possessed at these positions (Table 14.2).

T-2 toxin is the major mycotoxin of class A-trichothecenes, whereas deoxynivalenol is the main class B-trichothecene (Foroud and Eudes 2009; Sudakin 2003). Farnesyl pyrophosphate is a primary metabolite of all kinds of trichothecenes, and formation of the final product depends on the expression of different genes located in trichothecene biosynthetic gene cluster known as *tri5* gene cluster (Hammond-Kosack et al. 2004; Kimura et al. 2007).

Class B-trichothecenes are also known as phytotoxins. Class B-trichothecenes are accumulated on cereals, and they are resistant to degradation by high temperature conditions; they protect their stable structure even in food processing conditions. They are harmful to both plants and also for human and animals. They are potential eukaryotic protein synthesis inhibitors by binding ribosomes. Moreover, mortal and chronic toxicoses are caused by feeding with these mycotoxins (Gutleb et al. 2002; Rotter et al. 1996; Sudakin 2003). These toxins are produced by several species including *F. graminearum*, *F. culmorum*, *F. poae*, etc. *Fusarium* isolates and strains producing class B-trichothecenes are divided into two main chemotypes as deoxynivalenol (DON) and nivalenol (NIV) profiles. This phenomenon on class B-trichothecene classification is also detailed in three chemotypes and/or sub-chemotypes as (1) 3-acetylateddeoxynivalenol (3-ADON) and DON, (2) 15-acetylated deoxynivalenol (15-ADON) and DON and (3) NIV and 4-ANIV (Brown et al. 2002; Chandler et al. 2003; Jennings et al. 2004a, b; Ji et al. 2007;

Table 14.2 Five side chains of class B-trichothecenes

	C position				
Class B-trichothencenes	C-3	C-4	C-15	C-7	C-8
DON	-OH	-H	-OH	-OH	=O
3-ADON	-OA	-H	-OH	-OH	=O
15-ADON	-OH	-H	-OA	-OH	=O
NIV	-OH	-OH	-OH	-OH	=O

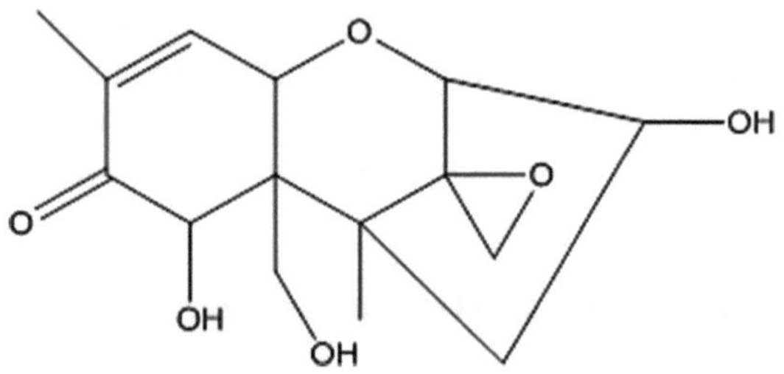

Fig. 14.2 Schematization of DON molecule generated by ChemOffice software V14 (Cambridge, England)

Lee et al. 2002; Miller et al. 1991; Wang et al. 2008; Yörük and Albayrak 2012). Deoxynivalenol mycotoxin is a characteristic mycotoxin in class B-trichothecenes, and this molecule deoxy is at C – 4 position (Fig. 14.2).

Class B-trichothecenes and the remaining trichothecenes are produced by fungus as a result of the expression of genes located in *tri5* gene cluster. Several research groups and laboratories have defined this gene cluster in class B-trichothecene producers *F. graminearum* strains and in class A-trichothecenes producer *F. sporotrichioides* strains (Brown et al. 2004; Hammond-Kosack et al. 2004; Isebaert et al. 2009; Kimura et al. 2003; Lee et al. 2002).

F. graminearum strains H-11 (15-ADON producer) and F15 (3-ADON producer), *F. graminearum* 88–1 strain (NIV producer) and *F. sporotrichioides* strain NRRL 3299 (T-2 toxin producer) were subjected to *tri5* gene cluster characterization studies (Brown et al. 2001; Kimura et al. 2003; Lee et al. 2001; Lee et al. 2002) (Table 14.3). Moreover, in the following years, *tri5* gene cluster in many strains of *F. graminearum* was sequenced totally, and interesting results were obtained in particular gene cluster size and number of genes located (see http://www.ncbi.nlm.nih.gov/nuccore/?term=tri5+gene+cluster). Three major differences were reported in *tri5* gene cluster of totally sequenced reference strains: (1) total size of the gene cluster, (2) full deletion of some genes and (3) functional activity of definite genes in specific chemotype profiles. The complete size of *tri5* gene cluster in three *F. graminearum* strains H-11, 88–1 and GZ3639 is 27,022, 30,159 and 57,840 bp, respectively. This situation is directly associated with specific chemotype profiles, and it is due to the presence of pseudogenes, insertion/deletions and completely deletion in genes (Brown et al. 2004; Chandler et al. 2003). Besides, the number of genes essential in the last metabolite production is variable in strains.

However, essentially expressing genes for all kinds of different trichothecenes could be accepted as stable. *tri3*, *tri4*, *tri6*, *tri5*, *tri10*, *tri9*, *tri11*, *tri12* and *tri14* genes (Fig. 14.3) were reported to critical genes in 3-ADON, 15-ADON and NIV chemotypes (Kimura et al. 2003, 2007). Figure 14.3 shows locations and functionality of genes. All genes located in essential 3-ADON, 15AcDON and NIV production are illustrated as filled boxes; *tri13* and *tri7* which are not essential in DON production are shown as null boxes (Fig. 14.3).

But Alexander et al. (2011) also reported that even if *tri3* is expressed in all three chemotypes, 3-ADON producers do not require *tri3* expression. *tri5* gene cluster is comprised of totally 12 genes including transporter, pathway and regulatory genes. Properties of genes located in the core gene cluster are mentioned in Table 14.3.

Table 14.3 Genes located in core tri5 gene cluster and their activities in *F. graminearum* strains H-11 (LEE et al. 2001), 88–1 (LEE et al. 2002) and F15 (Kimura et al. 2003)

Gene	Product	Size in bp	Exon/ intron	Activity in three *F. graminearum* strains		
				H-11	88–1	F15
tri8	Trichothecene C-3 deacetylase	1334	1/0	+[b]	+[b]	+[b]
tri7	3-Acetyltrichothecene 4-*O*-acetyltransferase	1376	2/1	X[c]	–[d]	+[b]
tri3	3-Acetyltrichothecene 15-*O*-acetyltransferase	1762	4/3	+[b]	+[b]	![e]
tri4	Multifunctional (in four steps) oxygenase	1741	4/3	+[b]	+[b]	+[b]
tri6	Transcription factor	656	1/0	+[b]	+[b]	+[b]
tri5	Trichodiene synthase	1179	2/1	+[b]	+[b]	+[b]
tri10	Regulatory protein	1352[a]	2/1	+[b]	+[b]	+[b]
tri9	*TRI9* with unknown function	131	1/0	![e]	![e]	![e]
tri11	C-15 hydroxylase	1740	5/4	+[b]	+[b]	+[b]
tri12	Final product transporter/efflux pump	1904	3/2	+[b]	+[b]	+[b]
tri13	3-Acetyltrichothecene C-4 hydroxylase	1801	2/1	X[c]	X[c]	+[b]
tri14	Hypothetical protein	1176	2/1	![e]	![e]	![e]

[a]means the information obtained from AF326571.1 and AY102604.1
[b]+means expressed gene
[c]xmeans pseudogene
[d]–means deleted gene
[e]!means gene with no crucial expression

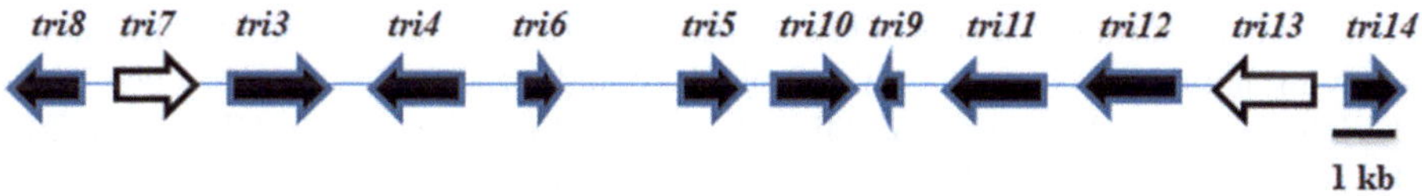

Fig. 14.3 Genes located in *tri5* gene cluster

The information given in Table 14.2 is generated from data currently deposited in NCBI database and also former literature (Goswami and Kistler 2004; Kimura et al. 2003, 2007). *tri101* gene and *tri1/tri6* gene cluster are independently located outside of the core gene cluster, and they are also involved in trichothecene production (Hammond-Kosack et al. 2004; Kimura et al. 2003). The third important subject is the presence of pseudogenes in different chemotypes. Several researchers confirmed that *tri7* and *tri13* genes are possessed as functional copies on genomes of *Fusarium* isolates of NIV chemotype, whereas isolates belonging to DON chemotypes carry non-functional copy (mainly 15-ADON chemotype isolates) or do not contain the genes (mainly 3-ADON chemotype isolates, Chandler et al. 2003; Jennings et al. 2004a; Kimura et al. 2003; Lee et al. 2001; Yli-Mattila 2010; Yli-Mattila and Gagkaeva 2010).

14.6 Review of Developments So Far

Class B-trichothecene chemotype distribution in *Fusaria* was more frequently investigated in *F. graminearum* and *F. culmorum* species (Pasquali and Migheli 2014). Chemotyping studies could be viewed in three approaches. The first approach is the identification of fungus' potential to produce toxin. The second is showing the presence of the toxin type. The third is the determination of presence and quantity of trichothecene molecules. Mycotoxin production potential could be assessed by polymerase chain reaction (PCR) and southern blot (SB) analysis. Precise detection of mycotoxin presence is provided by thin-layer chromatography (TLC) analysis. Quantity of mycotoxins is also analysed by semi-quantitative thin-layer chromatography, gas chromatography/mass spectrophotometer (GC/MS), high pressure liquid chromatography (HPLC) and enzyme-linked immunosorbent assay (ELISA) methods (Haratian et al. 2008; Jennings et al. 2004a; Kim et al. 2003; Kimura et al. 2003; Lee et al. 2001, 2002; Leisova et al. 2006; Naef et al. 2006; Tóth et al. 2004). Overall these methods have advantages and disadvantages in comparison by themselves. PCR technique gives a fast and reliable result. However, the potential to produce toxin does not always mean that the fungus produces mycotoxin since toxin production is associated with more than one factor. Similarly, a toxigenic fungus which produces mycotoxin in natural conditions may not secrete mycotoxin in in vitro laboratory conditions (Alexander et al. 2011). Carter et al. (2002), Tóth et al. (2004), and Llorens et al. (2006) reported *Fusarium* sp. with no or very low amount of trichtohecene production. Thus, researchers focused on the combination of two or more methods giving reliable results. In general, two chemotype-specific methods, PCR and GC/MS, are commonly used in trichothecene profiling (Haratian et al. 2008; Kim et al. 2003; Lee et al. 2002, 2009). Combined usage of two or more methods is important in terms of presenting of more than one mycotoxin production since in nature, *Fusarium* strains/isolates could produce more than one class B-trichothecenes (Pasquali and Migheli 2014). However, PCR + HPLC+GC/MS (Li et al. 2005), PCR + HPLC+TLC (Kim et al. 2003, 2008) and PCR + GC/MS + southern blot combinations were also used in chemotyping analysis of fusaria. Thus, PCR and chromatographic methods which are fast, reliable and promising are currently used in chemotyping analysis alone or together.

Hue et al. (1999) developed PCR assay based on amplification of *tri5* gene distinguishing trichothecene-producing *Fusarium* spp. However, polymerase chain reaction-based or genetic chemotyping has become popular since the year 2001, which Lee et al. (2001) first described as genic PCR approach. Great variation in core *tri5* gene cluster is present in *Fusarium* sp. strains with different chemotypes. Ward et al. (2002) published detailed study including polymorphism among strains of *F. graminearum* species complex dealing with 15-ADON, 3-ADON and NIV chemotypes. They reported that polymorphism in specific genes directly associated with modification and formation of final trichothecene product is responsible for 3 class B-trichothecene profiles. As an example, Fig. 14.4 shows CLUSTALW analysis of *tri13* gene which is most commonly targeted in chemotype determination studies.

```
H-11        TGCGTGTCAATCTTGGACAAGTGCTGGAGTGAGGATGTAGCTCCACCGCATCGAAGAGTC 531
F15         TGCGTGTCGCCCTTGGACAAGTGTTGAAGTGCGGGTGTAGCTCGACCGCATCGAAGAGTC 540
88-1        TGCGTGTCAACCATGGCCAAGCGTTGAAGTGCCGATGTAGCTCCACCGCATCGAAGACTC 529
            ********.. *:***.**** * **.****. *.******** ************* **
H-11        TCTTCAAAGACTGCTTCAAGGTGGAGGATTCGCGCAAAG--TAGCTCCAGGAAAGT---- 585
F15         TCTTCAATGGCTGTTTCAAGGTAGGGGATTCGCGCAAGG--TAGCTCTAGGAAAGT---- 594
88-1        TCTTCAATGACTGCTTCAAGGTAGGGGATTCGAGCAGCTGCTAGCTCTTGGAAAGTGGGA 589
            *******:*.*** ********.*.*******.***.     ****** :*******
H-11        ------------------------------------------------------------
F15         ------------------------------------------------------------
88-1        CTGCGGTTTTCGGATTTGGCCTTGGGGAACGCTGACTGCAAAGCCTGTCGGAGTCGCTTC 649

H-11        ------CCACTATTGCG------------------------------------------- 596
F15         ------CCACTGTTGCA------------------------------------------- 605
88-1        TGGGCGCCAGCATTGCCTGTAAGGAACTTCAAGCCCCACAGAGTGGTGGTGCTCGTTGTG 709
                  ***  .****
H-11        ------------------------------------------------------------
F15         ------------------------------------------------------------
88-1        TCATAACCTCCGAGGATCAGTCCGAATAGCTAGAATAACCCAACGTATCAGCTTCCATCA 769

H-11        -----ATCTGTGATTGAGTCTTTGTGCTTACCTCTCCTTGTATCATCGTTGACAATTAAT 651
F15         -----ATCTGTGATTGAGTCTTTGTGCTTACCTCTTCTTGCATCATCGTTGATACATAAT 660
88-1        TCTTAATATGTCATTGAGTCGTTGTGCTTACCTCTCCTTGCATCATTGTTGACAAATAGT 829
                 **.*** ******** ************** **** ***** ***** *.:**.*
H-11        TTCGGCTCCGAATTTTCGCTTCAGCCAACAATGTTTCACGTAGCATCATCTGTTCCAACG 711
F15         TCAAGCTCCGATTTTTTCGTTTCAGCCAGTAATGTTTCACGTAGCACCATCTGTTCCAACG 720
88-1        TTGGACTCCGATTTTTCCGCTTTGGCCAACAATGTTTCACGTAGCACCATCTGTTCCAAAG 889
            *   ..******:*** ** ** .****. **************** ************.*
```

Fig. 14.4 CLUSTALW analysis of *tri13* gene of *F. graminearum* 88–1 (NIV chemotype), H-11 (15-ADON chemotype) and F15 strains (3-ADON chemotype). '-' means deletions, '*' means conserved nucleotide sequences

According to the CLUSTALW analysis, functional or non-functional copies could be easily predicted in of 88–1, H-11 and F15 strains of *F. graminearum* deletion/insertion and single nucleotide polymorphisms, which led to functional or non-functional copies is clearly predicted. The moderate range of average genetic similarity values ranging from 66.62 to 75.23% of *tri13* gene in 88–1, H-11 and F15 strains of *F. graminearum* is an important example of these polymorphisms (Fig. 14.4). In addition to studies carried out by Lee et al. (2001, 2002), Chandler et al. (2003) and Kimura et al. (2003) reported complete nucleotide sequence information of *tri5* gene cluster. After the release of nucleotide sequence information of genes located in *tri5* gene cluster, several authors developed strategies for distinguishing three chemotypes in *Fusarium* isolates belonging to more than five species. Lee et al. (2001, 2002) developed generic PCR assays linked to *tri7* and *tri13* genes for differentiation of DON and NIV chemotypes. Functional copies of *tri7* and *tri13* genes were possessed on genomes of isolates with NIV chemotype. However, isolates belonging to 3-ADON and 15-ADON chemotypes have non-functional (disrupted or deleted) copies of *tri7* and *tri13* genes. Thus, generic PCR assay has been efficiently used.

Similarly, Chandler et al. (2003) and Waalwijk et al. (2003) differentiated two chemotypes since isolates of DON chemotypes have *tri7* and *tri13* genes as disrupted or deleted. Several specific PCR primer sets were developed and two strategies were aimed. First, differences in PCR amplicon sizes yielded two chemotypes, DON or NIV. Second, presence or absence of a PCR band determined two chemotypes. *tri3* have been also targeted by generic PCR assays in further analysis of two sub-chemotypes of DON profiles, 3-ADON and 15-ADON (Jennings et al. 2004a, b). Generic PCR with two primer sets results in two amplicons with different sizes which are signs of 3-ADON and 15-ADON chemotypes. Additionally, Wang et al. (2008) developed another generic PCR assay including amplification of *tri3* gene resulting with three amplicons with different sizes. Three PCR profiles differentiated three chemotypes NIV, 3-ADON and 15-ADON. Wang et al. (2008) suggested that former studies including the presence and absence of a band could give false-negative results. Moreover, Lee et al. (2002) suggested that in nature, some DON-producer isolates could carry a functional copy of *tri7* gene even if they did not determine any isolate belonging to DON chemotype with functional *tri7* gene. Also, some *Fusarium* sp. isolates can produce mycotoxins of two chemotypes or two sub-chemotypes (Castañares et al. 2014; Gilbert et al. 2001; Salas et al. 1999). Thus, a PCR-based approach requires confirmation by secondary methods in chemotyping investigations.

Thin-layer chromatography (TLC) is a basic and fast method for analysis of many kinds of secondary metabolites including mycotoxins produced by fungi. The method is constructed on the movement of a metabolite in the chromatographic system. The system is based on mobile phase including solvent/solvent mixture and stationary phase comprising of solid adsorbent (silica or alumina plate). Trichothecene chemotypes of *Fusarium* species including *F. graminearum*, *F. culmorum*, *F. cerealis* and *F. pseudograminearum* were detected by thin-layer chromatography on silica plates with different mixtures of mobile phase (Haratian et al. 2008; Kimura et al. 2003; Rahmani et al. 2009; Tóth et al. 2004; Trigo-Stockli et al. 1994; Völkl et al. 2004). Prompt results on the presence of NIV, 3-ADON and 15-ADON are obtained with suitable TLC markers in these investigations. Further chromatographic analysis or ELISA assays are currently preferred in chemotype determination analysis. HPLC and GC/MC methods are reliable and suitable chemical approaches for the class B-trichothecene determination as qualitative and quantitative. These methods present precise results, and they have been widely used and most frequently preferred in trichothecene profiling studies in *Fusarium graminearum*, *Fusarium culmorum*, *Fusarium cerealis*, *F. pseudograminearum* and *F. poae* species (Carter et al. 2002; Jennings et al. 2004a, b; Kammoun et al. 2010; Kim et al. 2003; Lee et al. 2002, 2009; Li et al. 2005; Llorens et al. 2006; Salas et al. 1999; Tóth et al. 2004; Walker et al. 2000; Wang et al. 2008). Three strain-specific class B-trichothecene profiles were detected by GC/MS and HPLC analysis in *Fusarium* spp., and precise quantification of mycotoxins produced has been also detected by these investigations. Bily et al. (2004) have developed the HPLC-based method, which is widely used in chemotyping investigations. However, these high-quality methods gave rise to the formation of an extraordinary dilemma about 'a *Fusarium* spp. strain could belong to two chemotypes' (Carter et al. 2002; Gilbert et al. 2001; Pasquali and Migheli 2014; Salas et al. 1999). In these researches, the presence of both 15-ADON and also 3-ADON mycotoxins had been detected in *Fusarium* spp. Another theoretical problem is that isolate or strain which is detected as non-trichothecene producer in laboratory conditions can be indeed a potential producer in nature. Thus, even if more than two methods are used in chemotype determination studies, isolates detected as non-producers could produce trichothecene toxins in natural conditions, or, in other words, false-negative identification could be applied to toxin producer isolates. Nevertheless, this situation is rarely presented.

In general, usage of just one of these methods or combination of more than one method gave information about class B-trichothecene profiles of *Fusarium* isolates worldwide. Especially, fungal samples isolated from diseased kernels planted in Asian, European and American countries have been subjected to chemotype identification studies. In Asia, predominantly *Fusarium graminearum* species complex and also *F. culmorum* are widely studied in Japan, Iran, Turkey, South Korea, Nepal and Syria (Table 14.4).

According to the investigations surveyed on Asian countries, wheat and *F. graminearum* species complex are the major host and causal agents. Fernández-Ortuño et al. (2013) reported *F. asiaticum*, *F. ussurianum* and *F. vorosii* are of Asian clade of species complex. These three species have been isolated from wheat and barley. Zhang et al. (2012) determined chemotypes of 457 *F. graminearum* species complex, including *F. meridionale*, *F. graminearum s.s.* and *F. asiaticum*, 10 *F. cerealis* and 1 *F. culmorum* isolate from China by PCR and HPLC analysis. 3-ADON was the prevalent chemotype in Chinese isolates. Similarly, Karugia et al. (2009) determined *F. asiaticum* as predominant pathogen in wheat samples with chemotypes of NIV (44.7%) and 3-ADON (53.3%). However, it is clear that chemotype distribution is not clearly associated only with geographic region, species, host and the year survey (sample collection year). In South Korea, both major chemotypes were detected as predominating in different years (Lee et al. 2009, 2012). But 15-ADON chemotype seems to be prevalent in *F. graminearum* sensu stricto in eastern Asian countries, except in Russian Far East (Table 14.4).

Slightly similar results were obtained in studies carried out in North and South America countries (Kim et al. 2003; Llorens et al. 2006; Qu et al. 2008; Reynoso et al. 2011; Scoz et al. 2009; Carter et al. 2002; Wang et al. 2008). Table 14.5 shows different *Fusarium* species associated with FHB in North and South American continents. Wheat, *F. graminearum* and DON/15-ADON seem to be predominating host plant, causal agent and chemotype, respectively. However, Salas et al. (1999) reported that *F. graminearum*, *F. aveneceum* and *F. poae* have almost equally account for the disease, and NIV chemotype is dominant in *F. poae*. Also, Malihipour et al. (2012) proved that NIV chemotype is dominant in *F. cerealis* isolates from Mexico.

In European countries, predominance issue of host plant species, causal agent and chemotype is more complicated. Jennings et al. (2004a, b) showed the presence of two major causal agents, *F. graminearum* and *F. culmorum*, with two main class B-trichothecene chemotypes in the UK (Table 14.6). Especially for these two species, predominancy could be variable according to/associated with agro-ecological regions, climatic conditions, crop rotations, etc. Recently, *F. graminearum* has been spreading to the north in Europe and displacing *F. culmorum* (Ward et al. 2002; Yli-Mattila 2010). In addition to wheat, winter rye, maize, rice and barley are also currently infected with *F. graminearum*, *F. culmorum*, *F. poae*, *F. cerealis* and *F.*

Table 14.4 Class B-trichothecene profile distribution of *Fusarium* spp. in Asian countries

Author(s)	Country(ies)	Species	Host(s)	Method(s)	No of isolates		
					3-ADON	15-ADON	NIV
Lee et al. (2001)	South Korea	*Fg*	Rice	PCR, GC/MS	31[a]		321
Carter et al. (2002)	Nepal	*Fg*	Maize, rice, wheat	GC/MS	26[a]		10
Kim et al. (2003)	South Korea	*Fg*	Wheat, barley, corn	PCR, SB GC/MS	34[a]		76
Li et al. (2005)	China	*Fg*	Wheat	PCR, HPLC GC/MS	310[a]		54
Ji et al. (2007)	China	*Fg*	Wheat, maize, barley	PCR	48	203	3
Zhang et al. (2007)	China	*Fg, Fa*	Wheat	PCR	155	91	53
Haratian et al. (2008)	Iran	*Fg*	Wheat	PCR, TLC	11[a]		46
Qu et al. (2008)	China, Nepal Japan	*Fgsc*	Wheat, maize, rice	PCR	26	3	10
Suga et al. (2008)	Japan	*Fg, Fa*	Wheat, barley	PCR, GC/MS	100	13	170
Wang et al. (2008)	Nepal, China	*Fgsc*	Rice, maize, Wheat	PCR, HPLC GC/MS	6	13	8
Karugia et al. (2009)	Japan	*Fa, Fgss*	Wheat	PCR, HPLC	99	0	84
Yli-Mattila and Gagkaeva (2010) Yli-Mattila et al. (2009)	Russia	*Fg, Fc, Fu, Fv, Fce*	Wheat, barley, oats	PCR	20/55	26/45	6/0
Yli-Mattila and Gagkaeva (2010)	China	*Fg, Fce*	Wheat	PCR	2	9	4
Desjardins and Proctor (2011)	Nepal	*Fgsc*	Maize	PCR, HPLC	48[a]		148
Lee et al. (2012)	South Korea	*Fgsc*	Maize	PCR, TLC	438[a]		54
Ndoye et al. (2012)	China	*Fg, Fa*	Maize		1	269	350
Shen et al. (2012)	China	*Fg, Fa*	Wheat	PCR	300	182	48
Yörük and Albayrak (2012)	Turkey	*Fg, Fc*	Wheat, maize, barley	PCR	20	11	1

(continued)

Table 14.4 (continued)

					No of isolates		
Author(s)	Country(ies)	Species	Host(s)	Method(s)	3-ADON	15-ADON	NIV
Zhang et al. (2012)	China	*Fmer, Fgss, Fa, Fc, Fce*	Wheat	PCR, HPLC	172	176	109
Alkadri et al. (2013)	Syria	*Fg, Fc*	Wheat	PCR, HPLC-MS/MS	4	0	10
Davari et al. (2013)	Iran	*Fg*	Wheat	PCR	3	121	11
Mert-Türk and Gencer (2013)	Turkey	*Fc*	Wheat	PCR	16	39	2
Venkataramana et al. (2013)	India	*Fc*	Maize	PCR, HPLC	54[a]		34
Yörük et al. (2014)	Turkey, Iran	*Fg, Fc, Fps, Fpo*	Wheat, barley	PCR	11	6	0

[a]means no further differentiation of DON chemotype
Fg, *F. graminearum*; Fc, *F. culmorum*; Fgss, *F. graminearum* sensu stricto; Fmer, *F. meridionale*; Fps, *F. pseudograminearum*; Fce, *F. cerealis*; Fa, *F. asiaticum*; Fu, *F. ussurianum*; Fv, *F. vorosii*; Fgsc, *F. graminearum* species complex

pseudograminearum (Miedaner et al. 2001, 2008; Pasquali and Migheli 2014; Yli-Mattila et al. 2013). In Finland, Norway and many parts of Sweden, the main host of *F. graminearum* is oats (Yli-Mattila et al. 2013; Yli-Mattila and Gagkaeva 2016). A limited number of *Fusarium* isolates from France, Italy and Germany have been investigated in chemotyping studies (Carter et al. 2002; Kimura et al. 2003; Qu et al. 2008; Tomczak et al. 2002). Nevertheless, Llorens et al. (2006) reported NIV was predominating chemotype in *F. graminearum*, *F. culmorum*, *F. cerealis*, *F. poae* and *Gibberella fujikuori* isolates from Spain. On contrary to global distribution of 15-ADON, 3-ADON has been reported as prevalent chemotype in some European countries including northern Europe (Finland, Sweden, Norway and Northwestern Russia) (Yli-Mattila et al. 2009; Yli-Mattila and Gagkaeva 2010, 2016). In Denmark 3-ADON of *F. graminearum* is the predominating chemotype in oats, while in other cereals the 15-ADON chemotype of *F. graminearum* is predominating (Nielsen et al. 2011, 2012). Thus, it has been suggested that there are two main populations of *F. graminearum* in Europe, of which the 3-ADON population is dominant in Central and Southern Europe and has been spreading, e.g. to Denmark, while 3-ADON chemotype has been dominant in Northern Europe and has recently spread from Finland to Northwestern Russia (Yli-Mattila et al. 2013). It may be that 3-ADON chemotype is also more specialized to oats, which is more common in Northern Europe.

In addition to studies conducted on three major continents, Asia, Europe and America, North Africa and Oceania countries have been also subjected to

Table 14.5 Class B-trichothecene profile distribution of *Fusarium* spp. in North and South American countries

Author(s)	Country(ies)	Species	Host(s)	Method(s)	No of isolates		
					3-ADON	15-ADON	NIV
Pineiro et al. (1996)	Uruguay	*Fg*	Barley	GC/MS	0	11	0
Salas et al. (1999)	USA	*Fg, Fpo, Fave, Fsp*	Wheat	GC/MS	11[a]		9
Walker et al. (2000)	USA	*Fg*	Wheat	GC	66[a]		0
Gilbert et al. (2001)	Canada	*Fg*	Wheat, barley, corn, weeds	GC	3	14[b]	2
Clear et al. (2006)	Canada	*Fps*	Wheat	GC	122	2	1
Qu et al. (2008)	USA	*Fgsc*	Wheat	PCR	0	7	0
Wang et al. (2008)	USA	*Fgsc*	Wheat	PCR, HPLC GC/MS	0	5	0
Alvarez et al. (2009)	Argentina	*Fgss*	Wheat	GC	75	115	34
Scoz et al. (2009)	Brazil	*Fgss, Fmer*	Wheat	PCR	0	76	6
Delgado et al. (2010)	USA	*Fg*	Potato, wheat, sugar beet	PCR, TLC, GC/MC	12[a]		2
Ohe et al. (2010)	Canada	*Fg*	Wheat	ELISA	12	12	0
Puri and Zhong (2010)	USA	*Fgss*	Wheat	PCR, GC/MS	4	111	5
Astolfi et al. (2011)	Brazil	*Fgss,Fmer, Faust*	Barley	PCR	4	61	27
Gale et al. (2011)	Louisiana	*Fa, Fgss*	Wheat	PCR, GC/MS	6	253	48
Sampietro et al. (2011)	Argentina	*Fmer, Fb*	Maize	PCR	0	10	56
Pan et al. (2013)	Uruguay	*Fgss*	Wheat	PCR	0	110	1
Castañares et al. (2014)	Argentina	*Fgss*	Barley	PCR, GC/MS	16[b]	110	1[b]
Malbran et al. (2014)	Argentina	*Fg*	Wheat	PCR, ELISA	0	112	0

[a]means no further differentiation of DON chemotype
[b]means co-production of two or three mycotoxins
Fg, *F. graminearum*; Fgss, *F. graminearum* sensu stricto; Faust, *F. austroamericanum*; Fb, *F. boothii*; Fmer, *F. meridionale*; Fa, *F. asiaticum*; Fave, *F. aveneceum*; Fps, *F. pseudograminearum*; Fsp, *F. sporotrichioides*; Fgsc, *F. graminearum* species complex

Table 14.6 Class B-trichothecene profile distribution of *Fusarium* spp. in European countries

					No of isolates		
Author(s)	Country(ies)	Species	Host(s)	Method(s)	3-ADON	15-ADON	NIV
Bakan et al. (2002)	France	*Fc*	Wheat	GC/MS	10	3[a]	0
Carter et al. (2002)	France, Germany, UK,	*Fg*	Wheat	GC/MS	13[b]		1
Hestbjerg et al. (2002)	Austria, Denmark, Norway	*Fc*	Wheat	HPLC	77	0	9
Jennings et al. (2004a)	England, Wales	*Fg*	Wheat	PCR, HPLC	4	76	21
Jennings et al. (2004b)	England, Wales	*Fc*	Wheat	PCR	88	0	65
Tóth et al. (2005)	Hungary, Austria	*Fgsc*	Wheat	PCR, HPLC	2[a]	21	0
Qu et al. (2008)	France, Germany, Italy, UK, Sweden	*Fgsc*	Wheat	PCR	5	17	2
Yli-Mattila et al. (2009)	Russia, Finland	*Fg*	Wheat, barley, oat	PCR	43	43	0
Yli-Mattila and Gagkaeva (2010)	Finland	*Fg*	Wheat, barley, oat	PCR	12	0	0
		Fc	Wheat, barley	PCR	13	0	0
	Russia	*Fc, Fce*	Potato, cirsium, wheat, barley	PCR	10	0	1
	Germany	*Fg*	Oat	PCR	0	4	1
Wang et al. (2008)	France, Germany, Italy, UK	*Fgsc*	Wheat, barley	PCR, HPLC GC/MS	0	18	3
Talas et al. (2011)	Germany	*Fgss*	Wheat	PCR	23	311	4
Nielsen et al. (2012)	Denmark	22 *Fusarium* sp. with 6 members of Fgsc	Wheat, barley, triticale, rye, oats	PCR	36	7	11[a]

(continued)

Table 14.6 (continued)

Author(s)	Country(ies)	Species	Host(s)	Method(s)	No of isolates		
					3-ADON	15-ADON	NIV
Cornea et al. (2013)	Romania	*Fg, Fc*	Wheat	PCR, HPLC	8	71	0
Purahong et al. (2013)	Italy	*Fg*	Wheat	PCR	5	21	6
Boutigny et al. (2014)	France	*Fgss, Fcor, Fb*	Wheat, maize, barley	PCR	1	255	38

[a]means co-production of two or three mycotoxins
[b]means no further differentiation of DON chemotype
Fg, *F. graminearum*; Fcor, *F. cortaderiae*; Fc, *F. culmorum*; Fgss, *F. graminearum* sensu stricto; Fb, *F. boothii*; Fgsc, *F. graminearum* species complex

Fusarium spp. studies with limited numbers. Obanor et al. (2010, 2013) reported that *F. graminearum* and *F. pseudograminearum* isolates from diseased wheat produced equally DON and NIV in Australia. *F. culmorum* and wheat are reported as a major causal agent and host plant in Tunisia (Kammoun et al. 2010; Rebib et al. 2014). 3-ADON was reported as major mycotoxin in the country among 183 *F. culmorum* isolates. However, the number of fungal isolates studied on chemotype determination assays seems to be limited in comparison with Asian, American and European samples.

14.7 Future Perspectives and Conclusions

Studies associated with class B-trichothecene producer *Fusarium* species have been increased in recent years. In particular, researchers focused on the declaration of the novel species in phytopathogenic *Fusarium* spp. Aoki and O'donnell (1999) distinguished former *F. graminearum* group as *F. pseduograminearum*. Moreover, lineage number of *F. graminearum* species complex changed few times after discovery of new species (O'Donnel let al., 2000; Przemieniecki et al. 2014; Sarver et al. 2011; Starkey et al. 2007), and the current number has reached to 15. Characterization studies on these novel species are still in progress, and output on agro-ecological region and chemotypes is increasing day by day. Recently, Aoki et al. (2015) recognized novel nivalenol producer species *Fusarium dactylidis* sp. nov., which is a species sister of *F. pseudograminearum*.

In addition to investigations on the identification of new trichothecene producer species, studies including inhibition or promotion of trichothecene production became popular. Kulik et al. (2014) showed that pinoresinol and secoisolariciresinol lignans produced by wheat decreased trichothecene level. Similarly, phenolic acids obtained from *Spirulina* sp. inhibited class B-trichothecene production (Pagnussatt et al. 2014). Lancovaa et al. (2008) detected the DON-3-Glc in naturally infected barley as well as in malt and beer samples. Nagl and Schatzmayr (2015) reported that the relative portion of these phase II metabolites of DON to free toxin form (DON) is of 20% on average. Analysis of masked forms of deoxynivalenol including deoxynivalenol-3-glucoside (DON-3-Glc), deoxynivalenol-3-glucuronide (DON-3-GlcA) and deoxynivalenol-15-glucuronide (DON-15-GlcA) has become popular in recent years (Michlmayr et al. 2015; Generotti et al. 2015). Vector-based RNA interference strategy was also used to inhibit trichothecene production in *Fusarium* sp. (Mcdonald et al. 2005; Scherm et al. 2011). On the contrary, acidic pH and H_2O_2 and amine supplements were recorded as trichothecene production indicators (Kawakami et al. 2014; Merhej et al. 2010; Ponts et al. 2009). Species-specific identification and chemotype determination have also been studied during the last two decades. Zhang et al. (2014) and Pallez et al. (2014) developed novel efficient PCR-based approaches to detect members of *F. graminearum* species complex at the species level and to determine the chemotypes.

Determination of class B-trichothecene profiles of *Fusarium* isolates worldwide could directly and/or indirectly contribute to the development of novel control strategies for FHB disease. First, pre-knowledge about the epidemiological distribution of fungal isolates would be gained by chemotyping studies. Since the toxicity of three trichothecene profiles has variation, potential toxigenic risk could be determined by chemotyping in local or global areas where FHB is present. Also, since additional risk factors including crop rotations, climatic changes, etc. could directly affect the spread of disease, fast, reliable and precise results become critical and crucial in chemotype determination. For these purposes, researchers have investigated global *Fusarium* spp. samples by modern genetic and chemical techniques in particular PCR and chromatography. As a result of output data originated from global chemotype of *Fusarium* spp., characterization of field populations of fungi, genotypic diversity and aggressiveness of fungus and severity levels of diseased host plants would be clearly understood.

Acknowledgements The authors express their thanks to the members of Molecular Biology and Genetics Department of Istanbul University including Dr. Gülruh Albayrak; M.Sc. Aylin Gazdagli; Dr. Berna Tunali, a member of Samsun 19 Mayis University Department of Plant Protection; Drs. Kerry O'Donnell and Todd Ward from USDA-ARS in Peoria; and Dr. Tatiana Gagkaeva from All-Russian Institute of Plant Protection in St. Petersburg for sharing their experience in *Fusarium* research area. Our research on chemotype distribution in *Fusarium* species was supported by Research Foundation of Istanbul University, Olvi Foundation, Finnish Cultural Foundation and the Academy of Finland.

References

Adejumo TO, Hettwer U, Karlovsky P (2007) Occurrence of *Fusarium* species and trichothecenes in Nigerian maize. Int J Food Microbiol 116(3):350–357

Alexander NJ, Mccormick SP, Waalwijk C, Lee TVD, Proctor RH (2011) The genetic basis for 3-ADON and 15-ADON trichothecene chemotypes in *Fusarium*. Fungal Genet Biol 48:485–495

Alkadri D, Nipoti P, Döll K, Karlovsky P, Prodi A, Pisi A (2013) Study of fungal colonization of wheat kernels in Syria with a focus on *Fusarium* species. Int J Mol Sci 14:5938–5951

Alvarez CL, Azcarate MP, Pinto VF (2009) Toxigenic potential of *Fusarium graminearum* sensu stricto isolates from wheat in Argentina. Int J Food Microbiol 135:131–135

Anand A, Zhou T, Trick HN, Gill BS, Bockus WW, Muthukrishnanl S (2003) Greenhouse and field testing of transgenic wheat plants stably expressing genes for thaumatin-like protein, chitinase and glucanase against *Fusarium graminearum*. J Exp Biol 54(384):1101–1111

Aoki T, O'Donnell K (1999) Morphological and molecular characterization of *Fusarium pseudograminearum* sp. nov., formerly recognized as the Group 1 population of *F. graminearum*. Mycologia 91(4):597–609

Aoki T, Vaughan MM, Mccormick SP, Busman M, Ward TJ, Kelly A, O'Donnell K, Johnston PR, Geiser DM (2015) *Fusarium dactylidis* sp. nov., a novel nivalenol toxin-producing species sister to *F. pseudograminearum* isolated from orchard grass (*Dactylis glomerata*) in Oregon and New Zealand. Mycologia 107(2):409–418

Astolfi P, Dos Santos J, Schneider L, Gomes LB, Silva CN, Tessmann DJ, Del Ponte EM (2011) Molecular survey of trichothecene genotypes of *Fusarium graminearum* species complex from barley in Southern Brazil. Int J Food Microbiol 148:197–201

Bai G, Shaner G (2004) Management and resistance in wheat and barley to *Fusarium* head blight. Annu Rev Phytopathol 42:135–161

Bakan B, Delville CG, PinsonL R-MD, Fournier E, Brygoo Y (2002) Identification of *Fusarium culmorum* strains producing large and small amounts of deoxinivalenol. Appl Environ Microbiol 68(11):5472–5479

Baldwin TK, Urban M, Brown N, Hammond-Kosack KE (2010) A role for topoisomerase I in *Fusarium graminearum* and *F. culmorum* pathogenesis and sporulation. Mol Plant Microbe Interact 23:566–577

Bernardo A, Bai G, Guo P, Xiao K, Guenzi AC, Ayoubi P (2007) *Fusarium graminearum*-induced changes in gene expression between *Fusarium* head blight-resistant and susceptible wheat cultivar. Funct Integr Genomics 7:69–77

Bily AC, Reid LM, Savard E, Reddy R, Blackwell BA, Campbell CM, Krantis A, Durst T, Philogene BJR, Arnason JT, Regnault-Roger C (2004) Analysis of *Fusarium graminearum* mycotoxins in different biological matrices by LC/MS. Mycopathologia 157:117–126

Bottalico A, Perrone G (2002) Toxigenic *Fusarium* species and mycotoxins associated with head blight in small-grain cereals in Europe. Eur J Plant Pathol 108:611–624

Boutigny AL, Ward TJ, Van Coller GJ, Flett B, Lamprecht SC, O'Donnell K, Viljoen A (2011) Analysis of the *Fusarium graminearum* species complex from wheat, barley and maize in South Africa provides evidence of species-specific differences in host preference. Fungal Genet Biol 48:914–920

Boutigny AL, Ward TJ, Ballois N, Iancu G, Ioos R (2014) Diversity of the *Fusarium graminearum* species complex on French cereals. Eur J Plant Pathol 138:133–148

Bowden RL, Leslie JF (1999) Sexual recombination in *Gibberella zeae*. Phytopathology 89:182–188

Brown DW, Mccormick SP, Alexander NJ, Proctor RH, Desjardins AE (2001) A genetic and biochemical approach to study trichothecene diversity in *Fusarium sporotrichioides* and *Fusarium graminearum*. Fungal Genet Biol 32(2):121–133

Brown DW, Mccormick SP, Alexander NJ, Proctor RH, Desjardins AE (2002) Inactivation of a cytochrome P-450 is a determinant of trichothecene diversity in Fusarium species. Fungal Genet Biol 36:224–233

Brown DW, Dyer RB, McCormick SP, Kendra DF, Plattner RD (2004) Functional demarcation of the *Fusarium* core trichothecene gene cluster. Fungal Genet Biol 41:454–462

Carter JP, RezanoorHN HD, Desjardins AE, Plattner RD, Nicholson P (2002) Variation in pathogenicity associated with the genetic diversity of *Fusarium graminearum*. Eur J Plant Pathol 108:573–583

Castañares E, Albuquerque DR, Dinolfo MI, Pinto VF, Patriarc A, Stenglein SA (2014) Trichothecene genotypes and production profiles of *Fusarium graminearum* isolates obtained from barley cultivated in Argentina. Int J Food Microbiol 179:57–63

Chandler EA, Simpson DR, Thomsett MA, Nicholson P (2003) Development of PCR assays to *tri7* and *tri13* trichothecene biosynthetic and characterisation of chemotypes of *Fusarium graminearum*, *Fusarium culmorum* and *Fusarium cerealis*. Physiol Mol Plant Pathol 62:355–367

Chehri K, Salleh B, Yli-Mattila T, Reddy KRN, Abbasi S (2011) Molecular characterization of pathogenic *Fusarium* species in cucurbit plants from Kermanshah province, Iran. Saudi J Biol Sci 18:341–351

Chung WH, Ishii H, Nishimura K, Ohshima M, Iwama T, Yoshimatsu H (2008) Genetic analysis and PCR-based identification of major *Fusarium* species causing head blight on wheat in Japan. J Gen Plant Pathol 10:110–118

Clear RM, Patrick SK, Gaba D, Roscoe M, Turkington TK, Demeke T, Pouleur S, Couture L, Ward TJ, O'Donnell K (2006) Trichothecene and zearalenone production, in culture, by isolates of *Fusarium pseudograminearum* from western Canada. Can J Plant Pathol 28:131–136

Cornea C, Israel-Roming F, Ciuca M, Voaides C (2013) Natural occurrence of *Fusarium* species and corresponding chemotypes in wheat scab complex from Romania. Rom Biotech Lett 18:8787–8795

Cuomo CA, Güldener U, Xu JR, Trail F, Turgeon BG, Di Pietro A, Adam G (2007) The Fusarium graminearum genome reveals a link between localized polymorphism and pathogen specialization. Science 317(5843):1400–1402

Davari M, Wei SH, Babay-Ahari A, Arzanlou M, Waalwijk C, Van Der Lee TAJ, Zare R, Van Den Ende AHGG, De Hoog GS, Van Diepeningen AD (2013) Geographic differences in trichothecene chemotypes of *Fusarium graminearum* in the Northwest and North of Iran. World Mycotoxin J6:137–150

Delgado JA, Schwarz PB, Gillespie J, Rivera-Varas V, Secor GA (2010) Trichothecene mycotoxins associated with potato dry rot caused by *Fusarium graminearum*. Phytopathology 100:290–296

Desjardins AE, Beremand M (1987) A genetic system for trichothecene toxin production in *Gibberella pulicaris* (*Fusarium sambucinum*). Phytopathology 77:678–683

Desjardins AE, Proctor RH (2007) Molecular biology of *Fusarium* mycotoxins. Int J Food Microbiol 119:47–50

Desjardins AE, Proctor RH (2011) Genetic diversity and trichothecene chemotypes of the *Fusarium graminearum* clade isolated from maize in Nepal and identification of a putative new lineage. Fungal Biol 115:38–48

Dinolfo MI, Stenglein SA, Moreno MV, Nicholson P, Jennings P, Salerno GL (2010) ISSR markers detect high genetic variation among *Fusarium poae* isolates from Argentina and England. Eur J Plant Pathol 127:483–491

Doohan FM, Weston G, Rezanoor HN, Parry DW, Nicholson P (1999) Development and use of a reverse transcription-PCR assay to study expression of *tri5* by *Fusarium* species *in vitro* and in planta. Appl Environ Microbiol 65:3850–3854

Dyakov TY, Dzhavakhlya VG, Korpela T (2007) Comprehensive and Molecular phytopathology. Elsevier, Amsterdam, p 469

Fernández-Ortuño D, Waalwijk C, Van Der Lee T, Fan J, Atkins S, West JS, Fraaije BA (2013) Simultaneous real-time PCR detection of *Fusarium asiaticum*, *F. ussurianum* and *F. vorosii*, representing the Asian clade of the *F. graminearum* species complex. Int J Food Microbiol 166:148–154

Foroud NA, Eudes F (2009) Trichothecenes in cereal grains. Int J Mol Sci 10:147–173

Gagkaeva TY, Yli-Mattila T (2004) Genetic diversity of *Fusarium graminearum*in Europe and Asia. Eur J Plant Pathol 110:551–562

Gale LR, Harrison SA, Ward TJ, O'Donnell K, Milus EA, Gale SW, Kistler HC (2011) Nivalenol-type populations of *Fusarium graminearum* and *F. asiaticum* are prevalent on wheat in Southern Louisiana. Phytopathology 101(1):124–134

Geiser DM, Aoki T, Bacon CW, Baker SE, Bhattacharyya MK, Brandt ME et al (2013) One fungus, one name: defining the genus *Fusarium* in a scientifically robust way that preserves longstanding use. Phytopathology 103:400–408

Generotti S, Cirlini M, Malachova A, Sulyok M, Berthiller F, Dall'asta C, Suman M (2015) Deoxynivalenol and deoxynivalenol-3-glucoside mitigation through bakery production strategies: effective experimental design within industrial rusk-making technology. Toxins 7:2773–2790

Gilbert J, Abramson D, Mccallum B, Clear R (2001) Comparison of Canadian *Fusarium graminearum* isolates for aggressiveness, vegetative compatibility, and production of ergosterol and mycotoxins. Mycopathologia 153:209–215

Goswami RS, Kistler HC (2004) Heading for disaster: *Fusarium graminearum* on cereal crop. Mol Plant Pathol 5(6):515–525

Gürel F, Albayrak G, Diken O, Cepni E, Tunali B (2010) Use of REP-PCR for genetic diversity analyses in *Fusarium culmorum*. J Phytopathol 158:387–389

Gutleb AC, Morrison E, Murk AJ (2002) Cytotoxicity assay for mycotoxins produced by *Fusarium* strains. Environ Toxicol Pharm 11:309–320

Hammond-Kosack K, UrbanM BT, Daudi A, Rudd J, Keon J, Lucas J, Maguire K, Kornyukhin D, Jing HC, Bass C, Antoniw J (2004) Plant pathogens: how can molecular genetic information on plant pathogens assist in breeding disease resistant crops. New Directions For a Diverse Planet 44:1–18

Haratian M, Sharifnabi B, Alizadeh A, Safaie N (2008) PCR analysis of the tri13 gene to determine the genetic potential of *Fusarium graminearum* isolates from Iran to produce nivalenol and deoxynivalenol. Mycopathologia 166:109–116

Hawksworth DLA (2011) New dawn for the naming of fungi: impacts of decisions made in Melbourne in July 2011 on the future publication and regulation of fungal names. IMA Fungus 2:155–162

Hestbjerg H, Felding G, Elmholt S (2002) *Fusarium culmorum* infection of barley seedlings: correlation between aggressiveness and deoxynivalenol content. J Plant Pathol 150:308–312

Hornok L, Waalwijk C, Leslie JL (2007) Genetic factors affecting sexual reproduction in toxigenic *Fusarium* species. Int J Food Microbiol 119:54–58

Hue FX, Huerre M, Rouffault MA, Bievre CD (1999) Specific detection of *Fusarium* species in blood and tissues by a PCR technique. J Clin Microbiol 37:2434–2438

Isebaert S, Saeger SD, Devreese R, Verhoeven R, Maene P, Heremans B, Haesaert G (2009) Mycotoxin-producing *Fusarium* species occurring in winter wheat in Belgium (Flanders) during 2002–2005. J Phytopathol 157:108–116

Jennings P, Coates ME, Walsh K, Turner JA, Nicholson P (2004a) Determination of deoxinivalenol- and nivalenol-producing chemotypes of *Fusarium graminearum* isolated from wheat crops in England and Wales. Plant Pathol 53:643–652

Jennings P, Coates ME, Turner JA, Chandle EA, Nicholson P (2004b) Determination of deoxinivalenol and nivalenol chemotypes of *Fusarium culmorum* isolates from England and Wales by PCR assay. Plant Pathol 53:182–190

Ji L, Cao K, Hu T, Wang S (2007) Determination of deoxinivalenol and nivalenol chemotypes of *Fusarium graminearum* isolates from China by PCR assay. J Phytopathol 155:505–512

Jurado M, Vázquez C, Patińo B, González-Jaén MT (2005) PCR detection assays for the trichothecene-producing species *Fusarium graminearum*, *Fusarium culmorum*, *Fusarium poae*, *Fusarium equiseti* and *Fusarium sporotrichioides*. Syst Appl Microbiol 28:562–568

Kammoun LG, Gargouri S, Barreau C, Richard-Forge F, Hajlaoui MR (2010) Trichothecene chemotypes of *Fusarium culmorum* infecting wheat in Tunisia. Int J Food Microbiol 140:84–89

Karugia GW, Suga H, Gale LR, Nakajima T, Tomimura K, Hyakumachi M (2009) Population structure of the *Fusarium graminearum* species complex from a single Japanese wheat field sampled in two consecutive years. Plant Dis 93(2):170–174

Kawakami A, Nakajima T, Hirayae K (2014) Effects of carbon sources and amines on induction of trichothecene production by *Fusarium asiaticum* in liquid culture. FEMS Microbiol Lett 352:204–212

Kerényi ZA, Moretti C, Waalwijk B, Olah L, Hornok L (2004) Mating type sequences in asexually reproducing *Fusarium* species. Appl Environ Microbiol 70:4419–4423

Kim HS, Lee T, Dawlatana M, Yun SH, Lee YW (2003) Polymorphism of trichothecene biosynthesis genes in deoxynivalenol- and nivalenol-producing *Fusarium graminearum* isolates. Mycol Res 107:190–197

Kim JE, Jin J, Kim H, Kim JC, Yun SH, Lee YW (2008) Functional characterization of genes located at the aurofusarin biosynthesis gene cluster in *Gibberella zeae*. Plant Pathol J 24:8–16

Kimura M, Tokai T, O'Donnell K, Ward TJ, Fujimura M, Hamamoto H, Shibata T, Yamaguchi I (2003) The trichothecene biosynthesis gene cluster of *Fusarium graminearum* F15 contains a limited number of essential pathway genes and expressed non-essential genes. FEBS Lett 539:105–110

Kimura M, Tokai T, Takahashi-Ando N, Ohsato S, Fujimura M (2007) Molecular and genetic studies of *Fusarium* trichothecene biosynthesis: pathways, gene and evolution. Biosci Biotech Bioch 71:2105–2123

Kristensen R, Gauthier G, Berdal KG, Hamels S, Remacle J, Holst-Jensen A (2007) DNA microarray to detect and identify trichothecene- and moniliformin-producing *Fusarium* species. J Appl Microbiol 102(4):1060–1070

Kulik T, Buśko M, Pszczółkowska A, Perkowski J, Okorski A (2014) Plant lignans inhibit growth and trichothecene biosynthesis in *Fusarium graminearum*. Lett Appl Microbiol 59:99–107

Laday M, Juhasz A, Mule G, Moretti A, Szecsi A, Logrieco A (2004) Mitochondrial DNA diversity and lineage determination of European isolates of *Fusarium graminearum* (*Gibberella zeae*). Eur J Plant Pathol 110:545–550

Lancovaa K, Hajslovaa J, Poustkaa J, Krplovaa A, Zachariasovaa M, Dostalekb P, Sachambula L (2008) Transfer of *Fusarium* mycotoxins and 'masked' deoxynivalenol (deoxynivalenol-3-glucoside) from field barley through malt to beer. Food Addit Contam 25(6):732–744

Larone DH (2011) Medically important fungi: a guide to identification. ASM Press, Washington, DC, 485 pp

Lauren DR, Smith WA (2001) Stability of *Fusarium* mycotoxins nivalenol, deoxynivalenol and zearelenone in ground maize under typical cooking conditions. Food Addit Contam 18:1011–1016

Lee T, Oh DW, Kim HS, Lee J, Kim YH, Yun SH, Lee YW (2001) Identification of deoxynivalenol- and nivalenol-producing chemotypes of *Gibberella zeae* by using PCR. Appl Environ Microbiol 67:2966–2972

Lee T, Han YH, Kim KH, Yun SH, Lee YW (2002) *Tri13* and *Tri7* determine deoxynivalenol- and nivalenol- producing chemotypes of *Gibberella zeae*. App Environ Microbiol 68:2148–2154

Lee J, ChangI Y, Kim H, Yun SH, Leslie JF, Lee YW (2009) Genetic diversity and fitness of *Fusarium graminearum* populations from rice in Korea. Appl Environ Microbiol 75:3289–3295

Lee J, Kim H, Jeon JJ, Kim HS, Zeller KA, Carter LLA, Leslie JF, Lee YW (2012) Population structure of and mycotoxin production by *Fusarium graminearum* from maize in South Korea. Appl Environ Microbiol 78:2161–2167

Leisova L, Kucera L, Chrpova J, Sykorova S, Sip V, Ovesna J (2006) Quantification of *Fusarium culmorum* in wheat and barley tissues using Real-Time PCR in comparison with DON content. J Phytopathol 154:603–611

Leslie JF, Summerell BA (2006) The *Fusarium* laboratuary manual. Blackwell, Manhattan, 387pp

Li HP, Wu AB, Zhao CS, Scholten O, Löffler H, Liao YC (2005) Development of a generic PCR detection of deoxynivalenol- and nivalenol-chemotypes of *Fusarium graminearum*. FEMS Microbiol Lett 243:505–511

Li RQ, He R, Zhang YB, Xu YM, Wang JM (2009) Establishment of ISSR reaction system of *Fusarium* and its analysis of genetic diversity. Sci Agric Sin 42(9):3139–3146

Llorens A, Hinojo MJ, Mateo R, Medina A, Valle-Algarre FM, Gonzalez-Jaen MT, Jimenez M (2006) Variability and characterization of mycotoxin producing *Fusarium spp.* isolates by PCR-RFLP analysis of the IGS-rDNA region. Anton Leeuw Int J G 89:465–478

Lori GA, Sisterna MN, Sarand6n SJ, Rizzo I, Chidichimo H (2009) Fusarium head blight in wheat: impact of tillage and other agronomic practices under natural infection. Crop Prot 28:495–502

Lysøe E, Harris LJ, Walkowiak S, Subramaniam R, Divon HH et al (2016a) The genome of the generalist plant pathogen *Fusarium avenaceum* is enriched with genes involved in redox, signaling and secondary metabolism. PLoS One 9(11):e112703. https://doi.org/10.1371/journal.pone.0112703

Lysøe E, Frandsen RJN, Divon HH, Terzi V, Orrù L, Lamontanara A, Kolseth AS-K, Nielsen KF, Thrane U (2016b) Draft genome sequence and chemical profiling of *Fusarium langsethiae*, an emerging producer of type A trichothecenes. Int J Food Microbiol 221:29–36

Malbran I, Mourelos CA, Girotti JR, Balatti PA, Lori GA (2014) Toxigenic capacity and trichothecene production by *Fusarium graminearum* isolates from Argentina and their relationship with aggressiveness and fungal expansion in the wheat spike. Phytopathology 104:357–364

Malihipour A, Gilbert J, Piercey-Normore M, Cloutier S (2012) Molecular phylogenetic analysis, trichothecene chemotype patterns, and variation in aggressiveness of *Fusarium* isolates causing head blight in wheat. Plant Dis 96:1016–1025

Mcdonald T, Brown D, Keller NP, Hammond TM (2005) RNA silencing of mycotoxin production in *Aspergillus* and *Fusarium* species. Mol Plant-Microbe Interact 18(6):539–545

Merhej J, Boutigny AL, Pinson-Gadais L, Richard-Forget F, Barreau C (2010) Acidic pH as a determinant of TRI gene expression and trichothecene B biosynthesis in *Fusarium graminearum*. Food Addit Contam 27:710–717

Mert-Türk F, Gencer G (2013) Distribution of the 3-ADON, 15-ADON, and NIV chemotypes of *Fusarium culmorum* in the North-West of Turkey. Plant Prot Sci 49(2):57–64

Michlmayr H, AlachováA VE, Kleinová J, Lemmens M, Newmister S, Rayment I, Berthiller F, Adam G (2015) Biochemical characterization of a recombinant UDP-glucosyltransferase from rice and enzymatic production of deoxynivalenol-3-*o*-β-d-glucoside. Toxins 7:2685–2700

Miedaner T, Schilling AG, Geiger HH (2001) Molecular genetic diversity and variation for aggressiveness in populations of *Fusarium graminearum* and *Fusarium culmorum* sampled from wheat fields in different countries. J Phytopathol 149:641–648

Miedaner T, Cumagun CJR, Chakraborty S (2008) Population genetics of three important head blight pathogens *Fusarium graminearum, F pseudograminearum* and *F culmorum.* J Phytopathol 156:129–139

Miller D, Mackenzie S (2000) Secondary metabolites of *Fusarium venenatum* strains with deletions in the Tri5 gene encoding trichodiene synthetase. Mycologia 92:764–771

Miller JD, Greenhalgh R, Wang YZ, Lu M (1991) Trichothecene chemotypes of three Fusarium species. Mycologia 83:121–130

Mishra PK, Tewari JP, Clear RM, Turkington TK (2004) Molecular genetic variation and geographical structuring in *Fusarium graminearum.* Ann Appl Biol 145:299–307

Mishra PK, Tewari JP, Clear RM, Turkington TK (2006) Genetic diversity and recombination within populations of *Fusarium pseudograminearum* from western Canada. Int Microbiol 9:65–68

Naef A, Senatore M, Défago GA (2006) Microsatellite based method for quantification of fungi in decomposing plant material elucidates the role of *Fusarium graminearum* DON production in the saprophytic competition with *Trichoderma atroviride* in maize tissue microcosms. FEMS Microbiol Lett 655:211–220

Nagl V, Schatzmayr G (2015) Deoxynivalenol and its masked forms in food and feed. Curr Opin Food Sci 5:43–49

National Centre for Biotechnology Information (NCBI) (2015). http://www.ncbi.nlm.nih.gov

Ndoye M, Zhang JB, Wang JH, Gong AD, Li HP, Qu B, Li SJ, Liao YC (2012) Nivalenol and 15-acetyldeoxynivalenol chemotypes of *Fusarium graminearum* clade species are prevalent on maize throughout China. J Phytopathol 160:519–524

Nielsen LK, Jensen JD, Nielsen GC, Jensen JE, Spliid NH, Thomsen IK, Justesen AF, Collinge DB, Jorgensen LN (2011) Fusarium head blight of cereals in Denmark: species complex and related mycotoxins. Phytopathology 101:960–969

Nielsen LK, Jensen JD, Rodríguez A, Jørgensen LN, Justesen AF (2012) TRI12 based quantitative real-time PCR assays reveal the distribution of trichothecene genotypes of *F. graminearum* and *F. culmorum* isolates in Danish small grain cereals. Int J Food Microbiol 157:384–392

Niessen L (2007) PCR-based diagnosis and quantification of mycotoxin producing fungi. Int J Food Microbiol 19:38–46

O'Donnell K, Kistler HC, Tacke BK, Casper HH (2000) Gene genealogies reveal global phylogeographic structure and reproductive isolation among lineages of *Fusarium graminearum*, the fungus causing wheat scab. Proc Natl Acad Sci U S A 97(14):7905–7910

O'Donnell K, Ward TJ, Aberra D, Kistler HC, Aoki T, Orwig N, Kimura M, Bjørnstad A, Klemsdal SS (2008) Multilocus genotyping and molecular phylogenetics resolve a novel head blight pathogen within the *Fusarium graminearum* species complex from Ethiopia. Fungal Genet Biol 45:1514–1522

Obanor F, Erginbas-Orakci G, Tunali B, Nicol JM, Chakraborty S (2010) *Fusarium culmorum* is a single phylogenetic species based on multilocus sequence analysis. Fungal Biol 114:753–765

Obanor F, Neate S, Simpfendorfer S, Sabburg R, Wilson P, Chakraborty S (2013) *Fusarium graminearum* and *Fusarium pseudograminearum* caused the 2010 head blight epidemics in Australia. Plant Pathol 62:79–91

Ohe CVD, Gauthier V, Tamburic-Ilincic L, Brule-Babel A, Fernando WGD, Clear R, Ward TJ, Miedaner T (2010) A comparison of aggressiveness and deoxynivalenol production between Canadian *Fusarium graminearum* isolates with 3-acetyl and 15-acetyldeoxynivalenol chemotypes in field-grown spring wheat. Eur J Plant Pathol 127:407–417

Osborne LE, Stein JM (2007) Epidemiology of *Fusarium* head blight on small-grain cereals. Int J Food Microbiol 119:103–108

Pagnussatt FA, Del Ponte EM, Garda-Buffon J, Badiale-Furlong E (2014) Inhibition of *Fusarium graminearum* growth and mycotoxin production by phenolic extract from *Spirulina* sp. Pestic Biochem Physiol 108:21–26

Pallez M, Pasquali M, Bohn T, Hoffmann L, Beyer M (2014) Validation of a quick PCR method suitable for direct sequencing: identification of *Fusarium* fungal species and chemotypes for preventive approaches in food safety. Food Technol Biotechnol 52:351–358

Pan D, Calero N, Mionetto A, Bettucci L (2013) Trichothecene genotypes of *Fusarium graminearum* from wheat in Uruguay. Int J Food Microbiol 162:120–123

Parry DW, Jenkinson P, Mcleod L (1995) *Fusarium* ear blight (scab) in small grain cereals-a review. Plant Pathol 44:207–238

Pasquali M, Migheli Q (2014) Genetic approaches to chemotype determination in type B-trichothecene producing *Fusaria*. Int J Food Microbiol 189:164–182

Pineiro MS, Scott PM, Kanhere SR (1996) Mycotoxin producing potential of *Fusarium graminearum* isolates from Uruguayan barley. Mycopathologia 132(3):167–172

Ponts N, Couedelo L, Pinson-Gadais L, Verdal-Bonnin MN, Barreau C, Richard-Forget F (2009) *Fusarium* response to oxidative stress by H2O2 is trichothecene chemotype dependent. FEMS Microbiol Lett 293:255–262

Przemieniecki SW, Kurowski TP, Korzekwa K (2014) Chemotypes and geographic distribution of the *Fusarium graminearum* species complex. Environ Biotechnol 10(2):45–54

Purahong W, Nipoti P, Pisi A, Lemmens M, Prodi A (2013) Aggressiveness of different *Fusarium graminearum* chemotypes within a population from Northern-Central Italy. Mycoscience 55(1):63–69

Puri KD, Zhong S (2010) The 3-ADON population of *Fusarium graminearum* found in North Dakota is more aggressive and produces a higher level of DON than the prevalent 15-ADON population in spring wheat. Phytopathology 100:1007–1014

Qu B, Li HP, Zhang JB, Xu YB, Huang T, Wu AB, Zhao CS, Carter J, Nicholson P, Liao YC (2008) Geographic distribution and genetic diversity of *Fusarium graminearum* and *F. asiaticum* on wheat spikes throughout China. Plant Pathol 57:15–24

Quellet T, Seifert KA (1993) Genetic characterization of *Fusarium graminearum* strains using RAPD and PCR amplification. Phytopathology 83:1003–1007

Rahmani A, Jinap S, Soleimany S (2009) Qualitative and quantitative analysis of mycotoxins. Compr Rev Food Sci Food Saf 8:202–251

Rebib H, Bouraoui H, Rouaissi M, Brygoo Y, Boudabbous A, Hajlaoui MR, Sadfi-Zouaoui N (2014) Genetic diversity assessed by SSR markers and chemotyping of *Fusarium culmorum* causal agent of foot and root rot of wheat collected from two different fields in Tunisia. Eur J Plant Pathol 139:481–493

Reynoso MM, Ramirez ML, Torres AM, Chulze SN (2011) Trichothecene genotypes and chemotypes in *Fusarium graminearum* strains isolated from wheat in Argentina. Int J Food Microbiol 145:444–448

Rotter B, Prelusky DB, Pestka JJ (1996) Toxicology of deoxynivalenol (vomitoxin). J Environ Sci Health 48:1–34

Ryu JC, Ohtsubo K, Izumiyama N, Nakamura K, Tanaka T, Yamamura H, Ueno Y (1988) The acute and chronic toxicities of nivalenol in mice. Fundam Appl Toxicol 11:38–47

Saharan MS, Kumar J, Nagarajan S (2004) *Fusarium* head blight (FHB) or head scab of wheat – a review. Proc Indian Natl Sci Acad 3:255–268

Salas B, Steffenson BJ, Casper HH, Tacke B, Prom LK Jr, Fetch TG, Schwarz PB (1999) *Fusarium* species pathogenic to barley and their associated mycotoxins. Plant Dis 83:667–674

Sampietro DA, Díaz CG, Gonzalez V, Vattuone MA, Ploper LD, Catalan CAN, Ward TJ (2011) Species diversity and toxigenic potential of *Fusarium graminearum* complex isolates from maize fields in northwest Argentina. Int J Food Microbiol 145:359–364

Sarver BAJ, Ward TJ, Gale LR, Broz K, Kistler HC, Aoki T, Nicholson P, Carter J, O'Donnell K (2011) Novel *Fusarium* head blight pathogens from Nepal and Louisiana revealed by multilocus genealogical concordance. Fungal Genet Biol 48:1096–1107

Scherm B, Orru M, Balmas V, Spanu F, Azara E, Delogu G, Hammond TM, Keller NP, Migheli Q (2011) Altered trichothecene biosynthesis in *TRI6*-silenced transformants of *Fusarium culmorum* influences the severity of crown and foot rot on durum wheat seedlings. Mol Plant Pathol 12(8):759–771

Scoz LB, Astolfi P, Reartes DS, Schmale III DG, Moraes MG, Del Ponte EM (2009) Trichothecene mycotoxin genotypes of *Fusarium graminearum* sensu stricto and *Fusarium meridionale* in wheat from southern Brazil. Plant Pathol 58:344–351

Shen CM, Hu TC, Sun HY, Li W, Guo JH, Chen HG (2012) Geographic distribution of trichothecene chemoptypes of *Fusarium graminearum* species complex in major winter wheat production area of China. Plant Dis 96:1172–1178

Starkey DE, WardTJ AT, Gale LR, Kistler HC, Geiser DM, Suga H, Tóth B, Varga J, O'Donnell K (2007) Global molecular surveillance reveals novel *Fusarium* head blight species and trichothecene toxin diversity. Fungal Genet Biol 44:1191–1204

Sudakin DL (2003) Trichothecenes in the environment: relevance to human health. Toxicol Lett 143:97–107

Suga H, Karugia GW, Ward T, Gale LR, Tomimura K, Nakajima T, Miyasaka A, Koizumi S, Kageyama K, Hyakumachi M (2008) Molecular characterization of the *Fusarium graminearum* species complex in Japan. Phytopathology 98:159–166

Talas F, Parzies HK, Miedaner T (2011) Diversity in genetic structure and chemotype composition of *Fusarium graminearum* sensu stricto populations causing wheat head blight in individual fields in Germany. Eur J Plant Pathol 131:39–48

Tomczak M, Wisniewska H, Stepien L, Kostecki M, Chelkowski J, Golinski P (2002) Deoxynivalenol, nivalenol and moniliformin in wheat samples with head blight (scab) symptoms in Poland (1998–2000). Eur J Plant Pathol 108:625–630

Tóth B, Mesterhazy A, Nicholson P, Teren J, Varga J (2004) Mycotoxin production and molecular variability of European and American isolates of *Fusarium culmorum*. Eur J Plant Pathol 110:587–599

Tóth B, Merterhazy A, Horvath Z, Bartok T, Varga M, Varga J (2005) Genetic variability of central European isolates of the *Fusarium graminearum* species complex. Eur J Plant Pathol 113:35–45

Trail F (2009) For blighted waves of grain: *Fusarium graminearum* in the postgenomics era. Plant Physiol 149:103–110

Trail F, Xu H, Loranger R, Gadoury D (2002) Physiological and environmental aspects of ascospore discharge in *Gibberella zeae*. Mycologia 94:181–189

Trigo-Stockli DM, Deyoe CW, Satumbaga RF, Pedersen JR (1994) Distribution of deoxynivalenol and zearalenone in milled fractions of wheat. Cereal Chem 73(3):388–391

Tunali B, Özsevenİ BO, Erdurmuş D, Demirci A (2006) *Fusarium* head blight and deoxynivalenol accumulation of wheat in Marmara region and reactions of wheat cultivars and lines to *F. graminearum* and *Fusarium culmorum*. Plant Pathol J 5(2):150–156

Venkataramana M, Shilpa P, Balakrishna K, Murali HS, Batra HV (2013) Incidence and multiplex PCR based detection of trichothecene chemotypes of *Fusarium culmorum* isolates collected from freshly harvested maize kernels in Southern India. Braz J Microbiol 44:401–406

Völkl A, Vogler B, Schollenberger M, Karlovsky P (2004) Microbial detoxification of mycotoxin deoxynivalenol. J Basic Microbiol 44(2):147–156

Waalwijk C, Kastelein P, De Vries I, Kerenyi Z, Van Der Lee T, Hesselink T (2003) Major changes in *Fusarium* spp. in wheat in the Netherlands. Eur J Plant Pathol 109:743–754

Walker SL, Leath S Jr, Hagler WM, Murphy JP (2000) Variation among isolates of *Fusarium graminearum* associated with fusarium head blight in North Carolina. Plant Dis 85:404–410

Wang JH, Li HP, Qu B, Zhang JB, Huang T, Chen FF, Liao YC (2008) Development of a generic PCR detection of 3-acetyldeoxy-nivalenol-, 15-acetyldeoxynivalenol- and nivalenol-chemotypes of *Fusarium graminearum* clade. Int J Mol Sci 9:2495–2504

Wang JH, Ndoye M, Zhang JB, Li HP, Liao YC (2011) Population structure and genetic diversity of the *Fusarium graminearum* species complex. Toxins 3(8):1020–1037

Ward TJ, Bielawski JP, Kistler HC, Sullivan E, O'Donnell K (2002) Ancestral polymorphism and adaptive evolution in the trichothecene mycotoxin gene cluster of phytopathogenic *Fusarium*. Proc Natl Acad Sci U S A 99:9278–9283

Windels CR (2000) Economic and social impacts of fusarium head blight: changing farms and rural communities in the Northern Great Plains. Phytopathology 90(1):17–21

Wisniewska H, Kowalczyk K (2005) Resistance of cultivars and breeding lines of spring wheat to *Fusarium culmorum* and powdery mildew. J Appl Genet 46(1):35–40

Yli-Mattila T (2010) Ecology and evolution of toxigenic *Fusarium* species in cereals in Northern Europe and Asia. J Plant Pathol 92:7–18

Yli-Mattila T (2012) Morphological and molecular taxonomyof highly toxigenic *Fusarium* species from small cereal grains in Northern Europe and Asia. In: Misra K, Tewari JP, Deshmukh K (eds) Systematics and evolution of fungi. Science publishers, New Hampshire, pp 277–303

Yli-Mattila T, Gagkaeva T (2010) Molecular chemotyping of *Fusarium graminearum, F. culmorum* and *F. cerealis* isolates from Finland and Russia. In: Gherbawy Y, Voigt K (eds) Molecular identification of fungi. Springer, Berlin/Heidelberg/New York, pp 159–177

Yli-Mattila T, Gagkaeva T (2016) *Fusarium* toxins in cereals in northern Europe and Asia. In: Deshmukh SK, Misra, JK, Tewari JP, Papp T (eds) Applications of fungi and their management strategies, CRC Press, pp 293–317

Yli-Mattila T, Gagkaeva T, Ward TJ, Aoki T, Kistler HC, O'Donnell K (2009) A novel Asian clade within the *Fusarium graminearum* species complex includes a newly discovered cereal head blight pathogen from the Russian far east. Mycologia 101:841–852

Yli-Mattila T, Ward TJ, O'Donnell K, Proctor RH, Burkin AA, Kononenko GP, Gavrilova OP, Aoki T, Mccormick SP, Gagkaeva T (2011) *Fusarium sibiricum* sp. nov, a novel type A trichothecene-producing *Fusarium* from northern Asia closely related to *F. sporotrichioides* and *F. langsethiae*. Int J Food Microbiol 147(1):58–68

Yli-Mattila T, Rämö S, Hietaniemi V, Hussien T, Carlobos-Lopez AL, Cumagun CJR (2013) Molecular quantification and genetic diversity of toxigenic *Fusarium* species in Northern Europe as compared to those in Southern Europe. Microorganisms 1:162–174

Yli-Mattila T, Garvrilova O, Hussien T, Gagkaeva T (2015) Identification of the first *Fusarium sibiricum* isolate in Iran and *Fusarium langsethiae* isolate in Siberia by morphology and species-specific primers. J Plant Pathol 97:183–187

Yörük E, Albayrak G (2012) Chemotyping of *Fusarium graminearum* and *F. culmorum* isolates from Turkey by PCR assay. Mycopathologia 173:53–61

Yörük E, Albayrak G (2013) Genetic characterization of *Fusarium graminearum* and *F. culmorum* isolates from Turkey by using random-amplified polymorphic DNA. Genet Mol Res 12(2):1360–1372

Yörük E, Albayrak G (2014) *Tri4* and *tri5* gene expression analysis in *Fusarium graminearum* and *F. culmorum* isolates by qPCR. Plant Pathol J 13(2):133–138

Yörük E, Gazdağli A, Albayrak G (2014) Class B trichothecene chemotyping in *Fusarium* species by PCR assay. Genetika 46(3):661–669

Zhang JB, Li HP, Dang FJ, Qu B, Xu JB, Zhao CS, Liao YC (2007) Determination of the trichothecene mycotoxin chemotypes and associated geographical distribution and phylogenetic species of the *Fusarium graminearum* clade from China. Mycol Res 111:967–975

Zhang H, Theo VDL, Waalwijk C, Chen W, Xu J, Xu J, Zhang Y, Feng H (2012) Population analysis of the *Fusarium graminearum* species complex from wheat in China show a shift to more aggressive isolates. PLoS One 7(2):1–13

Zhang X, Ma HX, Zhou YJ, Xing JC, Chen JH, Yu GH, Sun XB, Wang L (2014) Identification and genetic division of *Fusarium graminearum* and *Fusarium asiaticum* by species-specific SCAR markers. J Phytopathol 162:81–88

Aflatoxin and Ochratoxin A Detection: Traditional and Current Methods

15

Shraddha Rahi, Priyanka Choudhari, and Vandana Ghormade

Abstract

Mycotoxins such as aflatoxin and ochratoxin A are secondary metabolites secreted by *Aspergillus* and *Penicillium* species. These fungal species flourish in foodstuff and feeds under appropriate temperature and humidity conditions to produce mycotoxins. Aflatoxins are known carcinogens and ochratoxin A causes nephrotoxicity. The contamination of mycotoxins in food and feed, persistence during food processing, and toxicity make them a primary health hazard. Therefore, determination of aflatoxin and ochratoxin A contamination bears a critical importance. Classical methods like chromatographic separation including thin-layer chromatography, high-performance liquid chromatography, and mass spectroscopy are described. Detection of the causal organism by molecular approaches employing PCR and real-time PCR may contribute in early detection. Recently, immunochemical-based methods like enzyme-linked immunosorbent assay and electrical, optical, and piezoelectric immunosensors are being used for the screening purposes. Such detection platforms are portable, reducing the dependence on costly instrumentation. Current strategies to improve the mycotoxin detection involve nanotechnology-enabled sensors. One of the main challenges for the detection of mycotoxin contamination is the co-occurrence of two or more toxins in food and feed samples. The incorporation of novel recognition elements such as antibodies, peptides, or aptamers with nanoparticles for

S. Rahi (✉)
Nanobioscience, Agharkar Research Institute, Pune, Maharashtra, India

Savitribai Phule Pune University, Ganeshkhind, Pune, Maharashtra, India

P. Choudhari · V. Ghormade
Nanobioscience, Agharkar Research Institute, Pune, Maharashtra, India
e-mail: vandanaghormade@aripune.org

T. Satyanarayana et al. (eds.), *Advancing Frontiers in Mycology & Mycotechnology*,
https://doi.org/10.1007/978-981-13-9349-5_15

LFA and immunosensors has immense potential for simultaneously sensitive, specific, and cost-effective multitoxin analysis. Such devices will contribute to improved detection of toxic secondary fungal metabolites critical in food safety, human health, and food trade.

Keywords
Aflatoxin · Ochratoxin A · Chromatographic detection · Immunochemical methods · PCR · Nanotechnology · Lateral flow assay · Multitoxin detection

15.1 Introduction

Mycotoxins are secondary metabolites produced by fungi that contaminate food and feed leading to mycotoxicosis. FAO has estimated that 25% of food is contaminated with mycotoxins causing annual losses of around 1 billion metric tons of foods and food products globally (Smith et al. 2016). Mycotoxins are hepatotoxic and nephrotoxic causing diarrhea, vomiting, hemorrhage, and immune suppression, thus resulting in increased susceptibility to disease and possible death among animals and humans (Binder 2007; Bryden 2012). Mycotoxins are thermostable and generally resistant to sterilization. A recent study among patients suffering from kidney disease in Sri Lanka reported 93.5% of patients with presence of OTA in urine (Desalegn et al. 2011). The apathy toward the problem of mycotoxins is mainly due to low awareness about mycotoxins, their adverse health effects, and the approaches needed to control mycotoxin contamination at pre- and post-harvest stages. Mycotoxin detection and monitoring will contribute to better grain quality, health, and trade when carried out as a preliminary and routine activity in food testing. Remediation approaches incorporate strategies for prevention of mycotoxin contamination at pre- and post-harvest stages, detoxification of mycotoxins in food and feed, and inhibition of mycotoxin absorption in the gastrointestinal tract.

Major mycotoxins found in cereal grains such as wheat, maize, barley, oats, and rye and cereal-based products are aflatoxins (AF), ochratoxin A (OTA), deoxynivalenol, and zearalenone produced by the fungal genera of *Aspergillus*, *Penicillium*, and *Fusarium* (Smith et al. 2016). Mycotoxins are produced by growth of the fungus in adverse conditions of temperature, water activity, and oxygen content (Magan et al. 2002). Recent climate change with untimely rain and floods increases risk of mycotoxin contamination. According to hazard associated critical control point (HACCP) concept, mycotoxin contamination should be detected at every stage of production, harvesting, processing, and distribution; that warrants the requirement of toxin detection systems (FAO/WHO/UNEP 2001). Worldwide, HACCP is implemented for aflatoxin, while other mycotoxins are largely ignored. Worldwide, one or more mycotoxins may be found to contaminate cereal grains (Alshannaq and Yu 2017). In Europe, multi-mycotoxin studies reported that 75%–100% of animal feed samples contained more than one mycotoxin which could impact animal health and in turn the human health (Streit et al. 2013). In India 78% of feed is contaminated with more than one mycotoxin (https://en.engormix.com 2010). Aflatoxins and ochratoxins are secondary metabolites produced by the polyketide synthase pathway of the fungus. Aflatoxins produced by *A. flavus* and *A. parasiticus* in food and feedstuff are implicated in hepatocarcinogenesis. Aflatoxins have a difuranocoumarin structure, aflatoxins B1 and B2 exhibit blue fluorescence, and G1 and G2 display green fluorescence. Aflatoxins M1 and M2, found in milk and milk products, are the hydroxylated products of AFB1 and AFB2. The carcinogenicity of AFB1 is due to its interaction with DNA and alternation of the base sequence by transversion (Wacoo et al. 2014). Another mycotoxin, ochratoxin A, displaying a dihydroisocoumarin structure coupled to a phenylalanine amino acid by peptide linkage, is produced mainly by fungal strains belonging to *Aspergillus ochraceus*, *Penicillium verrucosum*, and *P. carbonarius*. OTA exposure leads to kidney toxicity due to the competitive binding of OTA with phenylalanine t-RNA synthase and interference in protein synthesis (Ha 2015).

Mycotoxins in human and animal food supply have been recognized as safety issue for many years and regulatory levels are defined by WHO and other national agencies. Wheat is a staple cereal consumed worldwide. Wheat and its by products are commonly contaminated by microbial organisms and their metabolites, contamination may survive and carry through cereal grain and their processed food. The European Commission has established regulatory limit of 5 μg/kg for OTA in processed cereal and baby foods (EC 2006). In India, a limit of 20 μg/kg of OTA and 15 μg/kg AFT has been defined by the Food Safety and Standards Authority of India in wheat and wheat-based products (Table 15.1).

Analytical methods for rapid, sensitive, and accurate determination of these mycotoxins in unprocessed food products are generally applied for the assessment of toxicological risk in tune with the regulatory levels. These methods generally involve toxin extraction from the matrix with an adequate extraction solvent, cleanup step designed to eliminate interference from the extract, and detection/determination by suitable analytical instruments/technologies. Often, commercial immunochromatographic assays, such as enzyme-linked immunosorbent assays (ELISA) are frequently used for screening purposes (Schneider et al. 2004). In addition, chromatographic methods such as high-performance liquid chromatography (HPLC) coupled with ultraviolet (UV); diode array (DAD), fluorescence, (FD) or

Table 15.1 Permissible regulatory limits for aflatoxin and ochratoxin A worldwide

Contaminant	Country	Maximum level (ppb)	Structure
Aflatoxin	USA Europe Japan China India	20 4–15 10 5–20 0.5–30	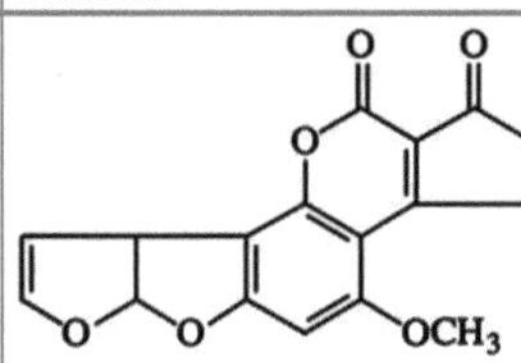
Ochratoxin	USA Europe Japan China India	0.5–2 2–10 0.5–20 5 20	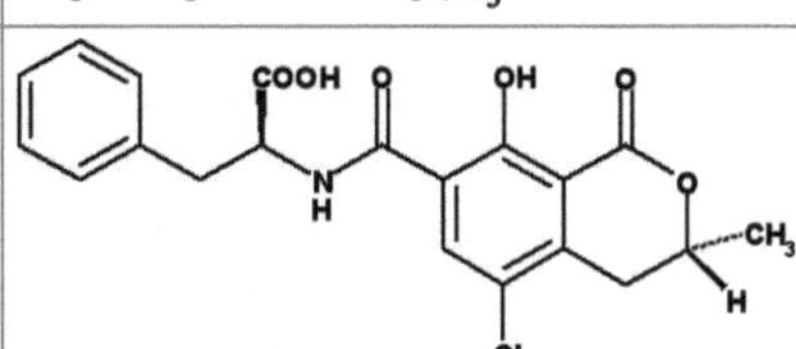

mass spectrometry (MS) detectors; and gas chromatography (GC) coupled with electron capture (ECD), flame ionization (FID), or MS detectors are used (Shephard 2008). These methods though available are expensive and require skilled personnel. A variety of rapid methods are proposed for mycotoxin analyses including infrared spectroscopy, immunochromatographic detection, electrochemical sensors, and biosensors (Wacoo et al. 2014).

Identification and characterization of fungal spoilage organisms by molecular methods such as polymerase chain reaction (PCR) or quantitative PCR (qPCR) can contribute to their early detection. The main efforts taken into ochratoxin and aflatoxin detection are directed toward causal organisms, i.e., *Aspergillus* and *Penicillium* spp. Molecular methods are used for validation with ITS regions of 18S RNA gene, conserved genes, and specific mycotoxigenic genes (Schmidt-Heydt et al. 2011; Sadhasivam et al. 2017).

Recently, several nanotechnology-enabled detection sensors are being developed for mycotoxin detection. Nanotechnology can contribute in rapid, sensitive detection of mycotoxins and their producers with nanoparticles as labels for signal detection. Nanoparticles can be used in monitoring food quality by detection of aflatoxins, OTA, as well as mycotoxin producers to provide safe and healthy food and improve food security regulations. Nanoparticles conjugated with antibodies, peptides, enzymes, or oligonucleotides are devised into sensitive detection sensors or lateral flow formats in diagnostic approaches (Baptista et al. 2005; Gao et al. 2009; Piro et al. 2016). Improved detection of toxic secondary fungal metabolites in food safety is of critical importance for human health and food trade.

15.2 Ochratoxin and Aflatoxin Detection

15.2.1 Spectroscopic Methods

Fluorescence spectrophotometry is used for recording the fluorescence emitted by ochratoxin and aflatoxins. The presence of aflatoxins and ochratoxin in grains and raw peanut was determined with this method. However, since crude samples show interference in fluorescence, may require derivatization, and often have a high LOD, these methods are limited to the initial screening for presence of toxins in food and feed samples (Wacoo et al. 2014). Fourier transform spectroscopy identifies organic compounds on basis of their specific infrared spectrum. FTIR analyses was used to identify the presence of aflatoxins in peanuts, peanut cake, and single corn kernels. Mirghani et al. (2001) used transmittance and reflectance spectroscopy to detect aflatoxin in maize kernels that were grouped with high (>100 ppb) or low (<10 ppb) aflatoxin concentrations.

15.2.1.1 Thin-Layer Chromatography

Thin-layer chromatography (TLC) is simple, robust, and qualitative method used for mycotoxin detection and can be optimized for their rapid screening. The samples are loaded on TLC plates, separated chromatographically by solvent mobile phase and analyzed by UV exposure by comparison to the standard. The presence of ochratoxin A (OTA) was analyzed in different cereals and feeds using TLC. Boudra et al. (1995) extracted OTA from *Aspergillus ochraceus* contaminated wheat using acidified chloroform and analyzed it by bidirectional TLC. Samples were loaded on silica gel aluminum sheet activated for 1 h at 100°C and developed first with mobile phase of anhydrous diethyl ether and later with toluene/ethyl acetate/formic acid (6:3:1) and exposed to UV for OTA detection (Boudra et al. 1995). Production of OTA by *Aspergillus ochraceus* on feed-grade wheat was confirmed by chloroform extraction followed by TLC detection using ethyl acetate and acetic acid for separation (Xiao et al. 1996). Pittet and Royer (2002) reported a rapid, low-cost method for detection of OTA in green coffee using bidirectional TLC with the detection limit of 10 μg/kg. Two varieties of green coffee obtained from Thailand, India, Kenya, Uganda, Salvador, and Guatemala were extracted using dichloromethane containing 0.1 M phosphoric acid and qualitatively screened using bidirectional TLC method that gave 98% OTA recovery.

A reliable, rapid, and inexpensive method to screen cereals for total and individual B1, B2, G1, G2 aflatoxins (AFs), as well as OTA, was developed by Braicu et al. (2008). After chloroform extraction, 43 cereal samples (wheat, maize, rye, and triticale) were separated by TLC with toluene/ethyl acetate/acetic acid and quantified using densitometry. Among them, 25 samples (58.14%) were contaminated with different mycotoxins: total aflatoxin, 11.2–10.8 mg/kg; individual AB1, AB2, AG1, and AG2, ranged from 0.89 to 5.7 mg/g; and ochratoxin A, 4.3–30 mg/kg. Wheat samples (62.5%) showed the highest contamination. Kushiro et al. (2017) carried out TLC-based rapid analysis of AFs, in presence of dichlorvos, an inhibitor of aflatoxin biosynthesis, using toluene/ethyl acetate/acetic acid (60:30:4) as a mobile phase. Inhibition of AF production was observed in aflatoxigenic *A. oryzae*, *A. parasiticus*, and *A. flavus* strains.

TLC determination of aflatoxin B1, B2, G1, and G2 was carried out for a total of 2668 brown rice, white rice, broken rice, Sella rice, and parboiled rice samples from Pakistan during 2006–2010 (Nisa et al. 2014). Anhydrous ether, chloroform, and acetone were used as the mobile phase. AFB1 contamination in these samples ranged from 22% to 39%, and AFG1 was low (3.6%), while AFG2 was absent in all samples. Aflatoxin B2 was found in 33% white rice, 23% brown rice, and 3.03% broken rice samples. In Ghana, maize kernels and groundnut seeds were studied for aflatoxin contamination, and the distribution of aflatoxin-producing potential of *Aspergillus* species were associated with both crops (Agbetiameh et al. 2018). Out of 326 maize and 183 groundnut samples, aflatoxin level exceeded the threshold limits of 15 and 20 ppb, respectively, set by the Ghana Standards Authority for 15% of maize and 11% of groundnut samples.

15.2.1.2 High-Performance Thin-Layer Chromatography

High-performance thin-layer chromatography (HPTLC) is an automated and sophisticated form of TLC. The efficiency of HPTLC is much enhanced due to the automated sampler, thinly layered sorbent HPLTC plates, and short migration distance in a controlled development chamber. The qualitative and quantitative scanning is recorded by an advanced densitometer. The semi-qualitative method is easy to use, robust and utilized for preliminary screening of food and feed for mycotoxin contamination.

Skarkova and Ostry (2000) employed HPTLC for the confirmation of aflatoxin M1 level in human urine samples using HPTLC after cleanup on immunoaffinity columns, with the mobile phase comprising of chloroform/acetone/2-propanol (85:10:5). The limit of quantification (LOQ) of aflatoxin M1 in urine was 5 ng/L. Twofold enhancement of the sensitivity of the HPTLC method was achieved by immersion of the chromatographic plate in a solution of paraffin oil in n-hexane.

OTA in wines was analyzed with HPTLC and HPLC after a dispersive liquid–liquid microextraction (Antep and Merdivan 2012). A linearity of 0.03–1.00 μg/L was established for HPTLC with a high correlation with the HPLC method. Additionally, the HPTLC method achieved the simultaneous analysis of different wine samples and standard of OTA on the same plate.

Welke et al. (2010) optimized the HPLTC method for wine that showed a mean recovery of 90.4% with a limit of quantification (LOQ) and limit of detection (LOD) of 0.1 μg/L and 0.016 μg/L, respectively. Quantification of OTA in 34 Brazilian red wine samples demonstrated that one sample exceeded (4.5 μg/L) the 2 μg/L limit of the Scientific Commission of the European Communities.

Kupski and Badiale-Furlong (2015) standardized the HPTLC method using hexane, ethyl acetate, and acetic acid (18:4:1.5) as mobile phase. The typical fluorescent spots of mycotoxin were observed under UV light plate after the development of plate with aluminum chloride in 15% methanol. The addition of dried magnesium sulfate and sodium chloride as salts to the extraction solvent acetonitrile/water (2:1) resulted in improved recovery of the toxin. The developed method could detect contamination of OTA from 0.22 to 0.85 μg/kg in 20 wheat flour samples.

Aflatoxin M1 (AFM1) levels in dairy products of South Korean origin were determined with HPTLC after immunoaffinity column purification (Yoon et al. 2016). The LOQ for milk, yoghurt, and cheese was 0.003, 0.07, and 0.05 μg/kg, respectively. Among 224 samples, the AFM1 contamination ranged from 0.001 to 0.1 μg/kg for milk and 0.015 to 0.136 μg/kg for yoghurt and cheese. The dairy products were safe for consumption according to the maximum limit of 0.5 μg/kg set by Korea Food and Drug Administration.

Contamination of livestock feed causes detrimental effects in both animal and human beings (Kotinagu et al. 2015). Analysis of 97 livestock feed and feed ingredient samples for aflatoxin B1 contamination showed that 33% of livestock feed and 24.5% of feed ingredients were positive.

15.2.1.3 High-Performance Liquid Chromatography

High-performance liquid chromatography (HPLC) allows detection and identification of aflatoxin and ochratoxin from food samples by recording the retention time through sensitive fluorescent, ultraviolet, or diode array detectors and mass spectroscopy (MS). In practice, the HPLC technique employs a chromatography column such as C-18 as the stationary phase, a mobile phase that moves through the column and a detector that displays the retention times of the separating molecules. Often, HPLC analysis for mycotoxin requires immunoaffinity column purification to eliminate other contaminants.

Commonly, the reversed phase HPLC method is used for separation and determination of aflatoxins. HPLC detection of aflatoxins with a sensitivity of 0.1 ng/kg using FLD (fluorescent detector) has been reported (Herzallah 2009). Occasionally, chemical derivatization using acid and halogens was employed to enhance the sensitivity of aflatoxins B1 and G1 detection when their natural fluorescence was not sufficient to reach the required detection limit (Kok et al. 1986). Recently, HPLC coupled to mass spectroscopy was able to overcome the challenges associated with derivatization processes in aflatoxins analysis. The HPLC–MS/MS uses low sample volumes to generate structural information (Rahmani et al. 2009). However, HPLC–MS/MS is a bulky, expensive equipment and requires skilled personnel for operation.

Aboul Enein et al. (2002) reported the use of HPLC-FLD for the determination of OTA from wheat, corn, red pepper, cheese, and wine. The retention time was 11.7 min using a mobile phase consisting of acetonitrile/water/acetic acid (99:99:2, v/v/v). The LOD and LOQ was 0.1 ng/ml and 3.3 ng/ml, respectively. Durguti et al. (2014) analyzed 54 wine samples by HPLC-FLD after sample cleanup using immunoaffinity columns. The OTA levels were below the EU limit of 2 ng/ml, and the LOD and LOQ were 0.05 and 0.1 ng/ml, respectively.

Infant formulas are an important food source for infants during the early stages. Ochratoxin A levels were analyzed from 150 samples of infant foods (50 infant formulas, 50 follow-on formulas, and 50 cereal-based supplementary foods for infants and children) from various supermarkets and pharmacies from Istanbul (Hampikyan et al. 2015). Among these, 52 (34.7%) analyzed samples were contaminated with OTA, though none exceeded the Turkish Food Codex maximum limit of OTA (0.5 μg/kg) in baby, infant, and young children foods.

Meuccia et al. (2010) surveyed the presence aflatoxin M1 (AFM1) and OTA in 14 leading brands of infant formulas marketed in Italy. Mycotoxin levels were determined by immunoaffinity column cleanup and HPLC-FLD. Among 185 samples, OTA was detected in 133 (72%) samples (range 35.1–689.5 ng/L). Ready-to-use preparations and powdered samples displayed 80% and 63% OTA contamination, respectively. The aflatoxin AFM1 was detected in two samples at levels below the European legislation limit of 25 ng/L.

Aflatoxin-contaminated livestock feeds are a potential health hazard as they can move up the food chain (Eun-mee et al. 2006). Screening of 249 feed samples collected in Korea by ELISA showed that ~10% contained AFB1. Among these, HPLC-FLD and LC/MS analysis confirmed that only one sample contained 11 ppb AFB1, while other samples did not contain AFB1. It was suggested that screening by ELISA should be followed by HPLC-FLD analysis for rapid and accurate detection of AFB1.

Yazdanpanah et al. (2013) analyzed AFB1 in rice, bread, puffed corn snack, wheat flour, and peanut samples with a HPLC method and obtained an average recovery of 94.4–100%. The LOD was 0.01 ng/g. Among 90 samples collected from Tehran retail market in June 2005, the bread and wheat flour samples were not contaminated, while rice, puffed corn snack, and peanut samples showed AFB1 level below 5 ng/g, the maximum tolerated level (MTL) in Iran. One rice sample and two peanuts samples tested positive for AFB1 contamination. In a study from Karaman, Turkey, aflatoxin analysis for 45 dried apricots, raisins, dried figs, nuts, peanuts, almonds, corn, red pepper, black pepper, bread, and moldy cheese samples was carried out by HPLC-FLD method after post-column derivatization (Kilicel et al. 2017). The total AF concentration in eight red pepper samples exceeded the MTL of 10 μg/kg that also exceeded the European MTL for AFB1 of 8 μg/kg. The amount of aflatoxin in other samples was negligible. AFs were present in 100% of raisins, dried figs, black pepper, red pepper, and corn, 75% of dried apricots, nuts, bread, and moldy cheese, and 50% of peanuts and almonds.

Kim et al. (2017) detected aflatoxins (B1, B2, G1, and G2) and OTA in animal feeds by a sensitive and robust HPLC-FLD method with a photochemical reaction device after immunoaffinity column purification. Aflatoxin and OTA were found in 44 of 496 samples with a concentration range of 1.76–162.69 μg/kg for AFB1 and 3.38–45.42 μg/kg for OTA for 2 years. The developed method was suitable for the routine analysis of aflatoxin in animal feed.

15.2.1.4 Liquid Chromatography Coupled Mass Spectroscopy

The technique of liquid chromatography coupled with tandem mass spectroscopy utilizing database searching has a high degree of sensitivity and accuracy that can be used in specific mycotoxin analysis. The accurate masses of protonated fungal metabolite molecule ions are obtained by optimizing the electrospray conditions, retention times, and UV spectra. Sulyok et al. (2006) used a single extraction step, without any cleanup step, followed by liquid chromatography with electrospray ionization triple quadrupole mass spectrometry (LC/ESI-MS/MS). They developed a validated method for the determination of 39 different mycotoxins in wheat and maize that included trichothecenes, zearalenone, fumonisins, ergot alkaloids, ochratoxins, aflatoxins, moniliformin, etc. The multianalyte analyses was carried out with positive as well as the negative ion ESI mode in two consecutive runs to accommodate the diverse mycotoxins. The limits of detection ranged from 0.03 to 220 mg/kg.

Jung et al. (2012) carried out the quantitative determination of aflatoxins (B1, B2, G, G2), OTA, deoxynivalenol, fumonisins, zearalenone, etc. in roasted and ground grains using LC–MS. The co-extraction of 11 mycotoxins was carried out by a double extraction using a phosphate buffer solution and methanol followed by cleanup with a multitoxin immunoaffinity column. The LODs of mycotoxins were 0.1–6.1 μg/kg, and LOQs were 0.3–18.4 μg/kg. Among 47 samples collected from Seoul, Korea, OTA was detected in 17% of samples, while aflatoxins were not detected in all samples.

LC–MS-based determination was carried out in a study on occurrence and levels of OTA in 98 infant formula (milk- and soy-based samples) and 155 infant cereal (barley-, rice-, oat-, wheat-, and mixed grain-based) products available in the US market. None of the infant formula samples tested positive, while 30% of infant cereals were contaminated with OTA in the range of 0.6–22.1 ng/g. Regulatory limits for OTA are lacking in the USA; however all positive samples were above the European Commission limits of 0.5 ng/g for OTA in baby foods. Oat-based infant cereals (59%) and mixed grain cereals (34%) showed the highest incidence and concentrations of ochratoxin contamination. Surveillance for OTA levels in grains used in infant foods can reduce exposure of infants and young children to OTA from cereal products.

The presence of aflatoxins B1, B2, G1, and G2 in four categories (oils, proteins, polysaccharides, and fatty oils) of traditional Chinese medicines was analyzed using ultra-performance liquid chromatography tandem mass spectrometry (Zhao et al. 2016). The aflatoxin concentrations ranged from 0.2 to 7.5 μg/Kg in 14 out of 22 samples. Fatty oils were most prone to contamination as compared to polysaccharides, proteins, and volatile oils.

15.2.1.5 Gas Chromatography

As other chromatographic methods, gas chromatography (GC) is based primarily on differential partitioning of analytes between the stationary phase consisting of inert particles coated with a layer of liquid in a column by the carrier gas as the mobile phase. The sample is vaporized into gaseous phase, and detection of the volatile products is carried out using either a flame ionization detector or an electron capture detector and mass spectrometer (MS). In case of aflatoxins, there is a need for derivatization in order to be detected, due to their nonvolatile nature (Scott 1995). Detection by GC-MS and an electronic nose did not find a correlation between odor and ochratoxin level, as samples with OTA levels both below and above 5 μg/Kg displayed pronounced or strong off-odors (Olsson et al. 2002).

However, due to the availability of other cheaper chromatographic methods, gas chromatography is less common in commercial analysis of mycotoxins (Wacoo et al. 2014). Besides, gas chromatography also requires a preliminary cleanup and derivatization before analysis, and it is therefore limited to analysis of a few mycotoxins, such as type A and B trichothecenes. Even in such analyses, the GC displays disadvantages of non-linearity of calibration curves, drifting responses, memory effects from previous samples, and high variation in reproducibility and repeatability.

15.2.1.6 PCR and Quantitative Real-Time PCR Detection

Mycotoxins are metabolites produced by the interaction of several biochemical pathways during later stages in fungal growth, while genes are involved in mycotoxin pathway expressed much earlier. Therefore, detection of the mycotoxin gene expression may contribute to early detection of mycotoxins. Conventionally, polymerase chain reaction (PCR)-based molecular methods are used for detection of mycotoxins producing fungi such as *Aspergillus* and *Penicillium* from food (Edwards et al. 2001). Gene-based methods reported in the literature are mainly based on unique ITS (internal transcribed spacer) regions of fungal species which are genus- and species-specific for identification purpose but do not confirm the production of toxins by the isolates. Genes of important enzymes in the mycotoxin biosynthetic pathway can be good targets for diagnostic tests as they can add to the specificity for detection of mycotoxin production and sensitivity of the test (Schmidt-Heydt et al. 2011).

Shapira et al. (1996) employed the genes coding for enzymes involved in *Aspergillus parasiticus* polyketide synthase pathway, namely, *apa-2*, *ver-1*, and *omt-1*, which are responsible for regulation, oxidation, and methylation, respectively, during aflatoxin synthesis. The presence of the fungus was detected by PCR method when 10^2 spores/g were inoculated in corn flour.

In a Korean study, multiplex PCR and high-performance liquid chromatography (HPLC) analyses were used to assess the ability of mycotoxin production in 32 *A. niger* isolates (Kim et al. 2014). Though multiplex PCR and HPLC analyses of *A. niger* isolates showed that OTA-producing strains exhibited positive PCR patterns for ochratoxin biosynthetic genes, it did not explain the presence of positive PCR products in the non-mycotoxin-producing strains.

A PCR method for differentiation and detection of OTA-producing *Penicillium* species was developed by utilizing polyketide synthase and non-ribosomal peptide synthetase genes of the ochratoxin A biosynthetic pathway (Bogs et al. 2006). An analysis of 62 strains showed that 11 PCR-positive (18%) strains produced ochratoxin A. However, some OTA-negative strains also tested PCR-positive, that was attributed to the presence of non transcribed biosynthetic gene.

Currently, ochratoxin A contamination in coffee beans was found to be associated with *Aspergillus carbonarius*, *A. niger*, and *A. ochraceus* (Sartori et al. 2006). A multiplex PCR method that detected PCR amplicons of 809, 372, and 260 bps for detection of *A. carbonarius*, *A. niger*, and *A. ochraceus* species, respectively, in coffee beans was developed.

A highly specific multiplex PCR method was developed using species-specific and mycotoxin metabolic pathway gene primers for detection of *Fusarium* and *Aspergillus* species (Sadhasivam et al. 2017). Stored wheat grain samples (34) were analyzed for the presence of mycotoxin-producing fungi by PCR and mycotoxin production by LC/MS/MS. The analyses had a strong correlation, and contamination of six samples with at least one mycotoxin, above EU regulatory limits was confirmed by both the methods.

As many of the genes involved in toxin biosynthetic pathways also contribute to other secondary metabolites, there is a lack of specificity for the PCR detection of toxin production. Detection of aflatoxin production by *Aspergillus versicolor* and *Aspergillus nidulans* is difficult as the genes involved in aflatoxin production are also involved in sterigmatocystin biosynthesis (Levin 2012). Aflatoxin and ochratoxin gene pathway provide an advantage in early monitoring of the mycotoxins producing *Aspergillus* and *Penicillium* spp. (Baptista et al. 2005). Dao et al. (2005) optimized the PCR amplification using primers toward the acyl transferase domain of a polyketide synthase gene involved in the OTA biosynthesis in OTA-producing fungi such as *A. carbonarius*, *A. melleus*, *A. sulfurous*, *A. ochraceus*, and *P. verrucosum*. Other studies indicate the role of several enzymes, such as polyketide synthase, non-ribosomal peptide synthase, halogenase, and P450 oxidase in OTA biosynthesis. Recently a multiplex PCR, based on genes of aflatoxin and sterigmatocystin biosynthesis metabolic pathways, was reported for the identification of AFT producers among *Aspergillus* spp. (Criseo et al. 2001). Such studies promote our understanding of the mechanisms of OTA production and regulation.

Quantitative real-time PCR systems may be useful for determining associations between detection of a gene at critical control points in food production and quantification of the mycotoxin contamination in the final product (Rodríguez et al. 2012). Though quantitative real-time PCR has been applied to monitor expression of mycotoxin biosynthetic genes, it is expensive.

The rapid and accurate differentiation of toxigenic and atoxigenic isolates was evaluated with PCR and real-time PCR in 22 *A. flavus* isolates from peanut kernels (Mahmoud 2015). The PCR amplification of genes did not correlate with aflatoxin production capability. Among the four aflatoxin biosynthetic pathway genes (*aflD*, *aflM*, *aflP*, and *aflQ*), the expression of *aflD* and *aflQ* was a good marker for differentiating the toxigenic from atoxigenic isolates. The real-time PCR and agar culture method for detection of *A. flavus* presence had 95% agreement. Aflatoxin production by these strains was confirmed by HPLC with 72% of isolates producing the toxin.

Rodríguez et al. (2012) proposed real-time PCR for the early detection and quantification of aflatoxin in peanut, spices, and sausages. The sensitivity and specificity of real-time quantitative PCR (qPCR)-based on SYBR Green and TaqMan were evaluated using the o-methyltransferase gene (*omt-1*) aflatoxin biosynthetic gene. Both qPCR methods gave a good linear correlation from 4 to 1 log cfu/g per reaction and a detection limit of 1–2 log cfu/g in the different food matrices tested.

The ochratoxin-producing fungus *Aspergillus carbonarius* is often associated with grapes and wine contamination (Atoui et al. 2007). Specific primers to the acyltransferase (AT) domain of polyketide synthase sequence of *A. carbonarius* amplified a 141 bp PCR product. These primers were also used for qPCR in 72 grape samples for direct quantification of the fungus. The expression of acyltransferase gene showed a positive correlation with the OTA concentration. Therefore, rapid detection of *A. carbonarius* by qPCR in grapes may offer an alternative to the traditional methods of OTA detection and culture identification.

15.2.2 Immunochemical Methods

Immunochemical techniques are based on the high affinity and specificity of antibody–antigen (Ab-Ag) binding. The binding event is recognized by signal amplification using labels such as enzymes, fluorophores, etc. Immunochemical methods such as immunosensors, immunoaffinity column assay (ICA), and enzyme-linked immunosorbent assay (ELISA) are sensitive, specific, and less labor-intensive, require less time, and allow simultaneous processing of several samples. Though radioimmunoassay technique was used for determination of aflatoxin B1 and aflatoxin M1 levels, the disadvantages such as requirement of antigen in a pure state and potential health hazards due to use of radioactive isotope have discouraged the method (Rauch et al. 1987). However, other immunochemical methods such as ELISA and immunosensors are in wide usage.

15.2.2.1 Enzyme-Linked Immunosorbent Assay (ELISA)

ELISA methods for mycotoxin detection rely on the recognition of specific mycotoxin by specific antibodies coated in microtiter plate wells. Mycotoxin capture is subsequently detected by antibodies, often labeled with horseradish peroxidase or alkaline phosphatase enzymes. The assay is rapid, simple, specific, and sensitive for the detection of mycotoxins in foods and feeds. This technology is available for more than two decades, and ELISA test kits require low sample volumes with less sample cleanup procedure as compared to conventional TLC and HPLC methods. This method has application for high throughput assays and simultaneous detection in several samples.

The ELISA technique is presently applied for detection of aflatoxins in food and feedstuffs, and a number of commercially available ELISA kits are widely used (Devi et al. 1999, 2002; Prestani et al. 2011). A commercial ELISA system and HPLC method were used to analyze the total aflatoxins in 178 foodstuff samples. The two methodologies had a high correlation for nuts, nut products, peanuts, and peanut butter samples (Azer and Cooper 1991).The presence of four types of aflatoxin (B1, B2, G1, and G2) in dairy cattle feeds and aflatoxin M1 in the milk samples was measured by HPLC and ELISA methods (Prestani et al. 2011). There was no significant difference by the two procedures, though the sensitivity and specificity of HPLC method were higher. Li et al. (2016) generated specific monoclonal antibodies that recognized G class aflatoxins, G1 and G2. A competitive indirect ELISA (CI-ELISA) was developed with a LOD of 0.06 ng/ml and low cross-reactivity toward aflatoxin B1. The CI-ELISA and HPLC methods were comparable for analysis of uncontaminated peanut samples spiked with different concentrations of aflatoxins G1 and G2.

Though antibodies are highly specific and sensitive toward the target mycotoxin, many a time, compounds with similar chemical groups also interact with the antibodies. Additionally, ELISA methods are restricted to certain food matrices for which they were validated. Sun et al. (2015) conducted an extensive study to evaluate the quality of five commercial ELISA kits from different suppliers for detecting aflatoxin B1 in 30 HPLC-verified feed samples such as corn, distillers dried grains, wheat, soybean meal, and poultry feed. The accuracy and precision of these kits were tested with positive controls. The authors reported that the qualities of five tested ELISA kits were significantly different and two kits had high false positive rates. Therefore, the complete validation for an ELISA method is critical for its application to a wide range of commodities.

ELISA technique for OTA was used for analyses of dry fruit samples from Central Iran that exhibited average concentrations of 6.7 ± 3.9 ng/g in 21% of samples (Rahimi and Shakerian 2013). Concentrations of OTA higher than the European Union maximum tolerance limit of 10 ng/g were detected in 7.9% of dried raisin samples and 2.1% of dried fig samples.

An anti-OTA monoclonal antibody-based indirect competitive ELISA was developed for OTA detection in coffee and coffee products with a detection limit of 3.73 ng/g (Fuji et al. 2007). The comparison between ELISA and HPLC methods resulted in high correlation for the green, roasted, and instant coffee samples.

A sandwich dot-ELISA method was developed for the simple, fast, and sensitive detection of OTA from contaminated food grain samples with a limit of detection of 5.0 ng/ml (Venkataramana et al. 2015). The ELISA method detected contaminations in 72 (19 maize, 38 wheat, 15 rice) of 195 samples, which was in agreement with the HPLC method. The mAb-based method was specific and did not show any cross-reactivity toward deoxynivalenol, fumonisin B1 fumonisin, or AFB1.

Similarly, Ekhtelat et al. (2018) established a good correlation between the ELISA and HPLC methods for determination of OTA levels in the medicinal herbs of *Zataria multiflora* and *Foeniculum vulgare* from Ahvaz, Iran.

A competitive direct ELISA format was developed and showed IC50 value of 0.07 ng/ml. A simple, rapid, and efficient method for extraction with 50% methanol and the ELISA method resulted in 74–110% recoveries of spiked samples. Dilution with PBS was applied for food samples such as barley, wheat, oat, corn, rice, and raisins, grape juice, and beer samples. The developed ELISA method and the HPLC method showed a good correlation.

15.2.2.2 Immunosensors

Immunosensors are based on a signal transducer which detects the binding of the antibody and antigen. The electrochemical, optical, and piezoelectric immunosensors detect changes in electrical current, mass, or optical signals (color or fluorescence), respectively, during Ab–Ag binding.

Electrochemical Immunosensors

An electrochemical immunosensor records the electroactive signals generated by the Ab–Ag interactions by differential pulse voltammetry, cyclic voltammetry, chronoamperometry, electrochemical impedance spectroscopy, or linear sweep voltammetry transducers (amplifiers). A potential difference is generated by the reaction conditions which is measured to establish the relationship between the potential difference and antigen concentration. Electrochemical immunosensors developed for aflatoxin detection mostly employ enzymes as the active biological component to generate signals.

Badea et al. (2016) developed an impedimetric immunosensor by immobilizing anti-OTA antibody on bovine serum albumin-modified gold electrodes for ochratoxin A (OTA) detection.

Modification of the impedance due to the specific antibody–antigen reaction at immunosensor surface was used in order to detect OTA. Linear proportionality of the charge transfer resistance to the OTA concentration allowed OTA detection in the range of 2.5–100 ng/ml.

Li et al. (2016) developed a portable immunosensor comprising of an impedance detector and a 3D-printed USB-compatible sensor chip for rapid and in situ detection of AFB1 in rice. The detection time was below an hour, and the limit of detection was 5 ng/ml. Such portable devices for detection offer the possibility of analyses of clinical and environmental samples.

A fast, sensitive, and efficient electrochemical immunoassay was developed by forming self-assembled monolayers by 2-aminoethanethiol on a gold electrode for

AFB1 capture (Kong et al. 2018). The noncompetitive immunoassays resulted in a significant current change at a minimum concentration of 0.01 ng/ml of AFB1. The immunosensor displayed excellent stability and sensitivity after storage for 7 days.

A simple, cost-effective, and portable antibody-modified screen-printed carbon working electrode with carbon counter was developed for direct aflatoxin M1 analysis with milk (Parker and Tothill 2009). An electrochemical detection scheme was constructed on the electrode surface in a competitive ELISA format with horseradish peroxidase (HRP) as the enzyme label. Milk samples were pre-treated with 18 mM calcium chloride to stabilize the whey proteins and eliminate the interfering signal. The sensor achieved a detection range up to 1000 ng/L and a LOD of 39 ng/L. The sensitivity of the immunosensor was comparable to the commercial ELISA kit and an a HPLC method.

Muchindu et al. (2011) produced an impedimetric immunosensor composed of a platinum disk electrode coated with polyaniline–polyvinyl sulfonate and anti-OTA antibody and calibrated it for OTA detection. When applied to the certified reference materials of corn, wheat, and roasted coffee, the sensor detected OTA concentrations of 21.1, 8.6, and 2.5 mg/kg, respectively. The immunosensor had a LOD of 10 pg/kg, and impedimetric estimation for corn was in agreement with the ELISA measurements.

Radi et al. (2009) modified an electrochemical immunosensor comprising of a screen-printed gold electrode with glutaraldehyde and tagged it with the anti-ochratoxin antibody. The competitive immunoassay between a horseradish peroxidase-labeled ochratoxin A (HRP-toxin) and ochratoxin A for antibody demonstrated a dynamic range up to 60 ng/ml and detection limit of 12 ng/ml. The binding of the antibody and HRP-toxin was measured by chronoamperometry in the presence of the substrate tetramethylbenzidine. The developed immunosensor displayed precision, accuracy, and stability.

Monoclonal antibodies were immobilized on disposable screen-printed electrodes to develop an electrochemical competitive enzyme-linked immunosorbent assay for quantitative determination of OTA (Alarcon et al. 2006). The assay had a working range from 0.05 to 2.5 and 0.1 to 7.5 g/L and detection limit of 60 and 100 g/L, respectively, in the direct and indirect assay formats. The immunosensor in the direct format was selected for the determination of OTA in wheat. Analysis of spiked and naturally contaminated wheat samples with the electrochemical assay and HPLC showed good correlation.

Optical Immunosensors

Optical immunosensors record the signals by surface plasmon resonance, optical waveguide light-mode spectroscopy, photoluminescence, etc. Surface plasmon resonance (SPR) measures the refractive index changes as a shift in the resonance angle during the Ab–Ag interactions. AFB1 detection and quantification was attempted with monoclonal as well as polyclonal antibodies using the SPR technology (Daly et al. 2000). Monoclonal antibodies encountered regeneration problems due to their high-affinity binding to the sensor surface. Immobilization of polyclonal anti-AFB1 antibodies on the sensor surface achieved a linear detection range of 3.0–98.0 ng/ml with good reproducibility. Regeneration was achieved using alkaline solution of methanolamine and acetonitrile.

Van der Gaag et al. (2003) utilized the SPR immunosensor for multiple detection of mycotoxins. Therefore, SPR immunosensors should offer label-free detection of aflatoxins if the current regeneration problems are overcome. Optical waveguide light-mode spectroscopy (OWLS) measures the resonance angle of polarized grating diffracted light, coupled into a thin waveguide. The Ab–Ag interaction on the surface of sensor chips results in multiple internal reflections within the waveguide layer which is detected by photodiodes. Aflatoxin and ochratoxin detection was carried out using OWLS in both competitive and direct immunoassays (Adanyi et al. 2007). A detection range of 0.5–10 ng/ml for aflatoxins was achieved when barley and wheat flour samples were analyzed.

Myndrul et al. (2018) fabricated a porous silicon (PSi) based photoluminescence immunosensor metal-assisted chemical etching procedure and modified by coating with protein A and binding with anti-OTA antibodies. The PL spectroscopy of PSi at room temperature resulted in an emission band at 680 ± 20 nm. The binding of OTA resulted in the photoluminescence quenching of the anti-OTA/protein A/PSi surface as compared to bare PSi. The response time of the immunosensor was in the range of 500–700 s.

Another immunosensor was developed by layering photoluminescent ZnO nanorods (ZnO-NRs) on glass substrate followed by coating for binding with protein A and binding anti-ochratoxin antibodies (glass/ZnO-NRs/protein A/anti-OTA) (Vitera et al. 2018). The immunosensor was integrated with portable fiber optic detection system and tested in a wide range of OTA concentrations, and the sensitivity and limit of detection were 0.1–1 ng/ml and 0.01 ng/ml, respectively. Response time of the immunosensor toward OTA was in the range of 500–800 s.

Piezoelectric Quartz Crystal Microbalances (QCMs)

This label-free direct detection method relies on changes in mass when the antibody immobilized on the quartz crystal surface interacts with the antigen (Janshoff et al. 2000). Spinella et al. (2013) immobilized anti-aflatoxin B1 antibody on gold-coated quartz crystals and achieved aflatoxin B1 detection in the range of 0.5–10 ppb. Antibody binding to the gold-coated surface was facilitated by 3,3'dithiodipropionic-acid-di-N hydroxysuccinimide ester. Jin et al. (2009) also detected 0.01–10.0 ng/ml aflatoxin B1 in spiked milk samples with the QCM-based sensor.

A label-free quartz crystal microbalance (QCM)-based immunosensor was developed by immobilization of OTA–bovine serum albumin conjugate on gold-coated quartz crystals (Vidal et al. 2009). An indirect competitive format was used to detect the binding of the excess of anti-OTA antibodies to the immobilized OTA that led to linear decrease in the resonant frequency at OTA concentration from 10 to 128 ng/ml, with a detection limit of 8 ng/ml. The quartz electrode was reusable after regeneration with a pepsin solution (pH 2.1).

Pirincci et al. (2018) developed OTA detection by direct immobilization of OTA to amine-bearing sensor surfaces using 1-ethyl-3-(3-dimethylaminopropyl) carbodiimide (EDC)/N-hydroxysuccinimide (NHS) chemistry. The QCM immunosensor was developed with a detection range between 17.2 and 200 ng/ml with the possible use for on-site detection of contamination in feedstuffs. The sensor was reusable up to 13 times without loss of performance after regeneration with 50 mM NaOH and 1% SDS.

15.3 Nanotechnology for Mycotoxin Detection

Nanotechnology can contribute to the detection of contaminated food and feed. Nanoparticles can be used in detection of mycotoxins and mycotoxin producers, monitoring food quality, and help in food security regulations to provide safe and healthy food. Nanoparticles, due to their size- and shape-dependent physical and chemical properties, have potential for development in various colorimetric, fluorometric enzymatic, and electrochemical diagnostic assays (Vo-Dinh 2007, Selvan et al. 2009, Gao et al. 2009, Kim and Park 2005). Semiconductors, noble metals, and metal oxide nanoparticles are used in various imaging and sensing applications (Fig. 15.1). Nanoparticles as labels can be conjugated with molecular recognition elements such as DNA, RNA, antibodies, or enzymes to offer a detection method by signal amplification (Fig. 15.1). Gold nanoparticles are often used in the lateral flow immunochromatographic devices that operate on a simple chromatographic flow separation to give a colorimetric visual signal based on the affinity of the gold nanoparticle- labeled atibodies to the antigen in either a direct or indirect format.

The early detection of mRNA expression was demonstrated in a cultured melanoma cells using nanoparticles carrying an imaging probe. The probe incorporated citrate-capped gold nanoparticles covalently bound to thiol-terminated hairpin oligonucleotides. The hairpin DNA-coated gold nanoparticles positively identified tyrosinase mRNA in melanoma cells (Harry et al. 2010). Gold nanoparticles functionalized with oligonucleotides which are complementary to unique sequences, present on the heat shock protein 70, were used for detection of the human parasite *Cryptosporidium parvum* (Javier et al. 2009). The use of oligonucleotide gold NPs for the molecular diagnosis of mycotoxins offers new opportunities for the further development of point of care diagnostic assays with low-cost, robust reagents and simple colorimetric detection (Baptista et al. 2005). Nanoparticle aided mRNA detection to identify the spoilage organism leading to early intervention by regulatory agencies. Therefore, the development of a diagnostic assay for rapid, sensitive detection of mycotoxin producers is the need for proper food safety in a country like India.

15.3.1 Nanoparticle-Assisted Electrochemical Immunoassay

Liu et al. (2013) fabricated an electrochemical sensor in the indirect competitive immunoassay format for detection of OTA by modifying a glassy carbon electrode with a composite film of polythionine and self-assembled gold nanoparticle monolayer. The sample OTA competed with the ochratoxin antibody–ovalbumin conjugates immobilized on the film for binding with the anti-ochratoxin

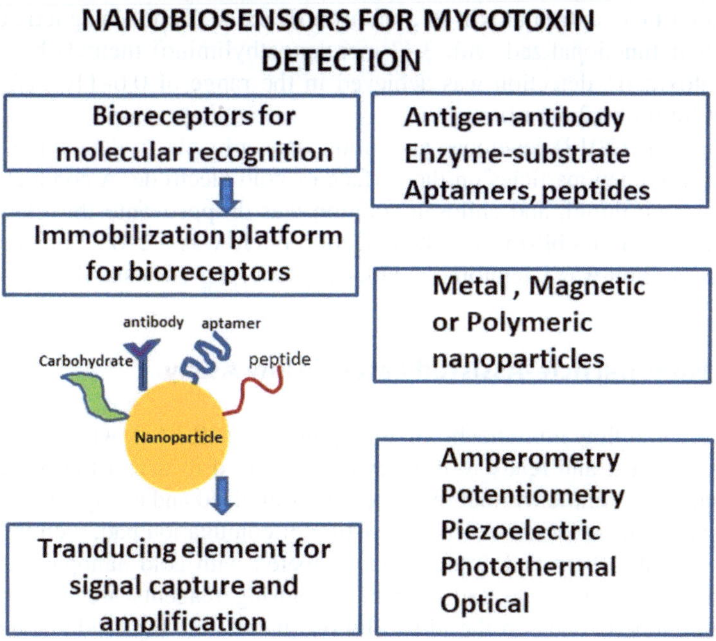

Fig. 15.1 Nanobiosensors for mycotoxin detection are devised with three main components: bioreceptors, an immobilization platform, such as nanoparticles, and a transducing element. The different variants for each component can be assembled for mycotoxin detection

monoclonal antibodies. The alkaline phosphatase-labeled secondary antibodies bound to the monoclonal antibody to give the electrochemical signal by oxidation of 1-naphthyl phosphate substrate. The electrochemical response was inversely proportional to the OTA concentration from 1 to 1000 ng/ml with a low detection limit of 0.2 ng/ml.

Rivas et al. (2015) developed an electrochemical impedance-based sensor by electropolymerizing a polythionine film on a screen-printed carbon electrode followed by the assembly of iridium oxide nanoparticles (IrO2 NPs). The aminated aptamer selective to OTA was bound by electrostatic interactions to the IrO2 NPs. Aptamers are single or double-stranded synthetic oligonucleotides that bind specifically to the target and are more advantageous as compared to antibodies due to their thermal and chemical stability and low-cost production. The sensor detected 14 pM OTA in white wine samples and exhibited high reproducibility.

Sharma et al. (2017) devised an electrochemical immunosensor by deposition of poly(3,4-ethylenedioxythiophene) functionalized gold nanoparticles and immobilization of monoclonal anti-aflatoxin antibodies. The immunosensor displayed a LOD of 0.0045 ng/ml and LOQ of 0.0156 ng/ml by amperometric measurement. Spiked maize samples were detected a high reproducibility.

A sandwich-type nonenzymatic electrochemical immunosensor was reported by Masoomi et al. (2013) by modification of glassy carbon electrodes with chitosan, gold nanoparticle, anti-aflatoxin B1, and iron III oxide (Fe_3O_4) magnetic core with a gold shell functionalized with 3-(2-mercaptoethylimino) methyl) benzene-1,2-diol. Aflatoxin B1 detection was achieved in the range of 0.6–110 ng/ml with a detection limit of 0.2 ng/ml.

Linting et al. (2012) developed an immunosensor by electrodepositing graphene oxide and gold nanoparticles on the surface of gold electrode. A conducting polymer film, ionic liquid, and chitosan solution was dropped onto this electrode for AFB1 antibody immobilization. The detection range was 3.2–0.32 picomoles, and the detection limit was 1 femtomole with excellent long-term stability.

15.3.2 Nanoparticle-Assisted Lateral Flow Assay

Recently, lateral flow immunochromatographic assays (LFAs), which are sensitive, simple, easy, fast, and ready-to-use devices, are used to detect the presence of a target analyte in sample without the need for specialized and costly equipment. The method uses a low-cost test device consisting of conjugation pad, membrane, sample pad, and absorbent pad. The antibody labeled with gold nanoparticles is pre-adsorbed on the conjugate pad, and the capturing reagents are immobilized in spatially confined zones on the nitrocellulose membrane (Quesada-Gonzalez and Merkoçi 2015). The detection is based upon either the competitive or direct formats (Moon et al. 2013). The high sensitivity and specificity of antibody–antigen reactions and the colorimetric visibility of the gold nanoparticles as labels are the basis for the rapid detection of analytes, as a red band on the membrane (Fig. 15.2).

An immunochromatographic strip test was developed and optimized for the rapid detection of aflatoxin B1 (AFB1) by nanocolloidal gold coupled with monoclonal antibody for AFB1 (Won-bo et al. 2007). The antibody showed cross-reactivity toward aflatoxin B2, G1, and G2. The test took 15 mins and the visual detection limit 0.5 ng/ml. Analyses of 172 grain samples by the immunochromatographic strip test was in agreement with the HPLC estimation. Gold nanoparticle coated with monoclonal antibodies was used as a detection agent for aflatoxins in soybean, a major Brazilian agricultural commodity (Santos et al. 2017). The monoclonal antibody displayed specificity toward aflatoxins B1, M1, G1, G2, and B2, and the strip test detected up to 0.5 μg/kg of aflatoxins in 10 min. Similarly Delmulle et al. (2005) developed a rapid lateral flow device for detecting aflatoxin B1 in pig feed rapidly within 10 min. Though most LFA platforms employ gold nanoparticles, recently Ren et al. (2014) employed quantum dots or semiconductor nanoparticles that show robust fluorescence as detection labels. Fluorescent CdSe/ZnS quantum dot beads (QBs) were synthesized for ultrasensitive detection of AFB1 in maize by polymer encapsulation of quantum dots (QDs) (Ren et al. 2014). The surface blocking method was used to prevent non-specific binding to the lateral flow membrane. The developed sensor had an LOD of 0.42 pg/ml in maize extract and twofold higher sensitivity than a gold nanoparticle-based immunoassay. This method compared favorably with the commercial ELISA and liquid chromatography tandem mass spectrometry measurements.

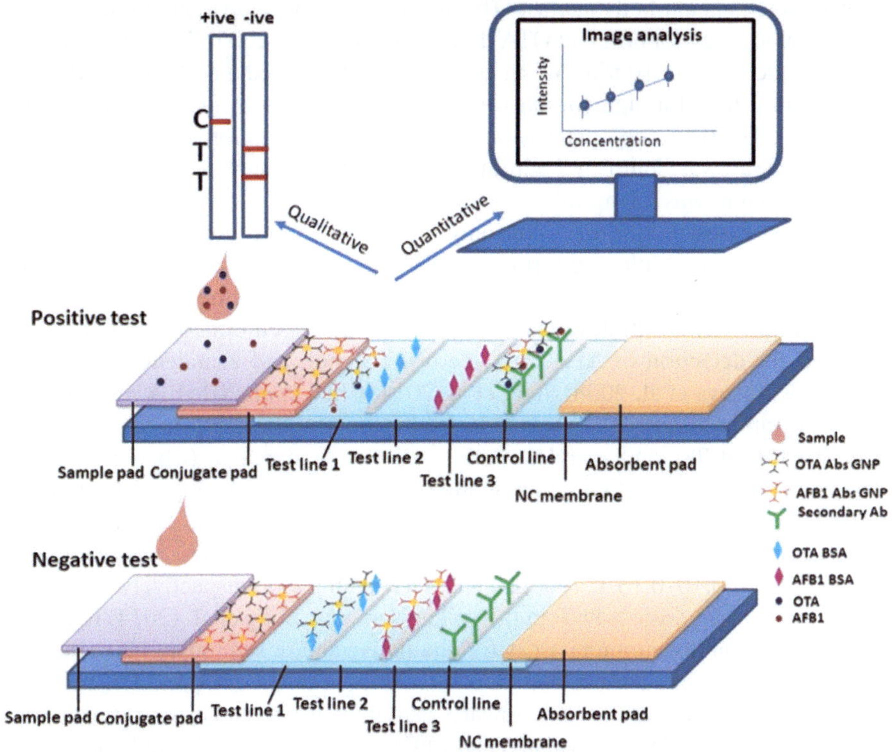

Fig. 15.2 Multitoxin detection by lateral flow assay. Aflatoxin and ochratoxin A can be simultaneously detected in the LFA visually and quantitatively. In the competitive format, the presence of red color bands due to gold nanoparticle on the control line indicates the presence of toxins. The presence of red bands on the test line indicates the absence of toxins. The color intensity can be compared to the standards for quantification

Anfossi et al. (2012) detected OTA semiquantitatively in wines and grape musts by a one-step lateral flow immunoassay. Matrix-matched calibration curves carried out in blank wines showed a detection limit of 1 μg/L and IC50 of 3.2 μg/L. The developed assay could detect OTA accurately and sensitively in 5 min. The assay was in agreement with the reference method of HPLC with 38 wines and 16 musts. A rapid immunochromatographic assay was combined with a simple, sample treatment of cereals for the quantitative determination of OTA with LOD as low as 1.5 μg/kg. Ochratoxin A was extracted from the cereal samples in presence of polyethylene glycol that reduced the matrix effects caused by different cereals (maize, wheat, and durum wheat) (Anfossi et al. 2011). The recoveries ranged from 87% to 119%. Fifteen maize, four wheat, and six durum wheat samples were analyzed by the developed assay that showed a good correlation when compared with a reference method.

A monoclonal antibody against OTA was raised to develop a rapid immunochromatographic assay for efficient OTA detection (Cho et al. 2005). Up to 500 ng/ml of OTA was detected in 10 min. Wang et al. (2007) developed a colloidal gold immunoassay in a flow-through format for the rapid detection of OTA in various food matrices. OTA–BSA conjugates were used to produce polyclonal antibodies in rabbits. The assay was completed within 10 min, and the visual detection limit of the developed assay was 1.0 ng/ml.

In most cases routine LFA can only detect one target molecule at a time. By comparison, the multiplexing format of LFA can simultaneously detect several target chemicals, thus it can further reduce operating cost and improve detection efficiency (Fig. 15.2). Multiplexed antibody-based immunochromatographic LFAs are reported for detection of AFs, OTA, deoxynivalenol, and zearalenone in various studies (Kolosova et al. 2008; Song et al. 2014; Chen et al. 2016). A multiplex lateral flow immunoassay (LFA) was reported for the simultaneous, on-site determination of AFB1, OTA, and zearalenone in corn, rice, and peanut (Chen et al. 2016). The size of gold nanoparticles, conjugation of antibody-gold nanoparticle, and location of capture antigen on the assay strip were optimized for the assay. The developed LFA visually detected 10, 15, and 50 μg/kg of AFB1, OTA, and zearalenone with a LOD of 0.10–0.13, 0.19–0.24, and 0.42–0.46 μg/kg, respectively. The recovery of mycotoxins were > 86% in the spiked samples. Song et al. (2014) reported a multiplex lateral flow immunoassay for qualitative and/or semiquantitative determination of AFB1, zearalenone, and deoxynivalenol in maize samples. The monoclonal antibodies used were specific toward each mycotoxin, and cross-reactivity was not reported. The LFA strip had a visual LOD of 0.03, 1.6, and 10 μg/kg and calculated LOD of 0.05, 1, and 3 μg/kg for AFB1, zearalenone, and deoxynivalenol, respectively. The mycotoxin recoveries ranged from 80% to 122%. The multiplex LFA and LC – MS/MS analyses of naturally contaminated maize samples were in a good agreement.

15.3.3 Aptamer-/Peptide-Coupled Nanoparticles for Mycotoxin Detection

Aptamers and peptides can enhance the sensitivity of the sensors (Piro et al. 2016). Aptamers are interesting short strand single or double oligonucleotides that are being incorporated into various assay formats to assist the mycotoxin detection. The aptamer-target recognition and the aptamer-controlled growth of gold nanoparticles (Au NPs) were used to develop a versatile, sensitive and visual colorimetric assay for rapid detection of OTA. The aptamer coverage on the Au NPs determined the development of the varied nanostructure morphologies that resulted in formation of different colored solutions. A red-colored solution was associated with spherical Au NPs with low aptamer coverage, whereas blue colored solutions produced branched Au NPs with high aptamer coverage. Thus, Soh et al. (2015) achieved visible colorimetric response for OTA detection at nanomolar levels (1 nM) for red wine samples, as well as cocaine. The sensitive and specific visual detection of contaminants without the need for sophisticated equipment makes such assays relevant for diagnostics and food sampling.

Zhang et al. (2018) described a competitive lateral flow strip format containing a fluorescent aptamer for the one-step determination of OTA in corn samples. In short, biotin-cDNA was immobilized on the test line on the nitrocellulose membrane. In the absence of OTA, the Cy5-labeled aptamer combined with the cDNA to form a stable double helix. The presence of OTA led to the formation of Cy5-aptamer/OTA complexes, therefore reducing the capture of free aptamer in the test zone and subsequently decreasing the fluorescent signals on the test line. A linear relationship from 1 to 1000 ng/ml with the LOD of 0.40 ng/ml and recoveries from 96.4% to 104.67% were reported for spiked corn samples.

Velu and DeRosa (2018) explored two different formats using 5′-biotin-modified OTA aptamer probes in combination with silver or gold nanoparticles in lateral flow colorimetric assays for the detection of OTA. First, in the "adsorption–desorption" approach, aptamers were adsorbed onto the metal nanoparticle surface and, in addition of OTA, lead to aptamer–ochratoxin binding, thereby releasing the NPs. A detection limit of 6.3 nM was achieved for both metal nanoparticles with this approach. The second approach involved a linkage inversion assembled nano-aptasensors (LIANAs) using a DNA linker containing a 5′-5′ linkage inversion (5′-5′ linker) to assemble biotinylated aptamer-functionalized metal nanoparticles. Briefly, OTA bound specifically with its aptamer triggered the release of the linker and disassembly of LIANA aggregates into dispersed nanoparticles. A LOD of 0.63 nM was achieved in the LFA format. A comparison of the LIANA-based LFA strips and the "adsorption–desorption" LFAs showed that the former were more sensitive.

On site rapid detection of OTA contamination in *Astragalus membranaceus*, used in Chinese medicine, was developed with a lateral flow strip in the competitive format using aptamers against OTA (Zhou et al. 2016). The sample extraction was optimized with 2.5 ml of methanol/water (80:20, v/v) for 1 g, followed by fourfold dilution with running buffer to eliminate the matrix and methanol interferences. A visual LOD of 1 ng/ml with no significant cross-reactivity with other similar toxins was achieved within 15 min. Analyses of *A. membranaceus* samples showed one OTA-positive sample among nine that were in agreement with the LC–MS/MS analysis.

In place of antibodies, which are expensive and unstable, peptides can also be used for recognizing ochratoxin in the LFA format. Ochratoxin-binding peptides were identified (Bazin et al. 2013; Giraudi et al. 2007). Using such peptides in lateral flow assay technology could provide a promising approach for semiquantitative, rapid, easy, and cost-effective mycotoxin detection. In an approach to develop a user-friendly lateral flow strip and avoid the direct use of ochratoxin in the assay due to its hazardous nature, a mimotope peptide to mimic OTA was screened from a random seven-peptide M13 phage-display library (Lai et al. 2009). The user-friendly lateral flow strip assay was developed by coating the mimotope peptide on the test line, replacing the ochratoxin–BSA conjugates, for capture of the gold nanoparticle-labeled anti-ochratoxin antibodies. The rapid, inexpensive, on-site ochratoxin testing LFA could detect up to 10 ppb ochratoxin in 10 min.

15.3.4 Nanoparticle-Aided Molecular Detection

The use of oligonucleotide gold nanoparticles for the molecular diagnosis of mycotoxins offers new opportunities for the further development of point of care diagnostic assays with low-cost robust reagents and simple colorimetric detection (Rosi and Mirkin 2005, Harry et al. 2010). The early detection of mycotoxins was correlated with mRNA expression of toxin gene; the detection of these mRNA by nanoparticles carrying oligonucleotide probe could contribute to rapid detection for mycotoxin (Baptista et al. 2005; Schmidt-Heydt et al. 2011). Gold nanoparticles can be conjugated to the biomolecules such as DNA and RNA to develop rapid diagnostic methods.

A qPCR aptasensor for sensitive detection of AFT (M1) was used through a strong interaction with biotin–streptavidin as a molecular recognition element (Guo and Wei 2005). The aptamer was used as a molecular recognition element for the complementary ssDNA. Real-time amplification showed a linear relationship from 0.0001 to 1 μg/L with an LOD of 0.03 ng/L.

15.4 Conclusions

Among the mycotoxin detection methods, traditional methods like HPLC–FLD and LC–MS are the gold standards. However, these methods lack field-level operational portability and ability for screening of large number of samples simultaneously, which is a challenge for mycotoxin detection in food and feed samples. The HACCP concept requires detection at each level to ensure safe and healthy food and feed. There is still a lack of worldwide awareness toward the health problems caused by mycotoxin consumption.

European Countries have better awareness with several food alert notifications on OTA occurrence being issued over the time. In 2008, 20 alerts concerning OTA in various cereal products were issued by the Rapid Alert System of Food and Feed of the European Union. In October 2014, notification of high OTA levels (14 μg/Kg) was issued in whole emmer wheat pasta from Italy that was distributed to other European countries. The Panel on Contaminants in the Food Chain (CONTAM) of the European Food Safety Authority (EFSA) has specified that consumers like infants and children are vulnerable and have specified a tolerable weekly intake (TWI) of 120 ng/Kg body weight for OTA.

It would be highly desirable if the same kind of alertness existed worldwide; however the sheer logistics is a great challenge considering the quantum of food and feed sources. In order to meet these challenges, portable screening methods like electrochemical sensors and LFA can be utilized. Newer strategies, involving nanotechnology, harness nanomaterials to increase the sensitivity and decrease the limit of detection. The use of simple, stable nanoparticle labels to enhance signal amplification will contribute in fabrication of sensitive biosensors. One of the main challenges for detection of mycotoxin contamination is the co-occurrence of two or more toxins in food and feed samples. The incorporation of novel recognition elements such as antibodies, peptides, or aptamers with nanoparticles for LFA and immunosensors has immense potential for simultaneous multitoxin analysis.

Acknowledgments VG and PC thank the Department of Biotechnology and Department of Science and Technology, Government of India for the funding (BT/PR10455/PFN/20/869/2013 & DST/INT/MECICO/P-06/2016). SR thanks the Department of Science and Technology for Junior Research Fellowship under INSPIRE program.

References

Aboul Enein HY, Kutluk OB, Altiokka G, Tuncel M (2002) A modified HPLC method for the determination of ochratoxin A by fluorescence detection. Biomed Chromatogr 16:470–474

Adanyi N, Levkovets IA, Rodriguez-Gil S, Ronald A, Varadi M, Szendro I (2007) Development of immunosensor based on OWLS technique for determining Aflatoxin B1 and Ochratoxin A. Biosens Bioelectron 22:797–802

Agbetiameh D, Ortega-Beltran A, Awuah RT, Atehnkeng J, Cotty PJ, Bandyopadhyay R (2018) Prevalence of aflatoxin contamination in maize and groundnut in ghana: population structure, distribution, and toxigenicity of the causal agents. Plant Dis 102:764–772

Alarcon SG, Palleschi G, Compagnonec D, Pascale M, Visconti A, Barna-Vetron I (2006) Monoclonal antibody based electrochemical immunosensor for the determination of ochratoxin A in wheat. Talanta 69:1031–1037

Alshannaq A, Yu JH (2017) Occurrence, toxicity and analysis of major mycotoxins in food. Int J Environ Res Public Health 14:632

Anfossi L, D'Arco G, Baggiani C, Giovannoli C, Giraudi G (2011) A lateral flow immunoassay for measuring ochratoxin A: Development of a single system for maize, wheat and durum wheat. Food Cont 22:1965–1970

Anfossi L, Giovannoli C, Giraudi G, Biagioli F, Passini C, Baggiani C (2012) A lateral flow immunoassay for the rapid detection of ochratoxin A in wine and grape must. J Agric Food Chem 60:11491–11497

Antep HM, Merdivan M (2012) Determination of ochratoxin A in grape wines after dispersive liquid–liquid microextraction using high performance thin layer and liquid chromatography–fluorescence detection. J Biol Chem 40:155–163

Atoui A, Mathieu F, Lebrihi A (2007) Targeting a polyketide synthase gene for Aspergillus carbonarius quantification and ochratoxin A assessment in grapes using real-time PCR. Int J Food Microbiol 115:313–318

Azer M, Cooper C (1991) Determination of aflatoxins in foods using HPLC and a commercial ELISA system. J Food Protect 54:291–294

Badea M, Floroian L, Restani P, Codruta S, Cobzac A, Moga M (2016) Ochratoxin A detection on antibody immobilized on bsa-functionalized gold electrodes. PLoS One 11:e0160021

Baptista P, Doria GA, Henriques D, Pereira E, Franco R (2005) Colorimetric detection of eukaryotic gene expression with DNA-derivatized gold nanoparticles. J Biotechnol 119:111–117

Bazin, I , Andreotti N, IbnHadjHassine A, DeWaard M, Sabatier JM, Gonzalez C (2013) Peptide binding to ochratoxin A mycotoxin: a new approach in conception of biosensors Biosens Bioelectron 40:240–246

Binder EM (2007) Managing the risk of mycotoxins in modern feed. Anim Feed Sci Tech 133:149–166

Bogs C, Battilani P, Geisen R (2006) Development of a molecular detection and differentiation system for ochratoxin A producing *Penicillium* species and its application to analyse the occurrence of *Penicillium nordicum* in cured meats. Int J Food Microbiol 107:39–47

Boudra H, Le Bars P, Le Bars J (1995) Thermostability of ochratoxin a in wheat under two moisture conditions. Appl Environ Microbiol 61:1156–1158

Braicu C, Puia E, Bodoki E, Socaciu C (2008) Screening and quantification of aflatoxins and ochratoxin a in different cereals cultivated in Romania using thin-layer chromatography-densitometry. J Food Qual 31:108–120

Bryden WL (2012) Mycotoxin contamination of the feed supply chain: implications for productivity and feed security. Anim Feed Sci Tech 173:134–158

Chen Y, Chen Q, Han M, Zhou J, Gong L, Niu Y, Zhang Y, He L, Zhang L (2016) Development and optimization of a multiplex lateral flow immunoassay for the simultaneous determination of three mycotoxins in corn, rice and peanut. Food Chem 213: 478–484

Cho YJ, Lee DH, Kim DO, Min WK, Bong KT, Lee GG, Seo JH (2005) Production of a monoclonal antibody against ochratoxin a and its application to immunochromatographic Assay. J Agric Food Chem 53:8447–8451

Criseo G, Bagnara A, Bisignano G (2001) Differentiation of aflatoxin-producing and non-producing strains of Aspergillus flavus group. Letts Appl Microbiol 33:291–295

Daly SG, Keating GJ, Dillon PP et al (2000) Development of surface plasmon resonance-based immunoassay for aflatoxin B1. J Agri Food Chem 48:5097–5104

Dao HP, Mathieu F, Lebrihi A (2005) Two primer pairs to detect OTA producers by PCR method. Int J Food Microbiol 104:61–67

Delmulle BS, de Saeger SMDG, Sibanda L, Barna-Vetro I, van Peteghem CH (2005) Development of an immunoassaybased lateral flow dipstick for the rapid detection of aflatoxin B1 in pig feed. J Agric Food Chem 53:3364–3368

Desalegn B, Nanayakkara S, Harada KH, Hitomi T, Chandrajith R, Karunaratne U, Abeysekera T, Koizumi A (2011) Mycotoxin detection in urine samples of patients with chronic kidney disease of uncertain etiology from Sri Lanka. Bull Environ Contam Toxicol 87:6

Devi KT, Mayo M, Reddy KLN, Delfosse P, Reddy G, Reddy SV, DVR R (1999) Production and characterization of monoclonal antibodies for aflatoxin B1. Lett Appl Microbiol 29:284–288

Devi KT, Mayo M, Hall AJ, Craufurd PQ, Wheeler TR, Waliyar F, Subrahmanyam A, Reddy K (2002) Development and application of an indirect competitive enzymelinked immunoassay for aflatoxin M1 in milk and milk-based confectionery. J Agri Food Chem 50:933–937

Durguti V, Georgieva A, Angelov A, Bajrami Z (2014) Quantitative determination of ochratoxin A in wine after the clarification and filtration. Croat J Food Sci Technol 6:79–83

Edwards SG, O'Callaghan J, Dobson ADW (2001) PCR-based detection and quantification of mycotoxigenic fungi. Mycol Res 106:1005–1025

Ekhtelat M, Badpa F, Khorasgani ZN, Azemi E (2018) High-performance Liquid Chromatography analysis of Ochratoxin A in *Zataria multiflora* and *Foeniculum vulgare* in Ahvaz (Iran). Asian J Pharmaceut 12:S523

Eun-mee H, Park HR, Hu SJ, Kwon KS, Lee H, Ha M, Kim K, Ko E, Ha S, Chun H, Chung D, Bae D (2006) Monitoring of Aflatoxin B1 in Livestock Feeds Using ELISA and HPLC. J Microbiol Biotechnol 16:643–646

Fuji S, Ono EYS, Ribeiro RMR, Assunção FGA, Takabayashi CR, Moreira de Oliveir TCR, Itano EN, Ueno Y, Kawamura O, Hirooka EY (2007) A Comparison between enzyme immunoassay and hplc for ochratoxin a detection in green, roasted and instant coffee. Braz Arch Biol Technol 50:349–359

Gao J, Gu H, Xu B (2009) Multifunctional magnetic nanoparticles: design, synthesis and biomedical applications. Accounts Chem Res 42:1097–1107

Giraudi G, Anfossi L, Baggiani C, Giovannoli C, Tozzi C (2007) Solid-phase extraction of ochratoxin A from wine based on a binding hexapeptide prepared by combinatorial synthesis. J Chromatogr A 1175:174–180

Guo P, Wei C (2005) Quantum dots for robust and simple assays using single particles in nanodevices. Nanomed Nanotechnol Biol Med 1:122–124

Ha TH (2015) Recent advances for the detection of ochratoxin A. Toxins 7:5276–5300

Hampikyan H, Bingol EB, Colak H, Cetin O, Bingol B (2015) Determination of ochratoxin a in baby foods by ELISA And HPLC. Acta Aliment 44:578–584

Harry SR, Hicks DJ, Amiri KI, Wright DW (2010) Hairpin DNA coated gold nanoparticles as intracellular mRNA probes for the detection of tyrosinase gene expression in melanoma cells. Chem Commun 46:5557–5559

Herzallah SM (2009) Determination of aflatoxins in eggs, milk, meat and meat products using HPLC fluorescent and UV detectors. Food Chem 114:1141–1146

Janshoff A, Galla HJ, Steinem C (2000) Piezoelectric mass-sensing devices as biosensors - an alternative to optical biosensors?Angew Chem. Int Ed 39:4004–4032

Javier DJ, Castellanos-Gonzalez A, Weigum SE, White AC, Richards-Kortum R (2009) Oligonucleotide-gold nanoparticle networks for detection of *Cryptosporidium parvum* heat shock protein 70 mRNA. J Clin Microbiol 47:4060–4066

Jin X, Liu X, Chen L,Jiang J, Shen G, Yu R (2009) Biocatalyzed deposition amplification for detection of aflatoxin B1 based on quartz crystal microbalance Anal Chim Acta 645:92–97

Jung S, Choe B, Shin G, Kim J, Chae Y (2012) Analysis of roasted and ground grains on the seoul (korea) market for their contaminants of aflatoxins, ochratoxin A and *Fusarium* toxins by LC-MS/MS. World Academy Sci Engineer Technol 6:12–23

Kilicel F, Karapinar HS, Cimen A (2017) Quantitation of aflatoxins in food materials using HPLC-FLD method. Sci J Analy Chem 5:90–97

Kim KS, Park JK (2005) Magnetic force-based multiplexed immunoassay using superparamagnetic nanoparticles in microfluidic channel. Lab Chip 5:657–664

Kim NY, Lee I, Ji GE (2014) Reliable and simple detection of ochratoxin and fumonisin production in black Aspergillus. J Food Protect 77:653–658

Kim HJ, Lee MJ, Kim HJ, Cho SK, Park HJ, Jeong MH (2017) Analytical method development and monitoring of aflatoxin B1, B2, G1, G2 and ochratoxin A in animal feed using HPLC with Fluorescence detector and photochemical reaction device. Cogent Food Agric 3:1419788

Kok WT, van Neer TCH, Traag WA, Tuinstra LGT (1986) Determination of aflatoxins in cattle feed by liquid chromatography and post-column derivatization with electrochemically generated bromine. J Chromat A 367:231–236

Kolosova AY, Sibanda L, Dumoulin F, Lewis J, Duveiller E, Van Peteghem C, De Saeger S (2008) Lateral-flow colloidal gold-based immunoassay for the rapid detection of deoxynivalenol with two indicator ranges. Anal Chim Acta 616:235–244

Kong Z, Wang H, Zou L, Ji Z (2018) Enhancement of aflatoxin B_1 detection using electrochemical immunoassay method and 2-aminoethanethiol. Mater Res Express 5:066414

Kotinagu K, Mohanamba T, Rathna Kumari N (2015) Assessment of aflatoxin B1 in livestock feed and feed ingredients by high-performance thin layer chromatography. Veterinary World. EISSN: 2231-0916

Kupski L, Badiale-Furlong E (2015) Principal components analysis: an innovative approach to establish interferences in ochratoxin a detection. Food Chem 177:354–360

Kushiro M, Hatabayashi H, Nakagawa H, Yabe K (2017) Improvement of mobile phase in thin-layer chromatography for aflatoxins and analysis of the effect of dichlorvos in aflatoxigenic fungi. JSM Mycotoxins 67:5–6

Lai W, Fung DYC, Xu Y, Liu R, Xiong Y (2009) Development of a colloidal gold strip for rapid detection of ochratoxin A with mimotope peptide. Food Control 20:791–795

Levin RE (2012) PCR detection of aflatoxin producing fungi and its limitations. Int J Food Microbiol 156:1–6

Li P, Zhou Q, Wang T, Zhou H, Zhang W, Ding X, Zhang Z, Chang PK, Zhang Q (2016). Development of an enzyme-linked immunosorbent assay method specific for the detection of g-group aflatoxins. Toxins 8:5. https://doi.org/10.3390/toxins8010005

Linting Z, Ruiyi L, Zaijun L, Qianfang X, Yinjun F, Junkang L (2012) An immunosensor for ultrasensitive detection of aflatoxin B1 with an enhanced electrochemical performance based on graphene/conducting polymer/gold nanoparticles/the ionic liquid composite film on modified gold electrode with electrodeposition. Sens Actuators B Chem 174:359–365

Liu X, Yang Z, Zhang Y, Yu R (2013) A novel electrochemical immunosensor for ochratoxin A with hapten immobilization on thionine/gold nanoparticle modified glassy carbon electrode. Anal Methods 5:1481–1486

Magan N, Hope R, Colleate A, Baxter ES (2002) Relationship between growth and mycotoxin production by *Fusarium* species, biocides and environment. Eur J Pl Pathol 108:685–690

Mahmoud MA (2015) Detection of *Aspergillus flavus* in stored peanuts using real-time pcr and the expression of aflatoxin genes in toxigenic and atoxigenic *A. flavus* isolates. Foodborne Pathog Dis 12:289–296

Maragos CM, Busman M (2010) Rapid and advanced tools for mycotoxin analysis: a review. Food Addit Contam 27:688–700

Masoomi L, Sadeghi O, Banitaba MH, Shahrjerdi A, Davarani SSH (2013) A non-enzymatic nano-magnetic electroimmunosensor for determination of Aflatoxin B1 as a model antigen. Sens Actuators B Chem 177:1122–1127

Meuccia V, Razzuolia E, Soldania G, Massart F (2010) Mycotoxin detection in infant formula milks in Italy. Food Addit Contam 27:64–71

Mirghani MES, Man YBC, Jinap S, Baharin BS, Bakar J (2001) A new method for determining aflatoxins in groundnut and groundnut cake using Fourier transform infrared spectroscopy with attenuated total reflectance. J Am Oil Chem Soc 78:985–992

Moon J, Kim G, Lee S (2013) Development of nanogold-based lateral flow immunoassay for the detection of ochratoxin a in buffer systems. J Nanosci Nanotechnol 13: 7245–7249

Muchindu M, Iwuoha E, Pool E, West N, Jahed N, Baker P, Waryo T, Williams A (2011) Electrochemical Ochratoxin A immunosensor system developed on sulfonated polyaniline. Electroanalysis 23:122–128

Myndrul V, Viter R, Savchuk M, Shpyrka N, Erts D, Jevdokimovs D, Silamiķelis V, Smyntyna V, Ramanavicius A, Iatsunskyi I (2018) Porous silicon based photoluminescence immunosensor for rapid and highly-sensitive detection of Ochratoxin A. Biosens Bioelectron 102:661–667

Nisa A, Zahra N, Hina S (2014) Detection of aflatoxins in rice samples. Bangladesh J Sci Ind Res 49: 189–194

Olsson J, Börjesson T, Lundstedt T, Schnürer J (2002) Detection and quantification of ochratoxin A and deoxynivalenol in barley grains by GC-MS and electronic nose. Int J Food Microbiol 72:203–214

Parker CO, Tothill IE (2009) Development of an electrochemical immunosensor for aflatoxin M1 in milk with focus on matrix interference. Biosens Bioelectron 24:2452–2457

Pirinçci SS, Ertekin O, Laguna DE,Özen FS, Öztürk ZZ , Öztürk S (2018) Label-free QCM immunosensor for the detection of ochratoxin A. Sensors 18:1161. https://doi.org/10.3390/s18041161

Piro B, Shi A, Reisberg S, Noël V, Anquetin G (2016) Comparison of electrochemical immunosensors and aptasensors for detection of small organic molecules in environment, food safety, clinical and public security. Biosensors 6:7

Pittet A, Royer D (2002) Rapid, low cost thin-layer chromatographic screening method for the detection of ochratoxin A in green coffee at a control level of 10íg/Kg. J Agric Food Chem 50:243–247

Prestani A, Tabatabaei SN, Fazeli MH, Antikchi M, Baabaei M (2011) Comparison of HPLC and Elisa for determination of aflatoxin concentration in the milk and feeds of dairy cattle. J Res Agri Sci 7:71–78

Quesada-González D, Merkoçi A (2015) Nanoparticle-based lateral flow biosensors Biosens Bioelectron. 73:47–63

Radi E, Muñoz-Berbel X, Cortina-Puig M, Marty JL (2009) An electrochemical immunosensor for ochratoxin A based on immobilization of antibodies on diazonium-functionalized gold electrode. Electrochim Acta 54:2180–2184

Rahimi E, Shakerian A (2013) Ochratoxin A in dried figs, raisings, apricots, dates on Iranian retail market. Health 5:2077–2080

Rahmani A, Jinap S, Soleimany F (2009) Qualitative and quantitative analysis of mycotoxins. Comprehen Rev Food Sci Food Saf 8:202–251

Rauch P, Fukal L, Prosek J, Brezina P, Kas J (1987) Radioimmunoassay of aflatoxin M1. J Radioanaly Nucl Chem 117:163–169

Ren M, Xu H, Huang X, Kuang M, Xiong Y, Xu H, Xu Y, Chen H, Wang A (2014) Immunochromatographic assay for ultrasensitive detection of aflatoxin b1 in maize by highly luminescent quantum dot beads. ACS Appl Mater Interfaces 6:14215–14222

Rivas L, Mayorga-Martinez CC, Quesada-Gonzalez D, Zamora-Galvez A, Escosura-Muñiz A, Merkoçi A (2015) Label-free impedimetric aptasensor for ochratoxin-A detection using iridium oxide nanoparticles. Anal Chem 87:5167–5172

Rodriguez A, Rodriguez M, Luque MI, Martin A, Cordoba JJ (2012) Real-time PCR assays for detection and quantification of aflatoxin-producing molds in foods. Food Microbiol 31:89e99

Rosi NL, Mirkin CA (2005) Nanostructures in biodiagnostics. Chem Rev 105:1547–1562

Sadhasivam S, Britzi M, Zakin V, Kostyukovsky M, Trostanetsky A, Quinn E, Ionov E (2017) Rapid detection and identification of mycotoxigenic fungi and mycotoxins in stored wheat grain. Toxins 9:302

Santos VO, Pelegrini PB, Mulinari F, Lacerda AF, Moura, RS, Cardoso LPV, Bührer-Sékula S, Miller RNG, Grossi-de-Sa MF (2017) Development and validation of a novel lateral flow immunoassay device for detection of aflatoxins in soy-based foods. Anal Methods 9:2715–2722

Sartori D, Furlaneto MC, Martins MK, de Paula MRF, Pizzirani-Kleiner AA, Taniwaki MH, Fungaro MHP (2006) PCR method for the detection of potential ochratoxin-producing *Aspergillus* species in coffee beans. Res Microbiol 157:350–354

Schmidt-Heydt M, Parra R, Geisen R, Magan N (2011) Modelling the relationship between environmental factors, transcriptional genes and deoxynivalenol mycotoxin production by strains of two *Fusarium* species. J. R. Soc Interf 8:117–126

Schneider E, Curtui V, Seidler C, Dietrich R, Usleber E, Märtlbauer E (2004) Rapid methods for deoxynivalenol and other tricothecenes. Toxicol Lett 153:113–121

Scott P (1995) Mycotoxin methodology. Food Addit Contam 12:395–403

Selvan ST, Tan TY, Yi DK, Jana NR (2009) Functional and multifunctional nanoparticles for bioimaging and biosensing. Langmuir 26:11631–11641

Shapira R, Paster N, Eyal O, Menasherov M, Mett A, Salomon R (1996) Detection of aflatoxinogenic molds in grains by PCR. Appl Environ Microbiol 62:3270–3273. PMid:8795215

Sharma A, Kumar A, Khan R (2017) Electrochemical immunosensor based on poly (3,4-ethylenedioxythiophene) modified with gold nanoparticle to detect aflatoxin B_1. Mater Sci Engineer C, Mater Biol Appl 76:802–809

Shephard GS (2008) Determination of mycotoxins in human food. Chem Soc Rev 37:2468–2477

Skarkova J, Ostry V (2000) An HPTLC method for confirmation of the presence of ultra, trace amounts of aflatoxin M1 in human urine. J Planar Chromat 13:42–49

Smith MC, Madec S, Coton E, Hymery N (2016) Natural co-occurrence of mycotoxins in foods and feeds and their *in vitro* combined toxicological effects. Toxins 8:94. https://doi.org/10.3390/toxins8040094

Soh JH, Lin Y, Rana S, Ying JY, Stevens MM (2015) Colorimetric detection of small molecules in complex matrixes via target-mediated growth of aptamer-functionalized gold nanoparticles. Anal Chem 87:7644–7652

Song S, Liu N, Zhao Z, Ediage EN, Wu S, Sun C, Saeger SD, Wu A (2014) Multiplex lateral flow immunoassay for mycotoxin determination. Anal Chem 86:4995–5001

Spinella K, Mosiello L, Palleschi G, Vitali F (2013) Development of a qcm (quartz crystal microbalance) biosensor to the detection of Aflatoxin B1. Open J Appl Biosensor 2:112–119

Streit E, Schwab C, Sulyok M, Naehrer K, Krska R, Schatzmayr G (2013) Multi-mycotoxin screening reveals the occurrence of 139 different secondary metabolites in feed and feed ingredients. Toxins 5:504–523

Sulyok M, Berthiller F, Krska R, Schuhmacher R (2006) Development and validation of a liquid chromatography/ tandem mass spectrometric method for the determination of 39 mycotoxins in wheat and maize. Rapid Commun Mass Spectrom 20:2649–2659

Sun D, Gu X, Li JG, Yao T, Dong YC (2015) Quality evaluation of five commercial enzyme linked immunosorbent assay kits for detecting Aflatoxin B1 in Feedstuffs. Asian Australas J Anim Sci 28:691–696

van der Gaag B, Spath S, Dietrich H et al (2003) Biosensors and multiple mycotoxin analysis. Food Cont 14:251–254

Velu R, DeRosa MC (2018) Lateral flow assays for Ochratoxin A using metal nanoparticles: comparison of "adsorption–desorption" approach to linkage inversion assembled nano-aptasensors LIANA. Analyst 143:4566–4574

Venkataramana M, Rashmi R, Uppalapati SR, Chandranayaka S, Balakrishna K, Radhika M, Gupta VK, Batra HV (2015) Development of sandwich dot-ELISA for specific detection of OchratoxinA and its application on to contaminated cerealgrains originating from India. Front Microbiol 6:article 511

Vidal JC, DuatoP BL, Castillo JR (2009) Use of polyclonal antibodies to ochratoxin a with a quartz – crystal microbalance for developing real-time mycotoxin piezoelectric immunosensors. Anal Bioanal Chem 394:575–582

Vitera R, Savchuk M, Iatsunskyic I, Pietralik Z, Starodub N, Shpyrka N, Ramanaviciene A, Ramanavicius A (2018) Analytical, thermodynamical and kinetic characteristics of photoluminescence immunosensor for the determination of Ochratoxin A. Biosens Bioelectron 99:237–243

Vo-Dinh T (2007). Nanotechnology for application in biology and medicine. CRC Press, Boca Raton

Wang XH, Liu T, Xu N, Zhang Y, Wang S (2007) Enzyme-linked immunosorbent assay and colloidal gold immunoassay for ochratoxin a: investigation of analytical conditions and sample matrix on assay performance. Anal Bioanal Chem 389:903–911

Won-bo S, Yang Z, Kim J, Kim JY, Kang SJ, Woo GJ, Chung YC, Eremin S, Chung DH (2007) Development of immunochromatography strip-test using nanocolloidal gold-antibody probe for the rapid detection of aflatoxin b1 in grain and feed samples. J Microbiol Biotechnol 17: 1629–1637

Wacoo AP, Wendiro D, Vuzi PC, Hawumba JF (2014) Methods for detection of aflatoxins in agricultural food crops. J Appl Chem 2014:Article ID 706291

Welke JE, Hoeltz M, Dottori HA, Noll IB (2010) Determination of ochratoxin A in wine by high-performance thin-layer chromatography using charged coupled device. J Braz Chem Soc 21:441–446

Xiao H, Marquardt RR, Abramson D, Frohlich AA (1996) Metabolites of Ochratoxins in rat urine and in a culture of *Aspergillus ochraceus*. Appl Environ Microbiol 62:648–655

Yazdanpanah H, Zarghi A, Shafaati AR, Foroutan SM, Aboul-Fathi F, Khoddam A, Nazari F, Shaki F (2013) Analysis of aflatoxin B1 in Iranian foods using hplc and a monolithic column and estimation of its dietary intake. Iranian J Pharmaceut Res 12:83–89

Yoon BR, Hong SY, Cho SM, Lee KR, Kim M, Chung SH (2016) Aflatoxin M1 levels in dairy products from South Korea determined by high performance liquid chromatography with fluorescence detection. J Food Nutrit Res 55:171–180

Zhang G, Zhu C, Huang Y, Yan J, Chen A (2018) A lateral flow strip based aptasensor for detection of ochratoxin A in corn samples. Molecules 23:291. https://doi.org/10.3390/molecules23020291

Zhao SP, Zhang D, Tan LH, Yu B, Cao WG (2016) Analysis of aflatoxins in traditional Chinese medicines: classification of analytical method on the basis of matrix variations. Sci Reports 6:30822

Zhou W, Kong W, Dou X, Zhao M, Ouyang Z, Yang M (2016) An aptamer based lateral flow strip for on-site rapid detection of ochratoxin A in *Astragalus membranaceus*. J Chromatogr B 1022:102–108

The Explosion of Brazilian Endophytic Fungal Diversity: Taxonomy and Biotechnological Potentials

16

Jadson Diogo Pereira Bezerra, Leticia Francisca da Silva, and Cristina Maria de Souza-Motta

Abstract

Brazil has the most immense biodiversity in the world, and therefore, it also harbors an incredible variety of fungal species. Over the past several decades, starting in the 1950s, Brazilian fungal diversity has been studied by several foreign groups as well as Brazilians not only because of its importance to taxonomy and ecology but also because of its potential as a source for biomolecules of economic interest. The study of endophytic fungi from Brazil has grown during the last 20–30 years, and its contribution to the estimation of global fungal diversity and its potential to produce several molecules have also contributed to the improvement of industrial procedures. Also, some studies investigating fungal endophytic diversity have contributed ecological information to programs for the protection of natural areas. In this chapter, selected publications about the taxonomy, ecology, and/or biotechnological potential of fungal endophytes from Brazil are summarized in order to show the importance of Brazilian endophyte mycodiversity.

Keywords

Bioproducts · Biotechnology · Brazilian diversity · Endophytes · Enzymes · Mycoendophytes

J. D. P. Bezerra (✉) · L. F. da Silva · C. M. de Souza-Motta
Departamento de Micologia Prof. Chaves Batista, Programa de Pós-Graduação em Biologia de Fungos (PPG-BF), CB, Universidade Federal de Pernambuco, Recife, PE, Brazil

T. Satyanarayana et al. (eds.), *Advancing Frontiers in Mycology & Mycotechnology*, https://doi.org/10.1007/978-981-13-9349-5_16

16.1 Introduction

The diversity of endophytic fungi in Brazil, if we consider its association with all plant species, may comprise an entire mycobiome still hidden in the Brazilian territory. Despite efforts to study this immense fungal diversity, there is a gap to be filled that requires more information about the association and ecological contribution of endophytes to plants and their environments.

In 2015, Maia et al. tried to estimate Brazilian fungal diversity, suggesting the occurrence of about 6000 species. These authors did not include in this estimation, among other things, the number of fungal species reported as endophytes from several environments in Brazil. Despite this, Maia et al. (2015) highlighted the importance of studies estimating fungal species in the country, demonstrating the gap of this knowledge considering the immense mycodiversity of endophytes that have been reported from several environments in Brazil (Rodrigues 1994; Camatti-Sartori et al. 2005; Magalhães et al. 2008; Abreu et al. 2010; Costa et al. 2012; Bezerra et al. 2013; Pádua et al. 2018).

One of the first studies of Brazilian fungal endophytic diversity was performed by Dreyfuss and Petrini (1984) who reported the community of endophytic fungi associated with the plants *Asplenium serratum*, *Anthurium* sp., *Philodendron* sp., *Guzmania* sp., *Epidendrum* sp., and *Maxiilaria* sp. Dreyfuss and Petrini (1984) presented a list of isolated endophytes (e.g., *Acremonium*, *Fusarium*, *Colletotrichum*, *Diaporthe*, etc.) and illustrated some of them. Other studies, such as Rodrigues (1994), Camatti-Sartori et al. (2005), Magalhães et al. (2008), Abreu et al. (2010), and Bezerra et al. (2013), used plants from different Brazilian biomes (e.g., the Amazon Forest, Atlantic Forest, Caatinga, and Cerrado) to show the diversity of these microorganisms in several habitats across the country.

Another way to verify the association of the endophytic fungal community is the use of plants that harbor fungi which can be used to produce some biomolecules with economic importance. These studies demonstrated fungal endophytes' capacity to produce several enzymes, such as xylanase, amylase, pectinase, cellulase, protease, lipase, and more recently L-asparaginase (Silva et al. 2006; Bezerra et al. 2012, 2015; Suryanarayanan et al. 2012; Santos et al. 2015a; Pádua et al. 2018). In addition, other surveys used endophytes to produce antimicrobial compounds against bacteria and/or fungi that are pathogenic to humans, other animals, and plants (Ferreira et al. 2015; Nascimento et al. 2015a; Pires et al. 2015; Silva et al. 2017a); other endophytic fungi demonstrated leishmanicidal, trypanocidal, cytotoxic, and antiviral activities (Campos et al. 2015; Ferreira et al. 2015) and were also reported as plant growth promoters (Silva et al. 2006; Luz et al. 2006). In this chapter, we aim to contribute to the knowledge of this field in Brazil in order to stimulate access to the endophytic fungal community and its biotechnological potential to produce several molecules. Because of the advance of anthropic destruction of natural environments, we also expect that protection programs for these environments can be developed to protect the Brazilian biodiversity and its biotechnological and ecological potential.

16.2 Diversity of Endophytic Fungi from Brazil

The known diversity of endophytic fungi from Brazil has increased owing to several publications investigating the mycobiome associated with several plant species (Ferreira et al. 2015; Nascimento et al. 2015a; Pádua et al. 2018; Silva et al. 2018). Since the first publication in the 1980s, the growing number of studies of Brazilian endophytes has created an expectation for the mycological community to yield new taxa and ecological patterns of species, e.g., *Corynespora subcylindrica* (Siqueira et al. 2008), *Bezerromyces* spp. and *Xiliomyces brasiliensis* (Bezerra et al. 2017a), *Toxicocladosporium* spp. (Bezerra et al. 2017b), and *Quambalaria fabacearum* (Bezerra et al. 2018).

Several studies from Brazil have demonstrated that the diversity of the Brazilian endophytic community is vastly greater than that of similar environments in other countries (Vaz et al. 2014; Oliveira et al. 2014; Bezerra et al. 2015, 2017c). However, the peculiarities associated with Brazilian plants, climate, and soil, among other characteristics, have also contributed to an explosion in the number of endophytic species, highlighting the country as a "hotspot" of fungal endophytic diversity.

For example, some studies performed with plants from the Amazon region have shown that these plants harbor a special fungal endophyte community which can also contribute to the survival of plants and its forest habitat (Rodrigues et al. 1993; Rodrigues 1994). When studying the plant *Euterpe oleracea* in areas of the Amazon Forest, Rodrigues (1994) reported the association of several endophytes belonging to 34 genera. Similar results were also obtained by Rodrigues and Dias-Filho (1996) and Rodrigues and Samuels (1999) who were studying the plants *Brachiaria brizantha* cv. Marandu, *B. humidicola* and *Spondias mombin* growing in Amazon regions. These authors published the first papers concerning endophytic fungal diversity associated with plants from the Amazon biome in Brazil. Later, other research used the diversity of plants in this region to verify its association with endophytic fungi (see Table 16.1).

Other examples are the medicinal plant species such as the endophytic fungi from leaves and stems of *Lippia sidoides* (Verbenaceae) that were isolated by Siqueira et al. (2011). A total of 203 endophytic fungi were recovered, representing 14 species belonging to *Ascomycota*. In that study, the most frequently isolated species were members of *Colletotrichum* [the most frequent], followed by *Alternaria*, *Phyllosticta*, and *Diaporthe*. Siqueira et al. (2011) also observed that nine species showed specificity for the host tissue; for example, *Curvularia pallescens*, *Drechslera dematioidea*, *Guignardia bidwellii*, *Microascus desmosporum*, *Paecilomyces variotii*, *Periconia byssoides*, and *Ulocladium oudemansii* were isolated only from leaves, while *Fusarium lateritium* and *Phoma tracheiphila* were isolated only from branches.

Because of the Brazilian biomes' diversity, new research studies have been developed using plant species from environments that are receiving serious anthropogenic pressure that can minimize their diversity and/or are considered high-stress habitats for plants that tolerate higher temperatures and water shortages. One of these examples is the Brazilian tropical dry forest, named Caatinga, which since

Table 16.1 Summary of selected studies on the diversity and/or biotechnological potential of endophytic fungi from Brazil

Plant species	Endophytic fungi (genera) isolated and/or tested	Environment and/or states	Biotechnological potential	References
Asplenium serratum *Anthurium* sp. *Philodendron* sp. *Guzmania* sp. *Epidendrum* sp. *Maxiilaria* sp.	*Acremonium, Fusarium, Hypoxylon, Nodulisporium, Phialophora, Rhizoctonia, Tubercularia, Xylaria, Colletotrichum, Cryptocline, Lasiodiplodia, Leptodothiorella, Pestalotia, Phomopsis*	Bacia Amazônica [Amazon Forest]	–	Dreyfuss and Petrini (1984)
Stylosanthes guianensis	*Alternaria, Colletotrichum, Curvularia, Drechslera, Glomerella, Guigniardia, Nigrospora, Nodulisporium, Periconia, Phomopsis, Sporormiella, Xylaria*	Animal Experiment Station of the Department of Agriculture in Nova Odessa, São Paulo and Genetics Department of the Agricultural School, São Paulo University, Piracicaba: São Paulo	–	Pereira et al. (1993)
Euterpe oleracea	*Xylaria*	River island of Combu: Floodplains of the Amazon	–	Rodrigues et al. (1993)
Baccharis coridifolia	*Ceratopycnidium*	Rio Grande do Sul	–	Bertoni (1994)
Euterpe oleracea	*Acrodictys, Acremonium, Anthostomella, Calonectria, Chloridium, Colletotrichum, Curvularia, Daldinia, Dendrodochium, Fusarium, Graphium, Hypoxylon, Idriella, Lasiodiplodia, Leiosphaerella, Letendraeopsis, Microdochium, Mycoleptodiscus, Neosartoria, Nigrospora, Nodulisporium, Oxydothis, Penzigia, Pestalotiopsis, Phoma, Phomatospora, Phomopsis, Physalacria, Physalospora, Thozetella, Trichoderma, Ustulina, Wardomyces, Xylaria*	Ilha do Combu: Estuário da Amazônia [Amazon Forest]	–	Rodrigues (1994)

Brachiaria brizantha cv. Marandu	*Acremonium*, *Colletotrichum*, *Curvularia*, *Fusarium*, *Leptosphaeria*, *Phoma*, *Phomopsis*, *Physalospora*, *Stagonospora*, *Trichoderma*	Pará: Amazonia [Amazon Forest]	–	Rodrigues and Dias-Filho (1996)
B. humidicola				
Spondias mombin	*Colletotrichum*, *Drechslera*, *Guignardia*, *Pestalotiopsis*, *Phoma*, *Phomatospora*, *Phomopsis*, *Physalospora*, *Trichoderma*, *Wiesneriomyces*, *Xylaria*	Amazon Forest: Pará and Rio de Janeiro Botanical Garden	–	Rodrigues and Samuels (1999)
Musa acuminata	*Xylaria*, *Colletotrichum*, *Cordana*, *Nigrospora*, *Epicoccum*, *Phylosticta*, *Sporomiella*, *Drechslera*, *Fusarium*, *Curvularia*, *Phomopsis*, *Periconia*, *Aspergillus*, *Alternaria*, *Humicola*, *Trichoderma*, Ascomycetes not identified, Yeasts not identified	Vale do Ribeira: São Paulo	–	Pereira et al. (1999)
Spondias mombin	*Guignardia*, *Phomopsis*, *Pestalotiopsis*	Amazon Forest: Pará and Rio de Janeiro Botanical Garden	Production of metabolites and antimicrobial activity	Rodrigues et al. (2000)
Citrus sinensis	*Colletotrichum*, *Guignardia*, *Cladosporium*	Center of Citriculture Sylvio Moreira, Agronomic Institute of Campinas: São Paulo	Antibacterial activity	Araújo et al. (2001)
C. limonia				
C. volkameriana				
C. reshni				
C. sunki				
Poncirus trifoliata				
C. sinensis × *P. trifoliate*				
C. paradisi × *C. reticulata*				
Zea mays	*Fusarium*, *Penicillium*, *Acremonium*	ESALQ/USP, Piracicaba: São Paulo	–	Pamphile and Azevedo (2002)

(continued)

Table 16.1 (continued)

Plant species	Endophytic fungi (genera) isolated and/or tested	Environment and/or states	Biotechnological potential	References
Anacardium giganteum *Myracrodruon urundeuva* *Spondias mombin* *Aspidosperma polyneuron* *Rhododendron sp.* *Bowdichia nitida* *Cassia occidentalis* *Citrus aurantium*	*Guignardia*	Park land of the Rio de Janeiro Botanical Garden, Fiocruz campus near Science Tent, Horticulture center on Fiocruz campus, Niterói city, Brazilian Amazon rain forest	–	Rodrigues et al. (2004)
Palicourea longiflora *Strychnos cogens*	*Colletotrichum*, *Guigniardia*, *Aspergillus*, *Phomopsis*, *Glomerella*, *Xylaria*, *Trichoderma* and other	Amazon Forest	Antimicrobial activity	Souza et al. (2004)
Malus domestica	*Alternaria*, *Botryosphaeria*, *Cladosporium*, *Colletotrichum*, *Epicoccum*, *Fusarium*, *Xylaria*, *Sporobolomyces*, *Sporodiobolus*, *Rhodotorula*, *Debaryomyces*, *Candida*, *Cryptococcus*, *Pichia*	Rio Grande do Sul	–	Camatti-Sartori et al. (2005)
Theobroma cacao	*Acremonium*, *Blastomyces*, *Botryosphaeria*, *Cladosporium*, *Colletotrichum*, *Cordyceps*, *Diaporthe*, *Fusarium*, *Geotrichum*, *Gibberella*, *Gliocladium*, *Lasiodiplodia*, *Monilochoetes*, *Nectria*, *Pestalotiopsis*, *Phomopsis*, *Pleurotus*, *Pseudofusarium*, *Rhizopycnis*, *Syncephalastrum*, *Trichoderma*, *Verticillium*, *Xylaria*	Orchards in nearest to Itabuna: Bahia	Antifungal activity	Rubini et al. (2005)

Palicourea marcgravii	*Xylaria*	Cerrado: São Paulo	Production of antifungal compounds (2-hexyl-3-methyl-butanodioic acid and cytochalasin D)	Cafêu et al. (2005)
Zea mays	*Beauveria*	Federal University of Paraná	Production of chitinases and proteases	Campos et al. (2005)
Vriesea gigantea	*Candida*, *Cryptococcus*, *Rhodotorula*, *Debaryomyces*	Itapuã Park (Pedreira Beach and Fora Beach), Viamão: Rio Grande do Sul	–	Landell et al. (2006)
Vriesea procera				
Tillandsia gardneri				
Passiflora edulis f. sp. *flavicarpa*	*Acremonium*, *Colletotrichum*, *Glomerella*, *Fusarium*, *Curvularia*, *Aspergillus*, *Alternaria* and other	–	Growth promotion and production of the enzyme lipase	Luz et al. (2006)
Annona squamosa	*Acremonium*, *Aspergillus*, *Chaetomium*, *Colletotrichum*, *Cylindrocladium*, *Fusarium*, *Glomerella*, *Nigrospora*, *Penicillium*, *Phomopsis* and other	Recife and Olinda: Pernambuco	Production of enzymes (lipase and protease) and growth promotion	Silva et al. (2006)
Annona muricata				
Eremanthus erythropappus	*Phomopsis*, *Xylaria*, *Dothiorella*, *Nigrospora*, *Fusarium*, *Cladosporium*, *Alternaria*, *Aspergillus* and other	BoqueirãoReserve, Ingaí: Minas Gerais	–	Magalhães et al. (2008)
Lippia sidoides	*Corynespora*	Experimental Station of the Agropecuary Search Company (IPA), Carpina: Pernambuco	–	Siqueira et al. (2008)
Baccharis dracunculifolia	*Cladosporium*, *Rizoctonia* and other	Ecological Station of the Federal University of Minas Gerais(Cerrado), Belo Horizonte: Minas Gerais.	–	Oki et al. (2009)
Citrus sinensis	*Pichia*, *Candida*, *Aureobasidium*, *Cryptococcus*, *Rhodotorula*	Catanduva, Colina, Elisiário and Novais: São Paulo	–	Gai et al. (2009)
C. reticulata				

(continued)

Table 16.1 (continued)

Plant species	Endophytic fungi (genera) isolated and/or tested	Environment and/or states	Biotechnological potential	References
Theobroma cacao	*Acremonium, Arthrinium, Aspergillus, Asteromella, Clonostachys, Colletotrichum,*	Instituto Nacional de Pesquisas da Amazônia (INPA),	Biological control	Hanada et al. (2010)
T. grandiflorum	*Coniothyrium, Curvularia, Cylindrocladium, Fusarium, Gliocladium, Lasiodiplodia, Myrothecium, Paecilomyces, Penicillium, Pestalotiopsis, Phoma, Septoria, Talaromyces, Tolypocladium, Trichoderma, Verticillium*	Manaus: Amazonas, and Almirante Cacau ltd., Itajuípe: Bahia		
Phoradendron perrottettii	*Aposphaeria, Acremonium, Alternaria, Amorphotheca, Aphanocladium, Arthrinium, Aspergillus, Beauveria, Bipolaris, Cladosporium, Clonostachys, Coleophoma, Curvularia, Cytospora, Diaporthe, Didymella, Didymosphaeria, Epicoccum, Eupenicillium, Fusarium, Geotrichum, Glomerella, Guignardia, Haematonectria, Lecanicillium, Leptosphaeria, Microsphaeropsis, Myrothecium, Monochaetinula, Monodictys, Nigrospora, Nodulisporum, Paecilomyces, Penicillium, Periconia, Phialocephala, Phialophora, Phomopsis, Phoma, Pseudocercospora, Pestalotiopsis, Paraconiothyrium, Pithomyces, Pleurophomopsis, Ramichloridium, Scolecobasidium, Sporothrix, Stagonospora, Torula, Trichoderma, Tubercularia, Verticillium*	Cerrado: Minas Gerais	–	Abreu et al. (2010)
Tapirira guianensis				
Zea mays	*Beauveria*	Canguiri Experimental Station, Federal University of Paraná	Biological control	Campos et al. (2010)
Bauhinia brevipes	*Phomopsis, Dothiorella, Acremonium, Pestalotopsis, Phoma*	Pirapitinga Ecological Station: Minas Gerais	–	Hilarino et al. (2011)

Baccharis dracunculifolia	*Penicillium, Aspergillus, Fusarium, Colletotrichum, Acremonium, Scopulariopsis, Cercospora, Cylindrocladium, Glomerella, Chaetomium, Phomopsis*	Micoteca of microbiology laboratory of UNIPAR - Campus Francisco Beltrão: Paraná	Enzyme productien (amylases, lipases and proteases)	Cuzzi et al. (2011)
Hevea brasiliensis	*Fusarium, Phomopsis, Glomerella, Myrothecium, Microsphaeropsis, Gibberella, Pestalotiopsis*	Igrapiúna: Bahia	Antagonistic effect	Rocha et al. (2011)
Psidium guajava	*Cladosporium*	São Carlos: São Paulo	Antibacterial activity	Medeiros et al. (2011)
Alibertia macrophylla *Piper aduncum*	*Xylaria, Penicillium*	Experimental Ecological Station of Mogi Guaçu, Farm Campininha, Mogi-Guaçu: São Paulo	Production of Dihydroisocoumarins, antifungal and acetylcholinesterase (AChE) inhibitory activities	Oliveira et al. (2011)
Mangifera indica *Stanhopea graveolens*	*Phyllosticta*	Pompéia and Lindóia: São Paulo	–	Glienke et al. (2011)
Gossypium	*Acremonium, Cladosporium, Colletotrichum, Curvularia, Fusarium, Glomerella, Guignardia, Lecanicillium, Nigrospora, Pestalotiopsis, Phoma, Phomopsis, Rhizopus, Rhodotorula, Talaromyces, Tritirachium, Xylaria*	Experimental field at the Department of Agronomy, Federal Rural University of Pernambuco (UFRPE), Recife: Pernambuco	–	Vieira et al. (2011)
Lippia sidoides	*Alternaria, Colletotrichum, Corynespora, Curvularia, Drechslera, Fusarium, Guignardia, Microascus, Peacilomyces, Periconia, Phoma, Phomopsis, Ulocladium*	Carpina: Pernambuco	Antibacterial and antifungal activity	Siqueira et al. (2011)

(continued)

Table 16.1 (continued)

Plant species	Endophytic fungi (genera) isolated and/or tested	Environment and/or states	Biotechnological potential	References
Spermacoce verticillata	*Aspergillus*, *Curvularia*, *Cladosporium*, *Guignardia*, *Penicillium*, *Rhinocladiella*, *Rhizomucor*, *Basidiomycota*	Recife: Pernambuco	Antibacterial activity	Conti et al. (2012)
Plectranthus barbatus	*Phomopsis*, *Colletotrichum*, *Nigrospora*, *Glomerella*, *Fusarium*, *Pestalotia*, *Trichoderma*	Unidade de Apoio à Pesquisa do Centro de Ciências e Tecnologias Agropecuárias, Universidade Estadual do Norte Fluminense Darcy Ribeiro: Rio de Janeiro	–	Mussi-Dias et al. (2012)
Vernonia condensata				
Pfaffia paniculata				
Foeniculum vulgare				
Cymbopogon citratus				
Cymbopogon nardus				
Cordia curassavica				
Maytenus ilicifolia				
Punica granatum				
Morus nigra				
Bauhinia forficata				
Sapindus saponaria	*Cochliobolus*, *Alternaria*, *Curvularia*, *Phomopsis*, *Diaporthe*, *Phoma*	Maringá: Paraná	–	García et al. (2012)
Combretum leprosum	*Fusarium*, *Hypocrea*, *Aspergillus*	Tropical dry forest (Caatinga): Brazilian semiarid region	Biological activity against cancer cell lines and human pathogenic fungi	Santos et al. (2012)
Trixis vauthieri	*Alternaria*	'RPPN Santuário do Caraça' Catas Altas: Minas Gerais	Antifungal activity	Johann et al. (2012)
Opuntia ficus-indica Mill.	*Acremonium*, *Aspergillus*, *Cladosporium*, *Fusarium*, *Monodictys*, *Nigrospora*, *Penicillium*, *Pestalotiopsis*, *Phoma*, *Phomopsis*, *Tetraploa*, *Xylaria*	Caatinga: Pernambuco	Production of enzymes (Pectinase, cellulase, xylanase, protease)	Bezerra et al. (2012)
Avicennia schaueriana	*Chloridium*, *Colletotrichum*, *Fusarium*, *Glomerella*, *Guignardia*, *Hormonema*, *Microsphaeropsis*, *Nodulisporium*, *Penicillium*, *Periconia*, *Phoma*, *Phomopsis*, *Phyllosticta*, *Preussia*, *Scopulariopsis*, *Sordaria*, *Sphaerosporium*, *Torula*, *Trichoderma*	Itamaracá Island (mangrove): Pernambuco	–	Costa et al. (2012)
Laguncularia racemosa				
Rhizophora mangle				

Solanum cernuum	*Arthrobotrys*, *Bipolaris*, *Botryosphaeria*, *Colletotrichum*, *Candida*, *Curvularia*, *Dothideomycetes*, *Edenia*, *Eutypella*, *Leptosphaeria*, *Chaetosphaeriaceae*, *Fusarium*, *Petriella*, *Phoma*, *Meyerozyma*, *Meruliaceae*, *Cercospora*, *Coprinellus*, *Oudemansiella*, *Phanerochaete*, *Phlebia*, *Polyporales*, *Coprinaceae*, *Cryptococcus*, *Diatrypella*, *Glomerella*, *Flavodon*, *Hohenbuehelia*, *Kwoniella*, *Phlebiopsis*, *Schizophyllum*, *Peniophora*, *Mucor* and other	Atlantic Forest: Minas Gerais	Antimicrobial activities	Vieira et al. (2012)
Vitis labrusca	*Aporospora*, *Aureobasidium*, *Bjerkandera*, *Colletotrichum*, *Diaporthe*, *Epicoccum*, *Flavodon*, *Fusarium*, *Guignardia*, *Lenzites*, *Paraphaeosphaeria*, *Phanerochaete*, *Phyllosticta*, *Pleurotus*, *Preussia*, *Tinctoporellus*, *Xylaria*	Salesópolis: São Paulo	Biological control	Brum et al. (2012)
Cereus jamacaru	*Acremonium*, *Aspergillus*, *Aureobasidium*, *Boeremia*, *Candida*, *Chrysonilia*, *Cladosporium*, *Cochliobolus*, *Curvularia*, *Cytospora*, *Debaryomyces*, *Fusarium*, *Gibberella*, *Guignardia*, *Nigrospora*, *Penicillium*, *Pestalotiopsis*, *Phoma*, *Phomopsis*, *Pseudocochliobolus*, *Purpureocillium*, *Redaellia*, *Sarocladium*, *Trichoderma*, *Rhodotorula*, *Sporobolomyces*, *Sterigmatomyces*, *Tritirachium*, *Cunninghamella*, *Syncephalastrum*	Caatinga: Paraíba	–	Bezerra et al. (2013)

(continued)

Table 16.1 (continued)

Plant species	Endophytic fungi (genera) isolated and/or tested	Environment and/or states	Biotechnological potential	References
Coffea arabica L.	*Rhodotorula*	Regions of Caatinga and Atlantic Forest: Pernambuco	–	Lima et al. (2013)
Vitis labrusca L. cv. Isabel				
Bauhinia guianensis	*Aspergillus*	Amazon Forest: Pará	Compound production (ergosterol, ergosterol peroxide, mevalolactone, monomethylsulochrin and trypacidin A) and antibacterial activity	Pinheiro et al. (2013)
Rhizophora mangle Avicennia schaueriana Laguncularia racemosa	*Diaporthe*, *Colletotrichum*, *Fusarium*, *Trichoderma*, *Xylaria*, *Arthothelium*, *Chrysoporthe*, *Coniothyrium*, *Coprinellus*, *Curvularia*, *Epicoccum*, *Eutipa*, *Gelasinospora*, *Lasiodiplodia*, *Neosartorya*, *Neurospora*, *Nigrospora*, *Periconia*, *Phaeoramularia*, *Phaeoseptoria*, *Fanerochaete*, *Pseudallescheria*, *Scolecobasidium*, *Valsa*, *Guignardia*, *Penicillium*, *Aspergillus*, *Alternaria*, *Botryosphaeria*, *Cylindrocladium*, *Endothia*, *Neofusicoccum*, *Pestalotiopsis*, *Pichia* and isolates representing the family *Glomerellaceae*, the class *Dothideomycetes*, the orders *Diaporthales*, *Xylariales*, *Hypocreales*, *Botryosphaeriales*, *Pleosporales*, *Trichosphaeriales*, *Pyrenulales*, and *Sordariales*	Mangroves (Cananeia and Bertioga): São Paulo	–	Sebastianes et al. (2013)
Eucalyptus benthamii Platanus orientalis	*Aspergillus*, *Trichoderma*, *Alternaria*, *Annulohypoxylon*, and others	Molecular Biology Laboratory of the Federal University of Paraná (LabMicro/UFPR)	Production of hemicellulases, xylanase, pectinase and β-glucosidase	Robl et al. (2013)
Glycine Max				
Solanum tuberosum Saccharum officinarum				

Myrceugenia ovata var. *nanophylla*	*Diaporthe*, *Colletotrichum*, *Annulohypoxylon*, *Xylaria*, *Guinardia*, *Cryptosporiopsis*, *Trametes*	Atlantic rain forest: Rio Grande do Sul	–	Vaz et al. (2014)
Eugenia neomyrtifolia				
Coffea arabica	*Aureobasidium*, *Cladosporium*, *Colletotrichum*, *Diaporthe*, *Drechslera*, *Khuskia*, *Lasiodiplodia*, *Nodulisporium*, *Pestalotiopsis*, *Phyllosticta*, *Phoma*, *Rhodotorula*, *Sarocladium*, *Xylaria*	Atlantic rain forest area: Pernambuco	–	Oliveira et al. (2014)
Syzygium cumini	*Pestalotiopsis*	Federal Institute of Education, Science and Technology of Pernambuco, Campus Barreiros: Pernambuco	Production of enzyme tannase	Sena et al. (2014)
Senna spectabilis	*Phomopsis*	Araraquara: São Paulo	Production of cytochalasins J and H, alternariol with anti-inflammatory potential and antifungal activity	Chapla et al. (2014)
Myrcia guianensis	*Pestalotiopsis, Phomopsis, Aspergillus, Xylaria, Nectria, Penicillium*, *Fusarium* and other	Amazonia, Santarém: Pará	Antimicrobial activity	Banhos et al. (2014)
Vernonia polyanthes	*Curvularia*, *Phomopsis*, *Guignardia*, *Diaporthe*, *Cochliobolus*, *Pestalotiopsis*, *Muscodor*, *Xylaria*	Itacolomi State Park, Ouro Preto: Minas Gerais	Anti-leishmanial activity	Nascimento et al. (2015b)

(continued)

Table 16.1 (continued)

Plant species	Endophytic fungi (genera) isolated and/or tested	Environment and/or states	Biotechnological potential	References
Glycine max	*Alternaria*, *Ampelomyces*, *Cercospora*, *Chaetomium*, *Cladosporium*, *Cochliobolus*, *Colletotrichum*, *Coniothyrium*, *Coprinellus*, *Curvularia*, *Diaporthe*, *Fusarium*, *Gibberella*, *Guignardia*, *Leptosphaeria*, *Leptospora*, *Macrophomina*, *Myrothecium*, *Neofusicoccum*, *Nigrospora*, *Peyronellaea*, *Pestalotiopsis*, *Phoma*, *Phomopsis*, *Saccharicola*, *Stagonosporopsis*	Experimental field Diogo Alves de Mello at the Universidade Federal de Viçosa, Viçosa: Minas Gerais	–	Fernandes et al. (2015)
Caesalpinia echinata	*Aspergillus*, *Epicoccum*, *Fusarium*, *Nectria*, *Taralomyces*, *Xylaria*	Zoo-Botanical Foundation, Belo Horizonte (FZB-BH): Minas Gerais	Leishmanicidal, trypanocidal, antimicrobial and cytotoxic activities	Campos et al. (2015)
Carapa guianensis	*Aspergilllus*, *Beltrania*, *Botryosphaeria*, *Colletotrichum*, *Diaporthe*, *Endomelanconiopsis*, *Fusarium*, *Guignardia*, *Pestalotiopsis*, *Phomopsis*, *Pilidiella*, *Trichoderma*, *Xylaria*	Amazon Forest	Antibacterial, trypanocidal and antiviral activities	Ferreira et al. (2015)
Calotropis procera	*Acremonium*, *Cercospora*, *Cladosporium*, *Colletotrichum*, *Curvularia*, *Diplodina*, *Glomerella*, *Guignardia*, *Microascus*, *Phaeoramularia*, *Rhodotorula*, *Xylaria* and other	Pernambuco	Antimicrobial activity	Nascimento et al. (2015a)
Bauhinia forficata	*Acremonium*, *Ascotricha*, *Aspergillus*, *Cochliobolus*, *Cladosporium*, *Diplococcium*, *Gibberella*, *Khuskia*, *Lasmenia*, *Myrmecridium*, *Myrothecium*, *Nodulisporium*, *Penicillium*, *Phoma*, *Phomopsis*, *Pithomyces*, *Spegazzinia*, *Talaromyces*, *Trichoderma*	Didatic Garden of the Center of Biological Sciences, Federal University of Pernambuco, Recife: Pernambuco	Production of enzymes (cellulase, protease, xylanase and lipase) and antibacterial activity	Bezerra et al. (2015)

Opuntia ficus-indica	*Acremonium*, *Aspergillus*, *Chrysonilia*, *Cladosporium*, *Cunninghamella*, *Curvularia*, *Fusarium*, *Mucor*, *Penicillium*, *Phialophora*, *Phoma*, *Rhinocladiela*	Caatinga: Pernambuco	–	Freire et al. (2015)
Cereus jamacaru	*Acremonium*, *Aspergillus*, *Aureobasidium*, *Candida*, *Cladosporium*, *Curvularia*, *Debaryomyces*, *Fusarium*, *Gibberella*, *Guinardia*, *Penicillium*, *Purpureocillium*, *Rhodotorula*, *Sterigmatomyces*, *Syncephalastrum*, *Trichoderma*, *Tritirachium*	Caatinga: Pernambuco	Production of enzyme L-asparaginase	Santos et al. (2015a)
Cereus jamacaru subsp. *jamacaru*	*Acremonium*, *Alternaria*, *Aspergillus*, *Aureobasidium*, *Beauveria*, *Curvularia*, *Cladosporium*, *Fusarium*, *Gibberella*, *Lecythophora*, *Nodulisporium*, *Penicillium*, *Pestalotiopsis*, *Phoma*, *Trichoderma*	Caatinga: Pernambuco	Antimicrobial activity	Pires et al. (2015)
Opuntia ficus-indica				
Pilosocereus gounellei subsp. *gounellei*				
Indigofera suffruticosa	*Colletotrichum*, *Pseudocochliobolus*, *Phomopsis*, *Curvularia*, *Pestalotiopsis*, *Khuskia*, *Chaetomella*, *Lasiodiplodia*	Caatinga and Atlantic Forest: Pernambuco	–	Santos et al. (2015b)
Piper hispidum	*Alternaria*, *Bipolaris*, *Colletotrichum*, *Phyllosticta*, *Lasiodiplodia*, *Marasmius*, *Phlebia*, *Phoma*, *Diaporthe*, *Schizophyllum* and one isolate from the order *Diaporthales*	Forest remnant, Maringá: Paraná	Antifungal activity and proteolytic activity	Orlandelli et al. (2015)
Solanum cernuum	*Mycosphaerella* and other	Atlantic Forest, Cerrado, Amazon rain forest	Production of compounds with antifungal activity	Pereira et al. (2015)
Eugenia bimarginata *Myrciaria floribunda* *Alchornea castaneifolia* *Stryphnodendron adstrigens*				
Carapa guianensis				

(continued)

Table 16.1 (continued)

Plant species	Endophytic fungi (genera) isolated and/or tested	Environment and/or states	Biotechnological potential	References
Tacinga inamoena	*Diaporthe*	Caatinga: Pernambuco	–	Crous et al. (2016)
Tacinga inamoena	*Bezerromyces*, *Xiliomyces*	Caatinga: Pernambuco	–	Bezerra et al. (2017a)
Schinus terebinthifolius	*Phoma* and other	Campus of Federal University of Pernambuco, Recife: Pernambuco	Antibacterial activity	Silva et al. (2017a)
Mandevilla catimbauensis *Lippia gracilis* *Sorghum bicolor*	*Phyllosticta*, *Achaetomium*, *Pseudophialophora*	Pernambuco and Rio Grande do Norte	–	Crous et al. (2017)
Melocactus zehntneri *Pilosocereus gounellei* subsp. *gounellei*	*Toxicocladosporium*	Caatinga: Pernambuco	–	Bezerra et al. (2017b)
Rhizophora mangle *Avicennia schaueriana* *Laguncularia racemosa*	*Aspergillus*, *Diaporthe*, *Fusarium*, *Hypocrea*, *Penicillium*, *Xylaria*	Mangroves (Cananeia and Bertioga): São Paulo	Production of organic acids	Dezam et al. (2017)
Piper hispidum	*Diaporthe*	Remnant of semideciduous forest, Maringá: Paraná	Production of _β-(1 → 3,1 → 6)-D-glucans and antiproliferative activity against human breast carcinoma (MCF-7) and hepatocellular carcinoma (HepG2-C3A) cells	Orlandelli et al. (2017)
Myracrodruon urundeuva Allemão *Anadenanthera colubrina* Vell. *Schinopsis brasiliensis* Engl. *Anacardium occidentale* L. *Caesalpinia pyramidalis* Tul	*Aspergillus* and other endophytes	Caatinga, Sumé: Paraíba	Production of enzyme tannase	Cavalcanti et al. (2017)

Hyptis suaveolens	*Aspergillus*	Pantanal	Production of compounds terrein, butyrolactone I and butyrolactone V with schistosomicidal activity, antioxidant effect, antitumor activity, antimicrobial potential and moderate leishmanicidal action	Silva et al. (2017b)
Myracrodruon urundeuva	*Alternaria*, *Colletotrichum*, *Diaporthe*, *Earliella*, *Exserohilum*, *Neofusicoccum*, *Penicillium*, *Phyllosticta*, *Rhinocladiella*, *Sarocladium*, *Talaromyces* and other	Caatinga: Pernambuco	Production of enzyme L-asparaginase	Pádua et al. (2018)
Mimosa tenuiflora *Poincianella pyramidalis* *Tillandsia catimbauensis*	*Quambalaria*, *Diaporthe*, *Umbelopsis*	Caatinga: Paraíba and Pernambuco	–	Bezerra et al. (2018)
Copaifera oblongifolia	–	Cerrado: Minas Gerais	–	Fernandes et al. (2018)
Poincianella pyramidalis *Citrullus lanatus*	*Diaporthe*, *Preussia*	Paraíba and Pernambuco	–	Crous et al. (2018)
Tillandsia catimbauensis	*Penicillium*, *Talaromyces*	Caatinga: Pernambuco	Production of enzyme L-asparaginase	Silva et al. (2018)
Syzygium cumini (L.) Skeels	*Aspergillus*	Federal Institute of Education, Science and Technology of Pernambuco, Campus Barreiros: Pernambuco	Production of enzyme tannase	Sena et al. (2018)

2012 has been studied for its fungal endophyte community. The first study to access the endophytic fungal community of plants from de Caatinga forest was performed by Bezerra et al. (2012) who studied the association of these microorganisms with the cactus *Opuntia ficus-indica*. The authors reported the isolation of members belonging to 12 genera and also the first record of 5 endophytic fungi species in Brazil. After this first publication, other research has been conducted by various groups in order to access the endophytic fungal community associated with plants from the Caatinga forest (Bezerra et al. 2013, 2015, 2017a, b; Freire et al. 2015; Pádua et al. 2018; Silva et al. 2018) (see Table 16.1).

Recently, Pádua et al. (2018) studied endophytic fungi from the leaves of *Myracrodruon urundeuva* in the Caatinga and *brejo de altitude* ecosystems in Brazil. The authors isolated 187 endophytic fungi belonging to *Ascomycota* (*Botryosphaeriales*, *Chaetothyriales*, *Diaporthales*, *Eurotiales*, *Glomerellales*, *Hypocreales*, and *Pleosporales*) and *Basidiomycota* (*Polyporales*). During this study, *Diaporthe* was the most frequent genus. Members of *Talaromyces* were indicative of Caatinga, while *Phyllosticta*, *Diaporthe*, and *Colletotrichum* were of the *brejo de altitude* ecosystem. Pádua et al. (2018) also demonstrated that the composition of endophytic fungi in *brejo de altitude* was higher than in the Caatinga forest. A similar survey was performed by Silva et al. (2018) who studied a bromeliad from the Caatinga forest and reported the isolation of 184 endophytes, of which 52 belonged to the *Penicillium* and *Talaromyces* genera.

Besides the remarkable diversity of endophytic fungi from the Caatinga forest, another Brazilian tropical dry forest, named Cerrado, has also been assessed for its diversity of endophytes. Studying *Palicourea marcgravii* growing in this environment, Cafêu et al. (2005) isolated an endophyte belonging to *Xylaria* and tested its capacity to produce antifungal compounds. Other authors, such as Magalhães et al. (2008), assessed the endophytic diversity of *Eremanthus erythropappus* growing in the Cerrado and related the isolation of species belonging to eight genera in *Ascomycota*. Similar results were obtained by Oki et al. (2009) and Abreu et al. (2010), who also used plants from the Cerrado forest to study its association with endophytic fungi species, reporting important fungal endophytic diversity (see Table 16.1).

The endophytic fungal diversity of the Atlantic Forest in Brazil has also been studied because of its important contribution to Brazilian biodiversity and is considered one of the global hotspots for biodiversity and endemism of species (Lima et al. 2013; Vaz et al. 2014; Oliveira et al. 2014). Some research studies in the Atlantic Forest have confirmed its important endophytic fungal diversity. Vieira et al. (2012) assessed the fungal endophyte community of *Solanum cernuum* and recorded the isolation of several endophytic fungi belonging to more than 33 taxa of *Ascomycota*, *Basidiomycota*, and *Mucoromycotina*. Similar work was performed by Vaz et al. (2014) who studied the plants *Myrceugenia ovata* var. *nanophylla* and *Eugenia neomyrtifolia* and isolated endophytic members of seven fungal genera. Other important plant species that have been studied in the Atlantic Forest areas include crops, such as *Coffea arabica*; its fungal endophytic community was assessed by Oliveira et al. (2014) who recorded the isolation of members of 14 genera of *Ascomycota* and several other isolates without sporulation in culture media. Other surveys are still in progress, which aim to assess the endophytic fungal diversity associated with other native and agricultural plants in the Atlantic Forest.

Another peculiar environment in Brazil is the mangrove forests, which are special because of their ecological contributions to the maintenance of biodiversity in coastal regions. Sebastianes et al. (2013) assessed the diversity of endophytes in this environment and reported isolation of about 4300 endophytes from *Rhizophora mangle*, *Avicennia schaueriana*, and *Laguncularia racemosa* in the Brazilian state of São Paulo. The endophytes isolated were classified as belonging to 34 genera in the orders *Diaporthales*, *Xylariales*, *Hypocreales*, *Botryosphaeriales*, *Pleosporales*, *Trichosphaeriales*, *Pyrenulales*, and *Sordariales*. The endophytic fungal diversity in mangroves was also verified by Costa et al. (2012) who studied the hosts *A. schaueriana*, *L. racemosa*, and *R. mangle* in an island in the Pernambuco province in Brazil. These authors highlighted the frequency of species from 19 genera, and these isolates were mainly pigmented fungal taxa. Due to the harsh conditions in the mangrove habitat, several pigmented fungal species can colonize plants in this environment, similar to plants living in dry forests as was discovered in the Caatinga (Bezerra et al. 2013, 2017c). The development of research in harsh environments is an important way to elucidate fungal diversity that is still unknown, and/or to understand the distribution of species only reported in a few substrates and places. Figure 16.1 shows the diversity of the fungal endophyte community in Brazil.

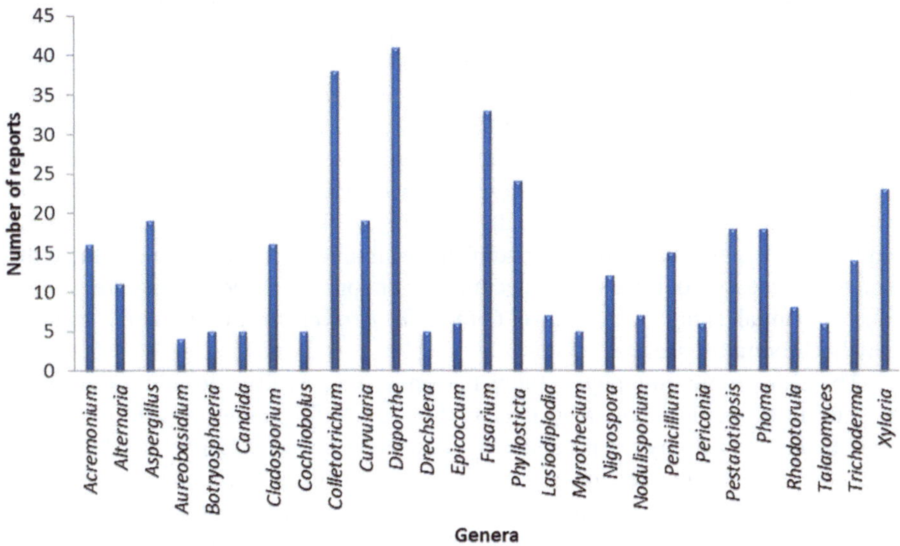

Fig. 16.1 Number of reported genera in the fungal endophyte community in Brazil

16.3 Biotechnological Potential of Endophytic Fungi from Brazil

The chemistry of endophytes has been seldom studied, especially if we consider the vast diversity of fungi and their specificity in the colonization of plant hosts. The study of interactions between microorganisms and plants, besides contributing to the understanding of chemical processes in ecology and nature, may result in the establishment of new sources of substances of human interest (Santos et al. 2008). Endophytic microorganisms are potential sources of new natural products that can be exploited in agriculture, industry, and medicine (Strobel and Daisy 2003; Chapla et al. 2013).

Several studies have demonstrated the potential of endophytic microorganisms in promoting plant growth (Santos and Varavallo 2011). For example, Silva et al. (2006) evidenced the potential of endophytic fungi to promote the growth of Annonaceae species seedlings, and Luz et al. (2006) verified that fungal endophytes presented potential to promote the growth of yellow passion fruit seedlings. Another study using the culture filtrate of the endophyte *Penicillium citrinum* showed that it positively influences the growth of the *Carex kobomugi* (Hwang et al. 2011). Other endophytic species, such as *Fusarium oxysporum*, have also been shown to be promising for banana growth (Ting et al. 2008). Endophytic microorganisms have the potential to stimulate plant growth due to mechanisms such as phytohormone production, nitrogen fixation, drug resistance, and pathogen antagonism (Peixoto Neto et al. 2002).

In addition, endophytes are reported to have potential for the control of plant pathogens. For example, Rocha et al. (2009) found an antagonist potential for endophytic fungi isolated from *Symphytum officinale* against *Sclerotinia sclerotiorum*, a plant pathogen of bean crops. Other examples of antagonist potential are the endophytic fungi from *Theobroma cacao* and *Theobroma grandiflorum* that showed promising biocontrol of *Phytophthora palmivora*, a phytopathogenic fungus of cocoa (Hanada et al. 2010). Among the endophytes tested during that experiment, members of *Trichoderma*, *Pestalotiopsis*, *Curvularia*, *Tolypocladium*, and *Fusarium* were reported as the best eco-friendly endophytes (Hanada et al. 2010). Studying endophytic fungi isolated from cacao plants, Rubini et al. (2005) demonstrated that some of these microorganisms, especially the endophyte *Gliocladium catenulatum*, were potential antagonists against *Crinipellis perniciosa*, the causative agent of witch's broom disease.

There are some endophytic fungi that may have evolved to allow transfer of genetic information to their host plant (Strobel 2002). Some organic substances with biological activity are produced by these endophytes, e.g., paclitaxel (Taxol®), the anticancer drug endogenously produced by *Taxus brevifolia* and its endophytic fungus *Taxomyces andreanae* (Stierle et al. 1993). Many endophytes isolated from medicinal plants produce secondary metabolites which are strong fungicides or bactericides (Tan and Zou 2001; Gunatilaka 2006). Siqueira et al. (2011) assessed the potential for antimicrobial compound production by endophytic fungi isolated from *L. sidoides*, a medicinal plant used as an antiseptic in the northeast of Brazil. In this study, *Staphylococcus aureus* ATCC6538, *Bacillus subtilis* UFPEDA-16, *Klebsiella pneumoniae* ATCC 29665, *Staphylococcus aureus* ATCC 6538, *Trichophyton rubrum* URM 4350, and *Malassezia furfur* URM4849 were inhibited by the endophytic fungi *Alternaria alternata*, *Phomopsis archeri*, *Colletotrichum gloeosporioides*, *Drechslera dematioidea*, and other endophytes without sporulation in culture. This study by Siqueira et al. (2011) indicated that endophytic fungi from *L. sidoides* have antifungal and antibacterial potential against microorganisms that are pathogenic to humans.

Natural products of endophytic fungi can act as antibiotics by inhibiting or eliminating disease-causing pathogens, such as protozoa, viruses, bacteria, and fungi, that affect humans, other animals, and plants (Strobel and Daisy 2003). An example of this potential is the study by Pires et al. (2015) that verified the antibacterial potential of endophytic fungi from the Brazilian tropical dry forest (Caatinga) and reported that of the 60 endophytes tested, 21 showed activity against bacteria that are pathogenic to humans. Another survey was performed by Souza et al. (2004) who used extracellular metabolic liquid from 79 endophytes isolated from Amazonian plants against microorganisms that are phytopathogenic and pathogenic to humans and found that 19 isolates were antagonistic against one or more pathogenic microorganisms. Additionally, Ferreira et al. (2015) reported that endophytic fungi isolated from *Carapa guianensis* demonstrated antiviral activity against the yellow fever virus.

Several other important medicinal properties have been reported for compounds produced by endophytic fungi. For example, Nascimento et al. (2015b) reported an anti-leishmanial activity in the endophyte of *Vernonia polyanthes*. Campos et al. (2015) reported similar results and reported anti-leishmanial activity, trypanocidal activity, and antitumor activity in the endophytes of *Caesalpinia echinata*. Orlandelli et al. (2017) identified additional important biotechnological potential and found antiproliferative activity against human breast carcinoma (MCF-7) and hepatocellular carcinoma (HepG2-C3A) cells in *Diaporthe* sp. isolates. Other compounds obtained from the endophyte *Aspergillus terreus* also demonstrated schistosomicidal activity, antitumor activity, and moderate leishmanicidal activity (Silva et al. 2017b). Moreover, Chapla et al. (2014) reported the anti-inflammatory potential of endophytic *Diaporthe* isolates.

Additionally, endophytic fungi are reported to be enzyme producers of major importance to industry. During studies verifying the capacity to produce extracellular enzymes using endophytic fungi, Cuzzi et al. (2011) verified proteolytic, lipolytic, and aminolytic activity of several endophytes. Bezerra et al. (2012) studied the potential to produce enzyme by fungal isolates from the cactus *O. ficus-indica* and reported the ability of these isolates to produce protease, cellulase, xylanase, and pectinase enzymes. In another study by Bezerra et al. (2015), the authors also verified the potential of endophytic fungi from different tissues of *Bauhinia forficata* to produce protease, cellulase, xylanase, and lipase. Silva et al. (2006) verified the production of protease and lipase by endophytic fungi isolated from *Annona squamosa* and *Annona muricata*. The production of other enzymes with industrial uses, such as tannase, chitinase, hemicellulases, and β-glucosidase, has also been reported in several studies of endophytic fungi (Campos et al. 2005; Robl et al. 2013; Sena et al. 2014, 2018; Cavalcanti et al. 2017). In addition, endophytic fungi have been reported to produce the enzyme L-asparaginase (Theantana et al. 2007, 2009; Chandra 2012; Kalyanasundaram et al. 2015; Lopes et al. 2015).

In Brazil, the demand for enzyme-producing fungi has contributed to the development of some projects searching for endophytes with various biotechnological potential. Recently, Azevedo and Quecine (2017) published the book *Diversity and benefits of microorganisms from the tropics* that includes information about the diversity and application of Brazilian endophytes in agriculture, renewable energy production, and environmental protection. Nevertheless, little information was included concerning the potential of endophytes to produce the enzyme L-asparaginase. L-asparaginase has been reported from endophytes in other countries (Theantana et al. 2007, 2009; Chandra 2012; Kalyanasundaram et al. 2015), and in recent years it also has been recorded from Brazilian endophytes (Santos et al. 2015a; Pádua et al. 2018; Silva et al. 2018). This enzyme is important as a medicine in the pharmacological industry and in food industries where it acts to inhibit the formation of acrylamide in foods treated with high temperature (Lopes et al. 2015). For example, the surveys by Santos et al. (2015a), Pádua et al. (2018), and Silva et al. (2018) emphasized that endophytes from the Caatinga forest in Brazil are an important source of strains with the potential to produce L-asparaginase.

The study by Santos et al. (2015a) was the first report of endophytic fungi isolated from a tropical dry forest plant with biotechnological potential to produce the enzyme L-asparaginase. The authors analyzed the ability of 44 endophytic fungi previously isolated from the cactus *C. jamacaru* to produce this enzyme. The researchers observed that 19 fungi were good producers and were from the genera *Aspergillus*, *Cladosporium*, *Curvularia*, *Fusarium*, *Gibberella*, *Penicillium*, and *Purpureocillium*. *Aspergillus japonicus* URM 6872, *A. ochraceus* URM 6885, *A. sydowii* URM 6866, *A. terreus* URM 6888, *F. oxysporum* URM 6815, *G. fujikuroi* var. *fujikuroi* URM 6816, and *P. brevicompactum* URM 6833 had the greatest ability to synthesize L-asparaginase and were indicated for further studies to optimize the production of the enzyme. Pádua et al. (2018) studied the potential of endophytic fungi from the leaves of *Myracrodruon urundeuva* in Brazil to produce L-asparaginase. In that study, *Diaporthe* sp. (URM 7793, URM 7779, and URM 7792) and *Talaromyces* sp. URM 7785 were the best producers of L-asparaginase. In addition, members of *Alternaria*, *Colletotrichum*, *Exserohilum*, *Penicillium*, *Phylosticta*, and *Sarocladium* were also reported to be producers of this enzyme and can also be studied for optimization of enzyme production processes. Similar results were obtained by Silva et al. (2018) studying endophytes from a bromeliad in the Caatinga forest in Brazil, who reported its potential to produce the enzyme L-asparaginase by species of *Penicillium* and *Talaromyces*, highlighting *T.* cf. *cecidicola* URM 7826 as the best producer. Other studies are still in progress to investigate the asparaginasic potential of several endophytes from different Brazilian environments (personal communication). Table 16.1 summarizes some selected papers on the diversity of the fungal endophyte community and/or the biotechnological potential of Brazilian endophytes.

16.4 Future Perspectives

It is believed that all plants have endophytic microorganisms yet to be studied, presenting genetic variants capable of producing several compounds of unknown bioactivity (Peixoto Neto et al. 2004). In Brazil, several studies have reported a great diversity of endophytic fungi (Rodrigues 1994; Rubini et al. 2005; Abreu et al. 2010; Vieira et al. 2012; Bezerra et al. 2013), including the description of new genera, such as *Bezerromyces* and *Xiliomyces* (Bezerra et al. 2017a) and several new species (Bezerra et al. 2017b, 2018; Crous et al. 2017). In addition, it is important to highlight the need for ex situ preservation of part of the Brazilian mycodiversity of endophytic fungi in mycological reference collections across the country in order to protect its diversity and contribute to current and future studies of taxonomic and biotechnological potential.

Coinciding with the diversity of endophytic fungi in the country, many studies also reported the biotechnological potential of these endophytes. For example, they are reported to have antimicrobial activity (Pires et al. 2015; Nascimento et al. 2015a; Ferreira et al. 2015; Silva et al. 2017a), leishmanicidal and trypanocidal activities (Campos et al. 2015; Ferreira et al. 2015), and potential to produce enzymes (Bezerra et al. 2012, 2015; Santos et al. 2015a; Pádua et al. 2018; Silva et al. 2018). According to Anbu et al. (2015), enzymes produced by microorganisms are more attractive to industries because they can be produced on a large scale and have short generation times.

Brazil holds the greatest botanical diversity in the world (Giulietti et al. 2005), which not only increases the possibility of research toward discovering endophytic fungal diversity and its relationship with hosts but also increases the chances of finding new bioactive molecules with different potentials and possibilities for industrial applications. This fact underscores the importance of conducting research that surveys the diversity and biotechnological potential of endophytic fungi associated with plants of different ecosystems in the country.

Acknowledgments The authors thank the Conselho Nacional de Desenvolvimento Científico e Tecnológico (CNPq), Coordenação de Aperfeiçoamento de Pessoal de Nível Superior (CAPES, Finance Code 001), Fundação de Amparo à Ciência e Tecnologia de Pernambuco (FACEPE), and Universidade Federal de Pernambuco (UFPE).

References

Abreu LM, Almeida AR, Salgado M, Pfenning LH (2010) Fungal endophytes associated with the mistletoe *Phoradendron perrottettii* and its host tree *Tapirira guianensis*. Mycol Progress 9:559–566

Anbu P, Gopinath SCB, Chaulagain BP, Tang TH, Citartan M (2015) Microbial enzymes and their applications in industries and medicine 2014. Biomed Res Int 2015:816419. https://doi.org/10.1155/2015/816419

Araújo WL, Maccheroni W Jr, Aguilar-Vildoso CI, Barroso PAV, Saridakis HO, Azevedo JL (2001) Variability and interactions between endophytic bacteria and fungi isolated from leaf tissues of citrus rootstocks. Can J Microbiol 147:229–236

Azevedo JL, Quecine MC (2017) Diversity and benefits of microorganisms from the tropics. Springer, Cham, p 439

Banhos EF, Souza AQL, Andrade JC, Souza ADL, Koolen HHF, Albuquerque PM (2014) Endophytic fungi from *Myrcia guianensis* at the Brazilian Amazon: distribution and bioactivity. Braz J Microbiol 45:153–161

Bertoni MD (1994) New record of *Ceratopycnidium baccharidicola* (endophyte on *Baccharis coridifolia*) in Brazil. Mycotaxon 52:193–196

Bezerra JDP, Santos MGS, Svedese VM, Lima DMM, Fernandes MJS, Paiva LM, Souza-Motta CM (2012) Richness of endophytic fungi isolated from *Opuntia ficus-indica* Mill.(Cactaceae) and preliminary screening for enzyme production. World J Microbiol Biotechnol 28:1989–1995

Bezerra JDP, Santos MGS, Barbosa RN, Svedese VM, Lima DMM, Fernandes MJS, Gomes BS et al (2013) Fungal endophytes from cactus *Cereus jamacaru* in Brazilian tropical dry forest: a first study. Symbiosis 60:53–63

Bezerra JDP, Nascimento CCF, Barbosa RN, Silva DCV, Svedese VM, Silva-Nogueira EB, Gomes BS et al (2015) Endophytic fungi from medicinal plant *Bauhinia forficata*: diversity and biotechnological potential. Braz J Microbiol 46:49–57

Bezerra JDP, Oliveira RJV, Paiva LM, Silva GA, Groenewald JZ, Crous PW, Souza-Motta CM (2017a) Bezerromycetales and Wiesneriomycetales ord. nov. (class Dothideomycetes), with two novel genera to accommodate endophytic fungi from Brazilian cactus. Mycol Progress 16:297–309

Bezerra JDP, Sandoval-Denis M, Paiva LM, Silva GA, Groenewald JZ, Souza-Motta CM, Crous PW (2017b) New endophytic *Toxicocladosporium* species from cacti in Brazil, and description of *Neocladosporium* gen. nov. IMA Fungus 8:77–97

Bezerra JDP, Azevedo JL, Souza-Motta CM (2017c) Why study endophytic fungal community associated with cacti species? In: Azevedo JL, Quecine MC (eds) Diversity and benefits of microorganisms from the tropics. Springer, Cham, pp 21–35

Bezerra JDP, Machado AR, Firmino AL, Rosado AWC, Souza CAF, Souza-Motta CM, Freire KTLS et al (2018) Mycological diversity description I. Acta Bot Bras 32:656–666

Brum MCP, Araújo WL, Maki CS, Azevedo JL (2012) Endophytic fungi from *Vitis labrusca* L. ('Niagara Rosada') and its potential for the biological control of *Fusarium oxysporum*. Genet Mol Res 11:4187–4197

Cafêu MC, Silva GH, Teles HL, Bolzani VS, Araújo AR, Young MCM, Pfenning LH (2005) Substâncias antifúngicas de *Xylaria* sp., um fungo endofítico isolado de *Palicourea marcgravii* (Rubiaceae). Quim Nova 28:991–995

Camatti-Sartori V, Silva-Ribeiro RT, Valdebenito-Sanhueza RM, Pagnocca FC, Echeverrigaray S et al (2005) Endophytic yeasts and filamentous fungi associated with southern Brazilian apple (*Malus domestica*) orchards subjected to conventional, integrated or organic cultivation. J Basic Microbiol 45:397–402

Campos RA, Arruda W, Boldo JT, Silva MV, Barros NM, Azevedo JL, Schrank A et al (2005) *Boophilus microplus* infection by *Beauveria amorpha* and *Beauveria bassiana*: SEM analysis and regulation of subtilisin-like proteases and chitinases. Curr Microbiol 50:257–261

Campos RA, Boldo JT, Pimentel IC, Dalfovo V, Araújo WL, Azevedo JL, Vainstein MH et al (2010) Endophytic and entomopathogenic strains of *Beauveria* sp to control the bovine tick *Rhipicephalus* (*Boophilus*) *microplus*. Genet Mol Res 9:1421–1430

Campos FF, Sales Junior PA, Romanha AJ, Araújo MSS, Siqueira EP, Resende JM, Alves TMA et al (2015) Bioactive endophytic fungi isolated from *Caesalpinia echinata* Lam. (Brazilwood) and identification of beauvericin as a trypanocidal metabolite from *Fusarium* sp. Mem Inst Oswaldo Cruz 110:65–74

Cavalcanti RMF, Ornela PHO, Jorge JA, Guimaraes LHS (2017) Screening, selection and optimization of the culture conditions for tannase production by endophytic fungi isolated from caatinga. J Appl Biol Biotechnol 5:1–9

Chandra S (2012) Endophytic fungi: novel sources of anticancer lead molecules. Appl Microbiol Biotechnol 95:47–59

Chapla VM, Biasetto CR, Araujo AR (2013) Fungos endofíticos: uma fonte inexplorada e sustentável de novos e bioativos produtos naturais. Rev Virtual Quim 5:421–437
Chapla VM, Zeraik ML, Ximenes VF, Zanardi LM, Lopes MN, Cavalheiro AJ, Silva DHS et al (2014) Bioactive secondary metabolites from *Phomopsis* sp., an endophytic fungus from *Senna spectabilis*. Molecules 19:6597–6608
Conti R, Cunha IGB, Siqueira VM, Souza-Motta CM, Amorim ELC, Araújo JM (2012) Endophytic microorganisms from leaves of *Spermacoce verticillata* (L.): diversity and antimicrobial activity. J Appl Pharm Sci 2:17–22
Costa IPMW, Maia LC, Cavalcanti MA (2012) Diversity of leaf endophytic fungi in mangrove plants of northeast Brazil. Braz J Microbiol 43:1165–1173
Crous PW, Wingfield MJ, Burgess TI, Hardy GESTJ, Crane C, Barrett S, Cano-Lira JF et al (2016) Fungal Planet description sheets: 469–557. Persoonia 37:218–403
Crous PW, Wingfield MJ, Burgess TI, Carnegie AJ, Hardy GESJ, Smith D, Summerell BA et al (2017) Fungal Planet description sheets: 625–715. Persoonia 39:270–467
Crous PW, Wingfield MJ, Burgess TI, Hardy GESJ, Gené J, Guarro J, Baseia IG et al (2018) Fungal Planet description sheets: 716–784. Persoonia 40:240–393
Cuzzi C, Link S, Vilani A, Onofre SB (2011) Enzimas extracelulares produzidas por fungos endofíticos isolados de *Baccharis Dracunculifolia* D.C. (Asteraeceae). Gl Sci Technol 4:47–57
Dezam APG, Vasconcellos VM, Lacava PT, Farinas CS (2017) Microbial production of organic acids by endophytic fungi. Biocatal Agric Biotechnol 11:282–287
Dreyfuss M, Petrini O (1984) Further investigations on the occurrence and distribution of endophytic fungi in tropical plants. Bot Helv 94:33–40
Fernandes EG, Pereira OL, Silva CC, Bento CBP, Queiroz MV (2015) Diversity of endophytic fungi in *Glycine max*. Microbiol Res 181:84–92
Fernandes EG, Valério HM, Duarte KLR, Capuchinho LMN, Fagundes M (2018) Fungi associated with Copaifera oblongifolia (Fabaceae) seeds: occurrence and possible effects on seed germination. Acta Bot Bras 33:179–182
Ferreira MC, Vieira MLA, Zani CL, Alves TMA, Junior PAS, Murta SMF, Romanha AJ et al (2015) Molecular phylogeny, diversity, symbiosis and discover of bioactive compounds of endophytic fungi associated with the medicinal Amazonian plant *Carapa guianensis* Aublet (Meliaceae). Biochem Syst Ecol 59:36–44
Freire KTLS, Araújo GR, Bezerra JDP, Barbosa RN, Silva DCV, Svedese VM, Paiva LM et al (2015) Fungos endofíticos de *Opuntia ficus-indica* (L.) Mill.(cactaceae) sadia e infestada por *Dactylopius opuntiae* (Cockerell, 1896) (Hemiptera: Dactylopiidae). Gaia Sci 9:104–110
Gai CS, Lacava PT, Maccheroni W Jr, Glienke C, Araújo WL, Miller TA, Azevedo JL (2009) Diversity of endophytic yeasts from sweet orange and their localization by scanning electron microscopy. J Basic Microbiol 49:441–451
García A, Rhoden SA, Rubin Filho CJ, Nakamura CV, Pamphile JA (2012) Diversity of foliar endophytic fungi from the medicinal plant *Sapindus saponaria* L. and their localization by scanning electron microscopy. Biol Res 45:139–148
Giulietti AM, Harley RM, Queiroz LP, Wanderley MGL, Van den Berg C (2005) Biodiversidade e conservação das plantas no Brasil. Megadiversidade 1:52–61
Glienke C, Pereira OL, Stringari D, Fabris J, Kava-Cordeiro V, Galli-Terasawa L, Cunnington J et al (2011) Endophytic and pathogenic *Phyllosticta* species, with reference to those associated with Citrus Black Spot. Persoonia 26:47–56
Gunatilaka AAL (2006) Natural products from plant-associated microorganisms: distribution, structural diversity, bioactivity and applications of their occurrence. J Nat Prod 69:509–526
Hanada RE, Pomella AWV, Costa HS, Bezerra JL, Loguercio LL, Pereira JO (2010) Endophytic fungal diversity in *Theobroma cacao* (cacao) and *T. grandiflorum* (cupuaçu) trees and their potential for growth promotion and biocontrol of black-pod disease. Fungal Biol 114:901–910
Hilarino MPA, Silveira FAO, Oki Y, Rodrigues L, Santos JC, Corrêa Junior A, Fernandes GW et al (2011) Distribution of the endophytic fungi community in leaves of *Bauhinia brevipes* (Fabaceae). Acta Bot Bras 25:815–821

Hwang JS, You YH, Bae JJ, Khan SA, Kim JG, Choo YS (2011) Effects of endophytic fungal secondary metabolites on the growth and physiological response of *Carex kobomugi* Ohwi. J Coast Res 27:544–548

Johann S, Rosa LH, Rosa CA, Perez P, Cisalpino PS, Zani CL, Cota BB (2012) Antifungal activity of altenusin isolated from the endophytic fungus *Alternaria* sp. against the pathogenic fungus *Paracoccidioides brasiliensis*. Rev Iberoam Micol 29:205–209

Kalyanasundaram I, Nagamuthu J, Srinivasan B, Pachayappan A, Muthukumarasamy S (2015) Production, purification and characterisation of extracellular L-asparaginase from salt marsh fungal endophytes. World J Pharm Pharm Sci 4:663–677

Landell MF, Mautone JN, Valente P (2006) Biodiversity of yeasts associated to bromeliads in Itapuã park, Viamão/RS. Biociencias 14:144–149

Lima TEF, Oliveira RJV, Neves RP, Bezerra JL, Cavalcanti MAQ (2013) Endophytic yeasts of *Coffea arabica* and *Vitis labrusca* cv. Isabel from Pernambuco, Brazil. Nova Hedwigia 96:463–469

Lopes AM, Oliveira-Nascimento L, Ribeiro A, Tairum CA Jr, Breyer CA, Oliveira MA, Monteiro G et al (2015) Therapeutic L-asparaginase: upstream, downstream and beyond. Crit Rev Biotechnol 37:82–99

Luz JS, Silva RLO, Silveira EB, Cavalcante UMT (2006) Atividade enzimática de fungos endofíticos e efeito na promoção do crescimento de mudas de maracujazeiro-amarelo. Caatinga 19:128–134

Magalhães WCS, Missagia RV, Costa FAF, Costa MCM (2008) Diversidade de fungos endofíticos em Candeia *Eremanthus erythropappus* (DC.) MacLeish. Cerne 14:267–273

Maia LC, Carvalho Júnior AA, Cavalcanti LH, Gugliotta AM, Drechsler-Santos ER, Santiago ALMA, Cáceres MES et al (2015) Diversity of Brazilian fungi. Rodriguésia 66:1033–1045

Medeiros LS, Murgu M, Souza AQL, Rodrigues-Fo E (2011) Antimicrobial depsides produced by *Cladosporium uredinicola*, an endophytic fungus isolated from *Psidium guajava* fruits. Helv Chim Acta 94:1077–1084

Mussi-Dias V, Araújo ACO, Silveira SF, Rocabado JMA, Araújo KL (2012) Fungos endofíticos associados a plantas medicinais. Rev Bras Pl Med 14:261–266

Nascimento TL, Oki Y, Lima DMM, Almeida-Cortez JS, Fernandes GW, Souza-Motta CM (2015a) Biodiversity of endophytic fungi in different leaf ages of *Calotropis procera* and their antimicrobial activity. Fungal Ecol 14:79–86

Nascimento AM, Soares MG, Torchelsen FKVS, Araujo JAV, Lage PS, Duarte MC, Andrade PHR et al (2015b) Antileishmanial activity of compounds produced by endophytic fungi derived from medicinal plant *Vernonia polyanthes* and their potential as source of bioactive substances. World J Microbiol Biotechnol 31:1793–1800

Oki Y, Soares N, Belmiro MS, Corrêa Junior A, Fernandes GW (2009) Influência dos fungos endofíticos sobre os herbívoros de *Baccharis dracunculifolia* (Asteraceae). Neotrop Biol Conserv 4:83–88

Oliveira CM, Regasini LO, Silva GH, Pfenning LH, Young MCM, Berlinck RGS, Bolzani VS (2011) Dihydroisocoumarins produced by *Xylaria* sp. and *Penicillium* sp., endophytic fungi associated with *Piper aduncum* and *Alibertia macrophylla*. Phytochem Lett 4:93–96

Oliveira RJV, Souza RG, Lima TEF, Cavalcanti MAQ (2014) Endophytic fungal diversity in coffee leaves (*Coffea arabica*) cultivated using organic and conventional crop management systems. Mycosphere 5:523–530

Orlandelli RC, Almeida TTD, Alberto RN, Polonio JC, Azevedo JL, Pamphile JA (2015) Antifungal and proteolytic activities of endophytic fungi isolated from *Piper hispidum* Sw. Braz J Microbiol 46:359–366

Orlandelli RC, Silva MDLC, Vasconcelos AFD, Almeida IV, Vicentini VEP, Prieto A, Hernandez MDD et al (2017) β-(1→3,1→6)-D-glucans produced by *Diaporthe* sp. endophytes: purification, chemical characterization and antiproliferative activity against MCF-7 and HepG2-C3A cells. Int J Biol Macromol 94:431–437

Pádua APSL, Freire KTLS, Oliveira TGL, Silva LF, Araújo-Magalhães GR, Agamez-Montalvo GS, Silva IR et al (2018) Fungal endophyte diversity in the leaves of the medicinal plant

Myracrodruon urundeuva in a Brazilian dry tropical forest and their capacity to produce L-asparaginase. Acta Bot Bras 33:39. https://doi.org/10.1590/0102-33062018abb0108

Pamphile JA, Azevedo JL (2002) Molecular characterization of endophytic strains of *Fusarium verticillioides* (= *Fusarium moniliforme*) from maize (*Zea mays*. L). World J Microbiol Biotechnol 18:391–396

Peixoto Neto PAS, Azevedo JL, Araújo WL (2002) Microrganismos endofíticos: Interação com plantas e potencial biotecnológico. Biotecnolog Cienc Desenvolv 29:62–76

Peixoto Neto PAS, Azevedo JL, Caetano LC (2004) Microrganismos endofíticos em plantas: status atual e perspectivas. BLACPMA 3:69–72

Pereira JO, Azevedo JL, Petrini O (1993) Endophytic fungi of *Stylosanthes*: a first report. Mycologia 85:362–364

Pereira JO, Carneiro Vieira ML, Azevedo JL (1999) Endophytic fungi from *Musa acuminata* and their reintroduction into axenic plants. World J Microbiol Biotechnol 15:37–40

Pereira CB, Oliveira DM, Hughes AFS, Kohlhoff M, Vieira MLA, Vaz ABM, Ferreira MC et al (2015) Endophytic fungal compounds active against *Cryptococcus neoformans* and *C. gattii*. J Antibiot 68:436–444

Pinheiro EAA, Carvalho JM, Santos DCP, Feitosa AO, Marinho PSB, Guilhon GMSP, Santos LS et al (2013) Chemical constituents of *Aspergillus* sp EJC08 isolated as endophyte from *Bauhinia guianensis* and their antimicrobial activity. An Acad Bras Cienc 85:1247–1252

Pires IMO, Silva AV, Santos MGS, Bezerra JDP, Barbosa RN, Silva DCV, Svedese VM et al (2015) Potencial antibacteriano de fungos endofíticos de cactos da Caatinga, uma Floresta Tropical Seca no Nordeste do Brasil. Gaia Sci 9:155–161

Robl D, Delabona SP, Mergel CM, Rojas JD, Costa PS, Pimentel IC, Vicente VA et al (2013) The capability of endophytic fungi for production of hemicellulases and related enzymes. BMC Biotechnol 13:94

Rocha R, Luz DE, Engels C, Pileggi SAV, Jaccoud Filho DS, Matiello RR, Pileggi M (2009) Selection of endophytic fungi from comfrey (*Symphytum officinale* L.) for in vitro biological control of the phytopathogen *Sclerotinia sclerotiorum* (Lib.). Braz J Microbiol 40:73–78

Rocha ACS, Garcia D, Uetanabaro APT, Carneiro RTO, Araújo IS, Mattos CRR, Góes-Neto A (2011) Foliar endophytic fungi from *Hevea brasiliensis* and their antagonism on *Microcyclus ulei*. Fungal Divers 47:75–84

Rodrigues KF (1994) The foliar fungal endophytes of the Amazonian palm *Euterpe oleracea*. Mycologia 86:376–385

Rodrigues KF, Dias-Filho MB (1996) Fungal endophytes in the tropical grasses *Brachiaria brizantha* cv. Marandu and *B. humidicola*. Pesq Agropec Bras 31:905–909

Rodrigues KF, Samuels GJ (1999) Fungal endophytes of *Spondias mombin* leaves in Brazil. J Basic Microbiol 39:131–135

Rodrigues KF, Leuchtmann A, Petrini O (1993) Endophytic species of *Xylaria*: cultural and isozymic studies. Sydowia 45:116–138

Rodrigues KF, Hesse M, Werner C (2000) Antimicrobial activities of secondary metabolites produced by endophytic fungi from *Spondias mombin*. J Basic Microbiol 40:261–267

Rodrigues KF, Sieber TN, Grünig CR, Holdenrieder O (2004) Characterization of *Guignardia mangiferae* isolated from tropical plants based on morphology, ISSR-PCR amplifications and ITS1-5. 8S-ITS2 sequences. Mycol Res 108:45–52

Rubini MR, Silva-Ribeiro RT, Pomella AW, Maki CS, Araújo WL, Santos DR, Azevedo JL (2005) Diversity of endophytic fungal community of cacao (*Theobroma cacao* L.) and biological control of *Crinipellis perniciosa*, causal agent of Witches' broom disease. Int J Biol Sci 1:24–33

Santos TT, Varavallo MA (2011) Aplicação de microrganismos endofíticos na agricultura e na produção de substâncias de interesse econômico. Semina: Ci Biol Saúde 32:199–212

Santos LS, Oliveira MN, Guilhon GMSP, Santos AS, Ferreira ICS, Lopes-Júnior ML, Arruda MSP et al (2008) Potencial herbicida da biomassa e de substâncias químicas produzidas pelo fungo endofítico *Pestalotiopsis guepinii*. Planta Daninha 26:539–548

Santos SN, Ferraris FK, Souza AO, Graças Henriques M, Melo IS (2012) Endophytic fungi from *Combretum leprosum* with potential anticancer and antifungal activity. Symbiosis 58:109–117

Santos MGS, Bezerra JDP, Svedese VM, Sousa MA, Silva DCV, Maciel MHC, Paiva LM et al (2015a) Screening of endophytic fungi from cactus of the Brazilian tropical dry forest according to their L-asparaginase activity. Sydowia 67:147–156

Santos IP, Bezerra JDP, Souza-Motta CM, Silva-Cavalcanti M, Menezes-Lima VL (2015b) Endophytic mycobiota from leaves of *Indigofera suffruticosa* Miller (Fabaceae): the relationship between seasonal change in Atlantic coastal Forest and tropical dry forest (Caatinga), Brazil. Afr J Microbiol Res 9:1227–1235

Sebastianes FLS, Romao-Dumaresq AS, Lacava PT, Harakava R, Azevedo JL, Melo IS, Pizzirani-Kleiner AA (2013) Species diversity of culturable endophytic fungi from Brazilian mangrove forests. Curr Genet 59:153–166

Sena AR, Santos ACB, Gouveia MJ, Mello MRF, Leite TCC, Moreira KA, Assis SA (2014) Production, characterization and application of a thermostable tannase from *Pestalotiopsis guepinii* URM 7114. Food Technol Biotechnol 52:459–467

Sena AR, Leite TCC, Nascimento TCES, Silva AC, Souza CS, Vaz AFM, Moreira KA et al (2018) Kinetic, thermodynamic parameters and *in vitro* digestion of tannase from *Aspergillus tamarii* URM 7115. Chem Eng Commun 205

Silva RLO, Luz JS, Silveira EB, Cavalcante UMT (2006) Fungos endofíticos em *Annona* spp.: isolamento, caracterização enzimática e promoção do crescimento em mudas de pinha (*Annona squamosa* L.). Acta Bot Bras 20:649–655

Silva GBPG, Silvino KF, Bezerra JDP, Farias TGS, Araújo JM, Stamford TLM (2017a) Antimicrobial activity of *Phoma* sp. URM 7221: an endophyte from *Schinus terebinthifolius* Raddi (Anacardiaceae). Afr J Microbiol Res 11:1–7

Silva IP, Brissow E, Kellner Filho LC, Senabio J, Siqueira KA, Vandresen Filho S, Damasceno JL et al (2017b) Bioactive compounds of *Aspergillus terreus*—F7, an endophytic fungus from *Hyptis suaveolens* (L.) Poit. World J Microbiol Biotechnol 33:62. https://doi.org/10.1007/s11274-017-2228-3

Silva LF, Freire KTLS, Araújo-Magalhães GR, Agamez-Montalvo GS, Sousa MA, Costa-Silva TA, Paiva LM et al (2018) *Penicillium* and *Talaromyces* endophytes from *Tillandsia catimbauensis*, a bromeliad endemic in the Brazilian tropical dry forest, and their potential for L-asparaginase production. World J Microbiol Biotechnol 34:162. https://doi.org/10.1007/s11274-018-2547-z

Siqueira VM, Braun U, Souza-Motta CM (2008) *Corynespora subcylindrica* sp. nov., a new hyphomycete species from Brazil and a discussion on the taxonomy of corynespora-like genera. Sydowia 60:113–122

Siqueira VM, Conti R, Araújo JM, Souza-Motta CM (2011) Endophytic fungi from the medicinal plant *Lippia sidoides* Cham. and their antimicrobial activity. Symbiosis 53:89–95

Souza AQL, Souza ADL, Astolfi Filho S, Belém Pinheiro ML, Sarquis MIM, Pereira JO (2004) Atividade antimicrobiana de fungos endofíticos isolados de plantas tóxicas da amazônia: *Palicourea longiflora* (aubl.) rich e *Strychnos cogens* Bentham. Acta Amaz 34:185–195

Stierle A, Strobel GA, Stierle D (1993) Taxol and taxen production by *Taxomyces andreanae* an endophytic fungus of pacific yew. Science 260:214–216

Strobel GA (2002) Rainforest endophytes and bioactive products. Crit Rev Biotechnol 22:315–333

Strobel G, Daisy B (2003) Bioprospecting for microbial endophytes and their natural products. Microbiol Mol Biol Rev 67:491–502

Suryanarayanan TS, Thirunavukkarasu N, Govindarajulu MB, Gopalan V (2012) Fungal endophytes: an untapped source of biocatalysts. Fungal Divers 54:19–30

Tan RX, Zou WX (2001) Endophytes: a rich source of functional metabolites. Nat Prod Rep 18:448–459

Theantana T, Hyde KD, Lumyong S (2007) Asparaginase production by endophytic fungi isolated from some Thai medicinal plants. KMITL Sci Tech J 7:13–18

Theantana T, Hyde KD, Lumyong S (2009) Asparaginase production by endophytic fungi from Thai medicinal plants: citoxicity properties. IJIB 7:1–8

Ting ASY, Meon S, Kadir J, Radu S, Singh G (2008) Endophytic microorganisms as potential growth promoters of banana. BioControl 53:541–553

Vaz ABM, Costa AGFC, Raad LVV, Góes-Neto A (2014) Fungal endophytes associated with three south American Myrtae (Myrtaceae) exhibit preferences in the colonization at leaf level. Fungal Biol 118:277–286

Vieira PDS, Souza-Motta CM, Lima D, Torres JB, Quecine MC, Azevedo JL, Oliveira NT (2011) Endophytic fungi associated with transgenic and non-transgenic cotton. Mycology 2:91–97

Vieira MLA, Hughes AFS, Gil VB, Vaz ABM, Alves TMA, Zani CL, Rosa CA et al (2012) Diversity and antimicrobial activities of the fungal endophyte community associated with the traditional Brazilian medicinal plant *Solanum cernuum* Vell.(Solanaceae). Can J Microbiol 58:54–66

Arbuscular Mycorrhizal Fungi in Alleviation of Cold Stress in Plants

17

Thokchom Sarda Devi, Samta Gupta, and Rupam Kapoor

Abstract

Cold stress is an important abiotic factor that adversely affects the growth and productivity of different agricultural crops globally. It leads to slower plant metabolism, cell membrane rigidification and loss of function, solute leakage, protein disintegration, depletion in sugar metabolism, and reproductive loss. The ever-increasing population and yet decrease in agricultural productivity is a global concern. So, there is a need for developing strategies that can help plants tolerate the harsh environmental condition and still not affect their productivity. In this context, utilization of arbuscular mycorrhizal fungi (AMF) for alleviation of cold stress in plants has gained much attention. Formation of AMF is reported to improve the performance of plants under both normal and stressful conditions. Although at low temperature (<15 °C) colonization of roots by AMF is often restrained, studies have reported improved tolerance in mycorrhizal plants to cold stress. Symbiotic association of plant roots with AMF improves cold tolerance through reduction of lipid peroxidation and maintenance of membrane integrity, enhancement of antioxidative potential, optimization of osmolytes accumulation and regulation of root hydraulic conductance, improvement of photosynthetic activity and respiration rate, and integrated transcriptional regulation of cold-responsive genes. This chapter discusses various mechanisms that AMF-colonized plants employ to mitigate the detrimental effects caused by low temperature.

Keywords

Arbuscular mycorrhiza fungi · Cold stress · Root hydraulic conductance · Osmotic adjustments · Transcriptional regulation

---

T. S. Devi · S. Gupta · R. Kapoor (✉)
Department of Botany, University of Delhi, New Delhi, Delhi, India

T. Satyanarayana et al. (eds.), *Advancing Frontiers in Mycology & Mycotechnology*,
https://doi.org/10.1007/978-981-13-9349-5_17

17.1 Introduction

Plants, being sessile, constantly interact with their immediate environment and are exposed to a myriad of changing environmental conditions. Temperature is a variable environmental factor, and exposure to extreme temperature is a common environmental stress experienced by plants. Temperature ranging from 0 to 15 °C is termed chilling and below 0 °C is called freezing temperature (Jan and Andrabi 2009). Chilling and freezing temperature both contribute to cold stress. Cold stress limits the growth and productivity in plants (Sanghera et al. 2011; Xin and Browse 2001). It directly affects the cellular macromolecules such as DNA, RNA, proteins, and membrane, leading to a slower metabolism, cell membrane rigidification, and loss of membrane's functions (Jewell et al. 2010).

In general, while plants from temperate climatic regions are chilling tolerant (Levitt 1980), plants of tropical and subtropical origins are sensitive to chilling stress and lack the mechanism for cold acclimatization (Hannah et al. 2005). Resistance to low-temperature stress in plants is a very complex trait and involves many different metabolic pathways in various organelles (Hannah et al. 2005).

Conventional breeding for improving cold tolerance in crop plants offers very limited success due to the lack of proficient selection criteria, complexity of the cold-tolerant traits, and low genetic variance (Sanghera et al. 2011). Although new strategies for developing transgenic crop plants with enhanced cold stress tolerance have been successfully reported (Byun et al. 2015; Wang et al. 2016; Wani et al. 2008), acceptance of transgenic crop plants is still a topic of global debate. It is therefore imperative to look for alternative strategies to help plants tolerate cold stress.

The symbiotic association between arbuscular mycorrhizal (AM) fungi (AMF) and roots of plants improve resistance of host plant to several abiotic stresses (Evelin et al. 2009; Miransari 2011; Sharma et al. 2017; Smith and Read 2010). About 70–90% of the terrestrial plant species are known to form AM (Smith and Read 2008; Zhu et al. 2010a). This symbiosis is the most ancient and ubiquitous interaction of plants (Parniske 2008) with fungi belonging to monophyletic phylum, the Glomeromycota (Schüßler et al. 2001). The fungal symbionts are obligate biotrophs and depend upon the host plants for acquiring carbon (Bago et al. 2000), and the plants are benefitted with enhanced nutrient and water uptake, increased metabolism, and also aboveground productivity (Sawers et al. 2008; Smith and Read 2008). Arbuscules, the specialized intracellular structures formed by the fungi, act as the site for exchange of carbon and nutrients between the two symbionts (Balestrini et al. 2015). The hyphae of AMF contribute extensively in acquiring minerals and water from the soil (Candido et al. 2015).

Arbuscular mycorrhiza can alleviate the effect of abiotic stresses, such as drought (Pavithra and Yapa 2018; Porcel and Ruiz-Lozano 2004; Quiroga et al. 2017), salinity (Evelin et al. 2012, 2013; Hajiboland et al. 2010; Kumar et al. 2015; Latef and Chaoxing 2011), heavy metals (Amir et al. 2014; Sharma et al. 2017), cold stress (Ma et al. 2015, 2018; Pedranzani et al. 2015; Zhu et al. 2010a, b), and biotic stresses (Marulanda-Aguirre et al. 2008; Pozo et al. 2010; Tayal et al. 2011).

The impact of cold stress on plants mainly comprises of two phases: the initial phase includes stress signal perception and plasma membrane rigidification (Beck

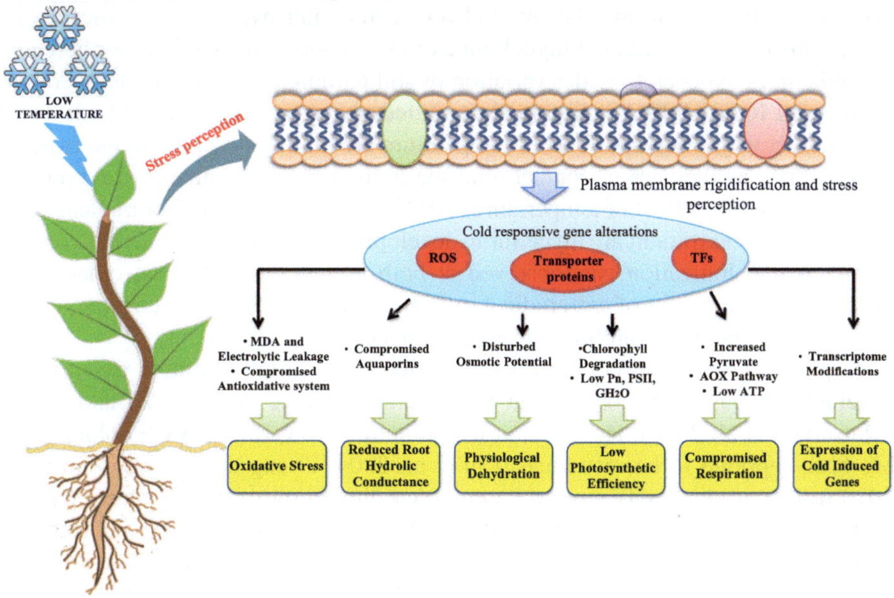

Fig. 17.1 Impact of cold stress on plants

et al. 2004) and the second phase involves multiple changes in cell morphology and metabolism (Fig. 17.1) (Ferullo and Griffith 2001; Xiong et al. 2002). In order to adapt to low temperature, mycorrhizal plants generally employ the second phase of response by altering various physiological activities and gene expression patterns (Guy 1990; Janicka-Russak et al. 2012; Thomashow 1999; Winfield et al. 2010; Wu and Zou 2010). In this chapter, the role played by AMF in alleviation of cold stress will be discussed.

The major cold-induced cellular responses following membrane rigidification disturbed plant-water relationship due to reduction in root hydraulic conductance, cellular dehydration resulting from changes in the osmotic potential, oxidative stress due to generation of large quantities of reactive oxygen species, lower respiration and photosynthetic efficiency as general metabolism of plants is hindered by low temperature, and activation of certain cold-inducible genes (Fig. 17.1). These responses altogether result in cellular damage, thereby affecting the performance of the plant exposed to cold stress.

17.2 Effect of Cold Stress on Arbuscular Mycorrhizal Fungi

The growth and development of AMF and formation of symbiotic association with plants is affected by the soil environmental conditions, light intensity, nutrient profile, and also cropping systems (Liu et al. 2004). Temperature is one of the major factors that influence the development and functioning of the symbiotic association (Gavito et al. 2005). The mycelium of AMF comprises of two linked mycelia, the extraradical mycelium and the internal root-colonizing hyphae (Heinemeyer and Fitter 2004). The extraradical mycelium explores a wider range of rhizosphere and hence is more exposed to wider variation of soil parameters, including temperature (Smith and Read 1997). Temperature can affect the growth of AMF in both direct and indirect ways. Direct effects of temperature are mainly observed in the extraradical hyphae as they are in contact with the immediate environment. Gavito et al. (2002) studied the effect of temperature (10 °C and 15 °C) on both intraradical colonization and extraradical mycelium development in *Pisum sativum* plants. Intraradical colonization was observed at both temperatures. On the other hand, there was development of extraradical mycelium only at 15 °C, thereby suggesting that lower temperature inhibits extraradical mycelium development.

In plants, growing in cool temperate climatic condition, optimum mycorrhizal development has been observed at a temperature range between 20 and 25 °C (Matsubara et al. 2000). However, at temperature below 15 °C, AMF colonization has been reported to be restrained (Liu et al. 2004; Zhang et al. 1995; Zhu et al. 2010b). Liu et al. (2004) reported a complete inhibition of AMF colonization in roots of *Sorghum bicolor* at 10 °C. Studies have also shown a positive correlation between mycorrhizal colonization and the increase in temperature (Gavito et al. 2002; Hetrick and Bloom 1984; Liu et al. 2004; Ma et al. 2015; Staddon et al. 2003; Zhu et al. 2010b). Gavito et al. (2002) reported that an increase in temperature from 10 to 15 °C resulted in twofold mycorrhizal colonization in *Pisum sativum*. Contrary to these findings, few studies reported no effect of cold temperature on AMF colonization rate (Charest et al. 1993; Zhu et al. 2010b). It may be due to the short duration of exposure to low temperature.

The reduced ability of AMF to colonize plant roots at low temperature could be due to the indirect effect of decreased carbohydrate supply from the host plant (Gavito et al. 2005). The symbiotic association is based on bidirectional transfer of nutrients between the participating symbionts (Kiers et al. 2011). As plant exposed to cold stress generally shows reduced growth, it can result in a decreased supply of carbohydrates to the fungal symbiont, leading to diminished symbiotic efficiency (Ma et al. 2015).

17.3 Arbuscular Mycorrhiza and Cold Stress Tolerance in Plants

Use of arbuscular mycorrhiza has been an effective approach to improve tolerance against cold stress in many plants such as *Zea mays* (Zhu et al. 2010a, b), *Solanum lycopersicum* (Latef and Chaoxing 2011), *Oryza sativa* (Liu et al. 2013a), *Cucumis sativus* (Chen et al. 2013; Ma et al. 2015), *Jatropha curcas* (Pedranzani et al. 2015), and *Elymus nutans* (Chu et al. 2016). These studies report significant contribution of AMF in improving plant's tolerance to cold stress. The mechanisms employed by AMF include efficient reactive oxygen species (ROS) scavenging system, osmotic adjustment and regulation of root hydraulic conductance, improved photosynthetic efficiency, improved respiration, increased nutrient uptake, and

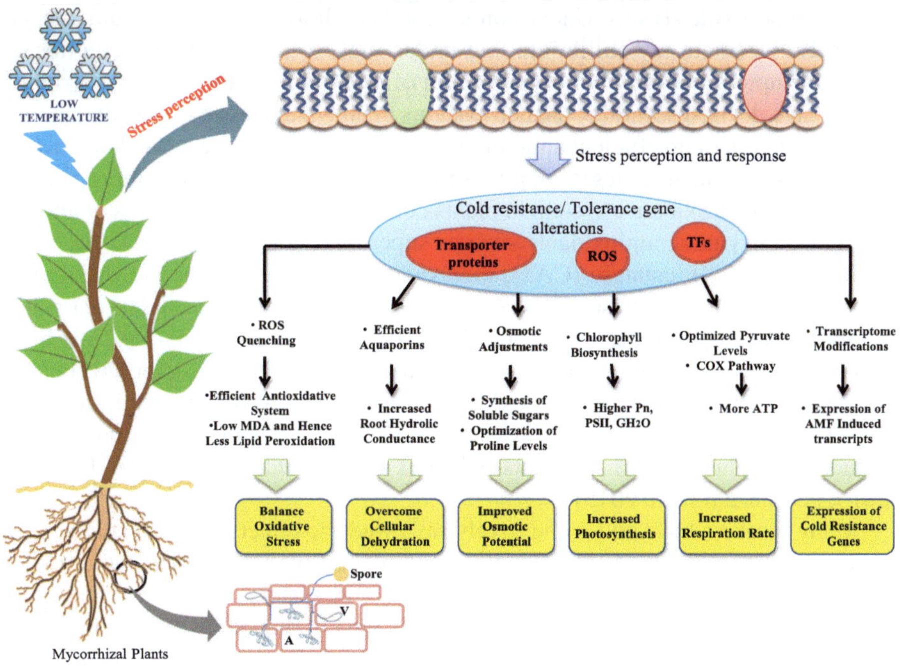

Fig. 17.2 Different mechanisms employed by arbuscular mycorrhizal (AM) plants to alleviate cold stress

adjustments at the transcriptomic level (Fig. 17.2) (Aroca et al. 2007; Chen et al. 2013; Latef and Chaoxing 2011; Liu et al. 2015; Ma et al. 2015, 2018).

Osmotic stress caused by low temperature is negated by mycorrhizal plants through osmotic adjustment brought by synthesis of osmolytes such as soluble sugars and optimization of proline level. Plant-water relationship is maintained by regulating the root hydraulic conductance and aquaporin functions. AM plants withstand cold-induced oxidative stress through enhanced production of antioxidant compounds that scavenge ROS and improve antioxidant enzyme activity. Both respiration and photosynthetic efficiency are also improved in plants upon mycorrhization. Molecular mechanisms regulated by AM to redress cold stress include activation of certain genes that code for functional proteins, such as aquaporins, sugar, and ion transporters. All these mechanisms work in synchrony to provide tolerance of mycorrhizal plants to cold stress.

17.3.1 Efficient Reactive Oxygen Species (ROS) Scavenging System

Low-temperature-mediated oxidative stress plays a crucial role in causing cold injury in plant cells (Bowler et al. 1992; Liu et al. 2016). Cold-induced oxidative damage is mediated by ROS such as superoxides [$O_2^{\bullet -}$], hydroxyl ion [$^{\bullet}OH$], and hydrogen peroxide [H_2O_2]. Under non-stressed conditions, production and quenching of ROS remain in equilibrium, but under stressed conditions this equilibrium gets interrupted causing the overproduction of ROS (Bowler et al. 1992). However, plants have devised several dedicated systems to alleviate these ROS-mediated injuries. Instigation of an efficient antioxidant system to reinstate the redox homeostasis in cell is among such strategies of plants to mitigate ROS-mediated damage (Lázaro et al. 2013). Antioxidative system in plants consists of three main classes of antioxidants: (1) enzymatic antioxidants such as superoxide dismutase (SOD), ascorbate peroxidase (APX), catalase (CAT), monodehydroascorbate reductase (MDHAR), dehydroascorbate reductase (DHAR), and glutathione reductase (GR); (2) water-soluble reductants such as ascorbate (AsA) and glutathione (GSH); and (3) lipid-soluble antioxidative molecules such as sugars (Foyer and Noctor 2011). AM improves the redox poise and antioxidant system in plants. Recent studies revealed the upregulation of antioxidant systems and hence efficient alleviation of cold-induced ROS in mycorrhizal plants (Chen et al. 2013; Chu et al. 2016; Liu et al. 2016, Pedranzani et al. 2015).

Many workers have comprehensively reviewed the effects of AMF colonization on enzymatic activities of antioxidants (Kapoor and Singh 2016; Wu et al. 2014). It is proposed that AM symbiosis boosts antioxidative system in plants by improving SOD activity. Plants inoculated with AMF showed higher SOD activity than their non-AM counterparts (Chu et al. 2016; Yang et al. 2015). This suggested that AMF-inoculated plants dismutate the superoxide radical to H_2O_2 more efficiently than the non-inoculated plants. Latef and Chaoxing (2011) reported that mycorrhizal inoculation of *Glomus mosseae* augments the activities of antioxidative enzymes such as SOD, POD, CAT, and APX in mycorrhizal plants exposed to low-temperature stress in *Solanum lycopersicum*. Chu et al. (2016) also suggested that *Glomus mosseae* inoculation can lessen the ROS accumulation and hence the oxidative stress in cold-sensitive and cold-tolerant genotypes of *Elymus nutans* by increasing the activity of a variety of antioxidant enzymes.

AM-mediated cold tolerance is also associated with efficient neutralization of H_2O_2 by regulating ascorbate and glutathione recycling (Liu et al. 2016). In ascorbate and glutathione cycle, APX uses AsA as an electron donor to eliminate H_2O_2 and oxidizes AsA to dehydroascorbate. Oxidized ascorbates such as monodehydroascorbate and dehydroascorbate, generated by the action of APX, are reduced back with the help of MDHAR and DHAR, respectively (Ordoñez et al. 2014). DHAR catalyzes the reduction of dehydroascorbate involving the oxidation of GSH to form ascorbate and glutathione disulfide. It is responsible for regenerating AsA from its oxidized state, regulating the cellular AsA redox state, which in turn shape cell responsiveness and tolerance to environmental ROS (Chen and Gallie, 2006). An increase in DHAR activity is a feedback regulation mechanism meant to recover AsA regeneration from DHA when AsA depletion occurs at its production sites due to enhanced ROS (Locato et al. 2009). Similarly, GR is another potential enzyme of ascorbate and glutathione cycle, which maintains the level of reduced glutathione inside the cell (Rao and Reddy 2008). Glutathione is crucial to maintain the normal redox status of the cells and neutralize the adverse effects of ROS-mediated oxidative damage in plants (Noctor et al. 2012). Liu et al. (2016) showed that AM plants showed higher expression levels of *APX*, *DHAR*, *MDHAR*, and *GR* genes and improved redox ratios of ascorbate and glutathione under cold stress than non-mycorrhizal plants.

17.3.2 Osmotic Adjustment and Regulation of Root Hydraulic Properties

Low temperature has a negative effect on cell membrane. The fatty acid tails of the phospholipid get saturated and hence become more rigid. The decrease in the overall fluidity of membrane results in reduced permeability, thereby restricting entry of important biomolecules into the cell. This disrupts the osmotic potential of plant cell affecting growth and development of the plant (Chen et al. 2013, Wu and Zou 2010). Plants adjust their cellular osmotic potential to tolerate cold stress and it has become an important mechanism for cold stress tolerance. Cold-tolerant plants produce a range of low molecular organic molecules called compatible osmolytes. These include soluble sugars, proline, polyamines, betaines, and acylated sterols (Chu et al. 2016; Ruelland and Zachowsk 2010; Theocharis et al. 2012). Synthesis and accumulation of these osmoprotectants lower the osmotic potential in the cytosol and maintain turgor pressure of the cells. AM is reported to alleviate cold stress in plants through adjustment of osmotic potential (Chu et al. 2016; Zhu et al. 2010a).

Soluble sugars, in addition to acting as osmoprotectants, also act as cryoprotectants to protect the plant cell membrane during cold stress (Welling and Palva 2006). Soluble sugars defend plant cell membranes during chilling-induced physiological dehydration by replacing water molecule in forming hydrogen bonds with lipid molecules present in cell membrane (Ruelland et al. 2009). They also act as signaling molecules and contribute in hormone-mediated stress responses in plants (Theocharis et al. 2012; Zeng et al. 2011). Chilling and AMF both are reported to influence the level of sugars in plants. Increased production of soluble sugars has been reported in mycorrhizal plants (Charest et al. 1993; Chu et al. 2016; Latef and Chaoxing 2011; Pedranzani et al. 2015; Zhu et al. 2010a). Studies have reported that mycorrhizal plants synthesize more sugar to cope with cold stress via hydrolysis of leaf starch (Liu et al. 2013a, b; Zhu et al. 2015).

Proline is a well-known and extensively reported osmo- and cryoprotectant that has been found to accrue in response to all type of abiotic stresses including chilling (Kishor et al. 2005; Theocharis et al. 2012). A positive correlation between proline content and improved chilling tolerance has been reported in most of the chilling-insensitive plants (Kaur et al. 2011; Szabados and Savoure 2010). However, chilling-sensitive plants don't attain cold tolerance, unless a high concentration of proline is applied prior to stress (Kushad and Yelenosky 1987; Xin and Li 1993), which indicates that proline possesses a potential to alleviate chilling injury in plants. It acts as a hydroxyl radical scavenger, a protein stabilizer, redox regulator, and ameliorates cytoplasmic acidosis (Hare and Cress 1997; Kishor et al. 2005; Theocharis et al. 2012). Studies have reported that under low-temperature stress, mycorrhizal plants were found to accumulate higher root and leaf proline content than non-mycorrhizal plants (Chen et al. 2014; Zhu et al. 2010a). However, Latef and Chaoxing (2011) reported that mycorrhizal plants accumulate lower proline in their leaves as compared to non-mycorrhizal plants when subjected to cold stress. This suggests that cold stress has lesser effect on mycorrhizal plants as compared to their non-mycorrhizal counterparts. As a result, mycorrhizal plant need not synthesize proline in larger amount for osmotic adjustment.

Cold stress can also induce dehydration of plant tissues due to formation of ice in the apoplast (Seki et al. 2003). The disparity in the amount of water lost in the leaves and water uptake rate by the roots can also lead to dehydration (Aroca et al. 2001). Thus, the availability of water and efficiency of plants to take up water determine the acclimatization ability of plants to cold stress (Thomashow 1999). Hydraulic conductance of the root regulates water uptake capacity of the plant under the guidance of aquaporins (Luu and Maurel 2005). Aquaporin, an integral membrane protein, is a channel that functions in the transfer of water and, at times, small solutes across the membrane of plant, animals, and bacteria (Takata et al. 2004).

Plants in symbiotic association with AMF have a higher ability of water uptake from the soil under water-deficit condition as compared to non-AM plants (Marulanda et al. 2003; Khalvati et al. 2005). However, there are very few reports on how AM symbiosis influences the root hydraulic properties under cold stress. Aroca and team (2007) reported that the hydraulic conductance capacity of AM *Phaseolus vulgare* under normal condition was approximately about half of that in non-AM plants. However, exposure to cold stress led to decreased conductance capacity in the non-AM plants while it remained unaltered in AM plants. This acquired advantage of AM plants gives them an edge over non-AM plants in tolerating cold stress. Plasma membrane aquaporin gene, *PIP1;3* (plasma membrane integral protein), showed increased expression in mycorrhizal *Phaseolus vulgare* and improved the root hydraulic conductance than the non-mycorrhizal plants under cold stress (Aroca et al. 2007). In *Lotus japonicus*, the expression of two putative aquaporin genes *NIP1* (NOD26-like intrinsic protein) and *XIP1* (X-intrinsic protein) was reported to be induced upon mycorrhization (Giovannetti et al. 2012). The role of aquaporins in improving root hydraulic conductivity for mitigation of cold stress needs to be explored further.

17.3.3 Improved Photosynthetic Efficiency

Photosynthesis is one of the many physiological aspects that are affected by low temperature (Adam and Murthy 2014; Ensminger et al. 2006; Paredes and Quiles 2015). Efficient photosynthesis is very important for chilling tolerance. A better photosynthetic rate provides energy in the form of photosynthates which is required for instigation of cold acclimation responses. Chlorophyll is a preeminent and vital biomolecule in plants that is involved in light harvesting and energy transformation during photosynthesis. Like any other cell component, chloroplast is also affected severely under cold stress (Ensminger et al. 2006). Low temperature causes alterations in chlorophyll biosynthesis and hence compromised chlorophyll content (Chu et al. 2016; Latef and Chaoxing 2011). Liu et al. (2013b) examined the effect of extremely low temperatures (1 °C and – 10 °C) on chlorophyll concentrations in cold-tolerant *Avena nuda*. The study reported that even with 5 days of chilling, chlorophyll content was reduced by 66% as compared to plants grown under ambient temperature. *De novo* transcriptomics and gene expression profiling studies in *Elymus nutans* grown under low temperature revealed that expression of genes involved in the biosynthesis of chlorophyll *a* decreased significantly in cold-tolerant as well as in cold-sensitive genotypes (Fu et al. 2016). This decrease in chlorophyll content due to chilling causes changes in the chlorophyll antenna complex, resulting in compromised PSII (photosystem II) and, hence, photosynthesis.

There are several studies where AMF help plants to combat the negative effects of chilling on chlorophyll content in plants, such as *Triticum aestivum* (Paradis et al. 1995), *Zea mays* (Zhu et al. 2010b; 2012), and *Solanum lycopersicum* (Latef and Chaoxing 2011). Chu et al. (2016) reported that in *Elymus nutans* cold temperature treatment decreased the content of photosynthetic pigments including chlorophyll *a* and chlorophyll *b* and total chlorophyll content. However, plants inoculated with *Glomus mosseae* showed improved concentration of photosynthetic pigments under cold stress. Porcel et al. (2015) suggested that the involvement of AMF in fortifying carbon sink may improve photosynthesis in AMF-colonized plants. As AMF colonization increases the chlorophyll content as well as the leaf gaseous exchange in host plants, it can be inferred that both the stomatal and non-stomatal factors are associated with AMF-mediated increase in photosynthesis. Also, the photosynthetic rate in plants is under austere control of two major biochemical processes, which are Rubisco carboxylation and RuBP regeneration (Amthor 1995). Both the processes are found to be positively affected by AMF colonization in host plants (Chen et al. 2017). Inoculation of plants with AMF increases the net photosynthetic rate, improves stomatal conductance and intercellular CO_2 concentration, maximizes Rubisco carboxylation and RuBP regeneration, and boosts quantum yield of PSII which consequently improves photosynthetic efficiency of plant (Chen et al. 2017; Yang et al. 2015).

17.3.4 Improved Respiration

Cold stress decreases the sink activity as well as the leaf carbohydrate exports in plants, which alters the level of assimilated carbon, and subsequently affects plant respiration (Bagnall et al. 1988; Labate and Leegood 1988). In addition to that, chilling also lowers the respiration rate and ATP production, which may be linked to cold-induced alterations in respiratory pathways (Karasawa et al. 2012; Kurimoto et al. 2004). Cytochrome C oxidase (COX) pathway and alternative oxidase (AOX) pathway are the two respiratory electron transport chain routes present in plant mitochondria. Under normal growth conditions, COX pathway is the chief route as it is directly involved in ATP synthesis (Maxwell et al. 1999; O'Leary and Plaxton 2016). However, it has been proposed that when plants are exposed to abiotic stress, such as chilling, heat, drought, and salinity, AOX route gets engaged in the optimization of respiratory metabolism (Li et al. 2013). This decreases the production of ATP per oxygen consumed and ultimately results in energetically less efficient respiratory system (Robinson et al. 1995). Several workers have also correlated the high AOX activity with high pyruvate levels (Dinakar et al. 2010; Oliver et al. 2008).

Recent studies have revealed that AMF colonization improves the respiration rate in plants exposed to low temperature. Liu et al. (2015) reported that *Oryza sativa* plants when inoculated with *Glomus intraradices* showed better respiration rates than non-inoculated plants under cold stress. Colonization of AMF augments levels of respiratory substrates such as soluble sugars and starch and also optimizes pyruvate. This consequently improves the efficiency of TCA cycle. AM plants are more efficient in ATP production than the non-inoculated plants during cold stress. AMF utilizes the COX pathway instead of AOX pathway, thereby assuring more ATP production (Atkin et al. 2009; Del-Saz et al. 2017a, b; Liu et al. 2015). High levels of pyruvate are associated with AOX pathway (Dinakar et al. 2010; Oliver et al. 2008), which is a less efficient respiratory pathway in terms of ATP production (Robinson et al. 1995). Liu et al. (2015) also reported that AMF inoculation results in increased expression levels of pyruvate kinase (*PK*) and pyruvate dehydrogenase (*PDHEα*) genes whose products are involved in pyruvate metabolism. Pyruvate dehydrogenase degrades the pyruvate and maintains its optimum level so that the respiratory pathway remains restricted to COX route as suggested by higher expression levels of *COX5c* gene in mycorrhizal plants. Additionally, AMF inoculation also results in overexpression of malate dehydrogenase (*MDH*) and isocitrate dehyhydrogenase *a* (*IDHa*) genes that ensures an efficient TCA cycle and hence respiration. Hence, it can be inferred that inoculating plants with AMF can boost plant respiration and thus help host plants to tolerate low temperatures more efficiently.

There are also some reports where AMF colonization has resulted in decreased root respiration in *Trifolium subterraneum* (Silsbury et al. 1983), *Ulmus glabra* (Rewald et al. 2015), and *Populus nigra* (Otgonsuren et al. 2016). The reasons for this difference might be specificity of the plant-fungus interactions, stage of AMF colonization, and other possible factors. Del-Saz et al. (2017b) have explained the respiratory ATP cost and benefit of AMF colonization in *Nicotiana tabacum* at different growth phases. The study reported that the early phases of symbiosis with *Rhizophagus irregularis* resulted in ATP cost for *Nicotiana tabacum* leaves and thus limited shoot growth. However at later stages, there is gain of ATP allowing better growth of the host plant. Smith et al. (2009) also reported that during the initial stages of colonization, synthesis of fungal structures imposes ATP cost, but if the interaction is positive, at mature associations, the symbiosis benefits the plant through enhanced ATP production (Evelin et al. 2009).

17.3.5 Increased Nutrient Uptake

Root zone temperature directly influences the uptake and translocation of essential nutrients (Cooper 1973). Tindall et al. (1990) reported reduced uptake of most of the macro- and micronutrients at low temperature (<25 °C). Low temperature impairs the ability of plants to take up both water and nutrients. Uptake of nutrient is influenced by low temperature by directly affecting ion absorption capacity of the plant by virtue of change in the root physiology (Gregory 1988).

Mycorrhizal symbiosis directly affects the mineral nutrition of host plants. Roots of AMF-colonized plants absorb nutrients either directly through root epidermis and root hairs or indirectly through fungal hyphae (Smith and Smith 2011). The hyphae of AMF greatly enhance mineral nutrient acquisition in host plants (Marschner and Dell 1994). AM plants are benefitted nutritionally with enhanced uptake of nutrients, particularly phosphorus and nitrogen (Smith and Read 2008). Enhanced uptake of other macro- and micronutrients in AM plants has also been reported (Raju et al. 1990; Karasawa et al. 2012).

Studies have reported that AMF can boost nutrient uptake by host plant under chilling temperatures. It is proposed that improved nutrient uptake and growth parameters in AM colonized plants could be an effective means to promote the plant performance under cold stress (Ma et al. 2015). Phosphorus uptake and its distribution in plants are mediated by different Pi transport systems (Versaw and Harrison 2002). Studies have shown that certain Pi transporter genes are strongly induced in cold-stressed mycorrhizal plants (Ma et al. 2015, 2018). Some high-affinity H^+-Pi cotransporters belonging to *Pht1* gene family, which show positive responses to AM symbiosis, are *PT3* in *Solanum tuberosum* (Rausch et al. 2001), *PT11* in *Oryza sativa* (Paszkowski et al. 2002), *PT4* in *Medicago truncatula* (Harrison et al. 2002), and *PT1–1* in *Cucumis sativus* (Ma et al. 2015). In addition to the *Pht1* gene family, there are several other predicted plant Pi transporters which are responsible for internal Pi cycling in the plant and release to subcellular compartments and organelles. These include low-affinity H^+-Pi cotransporter belonging to *Pht2* gene family such as *PT2–1*, mitochondrial phosphate transporter such as *MPT3*, and transporters involved in loading of Pi to the xylem in roots, such as *PHO1-H1* (Takabatake et al. 1999; Hamburger et al. 2002). Recent studies have reported high expression of *PT2-1*, *MPT3*, and *PHO1-H1* genes that are induced upon AM colonization under cold stress (Ma et al. 2015).

Potassium is an essential macronutrient and takes part in many physiological processes, such as photosynthesis, translocation of the photosynthates, turgidity maintenance, and activation of certain enzymes under stress (Marschner 1995). Deficiency of potassium stimulates ROS production resulting in degradation of chlorophyll and membrane damage (Waraich et al. 2011). Potassium alleviates cold stress in plants by increasing the antioxidant levels and reducing production of ROS (Cakmak 2005, Devi et al. 2012). Increase in potassium-mediated cold tolerance in plant is associated with increased phospholipids and membrane permeability (Hakerlerler et al. 1997). The positive effect of AMF in potassium uptake in plants subjected to low temperature has also been reported (Raju et al. 1990; Liu et al. 2016).

Nitrogen is another macronutrient whose uptake is affected by mycorrhizal colonization. It is a major limiting macronutrient, serving as a constituent of different cell components and takes part in synthesis of nucleic acids, proteins, and so on. Studies reported significant AM-mediated uptake and transfer of nitrogen (accounting to about 20–50% of total root N) to host plants (Govindarajulu et al. 2005). There are contradictory reports on role of AMF on nitrogen acquisition under temperature stress (Barrett et al. 2011; Liu et al. 2016; Raju et al. 1990). This disparity in the report of N uptake by AMF under temperature stress could be due to specific host-AMF combination. Thus, further exploration is required to find the appropriate host-AMF combination that benefits the plant under cold stress.

17.3.6 Adjustments at Transcriptomic Level

The successful symbiosis by AMF colonization causes modifications at transcription levels in the host plants (Bucher et al. 2014) and benefit plants by improving growth, development, and stress resistance/tolerance. Multifarious roles played by AMF in plant had recently gained the attention of researchers all around the world to utilize this mutualistic association to mitigate the adverse effect of environmental stress on agricultural plants. However, there are very scarce investigations on AMF-induced transcriptional modifications in plants. Furthermore, genome-wide alterations in gene expression of plants under the influence of AMF have not been fully characterized yet. This limits our understanding to answer the exact molecular mechanisms of AMF-induced cold tolerance in plants. Thus, appraisal of gene expression adjustments in plants and AMF exposed to low-temperature stress is required. Recently the genome resources of AMF such as *Rhizophagus irregularis* (Tisserant et al. 2012) and many agricultural plants have provided prospects to scrutinize the overall genome-wide alterations in plant and AMF gene expressions. Ma et al. (2018) have explored the global alterations in *Cucumis sativus* roots colonized by *Rhizophagus irregularis* under chilling stress and provided first report of transcription regulation. Extensive transcriptome analysis has identified several AMF-induced key cold-resistant genes that participate in AMF-mediated cold resistance in *Cucumis sativus* (Ma et al. 2018). The study revealed that cold temperature influences the expression of 2173 genes, out of which about 180 genes are specifically upregulated by AMF. The genome of *Rhizophagus irregularis* contains 28,232 genes (Tisserant et al. 2012); however out of these only 0.007% genes were AMF-induced differentially expressed genes (DEGs) when *Cucumis* plants were exposed to cold stress. This very minute difference in expression level suggests the high degree of AMF adaptability during stress, which may be the major factor in establishing the symbiosis with the variety of plant species. The functional enrichment studies of 2173 DEGs categorized these as genes involved in oxidative metabolism (efficient ROS scavenging) and genes involved in ion transport (P uptake, Ca^{2+} signaling, and N metabolism). Furthermore, the AMF-induced DEGs showed that most of the upregulated genes are involved in glycerol-3-phosphate metabolism and NADH oxidation, which crosstalk with host plants during cold stress by improving redox status. Also, the KEGG (Kyoto Encyclopedia of Genes and Genomes) enrichment analysis showed the involvement of these DEGs in multiple pathways involved in hormone signal transduction, glycolysis, metabolism (phenylalanine, pyruvate, glutathione, and ascorbate), and biosynthesis (zeatin, flavonoids, and diterpenoids). Crifo et al. (2011) also reported that cold stress increases the flavonoids biosynthesis, which in turn participate in AMF colonization (Antunes et al. 2006). Therefore, it can be proposed that AMF causes several transcriptiome modifications in host plants and induce the expression of several key metabolic genes, which can cause cold resistances in plants.

17.4 Future Perspectives and Conclusions

Thirty-three percent of the total land area is considered as possibly appropriate for arable agriculture. However, only 10% of the total arable area is under cultivation. This is on the grounds that abiotic stresses in one way or the other limit the agricultural productivity (Yadav 2011). Sustainable agriculture, less affected by environmental factors, is the need of the hour to feed the consistently expanding population in order to attain global food security (Basu et al. 2018). Terrestrial plant species have been naturally selected for mycorrhizal symbiosis. This symbiotic association has and will continue to play important roles in mitigating the adverse effects on plants caused by different abiotic stresses including cold stress. AM symbiosis plays an important role in governing terrestrial ecosystems and its presence has been documented since times immemorial. The potential of AMF inoculation in alleviation of cold stress experienced by plants has been convincingly realized. The association presents a superior road in plant's adaptation to adverse environmental conditions, which will be useful for future agriculture. Nonetheless, application of mycorrhizal technology in large-scale agricultural system still has innumerable challenges to meet. Most of the studies, done till date, have explored the effect of AMF on plant's performance under cold stress with limited attention on the physiological, biochemical, and molecular aspects (Zhu et al. 2017).

The role of hormones in regulation of cold stress response in plants is well established (Eremina et al. 2016). Plant hormones play crucial roles in various growth and developmental processes. In addition, they induce certain biotic and abiotic stress-responsive genes (Gomez-Roldan et al. 2008; Umehara et al. 2008). Jasmonic acid and abscisic acid are two plant hormones, whose role in cold tolerance in plants has been extensively studied (Eremina et al. 2016; Hu et al. 2013; Nair et al. 2015; Nakashima et al. 2014). Also, the biosynthesis of these hormones and their total content were reported to be regulated by AM (Danneberg et al. 1993; Hause et al. 2002; Mandal et al. 2015; Nair et al. 2015). Although there are independent studies on the role of hormones in cold tolerance and AMs' role in enhanced hormone production in plants, there are no studies yet that link both the factors. Hence, enhanced hormone production in AM plants as a mechanism to alleviate cold stress response in plants is an area of study that can be explored further. The present understanding of molecular basis of AM-mediated cold tolerance is very scarce and is limited to certain genes only. To decipher the exact mechanisms, the roles of chilling-tolerant genes with respect to AM symbiosis and signaling are needed to be deciphered. Applications of genomics, transcriptomics, and proteomic approaches can provide a better understanding of such responses.

Although the ability of AMF to improve cold tolerance in plants under experimental condition has been appreciated lately, their effectiveness in open-field condition is not yet known. Hence, it is necessary to check its efficiency in open-field condition to ensure sustainable agriculture. Future research may focus on (i) identification of mechanisms of the symbiotic interaction in natural field conditions, (ii) further exploration of cold alleviation mechanisms employed by host plants as well as the fungal symbionts, and (iii) screening of better fungal strains (cold-tolerant) and exploring other approaches that can be conveniently applied with mycorrhizal technology to strengthen their beneficial effect on host plants.

References

Adam S, Murthy SDS (2014) Effect of cold stress on photosynthesis of plants and possible protection mechanisms. In: Gaur RK, Sharma P (eds) Approaches to plant stress and their management. Springer, New Delhi, pp 219–226

Amir H, Jourand P, Cavaloc Y, Ducousso M (2014) Role of mycorrhizal fungi in the alleviation of heavy metal toxicity in plants. In: Solaiman ZM, Abott LK, Varma A (eds) Mycorrhizal Fungi: use in sustainable agriculture and land restoration. Springer, Berlin/Heidelberg, pp 241–258

Amthor JS (1995) Terrestrial higher-plant response to increasing atmospheric [CO_2] in relation to the global carbon cycle. Glob Change Biol. 1:243–274

Antunes PM, De Varennes A, Rajcan I, Goss MJ (2006) Accumulation of specific flavonoids in soybean (*Glycine max* (L.) Merr.) as a function of the early tripartite symbiosis with arbuscular mycorrhizal fungi and *Bradyrhizobium japonicum* (Kirchner) Jordan. Soil Biol Biochem 38:1234–1242

Aroca R, Tognoni F, Irigoyen JJ, Sánchez-Díaz M, Pardossi A (2001) Different root low temperature response of two maize genotypes differing in chilling sensitivity. Plant Physiol Biochem 39:1067–1073

Aroca R, Porcel R, Ruiz-Lozano JM (2007) How does arbuscular mycorrhizal symbiosis regulate root hydraulic properties and plasma membrane aquaporins in *Phaseolus vulgaris* under drought, cold or salinity stresses? New Phytol 173:808–816

Atkin OK, Sherlock D, Fitter AH, Jarvis S, Hughes JK, Campbell C, Hurry V, Hodge A (2009) Temperature dependence of respiration in roots colonized by arbuscular mycorrhizal fungi. New Phytol 182:188–199

Bagnall DJ, King RW, Farquhar GD (1988) Temperature-dependent feedback inhibition of photosynthesis in peanut. Planta 175:348–354

Bago B, Pfeffer PE, Shachar-Hill Y (2000) Carbon metabolism and transport in arbuscular mycorrhizas. Plant Physiol 124:949–958

Balestrini R, Lumini E, Borriello R, Bianciotto V (2015) Plant-soil biota interactions. Soil Microbiol Ecol Biochem:311–338

Barrett G, Campbell CD, Fitter AH, Hodge A (2011) The arbuscular mycorrhizal fungus *Glomus hoi* can capture and transfer nitrogen from organic patches to its associated host plant at low temperature. Appl Soil Ecol 48:102–105

Basu S, Rabara RC, Negi S (2018) AMF: the future prospect for sustainable agriculture. Physiol Mol Plant Pathol 102:36–45

Beck EH, Heim R, Hansen J (2004) Plant resistance to cold stress: mechanisms and environmental signals triggering frost hardening and dehardening. J Biosci 29:449–459

Bowler C, Montagu MV, Inzé D (1992) Superoxide dismutase and stress tolerance. Annu Rev Plant Biol 43:83–116

Bucher M, Hause B, Krajinski F, Küster H (2014) Through the doors of perception to function in arbuscular mycorrhizal symbioses. New Phytol 204:833–840

Byun MY, Lee J, Cui LH, Kang Y, Oh TK, Park H, Lee H, Kim WT (2015) Constitutive expression of *DaCBF7*, an Antarctic vascular plant *Deschampsia antarctica* CBF homolog, resulted in improved cold tolerance in transgenic rice plants. Plant Sci 236:61–74

Cakmak I (2005) The role of potassium in alleviating detrimental effects of abiotic stresses in plants. J Plant Nutr Soil Sci 168:521–530

Candido V, Campanelli G, D'Addabbo T, Castronuovo D, Perniola M, Camele I (2015) Growth and yield promoting effect of artificial mycorrhization on field tomato at different irrigation regimes. Sci Hortic 187:35–43

Charest C, Dalpé Y, Brown A (1993) The effect of vesicular-arbuscular mycorrhizae and chilling on two hybrids of *Zea mays* L. Mycorrhiza 4:89–92

Chen Z, Gallie DR (2006) Dehydroascorbate reductase affects leaf growth, development, and function. Plant Physiol 142:775–787

Chen S, Jin W, Liu A, Zhang S, Liu D, Wang F, Lin X, He C (2013) Arbuscular mycorrhizal fungi (AMF) increase growth and secondary metabolism in cucumber subjected to low temperature stress. Sci Hortic 160:222–229

Chen X, Song F, Liu F, Tian C, Liu S, Xu H, Zhu X (2014) Effect of different arbuscular mycorrhizal fungi on growth and physiology of maize at ambient and low temperature regimes. Sci World J 2014:1–7

Chen S, Zhao H, Zou C, Li Y, Chen Y, Wang Z, Jiang Y, Liu A, Zhao P, Wang M, Ahammed GJ (2017) Combined inoculation with multiple arbuscular mycorrhizal fungi improves growth, nutrient uptake and photosynthesis in cucumber seedlings. Front Microbiol 8:2516–2527

Chu XT, Fu JJ, Sun YF, Xu YM, Miao YJ, Xu YF, Hu TM (2016) Effect of arbuscular mycorrhizal fungi inoculation on cold stress-induced oxidative damage in leaves of *Elymus nutans* Griseb. S Afr J Bot 104:21–29

Cooper AJ (1973) Root temperature and plant growth. Commonwealth Agricultural Bureaux, Farnham Royal

Crifò T, Puglisi I, Petrone G, Recupero GR, Piero ARL (2011) Expression analysis in response to low temperature stress in blood oranges: implication of the flavonoid biosynthetic pathway. Gene 476:1–9

Danneberg G, Latus C, Zimmer W, Hundeshagen B, Schneider-Poetsch HJ, Bothe H (1993) Influence of vesicular-arbuscular mycorrhiza on phytohormone balances in maize (*Zea mays* L.). J Plant Physiol 141:33–39

Del-Saz NF, Romero-Munar A, Cawthray GR, Aroca R, Baraza E, Flexas J, Lambers H, Ribas-Carbó M (2017a) Arbuscular mycorrhizal fungus colonization in *Nicotiana tabacum* decreases the rate of both carboxylate exudation and root respiration and increases plant growth under phosphorus limitation. Plant Soil 416:97–106

Del-Saz NF, Romero-Munar A, Alonso D, Aroca R, Baraza E, Flexas J, Ribas-Carbó M (2017b) Respiratory ATP cost and benefit of arbuscular mycorrhizal symbiosis with *Nicotiana tabacum* at different growth stages and under salinity. J Plant Physiol 218:243–248

Devi BSR, Kim YJ, Selvi SK, Gayathri S, Altanzul K, Parvin S, Yang DU, Lee OR, Lee S, Yang DC (2012) Influence of potassium nitrate on antioxidant level and secondary metabolite genes under cold stress in *Panax ginseng*. Russ J Plant Physiol 59:318–325

Dinakar C, Raghavendra AS, Padmasree K (2010) Importance of AOX pathway in optimizing photosynthesis under high light stress: role of pyruvate and malate in activating AOX. Physiol Plant 139:13–26

Ensminger I, Busch F, Huner NP (2006) Photostasis and cold acclimation: sensing low temperature through photosynthesis. Physiol Plant 126:28–44

Eremina M, Rozhon W, Poppenberger B (2016) Hormonal control of cold stress responses in plants. Cell Mol Life Sci 73:797–810
Evelin H, Kapoor R, Giri B (2009) Arbuscular mycorrhizal fungi in alleviation of salt stress: a review. Ann Bot 104:1263–1280
Evelin H, Giri B, Kapoor R (2012) Contribution of *Glomus intraradices* inoculation to nutrient acquisition and mitigation of ionic imbalance in NaCl-stressed *Trigonella foenum-graecum*. Mycorrhiza 22:203–217
Evelin H, Giri B, Kapoor R (2013) Ultrastructural evidence for AMF mediated salt stress mitigation in *Trigonella foenum-graecum*. Mycorrhiza 23:71–86
Ferullo JM, Griffith M (2001) Mechanisms of cold acclimation. In: Basra AS (ed) Crop responses & adaptations to temperature stress. Food Products Press, New York, pp 109–150
Foyer CH, Noctor G (2011) Ascorbate and glutathione: the heart of the redox hub. Plant Physiol 155:2–18
Fu J, Miao Y, Shao L, Hu T, Yang P (2016) *De novo* transcriptome sequencing and gene expression profiling of *Elymus nutans* under cold stress. BMC Genomics 17:870–889
Gavito ME, Bruhn D, Jakobsen I (2002) Phosphorus uptake by arbuscular mycorrhizal hyphae does not increase when the host plant grows under atmospheric CO_2 enrichment. New Phytol 154:751–760
Gavito ME, Olsson PA, Rouhier H, Medina-Peñafiel A, Jakobsen I, Bago A, Azcón-Aguilar C (2005) Temperature constraints on the growth and functioning of root organ cultures with arbuscular mycorrhizal fungi. New Phytol 168:179–188
Giovannetti M, Balestrini R, Volpe V, Guether M, Straub D, Costa A, Ludewig U, Bonfante P (2012) Two putative-aquaporin genes are differentially expressed during arbuscular mycorrhizal symbiosis in *Lotus japonicus*. BMC Plant Biol 12:186–200
Gomez-Roldan V, Fermas S, Brewer PB, Puech-Pagès V, Dun EA, Pillot JP, Letisse F, Matusova R, Danoun S, Portais JC, Bouwmeester H (2008) Strigolactone inhibition of shoot branching. Nature 455:189–194
Govindarajulu M, Pfeffer PE, Jin HR, Abubaker J, Douds DD, Allen JW, Bücking H, Lammers PJ, Shachar-Hill Y (2005) Nitrogen transfer in the arbuscular mycorrhizal symbiosis. Nature 435:819–823
Gregory PJ (1988) Growth and functioning of plant roots. Russell's soil conditions and plant growth. Longman Scientific and Technical. Inglaterra, pp 113–67
Guy CL (1990) Cold acclimation and freezing stress tolerance: role of protein metabolism. Annu Rev Plant Biol 41:187–223
Hajiboland R, Aliasgharzadeh N, Laiegh SF, Poschenrieder C (2010) Colonization with arbuscular mycorrhizal fungi improves salinity tolerance of tomato (*Solanum lycopersicum* L.) plants. Plant Soil 331:313–327
Hakerlerler H, Oktay M, Eryuce N, Yagmur B (1997) Effect of potassium sources on the chilling tolerance of some vegetable seedlings grown in hotbeds. In: Johnston AE (ed) Food security in the WANA region, the essential need for balanced fertilization. International Potash Institute, Basel, pp 353–359
Hamburger D, Rezzonico E, Petétot JMC, Somerville C, Poirier Y (2002) Identification and characterization of the Arabidopsis PHO1 gene involved in phosphate loading to the xylem. Plant Cell 14:889–902
Hannah MA, Heyer AG, Hincha DK (2005) A global survey of gene regulation during cold acclimation in *Arabidopsis thaliana*. PLoS Genet 1:26–44
Hare PD, Cress WA (1997) Metabolic implications of stress-induced proline accumulation in plants. Plant Growth Regul 21:79–102
Harrison MJ, Dewbre GR, Liu J (2002) A phosphate transporter from *Medicagotruncatula* involved in the acquisition of phosphate released by arbuscular mycorrhizal fungi. Plant Cell 14:2413–2429
Hause B, Maier W, Miersch O, Kramell R, Strack D (2002) Induction of jasmonate biosynthesis in arbuscular mycorrhizal barley roots. Plant Physiol 130:1213–1220

Heinemeyer A, Fitter AH (2004) Impact of temperature on the arbuscular mycorrhizal (AM) symbiosis: growth responses of the host plant and its AM fungal partner. J Exp Bot 55:525–534
Hetrick BD, Bloom J (1984) The influence of temperature on colonization of winter wheat by vesicular-arbuscular mycorrhizal fungi. Mycologia 76:953–956
Hu Y, Jiang L, Wang F, Yu D (2013) Jasmonate regulates the inducer of CBF expression–c-repeat binding factor/DRE binding factor 1 cascade and freezing tolerance in Arabidopsis. Plant Cell 113:1–19
Jan N, Andrabi KI (2009) Cold resistance in plants: a mystery unresolved. EJB 12:14–15
Janicka-Russak M, Kabała K, Wdowikowska A, Kłobus G (2012) Response of plasma membrane H^+-ATPase to low temperature in cucumber roots. J Plant Res 125:291–300
Jewell MC, Campbell BC, Godwin ID (2010) Transgenic plants for abiotic stress resistance. In: Kole C, Michler CH, Abbott AG, Hall TC (eds) Transgenic crop plants. Springer, Berlin, Heidelberg, pp 67–132
Kapoor R, Singh N (2016) Arbuscular mycorrhiza and reactive oxygen species. In: Wu QS (ed) Arbuscular mycorrhizas and stress tolerance of plants. Springer Nature, Singapore, pp 225–243
Karasawa T, Hodge A, Fitter AH (2012) Growth, respiration and nutrient acquisition by the arbuscular mycorrhizal fungus *Glomus mosseae* and its host plant *Plantago lanceolata* in cooled soil. Plant Cell Environ 35:819–828
Kaur G, Kumar S, Thakur P, Malik JA, Bhandhari K, Sharma KD, Nayyar H (2011) Involvement of proline in response of chickpea (*Cicer arietinum* L.) to chilling stress at reproductive stage. Sci Hortic 128:174–181
Khalvati MA, Hu Y, Mozafar A, Schmidhalter U (2005) Quantification of water uptake by arbuscular mycorrhizal hyphae and its significance for leaf growth, water relations, and gas exchange of barley subjected to drought stress. Plant Biol 7:706–712
Kiers ET, Duhamel M, Beesetty Y, Mensah JA, Franken O, Verbruggen E, Fellbaum CR, Kowalchuk GA, Hart MM, Bago A, Palmer TM (2011) Reciprocal rewards stabilize cooperation in the mycorrhizal symbiosis. Science 333:880–882
Kishor PK, Sangam S, Amrutha RN, Laxmi PS, Naidu KR, Rao KRSS, Rao S, Reddy KJ, Theriappan P, Sreenivasulu N (2005) Regulation of proline biosynthesis, degradation, uptake and transport in higher plants: its implications in plant growth and abiotic stress tolerance. Curr Sci:424–438
Kumar A, Dames JF, Gupta A, Sharma S, Gilbert JA, Ahmad P (2015) Current developments in arbuscular mycorrhizal fungi research and its role in salinity stress alleviation: a biotechnological perspective. Crit Rev Biotechnol 35:461–474
Kurimoto K, Millar AH, Lambers H, Day DA, Noguchi K (2004) Maintenance of growth rate at low temperature in rice and wheat cultivars with a high degree of respiratory homeostasis is associated with a high efficiency of respiratory ATP production. Plant Cell Physiol 45:1015–1022
Kushad MM, Yelenosky G (1987) Evaluation of polyamine and proline levels during low temperature acclimation of citrus. Plant Physiol 84:692–695
Labate CA, Leegood RC (1988) Limitation of photosynthesis by changes in temperature. Planta 173:519–527
Latef AAHA, Chaoxing H (2011) Arbuscular mycorrhizal influence on growth, photosynthetic pigments, osmotic adjustment and oxidative stress in tomato plants subjected to low temperature stress. Acta Physiol Plant 33:1217–1225
Lázaro JJ, Jiménez A, Camejo D, Martí MC, Lázaro-Payo A, Barranco-Medina S, Sevilla F (2013) Dissecting the integrative antioxidant and redox systems in plant mitochondria. Effect of stress and S-nitrosylation. Front Plant Sci 460:1–20
Levitt J (1980) Responses of plants to environmental stress, chilling, freezing, and high temperature stresses. Academic:1–497
Li CR, Liang DD, Li J, Duan YB, Hao L, Yang YC, Qin RY, Li LI, Wei PC, Yang JB (2013) Unravelling mitochondrial retrograde regulation in the abiotic stress induction of rice alternative oxidase 1 genes. Plant Cell Environ 36:775–788
Liu A, Wang B, Hamel C (2004) Arbuscular mycorrhiza colonization and development at suboptimal root zone temperature. Mycorrhiza 14:93–101

Liu ZL, Li YJ, Hou HY, Zhu XC, Rai V, He XY, Tian CJ (2013a) Differences in the arbuscular mycorrhizal fungi-improved rice resistance to low temperature at two N levels: aspects of N and C metabolism on the plant side. Plant Physiol Biochem 71:87–95

Liu W, Yu K, He T, Li F, Zhang D, Liu J (2013b) The low temperature induced physiological responses of *Avena nuda* L., a cold-tolerant plant species. Sci World J 2013:1–8

Liu Z, Li Y, Wang J, He X, Tian C (2015) Different respiration metabolism between mycorrhizal and non-mycorrhizal rice under low-temperature stress: a cry for help from the host. J of Agric Sci 153:602–614

Liu A, Chen S, Wang M, Liu D, Chang R, Wang Z, Lin X, Bai B, Ahammed GJ (2016) Arbuscular mycorrhizal fungus alleviates chilling stress by boosting redox poise and antioxidant potential of tomato seedlings. J Plant Growth Regul 35:109–120

Locato V, De Pinto MC, De Gara L (2009) Different involvement of the mitochondrial, plastidial and cytosolic ascorbate-glutathione redox enzymes in heat shock responses. Physiol Plant 135:296–306

Luu DT, Maurel C (2005) Aquaporins in a challenging environment: molecular gears for adjusting plant water status. Plant Cell Environ 28:85–96

Ma J, Janoušková M, Li Y, Yu X, Yan Y, Zou Z, He C (2015) Impact of arbuscular mycorrhizal fungi (AMF) on cucumber growth and phosphorus uptake under cold stress. Funct Plant Biol 42:1158–1167

Ma J, Sun C, Bai L, Dong R, Yan Y, Yu X, He C, Zou Z, Li Y (2018) Transcriptome analysis of cucumber roots reveals key cold-resistance genes induced by AM fungi. Plant Mol Biol Report 36:135–148

Mandal S, Upadhyay S, Wajid S, Ram M, Jain DC, Singh VP, Abdin MZ, Kapoor R (2015) Arbuscular mycorrhiza increase artemisinin accumulation in *Artemisia annua* by higher expression of key biosynthesis genes via enhanced jasmonic acid levels. Mycorrhiza 25:345–357

Marschner H (1995) Mineral nutrition of higher plants. Academic, New York

Marschner H, Dell B (1994) Nutrient uptake in mycorrhizal symbiosis. Plant Soil 159:89–102

Marulanda A, Azcon R, Ruiz-Lozano JM (2003) Contribution of six arbuscular mycorrhizal fungal isolates to water uptake by *Lactuca sativa* plants under drought stress. Physiol Plant 119:526–533

Marulanda-Aguirre A, Azcón R, Ruiz-Lozano JM, Aroca R (2008) Differential effects of a *Bacillus megaterium* strain on *Lactuca sativa* plant growth depending on the origin of the arbuscular mycorrhizal fungus coinoculated: physiologic and biochemical traits. J Plant Growth Regul 27:10–18

Matsubara Y, Kayukawa Y, Fukui H (2000) Temperature-stress tolerance of asparagus seedlings through symbiosis with arbuscular mycorrhizal fungus. J Jpn Soc Hortic Sci 69:570–575

Maxwell DP, Wang Y, McIntosh L (1999) The alternative oxidase lowers mitochondrial reactive oxygen production in plant cells. Proc Natl Acad Sci U S A 96:8271–8276

Miransari M (2011) Hyperaccumulators, arbuscular mycorrhizal fungi and stress of heavy metals. Biotech Adv 29:645–653

Nair A, Kolet SP, Thulasiram HV, Bhargava S (2015) Systemic jasmonic acid modulation in mycorrhizal tomato plants and its role in induced resistance against *Alternaria alternata*. Plant Biol 17:625–631

Nakashima K, Yamaguchi-Shinozaki K, Shinozaki K (2014) The transcriptional regulatory network in the drought response and its crosstalk in abiotic stress responses including drought, cold, and heat. Front Plant Sci 5:170–176

Noctor G, Mhamdi A, Chaouch S, Han YI, Neukermans J, Marquez-Garcia BELEN, Queval G, Foyer CH (2012) Glutathione in plants: an integrated overview. Plant Cell Environ 35:454–484

O'Leary BM, Plaxton WC (2016) Plant respiration. eLS, pp 1–11

Oliver SN, Lunn JE, Urbanczyk-Wochniak E, Lytovchenko A, Van Dongen JT, Faix B, Schmälzlin E, Fernie AR, Geigenberger P (2008) Decreased expression of cytosolic pyruvate kinase in potato tubers leads to a decline in pyruvate resulting in an *in vivo* repression of the alternative oxidase. Plant Physiol 148:1640–1654

Ordoñez NM, Marondedze C, Thomas L, Pasqualini S, Shabala L, Shabala S, Gehring C (2014) Cyclic mononucleotides modulate potassium and calcium flux responses to H_2O_2 in *Arabidopsis* roots. FEBS Lett 588:1008–1015

Otgonsuren B, Rewald B, Godbold DL, Göransson H (2016) Ectomycorrhizal inoculation of *Populus nigra* modifies the response of absorptive root respiration and root surface enzyme activity to salinity stress. Flora 224:123–129

Paradis R, Dalpé Y, Charest C (1995) The combined effect of arbuscular mycorrhizas and short-term cold exposure on wheat. New Phytol 129:637–642

Paredes M, Quiles MJ (2015) The effects of cold stress on photosynthesis in *Hibiscus* plants. PLoS One 10:e0137472

Parniske M (2008) Arbuscular mycorrhiza: the mother of plant root endosymbioses. Nat Rev Microbiol 6:763–775

Paszkowski U, Kroken S, Roux C, Briggs SP (2002) Rice phosphate transporters include an evolutionarily divergent gene specifically activated in arbuscular mycorrhizal symbiosis. Proc Natl Acad Sci 99:13324–13329

Pavithra D, Yapa N (2018) Arbuscular mycorrhizal fungi inoculation enhances drought stress tolerance of plants. Groundw Sustain Dev 7:490–494

Pedranzani H, Tavecchio N, Gutiérrez M, Garbero M, Porcel R & Ruiz-Lozano JM (2015). Differential effects of cold stress on the antioxidant response of mycorrhizal and non-mycorrhizal *Jatropha curcas* (L.) plants. J Agric Sci 7:35

Porcel R, Ruiz-Lozano JM (2004) Arbuscular mycorrhizal influence on leaf water potential, solute accumulation, and oxidative stress in soybean plants subjected to drought stress. J Exp Bot 55:1743–1750

Porcel R, Redondo-Gómez S, Mateos-Naranjo E, Aroca R, Garcia R, Ruiz-Lozano JM (2015) Arbuscular mycorrhizal symbiosis ameliorates the optimum quantum yield of photosystem II and reduces non-photochemical quenching in rice plants subjected to salt stress. J Plant Physiol 185:75–83

Pozo MJ, Jung SC, López-Ráez JA, Azcón-Aguilar C (2010) Impact of arbuscular mycorrhizal symbiosis on plant response to biotic stress: the role of plant defence mechanisms. In: Koltai H, Kapulnik Y (eds) Arbuscular mycorrhizas: physiology and function. Springer, Dordrecht, pp 193–207

Quiroga G, Erice G, Aroca R, Chaumont F, Ruiz-Lozano JM (2017) Enhanced drought stress tolerance by the arbuscular mycorrhizal symbiosis in a drought-sensitive maize cultivar is related to a broader and differential regulation of host plant aquaporins than in a drought-tolerant cultivar. Front Plant Sci 8:1056

Raju PS, Clark RB, Ellis JR, Maranville JW (1990) Effects of species of VA-mycorrhizal fungi on growth and mineral uptake of sorghum at different temperatures. Plant Soil 121:165–170

Rao ASVC, Reddy AR (2008) Glutathione reductase: a putative redox regulatory system in plant cells. In: Khan NA, Singh S, Umar S (eds) Sulfur assimilation and abiotic stresses in plants. Springer, Dordrecht, pp 111–147

Rausch C, Daram P, Brunner S, Jansa J, Laloi M, Leggewie G, Amrhein N, Bucher M (2001) A phosphate transporter expressed in arbuscule-containing cells in potato. Nature 414:462–470

Rewald B, Holzer L, Göransson H (2015) Impact of arbuscular mycorrhiza inoculum on biomass accumulation and root respiration of salt-stressed *Ulmusglabra* seedlings. Urban For Urban Green 14:432–437

Robinson SA, Ribas-Carbo M, Yakir D, Giles L, Reuveni Y, Berry JA (1995) Beyond SHAM and cyanide: opportunities for studying the alternative oxidase in plant respiration using oxygen isotope discrimination. Funct Plant Biol 22:487–496

Ruelland E, Zachowsk A (2010) How plants sense temperature? EEB 69:225–232

Ruelland E, Vaultier MN, Zachowski A, Hurry V (2009) Cold signalling and cold acclimation in plants. Adv Bot Res 49:35–150

Sanghera GS, Wani SH, Hussain W, Singh NB (2011) Engineering cold stress tolerance in crop plants. Curr Genomics 12:30–43

Sawers RJ, Gutjahr C, Paszkowski U (2008) Cereal mycorrhiza: an ancient symbiosis in modern agriculture. Trends Plant Sci 13:93–97

Schüßler A, Schwarzott D, Walker C (2001) A new fungal phylum, the Glomeromycota: phylogeny and evolution. Mycol Res 105:1413–1421

Seki M, Kamei A, Yamaguchi-Shinozaki K, Shinozaki K (2003) Molecular responses to drought, salinity and frost: common and different paths for plant protection. Curr Opin Biotechnol 14:194–199

Sharma S, Anand G, Singh N, Kapoor R (2017) Arbuscular mycorrhiza augments arsenic tolerance in wheat (*Triticum aestivum* L.) by strengthening antioxidant defense system and thiol metabolism. Front Plant Sci 8:906–927

Silsbury JH, Smith SE, Oliver AJ (1983) A comparison of growth efficiency and specific rate of dark respiration of uninfected and vesicular-arbuscular mycorrhizal plants of *Trifolium subterraneum* L. New Phytol 93:555–566

Smith SE, Read DJ (1997) Mycorrhizal symbiosis. Academic, San Diego

Smith SE, Read DJ (2008) Mycorrhizal symbiosis. Academic, London

Smith SE, Read DJ (2010) Mycorrhizal symbiosis. Academic, London

Smith SE, Smith FA (2011) Roles of arbuscular mycorrhizas in plant nutrition and growth: new paradigms from cellular to ecosystem scales. Annu Rev Plant Biol 62:227–250

Smith FA, Grace EJ, Smith SE (2009) More than a carbon economy: nutrient trade and ecological sustainability in facultative arbuscular mycorrhizal symbioses. New Phytol 182:347–358

Staddon PL, Thompson K, Jakobsen I, Grime JP, Askew AP, Fitter AH (2003) Mycorrhizal fungal abundance is affected by long-term climatic manipulations in the field. Glob Change Biol 9:186–194

Szabados L, Savoure A (2010) Proline: a multifunctional amino acid. Trends Plant Sci 15:89–97

Takabatake R, Hata S, Taniguchi M, Kouchi H, Sugiyama T, Izui K (1999) Isolation and characterization of cDNAs encoding mitochondrial phosphate transporters in soybean, maize, rice and *Arabidopsis*. Plant Mol Biol 40:479–486

Takata K, Matsuzaki T, Tajika Y (2004) Aquaporins: water channel proteins of the cell membrane. Prog Histochem Cytochem 39:1–83

Tayal P, Kapoor R, Bhatnagar AK (2011) Functional synergism among *Glomus fasciculatum*, *Trichoderma viride* and *Pseudomonas fluorescens* on *Fusarium* wilt in tomato. J Plant Pathol 93:745–750

Theocharis A, Clément C, Barka EA (2012) Physiological and molecular changes in plants grown at low temperatures. Planta 235:1091–1105

Thomashow MF (1999) Plant cold acclimation: freezing tolerance genes and regulatory mechanisms. Annu Rev Plant Biol 50:571–599

Tindall JA, Mills HA, Radcliffe DE (1990) The effect of root zone temperature on nutrient uptake of tomato. J Plant Nutr 13:939–956

Tisserant E, Kohler A, Dozolme-Seddas P, Balestrini R, Benabdellah K, ColardA CD, Da Silva C, Gomez SK, Koul R, Ferrol N (2012) The transcriptome of the arbuscular mycorrhizal fungus *Glomus intraradices* (DAOM 197198) reveals functional tradeoffs in an obligate symbiont. New Phytol 193:755–769

Umehara M, Hanada A, Yoshida S, Akiyama K, Arite T, Takeda-Kamiya N, Magome H, Kamiya Y, Shirasu K, Yoneyama K, Kyozuka J (2008) Inhibition of shoot branching by new terpenoid plant hormones. Nature 455:195–202

Versaw WK, Harrison MJ (2002) A chloroplast phosphate transporter, PHT2; 1, influences allocation of phosphate within the plant and phosphate-starvation responses. Plant Cell 14:1751–1766

Wang H, Wang H, Shao H, Tang X (2016) Recent advances in utilizing transcription factors to improve plant abiotic stress tolerance by transgenic technology. Front Plant Sci 7:67–80

Wani SH, Sandhu JS, Gosal SS (2008) Genetic engineering of crop plants for abiotic stress tolerance. In: Malik CP, Kaur B, Wadhwani C (eds) Advanced topics in plant biotechnology and plant biology. MD Publications, New Delhi, pp 149–183

Waraich EA, Ahmad R, Ashraf MY, Saifullah, Ahmad M (2011) Improving agricultural water use efficiency by nutrient management in crop plants. Acta Agr Scan B-S P 61:291–304

Welling A, Palva ET (2006) Molecular control of cold acclimation in trees. Physiol Plant 127:167–181
Winfield MO, Lu C, Wilson ID, Coghill JA, Edwards KJ (2010) Plant responses to cold: transcriptome analysis of wheat. Plant Biotechnol J 8:749–771
Wu QS, Zou YN (2010) Beneficial roles of arbuscular mycorrhizas in citrus seedlings at temperature stress. Sci Hortic 125:289–293
Wu QS, Zou YN, Abd-Allah EF (2014) Mycorrhizal association and ROS in plants. In: Ahmad P (ed) Oxidative damage to plants. Elsevier, Amsterdam, pp 453–475
Xin Z, Browse J (2001) Cold comfort farm: the acclimation of plants to freezing temperatures. Plant Cell Environ 23:893–902
Xin Z, Li PH (1993) Relationship between proline and abscisic acid in the induction of chilling tolerance in maize suspension-cultured cells. Plant Physiol 103:607–613
Xiong L, Schumaker KS, Zhu JK (2002) Cell signaling during cold, drought, and salt stress. Plant Cell 14:165–183
Yadav SK (2011) Cold stress tolerance mechanisms in plants. In: Gaba S, Smith B, Lichtfouse E (eds) Sustainable agriculture. Springer, Dordrecht, pp 605–620
Yang Y, Han X, Liang Y, Ghosh A, Chen J, Tang M (2015) The combined effects of arbuscular mycorrhizal fungi (AMF) and lead (Pb) stress on Pb accumulation, plant growth parameters, photosynthesis, and antioxidant enzymes in *Robinia pseudoacacia* L. PLoS One 10:e0145726
Zeng Y, Yu J, Cang J, Liu L, Mu Y, Wang J, Zhang D (2011) Detection of sugar accumulation and expression levels of correlative key enzymes in winter wheat (*Triticum aestivum*) at low temperatures. Biosci Biotechnol Biochem 75:681–687
Zhang F, Hamel C, Kianmehr H, Smith DL (1995) Root-zone temperature and soybean [*Glycine max*.(L.) Merr.] vesicular-arbuscularmycorrhizae: development and interactions with the nitrogen fixing symbiosis. Environ Exp Bot 35:287–298
Zhu X, Song F, Xu H (2010a) Influence of arbuscular mycorrhiza on lipid peroxidation and antioxidant enzyme activity of maize plants under temperature stress. Mycorrhiza 20:325–332
Zhu XC, Song FB, Xu HW (2010b) Arbuscular mycorrhizae improves low temperature stress in maize via alterations in host water status and photosynthesis. Plant Soil 331:129–137
Zhu XC, Song FB, Liu SQ, Liu TD, Zhou X (2012) Arbuscular mycorrhizae improves photosynthesis and water status of *Zea mays* L. under drought stress. Plant Soil Environ 58:186–191
Zhu XC, Song FB, Liu FL, Liu SQ, Tian CJ (2015) Carbon and nitrogen metabolism in arbuscular mycorrhizal maize plants under low-temperature stress. Crop Pasture Sci 66:62–70
Zhu X, Song F, Liu F (2017) Arbuscular mycorrhizal fungi and tolerance of temperature stress in plants. In: Wu QS (ed) Arbuscular mycorrhizas and stress tolerance of plants. Springer, Singapore, pp 163–194

Challenges in Invasive Fungal Disease 18

Arunaloke Chakrabarti and Shreya Singh

Abstract

The term invasive fungal disease (IFD) incorporates severe systemic or deep-seated infections due to fungi. A rise in the prevalence of IFD has been observed in recent years, primarily due to the rising numbers of susceptible host population, increasing invasive therapeutic interventions and immunosuppressive treatments. In addition, the expanding spectrum of fungal pathogenicity, rise in infection due to previously obscure fungi, and development of fungal thermal adaptation are also main concerns. Challenge also lies in diagnostics due to the absence of validated, prompt, and accurate diagnostic test for IFD and also differentiating infection from mere colonization. Factors related to antifungal treatment, such as point of initiation, duration of treatment, lack of ideal drug with good pharmacokinetics and pharmacodynamics, drug toxicity, and difficulty in accessing antifungal drugs affect the patient outcome. The rise in antifungal resistance, which stems from incoherent antifungal use in clinical practice and the agriculture industry, is also a matter of concern. Although IFD continues to threaten humans worldwide, our understanding of the factors affecting its changing epidemiology is still not clear. Educating clinicians and laboratory personnel about the complex issues regarding IFD diagnosis and management is essential.

Keywords

Invasive fungal disease · Challenges · Antifungal

A. Chakrabarti (✉) · S. Singh
Department of Medical Microbiology, PGIMER, Chandigarh, India

T. Satyanarayana et al. (eds.), *Advancing Frontiers in Mycology & Mycotechnology*, https://doi.org/10.1007/978-981-13-9349-5_18

18.1 Introduction

An increase in the number of patients receiving intensive care, immunosuppression, instrumentation, and rise in multi-morbid patients, HIV infection and cancers has been paralleled by a rise in fungal infections globally. Classically, the term invasive fungal disease (IFD) is used to define deep-seated or systemic, severe, life-threatening infections due to fungi. Though invasive candidiasis, aspergillosis, mucormycosis, cryptococcosis, pneumocystosis, and endemic mycoses are the major IFD, recently new fungal pathogens and new susceptible hosts have increased the spectrum of fungal infections and have thrown various challenges related to their epidemiology, diagnosis, definition, and management of IFD.

18.2 The Current Scenario of IFD

The magnitude of IFD worldwide is depicted in Fig. 18.1 with invasive candidiasis as the commonest IFD. However, there is no universal methodology for computing the exact rate due to the lack of a unifying incidence denominator (Lamoth et al. 2018). While population-based surveillance using census population as denominator is used in some studies, others use patient days or hospital/intensive care unit (ICU) admissions as denominator. This makes it difficult to estimate the worldwide invasive candidiasis rate and to compare data across different regions. In a

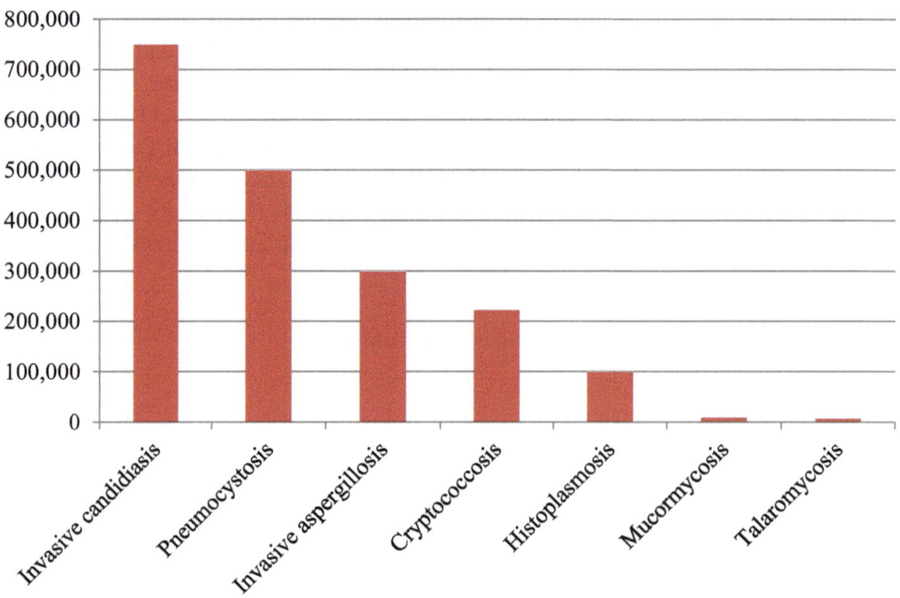

Fig. 18.1 The annual incidence of invasive fungal disease worldwide. [Source: Adapted from the estimates of global fungal disease burden published by Bongomin et al. (2017)]

point-prevalence study at 1265 ICUs in 75 countries, the incidence of invasive candidiasis was found to be 17% of all infections in critically ill patients (Chow et al. 2008; Vincent et al. 2009). In certain regions, a decline in the incidence of candidemia has been noted due to the improvement in healthcare practices [a significant decline from 14.1 and 30.9 per 100,000 person years in metropolitan Atlanta and Baltimore cities of United States to 9.5 and 14.4 per 100,000 persons in 2008, respectively (Cleveland et al. 2015)]. Additionally, a fall in the number of central venous line associated candidemia cases was also reported. Similarly, a nationwide survey of Denmark revealed a decrease in the incidence of fungemia from 2007 to 2009 (9.6 to 8.6/100,000) (Arendrup et al. 2011). On the other hand, the rate of 1.8 per 100,000 population in Australia in 2004 had gone up to 2.4 per 100,000 in 2015 (Chapman et al. 2016). In Norway too the incidence of 2.4 per 100,000 population in 2003 had gone up to 3.9 in 2012 (Hesstvedt et al. 2015). The incidence of invasive candidiasis in developing countries is difficult to predict due to absence of specific data. An upward trend of candidemia was noted at a tertiary care center from North India (Chakrabarti et al. 1996). Further, recent data reported high incidence of ICU-acquired candidemia in India (Chakrabarti et al. 2015). The epidemiology of invasive candidiasis continues to evolve with the progressive shift toward non-albicans *Candida* (NAC) species such as *C. tropicalis*, *C. parapsilosis*, and *C. glabrata* (Horn et al. 2009). The dramatic rise in the incidence of NAC may be attributed to various factors such as the increasing use of fluconazole prophylaxis in immunocompromised individuals, rise of in-hospital intervention practices, and use of central venous lines (Bajwa and Kulshrestha 2013). Various emerging yeast including *C. auris*, *Kodaemia ohmeri*, *Pichia anomala*, *C. haemulonii*, *C. famata*, and *C. rugosa* caused outbreaks in critical care settings (Chakrabarti et al. 2001, 2014). However, actual burden of infection due to these species is difficult to estimate due to the challenge in identifying the agents using conventional microbiological techniques. The emerging yeasts are less susceptible to the currently available antifungals, which is notable particularly in case of the multidrug-resistant species, *C. auris.* The lack of awareness regarding their pathogenic potential may further potentiate under-reporting of these new emerging yeasts.

Invasive aspergillosis (IA) is the second most common IFD with a proportion of 15% of all invasive fungal infections (IFIs), and the mortality due to the IA may be as high as 80% (Bulpa et al. 2001). The classical description of host factors for patients who develop IA is immunosuppressed individuals with congenital immunodeficiency, neutropenia or those undergoing aggressive immunosuppressive therapy. However, the spectrum of susceptible hosts has widened with the inclusion of patients with chronic obstructive pulmonary disease (COPD), previous history of tuberculosis, diabetes, and renal and liver failures. The emergence of various, previously obscure molds such as dematiaceous or black fungi like *Bipolaris*, *Curvularia*, *Alternaria*, and *Exophiala* species and hyaline fungi like *Fusarium* and *Scedosporium* species has also been observed.

The incidence of third most common IFD mucormycosis has also gone up over the past 20 years (Guinea et al. 2017; Kyvernitakis et al. 2016; Suh et al. 2012). The disease is commonly seen in patients with uncontrolled diabetes with or without

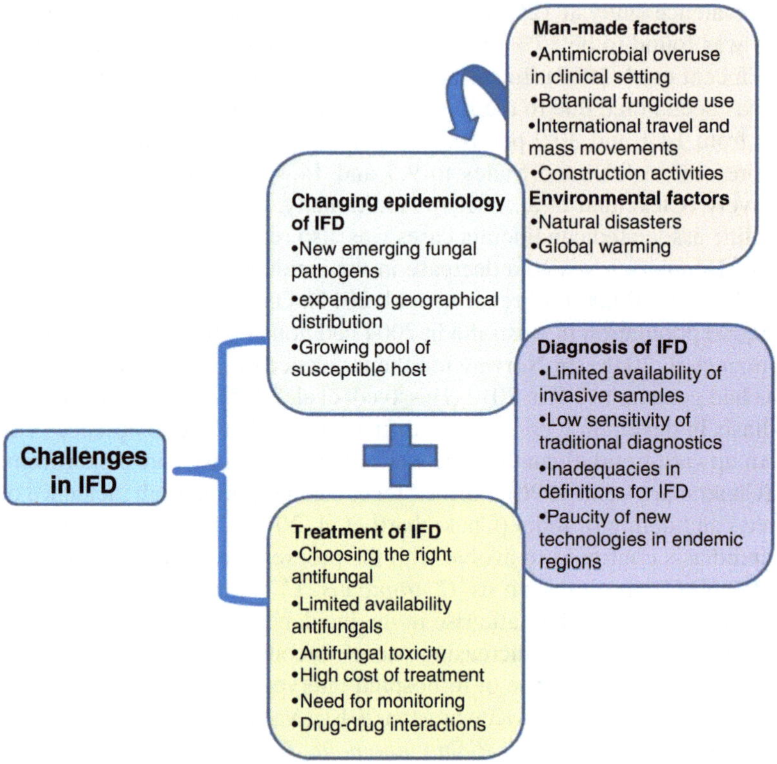

Fig. 18.2 The challenges associated with invasive fungal diseases

ketoacidosis, hematological malignancies, iron chelation therapy, immunosuppression, trauma, and burns, causing a mortality ranging from 50% to 90% (Lanternier et al. 2012; Jeong et al. 2019). Though the disease is known as community-acquired infection, it can also be acquired in the hospital setting (Perlroth et al. 2007). Iatrogenic outbreaks have also been described resulting from contaminated medical instruments, dressings, and hospital environment especially after construction activities (Krasinski et al. 1985).

18.3 The Challenges in IFD

The major challenges in IFD are change in epidemiology, diagnostic difficulty, and many issues in treatment. The factors are summarized in Fig. 18.2.

18.4 The Changing Epidemiology of IFD

The game changers in fungal disease happened with three recent developments:

1. The outbreak of *Cryptococcosis gattii* in Northwest USA and Canada: *Cryptococcosis gattii* infection was previously thought to be limited to semi-tropical and tropical regions, and its emergence in the temperate region of the Pacific Northwest and Canadian mainland in the late 1990s was unexpected. Although this pathogen was susceptible to all antifungal agents and diagnosis of the disease is not difficult, the mortality was over 30%. This fact possibly relates to the increased virulence of the organism. The patients presented with a primary pulmonary presentation as opposed to the classically described neurologic disease. Scientists investigating the presence of this predominantly tropical pathogen in temperate environments found clues by genomic dating, which suggests anthropogenic spread due to shipping and mass movements from South America many years ago (Firacative et al. 2018; Roe et al. 2018). Recently, international travel has shown to contribute to the spread of the infection from Vancouver to Japan (Kitaura et al. 2018).
2. *Sporothrix brasiliensis* outbreak in Brazil: Although subcutaneous sporotrichosis due to *Sporothrix schenckii* has been commonly seen in South America, thousands of cases due to *Sporothrix brasiliensis* have been reported from Brazil. The outbreak started from Rio de Janerio and spread across the entire country and various parts of the South American continent (Barros et al. 2008). The acquisition of the infection is contrary to basic dogma of endemic mycoses. Instead of acquiring the infection from environmental spores or conidia, the disease is spreading directly from infected cat to human (Bastos de Lima Barros et al. 2004). The zoonotic (feline) transmission has been associated with the long-lasting and large outbreak of this disease in this region.
3. Multidrug-resistant *Candida auris* behaving like bacteria and causing global outbreak: Contrary to slow development of antifungal resistance in fungi by mutation, *C. auris* develops resistance rapidly like bacteria. It can be transmitted easily in the hospital settings causing severe infections and high mortality (Jeffery-Smith et al. 2018). The identification of this organism is also difficult by traditional phenotypic methods; DNA sequencing or matrix-assisted laser desorption ionization (MALDI)-based approaches are needed.

These recent developments in the IFD scenario suggest that the epidemiology, pathophysiology, and factors affecting the emergence of fungi are quite complex. A better understanding of these parameters is integral to improve the outcome of IFD.

Experts have proposed various hypotheses for the sudden change in natural history of IFD. One interesting argument is better adaptation of fungi on human body by development of thermotolerance (Casadevall 2012). Mammals are naturally resistant to systemic mycoses by virtue of endothermy and homeothermy in addition to vertebrate level immunity (Robert and Casadevall 2009). Elevated body temperature, especially during fever contributes to the mammalian resistance to systemic fungal disease in contrast to the vulnerability of amphibians and other ectotherms to fungi. Temperature adaptation of fungi like *Cryptococcus neoformans* and *A. fumigatus* to elevated temperatures of the human host contributes to virulence (Cooney and Klein 2008). A recent analysis of the thermal susceptibility data for yeast has revealed that some fungi may be adapting to warmer temperatures (Robert et al. 2015). Although it is difficult to predict, global warming may help fungi to become more thermo-resistant, and this may be resulting in the emergence of new species of fungi causing infection in humans (Garcia-Solache and Casadevall 2010).

A link between natural disasters and development of IFI in affected population has been also recognized increasingly. Although the potential for adverse health-related events following natural disasters is well known, it poses significant challenges for public health. Infections due to *Mucorales*, *Aspergillus* spp., *Fusarium* spp., *Cladosporium* spp., *Cladophialophora bantiana*, and *Coccidioides immitis* have been reported following hurricanes, earthquakes, and Tsunami in Indian oceans (Benedict and Park 2014). Climate change could thus be affecting the ecology of fungi in ways that are not fully understood. The rising frequency coupled with greater severity of adverse climatic events could lead to more number of disaster-associated IFIs in the future.

Another major issue is the development of antifungal resistance among pathogenic fungi. Long-term use of antifungal in patients with chronic pulmonary aspergillosis and allergic broncho-pulmonary aspergillosis has led to development of itraconazole/voriconazole resistance in *Aspergillus* species, and indiscriminate use of antifungals as prophylaxis or empiric therapy has caused either emergence of intrinsically resistant *C. glabrata/C. krusei* or acquired azole resistance in common pathogenic *Candida* spp. (Chakrabarti 2011). The use of botanical fungicides for agriculture may also contribute to the rise of drug-resistant fungal infections (Ribas E Ribas et al. 2016). Azole fungicides are commonly used for the control of fungal phytopathogens. Since this group of agents is also widely used for the treatment of IFIs in humans, the environmental exposure to azole fungicides in agriculture can result in treatment failure even in azole naïve patients in the clinical setting (Berger et al. 2017). In view of the one health initiative, efforts to optimize surveillance and formulate guidelines regulating the use of such agents in the environment must be made.

18.5 Diagnosis of Invasive Fungal Disease

Though prompt initiation of antifungal therapy is crucial for the treatment of IFD, the diagnosis of IFD is often delayed by conventional techniques resulting in poor outcome of the patients. The initiation of empiric antifungal treatment when delayed more than 12 h after withdrawing the first positive blood sample is associated with a higher mortality in patients with invasive candidiasis (Morrell et al. 2005). Similar observations are also made in invasive aspergillosis and mucormycosis (Nivoix et al. 2008; Spellberg and Ibrahim 2010; Spellberg et al. 2012). Traditional mycological techniques of microscopy and culture lack sensitivity and are often time consuming, while invasive sampling needed for histopathology is usually difficult in patients with co-morbidities and thrombocytopenia (Clancy and Nguyen 2013). Even the best of laboratories cannot diagnose nearly 50% of IFD due to limitation of advancement of fungal diagnostics. Experts have commented that the current rates of IFD described are an underestimation of the problem due to the lack of sensitive diagnostic tests (Denning et al. 2017). However, researchers have brought out some better tests for diagnosis of fungal infections in recent years.

Recent progress in diagnostics has come in four specific areas:

1. Matrix-assisted laser desorption/ionization time-of-flight (MALDI-TOF) mass spectrometry: The introduction of MALDI-TOF MS for routine identification of fungal pathogens has profoundly minimized the turnaround time. In comparison to traditional techniques, it identifies yeasts and molds from culture isolates within a few minutes. But the identification of fungal pathogen by MALDI-TOF may be difficult as compared to bacteria, as it requires thorough standardization of extraction-preparation protocol and expansion of database. Nonetheless, the accuracy and reliability of MALDI-TOF for the identification of fungi has been demonstrated in several studies (Buchan and Ledeboer 2013; Huang et al. 2013). In addition, expanding the MS library of MALDI-TOF spectra in the database can provide accurate identification even when challenged by new emerging fungal species. New prospects using MALDI include antifungal susceptibility testing, direct identification of fungi in clinical samples and sub-species typing for outbreak investigation (Idelevich et al. 2017; Vella et al. 2017).
2. Nucleic acid detection tests: The use of multiplex polymerase chain reaction (PCR) has also improved turnaround time (Banerjee et al. 2015). The test is expected to be sensitive due to the scope of amplification of nucleic acid. But PCR-based screening tests for IFD have demonstrated a modest sensitivity and specificity. The test faces challenges of contamination from environment. Further, the test requires multi-center validation, which may be challenging.
3. Identification of fungi in tissue: The best method of diagnosis of IFD is the histopathological demonstration of fungal elements in tissue specimens. However, the common mycelia fungi are indistinguishable on histopathology. In this context, the in situ hybridization technique and extraction of nucleic acid from tissue followed by sequencing can assist in the rapid and accurate identification of such fungi (Hayden et al. 2002; Rickerts et al. 2012). The success rate of these nucleic acid detection-based methods is better in fresh tissue than paraffin-embedded tissue due to denaturation of nucleic acid in formalin.
4. Fungal biomarker detection: Advances in the development non-culture-based diagnostics including the detection of fungal cell wall components allow early diagnosis with a high sensitivity (Clancy and Nguyen 2013). The various biomarkers available for IFD diagnosis are summarized in Table 18.1. Galactomannan (GM) and 1,3-βeta-D-glucan (BDG) are both cell-wall components of fungi serving as biomarkers of IFD. Serum and BAL GM are used for the diagnosis of invasive aspergillosis, while BDG detection has been used for detecting infections due to *Candida*, *Pneumocystis*, *Aspergillus*, and *Fusarium* spp. Rather, the BDG may be raised in any fungal infection except mucormycosis and cryptococcosis. These tests have also been incorporated in the consensus group of European Organization for Research and Treatment of Cancer/Invasive

Table 18.1 Biomarkers available for IFD diagnosis

Biomarker	Use	Advantage	Disadvantage	Comment
Enolase	Invasive candidiasis	Produced by all *Candida* species. (except *C. glabrata*)	Low sensitivity	Enzyme in glycolytic pathway, produced by *Candida* species Commercial kit is withdrawn from the market by its manufacturer (Directigen; Becton Dickinson, Baltimore)
Mannan and antimannan antibodies	Invasive candidiasis	Combined detection of mannan and antimannan antibodies considerably improves the diagnosis of candidemia	Low sensitivity when used alone Not very useful in non-albicans *Candida* species infection May be present in superficial colonization	Cell wall component of *Candida species* Mannan may be rapidly removed from circulation by formation of immune complexes or phagocytoses by kupffer cells in the liver reducing the sensitivity
Secreted Aspartyl Proteinase (Sap)	Invasive candidiasis	Extracellular concentration correlates with invasive disease rather than simple colonization	Many sap proteins are there and detection may be difficult	Sap has been shown to be produced during active tissue invasion

(continued)

Table 18.1 (continued)

Biomarker	Use	Advantage	Disadvantage	Comment
1,3,Beta D glucan	Invasive fungal diseases except cryptococcosis and mucormycosis	Used in combination with clinical, radiological, and microbiological findings for early diagnosis of candidiasis and pneumocystosis	False-positive reactions are reported due to concurrent bacteremia (most commonly *Streptococcus* species); hemodialysis with cellulose membranes and filters; use of immunoglobulin products; food with glucans, e.g., mushrooms, oats, etc., and drugs	Cell wall polysaccharide, found in most fungi (exception: Cryptococci, *Mucorales*, and *Blastomyces dermatitidis*)
				A high sensitivity and good negative predictive value is reported for IFD diagnosis
				May be used to stop empiric antifungal treatment after repeated negative results
				Testing multiple samples is better than single sample

(continued)

Table 18.1 (continued)

Biomarker	Use	Advantage	Disadvantage	Comment
Galactomannan	Invasive aspergillosis	Screening tool for invasive aspergillosis in neutropenic patients, haematological malignancies, and/or HSCT recipients, solid organ transplant recipients, intensive care unit patients, autoimmune disorders/AIDS	Cross-reactivity reported with *Penicillium*, *Paecilomyces*, *Histoplasma*, *Alternaria*, and *Geotrichum* infection; bacterial sepsis due to *S. epidermidis*, *Enterococcus faealis*, *Corynebacterium jeikeium*, *Escerichia coli*; plasma lysate; antimicrobials such as piperacillin-tazobactum or amoxicillin-clavulanate	Polysaccharide found in the cell wall of molds: Released during hyphal growth
				Serum and bronchoalveolar lavage (BAL) samples can be tested
				BAL sample is better than serum
				Prior administration of antifungals as prophylaxis decreases test sensitivity
			False positivity is also reported in neonates and after yoghurt intake	Sample pretreatment with EDTA to dissociate immune complexes increases analytical sensitivity
				Testing multiple samples is better than single sample
Cryptococcal capsular polysaccharide antigen	Cryptococcosis	High sensitivity and specificity	False positive reported in disseminated trichosporonosis, *Capnocytophaga canimorsus* septicemia, and malignancy	Both serum and CSF samples can be tested
			False negative in case of prozone effect	Pre-treatment of sample with pronase or dilution can prevent false-positive/ false-negative reactions
Histoplasma antigen	Histoplasmosis	High sensitivity and specificity	Expensive and not available in majority of the laboratories	Noninvasive test and can be done in the urine

Fungal Infections Cooperative Group (EORTC) and the National Institute of Allergy and Infectious Diseases Mycoses Study Group (MSG) (EORTC/MSG) criteria for defining IFD in immunocompromised patients and are also useful in critically ill patients. However, the cautious interpretation of their results is essential and requires standardization in different patient groups. Additionally, false-positive GM results have been reported in patients receiving antibiotics like amoxicillin-clavulanate, piperacillin-tazobactam, carbapenems, ceftriaxone, etc. (Boonsarngsuk et al. 2010). Factors affecting the BDG values include hemodialysis with cellulose membranes, antibiotics, administration of albumin, immunoglobulins, etc. through BG-containing filters and concurrent bacterial sepsis (Bansal et al. 2018). Since most of these factors are virtually impossible to avoid in intensive care setting, a nuanced interpretation of the results of these biomarkers is needed while evaluating such patients. These tests are expected to be less sensitive than nucleic acid detection-based tests as cellular components have no scope of amplification and require sensitive platforms to detect. Among other biomarker tests, antigen detection tests for diagnosis of cryptococcosis and histoplasmosis are routinely used in the laboratories due to high sensitivity and specificity of the tests (Perfect et al. 2010). Unfortunately, these non-culture-based technologies are usually available only in the tertiary care centers and multi-specialty hospitals and may not be cost-effective in low-throughput laboratories.

5. Defining IFD: Since fungi are frequently present as colonizers or commensals, the discrimination of colonization from true infection decides the course of patient management. Clinical definitions have been developed to establish a formal framework for diagnosis of IFD with a reasonable level of certainty in immunosuppressed patients. A clear definition of IFD also helps in designing clinical trials and management strategies. In 2002, standard definitions for IFD in immunocompromised patients were published by the EORTC/MSG consensus group (Ascioglu et al. 2002), which were subsequently refined by De Pauw et al. (2008). This definition assigned three levels of probability for the diagnosis of IFD as “proven,” “probable,” and “possible” IFD. This definition is limited to patients with neutropenia and inherited immunodeficiency, recipients of hematopoietic stem cell transplantation, T-cell immunosuppressants or corticosteroid treatment. Since the above clinical definition has not been validated in other patients such as those in critical intensive care settings, a definitive diagnosis of IFD cannot be made in patients with renal and/or hepatic dysfunction and those with chronic airway diseases. The non-availability of specific criteria for the diagnosis of IFD in these patients makes it problematic to define such cases. To overcome this problem, IFD diagnostic algorithms have been proposed for different patient groups, such as chronic obstructive pulmonary disease patients (Bulpa et al. 2001), intensive care unit patients (Blot et al. 2012), and patients with acute chronic liver failure (Verma et al. 2018). However, those algorithms await validation in multi-center studies.
6. Distinction of colonization versus infection: Another point of contention is the detection of fungi in non-sterile samples such as urine, ET aspirate, and bronchoalveolar lavage (BAL). The presence of *Candida* in these samples is not

important as it may represent colonization even in a symptomatic patient. A careful evaluation of positive cultures from these samples is needed and they must not be treated hastily. A logical approach for the evaluation of *Candida* in urine and ET aspirates must precede therapeutic intervention. Although some studies have shown higher mortality in patients with respiratory tract *Candida* colonization, there is no definitive conclusion on its clinical relevance. The present data is insufficient to recommend the use of antifungals in those patients (De Pascale and Antonelli 2014).

18.6 Challenges in the Treatment of IFD

Successful management of IFDs continues to be an arduous task for clinicians. Management of IFD can be divided in four categories: prophylaxis (antifungal therapy in susceptible patients before developing symptoms and signs of fungal infections to prevent fungal diseases), empiric (antifungal therapy in susceptible patients after developing symptoms and signs of suspected fungal infections without definitive diagnosis), pre-emptive/diagnosis-driven (antifungal therapy in patients with biomarker positive for IFD and having symptoms and signs of IFD), targeted therapy (antifungal therapy in patients with specific diagnosis of fungal infections). An overview of the treatment options for IFD is shown in Table 18.2.

Following are the challenges in management of IFD:

1. Antifungal initiation and treatment strategy: Due to difficulty and delay in diagnosis, majority of patients with suspected fungal infections are on antifungal therapy by fever-driven approach (fever not responding to broad-spectrum antibacterial therapy for 3–5 days). It is not known how to monitor such patients on therapy, when to stop therapy, or when to modify therapy. In a recent study in critically ill patients, empiric antifungal treatment was seen to reduce the rates of IFD significantly; however, no overall improvement in mortality was noted (Wang et al. 2017). Antifungal prophylaxis is prescribed only in cohort with more than 10% IFD, as the NTT (number of persons to treat to save one patient with fungal infection) should not be high. Although there are validated studies and various proposals about defining IFI risk populations, there is still no unanimous agreement in management (Vazquez 2016). Antifungal prophylaxis is regularly used for selected patients with acute myeloid leukemia and sometimes in acute lymphoblastic leukemia and transplant recipients. In patients with multiple intestinal leak or acute pancreatitis in intensive care unit also, antifungal prophylaxis is recommended. Generally, a dynamic, risk-adapted antifungal strategy is needed, especially in high-risk individuals (Ko et al. 2018). The diagnosis-driven or pre-emptive strategy is increasingly preferred over the empiric strategy to minimize antifungal use. However, individualized decisions must be made in accordance with patient condition and local epidemiology.
2. Choice of antifungal agent: The choice of antifungal used for treatment or prophylaxis depends upon its spectrum of activity and target etiological agent. The

Table 18.2 Overview of antifungal drugs used in invasive fungal disease (IFD)

Regimen	Drug	IFD	Comment
Prophylaxis	Posaconazole	Any invasive fungal infections	In patients with acute myeloid leukemia and myelodysplastic syndrome, randomized clinical trial showed effective
	Fluconazole or itraconazole	Fluconazole when invasive candidiasis is the common disease, otherwise itraconazole	The drugs are less expensive and used frequently but with emergence of resistant or change of fungal species posaconazole/voriconazole/amphotericin B/echinocandin preferred
Empiric	Amphotericin B deoxycholate/lipid preparations	High risk of invasive candidiasis or mold infection (e.g., allogeneic HSCT recipient, patients undergoing chemotherapy for acute leukemia, patient with >3 weeks neutropenia)	In case of suspected yeast infection, amphotericin B deoxycholate 0.7 mg/kg/day (with the exception of *C. krusei* and *C. glabrata* where a dose of 1 mg/kg/day is recommended)—For a mold infection 1 mg/kg/day, liposomal amphotericin B 3–5 mg/kg/day
	Fluconazole	If mold infection is unlikely	Fluconazole dose: 400 mg/day
	Itraconazole	If high risk for mold infection in non-neutropenic patients	Itraconazole loading dose 200 mg IV twice daily for 2 days then 200 mg IV/day
Pre-emptive	Voriconazole	Invasive aspergillosis or invasive candidiasis	Voriconazole loading dose of 6 mg/kg/day IV twice daily for 2 doses, followed by 4 mg/kg twice daily orally
	Lipid preparations of amphotericin B	When mucormycosis is also a possibility	Lipid preparations of amphotericin B 3–5 mg/kg/day
Targeted	5 flucytosine	Cryptococcal meningitis	Used in combination of amphotericin B (deoxycholate or liposomal)
		Candida endocarditis	
	Amphotericin B (deoxycholate)	Invasive candidiasis	Allow serum creatinine to rise to 0.23 mmol/L before ceasing conventional amphotericin B except in patients where renal failure is expected (in them stop amphotericin B deoxycholate if serum creatinine reaches 0.15 mmol/L on 2 consecutive days or the measured creatinine clearance falls by 50% or if the rate of creatinine increase has been rapid
		Coccidioidomycosis	
		Severe pulmonary cryptococcosis	
		Induction-in disseminated histoplasmosis	
		Invasive mucormycosis	

(continued)

Table 18.2 (continued)

Regimen	Drug	IFD	Comment
	Amphotericin B (lipid preparations)	Mucormycosis	Monitor renal function regularly
		Fusariosis	
		Disseminated blastomycosis	
		Candidaemia (in neutropenic)	
		Candida osteomyelitis/ septic arthritis	
		Induction-cryptococcal meningitis	
		Disseminated sporotrichosis	
	Itraconazole	*Scedosporium* species infection in non-neutropenic patients	Monitor liver function test and other toxicity
		Follow up-disseminated histoplasmosis, sporotrichosis and blastomycosis	
	Fluconazole	Candidaemia (non-neutropenic)	In intensive care patients, dose adjustment may be required
		Candida osteomyelitis/ septic arthritis	
	Voriconazole	Invasive pulmonary aspergillosis	Therapeutic drug monitoring is essential
		Follow-up central nervous system blastomycosis	Monitor toxicity
			Take care of drug-interaction
		Scedosporium, *Fusarium* species and dematiaceous fungal infections	
	Anidulafungin, caspofungin, micafungin	Candidaemia	Cannot be used in central nervous system dissemination, retinitis; and in *Candida* urinary tract infections
		Candida endocarditis	

local epidemiology of pathogens implicated in IFD and its in vitro susceptibility pattern also guide patient management. However, the lack of such studies in developing countries makes the situation challenging. A systematic research investigating the epidemiology of IFD in targeted patient groups at different geographic locations is needed. This will also help in developing explicit and feasible guidelines facilitating the rational use of antifungal agents and facilitate decision-making in clinical practice.

The efficacy, toxicity profile, pharmacodynamics, and pharmacokinetic properties of antifungal agents are other factors deciding the antifungal choice. Amphotericin B deoxycholate had been the standard treatment for IFD for many years. However, renal toxicity and infusion associated side effects prompted the search for safer but equally efficacious alternatives. Lipid preparations of amphotericin B are preferred due to less toxicity, but the compounds are expensive and have limited use in developing countries. Fluconazole, the most widely prescribed antifungal agent, has been found to be the effective against yeast, cost-effective and well-tolerated agent for IFD prevention in critically ill patients (Wang et al. 2017). However, due to its limited activity against mycelial fungi and intrinsic resistance among some *Candida* species (e.g., *C. krusei, C. glabrata*), there is a need for other broad-spectrum options. Itraconazole has the limitations of erratic absorption or adverse gastrointestinal effects on oral dosing, though the drug is effective against mycelial fungi. Voriconazole is effective for treatment of invasive aspergillosis and fusariosis, while posaconazole is active against *Aspergillus* spp. and *Fusarium* spp. Unfortunately, this extended spectrum of action of antifungal agents comes at the cost of higher pharmacokinetic variability and increased drug interactions. Echinocandins are another class of antifungals with very little collateral toxicity. However, they still lack activity against *Cryptococcus* species, *Scedosporium, Fusarium* and Mucorales, which frequently develop as breakthrough infections in immunocompromised individuals. Thus, reformulation of existing drugs or development of new drugs with similar activity and better pharmacokinetic properties is still warranted. The list of available antifungal agents with spectrum of activity, mechanism of action, and pharmacokinetic details is shown in Table 18.3.

3. Pharmacokinetics and pharmacodynamics of antifungal drugs: Even if the correct drug and regimen are initiated, the antifungal exposure at the site of infection may be lower or in excess of the anticipated resulting in treatment failure or toxicity. Certain antifungal agents may not reach the tissue like echinocandin in central nervous stem or urinary tract infections. Inconsistencies in absorption, metabolism and/or elimination of antifungal agents can result in variability in blood drug levels. Thus, an insight into the pharmacokinetic and pharmacodynamics profile of antifungals is essential to optimize drug choice and dose. Monitoring the serum drug levels is particularly important for children due to wide alternations in the dose needed to achieve the target levels. Due to the correlation between subtherapeutic drug levels and poor patient outcome, therapeutic drug monitoring (TDM) of voriconazole and posaconazole has been widely accepted. However, a similar effectiveness for other antifungals is yet to be established and facilities for TDM in routine diagnostic laboratories need to be strengthened.
4. Drug interactions: Another challenge in the treatment of IFD in transplant patients and patients on chemotherapeutic agent is the presence of drug-drug interactions between antifungals (primarily azoles) and immunosuppressive agents like cyclosporine, tacrolimus, and chemotherapeutic agents (Lempers et al. 2015). Voriconazole alters the concentration of immunosuppressants resulting in a high risk of either transplant rejection or toxicity in such patients. These patients are at a high risk for developing IFDs and often need treatment with azoles; thus maintaining the adequate concentration of immunosuppressants with the desired therapeutic level of antifungal is needed which requires an

Table 18.3 Spectrum of activity, mechanism of action and pharmacokinetic properties of antifungal drugs

Antifungal agent	Mechanism of action	Spectrum of activity	Pharmacokinetic properties
Polyenes	Bind irreversibly to ergosterol, resulting in disruption of membrane integrity and ultimately cell death; some also predicts oxidative damage of the cell	*Candida* (except *C. lusitaniae*) and filamentous fungi (except. *Scedosporium* species, *A. terreus* and *Fusarium* species)	Concentration-dependent action
			Lipid formulations of amphotericin B are generally less potent in vivo compared with amphotericin B
			Penetration into CNS is best for liposomal amphotericin B
			Amphotericin B lipid complex formulation has higher concentration in lungs
Flucytosine	Metabolized to 5-fluorouracil within fungal cell and extensively incorporated into fungal RNA inhibiting synthesis of both DNA and RNA	*Candida* and *Cryptococcus*	Toxicity of flucytosine therapy has been associated with peak drug concentrations beyond 100 μg/ml; thus dosing to optimize the time > minimum inhibitory concentration (MIC) needed- so low doses very frequently (6 hourly) given
Azoles	Inhibition of 14α-lanosterol demethylase, a key enzyme in ergosterol biosynthesis, resulting in depletion of ergosterol and accumulation of toxic 14α-methylated sterols in membranes	Fluconazole on yeast; Itraconazole also on mycelial fungi; Voriconazole against majority of fungi except *Mucorales*; Posaconazole on majority of fungi; Isavuconazole can also against *Mucorales*	Concentration independent action
			Area under curve (AUC)/MIC ratio important
			Voriconazole and posaconazole trough concentrations ranging from 1–2 μg/ml are associated with optimal outcome for invasive aspergillosis
Echinocandin	Noncompetitively inhibit beta-1,3-D-glucan synthase enzyme complex fungi to disturb fungal cell glucan synthesis	*Candida* species (less in vitro potency against *C. parapsilosis*) and *Aspergillus* species (fungicidal activity against *Candida* and static against *Aspergillus* species)	Infrequent administration of large doses is optimal
			Wide therapeutic window—Potential to escalate drug doses

individualized approach toward each patient. Thus, TDM is also essential in such scenarios.

5. Monitoring response to antifungal therapy: The evaluation of patient response to antifungal treatment is a challenge faced by clinicians. The signs and symptoms of IFI are usually subtle in immunocompromised hosts, and radiological signs are non-specific. In addition, resolution of imaging findings requires a long time making it unsuitable for monitoring response. Biomarker tests may have a role in monitoring the therapy, though more studies are required to evaluate its effectiveness. The lack of effective clinical, radiological, and microbiological support in effectively monitoring the progress or control of IFD also creates difficulty in deciding the treatment duration.
6. Antifungal resistance: The challenge in the treatment of IFDs is further complicated with the emerging antifungal resistance in recent years, as it significantly limits the arsenal of systemically available treatment options. Some fungi are known to exhibit intrinsic resistance to some drugs (e.g., *C. krusei* to fluconazole, *C. lusitaniae* and *A. terreus* to amphotericin B). A rise in acquired resistance such as azole resistance in *A. fumigatus*, and echinocandins resistance in *C. glabrata* has also been reported (Vallabhaneni et al. 2015; Rivero-Menendez et al. 2016). In addition, new species demonstrating resistance to various antifungals are also emerging (e.g., *C. auris*) (Lockhart et al. 2017). The correct identification followed by antifungal susceptibility of the implicated fungal pathogen in IFD cases would certainly help in designing antifungal treatment. Unfortunately, such facilities are seldom available in resource-limited developing countries where IFDs are common. The rising antifungal resistance has been linked to prior exposure to antifungals and sub-optimal dosing, highlighting the need for antifungal stewardship (AFS) (Hamdy et al. 2017). Antifungal resistance refers to responsible planning with integrated interventions to audit and direct the judicious use of antifungals in order to preserve their future effectiveness and achieve the best clinical outcomes (Richardson 2016). Building a multidisciplinary team of experts in IFD management, suitable prescribing with post-prescription review of antifungals, TDM, de-escalation of drugs, and conversion from intravenous to oral formulations are necessary to implement AFS (Hamdy et al. 2017). Although the overall acceptance of AFS has been well demonstrated in literature, very few practical reports regarding AFS are present.
7. Cost of antifungal treatment: Although secondary, the cost of treatment is important to ensure the acceptance of any drug. Patient cannot afford antifungal drugs. An evaluation of the cost related to antifungal use has revealed that choosing pre-emptive therapy instead of empirical antifungal treatment and switching to oral form after intravenous formulations may reduce antifungal drug-related costs while treating IFD (Gedik 2015). Survival benefit with a low total drug cost has also been reported with the use of posaconazole while treating invasive aspergillosis (Herbrecht et al. 2010) and isavuconazole in mucormycosis compared to standard treatment (Bagshaw et al. 2018). Although validated assays for TDM are available, the pharmacoeconomic impact of using them has not been formally investigated. Since the antifungal treatment can be expensive, only

cost-effective protocols can be incorporated in clinical practice. Nonetheless, the expenditure on using these assays is certainly lower than even a single day of antifungal treatment with one of the new antifungal compounds and less than management charges of any complications resulting from inappropriate antifungal exposure (Andes et al. 2009).

8. Limited availability of antifungal drugs: Currently, the pace of antifungal drug development is inadequate to keep up with the clinical need. Unlike bacteria, fungal pathogens are more closely related to the host, as both are eukaryotes, and thus discovery of new antifungal molecule is difficult. The clinical evaluation of antifungal drugs is another challenge, in addition to the economic and regulatory challenges. The collaboration between pharmaceutical agencies and academic laboratories is essential to achieve success in antifungal drug development. Unfortunately, even amongst the currently available drugs, effective antifungal agents are inaccessible to a large number of developing countries due to regulatory constraints (Kneale et al.(2016).

18.7 Conclusions

Although IFD continue to pose a significant threat to the human health, our awareness and preparedness to face the challenge of fungal disease are still limited. The awareness of the fungal disease among clinicians is still limited to specialized units only, and there is a gross deficiency of fungal diagnostic laboratories especially in developing countries where the rate of IFD is high. Therefore, we need a) education and training of medical personnel, b) diagnostic mycology laboratory development, c) research for development of point of care test with good sensitivity and specificity, and d) development of low-cost, broad-spectrum, less toxic antifungal agent(s) with good bioavailability to overcome the present challenge due to systemic fungal infections.

References

Andes D, Pascual A, Marchetti O (2009) Antifungal therapeutic drug monitoring: established and emerging indications. Antimicrob Agents Chemother 53:24–34. https://doi.org/10.1128/AAC.00705-08

Arendrup MC et al (2011) National surveillance of fungemia in Denmark (2004–2009). J Clin Microbiol 49:325–334. https://doi.org/10.1128/JCM.01811-10

Ascioglu S et al (2002) Defining opportunistic invasive fungal infections in immunocompromised patients with cancer and hematopoietic stem cell transplants: an international consensus. Clin Infect Dis 34:7–14. https://doi.org/10.1086/323335

Bagshaw E et al (2018) Economic impact of treating invasive mold disease with isavuconazole compared with liposomal amphotericin B in the UK. Future Microbiol 13:1283–1293. https://doi.org/10.2217/fmb-2018-0119

Bajwa S, Kulshrestha A (2013) Fungal infections in intensive care unit: challenges in diagnosis and management. Ann Med Health Sci Res 3:238–244. https://doi.org/10.4103/2141-9248.113669

Banerjee R et al (2015) Randomized trial of rapid multiplex polymerase chain reaction–based blood culture identification and susceptibility testing. Clin Infect Dis 61:1071–1080. https://doi.org/10.1093/cid/civ447

Bansal N et al (2018) Experience with β-D-glucan assay in the management of critically ill patients with high risk of invasive candidiasis: an observational study. Indian J Crit Care Med 22:364–368. https://doi.org/10.4103/ijccm.IJCCM_4_18

Barros MBL et al (2008) An epidemic of sporotrichosis in Rio de Janeiro, Brazil: epidemiological aspects of a series of cases. Epidemiol Infect 136:1192–1196. https://doi.org/10.1017/S0950268807009727

Bastos de Lima Barros M et al (2004) Cat-transmitted sporotrichosis epidemic in Rio de Janeiro, Brazil: description of a series of cases. Clin Infect Dis 38:529–535. https://doi.org/10.1086/381200

Benedict K, Park BJ (2014) Invasive fungal infections after natural disasters. Emerg Infect Dis 20:349–355. https://doi.org/10.3201/eid2003.131230

Berger S et al (2017) Azole resistance in aspergillus fumigatus: a consequence of antifungal use in agriculture? Front Microbiol 8:1024. https://doi.org/10.3389/fmicb.2017.01024

Blot SI et al (2012) A clinical algorithm to diagnose invasive pulmonary aspergillosis in critically ill patients. Am J Respir Crit Care Med 186:56–64. https://doi.org/10.1164/rccm.201111-1978OC

Bongomin F et al (2017) Global and multi-national prevalence of fungal diseases—estimate precision. J Fungi 3:57. https://doi.org/10.3390/jof3040057

Boonsarngsuk V et al (2010) False-positive serum and bronchoalveolar lavage Aspergillus galactomannan assays caused by different antibiotics. Scand J Infect Dis 42:461–468. https://doi.org/10.3109/00365541003602064

Buchan BW, Ledeboer NA (2013) Advances in identification of clinical yeast isolates by use of matrix-assisted laser desorption ionization-time of flight mass spectrometry. J Clin Microbiol 51:1359–1366. https://doi.org/10.1128/JCM.03105-12

Bulpa PA et al (2001) Chronic obstructive pulmonary disease patients with invasive pulmonary aspergillosis: benefits of intensive care? Intensive Care Med 27:59–67. Available at: http://www.ncbi.nlm.nih.gov/pubmed/11280674. Accessed 14 Jan 2019

Casadevall A (2012) Fungi and the rise of mammals. PLoS Pathog 8:e1002808. https://doi.org/10.1371/journal.ppat.1002808

Chakrabarti A (2011) Drug resistance in fungi-an emerging problem. Available at: http://www.searo.who.int/publications/journals/regional_health_forum/media/2011/V15n1/rhfv15n1p97.pdf. Accessed 8 Jan 2019

Chakrabarti A et al (1996) Antifungal susceptibility pattern of non-albicans *Candida* species distribution of species isolated from candidaemia cases over a 5 year period. Indian J Med Res 104:171–176. Available at: http://www.ncbi.nlm.nih.gov/pubmed/8840655. Accessed 14 Jan 2019

Chakrabarti A et al (2001) Outbreak of Pichia anomala infection in the pediatric service of a tertiary-care center in northern India. J Clin Microbiol 39:1702–1706. https://doi.org/10.1128/JCM.39.5.1702-1706.2001

Chakrabarti A et al (2014) Epidemiological study of a large cluster of fungaemia cases due to Kodamaea ohmeri in an Indian tertiary care centre. Clin Microbiol Infect 20:O83–O89. https://doi.org/10.1111/1469-0691.12337

Chakrabarti A et al (2015) Incidence, characteristics and outcome of ICU-acquired candidemia in India. Intensive Care Med 41:285–295. https://doi.org/10.1007/s00134-014-3603-2

Chapman B et al (2016) Changing epidemiology of candidaemia in Australia. J Antimicrob Chemother 72:422. https://doi.org/10.1093/jac/dkw422

Chow JK et al (2008) Factors associated with candidemia caused by non- albicans candida species versus candida albicans in the intensive care unit. Clin Infect Dis 46:1206–1213. https://doi.org/10.1086/529435

Clancy CJ, Nguyen MH (2013) Finding the mMissing 50% of invasive candidiasis: how nonculture diagnostics will improve understanding of disease spectrum and transform patient care. Clin Infect Dis 56:1284–1292. https://doi.org/10.1093/cid/cit006

Cleveland AA et al (2015) Declining incidence of candidemia and the shifting epidemiology of Candida resistance in two US metropolitan areas, 2008–2013: results from population-based surveillance. PLoS One 10:e0120452. https://doi.org/10.1371/journal.pone.0120452

Cooney NM, Klein BS (2008) Fungal adaptation to the mammalian host: it is a new world, after all. Curr Opin Microbiol 11:511–516. https://doi.org/10.1016/j.mib.2008.09.018

De Pascale G, Antonelli M (2014) Candida colonization of respiratory tract: to treat or not to treat, will we ever get an answer? Intensive Care Med 40:1381–1384. https://doi.org/10.1007/s00134-014-3364-y

De Pauw B et al (2008) Revised definitions of invasive fungal disease from the European Organization for Research and Treatment of Cancer/Invasive Fungal Infections Cooperative Group and the National Institute of Allergy and Infectious Diseases Mycoses Study Group (EORTC/MSG) Consensus Group. Clin Infect Dis 46:1813–1821. https://doi.org/10.1086/588660

Denning DW et al (2017) Delivering on antimicrobial resistance agenda not possible without improving fungal diagnostic capabilities. Emerg Infect Dis 23:177–183. https://doi.org/10.3201/eid2302.152042

Firacative C et al (2018) The status of cryptococcosis in Latin America. Mem Inst Oswaldo Cruz 113:1–23. https://doi.org/10.1590/0074-02760170554

Garcia-Solache MA, Casadevall A (2010) Global warming will bring new fungal diseases for mammals. MBio 1:e00061-10. https://doi.org/10.1128/mBio.00061-10

Gedik H (2015) The expenditures related to the use of antifungal drugs in patients with hematological cancers: a cost analysis. Clinico-Econ Outcomes Res 7:537–543. https://doi.org/10.2147/CEOR.S92455

Guinea J et al (2017) Increasing incidence of mucormycosis in a large Spanish hospital from 2007 to 2015: epidemiology and microbiological characterization of the isolates. PLoS One 12(6):e0179136. https://doi.org/10.1371/journal.pone.0179136

Hamdy RF, Zaoutis TE, Seo SK (2017) Antifungal stewardship considerations for adults and pediatrics. Virulence 8:658–672. https://doi.org/10.1080/21505594.2016.1226721

Hayden RT et al (2002) In situ hybridization for the identification of filamentous fungi in tissue section. Diagn Mol Pathol 11:119–126. Available at: http://www.ncbi.nlm.nih.gov/pubmed/12045716. Accessed 14 Jan 2019

Herbrecht R et al (2010) Comparative survival and cost of antifungal therapy: posaconazole versus standard antifungals in the treatment of refractory invasive aspergillosis. Curr Med Res Opin 26:2457–2464. https://doi.org/10.1185/03007995.2010.516110

Hesstvedt L et al (2015) Twenty-two years of candidaemia surveillance: results from a Norwegian national study. Clin Microbiol Infect 21:938–945. https://doi.org/10.1016/j.cmi.2015.06.008

Horn DL et al (2009) Epidemiology and outcomes of candidemia in 2019 patients: data from the prospective antifungal therapy alliance registry. Clin Infect Dis 48:1695–1703. https://doi.org/10.1086/599039

Huang AM et al (2013) Impact of rapid organism identification via matrix-assisted laser desorption/ionization time-of-flight combined with antimicrobial stewardship team intervention in adult patients with bacteremia and candidemia. Clin Infect Dis 57:1237–1245. https://doi.org/10.1093/cid/cit498

Idelevich EA, Grünastel B, Becker K (2017) Rapid detection and identification of candidemia by direct blood culturing on solid medium by use of lysis-centrifugation method combined with matrix-assisted laser desorption ionization-time of flight mass spectrometry (MALDI-TOF MS). J Clin Microbiol 55:97–100. https://doi.org/10.1128/JCM.01787-16

Jeffery-Smith A et al (2018) Candida auris: a review of the literature. Clin Microbiol Rev 31:e00029–e00017. https://doi.org/10.1128/CMR.00029-17

Jeong W et al (2019) The epidemiology and clinical manifestations of mucormycosis: a systematic review and meta-analysis of case reports. Clin Microbiol Infect 25:26–34. https://doi.org/10.1016/j.cmi.2018.07.011

Kitaura T et al (2018) Cryptococcus gattii genotype VGIIa infection imported from Vancouver Island to Japan. J Infect Chemother 24:573–575. https://doi.org/10.1016/j.jiac.2017.12.014

Kneale M et al (2016) Global access to antifungal therapy and its variable cost. J Antimicrob Chemother 71:3599–3606. https://doi.org/10.1093/jac/dkw325
Ko BS et al (2018) 2016 guideline strategies for the use of antifungal agents in patients with hematological malignancies or hematopoietic stem cell transplantation recipients in Taiwan. J Microbiol 51(3):287–301. https://doi.org/10.1016/j.jmii.2017.07.005
Krasinski K et al (1985) Nosocomial fungal infection during hospital renovation. Infect Control 6:278–282. Available at: http://www.ncbi.nlm.nih.gov/pubmed/3847404. Accessed 14 Jan 2019
Kyvernitakis A et al (2016) Initial use of combination treatment does not impact survival of 106 patients with haematologic malignancies and mucormycosis: a propensity score analysis. Clin Microbiol Infect 22:811.e1–811.e8. https://doi.org/10.1016/j.cmi.2016.03.029
Lamoth F et al (2018) Changes in the epidemiological landscape of invasive candidiasis. J Antimicrob Chemother 73:i4–i13. https://doi.org/10.1093/jac/dkx444
Lanternier F et al (2012) Mucormycosis in organ and stem cell transplant recipients. Clin Infect Dis 54:1–8. https://doi.org/10.1093/cid/cis195
Lempers VJ et al (2015) Drug-interactions of azole antifungals with selected immunosuppressants in transplant patients: strategies for optimal management in clinical practice. Curr Opin Pharmacol 24:38–44. https://doi.org/10.1016/j.coph.2015.07.002
Lockhart SR et al (2017) Simultaneous emergence of multidrug-resistant candida auris on 3 continents confirmed by whole-genome sequencing and epidemiological analyses. Clin Infect Dis 64:134–140. https://doi.org/10.1093/cid/ciw691
Morrell M, Fraser VJ, Kollef MH (2005) Delaying the empiric treatment of *Candida* bloodstream infection until positive blood culture results are obtained: a potential risk factor for hospital mortality. Antimicrob Agents Chemother 49:3640–3645. https://doi.org/10.1128/AAC.49.9.3640-3645.2005
Nivoix Y et al (2008) Factors associated with overall and attributable mortality in invasive aspergillosis. Clin Infect Dis 47:1176–1184. https://doi.org/10.1086/592255
Perfect JR et al (2010) Clinical practice guidelines for the management of cryptococcal disease: 2010 update by the Infectious Diseases Society of America. Clin Infect Dis 50:291–322. https://doi.org/10.1086/649858
Perlroth J, Choi B, Spellberg B (2007) Nosocomial fungal infections: epidemiology, diagnosis, and treatment. Med Mycol 45:321–346. https://doi.org/10.1080/13693780701218689
Ribas E, Ribas AD et al (2016) Is the emergence of fungal resistance to medical triazoles related to their use in the agroecosystems? A mini review. Braz J Microbiol 47:793–799. https://doi.org/10.1016/j.bjm.2016.06.006
Richardson MD (2016) An introduction to antifungal stewardship. J Antimicrob Chemother 71:ii3–ii3. https://doi.org/10.1093/jac/dkw390
Rickerts V et al (2012) Enhanced fungal DNA-extraction from formalin-fixed, paraffin-embedded tissue specimens by application of thermal energy. Med Mycol 50:667–672. https://doi.org/10.3109/13693786.2012.665613
Rivero-Menendez O et al (2016) Triazole resistance in Aspergillus spp.: a worldwide problem? J Fungi 2(3):21. https://doi.org/10.3390/jof2030021
Robert VA, Casadevall A (2009) Vertebrate endothermy restricts most fungi as potential pathogens. J Infect Dis 200:1623–1626. https://doi.org/10.1086/644642
Robert V, Cardinali G, Casadevall A (2015) Distribution and impact of yeast thermal tolerance permissive for mammalian infection. BMC Biol 13:18. https://doi.org/10.1186/s12915-015-0127-3
Roe CC et al (2018) Dating the Cryptococcus gattii dispersal to the North American Pacific Northwest. mSphere 3:pii: e00499-17. https://doi.org/10.1128/mSphere.00499-17
Spellberg B, Ibrahim AS (2010) Recent advances in the treatment of mucormycosis. Curr Infect Dis Rep 12:423–429. https://doi.org/10.1007/s11908-010-0129-9
Spellberg B et al (2012) Risk factors for mortality in patients with mucormycosis. Med Mycol 50:611–618. https://doi.org/10.3109/13693786.2012.669502
Suh MK et al (2012) Molecular phylogenetics of exophiala species isolated from Korea. Ann Dermatol 24:287–294. https://doi.org/10.5021/ad.2012.24.3.287

Vallabhaneni S et al (2015) Epidemiology and risk factors for echinocandin nonsusceptible candida glabrata bloodstream infections: data from a large multisite population-based candidemia surveillance program, 2008–2014. Open Forum Infect Dis 2:ofv163. https://doi.org/10.1093/ofid/ofv163

Vazquez L (2016) Antifungal prophylaxis in immunocompromised patients. Mediterranean J Hematol Infect Dis 8:e2016040. https://doi.org/10.4084/MJHID.2016.040

Vella A et al (2017) Potential use of MALDI-ToF mass spectrometry for rapid detection of antifungal resistance in the human pathogen candida glabrata. Sci Rep 7(1):9099. https://doi.org/10.1038/s41598-017-09329-4

Verma N et al (2018) Invasive fungal infections amongst patients with acute-on-chronic liver failure at high risk for fungal infections. Liver Int 1:11. https://doi.org/10.1111/liv.13981

Vincent JL et al (2009) International study of the prevalence and outcomes of infection in intensive care units. JAMA 302:2323. https://doi.org/10.1001/jama.2009.1754

Wang Y et al (2017) Choosing optimal antifungal agents to prevent fungal infections in nonneutropenic critically ill patients: trial sequential analysis, network meta-analysis, and pharmacoeconomic analysis. Antimicrob Agents Chemother 61:piie00620-17. https://doi.org/10.1128/AAC.00620-17

Diversity of Endophytic Fungi and Their Role in Artificial Agarwood Production in *Aquilaria* Tree

19

Hemraj Chhipa and Sunil K. Deshmukh

Abstract

Endophytic microbial flora induces the plant immunity in response to pathogen attack. *Aquilaria* is the plant which has high value in diseased state due to production of agarwood after infection. Agarwood has enormous applications in fragrance and pharmaceutical industries. In the present chapter, we have reviewed the diversity of endophytic fungi in different species of the genus *Aquilaria*. Various fungi from *Ascomycota*, *Zygomycota* and *Basidiomycota* have been reported in the literature. Fungi from division *Ascomycota* were found to be dominant in *Aquilaria* and *Hypocreaceae*, *Nectriaceae*, *Pleosporaceae*, and *Trichocomaceae* families. Some of them have been reported as inducers of artificial infection. *Lasiodiplodia* sp. and *Penicillium polonicum* found most suitable endophytic fungi for the rapid production of agarwood within 3 months of infection period.

Keywords

Agarwood · *Aquilaria* · Endophyte · Oleoresin

19.1 Introduction

Aquilaria belongs to Thymelaeaceae family and is mainly found in evergreen rainforests. *Aquilaria* plant in its diseased state produces an oleoresin called agarwood. Oleoresin is one of the valuable raw materials in perfumery and pharmaceutical

---

H. Chhipa (✉)
College of Horticulture and Forestry, Agriculture University Kota, Jhalawar, Rajasthan, India

S. K. Deshmukh
Biotech & Management of Bioresources Div, The Energy and Resources Institute, New Delhi, Delhi, India

T. Satyanarayana et al. (eds.), *Advancing Frontiers in Mycology & Mycotechnology*,
https://doi.org/10.1007/978-981-13-9349-5_19

industries (Meng-Ling et al. 2005). The agarwood is known as aloe wood, Agar attar, gaharu, Jinkoh, Chen Xiang, and oud in different parts of the world.

In *Aquilaria*, the formation of agarwood is triggered by physical, chemical, or biological damages in natural or artificial ways (Chhipa et al. 2017). Chetpattananondh (2012) suggested that only 1–10% of *Aquilaria* trees produce agarwood naturally and usually high yield of agarwood is obtained from 50–80-year-old plants. Phytoalexin, terpenoids, glycosteroids, and alkaloids are produced as defense molecules (Novriyanti 2008). The presence of high quantity of resinous compounds determines the grade of agarwood, which may have price up to 1 million US$ per 10 kilogram in the international market (Akter et al. 2013). Exploitation of *Aquilaria* tree to fulfil market demand of agarwood made eight species of *Aquilaria* endangered in Appendix II of the Convention on International Trade in Endangered Species of Wild Fauna and Flora (CITES 2004).

Different methods have been reported for the artificial production of agarwood which include axe chopping, wounding by chisels and debarking for promoting infection, hammering to tree trunk, nail inserting and holing, burning chisel drilling, partly trunk pruning, chemical insertion into plant for wound formation, and artificial microbial inoculation method (Nobuchi and Siripatanadilok 1991; Pojanagaroon and Kaewrak 2003; Ito et al. 2005; Dai et al. 2010; Chen et al. 2011; Wei et al. 2012; Liu et al. 2013; Mohamed et al. 2014; Chhipa and Kaushik 2017). Among these methods, microbial infection is observed to reduce the time of agarwood formation and produced high-quality agarwood, which increased the interest of researchers to explore endophytic microbial flora of *Aquilaria* tree and their possible role in agarwood formation. Various endophytic fungi such as *Epicoccum granulatum*, *Botryosphaeria rhodina*, *Cladosporium edgeworthraem*, *Trichoderma* sp., *Lasiodiplodia* sp., *Colletotrichum gloeosporioides*, *Preussia* sp., and *Fusarium equiseti* have been reported in different *Aquilaria* species (Bhattacharyya et al. 1952; Gong and Guo 2009; Mohamed et al. 2010; Tian et al. 2013; Premalatha and Kalra 2013; Chhipa and Kaushik 2017). Tamuli et al. (2005) artificially infected *Aquilaria agallocha* with *Chaetomium globosum* and *Fusarium oxysporum* and found that chemical profile of the mixture of oil was similar to that of oil extracted from naturally infected agarwood. Sangareswari et al. (2016) reported the role of *Aspergillus*, *Lasiodiploidia*, *Chaetomium*, *Fusarium*, and *Penicillium* species in artificial production of agarwood. Different enzyme activities such as cellulases, ligninases, and laccases were involved in the infection process.

The following sections describe the diversity of endophytic fungi in various species of *Aquilaria*, their role in artificial production of agarwood, and future prospects of artificial production method

19.2 Distribution of *Aquilaria*

Aquilaria is mostly found in the tropical and sub-tropical regions of Southeast Asia (from Hong Kong to Papua New Guinea) and distributed in hardwood hill forests and rainforest of many countries like Australia, Bangladesh, Bhutan, Borneo,

Fig. 19.1 *Aquilaria malaccensis* plants in the field

Cambodia, India, Indonesia, Iran, Lao PRD, Malaysia, Myanmar, the Philippines, Singapore South China, South Africa, Thailand, and Vietnam (Fig. 19.1) (Burkill 1966; Gunasekera et al. 1981; Chakrabarty et al. 1994; Oldfield et al. 1998). Barden et al. (2000) reported that out of 27 species of *Aquilaria* found in all over the world, 23 species are found in Southeast Asian countries. *Aquilaria* species include *A. acuminata*, *A. apiculate*, *A. baillonii*, *A. banaensae*, *A. beccariana*, *A. brachyantha*, *A. citrinicarpa*, *A. crassna*, *A. cumingiana*, *A. filarial*, *A. grandiflora*, *A. hirta*, *A. malaccensis*, *Lam.*, *A. microcarpa Baill*, *A. ophispermum*, *A. parvifolia*, *A. pentandra*, *A. rostrate*, *A. rugose*, *A. sinensis*, *A. subintegra*, *A. urdanetensis*, and *A. yunnanensis*. Out of these, eight species, namely, *A. beccariana*, *A. crassna*, *A. cumingiana*, *A. hirta*, *A. malaccensis Lam*, *A. microcarpa Baill*, *A. sinensis*, and *A. subintegra*, have been reported as agarwood-producing species.

In different geographical areas, different *Aquilaria* species were found to be prominent, for instance, *A. apicultina* in the Philippines; *A. acuminata* in Papua New Guinea, Indonesia, and the Philippines; *A. baillonil* in Cambodia and Thailand; *A. baneonsis* only in Vietnam; *A. beccarain* mainly in Indonesia; *A. brachyantha* in Malaysia; *Aquilaria crassna* in Cambodia, Malaysia, Thailand, and Vietnam; *A. cumingiana* in Indonesia and Malaysia; *A. filarial* in the Philippines, Indonesia, and Papua New Guinea; *A. grandiflora* in China; *A. hirta* in Indonesia, Malaysia, and Thailand; and *A. khasiana* in North East India and Pakistan to name a few (Saikia and Khan 2012; Chua 2008; Chetpattananondh 2012; Tabin and Srivastava 2014).

19.3 Diversity of Endophytes in *Aquilaria*

It has been estimated that more than one million fungal species comprise the endophytic world and are present in 1:4 ratio with vascular plants (Sun et al. 2012). Endophytic fungi produce various types of bioactive compounds including flavonoids, alkaloids, terpenoids, and volatile compounds, which have shown a wide biological activities and also act as biocontrol agents (Gunatilaka 2006; Kharwar et al. 2011; Deshmukh 2018; Deshmukh et al. 2015, 2018). The role of endophytes in the production of bioactive compounds in *Aquilaria* has been explored by many researchers. The diversity of endophytes in *Aquilaria* has been explored from China, Malaysia and India.

In 1952, Bhattacharyya et al. reported the presence of endophytic fungus, namely, *Epicoccum granulatum* in *A. agallocha* species. Later on, Gong and Guo (2009) isolated endophytes *Botryosphaeria rhodina*, *Cephalosporium* sp., *Cladosporium edgeworthrae*, *Colletotrichum* sp., *Epicoccum* sp., *Fusarium oxysporum*, *Geotrichum* sp., and *Glomerularia* sp. from *A. sinensis* from Yunnan province, China. Wang et al. (2009) also reported presence of *Aspergillus* sp., *B. rhodina*, *Chaetomium* sp., *C. gloeosporioides*, *Fusarium* sp., *Penicillium* sp., *Pestalotiopsis*, *Trichoderma* sp., and *Xylaria* sp. in *A. sinensis*. In another study in West Malaysia, *Fusarium* sp., *Trichoderma* sp., *Curvularia* sp., *Cunninghamella* sp., and *Lasiodiplodia* sp. were observed as endophytes in *A. malaccensis* (Mohamed et al. 2010). Further, Cui et al. (2011) identified presence of *Cephalosporium* sp, *Cladosporium tenuissimum*, *Coniothyrium nitidae*, *Fusarium equiseti*, *Hypocrea lixii*, *Lasiodiplodia theobromae*, *Leptosphaerulina chartarum*, *Paraconiothyrium variabile*, *Phaeoacremonium rubrigenum*, *Pichia guilliermondii*, and *Rhizomucor variabilis* in *A. sinensis*. Tian et al. (2013) also investigated distribution of endophytic fungi in infected and non-infected *A. sinensis* tree. They found the presence of *C. gloeosporioides* in both types of wood. In another study, Premalatha and Kalra (2013) explored the diversity of endophytes in Indian *A. malaccensis* and reported presence of *Preussia* sp., *Alternaria*, *Cladosporium*, *Curvularia*, *Fusarium*, *Phaeoacremonium*, and *Trichoderma* sp. in resinous wood chips. Kee (2015) also reported presence of *T. harzanium*, *Trichoderma* sp., *T. reesei*, *Botryodiplodia theobromae*, and *Beauveria bassiana* in *Aquilaria* from Baram and Kota Samarahan area, Malaysia. Our group explored the diversity of endophytes in *A. malaccensis* in Assam, north eastern state of India. Chhipa and Kaushik (2017) isolated 340 fungal strains belonging to 13 fungal families including *Botryosphaeriaceae*, *Clavicipitaceae*, *Debaryomycetaceae*, *Diaporthaceae*, *Hypocreaceae*, *Mucoraceae*, *Nectriaceae*, *Pleosporaceae*, *Saccharomycetaceae*, *Syncephalastraceae*, *Trichocomaceae*, and *Trichosphaeriaceae*. It was observed that *Trichoderma virens* is the most dominating endophytic fungus followed by *Lasiodiplodia theobromae* and *Fusarium equiseti* (Fig. 19.2).

Most of the endophytic fungi belong to the class *Ascomycota*, except *Mucor circinelloides* and *Rhizomucor variabilis* which belong to class *Zygomycota* (Chhipa and Kaushik 2017; Cui et al. 2011); *Glomerularia* sp. and *Tylopilus* sp. belong to class *Basidiomycota* (Gong and Guo 2009; Li et al. 2018). Maximum number of

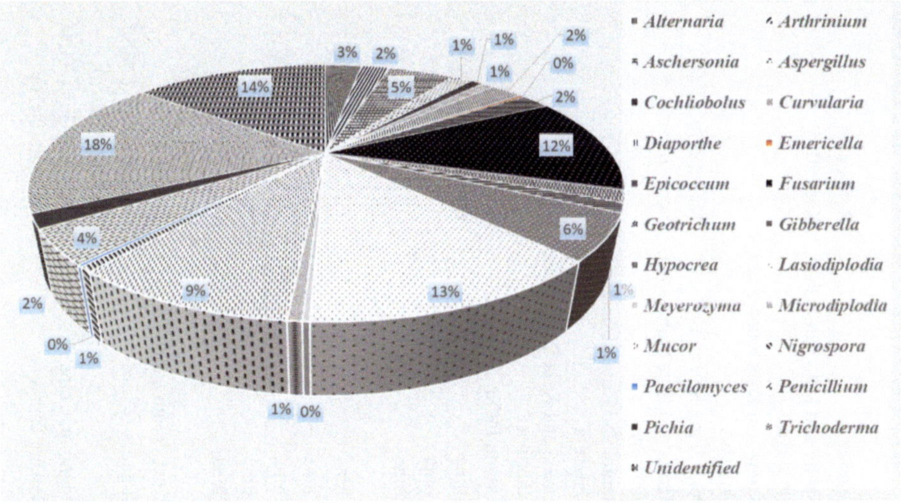

Fig. 19.2 Diversity of endophytes in *Aquilaria malaccensis* occurred in Assam, India

endophytic fungi was reported from the families *Hypocreaceae*, *Nectriaceae*, *Pleosporaceae*, and *Trichocomaceae* in *A. malaccensis*, *A. sinensis*, *A. crassna*, and *A. agallocha*. A large number of endophytic fungal genera were reported from *Aquilaria* species, which include *Acremonium*, *Alternaria*, *Aporospora*, *Arthrinium*, *Arthrobotrys*, *Aschersonia*, *Aspergillus*, *Beauveria*, *Botryodiplodia*, *Cephalosporium*, *Chaetomium*, *Cladosporium*, *Collectotrichum*, *Coniosporium*, *Corynespora*, *Cunninghamella*, *Curvularia*, *Cylindrocladium*, *Diaporthe*, *Epicoccum*, *Fimetariella*, *Fusarium*, *Geotrichum*, *Gibberella*, *Glomerularia*, *Guignardia*, *Hypocrea*, *Lasiodiplodia*, *Leptosphaerulina*, *Monilia*, *Mortierella*, *Mucor*, *Mycelia*, *Nigrospora*, *Nodulisporium*, *Ovulariopsis*, *Paraconiothyrium*, *Penicillium*, *Pestalotiopsis*, *Phaeoacremonium*, *Phoma*, *Phomopsis*, *Pichia*, *Pleospora*, *Rhinocladiella*, *Rhizomucor*, *Trichoderma*, *Tylopilus*, and *Xylaria*. The details of diversity of endophytes in various *Aquilaria* species are presented in Table 19.1.

19.4 Agarwood and Its Commercial Importance

Agarwood contains oleoresin which has different types of sesquiterpenoids, chromones, and aromatic compounds which determines the quality and grade of agarwood (Kalra and Kaushik 2017). Agarwood oleoresin is used in the production of incense and perfumes and in Chinese medicine. Agarwood is also used in different ancient practices such as curative ceremonies of Middle East, donation in Shinto-Buddhist temples by Japanese, and ceremonies by Mekong delta communities in Vietnamese temples (Akter et al. 2013). Agarwood oil is being used as essence in soap, shampoo, and perfumes. It is also used in curing asthma, vomiting arrest,

Table 19.1 Diversity of endophytic fungi in various species of *Aquilaria* plant

S.No.	Endophyte	Class	Family	Tree species	Country	Reference
1	*Beauveria bassiana*	*Sordariomycetes*	*Cordycipitaceae*	*Aquilaria* sp.	Baram and Kota Samarahan, Malasiya	Kee (2015)
2	*Botryodiplodia theobromae*	*Sordariomycetes*	Incertae sedis	*Aquilaria* sp.	Baram and Kota Samarahan, Malasiya	
3	*Fusarium trifosforium*	*Sordariomycetes*	*Nectriaceae*	*Aquilaria* sp.	Pekanbaru, Sumatra	Tabata et al. (2003)
4	*Trichoderma reesei*	*Sordariomycetes*	*Hypocreaceae*	*Aquilaria* sp.	Baram and Kota Samarahan, Malasiya	Kee (2015)
5	*Trichoderma harzanium*	*Sordariomycetes*	*Hypocreaceae*	*Aquilaria* sp.	Baram and Kota Samarahan, Malasiya	
6	*Trichoderma* sp.	*Sordariomycetes*	*Hypocreaceae*	*Aquilaria* sp.	Baram and Kota Samarahan, Malasiya	
7	*Aspergillus* sp.	*Eurotiomycetes*	*Trichocomaceae*	*Aquilaria* sp.	Assam, India	Bose (1934)
8	*Botryosphaeria rhodina*	*Dothideomycetes*	*Botryosphaeriaceae*	*A. sinensis*	China	Wang et al. (2009)
9	*Chaetomium* sp.	*Sordariomycetes*	*Chaetomiaceae*	*A. sinensis*	China	
10	*Colletotrichum gloeosporioides*	*Sordariomycetes*	*Glomerellaceae*	*A. sinensis*	China	
11	*Fusarium* sp.	*Sordariomycetes*	*Nectriaceae*	*A. sinensis*	China	
12	*Penicillium* sp.	*Eurotiomycetes*	*Trichocomaceae*	*A. sinensis*	China	
13	*Pestalotiopsis* sp.	*Sordariomycetes*	*Sporocadaceae*	*A. sinensis*	China	
14	*Trichoderma* sp.	*Sordariomycetes*	*Hypocreaceae*	*A. sinensis*	China	
15	*Xylaria* sp.	*Sordariomycetes*	*Xylariaceae*	*A. sinensis*	China	
16	*Alternaria* sp.	*Dothideomycetes*	*Pleosporaceae*	*A. sinensis*		Tian et al. (2013)
17	*Arthrobotrys* sp.	*Orbiliomycetes*	*Orbiliaceae*	*A. sinensis*	Xinyi, South China	Li et al. (2018)
18	*Botryosphaeria rhodina*	*Dothideomycetes*	*Botryosphaeriaceae*	*A. sinensis*	Jinghong, China	Gong and Guo (2009)
19	*Botryosphaeria dothidea*	*Dothideomycetes*	*Botryosphaeriaceae*	*A. sinensis*		Tian et al. (2013)
20	*Cephalosporium* sp.	*Ascomycetes*	Incertae sedis	*A. sinensis*	Jinghong, China	Gong and Guo (2009)
21	*Chaetomium globosum*	*Sordariomycetes*	*Chaetomiaceae*	*A. sinensis*	Hainan, China	Cui et al. (2011)

22	*Cladosporium edgeworthrae*	*Dothideomycetes*	*Davidiellaceae*	*A. sinensis*	Jinghong, China	Gong and Guo (2009)
23	*Cladosporium tenuissimum*	*Dothideomycetes*	*Davidiellaceae*	*A. sinensis*	Yunnan and Hainan province, China	Cui et al. (2011)
24	*Cladosporium tenuissimum*	*Dothideomycetes*	*Davidiellaceae*	*A. sinensis*	Yunnan, China	
25	*Collectotrichum* sp.	*Sordariomycetes*	*Glomerellaceae*	*A. sinensis*		Tian et al. (2013)
26	*Colletotrichum* sp.	*Sordariomycetes*	*Glomerellaceae*	*A. sinensis*	Jinghong, China	Gong and Guo (2009)
27	*Coniosporium* sp.	*Eurotiomycetes*		*A. sinensis*	Xinyi, South China	Li et al. (2018)
28	*Coniothyrium nitidae*	*Dothideomycetes*	*Leptosphaeriaceae*	*A. sinensis*	Yunnan, China	Cui et al. (2011)
29	*Corynespora* sp.	*Dothideomycetes*	*Corynesporascaceae*	*A. sinensis*	Xinyi, South China	Li et al. (2018)
30	*Cylindrocladium* sp.	*Sordariomycetes*	*Nectriaceae*	*A. sinensis*		Tian et al. (2013)
31	*Epicoccum* sp.	*Dothideomycetes*	*Pleosporaceae*	*A. sinensis*	Jinghong, China	Gong and Guo (2009)
32	*Epicoccum nigrum*	*Dothideomycetes*	*Pleosporaceae*	*A. sinensis*	Yunnan, China	Cui et al. (2011)
33	*Fimetariella rabenhorstii*	*Sordariomycetes*	*Lasiosphaeriaceae*	*A. sinensis*	China	Tao et al. (2011)
34	*Fusarium* sp.	*Sordariomycetes*	*Nectriaceae*	*A. sinensis*	Jinghong, China	Gong and Guo (2009)
35	*Fusarium equiseti*	*Sordariomycetes*	*Nectriaceae*	*A. sinensis*	Yunnan, China	Cui et al. (2011)
36	*Fusarium oxysporum*	*Sordariomycetes*	*Nectriaceae*	*A. sinensis*	Jinghong, China	Gong and Guo (2009)
37	*Geotrichum* sp.	*Saccharomycetes*	*Dipodascaceae*	*A. sinensis*	Jinghong, China	
38	*Glomerularia* sp.	*Pucciniomycetes*	*Platygloeaceae*	*A. sinensis*	Jinghong, China	
39	*Guignardia manqiferae*	*Dothideomycetes*	*Botryosphaeriaceae*	*A. sinensis*	Jinghong, China	
40	*Hypocrea lixii*	*Sordariomycetes*	*Hypocreaceae*	*A. sinensis*	Yunnan and Hainan, China	Cui et al. (2011)
41	*Lasiodiplodia theobromae*	*Dothideomycetes*	*Botryosphaeriaceae*	*A. sinensis*	Yunnan, China	Cui et al. (2011)
42	*Leptosphaerulina chartarum*	*Dothideomycetes*	Ncertae sedis	*A. sinensis*	Yunnan, China	
43	*Monilia* sp.	*Ascomycetes*	*Sclerotiniaceae*	*A. sinensis*	Jinghong, China	Gong and Guo (2009)
44	*Mortierella* sp.	*Mortierellomycetes*	*Mortierellaceae*	*A. sinensis*	Jinghong, China	
45	*Mycelia sterilia*			*A. sinensis*	Jinghong, China	

(continued)

Table 19.1 (continued)

S.No.	Endophyte	Class	Family	Tree species	Country	Reference
46	*Nigrospora oryzae*	*Sordariomycetes*	*Apiosporaceae*	*A. sinensis*	Xinyi. China	Li et al. (2014)
47	*Nigrospora* sp.	*Sordariomycetes*	*Apiosporaceae*	*A. sinensis*	Xinyi, South China	Li et al. (2018)
48	*Nodulisporium* sp.	*Sordariomycetes*,	*Xylariaceae*	*A. sinensis*	Guangdong province, China	Wu et al. (2010)
49	*Ovulariopsis* sp.	*Leotiomycetes*	*Erysiphaceae*	*A. sinensis*	Jinghong, China	Gong and Guo (2009)
50	*Paraconiothyrium variabile*	*Dothideomycetes*	*Didymosphaeriaceae*	*A. sinensis*	Yunnan and Hainan, China	Cui et al. (2011)
51	*Penicillium* sp.	*Eurotiomycetes*	*Trichocomaceae*	*A. sinensis*	Jinghong, China	Gong and Guo (2009)
52	*Phaeoacremonium rubrigenum*	*Sordariomycetes*	*Togniniaceae*	*A. sinensis*	Yunnan, China	Cui et al. (2011)
54	*Phoma herbarum*	*Dothideomycetes*	*Didymellaceae*	*A. sinensis*	Yunnan, China	
55	*Phoma* sp.	*Dothideomycetes*	*Didymellaceae*	*A. sinensis*		Tian et al. (2013)
56	*Phomopsis* sp.	*Sordariomycetes*	*Valsaceae*	*A. sinensis*		
57	*Pichia guilliermondii*	*Saccharomycetes*	*Saccharomycetaceae*	*A. sinensis*	Hainan, China	Cui et al. (2011)
58	*Pleospora* sp.	*Dothideomycetes*	*Pleosporaceae*	*A. sinensis*	Jinghong, China	Gong and Guo (2009)
59	*Rhinocladiella* sp.	*Eurotiomycetes*		*A. sinensis*	Jinghong, China	
60	*Rhizomucor variabilis*	*Zygomycetes*	*Mucoraceae*	*A. sinensis*	Yunnan, China	Cui et al. (2011)
61	*Tylopilus* sp.	*Agaricomycetes*	*Boletaceae*	*A. sinensis*	Xinyi, South China	Li et al. (2018)
62	*Xylaria mali*	*Sordariomycetes*	*Xylariaceae*	*A. sinensis*	Yunnan and Hainan, China	Cui et al. (2011)
63	*Acremonium* sp.	*Sordariomycetes*	*Hypocreaceae*	*A. microcarpa*	Malaysia	Rahayu and Juliarni (2007)
64	*Aporospora* sp.		Incertae sedis	*A. malaccensis*	Assam, India	Chhipa and Kaushik (2017)
65	*Arthrinium* sp.	*Sordariomycetes*	*Apiosporaceae*	*A. malaccensis*	Assam, India	
66	*Aschersonia* sp.	*Sordariomycetes*	*Clavicipitaceae*	*A. malaccensis*	Assam, India	
67	*Aspergillus flavus*	*Eurotiomycetes*	*Trichocomaceae*	*A. malaccensis*	Assam, India	
68	*Aspergillus* sp.	*Eurotiomycetes*	*Trichocomaceae*	*A. malaccensis*	Assam, India	
69	*Cunninghamella* sp.		*Cunninghamellaceae*	*A. malaccensis*		Mohamed et al. (2010)
70	*Curvularia* sp.	*Euascomycetes*	*Pleosporaceae*	*A. malaccensis*		

71	*Curvularia* sp.	*Euascomycetes*	*Pleosporaceae*	*A. malaccensis*	Assam, India	Chhipa and Kaushik (2017)
72	*Curvularia verruculosa*	*Euascomycetes*	*Pleosporaceae*	*A. malaccensis*	Assam, India	
73	*Curvularia lunata*	*Euascomycetes*	*Pleosporaceae*	*A. malaccensis*	Assam, India	
74	*Diaporthe phaseolorum*	*Sordariomycetes*	*Diaporthaceae*	*A. malaccensis*	Assam, India	
75	*Epicoccum sorghinum*	*Dothideomycetes*	*Pleosporaceae*	*A. malaccensis*	Assam, India	
76	*Epicoccum nigrum*	*Dothideomycetes*	*Pleosporaceae*	*A. malaccensis*	Assam, India	
77	*Epicoccum* sp.	*Dothideomycetes*	*Pleosporaceae*	*A. malaccensis*	Assam, India	
78	*Fusarium* sp.	*Sordariomycetes*	*Nectriaceae*	*A. malaccensis*		Mohamed et al. (2010)
79	*Fusarium incarnatum*	*Sordariomycetes*	*Nectriaceae*	*A. malaccensis*	Assam, India	Chhipa and Kaushik (2017)
80	*Fusarium* sp.	*Sordariomycetes*	*Nectriaceae*	*A. malaccensis*	Assam, India	
81	*Fusarium equiseti*	*Sordariomycetes*	*Nectriaceae*	*A. malaccensis*	Assam, India	
82	*Fusarium solani*	*Sordariomycetes*	*Nectriaceae*	*A. malaccensis*	Assam, India	
83	*Fusarium oxysporum*	*Sordariomycetes*	*Nectriaceae*	*A. malaccensis*	Assam, India	
84	*Gibberella intermedia*	*Sordariomycetes*	*Nectriaceae*	*A. malaccensis*	Assam, India	
85	*Hypocrea caerulescens*	*Sordariomycetes*	*Hypocreaceae*	*A. malaccensis*	Assam, India	
86	*Hypocreales* sp.	*Sordariomycetes*	*Hypocreaceae*	*A. malaccensis*	Assam, India	
87	*Lasiodiplodia pseudotheobromae*	*Dothideomycetes*	*Botryosphaeriaceae*	*A. malaccensis*	Assam, India	
88	*Lasiodiplodia theobromae*	*Dothideomycetes*	*Botryosphaeriaceae*	*A. malaccensis*	Assam, India	
89	*Lasiodiplodia* sp.	*Dothideomycetes*	*Botryosphaeriaceae*	*A. malaccensis*		Mohamed et al. (2010)
90	*Mucor* sp.	*Mucoromycotina*	*Mucoraceae*	*A. malaccensis*	Assam, India	Chhipa and Kaushik (2017)
91	*Mucor circinelloides*	*Mucoromycotina*	*Mucoraceae*	*A. malaccensis*	Assam, India	
92	*Nigrospora* sp.	*Sordariomycetes*	*Apiosporaceae*	*A. malaccensis*	Assam, India	
93	*Penicillium citrinum*	*Eurotiomycetes*	*Trichocomaceae*	*A. malaccensis*	Assam, India	
94	*Penicillium chrysogenum*	*Eurotiomycetes*	*Trichocomaceae*	*A. malaccensis*	Assam, India	
95	*Trichoderma* sp.	*Sordariomycetes*	*Hypocreaceae*	*A. malaccensis*		Mohamed et al. (2010)

(continued)

Table 19.1 (continued)

S.No.	Endophyte	Class	Family	Tree species	Country	Reference
96	*Trichoderma harzianum*	*Sordariomycetes*	*Hypocreaceae*	*A. malaccensis*	Assam, India	Chhipa and Kaushik (2017)
97	*Trichoderma longibrachiatum*	*Sordariomycetes*	*Hypocreaceae*	*A. malaccensis*	Assam, India	
98	*Trichoderma virens*	*Sordariomycetes*	*Hypocreaceae*	*A. malaccensis*	Assam, India	
99	*Alternaria* sp.	*Dothideomycetes*	*Pleosporaceae*	*A. malaccensis*	Assam, India	Premalatha and Kalra (2013)
100	*Cladosporium cladosporoides*	*Dothideomycetes*	*Davidiellaceae*	*A. malaccensis*	Assam, India	
101	*Cladosporium* sp.	*Dothideomycetes*	*Davidiellaceae*	*A. malaccensis*	Assam, India	
102	*Curvularia* sp.	*Euascomycetes*	*Pleosporaceae*	*A. malaccensis*	Assam, India	
103	*Fusarium* sp.	*Sordariomycetes*	*Nectriaceae*	*A. malaccensis*	Assam, India	
104	*Phaeoacremonium* sp.	*Sordariomycetes*	*Togniniaceae*	*A. malaccensis*	Assam, India	
105	*Trichoderma* sp.	*Sordariomycetes*	*Hypocreaceae*	*A. malaccensis*	Assam, India	
106	*Colletotrichum truncatum*	*Sordariomycetes*	*Glomerellaceae*	*A. crassna*	Southern Vietnam	Chi et al. (2016)
107	*Fusarium verticillioides*	*Sordariomycetes*	*Nectriaceae*	*A. crassna*	Southern Vietnam	
108	*Geotrichum candium*	*Saccharomycetes*	*Dipodascaceae*	*A. crassna*	Southern Vietnam	
109	*Aspergillus niger*	*Eurotiomycetes*	*Trichocomaceae*	*A. agallocha*		Tamuli et al. (2005)
110	*Epicoccum granulatum*	*Dothideomycetes*	*Pleosporaceae*	*A. agallocha*		Bhattacharyya et al. (1952)
111	*Fusarium* sp.	*Sordariomycetes*	*Nectriaceae*	*A. agallocha*		Tamuli et al. (2005)
112	*Penicillium* sp.	*Eurotiomycetes*	*Trichocomaceae*	*A. agallocha*		
113	*Trichoderma* sp.	*Sordariomycetes*	*Hypocreaceae*	*A. agallocha*		

carminative and pain reliever (Okugawa et al. 1993; Yao et al. 1995; Zhou et al. 2008). In Malaysia, it is used for jaundice and body ache treatment. Agarwood in powder form is used in religious ceremony of Buddhist, Muslims, and Hindus. All the above mentioned uses of agarwood made it most demanding non-timber wood (Naef 2011). The price of agarwood depends on grade of wood and oleoresin oil. First-grade agarwood is most expensive wood which is produced naturally and takes 20–30 years in development. Agarwood chips also have £20–£6000 per kilogram cost in the market. It has been estimated that global market of agarwood is approx. 6–8 billion US$ (Azah et al. 2013).

19.5 Application of Fungi in Artificial Production of Agarwood

It has been reported that biological method provides high-quality agarwood in comparison to physical and chemical methods (Mohamed et al. 2014; Tsan and Mohamed 2014; Chhipa and Kaushik 2017). Use of fungi in artificial inoculation can be more advantageous due to the fact that systemic growth of fungus in intermittent part of agarwood can induce agarwood formation continuously (Novriyanti et al. 2010). Initially, Bhattacharyya et al. (1952) reported association of *Epicoccum granulatum* with infected part of *Aquilaria* plant. Thereafter, *F. oxysporum*, *F. bulbigenium*, and *F. lateritium* found effective in agarwood formation (Nobuchi and Siripatanadilok 1991). Similarly, Tamuli et al. (2005) used *C. globosum* in artificial inoculation and found production of agarwood compounds. The production of oleoresin after infection changed the color of white wood into dark brown or black. Rahayu and Juliarni (2007) found the *Aquilaria* wood changed the color after inoculation of *Acremonium* sp. They observed the formation of terpenoid altered the wood coloration. Further, Tian et al. (2013) developed pinhole method by combing chemical and biological materials. They used formic acid and *B. dothidea* fungus for induction of agarwood formation. The succession pattern of fungi *C. bainieri*, *F. solani*, and *L. theobromae* in wounded *Aquilaria* tree was studied for 12 months by Mohamed et al. (2014). They found that succession of each fungus is tree specific and growth of inoculum influenced by microclimate conditions. In the initial first 2 weeks, fungi flourished in the tree, but after 6 months, the number of fungi decreased due to production of resin. It was assumed that increased amount of terpenes is responsible for reduction of fungal load level in infected wood (Tamuli et al. 2005; Naef 2011). Wong et al. (2013) reported that wounding in *Aquilaria* tree induced the expression of primary defense genes and found improved expression of genes encoding β-1,3-glucanase and phenylalanine ammonia lyase.

Production of agarwood by fungal inoculation has been reported by many researchers. Different fungi, including *Aspergillus* sp., *Acremonium*, sp. *B. dothidea*, *C. globosum*, *Colletotrichum* sp., *Diplodia* sp., *E. granulatum*, *F. lateritium*, *F. oxysporum*, *F. bulbiferum*, *F. lateritium*, *L. theobromae*, *Melanotus flavolivence*, *P. polonicum*, *Trichoderma* sp., and *Xylaria* sp., have been reported to induce

Fig. 19.3 Agarwood formation in *Aquilaria* plant and infected agarwood (dark in color)

agarwood formation (Chhipa and Kaushik 2017; Peng et al. 2015; Zhang et al. 2012, 2014; Mohamed et al. 2014; Akter et al. 2013; Cui et al. 2013; Lin et al. 2010; Yunita 2009; Rahayu and Juliarni 2007; Nobuchi and Siripatanadilok 1991; Bhattacharyya et al. 1952).

Fungus enters into plant through injury and induces defense system which consequently leads to the accumulation of secondary metabolites. Thus resin production is initiated and increases with infection (Cui et al. 2013). This process is known as "tylosis" and further wood coloration changed from white to dark brown or black (Fig. 19.3). Genetic variation, microenvironment, season, and age of *Aquilaria* tree are various parameters, which affect the formation of agarwood by fungi.

19.6 Future Prospects of Artificial Agarwood Production Technique

Artificial production of agarwood can be efficient technology, which can provide agarwood to fulfil the commercial demand in comparison to naturally generated agarwood. Some researchers have reported the production of agarwood in 3–6 months after artificial infection, whereas natural process takes 20–30 years. It has also been reported that the use of fungi in artificial production developed high quality of agarwood and similar terpenes and volatile compounds have been reported in chemical analysis, which are found in naturally produced agarwood.

19.7 Conclusions

Exploration of fungal diversity in *Aquilaria* tree and their role in artificial production of agarwood suggested that oleoresin accumulation depends on fungal species. Some of the endophytic fungal species get transformed into latent pathogen and induce the infection in *Aquilaria* tree. Various fungi from *Ascomycota* class such as *Trichoderma*, *Lasiodiplodia*, and *Penicillium* strains are appropriate in artificial production of high-quality agarwood in short time period. The development of artificial production technique using fungi can provide the new life to the threatened *Aquilaria* species by promoting interest in villagers for plantation. Biological approach of artificial production of agarwood will provide the sustainable source of economy to local communities. The prospect of artificial production of agarwood has been identified as small-scale industries for local villagers that mitigate the migration of local villagers for earning their livelihood.

References

Akter S, Islam MT, Zulkefeli M, Khan SI (2013) Agarwood production-a multidisciplinary field to be explored in Bangladesh. Int J Pharm Life Sci 2:22–32

Azah MN, Husni SS, Mailina J, Sahrim L, Majid JA, Faridz ZM (2013) Classification of agarwood (gaharu) by resin content. J Trop For Sci 25:213–219

Barden A, Anak NA, Mulliken T, Song M (2000) Heart of the matter: agarwood use and trade and CITES implementation for *Aquilaria malaccensis*. TRAFFIC International, Cambridge

Bhattacharyya B, Datta A, Baruah HK (1952) On the formation and development of agaru in *Aquilaria agallocha*. Sci Cult 18:240–243

Bose SR (1934) The nature of agar formation. Sci Cult 4:89–91

Burkill IH (1966) A dictionary of the economic products of the Malay Peninsula. A Dictionary of the Economic Products of the Malay Peninsula, 2(2nd edn)

Chakrabarty K, Kumar A, Menon V (1994) Trade in agarwood. TRAFFIC India and WWF-India, New Delhi

Chen H, Yang Y, Xue J, Wei J, Zhang Z, Chen H (2011) Comparison of compositions and antimicrobial activities of essential oils from chemically stimulated agarwood, wild agarwood and healthy *Aquilaria sinensis* (Lour.) Gilg trees. Molecules 16:4884–4896

Chetpattananondh P (2012) Overview of the agarwood oil industry. In: IFEAT international conference in Singapore, pp 131–138

Chhipa H, Kaushik N (2017) Fungal and bacterial diversity isolated from *Aquilaria malaccensis* tree and soil, induces agarospirol formation within 3 months after artificial infection. Front Microbiol 8:1286

Chhipa H, Chowdhary K, Kaushik N (2017) Artificial production of agarwood oil in Aquilaria sp. by fungi: a review. Phytochem Rev 16(5):835–860

Chi HK, Hang TTN, Luyen ND, Ha TTH (2016) Biological characterization of fungal endophytes isolated from agarwood tree *Aquilaria crassna* Pierre ex Lecomte. Vietnam J Biotechnol 14(1):149–156

Chua LSL (2008) Agarwood (*Aquilaria malaccensis*) in Malaysia. Forest Research Institute Malaysia, Kuala Lumpur, pp 1–17

CITES (2004) Convention on international trade in endangered species of wild fauna and flora. Amendment to appendices I and II of CITES. In: Thirteenth meeting of the conference of the parties, Bangkok, Thailand, 3–14 Oct 2004

Cui JL, Guo SX, Xiao PG (2011) Antitumor and antimicrobial activities of endophytic fungi from medicinal parts of Aquilaria sinensis. J Zhejiang Univ Sci B 12(5):385–392

Cui J, Wang C, Guo S, Yang L, Xiao P, Wang M (2013) Evaluation of fungus-induced agilawood from *Aquilaria sinensis* in China. Symbiosis 60(1):37–44

Dai HF, Liu J, Han Z, Zeng YB, Wang H, Mei WL (2010) Two new 2-(2-phenylethyl) chromones from Chinese eaglewood. J Asian Nat Prod Res 12:134–137

Deshmukh SK (2018) Translating endophytic fungal research towards pharmaceu-tical applications. Kavaka 50:1–13

Deshmukh SK, Verekar SA, Bhave S (2015) Endophytic fungi: an untapped source for antibacterials. Front Microbiol. https://doi.org/10.3389/fmicb.2014.00715

Deshmukh SK, Gupta MK, Prakash V, Saxena S (2018) Endophytic fungi: a source of potential antifungal compounds. J Fungi 4:77. https://doi.org/10.3390/jof4030077

Gong L, Guo S (2009) Endophytic fungi from Dracaena cambodiana and *Aquilaria sinensis* and their antimicrobial activity. Afr J Biotechnol 8(5):731–736

Gunasekera SP, Kinghorn AD, Cordell GA, Farnsworth NR (1981) Plant anticancer agents. XIX. Constituents of Aquilaria malaccensis. J Nat Prod 44(5):569–572

Gunatilaka AAL (2006) Natural products from plant associated microorganisms: distribution, structural diversity, bioactivity and implications of their occurrence. J Nat Prod 69:509–526

Ito M, Okimoto KI, Yagura T, Honda G, Kiuchi F, Shimada Y (2005) Induction of sesquiterpenoid production by methyl jasmonate in *Aquilaria sinensis* cell suspension culture. J Essent Oil Res 17:175–180

Kalra R, Kaushik N (2017) A review of chemistry, quality and analysis of infected agarwood tree (*Aquilaria* sp.). Phytochem Rev 16(5):1045–1079

Kee LK (2015) Isolation of endophytic fungi from agarwood. Thesis University of Malaysia Sarawak (UNIMAS)

Kharwar RN, Mishra A, Gond SK, Stierle A, Stierle D (2011) Anticancer compounds derivedfrom fungal endophytes: their importance and future challenges. Nat Prod Rep 28:1208–1228

Li D, chan Chen Y, Pan Q, hua Tao M, Zhang W (2014) A new eudesmane sesquiterpene from *Nigrospora oryzae*, an endophytic fungus of *Aquilaria sinensis*. Rec Nat Prod 8:330–333

Li T, Deng WQ, Li TH, Zhang WM, Hosen MI, Song B (2018) Endophytic fungal communities associated with leaves, stems and roots of four medicinal plants in South China. Stud Fungi 3:126–140

Lin F, Mei WL, Wu J, Dai HF (2010) GC-MS analysis of volatile constituents from Chinese eaglewood produced by artificial methods. J Chin Med Mater 33:222–225

Liu Y, Chen H, Yang Y, Zhang Z, Wei J, Meng H, Gao Z (2013) Whole-tree agarwood-inducing technique: an efficient novel technique for producing high-quality agarwood in cultivated *Aquilaria sinensis* trees. Molecules 18(3):3086–3106

Meng-Ling HE, Shu-yuan Q, Lan-juan HU (2005) Rapid in vitro propagation of medicinally important *Aquilaria agallocha*. J Zhejiang Univ Sci B(6):849–852

Mohamed R, Jong PL, Zali MS (2010) Fungal diversity in wounded stems of *Aquilaria malaccensis*. Fungal Divers 43:67–74

Mohamed R, Jong PL, Kamziah AK (2014) Fungal inoculation induces agarwood in young *Aquilaria malaccensis* trees in the nursery. J For Res 25(1):201–204

Naef R (2011) The volatile and semi-volatile constituents of agarwood, the infected heartwood of *Aquilaria* species: a review. Flavour Fragr J 26:73–87

Nobuchi T, Siripatanadilok S (1991) Preliminary observation of *Aquilaria crassna* wood associated with the formation of aloeswood. Bull Kyoto Univ For 63:226–235

Novriyanti E (2008) Role of extractive substances in formation agarwood *Aquilaria crassna* by Pierre ex Lecomte and *Aquilaria microcarpa* Baill. Ph.D. thesis, The Graduate School of Bogor Agricultural University. Bogor, Indonesia (Unpublished thesis)

Novriyanti E, Santosa E, Syafii W, Turjaman M, Sitepu IR (2010) Antifungal activity of wood extract of *Aquilaria crassna* Pierre ex Lecomte against agarwood-inducing fungi, *Fusarium solani*. Indones J For Res 15:155–165

Okugawa H, Ueda R, Matsumoto K, Kawanishi K, Kato A (1993) Effects of agarwood extracts on the central nervous system in mice. Planta Med 59:32–36

Oldfield S, Lusty C, MacKinven A (1998) The world list of threatened trees. World Conservation Press, Cambridge

Peng CS, Osman MF, Bahari N, Zakaria R, Rahim KA (2015) Agarwood inducement technology: a method for producing oil grade agarwood in cultivated *Aquilaria malaccensis* Lamk. J Agrobiotechnol 6:1–16

Pojanagaroon S, Kaewrak C (2003) Mechanical methods to stimulate aloes wood formation in *Aquilaria crassna* Pierre ex H. Lec. (Kritsana) trees. In: Jatisatienr A, Paratasilpin T, Elliott S, Anusarnsunthorn V, Wedge D, Craker LE, Gardner ZE (eds) III WOCMAP congress on medicinal andaromatic plants – volume 2: conservation, cultivation and sustainable use of medicinal and aromatic plants. ISHS Acta Horticulturae 676, Chiang Mai, pp. 161–166

Premalatha K, Kalra A (2013) Molecular phylogenetic identification of endophytic fungi isolated from resinous and healthy wood of *Aquilaria malaccensis*, a red listed and highly exploited medicinal tree. Fungal Ecol 6:205–211

Rahayu G, dan Juliarni AP (2007) *Acremonium* and methyl-jasmonate induce terpenoid formation in agarwood tree (*Aquilaria crassna*). In: Makalahdi presenta sikandalam 3rd Asian conference on crop protection, Jogyakarta, pp 22–24

Saikia P, Khan ML (2012) Seedling survival and growth of *Aquilaria malaccensis* in different microclimatic conditions of Northeast India. J For Res 23:569–574

Sangareswari M, Parthiban KT, Kanna SU, Karthiba L, Saravanakumar D (2016) Fungal microbes associated with agarwood formation. Am J Plant Sci 7:1445–1452

Sun X, Ding Q, Hyde KD, Guo LD (2012) Community structure and preference of endophytic fungi of three woody plants in a mixed forest. Fungal Ecol 5:624–632

Tabata Y, Widjaja E, Mulyaningsih T, Parman I, Wiriadinata H, Mandang YI, Itoh T (2003) Structural survey and artificial induction of aloeswood. Wood Res 90:11–12

Tabin T, Srivastava K (2014) Distribution and population status of critically endangered *Aquilaria malaccensis* Lamk. In The Forests of Arunachal Pradesh and Assam, India. Int J Innov Res Sci Eng Technol 3:17595–17604

Tamuli P, Boruah P, Nath SC, Leclercq P (2005) Essential oil of eaglewood tree: a product of pathogenesis. J Essent Oil Res 17:601–604

Tao MH, Yan J, Wei XY, Li DL, Zhang WM, Tan JW (2011) A novel sesquiterpene alcohol from *Fimetariella rabenhorstii*, an endophytic fungus of *Aquilaria sinensis*. Nat Prod Commun 6:763–766

Tian JJ, Gao XX, Zhang WM, Wang L, Qu LH (2013) Molecular identification of endophytic fungi from *Aquilaria sinensis* and artificial agarwood induced by pinholes-infusion technique. Afr J Biotechnol 12(21):3115–3131

Tsan PLJP, Mohamed R (2014) Gas chromatography-mass spectrometry analysis of agarwood extracts from mature and juvenile *Aquilaria malaccensis*. Int J Agric Biol 16:644–648

Wang L, Zhang WM, Pan QL, Li HH, Tao MH, Gao XX (2009) Isolation and molecular identification of endophytic fungi from Aquilaria sinensis. J Fungal Res 7:37–42

Wei JH, Zhang Z, Yang Y, Meng H, Gao ZH, Chen WP, Feng JD, Chen HQ (2012) An agarwood induction agent and its preparation method. Patent CN101731282B, 2012

Wong MT, Siah CH, Faridah QZ, Mohamed R (2013) Characterization of wound responsive genes in *Aquilaria malaccensis*. J Plant Biochem Biotechnol 22:168–175

Wu ZC, Li DL, Chen YC, Zhang WM (2010) A new isofuranonaphthalenone and benzopyrans from the endophytic fungus Nodulisporium sp. A4 from *Aquilaria sinensis*. Helv Chim Acta 93:920–924

Yao DM, Zhang JB, Chu L, Bao X, Shun Q, Qi P (1995) A coloured atlas of the Chinese materia medica specified in pharmacopoeia of the People's Republic of China. Guangdong Science and Technology Press, Guangdong

Yunita L (2009) Effectivity of *Acremonium* sp. and methyl Jasmonate to improve agarwood quality of *Aquilaria mcrocarpa*. IPB Scientific Repository, Bogor Agricultural University. Available online at: http://repository.ipb.ac.id/handle/123456789/14803

Zhang XL, Liu YY, Wei JH, Yang Y, Zhang Z, Huang JQ, Chen HQ, Liu YJ (2012) Production of high-quality agarwood in *Aquilaria sinensis* trees via whole-tree agarwood-induction technology. Chin Chem Lett 23(6):727–730

Zhang Z, Han XM, Wei JH, Xue J, Yang Y, Liang L, Li XJ, Guo QM, Xu YH, Gao ZH (2014) Compositions and antifungal activities of essential oils from agarwood of *Aquilaria sinensis* (Lour.) Gilg induced by *Lasiodiplodia theobromae* (Pat.) Griffon & Maubl. J Braz Chem Soc 25:20–26

Zhou M, Wang H, Kou J, Yu B (2008) Antinociceptive and anti-inflammatory activities of *Aquilaria sinensis* (Lour.) Gilg. Leaves extract. J Ethnopharmacol 117:345–350

Part IV

Bioprospects

Bioprospecting of Fungal Entomo- and Myco-Pathogens

20

E. K. Pathan, A. V. Patil, and M. V. Deshpande

Abstract

The pest and pathogen control in the field using fungi and their metabolites has indeed gone beyond 'proof of concept'. However, in view of the performances of these biocontrol fungi in the fields, moreover, acceptability by the end users regarding cost-effectiveness, shelf life, intellectual property rights (IPR), the additional roles, and possible applications are being explored. Dual pathogenicity, *viz.* entomopathogenesis and mycoparasitism, has added the advantage of wide-spectrum biocontrol in single-crop system. In addition to plant protection, these fungi can promote plant growth. Further, in integrated pest management, the bioremediation of pesticide residues, if any, can also be achieved by these organisms. For value addition, use of cuticle-degrading and mycolytic enzymes produced as killing components can be explored in food, detergent, organic synthesis, and pharmaceutical industries. While exploring the additional applications, the safety to beneficial insects and fungi, biodiversity, and humans, in general, is also addressed.

Keywords

Biocontrol · Bioremediation · Biotransformation · Entomopathogens · Healthcare · Industrial enzymes · Mycopathogens · Plant growth promotion

---

E. K. Pathan · A. V. Patil
Biochemical Sciences Division, CSIR-National Chemical Laboratory, Pune, Maharashtra, India
e-mail: mv.deshpande@ncl.res.in

M. V. Deshpande (✉)
Division of Biological Sciences, CSIR-National Chemical Laboratory, Pune, Maharashtra, India
e-mail: mv.deshpande@ncl.res.in

T. Satyanarayana et al. (eds.), *Advancing Frontiers in Mycology & Mycotechnology*,
https://doi.org/10.1007/978-981-13-9349-5_20

20.1 Introduction

The control of agricultural pests and pathogens constitutes a broad, highly technical and rapidly developing field of study. One of the reasons is the modern practice for large farming units that are devoted to growing only a single species of crop. The conventional practices such as crop rotation, selection of the resistant variety, and removal of the secondary hosts do not warrant complete control of pests and pathogens. As a result, chemicals are being used indiscriminately which create serious problems to human health such as blindness, asthma, cancer, skin disorders, enlargement of the liver, neural malfunction, and to some extent even psychological problems. Agrochemical consumption is maximum in Japan (11 kg/ha), while it is 4.5 kg/ha in the USA and 0.58 kg/ha in India. Still, by 2022 the market for chemical pesticides is expected to be USD 66 billion, which is ten times more than the market for biological control agents (https://www.marketsandmarkets.com/PressReleases/biopesticide.asp). Biological control of plant fungal pathogens and pests involves the use of microbial antagonists such as bacteria, viruses, fungi, and/or their products like enzymes, enzyme inhibitors, and antibiotics. The market for fungal biocontrol agents is considered to be 10–15% of the total biopesticide market, which is expected to go up to USD 0.66 billion by 2022.

20.2 Fungus-Host Interaction

20.2.1 Entomopathogenesis

Life cycles of insect host and entomopathogenic fungus usually synchronize with each other along with environmental conditions. The habitat influences insect population, which in turn affects the insect pathogens. As fungi are effective by contact, considerations such as adhesion of an infective propagule to the insect body, appressorium development for the entry of the organism into the host, and the killing are important factors in host-pathogen interaction. The conidia are the infective propagules for genera such as *Metarhizium*, *Beauveria*, *Nomuraea*, and *Verticillium*. Once the fungus breaks through the cuticle and underlying epidermis, it may grow profusely in the haemolymph, by forming the blastospores or hyphal bodies, in which case death is probably the result of starvation or physiological disruption brought about by the fungus. Alternately, insecticidal secondary metabolites such as destruxins and beauvericins may contribute to the killing process (Vey et al. 2001). Eventually, under high relative humidity, the fungus will emerge out through the body wall of the insect, producing aerial conidia. Indeed host specificity, virulence of the fungal pathogen, and/or the resistance of the host and finally killing of enzymic and non-enzymic components are the main concerns.

20.2.2 Mycoparasitism

The major findings regarding mycoparasitism of *Trichoderma lignorum* on plant-pathogenic fungi were made by Richard Weindling from the 1930s to 1940s (Harman and Bjorkman 1998). The active toxin such as gliotoxin was identified from *Gliocladium* as one of the killing components (Weindling 1941). Comprehensive details for various mycoparasitic interactions have been documented earlier (Chet et al. 1997; Jeffries 1997; Steyaert et al. 2003; Zeilinger and Omann 2007). There are two types of interactions, namely necrotrophic (destructive) and biotrophic (balanced). In the earlier one, the relationship results in the death and destruction of one or more components of the host while in the later development of the parasite is favoured by a living rather than dead host structure. The biotrophic interactions have a restricted host range, and the specialized structures are produced by the parasite to take up the nutrients from the host.

20.2.3 Dual Pathogenesis

Though host specificity is a major feature of fungus-host interactions, the common components of protective covers of insect pest (cuticle) and fungal pathogens (cell wall) make some of them dual specific mediated mostly through the production of hydrolytic enzymes. For instance, *Trichoderma harzianum*, a well-known mycoparasitic fungus, can parasitize the elm bark beetle, *Scolytus*, while *Verticillium lecanii*, a pathogen of potato aphid *Macrosiphum euphorbiae*, can also attack *Sphaerotheca fuliginea*, the causal agent of cucumber powdery mildew (Askary et al. 1998). Ownley et al. (2008) reported that endophytic colonization of tomato and cotton seedlings by entomopathogen *B. bassiana* was found to be effective in controlling root pathogens such as *Rhizoctonia solani* and *Pythium myriotylum*. The hyphae of *B. bassiana* showed mycoparasitism by coiling around host hyphae. Interestingly, the authors also reported the induction of systemic resistance in cotton against bacterial blight caused by *Xanthomonas axonopodis* pv. *malvecearum*.

20.3 Status of Mycopesticides in Biocontrol Industry

The total biopesticides market will be worth 6.60 billion USD by 2022, which was almost half in 2017. Furthermore, the market for mycoinsecticides is 10–15% of the total biopesticides market. The concept of use of the microbial (bacterial, viral and fungal) biocontrol agents to control insect pests is not new. The viral and the bacterial control agents are effective after ingestion by the insect pests while fungi are effective by contact. Moreover, the fungi can be produced easily on a large-scale, indigenous isolates are environmentally safe, and above all, there is no known resistance in the insect community against the fungal preparations as multiple virulence factors are involved in the killing of insect pests. Fungi such as *Metarhizium*, *Beauveria*, *Nomuraea*, *Verticillium*, and *Paecilomyces* have been studied on a large

Chart 20.1: Bioprospecting of an Entomo-and Myco-Pathogens

In agriculture

- Broad-spectrum biocontrol of pests and pathogens in single crop system
- Endophytes as bodyguards in agriculture crops
- To develop disease suppressive soils
- Plant growth promoting activities
- Transgenic entomo- and myco-pathogens
- One of the components in integrated pest management (IPM)

Value addition

- Isolation of cell wall polymers for variety of applications
- Bioremediation of chemicals used in IPM
- Use of mycelium for the synthesis of nanomaterials
- Enzymes in industries

Cellulases in

Paper and pulp
Textile

Lipases in

Dairy and food manufacture
Leather and detergent industry
Production of cosmetics and pharmaceuticals
Organic synthesis reactions in non-aqueous media
Biotransformation

Proteases in

Detergents
Leather dehairing
Hair waste processing

Chitinases and chitosanases in

Ethanol production
Single cell protein production
Chitooligosaccharide production for healthcare

Chitin deacetylases in

Enzymatic deacetylation of chitin/chitosan
Enzymatic *N*-acetylation of amino sugar

Healthcare

Malarial parasite control
Control of hypercholesterolaemia, hypoglycaemia, hypertension and gastric cancer
Treatment of hepatitis
Anti-TB activity

scale in the area of pest control (Butt et al. 2001). Butt and Copping (2000) discussed the issues that need to be dealt with to increase their market. For instance, the speed of action, virulence, large-scale production, shelf life and stability of formulation, targeting, etc. are the main concerns. From the industry point of view, low cost and strong incentives will be useful to increase the market. While Vega (2018) suggested the establishment of entomopathogenic fungi as endophytes can be a viable alternative for pest control. In the case of mycoparasitic *Trichoderma*, a main player in agriculture, Woo et al. (2014) suggested that in order to be economically competitive as well as effective, novel chemical formulations of *Trichoderma* are necessary. Keswani et al. (2014) discussed the bioprospecting potential of *Trichoderma* in plant growth promotion, bioremediation, and production processes.

The next sections will highlight the potential for bioprospecting and value addition to the fungal biocontrol agents (Chart 20.1).

20.4 Wide-Spectrum Biocontrol Agent for Pests and Pathogens in a Single-Crop System

20.4.1 Virulence

Habitat influences the population of entomopathogenic fungi qualitatively and quantitatively. Additionally, habitat affects the population of insects which in turn affects the population of entomopathogenic fungi (Bidochka et al. 2001). Vimala Devi et al. (2003) reported that *Nomuraea rileyi* isolates from different geographical locations showed different levels of virulence against two lepidopteran insect hosts, *Helicoverpa armigera* and *Spodoptera litura*. The higher virulence was attributed to the soil samples collected from the rhizosphere of crops where chemical insecticides are not routinely sprayed. Nahar et al. (2003) also observed that *Metarhizium* isolates from fields of tomato, okra, and other vegetables heavily sprayed with chemicals were less virulent to *H. armigera* than isolates from a custard apple field rarely sprayed with chemicals (Nahar et al. 2003). Interestingly, our studies highlight another facet to higher virulence of *Metarhizium* isolates from a custard apple field, which are discussed in the next section. The same strain of *Metarhizium* was reported to be effective to control the beet armyworm, *Spodoptera litura* (Fabricius), in a sugar beet field (Yadav and Deshpande 2012). Chavan et al. (2009) observed control of *Ceratovacuna lanigera* (woolly aphids) with conidia of *M. anisopliae* on sugarcane in the field. This was attributed to the cuticle-degrading enzymes, especially lipases produced extracellularly by the fungus. The same strain was also found to be useful in controlling mealy bugs, which have a hard, waxy coating, in a grape field (Tupe et al. 2017; Vidhate et al. 2015). While highlighting the wide-spectrum activities of the entomopathogen, one has to be careful about the non-target effects especially against useful insects. Ginsberg et al. (2002) while working with pathogenicity of *M. anisopliae* against different insects suggested that pathogenicity in laboratory bioassays alone was not sufficient to demonstrate activity under natural conditions. Field trials were needed to confirm the laboratory observations and to assess methods to minimize non-target exposure, if any. Garrido-Jurado et al. (2011) studied the effect of soil treatment with insect pathogenic fungi on soil-dwelling non-target arthropods such as ants, *Tapinoma nigerrimum*, at an olive orchard. According to them, ants have different defence mechanisms against entomopathogens, such as grooming, antimicrobial secretions, inducible antibiotic peptides, etc. Moreover, ants carry conidia which is one of the important dispersal mechanisms for fungus too. Hajek and Goettel (2007) reported guidelines to study the effect of entomopathogens on non-target organisms. Wu et al. (2014) observed that *Beauveria bassiana* strain showed high toxicity against thrips, *Frankliniella occidentalis*, but no pathogenicity to the predatory mite, *Neoseiulus barkeri*. According to the authors either self-grooming, different cuticular composition, or shrivelled conidia which could not adhere to the cuticle could be the possible mechanism for no pathogenicity of *B. bassiana* towards predatory mites.

20.4.2 Genetic Engineering

To increase the virulence of recombinant entomopathogens, the number of candidate genes were reported. For example, recombinant *M. anisopliae* strain with a gene coding protease *Pr1* showed less lethal time towards *Manduca sexta* as compared to wild strain. While genes such as *protein kinase A*, which controls the expression of some secreted virulence factors, an osmosensor *MOS1* that signals to penetrating hyphae that they have reached the hemocoel and a perilipin *MPL1* that regulates lipolysis, turgor pressure, and formation of infection structures were reported to be useful to increase the virulence. Interestingly, Wang and St. Leger (2007) modified *M. anisopliae* with a scorpion toxin (*AaIT*) gene. The engineered *M. anisopliae* achieved the same mortality rates as *M. sexta* at 22-fold lower spore doses than the wild type; with mosquitoes, LC_{50} was reduced 9-fold and for coffee berry borer beetle LC_{50} was 16-fold reduced. The co-expression of *Pr1* and *AaIT* in *B. bassiana* was attempted with the expectation of synergistic action. However, it was reported that the expressed *AaIT* was digested by *Pr1* produced in the haemolymph.

20.4.3 Natural Transformation

In view of the wide variation in the virulence of entomopathogens, it is necessary to isolate and select the highly virulent strain with reduced kill time for its application as a mycoinsecticide (Valero-Jiménez et al. 2016). Zhao et al. (2014) suggested the role of horizontal acquisition of genetic material by entomopathogens in their improved virulence. In a genome-wide analysis of seven different *Metarhizium* strains, Hu et al. (2014) suggested the possibility of horizontal gene transfer (HGT), where *Metarhizium* sp. acquired diverse genes from bacteria, archaea, arthropods, plants, and even from vertebrates. Out of 63 HGT genes acquired from soil-dwelling bacteria, only 8 genes including chymotrypsin, a pentose metabolizing phosphoketolase, a cold shock protein and chitinase, were common in all the *Metarhizium* sp., whereas most of the remaining genes were species specific. The genetic transfer between soil bacteria and *Metarhizium* sp. further suggested that the host environment could serve as a major source for HGT events. Such genetic transfers, especially from the host, were found to be involved in improved virulence of entomopathogens (Zhao et al. 2014). For instance, *M. robertsii* acquired virulence factor (Mr-NPC2a) gene from its insect host (Zhao et al. 2014), while Hu et al. (2014) observed selective acquisition of the gene coding for insecticidal toxin protein TcdB by *M. robertsii*. The gene cluster coding for destruxin, cyclic hexapeptides produced by *Metarhizium* sp. to kill insects, was suggested to be horizontally acquired by *Metarhizium* lineages 15 million years ago. This gene acquisition resulted in broadening the host range and also enhanced the pathogenicity of *Metarhizium* sp. (Hu et al. 2014; Wang et al. 2016; Xu et al. 2016; Zhao et al. 2014).

In our search for indigenous virulent strains of entomopathogens, we observed a wide variation in the virulence of *Metarhizium* isolates depending on their isolation source (Kulkarni et al. 2008; Nahar et al. 2003; Tupe et al. 2017). The *M. anisopliae* strains isolated from soils associated with *Annona squamosa* (custard apple) had higher virulence than strains isolated from the rhizosphere of other plants and from insect cadavers. This improved virulence was found to be due to horizontal transfer of insecticidal traits from *A. squamosa* to *M. anisopliae* strains. Proteomic analysis revealed two insecticidal cyclopeptides of *A. squamosa* origin in the *M. anisopliae* strains that led to higher virulence. Transcriptomic and genomic data indicated that the *M. anisopliae* strains and *A. squamosa* had more than 20 genes in common, including those for cyclic hexapeptide synthase (Tan and Zhou 2006), non-ribosomal peptide synthetase (Tanaka et al. 2005) and plant cyclotide genes (Pechy-Tarr et al. 2008), which are involved in the biosynthesis of insecticidal cyclopeptides. These genes were absent in *M. anisopliae* strains isolated from the rhizosphere of other crop plants. Phylogenetic analysis suggests that these genes were closer to those from *A. squamosa* than to those from fungi. Further, these strains could establish an endophytic relationship with *A. squamosa.* Thus, it was suggested that the higher virulence could be due to genes horizontally transferred to *M. anisopliae* during its endophytic existence in *A. squamosa*, which were eventually released into the soil. Our results suggested that during evolution *Metarhizium* spp. have acquired traits for improved virulence by HGT in a stepwise manner. Further, *Metarhizium* strains associated with *Azadirachta indica* (neem), *Capsicum annuum* (chili), and *Carica papaya* (papaya), plants with insecticidal properties, also had higher virulence. These findings gave a direction to the search for indigenous strains with higher virulence to save crops from insect pests. This can be an alternative to genetically modifying organisms in the lab to increase their potency (Pathan and Deshpande 2019). Given the social and cultural resistance against the acceptance of transgenic crops, as well as of genetically modified insect pathogens, the use of indigenous entomopathogens naturally transformed into more virulent strains is a viable and cost-effective strategy in agricultural biotechnology to save crops from insect pests.

20.5 Plant Growth-Promoting Activity

A number of entomopathogens, in addition to their ability to control insect pests, can provide plant growth-promoting benefits to their host plants (Vega et al. 2009). The plant growth-promoting potential of *Metarhizium*, *Beauveria*, *Trichoderma*, and others has been well documented. For instance, growth promotion by *M. anisopliae* has been reported in beans, corn, peanut, soybean, strawberry, and tomato to name a few where treated plants showed a significant increase in plant height, root and shoot lengths, and number of lateral roots (Garcia et al. 2011; Kabaluk and Ericsson 2007; Khan et al. 2012; Sasan and Bidochka 2012). In the case of *M. anisopliae*-treated peanut plants, increase in the number of lateral roots also increased nodule formation and eventually the yield as compared to untreated crop (Deshpande et al. unpublished data). *Trichoderma harzianum* treatment improved the strength as well as the length and surface area of roots in maize as compared to untreated plants (Harman et al. 2004), while the improved iron (Fe) nutrition of sorghum plants by *Metarhizium brunneum* was observed by Raya-DõÂaz et al. (2017). According to Liu et al. (2017), the method of application also influenced the plant growth-promoting activities. For instance, drenching of the *M. anisopliae* conidial suspension in peanut rhizosphere was more effective in promoting the growth of tap and lateral roots than seed dressing. In addition to controlling the plant pathogens in tomato, *Aureobasidium pullulans* and *Paraconiothyrium sporulosum* also promoted plant growth (Miles et al. 2012). Further, the endophytic colonization of *Piriformospora indica* stimulated the root growth of various crops including maize, tobacco, and parsley (Varma et al. 1999). Dara et al. (2017) found the positive effect of *B. bassiana* and *M. brunneum* application on growth, development, and health of cabbage plants under water stress conditions.

The number of studies have been carried out to understand the mechanism by which entomopathogens promote plant growth. Usuki and Narisawa (2007) suggested that *B. bassiana* and *Metarhizium* species provide insect-derived nitrogen to their host plants. According to Behie and Bidochka (2014), entomopathogens provide nutrients to the plants by stimulating the uptake of phosphorus and other minerals. *Lecanicillium psalliotae* stimulated the growth of cardamom by producing indole-3-acetic acid (IAA) and ammonia and also by making inorganic phosphate and zinc bioavailable to the plant (Senthil Kumar et al. 2018).

20.6 Bioremediation

Porto et al. (2011) extensively reviewed the microbial remediation of pesticides. For the degradation of organochlorine pesticides such as chlorinated derivatives of diphenyl ethane (dichlorodiphenyltrichloroethane—DDT) and its metabolites, hexachlorobenzene, lindane, aldrin, and others, soil bacteria are reported to be prominent, and fungi like *Trichoderma*, *Aspergillus*, *Penicillium*, *Pleurotus*, *Phanerochaete*, and others were also reported to be useful.

Organophosphorus insecticides are widely used to control soil-dwelling insect pests of corn such as rootworms (*Diabrotica* sp.) and cutworms (*Agrotis* sp.), while profenofos (PFF), an organophosphate (OP) insecticide, O-(4-bromo-2-chlorophenyl) O-ethyl S-propyl phosphorothioate, is a non-systemic foliar insecticide and acaricide effective against a wide range of insects including chewing and sucking insects and mites on various crop plants. Diazinon [*O*,*O*-diethyl *O*-(2-isopropyl-4-methyl-6-pyrimidinyl) phosphorothioate] and malathion [S-(1,2-dicarbethoxyethyl)-O,Odimethyldi-thiophosphate], are useful against agricultural as well as household pests. In view of the toxic effects of these pesticides, a number of clean-up physical and chemical technologies have been implemented. However, bioremediation using microorganisms is reported to be useful and cost-effective too. Abd El-Ghany and Masmali (2016) used *M. anisopliae* and *T. harzianum* mycelia to check the degradation of diazinon, malathion, and profenofos up to 20 days, which was almost 50–90% at 30–35 °C. In the soil experiment, the authors could find an increase in soluble phosphorus indicating the degradation of the pesticides. Harish et al. (2013) observed that *T. harzianum* and *Rhizopus nodosus* were capable of metabolizing the organophosphorous insecticides (chlorpyrifos and ethion, 100 ppm) almost 70–80% within 21 days.

Sylwia and co-workers (2013) used *Metarhizium* species to degrade 4-*n*-nonylphenol (4-NP), a degradation product of nonylphenol ethoxylate commercial surfactants. 4-NP is an endocrine-disrupting compound, a harmful contaminant of soil and water. *M. robertsii*, *M. brunneum*, and *M. lepidiotae* strains were reported to degrade 4-NP almost 50–90% to 4-hydroxybenzoic acid, 2-(4-hydroxyphenyl) acetic acid and 4-hydroxyphenylpentanoic acid in 24 h.

T. harzianum, *T. viride*, and *T. atroviride*, known biocontrol organisms, degraded benzimidazole (carbendazim) fungicide, one of the components of an integrated pest management system (Sharma et al. 2016). *T. harzianum* showed 85% degradation of carbendazim, while the other two species degraded it 20–50% within 5 days.

The compound, diuron, N-3,4-dichlorophenyl-N', N'-dimethylurea, is used for selective control of germinating grass and weeds in many crops, mainly cereals. After soil application, it usually persists for 4–8 months. Tixier et al. (2001) demonstrated the degradation and bio-transformation of diuron in a resting cell assay with fungi, *viz. A. niger*, *B. bassiana*, *Cunninghamella elegans*, and *Mortierella isabellina*. The hydroxylated compound of diuron was reported to be transformed completely by *B. bassiana* within 48 h.

20.7 Healthcare Applications

Vector-borne diseases such as malaria, filariasis, dengue, and chikungunya to name a few caused by mosquito genera such as *Anopheles stephensi* Liston, *Culex quinquifasciatus* Say, and *Aedes aegypti* (Linn.) affect significantly humans all over the world. The use of chemical pesticides like organochlorines, organophosphates, and pyrethroids was reported to show a quick knock-down effect on adults. While bacterial species such as *Bacillus thuringiensis* var. *israelensis* and *Bacillus sphaericus* are commercially used as larvicide exerting control by ingestion. Kapoor et al. (2013) observed effective control of both larvae and adults of *A. aegypti* with conidia of *M. anisopliae* strains. The survival of larvae was <5% in 7 days, while the survival of adults was decreased 5 times than untreated control.

More than 50% of the world population is facing the risk of malaria, and mostly in African countries thousands of people die of the disease every year. One of the reasons could be increased resistance of parasites and vectors to drugs and insecticides such as pyrethroid. Some entomopathogenic fungi, *B. bassiana*, *P. fumosoroseus*, and *F. moniliforme*, produce mosquito larvicidal compounds like cyclodepsipeptide, including beauvericin and the enniatin complex. Vyas et al. (2015) also studied the effect of extracellular metabolites of *M. anisopliae* against larvae of *A. stephensi*. The metabolites increased significantly the mortality of mosquito larvae.

Fang et al. (2011) constructed different recombinant *M. anisopliae* strains, which were tested for their ability to control *Plasmodium falciparum* development in *Anopheles gambiae*, Africa's main malaria vector. For example, recombinant strain of *M. anisopliae* expressing a 12 amino acid peptide 1 (SM1) that blocks the attachment of *Plasmodium* sporozoites to salivary glands of *Anopheles* was developed. Further, they introduced a synthetic gene [$(SM1)_8$] in *M. anisopliae* which expressed eight repeats of SM1 peptide. Earlier Wang and St. Leger (2007) showed that high-level expression of scorpion neurotoxin from the scorpion *Androctonus australis* in *M. anisopliae* increased fungal toxicity >20-fold against tobacco hornworm (*Manduca sexta*) caterpillars and almost tenfold against *A. aegypti*. Fang et al. (2011) expressed an $[SM1]_8$ scorpion fusion protein in *M. anisopliae* which reduced sporozoite counts by 98% and suggested that inhibition of *Plasmodium* development could be a powerful weapon for combating malaria.

Entomopathogenic fungi have opened a new area of natural structurally novel biologically active products which significantly contribute to human healthcare (Isaka et al. 2005a). In this regard, genus *Cordyceps*, with more than 300 species, is known for its use in Chinese traditional medicine for about 2000 years. Especially *C. sinensis* can be used to treat many diseases including hepatitis, hypercholesterolaemia, hypertension, and gastric cancer. In fact, the *Metarhizium* and *Beauveria* species are asexual states of some *Cordyceps* species (Shrestha et al. 2017). The extracts of *C. sinensis* containing polysaccharides and sterols were reported to show antioxidant, immunomodulatory, hypoglycemic, hypotensive, vasorelaxant, and antitumor activities. *Cordyceps unilateralis* was reported to produce naphthoquinones, which exhibited anti-malarial activity. The strains of *Beauveria* and *Paecilomyces* produced an ionophoric cyclodepsipeptide, beauvericin, which has proven insecticidal activity, and it also exhibited anti-malarial and anti-TB activities (Isaka et al. 2005a). Similarly, cyclohexadepsipeptide from *Hirsutella nivea* (Isaka et al. 2005b) and *Hirsutella kobayashii* (Vongvanich et al. 2002) were reported to have activity against *Mycobacterium tuberculosis* (Abedinzadeh et al. 2015). A cyclohexadepsipeptide Hirsutellide A containing isoleucine, sarcosine, and 2-hydroxy-3-phenylpropanoid acid from *H. kobayashii* exhibited anti-TB and anti-malarial activities but did not show any cytotoxic effect towards Vero cells

(Vongvanich et al. 2002). Later Isaka et al. (2005b) reported anti-mycobacterial hirsutellones A–E from the *H. nivea* mycelial methanol extract and hirsutatins A and B in ethyl acetate extract against the malarial parasite. Anti-tubercular hirsutellone F, a novel dimer, and monomers hirsutellones (A, B and C) were isolated from *Trichoderma* mycelial extract (Isaka et al. 2006). The entomopathogen, *Verticillium hemipterigenum*, was found to produce unique diketopiperazine dimers, vertihemiptellides (A and B) which exhibited anti-TB activity but did not show anti-malarial and anti-candida activities (Isaka et al. 2005c). Pruksakorn et al. (2010) isolated three amino-lipopeptides, designated trichoderins A (1), A1 (2), and B (3), from *Trichoderma* sp. as anti-mycobacterial against *Mycobacterium smegmatis*, *M. bovis*, and *M. tuberculosis*.

20.8 Industrial Enzymes

The main market for industrially important enzymes is governed by carbohydrases, proteases and lipases. Binod et al. (2013) have reviewed the present status of industrially important enzymes isolated from different sources. The entomo- and myco-pathogens produce extracellularly different hydrolytic enzymes as host-killing components which mainly include chitinolytic, proteolytic, and lipolytic enzymes (Chavan et al. 2009, Kulkarni et al. 2008, Silva et al. 2015, Yadav and Deshpande 2010). For instance, lipases are versatile as they catalyse various different reactions and have applications in dairy and food manufacture, leather and detergent industry, cosmetics and pharmaceuticals, and also biotransformations in non-aqueous media. While proteases are found to be useful in detergents, leather dehairing, and hair waste processing, the carbohydrases especially chitinases, cellulases, and others are being used extensively in paper and pulp, textile, single-cell protein, and ethanol production industries. Kuhad et al. (2011) extensively reviewed the industrial applications of cellulases produced by different microorganisms. The cellulases from mycoparasitic *T. harzianum* and entomopathogenic *Pacilomyces* sp. were found to be useful in different industries. For instance, in agriculture in addition to the control of plant pathogens, they can improve seed germination and root system and plant growth. The effective bioconversion of cellulosic materials to ethanol, other solvents, and single-cell protein for animal feed, in detergents, paper and pulp, textile, and food industry for the release of antioxidants from fruits and vegetables, and to improve maceration are other economically viable areas of applications. In addition to chitinases, chitin deacetylase and chitosanases contributing significantly to insect killing process, also can be used for a variety of applications. For instance, Vyas and Deshpande (1991) effectively used chitinolytic activities for the hydrolysis of the chitinous substrate and further hydrolysate for single-cell protein production using *S. cerevisiae*. Kulkarni et al. (2008) screened more than 50 strains of *M. anisopliae* for the production of protease, lipase, chitinase, chitosanase and chitin deacetylase. Earlier, Beys Silva et al. (2005) reported optimization studies for extracellular lipase production using *M. anisopliae*. Aranda-Martinez et al. (2017) observed that *M. anisopliae*, *B. bassiana*, and a nematophagous fungus *Pochonia chlamydosporia* produced chitosanases which can be used to utilize chitosan for ethanol production. Further, it was suggested that this can reduce the shellfish industry waste while providing an alternative for ethanol production. Aranda-Martinez et al. (2018) reported CDA from *P. chlamydosporia* which was involved in pathogenicity to nematodes. It was further observed that the enzyme exhibited a novel deacetylation pattern in chitosan and chito-oligomers. Therefore, it can also be explored for a variety of biotechnological applications.

20.9 Fungal Polymers

The galactomannans have the ability to form a viscous solution with a low concentration and resist drastic pH changes. While mechanical and barrier properties of galactomannan membranes and coatings enhance the safety and quality of food products. The biomass of *Lecanicillium muscarium*, *B. bassiana*, *Beauveria brongniartii,* and *Cordyceps sphingum* can be used to isolate galactomannans (Bernabé et al. 2011). Using gas liquid chromatography-mass spectrometry and nuclear magnetic resonance spectroscopy (NMR), the composition and structure of the galactomannans was suggested to be O-linked heterogalactomannans having an α- 1,6- mannose backbone. Furthermore, one of the strains of *L. muscarium* was reported to produce pullulan (α-1,4- and α-1,6-linked glucose polymer)-like exopolysaccharide also which has a lot of applications in the food industry (Bernabé et al. 2011). Mahapatra and Banerjee (2013) extensively reviewed the literature for reports on fungi which produce exopolysaccharides. The exopolysaccharides from entomopathogens such as *M. anisopliae*, *P. lilacinus*, *P. japonica*, and others were reported.

M. anisopliae mycelia biomass after removal of conidia for use in agriculture as mycoinsecticide was reported to be useful to isolate chitosan (Mane et al. 2017). The biophysical characterization using ^{1}H-NMR and viscometry of the isolated chitosan had 1.30 kDa molecular weight with >80% degree of deacetylation. Nahar et al. (2004) for the first time showed that *M. anisopliae* strains produce constitutively chitin deacetylase (CDA) and chitosanase while chitinases are induced. It was suggested that the CDA can modify the cell wall with more percentage of chitosan for self-defence from insect host chitinases during entomopathogenesis. Furthermore, as *M. anisopliae* produces chitin deacetylase extracellularly as one of the killing components in entomopathogenesis, the same was found to be effective to increase the deacetylation level of chitosan from marine as well as fungal sources (Ghormade et al. 2010).

20.10 Epilogue

Use of entomo- and myco-pathogens for the control of pests and pathogens is important in integrated pest management as well as to replace/supplement chemicals in the agriculture field. The cost-effective production and the shelf life are the two main concerns. To make this eco-friendly technology cost-effective, the value addition especially using biomass for the chitin and chitosan production is one of the main prospective ventures. Most importantly, there is no specific requirement of the downstream processing after harvesting conidia or commercially feasible hydrolytic enzymes, to get the biomass for further use.

Acknowledgements MVD is grateful to CSIR, New Delhi for Emeritus Scientist Scheme [21(0962)/13/EMR2] and Department of Biotechnology (DBT-BIRAC), New Delhi for financial support. We wish to thank Rajiv Gandhi Science and Technology Commission (RGSTC, Mumbai), Government of Maharashtra, for funding the project.

References

Abd El-Ghany TM, Masmali IA (2016) Fungal biodegradation of organophosphorus insecticides and their impact on soil microbial population. J Plant Pathol Microbiol 7:349

Abedinzadeh M, Gaeini M, Sardari S (2015) Natural antimicrobial peptides against *Mycobacterium tuberculosis*. J Antimicrob Chemother 70:1285–1289

Aranda-Martinez A, Ortiz MAN, García ISA, Zavala-Gonzalez EA, Lopez-Llorca LV (2017) Ethanol production from chitosan by the nematophagous fungus *Pochonia chlamydosporia* and the entomopathogenic fungi *Metarhizium anisopliae* and *Beauveria bassiana*. Microbiol Res 204:30–39

Aranda-Martinez A, Grifoll-Romero L, Aragunde H, Sancho-Vaello E, Biarnés X, Lopez-Llorca LV, Planas A (2018) Expression and specificity of a chitin deacetylase from the nematophagous fungus *Pochonia chlamydosporia* potentially involved in pathogenicity. Sci Rep 8:2170

Askary H, Carrière Y, Bélange RR, Brodeur J (1998) Pathogenicity of the fungus *Verticillium lecanii* to aphids and powdery mildew. Biocontr Sci Technol 8:23–32

Behie SW, Bidochka MJ (2014) Nutrient transfer in plant-fungal symbioses. Trends Plant Sci 19:734–740

Bernabé M, Salvachúa D, Jiménez-Barbero J, Leal JA, Prieto A (2011) Structures of wall heterogalactomannans isolated from three genera of entomopathogenic fungi. Fungal Biol 115:862–870

Beys Silva WO, Mitidieri S, Schrank A, Vainstein MH (2005) Production and extraction of an extracellular lipase from the entomopathogenic fungus *Metarhizium anisopliae*. Process Biochem 40:321–326

Bidochka MJ, Kamp AM, Lavender TM, DeKoning J, De Croos JNA (2001) Habitat association in two genetic groups of the insect-pathogenic fungus *Metarhizium anisopliae*: uncovering cryptic species? Appl Environ Microbiol 67:1335–1342

Binod P, Palkhiwala P, Gaikaiwari R, Nampoothiri KM, Duggal A, Dey K, Pandey A (2013) Industrial enzymes- present status and future perspectives for India. J Sci Ind Res 72:271–286

Butt TM, Copping LG (2000) Fungal biological agents. Pestic Outlook 11:186–191

Butt TM, Jackson C, Magan N (2001) Fungal biocontrol agents: progress, problems and potential. In: Butt TM, Jackson C, Magan N (eds) Fungi as biocontrol agents. CABI Publishing, Oxon, p 389

Chavan S, Kolomiets E, Kuptsov V, Mandrik M, Kulkarni M, Deshpande MV (2009) Significance of cuticle-degrading enzymes with special reference to lipase in biocontrol of sugarcane woolly aphids. J Mycol Plant Pathol 39:118–123

Chet I, Inbar J, Hadar I (1997) Fungal antagonists and mycoparasites. In: Wicklow DT, Söderström B (eds) The Mycota IV: environmental and microbial relationships. Springer, Berlin, pp 165–184

Dara SK, Dara SSR, Dara SS (2017) Impact of entomopathogenic fungi on the growth, development, and health of cabbage growing under water stress. Am J Plant Sci 8:1224–1233

Fang W, Vega RJ, Ghosh AK, Jacobs-LM KA, St. Leger RJ (2011) Development of transgenic fungi that kill human malaria parasites in mosquitoes. Science 331:1074–1077

Garcia JE, Posadas JB, Perticari A, Lecuona RE (2011) *Metarhizium anisopliae* (Metschnikoff) Sorokin promotes growth and has endophytic activity in tomato plants. Adv Biol Res 5:22–27

Garrido-Jurado I, Ruano F, Campos M, Quesada-Moraga E (2011) Effects of soil treatments with entomopathogenic fungi on soil-dwelling non-target arthropods at a commercial olive orchard. Biol Control 59:239–244

Ghormade V, Kulkarni S, Doiphode N, Rajamohanan PR, Deshpande MV (2010) Chitin deacetylase: a comprehensive account on its role in nature and its biotechnological applications. In: Mendez-Vilas A (ed) Current research, technology and education topics in applied microbiology and microbial biotechnology. Formatex research center, Badajoz, pp 1054–1066

Ginsberg HS, Lebrun RA, Heyer K, Zhioua E (2002) Potential nontarget effects of *Metarhizium anisopliae* (Deuteromycetes) used for biological control of ticks (Acari: Ixodidae). Environ Entomol 31:1191–1196

Hajek AE, Goettel MS (2007) Guidelines for evaluating effects of entomopathogens on non-target organisms. In: Lacey LA, Kaya HK (eds) Field manual of techniques in invertebrate pathology. Springer, Dordrecht, pp 816–833

Harish R, Supreeth M, Chauhan JB (2013) Biodegradation of organophosphate pesticide by soil fungi. Adv BioTech 9:4–8

Harman GE, Bjorkman T (1998) Potential and existing uses of *Trichoderma* and *Gliocladium* for plant disease control and plant growth enhancement. In: Harman G, Kubicek C (eds) *Trichoderma* and *Gliocladium*. Taylor and Francis, London, pp 229–265

Harman GE, Petzoldt R, Comis A, Chen J (2004) Interactions between *Trichoderma harzianum* strain T22 and maize inbred line M017 and effects of these interactions on disease caused by *Pythium ultimum* and *Colletotrichum graminicola*. Phytopathology 94:147–153

Hu X, Xiao G, Zheng P, Shang Y (2014) Trajectory and genomic determinants of fungal-pathogen speciation and host adaptation. Proc Natl Acad Sci U S A 111:16796–16801

Isaka M, Kittakoop P, Kirtikara K, Hywel-jones NL, Thebtaranonth Y (2005a) Bioactive substances from insect pathogenic fungi. Acc Chem Res 38:813–823

Isaka M, Palasarn S, Sriklung K, Kocharin K (2005b) Cyclohexadepsipeptides from the insect pathogenic fungus *Hirsutella nivea* BCC 2594. J Nat Prod 68:1680–1682

Isaka M, Palasarn S, Rachtawee P, Vimuttipong S, Kongsaeree P (2005c) Unique diketopiperazine dimers from the insect pathogenic fungus *Verticillium hemipterigenum* BCC 1449. Org Lett 7:2257–2260

Isaka M, Prathumpai W, Wongsa P, Tanticharoen M (2006) Hirsutellone F, a dimer of antitubercular alkaloids from the seed fungus *Trichoderma* species BCC 7579. Org Lett 8:2815–2817

Jeffries P (1997) In the Mycota IV, environmental and microbial relationships (ed. W. Soderstrom). Springer, Berlin, pp 149–163

Kabaluk JT, Ericsson JD (2007) *Metarhizium anisopliae* seed treatment increases yield of field corn when applied for wireworm control. Agron J 99:1377–1381

Kapoor M, Pawar PV, Joseph M, Sen A, Deshpande MV (2013) Evaluation of biocontrol potential of *Metarhizium anisopliae* strains against larvae and adults of *Aedes aegypti* (L.). J Biol Control 27:194–203

Keswani C, Mishra S, Sarma BK, Singh SP, Singh HB (2014) Unraveling the efficient applications of secondary metabolites of various *Trichoderma* spp. Appl Microbiol Biotechnol 98:533–544

Khan AL, Hamayun M, Khan SA, Kang SM, Shinwari ZK, Kamran M, Ur Rehman S, Kim JG, Lee IJ (2012) Pure culture of *Metarhizium anisopliae* LHL07 reprograms soybean to higher growth and mitigates salt stress. World J Microbiol Biotechnol 28:1483–1494

Kuhad RC, Gupta R, Singh A (2011) Microbial cellulases and their industrial applications. Enzyme Res 2011:280696. https://doi.org/10.4061/2011/280696

Kulkarni SA, Ghormade V, Kulkarni G, Kapoor M, Chavan SB, Rajendran A, Patil SK, Shouche Y, Deshpande MV (2008) Comparison of *Metarhizium* isolates for biocontrol of *Helicoverpa armigera* (Lepidoptera: Noctuidae) in chickpea. Biocontrol Sci Tech 18:809–828

Liu SF, Wang GJ, Nong XQ, Liu B, Wang MM, Li SL, Cao GC, Zhang ZH (2017) Entomopathogen *Metarhizium anisopliae* promotes the early development of peanut root. Plant Protect Sci 53:101–107

Mahapatra S, Banerjee D (2013) Fungal exopolysaccharide: production, composition and applications. Microbiol Insights 6:1–16

Mane SR, Ghormad V, Rajamohanan PR, Badiger MV, Deshpande MV (2017) Isolation of low molecular weight chitosan from agriculturally important ascomycetous fungi *Metarhizium anisopliae* and *Myrothecium verrucaria*. Asian Chitin J 13:31–38

Miles LA, Lopera CA, Gonzalez S, Cepero de Garcia MC, Franco AE, Restrepo S (2012) Exploring the biocontrol potential of fungal endophytes from an Andean Colombian Paramo ecosystem. Biol Control 57:697–710

Nahar PB, Kulye M, Yadav P, Hassani M, Tuor U, Keller S, Deshpandc MV (2003) Comparative evaluation of indigenous fungal isolates, *Metarhizium anisopliae* M34412, *Beauveria bassiana* B3301 and *Nomuraea rileyi* N812 for the control of *Helicoverpa armigera* (Hub) on Chickpea. J Mycol Plant Pathol. 33:372–377

Nahar PB, Ghormade V, Deshpande MV (2004) The extracellular constitutive production of chitin deacetylase in *Metarhizium anisopliae*: possible edge to entomo-pathogenic fungi in the biocontrol of insect pest. J Invertebr Pathol 85:80–88

Ownley BH, Griffin MR, Klingeman WE, Gwinn KD, Moulton JK, Pereira RM (2008) *Beauveria bassiana*: endophytic colonization and plant disease control. J Invertebr Pathol 98:267–270

Pathan EK, Deshpande MV (2019) The puzzle of highly virulent *Metarhizium anisopliae* strains from *Annona squamosa* fields against *Helicoverpa armigera*. J Basic Microbiol 59:392–401. https://doi.org/10.1002/jobm.201800631

Pechy-Tarr M, Bruck DJ, Maurhofer M, Fischer E (2008) Molecular analysis of a novel gene cluster encoding an insect toxin in plant-associated strains of *Pseudomonas fluorescens*. Environ Microbiol 10:2368–2386

Porto ALM, Melgar GZ, Kasemodel MC, Nitschke M (2011) Biodegradation of pesticides. In: Stoytcheva M (ed) Pesticides in the modern world – pesticides use and management. InTeck, Shangai, pp 407–438

Pruksakorn P, Arai M, Kotoku N, Vilchze C, Baughn AD, Moodley P, Jacobs WR, Kobayashi M (2010) Trichoderins, novel aminolipopeptides from a marine sponge-derived *Trichoderma* sp., are active against dormant mycobacteria. Bioorg Med Chem Lett 20:3658–3663

Raya-DõÂaz S, SaÂnchez-RodrõÂguez AR, Segura-FernaÂndez JM, del Campillo MC, Quesada-Moraga E (2017) Entomopathogenic fungi-based mechanisms for improved Fe nutrition in sorghum plants grown on calcareous substrates. PLoS One 12:e0185903

Sasan RK, Bidochka MJ (2012) The insect-pathogenic fungus *Metarhizium robertsii* (Clavicipitaceae) is also an endophyte that stimulates plant root development. Am J Bot 99:101–107

Senthil Kumar CM, Jacob TK, Devasahayam S, Thomas S, Geethu C (2018) Multifarious plant growth promotion by an entomopathogenic fungus *Lecanicillium psalliotae*. Microbiol Res 207:153–160

Sharma P, Sharma M, Raja M, Singh DV, Srivastava M (2016) Use of *Trichoderma* spp. in biodegradation of Carbendazim. Indian J Agri Sci 86:891–894

Shrestha B, Sung GH, Sunga JM (2017) Current nomenclatural changes in *Cordyceps sensu lato* and its multidisciplinary impacts. Mycol 8:293–302

Silva WOB, Mitidieri S, Schrank A, Vainstein MH (2015) Production and extraction of an extracellular lipase from the entomopathogenic fungus *Metarhizium anisopliae*. Process Biochem 40:321–326

Steyaert JM, Ridgway HJ, Elad Y, Stewart A (2003) Genetic basis of mycoparasitism: a mechanism of biological control by species of *Trichoderma*. New Zeal J Crop Hort 31:281–291

Sylwia R, Julia P, Marta W, Cezary T, Długoński J (2013) Utilization of 4-*n*-nonylphenol by *Metarhizium* sp. isolates. Acta Biochim Pol 60:677–682

Tan NH, Zhou J (2006) Plant cyclopeptides. J Chem Rev 106:840–895

Tanaka A, Tapper B, Popay A, Parker EJ (2005) A symbiosis expressed non-ribosomal peptide synthetase from a mutualistic fungal endophyte of perennial ryegrass confers protection to the symbiotum from insect herbivory. Mol Microbiol 57:1036–1050

Tixier C, Sancelme M, Bonnemoy F, Cuer A, Veschambre H (2001) Degradation products of a phenylurea herbicide, diuron: synthesis, ecotoxicity, and biotransformation. Environ Toxicol Chem 20:1381–1389

Tupe SG, Pathan EK, Deshpande MV (2017) Development of *Metarhizium anisopliae* as a mycoinsecticide: from isolation to field performance. J Vis Exp 125:e55272

Usuki F, Narisawa H (2007) A mutualistic symbiosis between a dark septate endophytic fungus, *Heteroconium chaetospira*, and a nonmycorrhizal plant, Chinese cabbage. Mycologia 99:175–184

Valero-Jiménez CA, Wiegers H, Zwaan BJ, Koenraadt CJ, van Kan JA (2016) Genes involved in virulence of the entomopathogenic fungus *Beauveria bassiana*. J Invertebr Pathol 133:41–49

Varma A, Verma S, Sudha Sahay N, Butehorn B, Franken P (1999) *Piriformospora indica*, a cultivable plant-growth promoting root endophyte. Appl Environ Microbiol 65:2741–2744

Vega FE (2018) The use of fungal entomopathogens as endophytes in biological control: a review. Mycologia 110:4–30

Vega FE, Goettel MS, Blackwell M, Chandler D, Jackson MA, Keller S, Koike M, Maniania NK, Monzón A, Ownley BH, Pell JK, Rangel DEN, Roy HE (2009) Fungal entomopathogens: new insights on their ecology. Fungal Ecol 2:149–159

Vey A, Hoagland RE, Butt TM (2001) Toxic metabolites of fungal biocontrol agents. In: Butt TM, Jackson CW, Magan N (eds) Fungi as biocontrol agents: progress, problems and potential. CAB Publishing Co., Oxon, pp 311–346

Vidhate R, Singh J, Ghormade V, Chavan S, Patil A, Deshpande MV (2015) Use of hydrolytic enzymes of *Myrothecium verrucaria* and conidia of *Metarhizium anisopliae*, singly and sequentially to control pest and pathogens in grapes and their compatibility with pesticides used in the field. Biopestic Int 11:48–60

Vimala Devi PS, Prasad YG, Chowdary DA, Rao LM, Balakrishnan K (2003) Identification of virulent isolates of the entomopathogenic fungus *Nomuraea rileyi* (F) Samson for the management of *Helicoverpa armigera* and *Spodoptera litura* (identification of virulent isolates of *N. rileyi*). Mycopathologia 156:365–373

Vongvanich N, Kittakoop P, Isaka M, Trakulnaleamsai S, Vimuttipong S, Tanticharoen M, Thebtaranonth Y (2002) Hirsutellide A, a new antimycobacterial cyclohexadepsipeptide from the entomopathogenic fungus *Hirsutella kobayasii*. J Nat Prod 65:1346–1348

Vyas PR, Deshpande MV (1991) Enzymatic hydrolysis of chitin by *Myrothecium verrucaria* chitinase complex and its utilization to produce SCP. J Gen Appl Microbiol 37:267–275

Vyas N, Dua KK, Prakash S (2015) Metabolites of *Metarhizium anisopliae* against malaria vector and non-target organisms. Entomol Ornithol Herpetol 4:147

Wang C, St Leger RJ (2007) A scorpion neurotoxin increases the potency of a fungal insecticide. Nat Biotechnol 25:1455–1456

Wang JB, St Leger RJ, Wang C (2016) Advances in genomics of entomopathogenic Fungi. Adv Genet 94:67–105

Weindling R (1941) Experimental consideration of the mold toxins of *Gliocladium* and *Trichoderma*. Phytopathology 31:991–1003

Woo SL, Ruocco M, Vinale F, Nigro M, Marra R, Lombardi N, Pascale A, Lanzuise S, Manganiello G, Lorito M (2014) *Trichoderma*-based products and their widespread use in agriculture. The Open Mycol J 8:71–126

Wu S, Gao Y, Zhang Y, Wang E, Xu X, Lei Z (2014) An entomopathogenic strain of *Beauveria bassiana* against *Frankliniella occidentalis* with no detrimental effect on the predatory mite *Neoseiulus barkeri*: evidence from laboratory bioassay and scanning electron microscopic observation. PLoS One 9:e84732

Xu Y-J, Luo F, Li B, Shang Y, Wang C (2016) Metabolic conservation and diversification of *Metarhizium* species correlate with fungal host-specificity. Front Microbiol 7:2020

Yadav P, Deshpande MV (2010) Fungus- fungus and fungus–insect interactions. Biopestic Int. 6:21–35

Yadav P, Deshpande MV (2012) Control of beet armyworm, Spodoptora litura (Fabricius) by entomopathogenic fungi, Metarhizium anisopliae M34412, Beauveria bassiana B3301 and Nomuraea rileyi N812. Biopestic Int 8:107–114

Zeilinger S, Omann M (2007) *Trichoderma* biocontrol: signal transduction pathways involved in host sensing and mycoparasitism. Gene Regul Syst Bio 1:227–234

Zhao H, Xu C, Lu H-L, Chen X, St. Leger RJ, Fang W (2014) Host-to-pathogen gene transfer facilitated infection of insects by a pathogenic fungus. PLoS Pathog 10(4):e1004009

Fungal Enzymes: Sources and Biotechnological Applications

21

Naveen Kango, Uttam Kumar Jana, and Ritumbhara Choukade

Abstract

Fungi, being obligate heterotrophs, are natural decomposers and elaborate a number of enzymes. Currently, more than half of the industrial enzymes are of fungal origin and are being used successfully in diverse industrial processes and products. Some of the well-known areas are pulp and paper, textiles, detergents, food, feeds, nutraceuticals, and therapeutics. Production of industrial enzymes utilizes different fungal genera, *Aspergillus* being the most exploited one. Apart from protease, phytase, L-asparaginase, and few others, most commercial fungal enzymes are glycosyl hydrolases (cellulases, xylanase, mannanase, amylase, pectinase, β-fructofuranosidase, and others).

Cellulase and amylase (including glucoamylase) from *Trichoderma* sp. and *Aspergillus* spp., respectively, are exploited for bio-ethanol, textiles, and detergent industries. Fungal proteases, including keratinases, find application in detergent, food, leather, pharmaceutical, and waste management sectors. The role of fungal acidic pectinases in bringing down the cloudiness and bitterness of fruit juices is well recognized, while fungal phytases are being explored in enriching the nutritive value of poultry diets. L-Asparaginases sourced from molds are being examined for cancer therapy and mitigation of acrylamide formation in food. With the advent of biotechnological interventions, heterologous overexpression in suitable hosts, immobilization on novel matrices, and tailoring of fungal enzymes are being pursued. In this chapter, some of the important fungal enzymes are explored from recent perspective of their biotechnological applications.

N. Kango (✉) · U. K. Jana · R. Choukade
Department of Microbiology, Dr. Harisingh Gour Vishwavidyalaya, Sagar, Madhya Pradesh, India

T. Satyanarayana et al. (eds.), *Advancing Frontiers in Mycology & Mycotechnology*,
https://doi.org/10.1007/978-981-13-9349-5_21

Keywords
Fungi · Enzymes · Cellulases · Xylanase · Mannanase · Amylase · Fructosyltransferase · Inulinase · L-asparaginase

21.1 Introduction

Although the term enzyme (Greek "en" meaning "in," and "zyme" meaning "east" or "leaven") derives its origin from the yeast-mediated fermentation of sugary syrups, the first reference to the successful application of fungal enzymes dates back to 1894 on account of a patent on Taka-diastase, α-amylase from *Aspergillus oryzae* cultivated on rice by Jokichi Takamine. Fungi are natural decomposers and therefore are bestowed with a number of enzymes required for bioconversion of a variety of complex substrates (Berbee et al. 2017). Owing to ease of culturing, amenability to genetic manipulation and amazing enzymatic spectra, fungi predominate the scenario of microbial enzyme producers. Moreover, fungal cultivation in a variety of traditional preparations (brewing, baking) dates back to time immemorial, thus providing a firm and safe background for their modern-day exploitation. The advent of industrial enzymes geared up with a better understanding of their nature and function. Among the six classes of enzymes, hydrolases belonging to Class 3 make most of the industrial enzymes with predominating alkaline protease and glycoside hydrolases (Murphy et al. 2011).

Fungi are the natural factories that produce versatile enzymes which are proficient catalysts for various chemical reactions. Enzymes offer a number of favorable and economic factors over chemical catalysts. As a conventional estimate, more than half of the enzymes known are of fungal origin. Advent in fungal genomics is unraveling more number of enzymes that may play important role (Peciulyte et al. 2017). Many of these have been screened for their ability to produce industrially sound products. Fungi have been important in both ancient and modern biotechnological processes. Processes and products that make use of fungi include production of antibiotics, enzymes, organic acids, baking, brewing, alcohols, and numerous pharmaceuticals. The industrial production of numerous enzymes utilizes different fungal species. The use of fungal cells for most of the industrial enzyme production is based on their characteristics such as pH tolerance, thermostability, high yield, low operational cost, easy and cheap downstream processing etc.

Fungi being obligate heterotrophs secrete a battery of extracellular enzymes to hydrolyze complex polymeric substrates around (Kües 2015). Many a times these enzymes are robust enough to survive harsh conditions including low water activity level and high temperature. Solid state fermentation of complex substrates, particularly agro-industrial wastes such as sugarcane bagasse, palm kernel cake, copra meal, wheat bran, rice hulls, orange peel etc., naturally suits them as molds thrive well in xerophilic conditions (Hölker et al. 2004; Diaz et al. 2016).

Over the past few decades, the worldwide market of enzyme has rapidly grown. It was valued at $7.082 billion in 2017 and is projected to reach $10.519 billion in 2024, amounting to a compound annual growth rate of 5.7% from 2018 to 2024. As per the Global Enzymes Market Report 2018, protease segment alone made one-fourth share of the global enzymes market in 2017, and it projects that lyase segment will grow at the fastest rate in the coming years (https://www.businesswire.com/news/home/20180628006408/en/Global-Enzymes-Market-report-2018). Surge in the demand of first- and second-generation biofuels has increased the demand of amylolytic and cellulolytic enzymes.

As per the Association of Manufacturers and Formulators of Enzyme Products (AMFEP 2009), out of about 260 commercial enzymes, 60% are sourced from about 25 fungal genera. The most dominating among microorganisms is the versatile genus *Aspergillus,* accounting for about 25% of total industrial enzymes. Enzymes sourced from *Trichoderma*, *Penicillium*, *Rhizopus,* and *Humicola* add up to another 20% of the industrial enzymes.

The industrial enzyme market is dominated by Class 3 (hydrolases), making 85% of the total, followed by Class 2 oxidoreductases (8%), Class 4 lyases (4%), Class 2 transferases (2%), and Class 1 isomerases (1%). Many enzymes find applications in more than one industry, especially hydrolases like cellulases, amylases, and proteases. Often fungi are known to produce a spectrum of enzymes desirable for efficient depolymerization of complex substrates like lignocelluloses. Some of the prominent industrial enzymes in the light of recent developments are discussed in the present chapter (Fig. 21.1).

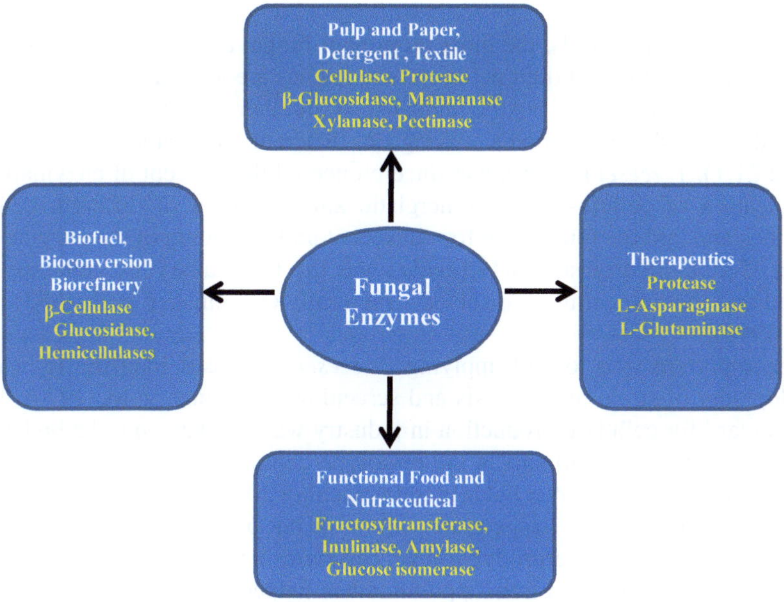

Fig. 21.1 Application of fungal enzymes

21.2 Cellulases

Cellulose is the most abundant renewable carbohydrate on the earth and the major constituent of plant cell wall. Cellulose is naturally embedded with lignin-hemicellulose matrix within plant cell wall. It is a homopolymer composed of glucose units linked by β-1,4-glycosidic bonds. Hydrogen bonding between individual cellulose fibrils gives rise to compact crystalline structure which is difficult to digest by a single hydrolase (Payne et al. 2015; Ghosh et al. 2019a).

Cellulases represent a complex group of synergistically acting enzymes. They principally contain endo-1,4-glucanase (EC 3.2.1.4) which cleave randomly at internal amorphous cellulose sites causing rapid reduction in the cellulose while liberating cello-oligomers in the process, cellobiohydrolases (EC 3.2.1.91) or exo-1,4-glucanases which act progressively on crystalline cellulose and primarily attack the reducing ends of polymer to produce cellobiose, and short-chain oligosaccharides and β-glucosidases (E.C. 3.2.1.21) which hydrolyze cellobiose to glucose monomers (Adlakha et al. 2011; Gastelum-Arellanez et al. 2014; Prajapati et al. 2018).

Biofuel generation from cellulosic biomass utilizes three steps, *viz*., pretreatment, enzymatic saccharification, and ethanolic fermentation. After pretreatment, generation of monosugars is catalyzed by cellulase, hemicellulases, and glucosidase. Alcoholic fermentation of the released sugars for bio-ethanol production is carried out by widely used yeasts like *Saccharomyces cerevisiae* (Huang et al. 2018).

Based on higher yields, the most prominent fungal cellulase producers belong to *Trichoderma* spp. and their cellulolytic enzymes are applied in food, feed, biofuel and biorefinery, and textile industry. A number of commercial cellulases sourced from different molds suiting to different applications are available (Table 21.1). *T. reesei* research has since pioneered the concept of enzymatic saccharification of cellulose by a synergistic amalgamation of different cellulase activities and laid the foundation for our recent understanding of the enzyme regulation. Cellobiohydrolase CBH1 (*cel7a*) was the first eukaryotic cellulase to be cloned and the first cellulase resolved structurally (Shoemaker et al. 1983; Divne et al. 1994).

An important step toward applying *T. reesei* cellulases industrially, was the development of strain mutagenesis and screening procedures in the 1970s. While the standard for cellulase production in industry was proclaimed to be higher than 100 g/L, strain RUT-C30 still is the prototype cellulase hyperproducer available with concentration of extracellular protein reaching 30 g/L (Bischof et al. 2016).

Over the past decades, the genome of *T. reesei* has been explored to help achieve overexpression and hyperproduction of cellulase heterologously. Better understanding of genomics and transcriptomics has helped in identifying constitutive and tunable promoters in the strain. All these lead to developing novel synthetic expression systems. Understanding of the gene expression mechanism and control will help in understanding gene function and enhance yields for biotechnological purposes (Fitz et al. 2018). *Trichoderma reesei* has been established

Table 21.1 Commercial cellulases and their fungal sources

Trade name	Manufacturer	Source	Composition	Application
Novozyme 188	Novozymes, Denmark	*A. niger*	Cellobiase	Biomass depolymerization
Multifect CL	Genencor, USA	*T. reesei*	Cellobiase	Bioblasting in textile mills
Biocellulase A	Quest Intl., USA	*A. niger*	Cellulase and xylanase	Improves the nutritional value substantially
GC 880	Genencor	*T. reesei* *T. longibrachiatum*	β-Glucanase and xylanase complex	Biomass hydrolysis
Accellerase® 1500	Genencor	*T. reesei*	Exo- and endoglucanases, hemicellulase, and β-glucosidase	Hydrolyzes lignocellulosic biomass into fermentable monosaccharides
Bio-feed beta L	Novozymes	*T. longibrachiatum* *T. reesei*	β-Glucanase and xylanase complex	Beta-Glucanase hydrolyzes the barley beta-glucans under formation of mono- or oligomers
Celluclast 1.5 LFG®	Novozymes	*T. reesei*	Cellobiase	To degrade oligosaccharides into glucose in order to investigate the biodegradability of bioabsorbable bacterial cellulose (BBC)
Rovabio	Excel Adisseo, France	*P. funiculosum*	Feedase	Improves the digestibility of feedstuffs from vegetal origin for animals, poultry, and swine; also contains arabinanase activities allowing to break down arabinose links
Cellic CTec2	Novozymes	*T. reesei*	Cellulase	Degrades cellulose to fermentable sugars

as a model organism for cellulase development and regulation machinery. In this context, many researchers have recently started working on the role of mitogen-activated protein kinases (MAPKs) in cellulase formation (Wang et al. 2017; de Paula et al. 2018). MAPKs, extremely conserved family of serine/threonine protein kinases, regulate diversity of essential cellular processes that help fungus differentiate in carbon sources, stress response, transport, proliferation, etc. (de Paula et al. 2018). The integration of light and nutrient signals has been used for strain improvement and adaptation of enzyme production in *T. reesei* (Schmoll 2018). Transcriptome and secretome analysis for cellulases is being explored in fungi such as *Aspergillus fumigatus* and *Aspergillus tamarii* grown on sugarcane bagasse (de Gouvêa et al. 2018; Midorikawa et al. 2018).

Heterologous expression of cellulases can also be triggered in rich growth media by utilization of inducible or auto-inducible promoters. Upward of 20 g and reportedly up to 100 g of crude cellulases per liter are reachable with engineered *Trichoderma reesei* strains (Cherry and Fidantsef 2003). Furthermore, other fungi, such as *Penicillium, Acremonium,* and *Chrysosporium,* are viewed as probable and promising alternatives to *Trichoderma* (Gusakov et al. 2005). For the alteration of biomass to biofuels on an industrial scale, several hurdles need to be overcome. For example, continued high production costs of cellulases, which comprise up to 20% of the total ethanol production costs as evaluated by the US National Renewable Energy Laboratory (NREL), reduced production efficiency on a commercial level. In addition, to achieve efficient biomass conversion, concerted action of a set of enzymes is required as per the composition of particular substrate. The use of traditional fungal host organisms for cellulose degradation is constrained by the need for special culturing and induction conditions. To triumph over these limitations, researchers are not only working on increasing the expression level of fungal cellulases to lower the fabrication costs but also on the optimization of recombinant expression systems in plants or microorganisms (Lambertz et al. 2014).

β-Glucosidases carry out hydrolysis of β-1,4-glycosidic bonds in aryl- and alkyl β-D-glucosides through non-reducing terminal and act in combination with endoglucanase for complete hydrolysis of cellulose to glucose (Maitan-Alfenas et al. 2015). The commercial β-glucosidase (Novozyme 188) is obtained mainly from *Aspergillus niger*, and other filamentous fungi such as *Penicillium decumbens* (Chen et al. 2010), *Phanerochaete chrysosporium* (Tsukada et al. 2006), *Paecilomyces thermophila* (Yang et al. 2009), *Aspergillus unguis* (Rajasree et al. 2013), and *Penicillium verruculosum* (Korotkova et al. 2009) are also reported to be potent β-glucosidase producers. Some of the fungal β-glucosidase producers are listed in Table 21.2. The hydrolysis carried out by glucosidase is a two-step process. The first step is nucleophilic addition reaction which results in an α-glycosyl enzyme intermediate which ultimately hydrolyzed to β-glucose in the presence of H_2O. Sawant et al. (2016) studied the two-way dynamics with the release of glucose from cellobiose and

Table 21.2 Some fungal sources of β-glucosidases

Source	References
Myceliophthora thermophila	Bonfa et al. (2018)
Clavispora sp.	Wang et al. (2016a)
Aspergillus unguis	Rajasree et al. (2013)
Penicillium piceum	Gao et al. (2013)
Trichoderma reesei	Nakazawa et al. (2011)
Neosartorya fischeri	Kalyani et al. (2011)
Humicola insolens	Souza et al. (2010)
Paecilomyces thermophila	Yang et al. (2009)
Daldinia eschscholzii	Karnchanatat et al. (2007)
Thermomyces lanuginosus	Lin et al. (1999)

cello-oligosaccharides by β-glucosidase. Soluble cellodextrin hydrolyzing β-glucosidase also helps to avoid cellulase inhibition by cellobiose (Karnaouri et al. 2013). Recently, two isoforms of β-glucosidase (50 and 200 kDa) were obtained when thermophilic *Myceliophthora thermophila* M.7.7 was grown on a mixture of sugarcane bagasse and wheat bran (1:1). The lower molecular weight β-glucosidase showed thermostability at higher temperature (60 °C) with half-life of 855.6 min (Bonfa et al. 2018).

21.3 Hemicellulases

Hemicelluloses, comprising a significant part of plant biomass, are a diverse group of structural polysaccharides. Xylan, mannan, arabinan, and other hemicelluloses make up to 30% of the lignocelluloses. Hemicelluloses have dissimilar compositions (heteropolymeric) as they contain both hexose and pentose sugars (Sjostrom 1993; Chaikumpollert et al. 2004; Bajpai 2014). Their enzymatic hydrolysis using fungal hemicellulases has found a number of applications, *viz.*, biobleaching; waste paper deinking; fruit juice maceration; upgradation of feed, fodder, and fibers; and saccharification of biomass. Hemicellulases include backbone hydrolyzing enzymes xylanase, mannanase, and arabinase and accessory enzymes α-glucuronidase, α-arabinofuranosidase, β-mannosidase, acetyl xylan esterase, and feruloyl xylan esterase (Saha 2003; Juturu and Wu 2012; Obeng et al. 2017).

Xylanases are hemicellulases which act upon β-1, 4-xylosidic bonds in xylan, a polymer of xylose, and include endo-1, 4-β-xylanase, and β-xylosidase (Walia et al. 2017). Complete hydrolysis of xylan requires combined action of endo-1, 4-β-xylanase, β-1,4-D-xylan-xylanohydrolase, β-xylosidase, and some accessory enzymes (Kango et al. 2003). Many microbes such as bacteria, yeast, fungi, and actinobacteria are known for their xylanase production; however, filamentous fungi are the most proficient and most explored xylanase producers among these (Kumar et al. 2018). *Thermomyces, Trichoderma,* and *Aspergillus* are the most exploited genera for xylanase production (Kango and Jain 2005). *Thermomyces lanuginosus* (previously known as *Humicola lanuginosa*) has gained considerate interest due to its ability to produce high titers of thermostable endoxylanase (Mchunu et al. 2013). Apart from being used in conjunction with cellulases for biofuel production, xylanases have numerous applications in various industries such as food and animal feed, paper and pulp processing, textiles, etc. (Cesar and Mrša 1996; Kang et al. 2004; Kango et al. 2017). Enzymatic hydrolysis of xylan and mannan is much relevant to biobleaching and efficient saccharification of lignocellulosic biomass (Viikari et al. 1994, Maijala et al. 2012). Cellulase-free xylanases are desirable for biobleaching where they replace chlorine-based bleaching agents, and thus, release of toxic organo-chloro compounds is avoided.

Commercial production of xylanases at industrial level is being done in several countries (Table 21.3). The main microorganisms used to obtain these enzymes are *Aspergillus niger*, *Trichoderma* sp. and *Humicola* (Bajpai 2014). The desirability of cellulase-free xylanase for biobleaching is to ascertain the selective removal of

Table 21.3 Some commercial hemicellulases and their manufacturers

Trade name	Company
Commercial xylanases	
Biobrite 100 series	Iogen, Canada
Sumizyme X	Shin Nihon, Japan
Ecopulp, Econase	AB Enzymes, Germany
Multifect XL Optimase CX 72 L	Genencor, USA
FibreZyme PBL 100, FibreZyme LBL	Dyadic International, USA
SEBrite BL 1	Advanced Enzyme, India
Bleachzyme P	Aumgene Biosciences, India
Ecozyme	Thomas Swan Co., UK
Sternzym HC 46	SternEnzym, Germany
Pulpzyme HC, NS 51024, NS 51025	Novozymes, Denmark
Commercial mannanases	
Gamanase	Novo Nordisk, Denmark
Mannaway	Novo Nordisk, Denmark
Hemicell	Elanco Animal Health, USA
Purabrite	Genencor, USA
CTCzyme	CTC Bio Inc., South Korea
DigeGrain	Advanced Enzyme, India
Mannazyme XP	Aumgene Biosciences, India

hemicellulose fraction from the pulp (Archana and Satyanarayana 2003). Fungi produce xylanase extracellularly into the medium and their titers are much higher than yeasts and bacteria (Polizeli et al. 2005).

Mannans, chiefly composed of mannose, are plant polysaccharides commonly known as gums and occur in a variety of forms. These being heteropolymeric require a number of enzymes for complete degradation (Suryawanshi et al. 2019). β-Mannanase (EC 3.2.1.78) and β-mannosidase (EC 3.2.1.25) act upon the β-1-4 mannopyranosyl linkages of the mannan backbone, while β-glucosidase (EC 3.2.1.21), α-galactosidase (EC 3.2.1.22) and acetyl esterase (EC 3.1.1.6) cleave the respective moieties from the side chains (Soni and Kango 2013). β-Mannanase, the main enzyme, and other accessory enzymes are synthesized by a variety of microorganisms. Fungal mannanases have been investigated by various workers (Moreira and Filho 2008; van Zyl et al. 2010). Some of the prominent fungal mannanase producers are reported from the genus *Aspergillus* (Soni et al. 2016; Jana et al. 2018) followed by *Penicillium* sp. (Blibech et al. 2011) and *Trichoderma* sp. (Chai et al. 2016). About 50% of commercial mannanase preparations are sourced from genetically engineered microorganisms (Dhawan and Kaur 2007). A 1345 bp gene encoding mannanase (ManN) from *Aspergillus sulphureus* was expressed in *Pichia pastoris* (Chen et al. 2007). Malherbe et al. (2014) have expressed *Aspergillus aculeatus* endo-β-mannanase (Man1) and *Talaromyces emersonii* α-galactosidase (Agal) genes in *S. cerevisiae* Y294. Mannans occurring in animal feeds made from soybean and legumes are anti-nutritive and elicit a Feed-Induced Immune Response (FIIR) in animals (Hsiao et al. 2006; Zhang and Tizard 1996). Commercial mannanase preparations specifically designed to mitigate the problem of immunogenicity, Hemicell digest the immunogenic mannan in feed and improve the poultry health (Korver 2006).

Rhizomucor miehei mannanase showed classical (β/α) 8-TIM barrel-fold structure which provides high specific activity and hydrolyzing property. Using directed evolution strategies such as error-prone polymerase chain reaction (error-prone PCR), DNA shuffling, site-directed mutagenesis (SDM), and site-saturation mutagenesis (SSM), the catalytic activity of mannanase in acidic and thermophilic conditions was further improved (Li et al. 2017). β-Mannanases have extensive applications in industries such as food and feed processing. For the enhancement in the activity of mannanase, rational design strategy was applied which included N-glycosylation in the loop area intern. Improved thermal stability, pH stability, and protease resistance of the *Armillaria tabescens* β-mannanase were noticed (Hu et al. 2017). Structure of *Rhizopus microsporus* endo-β-mannanase was elucidated and it showed different binding behaviors with different oligosaccharides (You et al. 2018). Recently, fungal β-mannnanases from *Malbranchea cinnamomea, Aspergillus oryzae,* and *A. terreus* that generate mannooligosaccharide (MOS) from locust bean gum, guar gum, and konjac gum have been reported (Ahirwar et al. 2016; Li et al. 2017; Jana et al. 2018).

21.4 Amylases

Starch is the most abundant storage polysaccharide on the earth and major component of potato, wheat, corn, and rice. Apart from being a staple food such as bread or rice, it also finds use as a thickener and a gelling agent in food industry. Starch consists of linear insoluble amylose and branched soluble amylopectin. In amylose, glucose is linked by β-1,4-glycosidic bonds in a linear fashion, while in amylopectin some of the chains are linked by α-1,6 linkages giving it a branched structure (Buléon et al. 1998). A number of enzymes are known to act upon starch, among which the α-amylases and glucoamylases are the prominent ones (Parashar and Satyanarayana 2017). Starch being the most common source of energy, amylases occur in a wide array of organisms including bacteria and fungi. As mentioned earlier, α-amylase bears historical relevance from the point of view of industrial application of enzymes. After application of Taka-diastase in 1894 from *A. oryzae* as a digestive enzyme, α-amylase was also used as a textile desizing agent in Japan in 1905. Later in 1959, *Rhizopus* sp. was used for production of glucoamylase. Amylolytic enzymes account for about 30% of total industrial enzymes (Vaidya et al. 2015). Due to enormous advantages of enzymatic processing of starch over chemical hydrolysis, amylases have replaced the harsh chemicals in industries.

α-Amylases (EC 3.2.1.1) are extracellular endo-acting enzymes that randomly hydrolyze α-1,4-glycosidic bonds in starch and produce maltose and dextrins. β-Amylases (E.C.3.2.1.2) are starch hydrolyzing enzymes that cleave α-1,4 linkages from non-reducing end of starch and cause inversion of maltose to its β-form (Zhang et al. 2017a). However, most industrial applications employ *α*-amylases for saccharification or liquefaction purposes. Fungal sources of industrial α-amylases are mostly confined to *Aspergillus, Penicillium,* and *Rhizopus* spp. (Li et al. 2011). *Aspergillus* is one of the prominent and notably the most explored genera for α-amylases. *Aspergillus oryzae* (Taka-diastase) and *Aspergillus niger* have been used extensively in starch industry (Kammoun et al. 2008; Porfirif et al. 2016; Avwioroko et al. 2018). Fungal amylases sourced from these two molds are preferred over other sources as they enjoy GRAS (generally regarded as safe) status. These molds are prolific producers of hydrolases and due to secretion of organic acids help avoid contamination. However, being mesophilic, the enzymes are not thermostable, and thus bacterial α-amylases replace them in the very first step of gelatinization (or cooking) at high temperature. Some workers have explored some thermophilic molds including *Thermomyces lanuginosus, Humicola griseus, Malbranchea pulchella, Rhizomucor pusillus,* and *R. miehei* for production of extracellular thermostable α-amylases (Arnesen et al. 1998; Jensen et al. 2002; Kumar and Satyanarayana 2003). Recently, Abdulaal (2018) has described occurrence of five α-amylases (A1-A5) from *Trichoderma pseudokoningii* and purified one A4 (Mr 30 kDa) stable at 80 °C.

The amylolytic enzyme to be discovered after α- and β-amylase is another glucose liberating enzyme referred as γ- or glucoamylase (Azzopardi et al. 2016). It is a very important enzyme for successive and complete degradation of starch into glucose. It is an exo-acting enzyme that cleaves α-1,4 linkages from the nonreducing ends, but can also cleave α-1, 6 linkages, thus leading to complete saccharification. Most commercial glucoamylases are sourced from *Aspergillus* or *Rhizopus* spp. (Carrasco et al. 2017). *Thermomucor indicae-seudaticae* produced thermostable glucoamylase optimally at 60 °C and pH 7.0 (Kumar and Satyanarayana 2003; Kumar and Satyanarayana 2007).

To achieve saccharification of starch in a single step, a chimeric biocatalyst (Amy-Glu) was prepared using α-amylase of *Bacillus acidicola* and glucoamylase of *A. niger* linked by a peptide. The chimeric enzyme (145 kDa) was expressed in *E. coli* (Parashar and Satyanaryana 2017). In an effort to co-immobilize alpha- and gluco- amylase, Salgaonkar et al. (2018) have used metal organic framework (MOF) by mixing zinc acetate and 2-methylimidazole with enzyme mixture in one pot. The product showed remarkable thermal stability (temperature ranges of 55–75 °C) and catalytic efficiency (V_{max}).

21.5 Protease

Proteases make a large class of enzymes that are involved in peptide bond (CO-NH) hydrolysis within a protein molecule. A wide variety of proteases are produced by different microbial sources. Generally, bacteria produce alkaline proteases and fungi are known to be good producers of acid proteases. Proteases have been utilized for a number of industrial applications and thus attracted attention of researchers to explore microbial diversity. Although bacterial proteases dominate the commercial scenario, fungal proteases have gained considerable interest due to their broad pH activity range and stability over diverse industrial conditions (Banerjee and Ray 2017). Proteases are the most important industrial enzymes that make approximately 60% of the total enzyme market (Budak et al. 2014). A number of endo- and exopeptidases belonging to different families are produced by aspergilli in protein-rich medium (Machida et al. 2005). Filamentous fungi produce peptidases with varying specificities which must be taken into account in choosing a peptidase to catalyze the protein hydrolysis for the desired application (Hamin Neto et al. 2017a).

Proteases are subcategorized into two major classes, exo- and endopeptidases, based on their site of action. Exopeptidases (also known as peptidases) are known to cleave off N- or C-terminal amino acid from the peptide chain (Jain et al. 2010). Endopeptidases (also called proteinases) hydrolyze internal peptide bond within the protein molecule. Endo- and exopeptidases are further subdivided into four major groups, *viz*., cysteine, aspartic, serine, and metalloproteases, based on the functional groups present on active site. Most of the metalloproteases act as virulent factors of pathogenic fungi to the plants (Barrett and Rawlings 1991). Aspartic proteases, having aspartic acid residue in their active site, are generally produced by a number of filamentous fungi such as *Aspergillus*, *Rhizopus*, *Mucor*, and *Rhizomucor*. They are industrially important because they are unaffected by serine protease inhibitors, reagents having thiol group, and various chelating agents. Most of the aspartic proteases share similarities with pepsin and rennin and therefore can be used in bakery and animal feed industries (Mandujano-González et al. 2016). Some of the proteases sourced from molds are listed in Table 21.4.

Keratinases (EC 3.4.99.11) are proteases with the unique ability to attack highly cross-linked, recalcitrant structural proteins such as keratin (Pawar et al. 2018). Unlike most proteins which are easily degraded by common proteolytic enzymes like papain, pepsin, or trypsin, feather keratin protein is not degraded by these enzymes. Feather keratin is stabilized by disulfide bonds, hydrogen bonds and hydrophobic interactions (Ghosh et al. 2019b). Although keratinolytic proteases are produced by many microorganisms, keratinophilic fungi deserve special mention for colonizing keratin and production of keratinolytic enzymes (Lange et al. 2016). Feathers are almost pure keratin protein and hence can be used as a cheap alternative for production of protein-rich animal feed. Among various agriculture segments in India, poultry is considered to be one of the fastest growing segments, increasing at a rate of 8–10% per year. Thus, in the approaching years, there will be substantial increase in the generation of poultry waste which, if not handled properly, can lead to environmental pollution and health hazard (Farag and Hassan 2004). Current methods to convert feathers into animal feed include physical and chemical processing requiring significant amounts of energy and chemicals. Further, these processes also cause destruction of certain essential amino acids, reducing nutritional value of the feed. Chemicals used in feather processing are responsible for environmental pollution as the bulk effluents are released into water bodies. In this context, biodegradation of feathers by keratinolytic fungi is seen as a potential eco-friendly alternative to chemical treatment. They have gained importance in various biotechnological and pharmaceutical applications, yet the commercial availability of keratinases is still

Table 21.4 Types of some fungal proteases and their applications

Protease	Trade name/fungus	Application	Source/references
Exopeptidases			
Aminopeptidases (EC 3.4.11)	Flavourzyme	Food industry	Novo Nordisk (Denmark)
	Neutrase	Dairy industry	
	Aspergillus oryzae	Fish industry	
	Corolase	–	Rohm enzyme (Finland)
	Aspergillus sojae		
Carboxypeptidase (EC 3.4.16–3.4.18)	*Aspergillus oryzae*	–	Salamin et al. (2017)
Serine-type protease (EC 3.4.16)	*Neurospora sitophila*	Fibrinolysis and plasminogen activation	Deng et al. (2018)
Metalloprotease (EC 3.4.17)	*Candida kefyr* 41PSB	–	Yavuz et al. (2017)
	Eupenicillium javanicum	–	Hamin Neto et al. (2017b)
Cysteine-type protease (EC 3.4.18)	*Aspergillus nidulans*	–	Futai et al. (2001)
Peptidyl-dipeptidases (EC 3.4.15)	*Aspergillus oryzae* (Flavourzyme)	Food processing	Merz et al. (2015)
Dipeptidases (EC 3.4.13)	*Aspergillus oryzae* (Flavourzyme)	Food processing	Merz et al. (2015)
Endopeptidases			
Serine protease (EC 3.4.21)	*T. harzianum* CECT 2413	Biocontrol	Suárez et al. (2007)
Aspartic protease (EC 3.4.23)	*Piptoporus soloniensis*	Dairy industry	El-Baky et al. (2011)

Table 21.5 Keratinases from some fungal sources

Fungus	Optimum pH	Reference
Chrysosporium articulatum	7.58	Bohacz (2016)
Onygena corvina	8.0	Huang et al. (2015)
Purpureocillium lilacinum	–	Cavello and Cavalitto (2014)
Aspergillus parasiticus	7.0	Anitha and Palanivelu (2013)
Myrothecium verrucaria	8.3	Moreira-Gasparin et al. (2009)
Trichoderma atroviride	8.0–9.0	Cao et al. (2008)
Paecilomyces marquandii	8.0	Gradisar et al. (2005)
Aspergillus oryzae	8.0	Farag and Hassan (2004)
Aspergillus fumigatus	8.0	Noronha et al. (2002)
Doratomyces microsporus	7.5	Gradisar et al. (2000)
Microsporum canis	9.0	Mignon et al. (1998)
Chrysosporium keratinophilum	9.0	Dozie et al. (1994)
Trichophyton schoenleinii	5.5	Qin et al. (1992)

limited (Noronha et al. 2002). Keratinophilic fungi include hyphomycetes and several other taxa (Table 21.5). Hyphomycetes include dermatophytic (e.g. *Microsporum* sp.) and non-dermatophytic (e.g. *Chrysosporium* sp.) keratinophilic molds (Gopinath et al. 2015).

21.6 L-Asparaginase

L-Asparaginase (L-asparagine amidohydrolase, E.C.3.5.1.1) hydrolyzes L-asparagine (essential amino acid) to aspartic acid and ammonia. Since several types of tumor cells require L-asparagine for protein synthesis, they are deprived of an essential growth factor in the presence of L-asparaginase, thus resulting in starvation and death of leukemic cells. Low levels of the non-essential amino acid asparagine only affect the viability of abnormal cells as these cells have abnormally high requirement for asparagine. This is because normal cells produce enzyme asparagine synthetase, which is able to synthesize asparagine, whereas, in cancer and tumor cells, enzyme is present in low levels. L-Asparaginase enzyme is being used effectively in the treatment of acute lymphoblastic and myelocytic leukemia, Hodgkin's lymphoma, lymphocytic leukemia, and lymphosarcoma treatment (Saxena et al. 2015; Agrawal and Kango 2019).

Commercially available L-asparaginase from *Escherichia coli* and *Erwinia chrysanthemi* elicits a relatively high rate of immune response including silent hypersensitivity, thrombosis, pancreatitis, and hyperglycemia (Li et al. 2018). Hence, efforts are underway to find newer sources of L-asparaginase, and fungal L-asparaginases may be a promising alternative due to their eukaryotic origin. Vala et al. (2018) have reported a marine-derived *Aspergillus niger* AKV-MKBU L-asparaginase with anticancer properties. *A. terreus* L-asparaginase gene encoding a protein of 376 amino acids (42.0 kDa) was expressed in *E. coli* (Saeed et al. 2018). The gold nano-biocomposite was also prepared by immobilizing fungal L-asparaginase (*Aspergillus terreus* MTCC 1782) onto gold nanoparticles which showed anticancer activity against lung cancer cell line A549 (Baskar et al. 2018). Microbial production of L-asparaginase depends on a variety of environmental factors such as temperature, pH, oxygen availability, nutrient type and availability, etc. Commercial production of enzyme requires complete analysis of various optimum conditions and genetic makeup for highest yield. Native microbial strains produce asparaginase either constitutively or after induction by asparagine. Sarquis et al. (2004) reported L-asparaginase production by filamentous fungal species *Aspergillus tamarii* and *Aspergillus terreus*. Some fungal L-asparaginases are listed in Table 21.6.

Besides being an anticancer agent, L-asparaginase has application in food industry as well. Acrylamide, a potent carcinogen, is formed by Maillard reaction between reducing sugars and asparagine present in starchy foodstuff (Agrawal et al. 2018). In a report, L-asparaginase from *Aspergillus terreus* was used for the pretreatment of banana slices before frying to mitigate acrylamide formation during frying. The soaking and frying conditions were optimized using free and chitosan-immobilized fungal L-asparaginase (Aiswarya and Baskar 2017). L-asparaginases sourced from fungi with GRAS status are more suitable for application in food industry.

Table 21.6 Production of some fungal L-asparaginases

Fungus	Method	Yield	References
Aspergillus niger	SmF, cloned	15.78 U/ml	Vala et al. (2018)
Aspergillus terreus	SmF, cloned	42.46 U/mg	Saeed et al. (2018)
Saccharomyces cerevisiae	SmF, cloned	–	Costa et al. (2016)
Aspergillus terreus	SSF	273 U/gds	Varalakshmi and Raju (2013)
Penicillium digitatum	SmF	363.80 U/ml	Shrivastava et al. (2012)
Aspergillus terreus	SmF	24.10 U/ml	Gurunathan and Sahadevan (2011)
Fusarium equiseti	SSF	8.51 U/ml	Hosamani and Kaliwal (2011)
Emericella nidulans		1.1 U	Jayaramu et al. (2010)
Aspergillus niger	SSF	12.52 U/ml	Mishra (2006)

21.7 Inulinase and Fructosyltransferase (FTase)

Prebiotics have attracted eager interest of people as well as nutraceutical industries to process food due to their high therapeutic and nutritional properties (Rawat et al. 2017; Choukade and Kango 2019). Prebiotics contain short-chain non-digestible carbohydrates (NDC) which selectively nourish healthy gut microbiota and ultimately facilitate better health. Fructooligosacharides consist of 1-kestose (GF2), nystose (GF3), and β-fructofuranosyl nystose (GF4) produced from sucrose upon action of fructosyltransferase (FTase) from plants, bacteria, yeasts, and fungi (Flores-Maltos et al. 2014). FOS, a leading prebiotic, has various health-promoting properties as it is bifidogenic, non-cariogenic, and hypolipidemic and helps in ion absorption through gut. Inulinases hydrolyze plant fructan, inulin into inulooliogsaccharides (endoinulinase) and fructose (exoinulinase) by breaking on glycosidic linkages (Kango 2008; Kango and Jain 2011; Rawat et al. 2016).

Fructosyltransferase (FTase; EC 2.4.1.9) is known to hydrolyze sucrose and transfer fructosyl group to an acceptor molecule to generate fructooligosaccharides (FOS) along with glucose and fructose (Ganaie et al. 2013, 2014). FTase cleaves the β-1,2 linkage of sucrose and transfers fructosyl group to an acceptor molecule leading to the formation of fructooligosaccharides and release of glucose. β-Fructofuranosidase (FFase, EC 3.2.1.26) catalyzes both hydrolytic and transfructosylating reactions; however, the latter is evidenced only with higher sucrose concentrations (Rawat et al. 2015a, b).

Bali et al. (2015) have reviewed microbial production of FOS and mentioned fungi such as *A. niger*, *Aspergillus japonicus*, *A. sydowii*, *A. foetidus*, *A. oryzae*, *Aureobasidium pullulans*, *Penicillium citrinum*, *P. frequentans,* and *Fusarium oxysporum* as the prominent producers. Rawat et al. (2015a, b) have also provided a comparative account of fructosyltransferase, inulinase, and sucrase activities in some aspergilli and penicillia. Jiang et al. (2016) isolated a novel yeast *Aureobasidium* sp. P6 from a mangrove ecosystem and cloned inulinase gene. It produced inulin hydrolyzing enzyme (30.98 ± 0.8 U/ml) that showed transfructosylating activity at 30.0% sucrose concentration and generated fructooligosaccharides (FOS).

Zhang et al. (2017b) have used an industrial strain, *Aspergillus niger* ATCC 20611, to enhance the production of FOS wherein they have used polyethylene glycol (PEG)-mediated protoplast transformation system for strain improvement. The transformed *A. niger* ATCC 20611 displayed a 58% increase in β-fructofuranosidase production (507 U/g), compared to the parental strain *A. niger* ATCC 20611 (320 U/g). Production of an extracellular, thermostable inulinase was carried out by *Aspergillus tubingensis* CR16 using wheat bran and corn steep liquor (CSL) under solid state fermentation (SSF). The fungus produced 1358.6 U/g inulinase after parametric optimization which was fivefolds higher (Trivedi et al. 2012).

Tanriseven and Aslan (2005) have immobilized commercially available *Aspergillus aculeatus* FTase (Pectinex Ultra SP-L) in Eupergit C with 96% efficiency and maintained the recycling up to 20 days effectively to obtain GF4, GF3, GF2, glucose, and fructose. Immobilized enzyme also showed a higher temperature optimum at 65 °C. Heteroexpression of endoinulinase encoding gene from *Aspergillus ficuum* in *E. coli* with high inulooligosaccharide (IOS) yield of 94.41% has been reported by Wang et al. (2016b). Some heterologously expressed FTases and inulinases are listed in Table 21.7.

Table 21.7 Cloning and heterologous expression of fungal FTase, FFase, and inulinase

Organism	Enzyme	GenBank accession no.	Cloning host	Expression host	References
A. niger ATCC 20611	Inulinase	–	*Escherichia coli* DH5α	*A. niger* ATCC 20611 protoplast	Zhang et al. (2017b)
A. oryzae FS4	Inulinase	bfrAFS4 (CGMCC no. 9087)	*Escherichia coli* DH5α	BL21, *Pichia pastoris*	Xu et al. (2014)
Aspergillus oryzae	FTase	EU130944	*Escherichia coli* DH5α	BL21	Wang et al. (2016c)
A. terreus NIH2624	FFase ATEG 04996	XP 001214174.1	*Escherichia coli* strain GB05	*K. lactis* GG799	Spohner and Czermak (2016)
Aspergillus sp.	Endoinulinase	–	*Escherichia coli*	*Escherichia coli*	Raba'atun Adawiyah et al. (2011)
Aspergillus awamori	Exoinulinase	AJ315793	*Escherichia coli*	*Escherichia coli*	Arand et al. (2002)
Aspergillus fumigatus Cl1	Endoinulinase	AFUA 5G00530	*Escherichia coli* DH5α	*P. pastoris* GS115	Chen et al. (2014)
Aspergillus kawachii	Exoinulinase	CAC44220	*Escherichia coli* TOP10F	*P. pastoris* GS115 and X33	Chesini et al. (2018)
A. oryzae	FTase	–	*Escherichia coli* DH5α	*Y. lipolytica* CGMCC7326	Zhang et al. (2016)

High-yielding strain of *Aspergillus oryzae* was developed using strains with high fructosyltransferase (FTase) activity for intraspecific protoplast fusion via genome shuffling. The resulting strain produced 353 U/g FTase activity (Wang et al. 2016c). More recently, Wang et al. (2016d) have cloned endoinulinase in *Saccharomyces cerevisiae* and deleted its sucrase gene, resulting in high-content FOS production (90%) from inulin in a single step.

21.8 Future Perspectives and Conclusions

Fungi produce a number of industrial enzymes which find multifarious applications in a variety of industrial processes. Owing to their ability to utilize low-value substrates, amenability to manipulation, and ability to produce high enzyme titers, fungi are being explored extensively for industrial enzymes. Often, fungal species are noticed to elaborate spectra of hydrolases including main and accessory enzymes that can be used as consortia for efficient and complete depolymerization of complex substrates. Out of about 260 commercial enzymes, 60% are sourced from about 25 fungal genera. The enzyme market is projected to grow up to $10.5 billion by 2024. The rapid growth in enzyme market is indicative of the ever-increasing demand of enzymes in various sectors like biofuel, food, detergents, pharmaceuticals, etc. To realize the aim of replacing harmful toxic chemicals in industries, enzymes should be able to work under harsh or extreme conditions. This is one bottleneck where fungal enzymes lag behind bacterial extremozymes. Development of strains expressing robust and multifunctional (chimeric) enzymes using recombinant DNA technology, high-throughput screening of novel isolates, metagenomic screening, *in silico* enzyme engineering, site-directed mutagenesis, and directed evolution will pave a way to cater future demands.

References

Abdulaal WH (2018) Purification and characterization of α-amylase from *Trichoderma pseudokoningii*. BMC Biochem 19:4

Adlakha N, Rajagopal R, Kumar S, Reddy VS, Yazdani SS (2011) Synthesis and characterization of chimeric proteins based on cellulase and xylanase from an insect gut bacterium. Appl Environ Microbiol 77:4859–4866

Agrawal S, Kango N (2019) Development and catalytic characterization of L-asparaginase nanobioconjugates. Int J Biol Macromol 145:1145–1150

Agrawal S, Sharma I, Prajapati BP, Suryawanshi RK, Kango N (2018) Catalytic characteristics and application of L-asparaginase immobilized on aluminum oxide pellets. Int J Biol Macromol 114:504–511

Ahirwar S, Soni H, Rawat HK, Ganaie MA, Pranaw K, Kango N (2016) Production optimization and functional characterization of thermostable β-mannanase from *Malbranchea cinnamomea* NFCCI 3724 and its applicability in mannotetraose (M4) generation. J Taiwan Inst Chem Eng 63:344–353

Aiswarya R, Baskar G (2017) Microbial production of L-asparaginase and its immobilization on chitosan for mitigation of acrylamide in heat processed carrot slices. Indian J Exp Biol 56:504–510

AMFEP (2009) List of enzymes. In: Association of Manufacturers and Formulators of enzyme products. http://www.amfep.org

Anitha TS, Palanivelu P (2013) Purification and characterization of an extracellular keratinolytic protease from a new isolate of *Aspergillus parasiticus*. Protein Expr Purif 88:214–220

Arand M, Golubev AM, Neto JR, Polikarpov I, Wattiez R, Korneeva OS, Eneyskaya EV, Kulminskaya AA, Shabalin KA, Shishliannikov SM, Chepurnaya OV, Neustroev KN (2002) Purification, characterization, gene cloning and preliminary X-ray data of the exo-inulinase from *Aspergillus awamori*. Biochem J 362:131–135

Archana A, Satyanarayana T (2003) Purification and characterization of a cellulase-free xylanase of a moderate thermophile *Bacillus licheniformis* A99. World J Microbiol Biotechnol 19:53–57

Arnesen S, Havn Eriksen S, Olsen JO, Jensen B (1998) Increased production of α-amylase from *Thermomyces lanuginosus* by the addition of tween 80. Enzym Microb Technol 23:249–252

Avwioroko OJ, Anigboro AA, Unachukwu NN, Tonukari NJ (2018) Isolation, identification and in silico analysis of alpha-amylase gene of *Aspergillus niger* strain CSA35 obtained from cassava undergoing spoilage. Biochem Biophys Rep 14:35–42

Azzopardi E, Lloyd C, Teixeira SR, Conlan RS, Whitaker IS (2016) Clinical applications of amylase: novel perspectives. Surgery 160:26–37

Bajpai P (2014) Introduction. In: Xylanolytic enzymes. Academic, Burlington, pp 1–7

Bali V, Panesar PS, Bera MB, Panesar R (2015) Fructo-oligosaccharides: production, purification and potential applications. Crit Rev Food Sci Nutr 55:1475–1490

Banerjee G, Ray AK (2017) Impact of microbial proteases on biotechnological industries. Biotechnol Genet Eng Rev 33:119–143

Barrett AJ, Rawlings ND (1991) Types and families of endopeptidases. Biochem Soc Trans 19:707–715

Baskar G, Garrick BG, Lalitha K, Chamundeeswari M (2018) Gold nanoparticle mediated delivery of fungal asparaginase against cancer cells. J Drug Delivery Sci Technol 44:498–504

Berbee M, James TY, Strullu-Derrien C (2017) Early diverging fungi: diversity and impact at the dawn of terrestrial life. Annu Rev Microbiol 71:41–60

Bischof RH, Ramoni J, Seiboth B (2016) Cellulases and beyond: the first 70 years of the enzyme producer *Trichoderma reesei*. Microb Cell Factories 15:106

Blibech M, Ellouz Ghorbel RE, Chaari F, Dammak I, Bhiri F, Neifar M, Ellouz Chaabouni SE (2011) Improved mannanase production from *Penicillium occitanis* by fed-batch fermentation using acacia seeds. ISRN Microbiol 2011:1–5

Bohacz J (2016) Biodegradation of feather waste keratin by a keratinolytic soil fungus of the genus *Chrysosporium* and statistical optimization of feather mass loss. World J Microbiol Biotechnol 33:13

Bonfá EC, de Souza Moretti MM, Gomes E, Bonilla-Rodriguez GO (2018) Biochemical characterization of an isolated 50 kDa beta-glucosidase from the thermophilic fungus *Myceliophthora thermophila* M.7.7. Biocatal Agric Biotechnol 13:311–318

Budak SO, Zhou M, Brouwer C, Wiebenga A, Benoit I, Di Falco M, Tsang A, de Vries RP (2014) A genomic survey of proteases in aspergilli. BMC Genomics 15:523

Buléon A, Colonna P, Planchot V, Ball S (1998) Starch granules: structure and biosynthesis. Int J Biol Macromol 23:85–112

Cao L, Tan H, Liu Y, Xue X, Zhou S (2008) Characterization of a new keratinolytic *Trichoderma atroviride* strain F6 that completely degrades native chicken feather. Lett Appl Microbiol 46:389–394

Carrasco M, Alcaíno J, Cifuentes V, Baeza M (2017) Purification and characterization of a novel cold adapted fungal glucoamylase. Microb Cell Factories 16:75

Cavello IA, Cavalitto SF (2014) Kinetic modelling of thermal inactivation of a keratinase from *Purpureocillium lilacinum* LPSC # 876 and the influence of some additives on its thermal stability. Appl Biochem Biotechnol 173:1927–1939

Cesar T, Mrša V (1996) Purification and properties of the xylanase produced by *Thermomyces lanuginosus*. Enzym Microb Technol 19:289–296

Chai SY, Abu Bakar FD, Mahadi NM, Murad AMA (2016) A thermotolerant Endo-1,4-β-mannanase from *Trichoderma virens* UKM1: cloning, recombinant expression and characterization. J Mol Catal B Enzym 125:49–57

Chaikumpollert O, Methacanon P, Suchiva K (2004) Structural elucidation of hemicelluloses from Vetiver grass. Carbohydr Polym 57:191–196

Chen X, Cao Y, Ding Y, Lu W, Li D (2007) Cloning, functional expression and characterization of *Aspergillus sulphureus* β-mannanase in *Pichia pastoris*. J Biotechnol 128:452–461

Chen M, Qin Y, Liu Z, Liu K, Wang F, Qu Y (2010) Isolation and characterization of a β-glucosidase from *Penicillium decumbens* and improving hydrolysis of corncob residue by using it as cellulase supplementation. Enzym Microb Technol 46:444–449

Chen M, Lei X, Chen C, Zhang S, Xie J, Wei D (2014) Cloning, overexpression, and characterization of a highly active endoinulinase gene from *Aspergillus fumigatus* Cl1 for production of inulo-oligosaccharides. Appl Biochem Biotechnol 175:1153–1167

Cherry JR, Fidantsef AL (2003) Directed evolution of industrial enzymes: an update. Curr Opin Biotechnol 14:438–443

Chesini M, Wagner E, Baruque DJ, Vita CE, Cavalitto SF, Ghiringhelli PD, Rojas NL (2018) High level production of a recombinant acid stable exoinulinase from *Aspergillus kawachii*. Protein Expr Purif 147:29–37

Choukade R, Kango N (2019) Characterization of a mycelial fructosyltransferase from *Aspergillus tamarii* NKRC 1229 for efficient synthesis of fructooligosaccharides. Food Chem 286:434–440

Costa IM, Schultz L, de Araujo Bianchi Pedra B, Leite MSM, Farsky SHP, de Oliveira MA, Pessoa A, Monteiro G (2016) Recombinant L-asparaginase 1 from *Saccharomyces cerevisiae*: an allosteric enzyme with antineoplastic activity. Sci Rep 6:36239

de Gouvêa PF, Bernardi AV, Gerolamo LE, de Souza SE, Riaño-Pachón DM, Uyemura SA, Dinamarco TM (2018) Transcriptome and secretome analysis of *Aspergillus fumigatus* in the presence of sugarcane bagasse. BMC Genomics 19:232

de Paula RG, Antoniêto ACC, Carraro CB, Lopes DCB, Persinoti GF, Peres NTA, Martinez-Rossi NM, Silva-Rocha R, Silva RN (2018) The duality of the MAPK signaling pathway in the control of metabolic processes and cellulase production in *Trichoderma reesei*. Sci Rep 8:14931

Deng Y, Liu X, Katrolia P, Kopparapu NK, Zheng X (2018) A dual-function chymotrypsin-like serine protease with plasminogen activation and fibrinolytic activities from the GRAS fungus, *Neurospora sitophila*. Int J Biol Macromol 109:1338–1343

Dhawan S, Kaur J (2007) Microbial mannanases: an overview of production and applications. Crit Rev Biotechnol 27:197–216

Diaz AB, Blandino A, Webb C, Caro I (2016) Modelling of different enzyme productions by solid-state fermentation on several agro-industrial residues. Appl Microbiol Biotechnol 100:9555–9566

Divne C, Stahlberg J, Reinikainen T, Ruohonen L, Pettersson G, Knowles JK, Teeri TT, Jones TA (1994) The three-dimensional crystal structure of the catalytic core of cellobiohydrolase I from *Trichoderma reesei*. Science 265:524–528

Dozie INS, Okeke CN, Unaeze NC (1994) A thermostable, alkaline-active, keratinolytic proteinase from *Chrysosporium keratinophilum*. World J Microbiol Biotechnol 10:563–567

El-Baky HA, Linke D, Nimtz M, Berger RG (2011) PsoP1, a milk-clotting aspartic peptidase from the basidiomycete fungus *Piptoporus soloniensis*. J Agric Food Chem 59:10311–10316

Farag AM, Hassan MA (2004) Purification, characterization and immobilization of a keratinase from *Aspergillus oryzae*. Enzym Microb Technol 34:85–93

Fitz E, Wanka F, Seiboth B (2018) The promoter toolbox for recombinant gene expression in *Trichoderma reesei*. Front Bioeng Biotechnol 6:135

Flores-Maltos DA, Mussatto SI, Contreras-Esquivel JC, Rodríguez-Herrera R, Teixeira JA, Aguilar CN (2014) Biotechnological production and application of fructooligosaccharides. Crit Rev Biotechnol 36:259–267

Futai E, Kubo T, Sorimachi H, Suzuki K, Maeda T (2001) Molecular cloning of PalBH, a mammalian homologue of the *Aspergillus* atypical calpain PalB. Biochim Biophys Acta Gene Struct Expr 1517:316–319

Ganaie MA, Gupta US, Kango N (2013) Screening microorganisms for fructosyltransferase (FTase) activity for generation of fructo-oligosaccharides (FOS). J Mol Catal B Enzym 97:12–17

Ganaie MA, Rawat HK, Wani OA, Gupta US, Kango N (2014) Immobilization of fructosyltransferase by chitosan and alginate for efficient production of fructo-oligosaccharides. Process Biochem 49:840–844

Gao L, Gao F, Zhang D, Zhang C, Wu G, Chen S (2013) Purification and characterization of a new β-glucosidase from *Penicillium piceum* and its application in enzymatic degradation of delignified corn stover. Bioresour Technol 147:658–661

Gastelum-Arellanez A, Paredes-López O, Olalde-Portugal V (2014) Extracellular endoglucanase activity from *Paenibacillus polymyxa* BEb-40: production, optimization and enzymatic characterization. World J Microbiol Biotechnol 30:2953–2965

Ghosh M, Prajapati BP, Suryawanshi RK, Dey KK, Kango N (2019a) Study of the effect of enzymatic deconstruction on natural cellulose by NMR measurements. Chem Phys Lett 727:105–115

Ghosh M, Prajapati BP, Kango N, Dey KK (2019b) A comprehensive and comparative study of the internal structure and dynamics of natural β-keratin and regenerated β-keratin by solid state NMR spectroscopy. Solid State Nucl Mag 101:1–11

Gopinath SCB, Anbu P, Lakshmipriya T, Tang TH, Chen Y, Hashim U, Ruslinda AR, Arshad MKM (2015) Biotechnological aspects and perspective of microbial keratinase production. Biomed Res Int 2015:1–10

Gradisar H, Kern S, Friedrich J (2000) Keratinase of *Doratomyces microsporus*. Appl Microbiol Biotechnol 53:196–200

Gradisar H, Friedrich J, Krizaj I, Jerala R (2005) Similarities and specificities of fungal keratinolytic proteases: comparison of keratinases of *Paecilomyces marquandii* and *Doratomyces microsporus* to some known proteases. Appl Environ Microbiol 71:3420–3426

Gurunathan B, Sahadevan R (2011) Design of experiments and artificial neural network linked genetic algorithm for modeling and optimization of L-asparaginase production by *Aspergillus terreus* MTCC 1782. Biotechnol Bioprocess Eng 16:50–58

Gusakov AV, Sinitsyn AP, Salanovich TN, Bukhtojarov FE, Markov AV, Ustinov BB, Zeijl CV, Punt P, Burlingame R (2005) Purification, cloning and characterisation of two forms of thermostable and highly active cellobiohydrolase I (Cel7A) produced by the industrial strain of *Chrysosporium lucknowense*. Enzym Microb Technol 36:57–69

Hamin Neto YAA, da Rosa Garzon NG, Pedezzi R, Cabral H (2017a) Specificity of peptidases secreted by filamentous fungi. Bioengineered 9:30–37

Hamin Neto YAA, de Oliveira LCG, de Oliveira JR, Juliano MA, Juliano L, Arantes EC, Cabral H (2017b) Analysis of the specificity and biochemical characterization of metalloproteases isolated from *Eupenicillium javanicum* using fluorescence resonance energy transfer peptides. Front Microbiol 7:2141

Hölker U, Höfer M, Lenz J (2004) Biotechnological advantages of laboratory-scale solid-state fermentation with fungi. Appl Microbiol Biotechnol 64:175–186

Hosamani R, Kaliwal BB (2011) L-Asparaginase- an antitumor agent production by *Fusarium equiseti* under solid state fermentation. Int J Drug Discov 3:88–99

Hsiao HY, Anderson DM, Dale NM (2006) Levels of β-mannan in soybean meal. Poult Sci 85:1430–1432

Hu W, Liu X, Li Y, Liu D, Kuang Z, Qian C, Yao D (2017) Rational design for the stability improvement of *Armillariella tabescens* β-mannanase MAN47 based on N-glycosylation modification. Enzym Microb Technol 97:82–89

Huang Y, Busk PK, Lange L (2015) Production and characterization of keratinolytic proteases produced by *Onygena corvina*. Fungal Genom Biol 5:119

Huang C, Ragauskas AJ, Wu X, Huang Y, Zhou X, He J, Huang C, Lai C, Li X, Yong Q (2018) Co-production of bio-ethanol, xylonic acid and slow-release nitrogen fertilizer from low-cost straw pulping solid residue. Bioresour Technol 250:365–373

Jain R, Kango N, Jain PC (2010) Proteases: significance and applications. In: Maheshwari DK, Dubey RC, Saravanamuthu R (eds) Industrial exploitation of microorganisms. I.K International Publishers, New Delhi, pp 228–254

Jana UK, Suryawanshi RK, Prajapati BP, Soni H, Kango N (2018) Production optimization and characterization of mannooligosaccharide generating β-mannanase from *Aspergillus oryzae*. Bioresour Technol 268:308–314

Jayaramu M, Hemalatha N, Rajeshwari C, Siddalingeshwara K, Mohsi S (2010) A novel approach for detection, confirmation, and optimization of L-asparaginase from *Emericella nidulans*. Curr Pharm Res 1:20–24

Jensen B, Nebelong P, Olsen J, Reeslev M (2002) Enzyme production in continuous cultivation by the thermophilic fungus, *Thermomyces lanuginosus*. Biotechnol Lett 24:41–45

Jiang H, Ma Y, Chi Z, Liu GL, Chi ZM (2016) Production, purification, and gene cloning of a β-fructofuranosidase with a high inulin-hydrolyzing activity produced by a novel yeast *Aureobasidium* sp. p6 isolated from a mangrove ecosystem. Mar Biotechnol 18:500–510

Juturu V, Wu J (2012) Insight into microbial hemicellulases other than xylanases: a review. J Chem Technol Biotechnol 88:353–363

Kalyani D, Lee KM, Tiwari MK, Ramachandran P, Kim H, Kim IW, Jeya M, Lee JK (2011) Characterization of a recombinant aryl β-glucosidase from *Neosartorya fischeri* NRRL181. Appl Microbiol Biotechnol 94:413–423

Kammoun R, Naili B, Bejar S (2008) Application of a statistical design to the optimization of parameters and culture medium for α-amylase production by *Aspergillus oryzae* CBS 819.72 grown on gruel (wheat grinding by-product). Bioresour Technol 99:5602–5609

Kang S, Park YS, Lee JS, Hong SI, Kim SW (2004) Production of cellulases and hemicellulases by *Aspergillus niger* KK2 from lignocellulosic biomass. Bioresour Technol 91:153–156

Kango N (2008) Production of inulinase using tap roots of dandelion (*Taraxacum officinale*) by *Aspergillus niger*. J Food Eng 85:473–478

Kango N, Jain PC (2005) Production and application of fungal xylanases. In: Rai MK, Deshmukh SK (eds) Fungi: diversity and biotechnology. Scientific Publishers, New Delhi, pp 251–281

Kango N, Jain SC (2011) Production and properties of microbial inulinases: recent advances. Food Biotechnol 25:165–212

Kango N, Agrawal SC, Jain PC (2003) Production of xylanase by *Emericella nidulans* NK-62 on low-value lignocellulosic substrates. World J Microbiol Biotechnol 19:691–694

Kango N, Soni H, Rawat H (2017) Extremophilic xylanases. In: Sani RK, Navanietha R (eds) Extremophilic bioprocessing of lignocellulosic feedstocks to biofuels, value-added products, and usable power. Springer, Cham, pp 73–88

Karnaouri A, Topakas E, Paschos T, Taouki I, Christakopoulos P (2013) Cloning, expression and characterization of an ethanol tolerant GH3 β-glucosidase from *Myceliophthora thermophila*. PeerJ 1:e46

Karnchanatat A, Petsom A, Sangvanich P, Piaphukiew J, Whalley AJS, Reynolds CD, Sihanonth P (2007) Purification and biochemical characterization of an extracellular beta-glucosidase from the wood-decaying fungus *Daldinia eschscholzii* (Ehrenb.:Fr.) Rehm. FEMS Microbiol Lett 270:162–170

Korotkova OG, Semenova MV, Morozova VV, Zorov IN, Sokolova LM, Bubnova TM, Okunev ON, Sinitsyn AP (2009) Isolation and properties of fungal β-glucosidases. Biochemistry 74:569–577

Korver DR (2006) Overview of the immune dynamics of the digestive system. J Appl Poult Res 15:123–135

Kües U (2015) Fungal enzymes for environmental management. Curr Opin Biotechnol 33:268–278

Kumar S, Satyanarayana T (2003) Purification and kinetics of a raw starch-hydrolyzing, thermostable, and neutral glucoamylase of the thermophilic mold *Thermomucor indicae-seudaticae*. Biotechnol Prog 19:936–944

Kumar P, Satyanarayana T (2007) Optimization of culture variables for improving glucoamylase production by alginate-entrapped *Thermomucor indicae-seudaticae* using statistical methods. Bioresour Technol 98:1252–1259

Kumar V, Dangi AK, Shukla P (2018) Engineering thermostable microbial xylanases toward its industrial applications. Mol Biotechnol 60:226–235

Lambertz C, Garvey M, Klinger J, Heesel D, Klose H, Fischer R, Commandeur U (2014) Challenges and advances in the heterologous expression of cellulolytic enzymes: a review. Biotechnol Biofuels 7:135

Lange L, Huang Y, Busk PK (2016) Microbial decomposition of keratin in nature-a new hypothesis of industrial relevance. Appl Microbiol Biotechnol 100:2083–2096

Li S, Sing S, Wang Z (2011) Improved expression of *Rhizopus oryzae* α-amylase in the methylotrophic yeast *Pichia pastoris*. Protein Expr Purif 79:142–148

Li Y, Yi P, Yan Q, Qin Z, Liu X, Jiang Z (2017) Directed evolution of a β-mannanase from *Rhizomucor miehei* to improve catalytic activity in acidic and thermophilic conditions. Biotechnol Biofuels 10:143

Li X, Zhang X, Xu S, Zhang H, Xu M, Yang T, Wang L, Qian H, Zhang H, Fang H, Osire T, Rao Z, Yang S (2018) Simultaneous cell disruption and semi-quantitative activity assays for high-throughput screening of thermostable L-asparaginases. Sci Rep 8:7915

Lin J, Pillay B, Singh S (1999) Purification and biochemical characteristics of β-D-glucosidase from a thermophilic fungus, *Thermomyces lanuginosus*–SSBP. Biotechnol Appl Biochem 30:81–87

Machida M, Asai K, Sano M, Tanaka T, Kumagai T et al (2005) Genome sequencing and analysis of *Aspergillus oryzae*. Nature 438:1157–1161

Maijala P, Kango N, Szijarto N, Viikari L (2012) Characterization of hemicellulases from thermophilic fungi. Antonie van Leeuwenhoek 101:905–917

Maitan-Alfenas GP, Visser EM, Guimarães VM (2015) Enzymatic hydrolysis of lignocellulosic biomass: converting food waste in valuable products. Curr Opin Food Sci 1:44–49

Malherbe AR, Rose SH, Viljoen-Bloom M, van Zyl WH (2014) Expression and evaluation of enzymes required for the hydrolysis of galactomannan. J Ind Microbiol Biotechnol 41:1201–1209

Mandujano-González V, Villa-Tanaca L, Anducho-Reyes MA, Mercado-Flores Y (2016) Secreted fungal aspartic proteases: a review. Rev Iberoam Micol 33:76–82

Mchunu NP, Permaul K, Abdul Rahman AY, Saito JA, Singh S, Alam M (2013) Xylanase superproducer: genome sequence of a compost-loving thermophilic fungus, *Thermomyces lanuginosus* strain SSBP. Genome Announc 1(3):pii: e00388-13

Merz M, Eisele T, Berends P, Appel D, Rabe S, Blank I, Stressler T, Fischer L (2015) Flavourzyme, an enzyme preparation with industrial relevance: automated nine-step purification and partial characterization of eight enzymes. J Agric Food Chem 63:5682–5693

Midorikawa GEO, Correa CL, Noronha EF, Filho EXF, Togawa RC, Costa MM d C, Silva-Junior OB, Grynberg P, RNG M (2018) Analysis of the transcriptome in *Aspergillus tamarii* during enzymatic degradation of sugarcane bagasse. Front Bioeng Biotechnol 6:123

Mignon B, Swinnen M, Bouchara JP, Hofinger M, Nikkels A, Pierard G, Gerday CH, Losson B (1998) Purification and characterization of a 315 kDa keratinolytic subtilisin-like serine protease from *Microsporum canis* and evidence of its secretion in naturally infected cats. Med Mycol 36:395–404

Mishra A (2006) Production of L-Asparaginase, an anticancer agent, from *Aspergillus niger* using agricultural waste in solid state fermentation. Appl Biochem Biotechnol 135:33–42

Moreira LRS, Filho EXF (2008) An overview of mannan structure and mannan-degrading enzyme systems. Appl Microbiol Biotechnol 79:165–178

Moreira-Gasparin FG, de Souza CGM, Costa AM, Alexandrino AM, Bracht C, Boer CK, Peralta RM (2009) Purification and characterization of an efficient poultry feather degrading-protease from *Myrothecium verrucaria*. Biodegradation 20:727–736

Murphy C, Powlowski J, Wu M, Butler G, Tsang A (2011) Curation of characterized glycoside hydrolases of fungal origin. Database 2011:bar020-bar020

Nakazawa H, Kawai T, Ida N, Shida Y, Kobayashi Y, Okada H, Tani S, Sumitani JI, Kawaguchi T, Morikawa Y, Ogasawara W (2011) Construction of a recombinant *Trichoderma reesei* strain expressing *Aspergillus aculeatus* β-glucosidase 1 for efficient biomass conversion. Biotechnol Bioeng 109:92–99

Noronha EF, de Lima BD, de Sá CM, Felix CR (2002) Heterologous production of *Aspergillus fumigatus* keratinase in *Pichia pastoris*. World J Microbiol Biotechnol 18:563–568

Obeng EM, Adam SNN, Budiman C, Ongkudon CM, Maas R, Jose J (2017) Lignocellulases: a review of emerging and developing enzymes, systems, and practices. Bioresour Bioprocess 4:16

Parashar D, Satyanarayana T (2017) Engineering a chimeric acid-stable α-amylase-glucoamylase (Amy-Glu) for one step starch saccharification. Int J Biol Macromol 99:274–281

Pawar VA, Prajapati AS, Akhani RC, Patel DH, Subramanian RB (2018) Molecular and biochemical characterization of a thermostable keratinase from *Bacillus altitudinis* RBDV1. 3 Biotech 8:107

Payne CM, Knott BC, Mayes HB, Hansson H, Himmel ME, Sandgren M, Ståhlberg J, Beckham GT (2015) Fungal cellulases. Chem Rev 115:1308–1448

Peciulyte A, Pisano M, de Vries RP, Olsson L (2017) Hydrolytic potential of five fungal supernatants to enhance a commercial enzyme cocktail. Biotechnol Lett 39:1403–1411

Polizeli MLTM, Rizzatti ACS, Monti R, Terenzi HF, Jorge JA, Amorim DS (2005) Xylanases from fungi: properties and industrial applications. Appl Microbiol Biotechnol 67:577–591

Porfirif MC, Milatich EJ, Farruggia BM, Romanini D (2016) Production of alpha-amylase from *Aspergillus oryzae* for several industrial applications in a single step. J Chromatogr B 1022:87–92

Prajapati BP, Suryawanshi RK, Agrawal S, Ghosh M, Kango N (2018) Characterization of cellulase from Aspergillus tubingensis NKBP-55 for generation of fermentable sugars from agricultural residues. Bioresour Technol 250:733–740

Qin LM, Dekio S, Jidoi J (1992) Some biochemical characteristics of a partially purified extracellular keratinase from *Trichophyton schoenleinii*. Zentralbl Bakteriol 277:236–244

Raba'atun Adawiyah S, Shuhaimi M, Mohd Yazid AM, Abdul Manaf A, Rosli N, Sreeramanan S (2011) Molecular cloning and sequence analysis of an inulinase gene from an *Aspergillus* sp. World J Microbiol Biotechnol 27:2173–2185

Rajasree KP, Mathew GM, Pandey A, Sukumaran RK (2013) Highly glucose tolerant β-glucosidase from *Aspergillus unguis*: NII 08123 for enhanced hydrolysis of biomass. J Ind Microbiol Biotechnol 40:967–975

Rawat HK, Ganaie MA, Kango N (2015a) Production of inulinase, fructosyltransferase and sucrase from fungi on low-value inulin-rich substrates and their use in generation of fructose and fructooligosaccharides. Antonie Van Leeuwenhoek 107:799–811

Rawat HK, Jain SC, Kango N (2015b) Production and properties of inulinase from *Penicillium* sp. NFCC 2768 grown on inulin containing vegetal infusions. Biocatal Biotransformation 33:61–68

Rawat HK, Soni H, Treichel H, Kango N (2016) Biotechnological potential of microbial inulinases: recent perspective. Crit Rev Food Sci Nutr 57:3818–3829

Rawat H, Soni H, Kango N (2017) In: Satyanarayana T, Deshmukh SK, Johri BN (eds) Fungal Inulinolytic enzymes: a current appraisal in developments in fungal biology and applied mycology. Springer, Singapore, pp 279–293

Saeed H, Ali H, Soudan H, Embaby A, El-Sharkawy A, Farag A, Hussein A, Ataya F (2018) Molecular cloning, structural modeling and production of recombinant *Aspergillus terreus* L-asparaginase in *Escherichia coli*. Int J Biol Macromol 106:1041–1051

Saha BC (2003) Hemicellulose bioconversion. J Ind Microbiol Biotechnol 30:279–291

Salamin K, Eugster PJ, Jousson O, Waridel P, Grouzmann E, Monod M (2017) AoS28D, a proline-Xaa carboxypeptidase secreted by *Aspergillus oryzae*. Appl Microbiol Biotechnol 101:4129–4137

Salgaonkar M, Nadar SS, Rathod VK (2018) Combi-metal organic framework (Combi-MOF) of α-amylase and glucoamylase for one pot starch hydrolysis. Int J Biol Macromol 113:464–475

Sarquis MI d M, EMM O, Santos AA, da Costa GL (2004) Production of L-asparaginase by filamentous fungi. Mem Inst Oswaldo Cruz 99:489–492

Sawant S, Birhade S, Anil A, Gilbert H, Lali A (2016) Two-way dynamics in β-glucosidase catalysis. J Mol Catal B Enzym 133:161–166

Saxena A, Upadhyay R, Kango N (2015) Isolation and identification of actinomycetes for production of novel extracellular glutaminase free L-asparaginase. Indian J Exp Biol 53:786–793

Schmoll M (2018) Regulation of plant cell wall degradation by light in *Trichoderma*. Fungal Biol Biotechnol 5:10

Shoemaker S, Schweickart V, Ladner M, Gelfand D, Kwok S, Myambo K, Innis M (1983) Molecular cloning of exo-cellobiohydrolase I derived from *Trichoderma reesei* strain L27. Nat Biotechnol 1:691–696

Shrivastava A, Khan AA, Shrivastav A, Jain SK, Singhal PK (2012) Kinetic studies of l-asparaginase from *Penicillium digitatum*. Prep Biochem Biotechnol 42:574–581

Sjostrom E (1993) Wood chemistry, fundamentals and application. Academic, San Diego, pp 12–23

Soni H, Kango N (2013) Microbial mannanases: properties and applications. In: Shukla P, Pletscke BI (eds) Advances in enzyme biotechnology. Springer, New Delhi, pp 41–56

Soni H, Rawat HK, Pletschke BI, Kango N (2016) Purification and characterization of β-mannanase from *Aspergillus terreus* and its applicability in depolymerization of mannans and saccharification of lignocellulosic biomass. 3 Biotech 6:136

Souza FHM, Nascimento CV, Rosa JC, Masui DC, Leone FA, Jorge JA, Furriel RPM (2010) Purification and biochemical characterization of a mycelial glucose- and xylose-stimulated β-glucosidase from the thermophilic fungus *Humicola insolens*. Process Biochem 45:272–278

Spohner SC, Czermak P (2016) Heterologous expression of *Aspergillus terreus* fructosyltransferase in *Kluyveromyces lactis*. New Biotechnol 33:473–479

Suárez MB, Vizcaíno JA, Llobell A, Monte E (2007) Characterization of genes encoding novel peptidases in the biocontrol fungus *Trichoderma harzianum* CECT 2413 using the TrichoEST functional genomics approach. Curr Genet 51:331–342

Suryawanshi RK, Jana UK, Prajapati BP, Kango N (2019) Immobilization of *Aspergillus quadrilineatus* RSNK-1 multi-enzymatic system for fruit juice treatment and mannooligosaccharide generation. Food Chem 289:95–102

Tanriseven A, Aslan Y (2005) Immobilization of pectinex ultra SP-L to produce fructooligosaccharides. Enzym Microb Technol 36:550–554

Trivedi S, Divecha J, Shah A (2012) Optimization of inulinase production by a newly isolated *Aspergillus tubingensis* CR16 using low cost substrates. Carbohydr Polym 90:483–490

Tsukada T, Igarashi K, Yoshida M, Samejima M (2006) Molecular cloning and characterization of two intracellular β-glucosidases belonging to glycoside hydrolase family 1 from the basidiomycete *Phanerochaete chrysosporium*. Appl Microbiol Biotechnol 73:807–814

Vaidya S, Srivastava PK, Rathore P, Pandey AK (2015) Amylases: a prospective enzyme in the field of biotechnology. J Appl Biosci 41:1–18

Vala AK, Sachaniya B, Dudhagara D, Panseriya HZ, Gosai H, Rawal R, Dave BP (2018) Characterization of L-asparaginase from marine-derived *Aspergillus niger* AKV-MKBU, its antiproliferative activity and bench scale production using industrial waste. Int J Biol Macromol 108:41–46

van Zyl WH, Rose SH, Trollope K, Görgens JF (2010) Fungal β-mannanases: mannan hydrolysis, heterologous production and biotechnological applications. Process Biochem 45:1203–1213

Varalakshmi V, Raju KJ (2013) Optimization of L-asparaginase production by *Aspergillus terrus* MTCC1782 using bajra seed flour under solid state fermentation. Int J Res Eng Technol 2:121–129

Viikari L, Kantelinen A, Sundquist J, Linko M (1994) Xylanases in bleaching: from an idea to the industry. FEMS Microbiol Rev 13:335–350

Walia A, Guleria S, Mehta P, Chauhan A, Parkash J (2017) Microbial xylanases and their industrial application in pulp and paper biobleaching: a review. 3 Biotech 7:11

Wang X, Liu ZL, Weber SA, Zhang X (2016a) Two new native β-glucosidases from *Clavispora* NRRL Y-50464 confer its dual function as cellobiose fermenting ethanologenic yeast. PLoS One 11:e0151293

Wang P, Ma J, Zhang Y, Zhang M, Wu M, Dai Z, Jiang M (2016b) Efficient secretory overexpression of endoinulinase in *Escherichia coli* and the production of inulooligosaccharides. Appl Biochem Biotechnol 179:880–894
Wang S, Duan M, Liu Y, Fan S, Lin X, Zhang Y (2016c) Enhanced production of fructosyltransferase in *Aspergillus oryzae* by genome shuffling. Biotechnol Lett 39:391–396
Wang D, Li FL, Wang SA (2016d) A one-step bioprocess for production of high-content fructooligosaccharides from inulin by yeast. Carbohydr Polym 151:1220–1226
Wang M, Zhang M, Li L, Dong Y, Jiang Y, Liu K, Zhang R, Jiang B, Niu K, Fang X (2017) Role of *Trichoderma reesei* mitogen-activated protein kinases (MAPKs) in cellulase formation. Biotechnol Biofuels 10:99
Xu L, Wang D, Lu L, Jin L, Liu J, Song D, Guo Z, Xiao M (2014) Purification, cloning, characterization and n-glycosylation analysis of a novel β-fructosidase from *Aspergillus oryzae* FS4 synthesizing levan- and neolevan-type fructooligosaccharides. PLoS One 9:e114793
Yang S, Wang L, Yan Q, Jiang Z, Li L (2009) Hydrolysis of soybean isoflavone glycosides by a thermostable β-glucosidase from *Paecilomyces thermophila*. Food Chem 115:1247–1252
Yavuz S, Kocabay S, Çetinkaya S, Akkaya B, Akkaya R, Yenidunya AF, Bakıcı MZ (2017) Production, purification, and characterization of metalloprotease from *Candida kefyr* 41 PSB. Int J Biol Macromol 94:106–113
You X, Qin Z, Li YX, Yan QJ, Li B, Jiang ZQ (2018) Structural and biochemical insights into the substrate-binding mechanism of a novel glycoside hydrolase family 134 β-mannanase. Biochim Biophys Acta 1862:1376–1388
Zhang L, Tizard IR (1996) Activation of a mouse macrophage cell line by acemannan: the major carbohydrate fraction from Aloe vera gel. Immunopharmacology 35:119–128
Zhang L, An J, Li L, Wang H, Liu D, Li N, Cheng H, Deng Z (2016) Highly efficient fructooligosaccharides production by an erythritol-producing yeast *Yarrowia lipolytica* displaying fructosyltransferase. J Agric Food Chem 64:3828–3837
Zhang Q, Han Y, Xiao H (2017a) Microbial α-amylase: a biomolecular overview. Process Biochem 53:88–101
Zhang J, Liu C, Xie Y, Li N, Ning Z, Du N, Huang X, Zhong Y (2017b) Enhancing fructooligosaccharides production by genetic improvement of the industrial fungus *Aspergillus niger* ATCC 20611. J Biotechnol 249:25–33

22 Fungi in Hypogean Environment: Bioprospection Perspective

S. R. Joshi and Upashna Chettri

Abstract

Fungi are ubiquitous in their presence almost similar to bacterial distribution. With regard to the products and metabolites that have been bioprospected and utilized for human welfare, fungi are always the first among the living organisms that have benefitted mankind. The exploration and distribution of fungi from the epigean environment is well known along with the targeted species of interest which have been manipulated for optimizing the production of metabolic by-products. However, when it comes to the hypogean or subterranean habitats, literature on the nature, type, and distribution of fungi is still scarce. Considering the benefits that can be offered by fungi from extreme environments as reported in literature, it becomes more relevant to explore and document fungal diversity from hypogean habitats such as the caves. This niche may provide vistas for discovery of products and metabolites hitherto unknown to mycoscience and mycotechnology.

Keywords

Caves · Hypogean · Fungi · Diversity · Metabolites · Bioprospection

22.1 Introduction

The biosphere is divided into the epigean (epi-upon, gean-earth) (spaces on the surface of the earth) and hypogean (hypo-under, gean-earth) (spaces below the surface of the earth) or subterranean environments. The former is exposed to direct light and the latter is in perpetual darkness. One of the striking features of the

S. R. Joshi (✉) · U. Chettri
Microbiology Laboratory, Department of Biotechnology and Bioinformatics, North-Eastern Hill University, Shillong, Meghalaya, India

T. Satyanarayana et al. (eds.), *Advancing Frontiers in Mycology & Mycotechnology*,
https://doi.org/10.1007/978-981-13-9349-5_22

hypogean environment is its exposure to the amount of light (Romero 2011). Hypogean environments can be best described as stable environments with relatively lower temperature, high humidity, low nutrient, and lack of photosynthetic process resulting in low organic carbon, a vital component of life (Poulson and White 1969). Therefore, chemoautotrophy becomes one of the drivers of ecosystem in such cave habitats (Sarbu et al. 1996; Kinkle and Kane 2000).

The hypogean environment constitutes natural (phreatic or artesian and cave) and man-made underground structures like tombs and temples (Romero 2011; Agrawal et al. 1988). Caves serve a suitable example of hypogean environment. They are buffered habitats having no natural light added to the presence of physical obstacles (Kambesis 2007). A "cave" was first described as areas accessible to humans and extending beyond the reach of twilight (Gillieson 1996; Hill and Forti 1997). Caves are mainly classified based on the formations and the nature of rocks (Palmer 1991).

Karst region caves mostly have limestone and calcareous rocks as major constituents, while lava tubes are prevalent in basaltic rock. However, the caves formed in gypsum, granite, talus, quartzite, ice, and sandstone are usually less extent (Zhou et al. 2007). Caves offer extreme environments for life to grow, and specialized microorganisms tend to grow in such niches which are characterized by limited organic matter with consistently low temperature, high humidity, and mineral-enriched conditions (Schabereiter-Gurtner et al. 2002). The prevailing internal environmental conditions of a cave are invariably regulated by the proximate surrounding environments and climatic conditions (Banerjee and Joshi 2013). A cave can have several biotic zones based on the amount and presence of light as well as other environmental conditions, thus making this hypogean habitat a zonal environment (Poulson and Lavoie 2000). The entrance zone that is closest to the surface which can have photosynthetic activities is dependent on the diurnal variations in light and environmental conditions. The twilight zone follows the entrance and is characterized by the presence of moderate light and also lesser fluctuations in temperatures as compared to surface zone. Photosynthetic organisms like lichens and algae having capacity to use low light intensity are present in the twilight zone. The aphotic zone marks the end of twilight zone and the beginning of total darkness. This is the deepest part of the cave with complete absence of light, stable temperature, little to no airflow with high CO_2 concentrations, and nearly water-saturated environments creating a unique niche (Howarth and Stone 1990). Organisms which can tolerate and exploit the hostile conditions of the subterranean cave ecosystem are present here, and such habitats can be a good source of extremophilic microorganisms like fungi (Ogorek et al. 2017).

Organisms in the subterranean environments can be temporary inhabitants called stygoxene or trogloxene, communities with strong hypogean affinities known as stygophile or troglophile, and obligatory hypogean communities called stygobite or troglobite. The communities belonging to the stygobites and troglobites are specialized morphologically and physiologically for underground life. Studies conducted on the deep pristine karstic aquifer of the Lez basin of France revealed that out of a total of 53 taxa identified, 13 were observed to be stygoxene, 8 to be stygophile, and

32 to be stygobite species (Malard and Gibert 1997). About 60% stygobite species were reported to constitute the subterranean community and the percentage could be higher up to 80% and 100% where the stygobites do not have direct contact with the environment of the surface (Gibert and Deharveng 2002). The same researchers listed around 6634 stygobite species and found that some crustacean species were strictly restricted to the subterranean habitats. These figures still remain underestimated as most of the caves worldwide remain to be explored. The species distribution and diversity in the borderline of epigean and hypogean environments could vary depending upon the immediate external environment of the cave. A comprehensive documentation of caves when carried out can provide the worldwide scenario of species' richness and diversity transition in hypogean-epigean cave ecosystems.

An interesting case in biospeleologic environment is the presence of caves with overhead openings that permit the penetration of light making the environment similar to epigean surroundings. Photosynthesis is possible in such areas increasing or modifying the local flora and fauna. In such areas of caves, organisms possessing eyes and pigmented morphology are found as in epigean habitat (Romero 2011).

22.2 Biospeleogenesis: Microbial Perspective

The metabolic flexibility of microbes makes them possible to inhabit virtually all habitats on the earth which may be hostile for other living organisms. Life in the underground habitats like caves is encountered with extreme environments due to the lack of nutrient input through primary production by photosynthetic activities, and hence, the caves are often oligotrophic. In terms of the physical parameters prevalent in such habitats, they are relatively predictable, mild, and invariably constant (Northup and Lavoie 2001). Caves render a microclimatic condition that is distinguished not only by constant temperature and relative humidity but also by a luminous intensity that varies from the entrance to the back of the cave (Martinez and Asencio 2009). The light intensity attenuates as we go further into the cave and twilight zone prevails subsequently in the deeper parts. The caves become aphotic in the deeper depths characterized by the complete absence of light where the temperature is at or near the Mean Annual Surface Temperature (MAST) and consistent with high humidity. Earliest insight into the microbial presence on such habitats was provided by Hoeg (1946) from the walls of the Norwegian cave. Further reviews on microbes from caves were provided by Caumartin (1963), Dyson and James (1973), Rutherford and Huang (1994), and Vanderwolf et al. (2013), primarily focusing on fungi, yeast, and slime molds.

Microorganisms prevalent in the deep caves are found to be mostly identical to forms present in the surface and are opportunistic in nature as they are active only under favorable environmental conditions (Dickson and Kirk 1976; Jones and Motyka 1987; James 1994). Some microbes may be nonresidents in caves, and external forces like water current, animals, and air normally carry them into the caves (Northup and Lavoie 2001). Lately, microbes have grabbed the attention of the geomicrobiologists worldwide. Till date, substantial work has been done on microbes from caves and their role in biospeleogenesis. Microorganisms are always found to interact with their geological surroundings owing to their ability to generate energy by consumption of inorganic material (Barton 2006). In the subterranean voids where there is no light for primary production by photosynthesis, chemoautotrophic microorganisms produce organic matter from reduced compounds like hydrogen sulfide and methane that provides energy sources for microbial communities which in turn becomes the food base for rich and abundant communities of invertebrates that inhabit hypogean caves, therefore becoming important components from a microbiological perspective (Forti et al. 2002; Kumaresan et al. 2014). Geochemistry of caves like the rock fabric and mineralogy may often be the important drivers of microbial diversity and community. Very low microbial community belonging to the members of *Actinobacteria* was found to dominate a geochemically simple cave as compared to the geochemically complex site where more diverse bacterial communities representing *Alpha-*, *Beta-*, and *Gammaproteobacteria* have been reported (Barton et al. 2007). Modern-day approaches involving culture-independent protocols have been used to study the microbial diversity in different cave systems around the world. Common bacterial groups such as the *Actinobacteria, Alpha-*, *Beta-*, and *Gammaproteobacteria* (Schabereiter-Gurtner et al. 2002; Barton et al. 2004; Barton et al. 2007; Zhou et al. 2007), which have been characterized using culture-independent molecular techniques applied to study the karstic subterranean habitats, have provided newer evidences for the occurrence of groups such as *Gemmatimonadetes*, *Nitrosomonadales*, *Oceanospirillales*, and *Rubrobacterales* in caves. Modern approaches which encompass culture-independent methods and phylogenetic analysis using molecular markers have become very handy in description and discovery of novel organisms in caves that are not culturable in defined formulations (Northup and Lavoie 2001).

Fungi which form the second largest group in the eukaryotes are ubiquitous and are important microflora that are found in various diversified ecological niches comprising ecosystems in water and land (Ritz and Young 2004). They are reported from lakes and deep sea (Shearer et al. 2007; Nagahama and Nagano 2012) as well as caves and rock surfaces (Sterflinger 2000; Gorbushina 2007; Vanderwolf et al. 2013; Man et al. 2018). Extensive efforts have been put to unfold the fungal diversity in caves through Next Generation Sequencing (NGS) technologies (Man et al. 2015; Jiang et al. 2017; Zhang et al. 2017). Zhang et al. (2017) have reviewed fungi that have been discovered from caves and mines around the world and reported more than 1150 fungal species in 550 genera. From among the various fungal taxa that have been reported from subterranean environments, species and genera belonging to *Ascomycota*, *Basidiomycota*, *Zygomycota*, Mycetozoa, Oomycota, and others such as Amoebozoa, *Chytridiomycota*, *Microsporidiomycota*, and Percolozoa have been present to the tune of 69.1%, 20%, 6.6%, 2.6%, 1%, and 0.8%, respectively (Vanderwolf et al. 2013). Novel species belonging to oligotrophic fungi such as *Cephalotrichum* was isolated using oligotrophic carbon-free silica gel medium (SGM), indicating that fungi can grow on carbon-free media (Jiang et al. 2017). Even though fungi are well distributed in cave habitats, their role in mineral precipitation and the process of biospeleogenesis remain scanty. Studies are focused more on the role of bacterial mineral precipitation, emphasizing that the caves are sites of active mineralization and sediment deposition (Jones 2001).

Chemolithotrophs and chemoautotrophs are dominant microbial biota that are found to be inhabiting the deep caves which are devoid of organic matter (Cunningham et al. 1995; Northup et al. 1997). Cave environments where organic carbon gets transported by water flow and wind support the growth of heterotrophic microbes (Northup et al. 1997). Heterotrophic microorganisms prevalent in some caves are the resultant effect of the changing microclimatic conditions affected by the animals, cave visitors, and dripping water from the surface which contribute to the total organic matter input (Groth and Saiz-Jimenez 1999). Microbes in the cave ecosystem interact with minerals mobilizing inorganic phosphate, oxidizing methane, hydrogen, and sulfide and also derive energy by hydrolyzing macromolecules from other cave microbial communities (Barton and Jurado 2007). Barton and Luiszer (2005) studied the microbial metabolic structure in sulfidic cave, suggesting that the sulfuric acid production by the sulfate-reducing and sulfide-oxidizing bacteria is the primary component of cavern enlargement in sulfuric acid caves. Microbes' role in the genesis of secondary cave minerals can also be attributed to the ability of microorganisms to cause biomineralization through different enzymes or by the production of substances that may lead to changes in pH causing precipitation of minerals (Forti 2001). Therefore, there is substantial amount of acidity generated by the metabolic activities through oxidation of sulfur, iron, and manganese by microbes which contribute to the development of wall formations in caves (Jones 1992; Sarbu et al. 1994; Provencio and Polyak 2001; Engel et al. 2004). Many microbes are identified which are involved in these cave wall formations and are composed of carbonates, sulfur, salpeter, clays, manganese, and iron oxides, collectively described in such cases as biogenic materials (Northup and Lavoie 2001). Evidence of biogenic activities due to the occurrence of *Bacillus cereus, B. licheniformis, and B. mycoides* has been studied in the Mawsmai Cave and Krem Phyllut, India (Baskar et al. 2009).

Calcium carbonate formations are observed to be the dominant structures in most caves of the world, and microbes have been attributed to calcium carbonate precipitation which is controlled by various factors influencing the growth of microbes (Banerjee and Joshi 2013). Microorganisms characterized from cave deposits have been attributed to precipitation of calcium carbonate (Danielli and Edington 1983). Metabolic processes such as photosynthesis, ammonification, denitrification, sulfate reduction, and anaerobic sulfide oxidation by bacteria and fungi can lead to extracellular calcium carbonate precipitation (Castanier et al. 2000; Riding 2000).

The fact that microbes are associated with carbonates and speleothems has already been established through in vitro studies in the laboratory. *Bacillus* species isolated from saline soil was reported to precipitate carbonate (Rivadeneyra et al. 1993). Microorganisms produce crystals from organic calcium salts and it has been shown that bacteria can form calcium carbonate crystals in the laboratory conditions, thereby proving the geomicrobiological processes in stalactite formation (Baskar et al. 2006). Organisms such as bacteria, cyanobacteria, microalgae, and fungi which form biofilms or microbial mats are the primary organisms involved in such formations. The extracellular polymeric substances produced by microbes can bring about adherence of sediments which become not only a critical factor in the formation of microbial carbonates but also provide sites for nucleation of carbonate minerals (Riding 2000; Banerjee and Joshi 2013).

Investigations carried out in different caves from Meghalaya in India showed that biofilm bacteria present in the calcareous speleothem fragments could be cultivated on B-4 medium in the laboratory, mimicking the ideal cave conditions and parameters. This demonstrated a close association between the bacterial cells as biofilms growing in the cave walls and the nature of the polymorphic crystals. In addition, the ultrastructural revelations of the polymorphs of crystals suggest that the identified strains were capable of depositing $CaCO_3$ that varied under different cultural conditions. Majority of the calcifying bacteria isolated from the cave environment were identified by 16SrRNA-based molecular techniques and were found to belong to *Bacillus* and *Lysinibacillus*, while a few of these microbes belonged to genera *Acinetobacter, Kocuria, and Brevibacillus* (Banerjee and Joshi 2014).

The microbial sulfur cycling by different sulfur-oxidizing and sulfate-reducing microbes has been the major factor in the development of cave minerals and gypsum as well as iron oxides and hydroxides (Hill and Forti 1997). Speleothems in gypsum caves are mainly constituted by calcium carbonate (calcite) or calcium sulfate (gypsum), and the abundance of these deposits is believed to differ with climatic conditions where gypsum speleothems are found dominant in arid regions and humid conditions support carbonate deposition (Calaforra et al. 2007).

Hubbard et al. (1986) undertook a pioneering work on cave formations and reported the contribution of sulfur-oxidizing microbes in cave formation resulted by the generation of sulfuric acid in Cesspool Cave, Virginia. They reported the occurrence of morphologically distinct sulfur bacteria *Beggiatoa, Thiothrix,* and *Achromatium.* Engel et al. (2001), while working on the same cave, concluded that the chemoautotrophic processes dominate the microbial mat community, while chemoautotrophic energy production was found to ecologically limit the occurrence of higher trophic levels. The sulfur-oxidizing and acid-producing bacteria and their metabolic activity bear extensive geological implications in the subterranean ecosystems.

The coating and the crusts in caves are due to the iron oxides and hydroxides which also are seen to occur in typical stalactites. A wide variety of aqueous and terrestrial habitats consist of microbial iron mineral in the form of ferrihydrite, siderite, and magnetites as well as oxides, sulfates, sulfides, silicates, and phosphates of iron (Konhauser 1997, 1998). In one of the studies, spherical bodies in stalactites of Grand Cayman Island in British West Indies have been reported to contain considerable amounts of iron or manganese (Jones and Motyka 1987). Peck (1986) demonstrated the presence of iron oxidizers in cave sediment. In his experiment, he reported *Gallionella ferruginea* showing precipitates of iron hydroxide and iron-impregnated sheath formations by *Leptothrix* sp. in enrichment media containing inoculation of mud from cave pools and sumps in Level Crevice Cave near Dubuque, Iowa. Soft deposits of manganese compounds present on speleothems or as coatings on walls of caves were found in caves (Gascoine 1982; Rogers and Williams, 1982). One of the most common manganese minerals prevalent in caves is Birnessite (Hill and Forti 1997). *Leptothrix*, a manganese oxidizing bacterium studied by Moore (1981) in Matt Black Cave, West Virginia, was found to be involved in the deposition of Birnessite. The fossil remains of precipitated manganese revealed the occurrence of rods, strands, spheroid morphologies, and sheets in Grand Cayman Caves and were attributed to be biogenic in nature (Jones 1992).

22.3 Fungi in Extreme Habitats

Fungi are important in the functioning of ecosystems. They are cosmopolitan and play different roles from saprotrophs/decomposers or parasites to the symbionts. Fungi are specialized in adapting to harsh environments due to their ecological plasticity (Selbmann et al. 2013). Fungi have been reported from extreme cold conditions (Gunde-Cimerman et al. 2003; Zalar et al. 2008), dry and salty habitats (Butinar et al. 2005; Zalar et al. 2007), acidic and basic environments (Ranta 1990; Shiomi et al. 2004), deep-sea extreme niches (Nagano and Nagahama 2012), and oligotrophic ecosystems (Onofri 1999). Similarly, hypogean caves represent an extreme environment where life is hardly possible. Fungi play a critical role in the underground ecology like caves by way of contributing to the process of decomposition of organic material and also constituting the fundamental source of food for other cave inhabitants (Sustr et al. 2005; Walochnik and Mulec 2009; Bastian et al. 2010).

A large area of the hypogean environment is partly dark with darkness increasing toward the interior which makes the fungi to carry out the role of decomposer or parasites that could derive nourishment from dead organic matter, whereas in the parasitic mode of nourishment it may grow as an obligate or the true parasite which cannot grow in the absence of a suitable host and the facultative parasites which exhibit parasitic mode but under certain conditions. A very specific microfungal community exists in the underground environment, particularly when there are organic matter inputs (Novakova 2009). Inside the cave, fungi grow on organic matter like bat guano or animals (Mulec 2008; Novakova 2009). The greatest species diversity of fungi has been on bat guano when compared to other cave substrates as studied by Novakova (2009) in a Slovakian cave. Bat guano can determine the mycoflora in caves (Min 1988). Around 50 species of bat guano-loving fungi known as guanophilic fungi were characterized from three caves in southwestern Puerto Rico. *Circinella umbellata* was found to dominate the bat guano (Nieves-Rivera et al. 2009).

Cadavers of insects also represent a rich source of food for primary and secondary saprophytic cavernicolous mycobiota (Gunde-Cimerman et al. 1998). Many entomopathogenic and opportunistic fungi were discovered from the larval as well as adult stage of the cave cricket *Troglophilus neglectus* cadavers (Gunde-Cimerman et al. 1998). The frequently isolated fungi belonged to the genus *Mucor* with the prevailing one being a new species *Mucor troglophilus*, and *Beauveria bassiana*, a well-known entomopathogen, was the dominating fungus isolated from the adult stage (Gunde-Cimerman et al. 1998). Fungal spores might enter the cave with water and air currents, animals, and also humans in case of show caves (Chelius et al. 2009; Griffin et al. 2014). Speleomycological study and research related to this domain is mainly focused on the fungal diversity. Various approaches comprising culture-dependent and culture-independent methods have been used worldwide in the microbial studies of caves and their mycobiota. Recently, 20 novel species of fungi were enumerated and characterized from the karst cave in China by Zhang et al. (2017). They obtained 563 fungal strains which belonged to *Ascomycota*, *Basidiomycota*, and *Mucoromycotina*, and 59% of the species reported were enlisted as discovered for the first time from Karst caves (Zhang et al. 2017). Recently, Zhang et al. (2018a, b) in their review article reported more than 1150 fungal species in 550 genera which were among the ones discovered from caves and mines worldwide. Some obligate troglobitic fungi such as *Acaulium caviariforme*, *Aspergillus baeticus,* and *A. thesauricus* have been reported by Novakova et al. (2012). Mycologists have discovered many new species till date from different cave habitats (Novakova et al. 2012; Man et al. 2015; Jiang et al. 2017; Zhang et al. 2017), but the existence and origin of troglobitic fungi still remains unclear. It is suggested that the novel species reported from the caves could be travellers from other environments and not a true troglobitic fungus (Zhang et al. 2018a, b). To understand the origin of subterranean fauna, adaptive shift hypothesis and climate relict hypothesis have been put forward. The climate relict hypothesis explains how the epigean species pre-adapted to the hypogean life may take refuge in the underground environment when environment in the surface habitats becomes unfavorable, whereas under the adaptive shift hypothesis, a pre-adapted epigean species may enter the below-ground habitats to explore the unexplored resources (Belles 1992; Leys et al. 2003; Juan and Emerson 2010; Ribera et al. 2010). Likewise, the origin of hypogean fungi has always been a subject of interest for mycologists. The origin of cave fungi has been inferred by making a comparative analysis of the divergence time of the suspected troglobitic fungi with the geologic age of the cave development by Zhang et al. (2018a, b) who have put an effort to study the evolution and origin of cave fungi by molecular clock method calibrated by fossil records. The obtained result revealed an interesting observation where the divergence time of 20 suspected troglobitic fungi was found to lie between late Miocene and late Jurassic, but the caves developed much later in the then middle Pliocene when viewed from the perspectives based on the historical geological movement. Therefore, the new species of fungi under study may not be true troglobites (Zhang et al. 2018a, b). Similar kind of studies can be carried out with the new fungus discovered in caves from other parts of the world in order to have a clear understanding on the origin of troglobitic fungi.

The fungal community structure is vulnerable to changes. The movement of animals into caves from the outside environment is thought to be one factor that affects cave mycobiota. Arthropods are associated with the dispersal of fungal spores into the cave environment. Adult females of the cave orb weaver spider, *Meta ovalis*, have entomopathogenic fungi on their surfaces and have the potential to disperse fungal spores (conidia) throughout the cave environment (Yoder et al. 2009). The microclimate of a cave is also an important factor in determining the mycobiota that prevails inside caves. Visitors and tourists leading to anthropogenic contamination can lead to changes in the microclimatic conditions of a cave (Hoyos et al. 1998). This can be in the form of increase in air temperatures and CO_2 concentration resulting from visits of tourists/visitors. The visitors can also lead to the organic matter input and new microorganisms (Ogorek et al. 2014). Anthropogenic influences on the fungal diversity have been studied by Shapiro and Prinjle (2010) in caves in Kentucky and Tennessee. Undisturbed habitats support higher fungal diversity which decreases in moderately visited sites, and the lowest diversity is observed in heavily human trafficked sites (Shapiro and Prinjle 2010).

Fungal diversity in areas of Naracoorte cave was studied using culture-dependent and culture-independent methods, and significant variations in fungal distribution as well as diversity were observed in areas accessed by tourists and areas which were inaccessible to tourists (Adetutu et al. 2011). Analysis of fungal diversity using molecular approaches revealed higher diversity in tourist-accessible areas of the caves when compared to culturable methods with regard to Stick-Tomato and Alexandra caves which were compared to inaccessible areas and the Strawhaven Cave considered as control. However, the differing conclusions derived using different methods can be attributed to the biases associated with each method in terms of phylogenetic analysis (Adetutu et al. 2011). The environment and the vegetation around the cave also influence the air-borne fungal diversity in caves. Air currents are an important mode for spreading the inocula, but also depend on seasonal air circulation (Ogorek et al. 2014).

More air-borne fungal diversity is reported from air outside cave than its interior which may be due to the internal microclimate of the cave as reported in the case study of Niedzwiedzia Cave (Ogorek et al. 2014) and Demanovska Ice Cave (Ogorek et al. 2017).

Fungi in hypogean environments are an emerging threat to the cave organism and also to the cultural heritage in some show caves. The appearance of white-nose syndrome disease that hit the bat population in the USA has been a devastating one (Zhang et al. 2014). The white-nose syndrome is described after the snowy-white fungus, *Geomyces destructans,* which grows on the muzzle and even on the wings of afflicted hibernating bats. Animals such as flies, ants, spiders, and moths have been reported to be the hosts to fungi like *Verticillium lamellicola* (Greif and Currah 2007), and the presence of such fungus in the hypogean environments like caves is related to the occurrence of arthropod population. *Histoplasma capsulatum* is a pathogenic species commonly found in caves and associated with bat and bird guano and was isolated from bats living in the Aguas Buenas Caves, Puerto Rico (Zamora 1977).

Fungi are also known for their biodeteriorative properties in other subterranean habitats like historical caves and catacombs. Several fungi have been associated with the deterioration of Paleolithic paintings affecting the integrity of rock arts and mural paintings. One of the priceless heritages and cultures which have been of great interest to mankind has been the conservation of Paleolithic paintings in caves

(Bastian et al. 2010). Species of *Fusarium solani* (Dupont et al. 2007), *Ochroconis lascauxensis* and *O. anomala* (De la Rosa et al. 2017), *Acremonium nepalense* (Saiz-Jimenez 2012), and *Penicillium glandicola* (Ogorek et al. 2015) have been attached to the occurrence of black stains on the walls of the Lascaux Cave in France and Driny Caves in Slovakia.

Fungi are well-known geologic agents, and they were first considered to be the agents of carbonate deterioration by Krumbein (1968). Production of various organic acids such as acetic, oxalic, citric, formic, gluconic, and tartaric acid has been attributed as the major cause for the dissolution of rocks (Braams 1992). The hypha of *Verticillium lamellicola* suspended from the stalactite wall in Lehman Caves, USA, was found to show the crystallization of calcite which occurred in the terminal drop as reported by Went (1969). Barton and Northup (2007) in their study from Hidden Cave, USA, reported a similar observation in a stalactite formation. Fungal communities in the speleothemic surfaces from Kartchner Caverns, USA, have been reported by Vaugan et al. (2011) who reported *Penicillium*, *Paecilomyces*, *Phialophora,* and *Aspergillus* as the predominantly occurring fungi from the studied caverns. In spite of such studies, the fungal role in biospeleogenesis remains poorly understood. Further, secondary mineral deposit in caves by prokaryotes has been extensively studied in comparison with fungi. Hence, it is essential to study the role of fungus in such environment as these members are capable of producing wide range of metabolites which could be harnessed for bioprospecting purpose.

22.3.1 Extremophilic Fungi and Advancing Frontiers

Species colonizing unique niches experiencing extreme environmental conditions have attracted the attention of researchers as they invariably provide the best resources for novel metabolites and compounds (Jensen and Fenical 1996). Extreme conditions play a major role in the physiological activity of organisms thriving in extreme habitats and offer an environment favorable for the production of molecules of biological interest. Fungi offer huge metabolic potential, and different fungi have been the resource base of bioactive metabolites, majority of which are regarded as fundamental prototype molecules explored for development and synthesis of new drugs (Goncalves et al. 2016). Extreme habitats are possible niches for search of novel bioactive compounds which open vistas and newer paradigm for bioprospecting research.

Fungal enzymes and secondary metabolites have already grabbed particular attention for biotechnology and pharmaceutical purposes. A vast spectrum of enzymes produced by fungi are known to have huge industrial applications. The cold active enzymes and their application from psychrophilic and psychrotolerant fungi have been reviewed by Hassan et al. (2016). The different enzymes produced by psychrophilic fungi are α-amylases, cellulases, glucose oxidases, lipases, phosphatases, proteases, and xylanases which have wider applicability in detergent and textile, food and alcohol, and paper and pulp industries (Hassan et al. 2016). Deep-sea sediments inhabiting fungi have been explored for the production of proteases

(Damare et al. 2006). The yeast cells from supraglacial sediments were found to produce significant titers of extracellular starch degrading, lipolytic, esterolytic, and pectinolytic enzymes (Turchetti et al. 2008). Thermophilic fungi potentially produce large number of thermostable enzymes like β-glucosidases, cellobiohydrolases, and endo-1,4-β-glucanases (Bayer et al. 1998; Baldrian and Valaskova 2008), which find a wide use in industry for the conversion of biomass into sugars. Furthermore, thermophilic fungi producing biomass-degrading enzymes consistently have superior hydrolytic capacity (Wojtczak et al. 1987). Some of the other thermostable enzymes from thermophilic fungi are laccase and phytase. Laccases can degrade both phenolic and non-phenolic compounds and find use in the remediation of xenobiotics (Upadhyay et al. 2016), while phytases find use in animal feeds, since they increase the availability of phosphorus in the feeds (Maheshwari et al. 2000). Fungi from marine habitats are also great reservoirs of industrially important enzymes. They have been isolated from different substrates like the decaying mangrove woods and mangrove leaf detritus (Raghukumar et al. 2004; D-Souza-Ticlo et al. 2009), marine soil and sediments (Huang et al. 2004; Chi et al. 2007; Elyas et al. 2010), and sponges (Bonugli-Santos et al. 2010), which are the sources of different hydrolytic and oxidative enzymes like chitinase, keratinase, inulinase, ligninase, and xylanase (Bonugli-Santos et al. 2015), having widespread usage in food beverage, textile, cosmetics, and drug manufacturing industries. High salt tolerance, thermostability, cold activity, and barophilicity are novel features observed in enzymes produced by marine fungi (Velmurugan and Lee 2012).

The search for natural compounds has taken an upturn after an era dominated by chemically synthesized drugs. Fungi are also the prolific producers of natural compounds with bioactivities and thus have huge pharmaceutical potential. In fact, natural products such as drugs have been mainly produced by microbes. Various active metabolites and compounds used as antimicrobials, antivirals, and cytotoxic and immunosuppressive drugs comprising about 23,000 active compounds from microorganisms constitute about 42% derived from fungi and about 32 % from actinobacteria (Demain 2014). Furthermore, the adaptational features and evolutionary changes encountered by extremophilic fungi have enabled them to acquire specialized defense strategies to survive extremes of pressure, temperature, salinity, desiccation, and pH, which have been the result of their capacity to biosynthesize new natural products and metabolites with diverse biological activities (Zhang et al. 2018a, b). Striking examples of various antibacterial agents are the penicillins produced by *Penicillium* species; cephalosporins produced by *Cephalosporium acremonium*; aminoglycosides, tetracyclines and other polyketides produced by *Actinobacteria*; immunosuppressive drugs such as the cyclosporins from *Tolypocladium* species and rapamycin from *Streptomyces* species; cholesterol-lowering agents like mevastatin from *Penicillium* spp. and lovastatin from *Aspergillus* spp.; and anthelmintics and antiparasitic drugs such as the ivermectins from *Streptomyces* spp. (Cragg and Newman 2013). Recently, Zhang et al. (2018a, b) reviewed 314 novel natural compounds from extremophilic fungi which belonged to various classes of compounds such as terpenoids/steroids, alkaloids/peptides/amides, quinones/phenols, esters/lactones, xanthones, polyketides, and other compounds, majority being terpenoids and alkaloids. All these compounds are produced by fungal strains especially the asexual stages of ascomycetes, e.g., *Penicillium* spp., *Aspergillus* spp., etc. Moreover, these compounds are found to be antibacterial, antifungal, anti-inflammatory, cytotoxic, and antiviral. Some compounds are found to be radical scavenging inhibitors of PTP1B, BRD4, MMP-3, and NF-κB and activators of Nrf2 (Zhang et al. 2018a, b).

Other studies on the fungal extrolites include those of Gomes et al. (2018) who studied fungal diversity in the Antarctic soil from their bioprospection perspective. The crude extracts of fungi like *Pseudogymnoascus destructans*, *Mortierella parvispora, M. amoeboidea, Penicillium chrysogenum,* and *P. tardochrysogenum* grown in solid state fermentation showed antiparasitic and herbicidal activities. In addition, *Pseudogymnoascus destructans*, phylogenetically close to the strains that infect bats, exhibited trypanocidal activities and may be a source of bioactive metabolites. Similarly, *P. destructans* responsible for the white-nose syndrome prevalent in cave bats could be screened for such molecules which could be considered for chemotherapeutic purpose. In another study by Godinho et al. (2015), the extracts of fungal isolates (*Aspergillus sydowii*, *Penicillium allii-sativi*, *P. brevicompactum*, *P. chrysogenum,* and *P. rubens)* from the oligotrophic soils of Antarctica showed antiviral, antibacterial, antifungal, antitumor, herbicidal, and antiprotozoal activities.

Natural pigments are also gaining interest owing to their use in cosmetics, textiles, food industry, and other important biological activities. Souza et al. (2016) studied pigment production from different strains of cave fungi obtained from the collection at the Laboratory of Bioprospecting and Genetics, Brazil. The fungi produced various kinds of pigments, and the production was influenced by the media composition wherein complex media favored more pigment production. Some of the pigment compounds identified by the authors were oosporein, orevactaene, and dihydrotrichodimerol which have been reported to be phytotoxic (Cole et al. 1974), mycotoxic (Kogl & van Wessem 1944), antitumor (Lee et al. 2005; Liu et al. 2005; Mao et al. 2010), and antiviral (Shu et al. 1997), besides their use as food colorants (Mapari et al. 2010).

The hypogean environment which constitutes an important habitat for fungi is a storehouse of fungal biodiversity, boasting a large number of biotechnologically important, rare, and novel species of fungi (Fig. 22.1-unpublished data). The hypogean habitat exhibits a wide range of environmental variations that is instrumental in creating strain variations ranging from the oligotrophic to the extreme inhospitable aphotic type which consequently leads to genetic variations. The habitat, however, shows predominantly extreme conditions, and the members of the hypogean fungi have adapted themselves to environment where the nutritional factors remain low beyond which the normal fungi cannot tolerate. Some of the major adaptive features exhibited by the members of this group are usually found in environments where humidity remains much higher and the temperature mostly remains optimal for maximal growth.

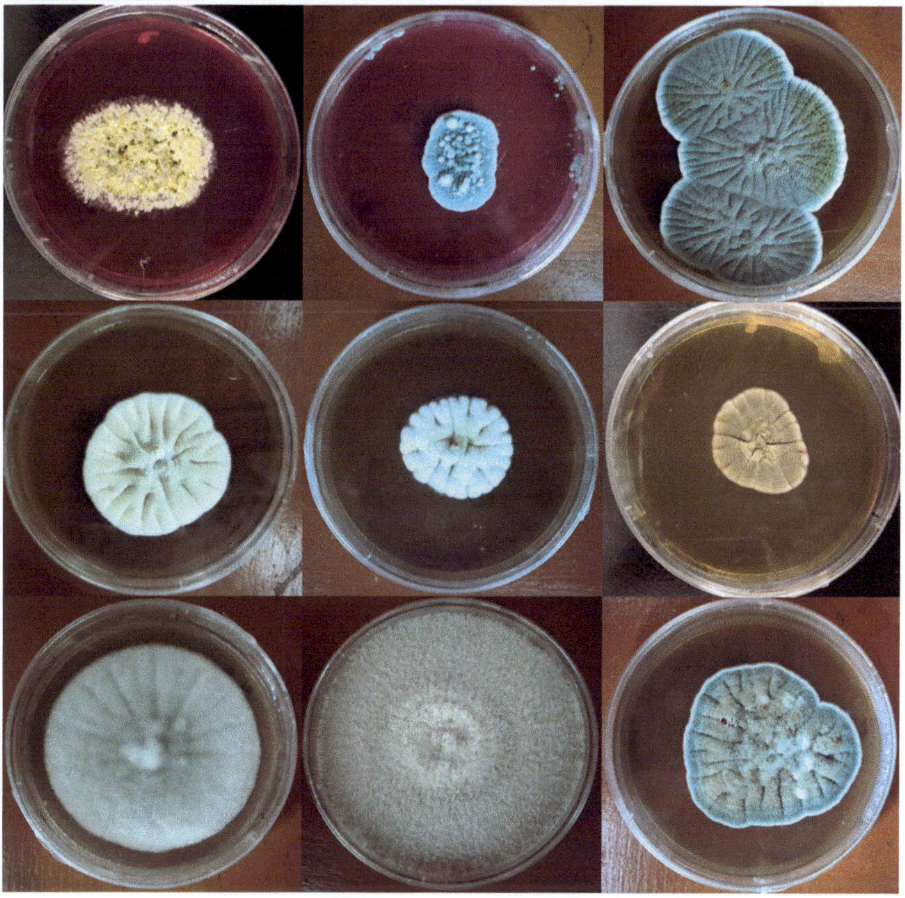

Fig. 22.1 Some representative isolates of fungi from the caves of Meghalaya, India. (Unpublished data)

Studies have shown that the extreme environmental conditions play a vital role in the physiological activity of the members belonging to fungi which have evolved highly specific adaptive features that permit the growth under stress conditions. The harsh oligotrophic conditions coupled with no light penetration may account for the evolution chiefly with respect to the ability to produce novel metabolites of profound biological significance. The majority of the members of the fungi belonging to hypogean region face environmental and nutritional challenges deployed by the habitat which are considered to be the most important sources of untapped secondary metabolites. However, studies have shown that there has been scanty work done on hypogean fungal communities, which is one of the less frequently explored members of the kingdom fungi. Furthermore, members of different groups are in existence in this habitat, and these groups can be readily recognized if they are cultured in growth media and allowed to grow to harness the fungal pool, which are promising and potential source for the discovery of economically profitable novel pharmaceutical substance. So, strategies for the studies for a variety of reasons, including the preparation of inventory, taxonomic diversities across habitat, and exploitation of the selected species with high potentiality for commercialization of the novel chemicals with bioactive properties, have to be carried out. At a time when rampant uses of antimicrobial agents are ravaging the normal microflora, understanding the indispensable worth of the hypogean fungi is urgent. This is particularly important and it has to be done at the earliest to screen drugs which could replace the already resistant groups of antimicrobial agents. The novel chemical substance produced by the members of these fungi could be used as a reserve antimicrobial agent in the near future.

In addition, many of the fungi show the presence of diverse class of enzymes which could have evolved to overcome the defense mechanism exhibited by the insect host along with the biosynthesis of certain enzymes that make up the central part of the cellular processes required for growth. Further, it becomes essential that these enzymes must be characterized as the presence of these bioactive molecules may be essential for their survival in such a hostile environment. However, with gradual advancement in techniques related to microbiology and with the understanding of various biological and physiological differences exhibited by these fungi, the studies in relation to functional biomolecules should be done without much difficulty (Table 22.1).

22.4 Conclusions

Majority of the members of the hypogean fungi are recognized with high medicinal value, which are being chiefly studied for metabolites or products for their biological activities. Members belonging to this unique habitat belong to a mixed group of genera representing a number of taxonomic sections of the kingdom fungi. A large number of bioactive metabolites along with economically important enzymes have been reported, yet only a certain percentage of these fungi have been accessed as a source of novel therapeutically important substances with lesser number of species tested for biotechnological significance. Moreover, as the hypogean environment is a challenging one and hostile toward the survival and growth of organisms occupying this habitat, it may serve as a strong selective force for eliciting the novel biosynthetic pathways. With more exploration of microbes especially the ones belonging to fungal domain, newer molecules of human significance can be expected to be available from extreme habitats. This opens up new vistas for deeper exploration and bioprospection of fungi from hypogean environment which could be harnessed for untapped source of bio-metabolites awaiting the discovery for use in human welfare.

Table 22.1 Utility metabolites/products recovered from extremophilic fungi

Habitat	Fungi	Compounds/products recovered	Utility	Reference
Antarctica (oligotrophic soil)	*Aspergillus, Debaryomyces, Cladosporium, Pseudogymnoascus, Penicillium, Hypocreales*	Aromatic compounds, fatty acids, triglyceride	Antibacterial, antifungal, Antitumoral, antiprotozoal, herbicidal	Godinho et al. (2015)
Atacama Desert (rock-associated fungus)	*Cladosporium halotolerans, Penicillium chrysogenum, P. citrinum*	α-Linolenic acid, ergosterol, endoperoxide	Antibacterial, antimycobacterial, antimalarial	Goncalves et al. (2016)
Antarctica (soil)	*Pseudogymnoascus destructans, Mortierella parvispora, M. amoeboidea, Penicillium chrysogenum, P. tardochrysogenum*	Not identified	Antiparasitic, herbicidal	Gomes et al. (2018)
Brazilian caves	*Aspergillus syndowii, A. aureolatus, A. keveii, Penicillium flavigenum, P. chermesinum, Epicoccum nigrum, Lecanicillium aphanocladii, Fusarium* sp.	Oosporein, orevactaene, dihydrotrichodimerol	Food colorants, antitumor, mycotoxic, phytotoxic	Souza et al. (2016)
Sonoran Desert (endophytes)	*Chaetomium globosum*	Globosumones A-C	Moderately anticancerous	Bashyal et al. (2005)
Deep-sea sediment	*Aspergillus westerdijkiae*	Circumdatin K&L, 5-chlorosclerotiamide, 10-epi-sclerotiamide, aspergillamide B, circumdatin F	Non-cytotoxic; other activities unreported	Peng et al. (2013)
Hot spring	*Malbranchea sulfurea*	Malbranpyrroles A-F	Pharmacophore	Yang et al. (2009)
Antarctic deep-sea sediments	*Penicillium crustosum* PRB-2	Penilactones A and B	Activities unreported	Wu et al. (2012)
Antarctic soil	*Aspergillus ochraceopetaliformis*	Merosesquiterpenoids	Antiviral	Wang et al. (2016a)
Deep-sea sediment	*Aspergillus versicolor*	Alkaloids and xanthones	Antimicrobial	Wang et al. (2016b)

(continued)

Table 22.1 (continued)

Habitat	Fungi	Compounds/products recovered	Utility	Reference
Deep ocean sediment	*Penicillium* sp. *P. granulatum*	Diketopiperazine alkaloids, spiro-tetracyclic diterpene	Anti-inflammatory	Du et al. (2009), Niu et al. (2017)
Deep-sea sediment	*Penicillium commune*	Oxindole alkaloids	Insecticidal	Xu et al. (2015)
Antarctic moss-derived fungus	*Penicillium funiculosum*	Pyridone alkaloids	Anti-hypercholesterol	Zhou et al. (2015)
Salt field sediment	*Penicillium citrinum*	Citrinin derivatives	Radical scavenging	Lu et al. (2008)
Hot spring	*Talaromyces thermophilus*	Macrolides	Nematocidal	Guo et al. (2012)
Alpine glaciers	*Aureobasidium pullulans*, *Cryptococcus gilvescens*, *C. terricolus*, *Rhodotorula glacialis*, *R. phenolica*	Industrial enzymes	Food, textile, chemical industries	Turchetti et al. (2008)
Marine soil, sediments, algae, sponges	*Aureobasidium pullulans*, *Aspergillus* sp., *Mucor racemosus*	Industrial enzymes	Food, paper, textile, chemical, Agricultural, cosmetic industries	Chi et al. (2007), Elyas et al. (2010), Mohapatra et al. (1998), and Bonugli-santos et al. (2010)

References

Adetutu EM, Thorpe K, Bourne S, Cao X, Shahsavari E, Kirby G, Ball AS (2011) Phylogenetic diversity of fungal communities in areas accessible and not accessible to tourists in Naracoorte Caves. Mycologia 103(5):959–968

Agrawal OP, Dhawan S, Garg KL, Shaheen F, Pathak N, Misra A (1988) Study of biodeterioration of the Ajanta wall paintings. Int Biodeterior 24:121–129

Baldrian P, Valaskova V (2008) Degradation of cellulose by basidiomycetous fungi. FEMS Microbiol Rev 32:501–521

Banerjee S, Joshi SR (2013) Insights into cave architecture and the role of bacterial biofilm. Proc Natl Acad Sci India, Sect B Biol Sci 83(3):277–290

Banerjee S, Joshi SR (2014) Ultrastructural analysis of calcite crystal patterns formed by biofilm bacteria associated with cave speleothems. J Microsc Ultrastruc 2(4):217–223

Barton HA (2006) Introduction to cave microbiology: a review for the non-specialists. J Cave Karst Stud 68:43–64

Barton HA, Jurado V (2007) What's up down there: microbial diversity in starved cave environments. Microbe 2:132–138
Barton HA & Northup DE (2007). Geomicrobiology in cave environments: Past, current, and future perspectives. J Cave Karst Stu 69:163–178.
Barton HA, Luiszer F (2005) Microbial metabolic structure in a sulfidic cave hot spring: potential mechanisms of Biospeleogenesis. J Cave Karst Stud 67:28–38
Barton HA, Taylor MR, Pace NR (2004) Molecular phylogenetic analysis of a bacterial community in an Oligotrophic cave environment. Geomicrobiol J 21:11–20
Barton HA, Taylor NM, Kreate MP, Springer AC, Oehrle SA, Bertog JL (2007) The impact of host rock geochemistry on bacterial community structure in oligotrophic cave environments. Int J Speleol 36:93–104
Bashyal BP, Wijeratne EMK, Faeth SH, Gunatilaka AAL (2005) Globosumones A-C, cytotoxic orsellinic acid esters from the Sonoran Desert endophytic fungus *Chaetomiumglobosum*. J Nat Prod 68:724–728
Baskar S, Baskar R, Mauclaire L, McKenzie JA (2006) Microbially induced calcite precipitation in culture experiments: possible origin for stalactites in Sahastradhara caves, Dehradun, India. Curr Sci 90:58–64
Baskar S, Baskar R, Lee N, Theophilus PK (2009) Speleothems from Mawsmai and Krem Phyllut caves, Meghalaya, India: some evidences on biogenic activities. Environ Geol 57:1169–1186
Bastian F, Jurado V, Novakova A, Alabouvette C, Saiz-Jimenez C (2010) The microbiology of *Lascaux cave*. Microbiology 156:644–652
Bayer EA, Morag E, Lamed R, Yaron S, Shoham Y (1998) Cellulosome structure: four-pronged attack using biochemistry, molecular biology, crystallography and bioinformatics. In: Claeyssens M, Nerinckx W, Piens K (eds) Carbohydrases from *Trichoderma reesei* and other microorganisms. Royal Society of Chemistry, London, pp 39–65
Belles X (1992) Survival, opportunism and convenience in the processes of cave colonization by terrestrial fauna. Oecol Aquat 10:325–335
Bonugli-Santos RC, Durrant LR, Sette LD (2010) Laccase activity and putative laccase genes in marine-derived basidiomycetes. Fungal Biol 114:863–872
Bonugli-Santos RC, dos Santos Vasconcelos MR, Passarini MR et al (2015) Marine-derived fungi: diversity of enzymes and biotechnological applications. Front Microbiol 6:1–15
Braams J (1992) Ecological studies of the fungal microflora inhabiting historical sandstone monuments. PhD thesis, University of Oldenburg, Germany, 104 p
Butinar L, Sonjak S, Zalar P, Plemenitas A, Gunde-Cimerman N (2005) Melanized halophilic fungi are eukaryotic members of microbial communities in hypersaline waters of solar salterns. Bot Mar 48:73–79
Calaforra JM, Forti P, Fernandez-Cortes A (2007) Speleothems in gypsum caves and their paleoclimatological significance. Environ Geol 53:1099–1105
Castanier S, Le Metayer-Levrel G, Perthuisot JP (2000) Bacterial roles in the precipitation of carbonate minerals. In: Riding RE, Awramik SM (eds) Microbial sediments. Springer, Heidelberg, pp 32–39
Caumartin V (1963) Review of the microbiology of underground environments. Bull Nat Speleol Soc 25:1–14
Chelius MK, Beresford G, Horton H, Quirk M, Selby G, Simpson RT, Horrocks R, Moore JC (2009) Impacts of alterations of organic inputs on the bacterial community within the sediments of Wind Cave, South Dakota, USA. Int J Speleol 38:1–10
Chi Z, Ma C, Wang P, Li HF (2007) Optimization of medium and cultivation conditions for alkaline protease production by the marine yeast *Aureobasidium pullulans*. Bioresour Technol 98:534–538
Cole RJ, Kirksey JW, Cutler HG, Davis EE (1974) Toxic effects of oosporein from *Chaetomium trilaterale*. J Agric Food Chem 22:517–520
Cragg GM, Newman DJ (2013) Natural products: a continuing source of novel drug leads. Biochim Biophys Acta 1830:3670–3695

Cunningham KI, Northup DE, Pollastro RM, Wright WG, LaRock EJ (1995) Bacteria, fungi and biokarst in Lechuguilla cave, Carlsbad Caverns National Park. New Mexico Environ Geol 25:2–8

D'Souza-Ticlo D, Sharma D, Raghukumar C (2009) A thermostable metal- tolerant laccase with bioremediation potential from a marine-derived fungus. Mar Biotechnol 11:725–737

da Costa Souza PN, Grigoletto TLB, Moraes LAB, Abreu LM, Guimaraes LHS, Santos C, Galvao LR, Cardoso G (2016) Production and chemical characterization of pigments in filamentous fungi. Microbiology 162:12–22

Damare S, Raghukumar C, Raghukumar S (2006) Fungi in deep-sea sediments of the Central Indian Basin. Deep Sea Res 53:14–27

Danielli HMC, Edington MA (1983) Bacterial calcification in limestone caves. Geomicrobiol J 3:1–16

De la Rosa JM, Martin-Sanchez PM, Sanchez-Cortes S, Hermosin S, Knicker H, Saiz-Jimenez C (2017) Structure of melanins from the fungi *Ochroconis lascauxensis* and *Ochroconis anomala* contaminating rock art in the Lascaux Cave. Sci Rep 7:13441

Demain AL (2014) Valuable secondary metabolites from fungi. In: Martin JF, Garcia-Estrada C, Zeilinger S (eds) Biosynthesis and molecular genetics of fungal secondary metabolites. Springer, New York, pp 1–15

Dickson GW, Kirk PW Jr (1976) Distribution of heterotrophic microorganisms in relation to detritivores in Virginia caves (with supplemental bibliography on cave mycology and microbiology). In: Parker BC, Roane MK (eds) The distributional history of the biota of the Southern Appalachians. Part IV Algae and fungi. University Press of Virginia, Charlotteville, pp 205–226

Du L, Yang X, Zhu T, Wang F, Xiao X, Park H, Gu Q (2009) Diketopiperazine alkaloids from a deep ocean sediment derived fungus *Penicillium* sp. Chem Pharm Bull 57:873–876

Dupont J, Jacquet C, Dennetiere B, Lacoste S, Bousta F, Orial G, Cruaud C, Couloux A et al (2007) Invasion of the French Paleolithic painted cave of Lascaux by members of the *Fusarium solani* species complex. Mycologia 99:526–533

Dyson HJ, James JM (1973) A preliminary study of cave bacteria. J Sydney Speleol Soc 17:221–230

Elyas KK, Mathew A, Sukumaran RK, Ali PPM, Sapna K, Kumar SR et al (2010) Production optimization and properties of beta glucosidases from a marine fungus *Aspergillus*-SA58. N Biotechn 27:347–351

Engel AS, Porter ML, Kinkle BK, Kane TC (2001) Ecological assessment and geological significance of microbial communities from cesspool cave, Virginia. Geomicrobiol J 18(3):259–274

Engel AS, Stern LA, Bennett PC (2004) Microbial contributions to cave formation: new insights into sulfuric acid speleogenesis. Geology 32:369–372

Forti P (2001) Biogenic speleothems: an overview. Int J Speleol 30:39–56

Forti P, Galdenzi S, Sarbu SM (2002) The hypogenic caves: a powerful tool for the study of seeps and their environmental effects. Cont Shelf Res 22:2373–2386

Gascoine W (1982) The formation of black deposits in some caves of South East Wales. Cave Sci 9:165–175

Gibert J, Deharveng L (2002) Subterranean ecosystems: a truncated functional Biodiversity. BioScience 52:473–481

Gillieson D (1996) Caves, processes, development, and management. Blackwell, Oxford, p 324

Godinho VM, Gonçalves VN, Santiago IF, Figueredo HM, Vitoreli GA, Schaefer CEGR et al (2015) Diversity and bioprospection of fungal community present in oligotrophic soil of continental Antarctica. Extremophiles 19:585–596

Gomes ECQ, Godinho VM, Silva DAS, De Paula MD, Vitoreli GA, Zani CL et al (2018) Cultivable fungi present in Antarctic soils: taxonomy, phylogeny, diversity, and bioprospecting of antiparasitic and herbicidal metabolites. Extremophiles 22:381–393

Goncalves VN, Cantrell CL, Wedge DE, Ferreira MC, Soares MA, Jacob MR et al (2016) Fungi associated with rocks of the Atacama Desert: taxonomy, distribution, diversity, ecology and bioprospection for bioactive compounds. Environ Microbiol 18(1):232–245

Gorbushina AA (2007) Life on the rocks. Environ Microbiol 9:1613–1631

Greif MD, Currah RS (2007) Patterns in the occurrence of saprophytic fungi carried by arthropods caught in traps baited with rotted wood and dung. Mycologia 99:7–19

Griffin DW, Gray MA, Lyles MB, Northup DE (2014) The transport of nonindigenous microorganisms into caves by human visitation: a case study at Carlsbad Caverns National Park. Geomicrobiol J 31:175–185

Groth I, Saiz-Jimenez C (1999) Actinomycetes in Hypogean environments. Geomicrobiol J 16:1–8

Gunde-Cimerman N, Zalar P, Jeram S (1998) Mycoflora of cave cricket *Troglophilus neglectus* cadavers. Mycopathologia 141:111–114

Gunde-Cimerman N, Sonjak S, Zalar P, Frisvad JC, Diderichsen B, Plemenitas A (2003) Extremophilic fungi in Arctic ice: a relationship between adaptation to low temperature and water activity. Phys Chem Earth 28:1273–1278

Guo JP, Zhu CY, Zhang CP, Chu YS, Wang YL, Zhang JX, Wu DK, Zhang KQ, Niu XM (2012) Thermolides, potent nematocidal pks-nrps hybrid metabolites from thermophilic fungus *Talaromyces thermophilus*. J Am Chem Soc 134:20306–20309

Hassan N, Rafiq M, Hayat M, Shah AA, Hasan F (2016) Psychrophilic and psychrotrophic fungi: a comprehensive review. Rev Environ Sci Biotechnol 15:147–172

Hill CA, Forti P (1997) Cave minerals of the world, 2nd edn. Huntsville, Alabama, p 463

Hoeg O (1946) Cyanophyceae and bacteria in calcareous sediments in the interior of limestone caves in Nord-Rana. Norway Nytt Mag Naturvidensk 85:99–104

Howarth FG, Stone FD (1990) Elevated carbon dioxide levels in Bayliss Cave, Australia: implications for the evolution of Obligate Cave species. Pac Sci 44:207–218

Hoyos M, Soler V, Canaveras JC, Sanchez-Moral S, Sanz-Rubio E (1998) Microclimatic characterization of a karstic cave: human impact on microenvironmental parameters of a prehistoric rock art cave (Candamo Cave, northern Spain). Environ Geol 33:231–242

Huang YL, Locy R, Weete JD (2004) Purification and characterization of an extracellular lipase from *Geotrichum marinum*. Lipids 39:251–257

Hubbard D, Herman J, Bell P (1986) The role of sulfide oxidation in the genesis of Cesspool Cave, Virginia, USA. Proc Int Congr Speleol Barcelona Spain 9:255–257

James JM (1994) Microorganisms in Australian caves and their influence on speleogenesis. In: Sasowsky ID, Palmer MV (eds) Breakthroughs in karst geomicrobiology and redox geochemistry. Karst Waters Institute, Colorado, pp 31–34

Jensen PR, Fenical W (1996) Marine bacterial diversity as a source for novel microbial products. J Ind Microbiol 17:346–351

Jiang JR, Cai L, Liu F (2017) Oligotrophic fungi from a carbonate cave, with three new species of *Cephalotrichum*. Mycology 8:164–177

Jones B (1992) Manganese precipitates in the karst terrain of grand Cayman, British West Indies. Can J Earth Sci. 29:1125–1139

Jones B (2001) Microbial activity in caves – a geological perspective. Geomicrobiol J 18:345–357

Jones B, Motyka A (1987) Biogenic structures and micrite in stalactites from grand Cayman Island, British West Indies. Can J Earth Sci 24:1402–1411

Juan C, Emerson BC (2010) Evolution underground: shedding light on the diversification of subterranean insects. J Biol 9:17

Kambesis P (2007) The importance of cave exploration to scientific research. J Cave Karst Stud 69:46–58

Kinkle B, Kane TC (2000) Chemolithoautotrophic microorganisms and their potential role in subsurface environments. In: Wilkens H, Culver DC, Humpreys WF (eds) Ecosystems of the world 30. Elsevier, Amsterdam, pp 309–318

Kogl F, van Wessem GC (1944) Analysis concerning pigments of fungi XIV. Concerning oosporein, the pigment of *Oospora colorans* van Beyma. Recl Trav Chim Pays Bas Belg 63:5–24

Konhauser KO (1997) Bacterial iron mineralization in nature. FEMS Microbiol Rev 20(315):326

Konhauser KO (1998) Diversity of bacterial iron mineralization. Earth Sci Rev 43:91–121

Krumbein WE (1968) Über den Einfluß der Mikroflora auf die exogene Dynamik (Verwitterung und Krustenbildung). Geol Rundsch 58(2):333–363

Kumaresan D, Wischer D, Stephenson J, Hillebrand-Voiculescu A, Murrell JC (2014) Microbiology of Movile Cave-A Chemolithoautotrophic ecosystem. Geomicrobiol J 31:186–193

Lee D, Lee JH, Cai XF, Shin JC, Lee K, Hong YS, Lee JJ (2005) Fungal metabolites, sorbicillinoid polyketides and their effects on the activation of peroxisome proliferator-activated receptor c. J Antibiot (Tokyo) 58:615–620

Leys BR, Watts CH, Cooper SJ, Humphreys WF (2003) Evolution of subterranean diving beetles (Coleoptera: dytiscidae: hydroporini, Bidessini) in the arid zone of Australia. Evolution 57:2819–2834

Liu W, Gu Q, Zhu W, Cui C, Fan G (2005) Dihydrotrichodimerol and tetrahydrotrichodimerol, two new bisorbicillinoids, from a marine derived *Penicillium terrestre*. J Antibiot (Tokyo) 58:621–624

Lu ZY, Lin ZJ, Wang WL, Du L, Zhu TJ, Fang YC, Gu QQ, Zhu WM (2008) Citrinin dimers from the halotolerant fungus *Penicillium citrinum* B-57. J Nat Prod 71:543–546

Maheshwari R, Bharadwaj G, Bhat MK (2000) Thermophilic Fungi: their physiology and enzymes. Microbiol Mol Biol Rev 64:461–488

Malard F, Gibert J (1997) Stygobitic fauna of the north Montpellierains karsts with special emphasis on the Lez karst system. From first annual top ten list of endangered karst ecosystems. Karst Waters Institute, Charles Town

Man B, Wang H, Xiang X, Wang R, Yun Y, Gong L (2015) Phylogenetic diversity of culturable fungi in the Heshang Cave, Central China. Front Microbiol 6:1158

Man B, Wang H, Yun Y, Xiang X, Wang R, Duan Y, Cheng X (2018) Diversity of fungal communities in Heshang Cave of Central China revealed by mycobiome-sequencing. Front Microbiol 9:1400

Mao BZ, Huang C, Yang GM, Chen YZ, Chen SY (2010) Separation and determination of bioactivity of oosporein from *Chaetomium cupreum*. Afr J Biotechnol 9:5955–5961

Mapari SAS, Thrane U, Meyer AS (2010) Fungal polyketide azaphilone pigments as future natural food colorants? Trends Biotechnol 28:300–307

Martinez A, Asencio AD (2009) Distribution of cyanobacteria at the Gelada Cave (Spain) by physical parameters. J Cave Karst Stud 72:11–20

Min KH (1988) Fungus flora of Seongrya cave in Korea. Trans Mycol Soc Jpn 29:479–487

Mohamed R, Jong PL, Kamziah AK (2014) Fungal inoculation induces agarwood in young *Aquilaria malaccensis* trees in the nursery. J For Res 25(1):201–204

Mohapatra BR, Banerjee UC, Bapuji M (1998) Characterization of a fungal amylase from *Mucor* sp. associated with the marine sponge *Spirastrella* sp. J Biotechnol 60(1–2):113–117

Moore GW (1981) Manganese deposition in limestone caves. In: Beck BF (ed) Proceedings of the Eighth International Congress of Speleology. Georgia Southwestern College, Americus, pp 642–645

Mulec J (2008) Microorganisms in hypogen: examples from Slovenian karst caves. Acta Carsologica 37:153–160

Nagahama T, Nagano Y (2012) Cultured and uncultured fungal diversity in deep-sea environments. In: Raghukumar C (ed) Biology of marine fungi. Progress in molecular and subcellular biology. Springer, Berlin, pp 173–187

Nagano Y, Nagahama T (2012) Fungal diversity in deep-sea extreme environments. Fungal Ecol 5:463–471

Nieves-Rivera AM, Santos-Flores CJ, Dugan FM, Miller TE (2009) Guanophilic fungi in three caves of southwestern Puerto Rico. Int J Speleol 38:61–70

Niu S, Fan ZW, Xie CL, Liu Q, Luo ZH, Liu G, Yang XW (2017) Spirograterpene a, a tetracyclic spiro-diterpene with a fused 5/5/5/5 ring system from the deep-sea-derived fungus *Penicillium granulatum* MCCC 3A00475. J Nat Prod 80:2174–2177

Northup DE, Lavoie KH (2001) Geomicrobiology of caves: a review. Geomicrobiology 18:199–220

Northup DE, Reysenbach AL, Pace NR (1997) Microogranisms and speleothems. In: Hill C, Forti P (eds) Cave minerals of the world. Speleological Society, Huntsville, pp 261–266

Novakova A (2009) Microscopic fungi isolated from the Domica Cave system (Slovak Karst National Park, Slovakia). A review. Int J Speleol 38(1):71–82

Novakova A, Hubka V, Saiz-Jimenez C, Kolarik M (2012) *Aspergillus baeticus* sp. nov. and *Aspergillus thesauricus* sp. nov., two species in section Usti from Spanish caves. Int J Syst Evol Microbiol 62:2778–2785

Ogorek R, Lejman A, Matkowski K (2014) Influence of the external environment on airborne fungi isolated from a cave. Pol J Environ Stud 23(2):435–440

Ogorek R, Dylag M, Kozak M, Visnovska Z, Tancinova D, Lejman A (2015) Fungi isolated and quantified from bat guano and air in Harmanecka and Driny Caves (Slovakia). J Cave Karst Stud 78(1):41–49

Ogorek R, Pusz W, Zagozdzon PP, Kozak B, Bujak H (2017) Abundance and diversity of psychrotolerant cultivable mycobiota in winter of a former aluminous shale mine. Geomicrobiol J 34:823–833

Onofri S (1999) Antarctic microfungi. In: Seckbach J (ed) Enigmatic microorganisms and life in extreme environments. Kluwer Academic, Dordrecht, pp 323–336

Palmer AN (1991) Origin and morphology of limestone caves. Geol Soc Am Bull 103:1–21

Peck SB (1986) Bacterial deposition of iron and manganese oxides in North American caves. NSS Bull 48:26–30

Peng J, Zhang XY, Tu ZC, Xu XY, Qi S-H (2013) Alkaloids from the deep-sea-derived fungus *Aspergillus westerdijkiae* DFFSCS013. J Nat Prod 76:983–987

Poulson TL, Lavoie KH (2000) The trophic basis of subsurface ecosystems. In: Wilkins H, Culver DC, Humphreys WF (eds) Subterranean ecosystems, Ecosystems of the World 30. Elsevier Press, Amsterdam, pp 231–249

Poulson TL, White WB (1969) The cave environment. Science 165:971–981

Provencio PP, Polyak VJ (2001) Iron-oxide-rich filaments: possible fossil bacteria in Lechuguilla cave, New Mexico. Geomicrobiol J 18:297–309

Raghukumar C, Muraleedharan U, Gaud VR, Mishra R (2004) Xylanases of marine fungi of potential use for biobleaching of paper pulp. J Ind Microbiol Biotechnol 31:433–441

Ranta HM (1990) Effect of simulated acid rain on quantity of epiphytic microfungi on Scots pine (*Pinus sylvestris* L.) needles. Environ Pollut 67:349–359

Ribera I, Fresneda J, Bucur R, Izquierdo A, Vogler AP, Salgado JM et al (2010) Ancient origin of a Western Mediterranean radiation of subterranean beetles. BMC Evol Biol 10:29

Riding R (2000) Microbial carbonates: the geological record of calcified bacterial–algal mats and biofilms. Sedimentology 47:179–214

Ritz K, Young IM (2004) Interactions between soil structure and fungi. Mycologist 18:52–59

Rivadeneyra MA, Delgado R, Delgado G, Del Moral A, Ferrer MR, Ramos-Cormenza A (1993) Precipitation of carbonate by *Bacillus* sp. isolated from saline soils. Geomicrobiol J 11:175–184

Rogers BW, Williams KM (1982) Mineralogy of Lilburn Cave, Kings Canyon National Park, California. NSS Bull 44:23–31

Romero A (2011) The evolution of cave life: new concepts are challenging conventional ideas about life underground. Am Sci 99:144–151

Rutherford JM, Huang LH (1994) A study of fungi of remote sediments in West Virginia caves and a comparison with reported species in the literature. NSS Bull 56:38–45

Saiz-Jimenez C (2012) Microbiological and environmental issues in show caves. World J Microbiol Biotechnol 28:2453–2464

Sarbu SM, Kinkle BK, Vlasceanu L, Kane TC, Popa R (1994) Microbiological characterization of a sulfide-rich groundwater ecosystem. Geomicrobiol J 12:175–182

Sarbu SM, Kane TC, Kinkle BK (1996) A chemoautotrophically based cave ecosystem. Science 272:1953–1955

Schabereiter-Gurtner C, Saiz-Jimenez C, Pinar G, Lubitz W, Rolleke S (2002) Phylogenetic 16S rRNA analysis reveals the presence of complex and partly unknown bacterial communities in Tito Bustillo cave, Spain, and on its Paleolithic paintings. Environ Microbiol 4:392–400

Selbmann L, Egidi E, Isola D, Onofri S, Zucconi Z, de Hoog GS, Chinaglia S, Testa L, Tosi S, Balestrazzi A, Lantieri A, Compagno R, Tigini V, Varese G (2013) Biodiversity, evolution and adaptation of fungi in extreme environments. Plant Biosyst 147:237–246

Shapiro J, Prinjle A (2010) Anthropogenic influences on the diversity of Fungi isolated from caves in Kentucky and Tennessee. Am Midl Nat 163:76–86
Shearer CA, Descals E, Kohlmeyer B, Kohlmeyer J, Marvanova L, Padgett D et al (2007) Fungal biodiversity in aquatic habitats. Biodivers Conserv 16:49–67
Shiomi N, Yasuda T, Inoue Y, Kusumoto N, Iwasaki S, Katsuda T, Katoh S (2004) Characteristics of neutralization of acids by newly isolated fungal cells. J Biosci Bioeng 97:54–58
Shu YZ, Ye Q, Li H, Kadow KF, Hussain RA, Huang S, Gustavson DR, Lowe SE, Chang LP et al (1997) Orevactaene, a novel binding inhibitor of HIV-1 rev protein to Rev response element (RRE) from *Epicoccum nigrum* WC47880. Bioorg Med Chem Lett 7:2295–2298
Sterflinger K (2000) Fungi as geologic agents. Geomicrobiol J 17:97–124
Sustr V, Elhottova D, Kristufek V, Lukesova A, Novakova A, Tajovsky K, Triska J (2005) Ecophysiology of the cave isopod Mesoniscus graniger (Frivaldszky 1865) (Crustacea: Isopoda). Eur J Soil Biol 41:69–75
Turchetti B, Buzzini P, Goretti M, Branda E, Diolaiuti G, D'Agata C et al (2008) Psychrophilic yeasts in glacial environments of Alpine glaciers. FEMS Microb Ecol 63:73–83
Upadhyay P, Shrivastava R, Agrawal PK (2016) Bioprospecting and biotechnological applications of fungal laccase. 3 Biotech 6:15
Vanderwolf KJ, Malloch D, McAlpine DF, Forbes GJ (2013) A world review of fungi, yeasts, and slime molds in caves. Int J Speleol 42:77–96
Vaugan MJ, Maier RM, Pryor BM (2011) Fungal communities on speleothem surfaces in Kartchner caverns, Arizona, USA. Int J Speleol 40:65–77
Velmurugan N, Lee YS (2012) Enzymes from marine fungi: current research and future prospects. In: Jones EBG (ed) Marine Fungi and Fungal-like Organisms, Marine and fresh water botany. Walter de Gruyter, Berlin, pp 441–474
Walochnik J, Mulec J (2009) Free-living amoebae in carbonate precipitating microhabitats of karst caves and a new vahlkampfiid amoeba, *Allovahlkampfia spelaea gen. nov., sp. nov.* Acta Protozool 48:25–33
Wang J, Wei X, Qin X, Tian X, Liao L, Li K, Zhou X, Yang X, Wang F, Zhang T et al (2016a) Antiviral merosesquiterpenoids produced by the Antarctic fungus *Aspergillus ochraceopetaliformis* SCSIO 05702. J Nat Prod 79:59–65
Wang J, He W, Huang X, Tian X, Liao S, Yang B, Wang F, Zhou X, Liu Y (2016b) Antifungal new oxepine-containing alkaloids and xanthones from the deep-sea-derived fungus *Aspergillus versicolor* SCSIO 05879. J Agric Food Chem 64:2910–2916
Went FE (1969) Fungi associated with stalactite growth. Science 166:385–386
Wojtczak G, Breuil C, Yamuda J, Saddler JN (1987) A comparision of the thermostability of cellulose from various thermophilic fungi. Appl Miocrobiol Biotechnol 27:82–87
Wu G, Ma H, Zhu T, Li J, Gu Q, Li D (2012) Penilactones A and B, two novel polyketides from Antarctic deepsea derived fungus *Penicillium crustosum* PRB-2. Tetrahedron 68:9745–9749
Xu X, Zhang X, Nong X, Wei X, Qi S (2015) Oxindole alkaloids from the fungus *Penicillium commune* DFFSCS026 isolated from deep-sea-derived sediments. Tetrahedron 71:610–615
Yang YL, Liao WY, Liu WY, Liaw CC, Shen CN, Huang ZY, Wu SH (2009) Discovery of new natural products by intact-cell mass spectrometry and LC-SPE-NMR: malbranpyrroles, novel polyketides from thermophilic fungus *Malbranchea sulfurea*. Chem Eur J 15:11573–11580
Yoder JA, Benoit JB, Christensen BS, Croxall TJ, Hobbs HH III (2009) Entomopathogenic fungi carried by the cave orb weaver spider, *Meta ovalis* (Araneae, Tetragnathidae), with implications for mycoflora transfer to cave crickets. J Cave Karst Stud 71:116–120
Zalar P, de Hoog GS, Schroers HJ, Crous PW, Groenewald JZ, Gunde-Cimerman N (2007) Phylogeny and ecology of the ubiquitous saprobe *Cladosporium sphaerospermum*, with descriptions of seven new species from hypersaline environments. Stud Mycol 58:157–183
Zalar P, Frisvad JC, Gunde-Cimerman N, Varga J, Samson RA (2008) Four new species of Emericella from the Mediterranean region of Europe. Mycologia 100:779–795
Zamora JRC (1977) Isolation of Histoplasma capsulatum from the air in the Aguas Buenas caves, Aguas Buenas, Puerto Rico. Mycopathologia 60:163–165

Zhang T, Victor TR, Rajkumar SS, Li X, Okoniewski JC et al (2014) Mycobiome of the bat White nose syndrome affected caves and mines reveals diversity of Fungi and local adaptation by the fungal pathogen *Pseudogymnoascus (Geomyces) destructans*. PLoS One 9(9):e108714

Zhang ZF, Liu F, Zhou X, Liu XZ, Liu SJ, Cai L (2017) Culturable mycobiota from Karst caves in China, with descriptions of 20 new species. Persoonia 39:1–31

Zhang X, Li SJ, Li JJ, Liang ZZ, Zhao CQ (2018a) Novel natural products from Extremophilic Fungi. Mar Drugs 16:194

Zhang ZF, Zhao P, Cai L (2018b) Origin of cave Fungi. Front Microbiol 9:1407

Zhou JP, Gu YQ, Zou CS, Mo MH (2007) Phylogenetic diversity of bacteria in an earth-cave in Guizhou province southwest of China. J Microbiol 45:105–112

Zhou H, Li L, Wu C, Kurtan T, Mandi A, Liu Y, Gu Q, Zhu T, Guo P, Li D (2015) Penipyridones a–f, pyridone alkaloids from *Penicillium funiculosum*. J Nat Prod 79(1783):1790

Secondary Metabolites of Mushrooms: A Potential Source for Anticancer Therapeutics with Translational Opportunities

23

Sudeshna Nandi, Rimpa Sikder, and Krishnendu Acharya

Abstract

Across the globe cancer is emerging as an overriding source of death raising enduring consequences all through the life period of patients. Chemotherapeutic drugs adopted for cancer therapy have grievous after-effects, and growth of resistance is a major downside for these agents. For decades natural products and medicine have been intimately connected through the usage of conventional medicines. Mushrooms own substantial antiquity of use in conventional medicine with no or minimal side effects. Mushrooms are considered as superfoods due to the presence of bioactive compounds and origin of medical drug and nutraceutical development for enhancing longevity of people. Recently there is an elevated interest found in people for the secondary metabolites of higher fungi in order to explore novel medical substance or lead compounds for cancer treatment. A number of novel fungal metabolites have been extracted from higher fungi which are likely to provide drugs with chemopreventive property. The researches involved the drug discovery from medicinal fungi mainly implicating multifaceted approach. Most of the isolated compounds have exhibited prominent in vitro cytotoxic effects in cancer cell lines from human tissue, and specifically selected compounds were used for in vivo experiments. This chapter cautiously deals with the review of low-molecular-weight compounds obtained from higher fungi with anticancer potential identified so far. In the near future, among these novel compounds many are presumed to enter human clinical trials.

Authors Sudeshna Nandi, Rimpa Sikder have been equally contrbuted to this chapter.

S. Nandi · R. Sikder · K. Acharya (✉)
Molecular and Applied Mycology and Plant Pathology Laboratory, Department of Botany, University of Calcutta, Kolkata, West Bengal, India

T. Satyanarayana et al. (eds.), *Advancing Frontiers in Mycology & Mycotechnology*,
https://doi.org/10.1007/978-981-13-9349-5_23

Keywords
Mushrooms · Medicinal value · Chemotherapy · Anticancer · Biological activity · Low-molecular-weight compounds · Cytotoxicity · Molecular insight

23.1 Introduction

Cancer, a collective term for invasive neoplasms, is the foremost reason of mortality and morbidity around the world (Joseph et al. 2018). This is a protracted disease that religiously trenches the patients along with their family (Kim et al. 2004). As per the survey reports of World Health Organization (WHO), if cancer remains untreated, especially in lower- and middle-earning countries, millions of people will die of it. In 1942, the current era of cancer therapy has begun; from there onwards the chemotherapeutic agents were utilized enormously as solitary agents, in combination (Li et al. 2014) or as an adjuvant to other therapies (Liao et al. 2007). Although developing resistance to chemotherapy (Thomas and Coley 2003) and molecular targeted therapies (Shervington and Lu 2008) are overly toxic to an ample range of normal cells and tissues. The resuming enormity of the cancer problems and the defeat of traditional chemotherapy against the foster invasive disease affected the considerable drop in the survival rates for the prevalent forms of malignancies inclusive of carcinoma of lung, colon, breast, prostate, pancreas, etc., which pointed out that modern outlook to the curb of cancer is highly needed. In spite of progress made in fundamental scientific knowledge connected with cancer-related clinical treatment, morbidity rates of few of the customary cancers proceed to advance (Sporn and Suh 2002). Over these years, our perception over cancer has grown exceedingly which allows the reconnaissance of advanced molecular pathways as well as cancer-related tissue architecture, genes and proteins (Kellof and Sigman 2000). Cancerous cells mature when normal healthy cells start growing anomalously and become malevolent and it's not workable for any sort of drug to block carcinogenesis and suppress the advancement of cancerous cells without harming the normal cells. Recently attention has been focussed on the development of immunopotentiators and biological response modifiers (Mizuno 1999b) or the adjuvant utilization of natural bioactive compounds with no or minimal toxicity in conventional chemotherapy that would check carcinostasis and carcinogenesis (Petrova et al. 2008). This sort of research is now gaining more attentions.

The substance or chemical compound which is formed in nature and produced by living organism is termed a 'natural product'. Investigation of natural products for novel bioactive agents may furnish resolutions for discovering drugs and their evolution (Debbab et al. 2011; Debbab et al. 2012). Historically, natural products or chemical substances procured from microorganisms and plants have conventionally been a chief source of novel anticancer agents (Harvey 2008; Dias et al. 2012). Later, several compounds or drugs were obtained from natural reserves, which include functional chemotherapeutic agents such as mitomycin C, doxorubicin, daunomycin, bleomycin or etoposide, paclitaxel, teniposide, topotecan and the vinca alkaloids derived from plant-based natural products (Evidente et al. 2014).

In drug discovery, entertaining the capacity and involvement of natural resources, one cannot disregard the therapeutic prospective and input of edible as well as medicinally important mushrooms as fungi-derived natural products have been a magnificent source of pharmaceutical as well as medicine. On earth out of 14,000 discrete species of mushrooms, around 700 showed medicinal properties and 7000 remain undiscovered (Wasser 2011). Mushrooms comprising of enormous and yet mainly untapped source new dynamic pharmaceutical products. For centuries mushrooms have been used for medicinal purposes and food and recently considered as superfood, and their intake has useful results on human well-being due to its therapeutic properties (Figueiredo and Regis 2017; Zaidman et al. 2005). Further present-day scientific researches on medicinally important mushrooms have amplified rapidly through the last two decades (Zaidman et al. 2005). A unique term 'mycotherapy' was introduced for the study of bioactive compounds and extracts isolated from mushroom as health-stimulating factor as mycotherapy of cancer is comparatively new and promising research area (Popovic et al. 2013). Most of the mushrooms of basidiomycetes have prevalently established in traditional Chinese medicine, and recently focus has been expanded in the isolation and characterization of secondary metabolic product from higher basidiomycetes for finding out novel drugs or lead compounds. An array of bioactive comixture and composite fractions were extracted from edible and medicinal mushrooms with notable value regarding anticancer or growth preventive activity (Quang et al. 2006; Xu et al. 2012; Wasser 2014). For present-day medicine, mushroom illustrated an unbound resource of bioactive compounds which includes high-molecular-weight (HMW) compounds including hetero- and homopolysaccharides, proteins, glycopeptides, RNA protein complexes, lipids and an array of complex low-molecular-weight (LMW) compound with multiple chemical compositions such as terpenoids, quinones, catechols, cerebrosides, isoflavones, amines, steroids, polyketides, alkaloids and non-ribosomal peptides (Ferreira et al. 2010; De Silva et al. 2013).

Our study principally deals with the low-molecular-weight compounds obtained from various mushrooms. Low-molecular-weight compounds are the secondary metabolites which are not included in the primary metabolic mechanism of the fungi, i.e. production of energy and the generation of the elementary unit of nucleic acids, cell membranes and proteins (Kozlovskii et al. 2013). The particular secondary metabolites emerge as transition of primary metabolic products, and classified into four chief metabolic sources: (a) amino acid-origin pathways, comprising non-ribosomal proteins; (b) the shikimic acid pathway, producing aromatic compounds; (c) the acetate–malonate pathway, leading to polyketides; and (d) the mevalonic acid pathway, for the biosynthesis of terpenoids (Lung and Hsieh 2011; Silva et al. 2012). Among these polyketides and terpenoids were often been described from basidiomycota, and polyketides were also predominantly produced from ascomycota (Brakhage 2013). Many such metabolites were being utilized in the medication of various health-related issues, including cancer (Paterson 2001). In the present chapter, we tried to group all the studies of literature that inspect positive impression of all mushroom-related compounds on cancer treatment because of their antitumour and anticarcinogenic effects and feasible empathy with chemotherapy management. This review provides an overview of almost all the reported mushroom with their secondary metabolites possessing property of apparently reducing the risk of developing cancer with defined molecular mechanistic pathway. However, a detailed and clear interpretation of how mushrooms' bioactive principles might influence adjuvant treatments demand additional investigation. Bioactive compounds which are obtained from specific mushrooms with their effectivity towards cancer are listed in Table 23.1.

23.2 Mushroom's Secondary Metabolites with Anticancer Property

23.2.1 *Antrodia cinnamomea*

Antrodia cinnamomea, previously known as *Antrodia camphorata*, is a scarce medicinally valuable mushroom popular for its inexplicable response on a variety of health-related issues, largely those relevant to the liver and cancerous growths. *Antrodia cinnamomea* solely observed in the islands of Taiwan usually breed in the decayed inner wall of the indigenous, unfamiliar, endangered camphor tree *Cinnamomum kanehirae*. It has been believed to boost body metabolism and strength and promote longevity (Levin et al. 2012). Five lanostane-dehydroeburicoic acid (Fig. 23.1c), 3β,15a-dihydroxy lanosta-7,9(11), 24-triene-21-oic acid, 15a-acetyl dehydrosulphurenic acid, sulphurenic acid, dehydrosulphurenic acid and three ergostane-type (methyl zhankuic acid A, zhankuic acid A and zhankuic acid C) triterpenes were procured from basidiocarp of *A. camphorata* and examined for their in vitro cytotoxicity against colon and breast cancer cell groups. The above-mentioned three zhankuic acids, zhankuic acid A, methyl zhankuic acid A and zhankuic acid C (Fig. 23.1a), manifested the best cytotoxic outcome with LD_{50} value of 22.3–75.0 μM. 3β,15a-Dihydroxy lanosta-7,9 (11), 24-triene-21-oic acid and 15a-acetyl dehydrosulphurenic acid were found to be selectively cytotoxic in breast cancer MDA-MB-231 model and three colon cancer cell lines (HCT-116, HT-29 and SW-480), whereas sulphurenic acid only expressed its cytotoxicity for MDA-MB-231. Interestingly, neither of these isolates have toxicity towards normal cell lines, i.e. mammary epithelial (MCF10A) and primary foreskin fibroblast (HS68) cells (Yeh et al. 2009). Later, ten new triterpenoids, namely, camphoratins A–J (Fig. 23.1d), were also obtained from this species, and camphoratins (B–F) showed moderate cytotoxicity in KB-VIN and KB human cancer cell lines (Wu et al. 2010). Later Huang et al. (2012) isolated four new lanostane triterpenes (methyl 11α-3,7-dioxo-5α-lanosta-8,24(E)-dien-26-oate, 3,7,11-trioxo-5α-lanosta-8,24(E)-dien-26-oic acid, methyl 3,7,11,12,15,23-hexaoxo-5α-lanost-8-en-26-oate and ethyl 3,7,11,12,15,23-hexaoxo-5α-lanost-8-en-26-oate) and two sterols, named as camphosterol A and (14α,22E)-14-hydroxyergosta-7,22-diene-3,6-dione, from a concoction of mycelia and mature basidiocarps of *A. camphorata*. All these six compounds were tested for cytotoxicity showing moderate activity against various

Table 23.1 Bioactive compounds from mushrooms and their antitumour potential

Mushrooms (scientific name)	Secondary metabolite type	Cell type/cancer type/ model type	References
Agaricus bisporus	490 quinone (γ-L-glutaminyl-4-hydroxy-2,5-benzoquinone)	L1210 leukaemia cells	Ferreira et al. (2010)
Agaricus blazei	Ergosterol	Solid tumours	Takaku et al. (2001) and Didukh et al. (2004)
Albatrellus confluens	Neogrifolin, grifolin	Osteosarcoma(U2OS, MG63 cell line)	Chen et al. (2015)
Amauroderma rude	Ergosterol	Breast cancer (MDA-MB-231, SK-BR-3, MDA-MB-468, MCF-7, 4T1)	Li et al. (2015)
Antrodia camphorata/ Antrodia cinnamomea	Dehydroeburicoic acid, 15a-acetyl dehydrosulphurenic acid, 3β,15a-dihydroxy lanosta-7,9 (11), 24-triene-21-oic acid, Dehydrosulphurenic acid, Sulphurenic acid, Methyl zhankuic acid A, Zhankuic acid A Zhankuic acid C	HT-29, HCT-116, SW-480, MDA-MB-231	Yeh et al. (2009)
	Camphoratins (A–J)	KB and KB-VIN cells	Wu et al. (2010)
	3,7,11-trioxo-5α-lanosta-8,24(E)-dien-26-oic acid, methyl 11α-3,7-dioxo-5α-lanosta-8,24(E)-dien-26-oate, methyl 3,7,11,12,15,23-hexaoxo-5α-lanost-8-en-26-oate, and ethyl 3,7,11,12,15,23-hexaoxo-5α-lanost-8-en-26-oate, (14α,22E)-14-hydroxyergosta-7,22-diene-3,6-dione and Camphosterol A	Several human cancer cell lines	Huang et al. (2012)
	Anticin K and Anticin C	Leukaemia cell lines	Du et al. (2012)
Astraeus odoratus	Lanostane triterpenes: Astraodoric acids A–D Astraeusins A–L	MCF-7, NCI-H187 and KB	Arpha et al. (2012) and Isaka et al. (2016)

(continued)

Table 23.1 (continued)

Mushrooms (scientific name)	Secondary metabolite type	Cell type/cancer type/ model type	References
Clitocybe clavipes	Clavilactones	A431 cells	Zaidman et al. (2005)
Cordyceps militaris	Cordycepin (3′-deoxyadenosine)	Human leukaemia, prostate cancer cells, follicular thyroid carcinoma, colon cancer cells, MDA-MB-435, Hep3B, NIH3T3	Jeong et al. (2011), Lee et al. (2013a, b), Chen et al. (2010a), Lee et al. (2010), Lee et al. (2012), and Wong et al. (2010)
	Compound 1–3	PC-3, colon 205 and HepG2 cells	Rao et al. (2010)
Cordyceps ophioglossoides	Cordycepol A–C Cordycol	HepG2, MCF-7, HeLa and A549	Sun et al. (2013)
Cordyceps sinensis	5,8-Epidioxy-24(*R*)-methylcholesta-6,22-dien-3β-ol (steroids), 5,8-Epidioxy-24(*R*)-methylcholesta-6,22-dien-3β-D-glucopyranoside, 5,6-Epoxy-24(*R*)-methylcholesta-7,22-dien-3β-ol	RPMI-8226, K562, HL-60, Jurkat, WM-1341	Bok et al. (1999)
Daedalea dickinsii	Polyporenic acid C	Solid tumour cells	Zaidman et al. (2005)
Flammulina velutipes	Genistein	Human prostate carcinoma cells (LNCaP; PC-3-M)	Rao et al. (2004)
		Human ovarian carcinoma cell line (SKOV-3)	Chang et al. (2004)
		Human hepatocellular carcinoma cells (HepG2)	Choi et al. (2000)
Fomitella fraxinea	Fomitellic acids A and B (steroids)	Gastric cancer cells (NUGC)	Zaidman et al. (2005)
Fomitopsis nigra	Fomitoside-K	Oral squamous carcinoma cells (YD-10B)	Bhattarai et al. (2012) and Lee et al. (2012)
Ganoderma applanatum	Ergosterol (steroids)	COX-1 and COX-2 inhibitor	Ferreira et al. (2010)
	Ergosta-4,6,8,22-tetraen-3-one (steroids)		

(continued)

Table 23.1 (continued)

Mushrooms (scientific name)	Secondary metabolite type	Cell type/cancer type/ model type	References
Ganoderma hainanense	Ganoderone A Lucidadiol Ganodermanontriol 4, 4, 14a-trimethyl-3,7-dioxo-5a-chol-8-en-24-oic acid	MCF-7 cells	Peng et al. (2015)
Ganoderma japonicum	Ergosterol (steroids) Ergosta-4,6,8,22-tetraen-3-one (steroids)	COX-1 and Cox-2 inhibitor	Gan et al. (1998) and Ferreira et al. (2010)
Ganoderma lucidum	Ganoderic acid T (GA-T)	Colon tumour cell line (HCT-116), SMMC-7721, lung cancer cell line (95-D), xenographed in athymic mice	Chen and Zhong et al. (2011), Tang et al. (2006), Chen et al. (2010b), and Liu et al. (2012a)
	Ganoderic acid me (GA-Me)	HCT-116, breast cancer cell line	Zhou et al. (2011), Li et al. (2012), and Chen et al. (2008a)
	Ganoderic acid mf (GA-Mf) and ganoderic acid S (GA-S)	Human cervical cell line-HeLA	Liu and Zhong (2010)
	Ganoderic acid Mk (GA-Mk)	HeLA cell line	Liu et al. (2012b)
	Ganoderic acid V–Z	HTC cell line	Ivanova et al. (2014)
	Ganoderic acid F	Human carcinoma cells	Kimura et al. (2002)
	Lucialdehydes A–C	Sarcoma 180, sarcoma meth-A, LLC and T-47D cell lines	Gao et al. (2002)
	20-Hydroxylucidenic acid N	Mouse skin two-stage model	Akihisa et al. (2007)
	Ganoderic acid A and H	MDA-MB-231 cell line	Jiang et al. (2008)
	Lucidenic acids A, B, C, N	HL-60, HepG2, Calf	Weng et al. (2007) and Hsu et al. (2008)
	Lucidenic acid O and lucidenic lactone	DNA polymerase α, rat DNA polymerase β	Mizushina et al. (1997)
	Ganodermanontriol	Human breast cancer cell lines-MDA-MB 231 and MCF-7	Jiang et al. (2011)

(continued)

Table 23.1 (continued)

Mushrooms (scientific name)	Secondary metabolite type	Cell type/cancer type/ model type	References
	Ganoderic acid DM	Prostate cancer cells, LnCaP, MCF-7, MDA-MB-231	Johnson et al. (2010), Liu et al. (2009), Wu et al. (2012), and Liu et al. (2012c)
	Ganoderiol F	Lewis lung carcinoma cell	Gao et al. (2006)
	Ganoderic acid X (GA-X)	Human hepatoma HuH-7 cells	Li et al. (2005)
	Cerebrosides [(4E,8E)-ND-2′-hydroxypalmitoyl-1-O-β-D-glucopyranosyl-9-methyl-4,8 sphingadienine and (4E,8E)-N-D-2′-hydroxystearoyl-1-O-β-Dglucopyranosyl-9-methyl-4,8-sphingadienine]	Inhibition of replicative DNA polymerase	Zaidman et al. (2005)
	Cerevisterol		Mizushina et al. (1997)
	Lucidumol A and B	Inhibition of replicative DNA polymerase α Human tumour cell lines	Moradali et al. (2007)
Ganoderma zonatum	Ganoderic acid Y	A549, SMMC-7221	Kinge and Mih (2011)
Grifola frondosa	Ergosterol Ergosta-4,6,8(14),22-tetraen-3-one	Human bladder tumour	Poucheret et al. (2006), Zaidman et al. (2005), and Yazawa et al. (2000)
Gymnopilus marginatus	6-(3,4-dihydroxystyryl)-4-hydroxy-2-pyrone (hispidin)	SCL-1, Capan-1 cell line	Gonindard et al. (1997)
Gymnopilus parvisporus	6-(3,4-dihydroxystyryl)-4-hydroxy-2-pyrone (hispidin)	SCL-1, Capan-1 cell line	Zaidman et al. (2005) and Gonindard et al. (1997)
Gymnopilus patriae	6-(3,4-dihydroxystyryl)-4-hydroxy-2-pyrone (hispidin)	SCL-1, Capan-1 cell line	Zaidman et al. (2005) and Gonindard et al. (1997)

(continued)

Table 23.1 (continued)

Mushrooms (scientific name)	Secondary metabolite type	Cell type/cancer type/ model type	References
Hericium erinaceus	Hericenones	Human leukaemia (K562) and human prostate cancer (LANCAP) cells	Mizuno (1999a)
	Erinacines		
	Erinacerins C – L, Hericerins, Resorcinols, Ergosterol, Erinarols A – J		Friedman (2015)
Hexagonia speciosa	Cyclohexanoids—Speciosin B	Various tumour cell lines	De Silva et al. (2013)
Inonotus obliquus	Lanosterol, 3β-hydroxylanosta-8,24-dien-21-al, Inotodiol, Trametenolic acid, lanosta-8,25-dien-3,22,24-triol, β-Sitosterol, Ergosterol peroxide, Betulin, 2,5-dihydroxy-benzaldehyde, 3,4-dihydroxybenzalacetone	A549, MCF-7 L1210, COLO 205 and HL-60 cancer cell lines	Kim et al. (2011)
	Spiroinonotsuoxodiol	P388, L1210, HL-60 and KB cells	Handa et al. (2010)
	Inonotsuoxodiol A		
	Inonotsudiol A		
	Inonotsuoxides A and B	Epstein–Barr virus early antigen (EBV-EA)	Nakata et al. (2007) and Nomura et al. (2008)
Inonotus hispidus	6-(3,4-dihydroxystyryl)-4-hydroxy-2-pyrone (hispidin)	Human cancerous keratinocytes (SCL-1), human cancerous pancreatic duct cells (Capan-1)	Zaidman et al. (2005), Berns et al. (2000), Yoshiji et al. (1999), and Gonindard et al. (1997)
Lampteromyces japonicus	Illudin S and M and derivatives	Pancreatic, ovarian, colorectal carcinoma, malignant glioma and non-small cell lung cancer	Poucheret et al. (2006) and Zaidman et al. (2005)
Lentinus crinitus	Panepoxydone (PP)	Breast cancer (MDA-MB-453, MCF-7, MDA-MB-468 and MDA-MB-231 cell lines)	Erkel et al. (1996); Erkel et al. (2007), Arora et al. (2014)

(continued)

Table 23.1 (continued)

Mushrooms (scientific name)	Secondary metabolite type	Cell type/cancer type/ model type	References
Lepiota americana	5,8-Epidioxy-24(*R*)-methylcholesta-6,22-dien-3β-ol	Breast cancer	Zaidman et al. (2005) and Kim et al. (2000)
Macrolepiota procera	Lepiotaprocerins A – L	HL-60, A-549, SMMC-7721, MCF-7, SW480	Chen et al. (2018)
Neonothopanus nambi	Nambinones A–C Nambinone D 1-epi-nambinone B Aurisin A Aurisin K	NCI-H187, cholangiocarcinoma	Kanokmedhakul et al. (2012)
Omphalotus illudens	Illudin S and M and derivatives	Pancreatic, ovarian, colorectal carcinoma, malignant glioma and non-small-cell lung cancer	Poucheret et al. (2006) and Zaidman et al. (2005)
Panus conchatus	Panepoxydone	Five human cancer cell lines	Zaidman et al. (2005) and Ding et al. (2018)
	Panutorulon		
Panus rudis	Panepoxydone	Regulate NF-κB	Zaidman et al. (2005)
Phellinus igniarius	Hispolon	Hep3B cell line	Huang et al. (2011)
Phellinus linteus	Hispolon	Human epidermoid KB cell, human gastric cancer cells (SGC-7901, MGC-803, MKN-45), NB-4 human leukemia cells, breast and bladder cancer cells	Chen et al. (2006, 2008, 2013), Lu et al. (2009)
Piptoporous betulinus/ Fomitopsis betulina	(E)-2-(4-hydroxy-3-methyl-2-butenyl)-hydroquinone, Polyporenic acid C	Rectal cancer cells	Ferreira et al. (2010)
Polyporus umbellatus	Ergone (22E,24R)-ergosta-7,22-dien-3β-ol, 5 α,8α-epidioxy-(22E,24R)-ergosta-6,22-dien-3β-ol, Ergosta-6,22-dien-3β,5α,6β-triol polyporusterone A–G	HepG2, HEp-2, HeLa cells	Zhao et al. (2010) and Zhao et al. (2011)
	Ergosta-4, 6, 8(14), 22-tetraen-3-one	Human carcinoma cells	Lee et al. (2005)

(continued)

Table 23.1 (continued)

Mushrooms (scientific name)	Secondary metabolite type	Cell type/cancer type/ model type	References
Poria cocos	Poricotriol A	HL-60, A549, CRL1579, NIH:OVCAR-3, SK-BR-3, DU145, AZ521, PANC-1	Kikuchi et al. (2011)
	Dehydroebriconic acid Dehydrotrametenonic acid	Gastric cancer cells	Akihisa et al. (2004) and Mizushina et al. (2004)
Sarcodon aspratus	5α,8α-epidioxy-22E-ergosta-6,22-dien-3β-ol 5α,8α-epidioxy-22E-ergosta-6,9(11),22-trien-3β-ol	HT29 cells	Kobori et al. (2006)
Tricholoma terreum	Terreumols A – D	MCF-7, SW480, hepatocellular carcinoma SMMC-7721, HL-60 and A-549 cells	Yin et al. (2013)

tumorous cell lines of humans. Furthermore, two new triterpenoids were obtained, namely, anticin K and anticin C, along with few known compounds from the same fruiting body which were observed to inhibit leukaemia cell lines (Du et al. 2012).

23.2.2 Genus *Cordyceps*

Species belonging to the genus *Cordyceps* were found to grow on mature insects, insect larvae or fruiting bodies of other mushroom. *Cordyceps*, natively known as caterpillar fungus, is a costlier ascomycete medicinal mushroom which had a giant history as exotic and rare medicinal fungi used by the people worldwide. For centuries it has been hugely recognized as cornerstone of Chinese medicine. Despite its rarity and overprice, the unparalleled importance of medicinal feasibility for *Cordyceps* species has moulded it an overly valued staple of traditional Chinese medicine (Holliday and Cleaver 2008). ***C.*** sinensis and C. militaris are the well-known species of *Cordyceps* with huge medicinal value. Apart from these two, a new addition is *C. ophioglossoides* with potential medicinal importance.

Cordyceps militaris is a very well-known medicinally used mushroom. Previous reports had provided that pharmacological activity of both natural and cultivated *C. militaris* is comparable. The cultivated products of *C. militaris* are customarily vended in markets of China and Southeast Asia as medicated and healthy food material. Rao et al. (2010) had isolated and purified ten pure biologically active compounds from extracts of *C. militaris* through silica gel column chromatography. The compounds 1 and 2 manifested dynamic anti-proliferative property against colon 205 and PC-3 cells, whereas compound 3 only showed inhibitory effect

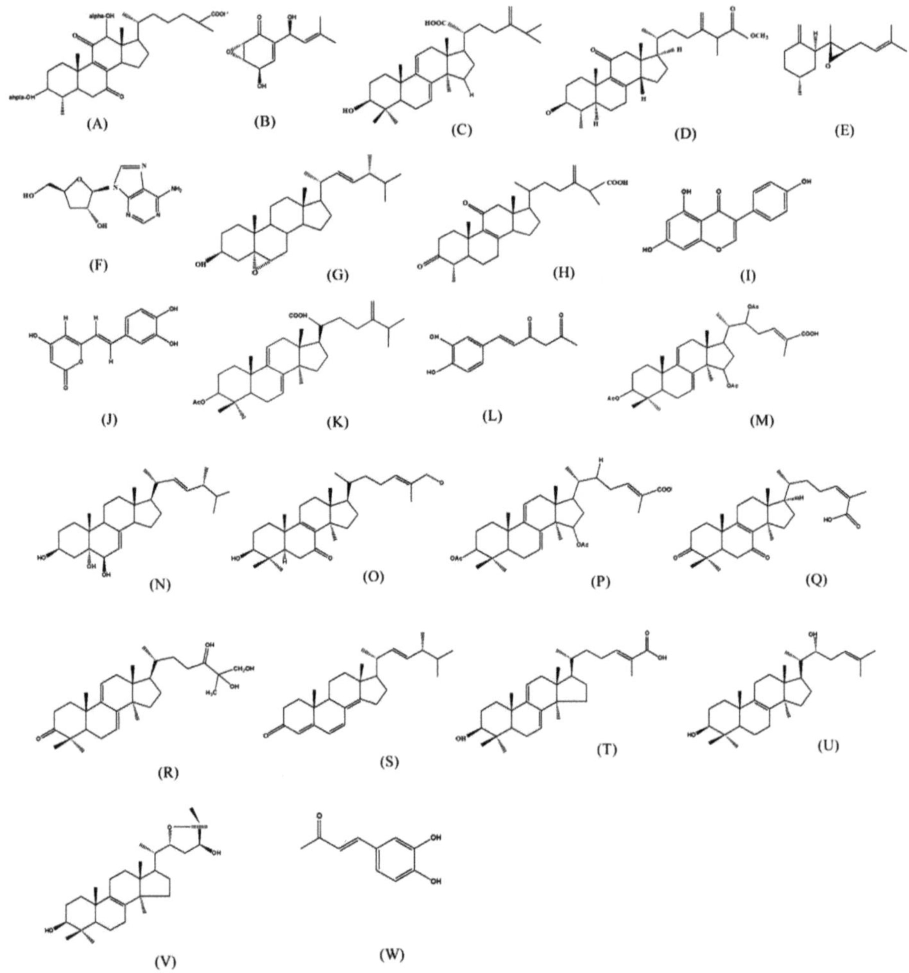

Fig. 23.1 Chemical structure of some important bioactive metabolites obtained from mushroom with potent anticancer property. (**a**) Zhankuic acid C, (**b**) Panepoxydone, (**c**) Dehydroeburicoic acid, (**d**) Camphoratin J, (**e**) Cordycol, (**f**) Cordycepin, (**g**) 5,6-Epoxy-24(R)- methylcholesta-7,22-dien-3β-ol, (**h**) Antcin A, (**i**) Genistein, (**j**) Hispidin, (**k**) Dehydroebriconic acid, (**l**) Hispolon, (**m**)-Ganoderic acid T, (**n**) Cerevisterol, (**o**) Lucialdehyde C, (**p**) Ganoderic acid Me, (**q**) Ganoderic acid DM, (**r**) Ganodermanontriol, (**s**) Ergosta-4,6,8(14),22-tetraen-3-one (Ergone), (**t**) Ganoderic acid Y, (**u**) Inotodiol, (**v**) Inonotsuoxides A, (**w**) 3,4- dihydroxybenzalacetone

against liver cancer HepG2 cells. Another active component was obtained from the fungus *C. militaris*, polyadenylation inhibitor cordycepin (3′-deoxyadenosine) (Ito et al. 1981). Treating human leukaemia cells and prostate cancer cells with cordycepin (Fig. 23.1f) has notably supressed the proliferation of both the cell types through inducing apoptosis by a signalling cascade including mitochondria-mediated caspase pathway (Lee et al. 2013b; Jeong et al. 2011). With cordycepin it has also been observed to decrease the cell viability of follicular thyroid carcinoma cells and resulted in apoptosis via calcium-calpain-caspase 7-poly(ADP-ribose) polymerase (PARP) pathway (Chen et al. 2010a, b). This polyadenylation inhibitor has also established its anticancer property in human colon cancer cells via arresting the cell cycle at G2/M phase through p21WAF1 expression (Lee et al. 2010) and by inhibiting RNA synthesis thereby inducing double strand breaks in DNA of MDA-MB-435 cells (Lee et al. 2012). Lately Lee et al. (2013a) claimed that cordycepin can also sensitize TRAIL-resistant Hep3B human hepatocellular carcinoma cells to tumour necrosis factor-related apoptosis-inducing ligand (TRAIL)-mediated apoptosis and inhibiting the activation of c-Jun N-terminal kinase (JNK). Wong et al. (2010) confirmed that lower dose of cordycepin has reduced length of poly A tail and proliferation of NIH3T3 fibroblasts whereas at higher doses a marked retardation of cell attachment as well as depletion of focal adhesions was seen. Moreover, a strong impedetion of mechanistic target of rapamycin (mTOR) and adenosine monophosphate-activated protein kinase (AMPK) signalling was also observed on treatment with the drug. Furthermore, investigations were also made to check the efficacy of cordycepin on the survival rate of cells which are vulnerable to tumour necrosis factor (TNF)-α. At subtoxic dose of cordycepin, it has been observed that it enhanced sensitivity of cells to TNF-α-induced apoptosis and correlated with inhibition of pro-survival component nuclear factor-κB (NF-κB) which is associated with TNF-α signalling. It has been scrutinized that cordycepin had potentially phosphorylated eukaryotic translation initiation factor 2α (eIF2α) and phosphorylated eIF2α further imitated the restrictive activity of cordycepin on the NF-κBpathway. Downstream molecular events of eukaryotic translation initiation factor 2α were identified by examining the importance of mammalian target of rapamycin complex 1 (mTORC1). Thus it can be stated that cordycepin has the potentiality of sensitizing the cells to TNF-α-mediated apoptosis, partially by inducing the eIF2α-mTORC1 pathway and proceeding in the repression of NF-κB (Kadomatsu et al. 2012). The genotoxic efficacy of cordycepin has made it a potential curative drug for both the prevention as well as treatment of cancer.

Cordyceps sinensis has been well described as a medicinal mushroom in old Chinese medical and Tibetan medicinal books. Common people of North Sikkim has approved *Cordyceps sinensis* for all diseases either as a singly or in combination with other herbs (Panda and Swain 2011). 5,8-Epidioxy-24(*R*)-methylcholesta-6,22-dien-3β-ol (Fig. 23.2h) and two more steroids 5,6-epoxy-24(*R*)-methylcholesta-7,22-dien-3β-ol (Fig. 23.1g) and 5,8-epidioxy-24(*R*)-methylcholesta-6,22-dien-3β-D-glucopyranoside were characterized from *C. sinensis* which prevented the proliferation of RPMI-8226, K562, HL-60, Jurkat and WM-1341, tumour cell lines by 10–40% at 10 μg/mL than its previously identified aglycone (Bok et al. 1999).

The cultured mycelia of *Cordyceps ophioglossoides* have yielded three unique sesquiterpenes, namely, cordycepol A–C and one fumagillol analogue cordycol (Fig. 23.1e). All four isolated compounds were experimented for their cytotoxicity against MCF-7, HeLa, HepG2 and A549 cell lines. Among these four, cordycepol C and cordycol both dose and time dependently restarted the number of viable carcinoma cells, with comparatively low effect on normal liver LO2 cell line (Sun et al. 2013).

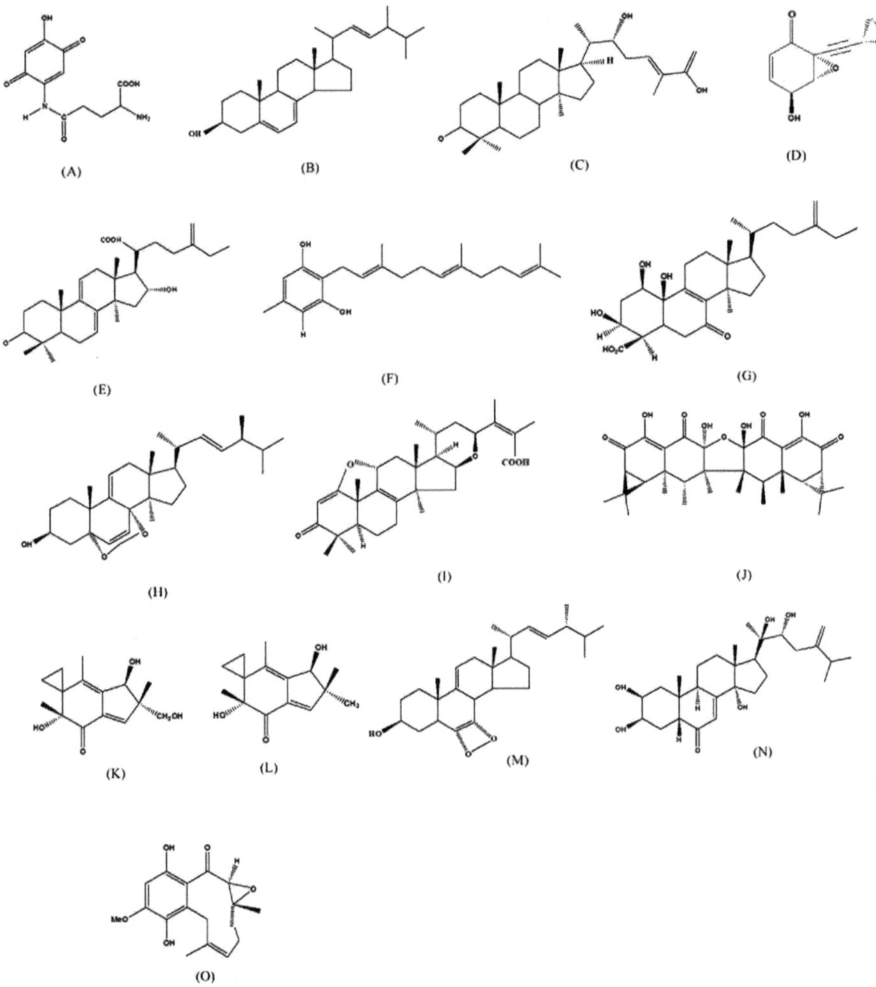

Fig. 23.2 Chemical structure of some less studied important low molecular weight compounds with anticarcinogenic activity. (**a**) 490 Quinone (γ-L-glutaminyl-4-hydroxy-2,5-benzoquinone), (**b**) Ergosterol, (**c**) Astraodoric acid B, (**d**) Speciosin B, (**e**) Polyporenic acid C (**f**) Grifolin, (**g**) Fomitellic acids A, (**h**) 5,8-Epidioxy-24(R)-methylcholesta-6,22-dien-3β-ol, (**i**) Lepiotaprocerins C, (**j**) Aurisin A, (**k**) Iludin S, (**l**) Iludin M, (**m**) 5α,8α-epidioxy-22E-ergosta-6,9 (11),22-trien-3β-ol, (**n**) Polyporusterone B, (**o**) Terreumol A

23.2.3 *Flammulina velutipes*

Flammulina velutipes mostly called as golden needle mushroom has traditionally being used as oriental medicine in many countries of Asia like Japan, China and Korea. Genistein (4′,5,7-trihydroxyisoflavone) is a natural isoflavonoid phytoestrogen, which has been isolated from *F. velutipes*, and showed strong inhibitory effects on DNA topoisomerase II activity and protein tyrosine kinase. Protein kinase is a key regulator of neoplasia that includes cell proliferation, invasion, angiogenesis and metastasis. In human prostate adenocarcinoma LNCaP cells, genistein (Fig. 23.1i) was found to upregulate the levels of Cdc2 active kinase, Weel and $p21^{WAF1/CIP1}$ which are crucial for significant growth suppression (Rao et al. 2004). It also suppressed the proliferation of PC-3-M, a p53-null human prostate carcinoma cell (Choi et al. 2000). Human ovarian carcinoma cell line (SKOV-3) growth proliferation was inhibited by treating with genistein, through upregulation of $p21^{WAF1/CIP1}$ and the downregulation of PCNA and cyclin B1 protein (Li and Mi 2003). Genistein was also found to be active in cells of human hepatocellular carcinoma (HepG2) through arresting their proliferation at G2/M phase by modulating Cdc2 kinasc activity (Chang et al. 2004).

23.2.4 Genus *Ganoderma*

Ganoderma is admirably considered as a genus of traditionally used remedial mushrooms (Baby et al. 2015). Individual bioactive compounds or entire mushroom extracts of *Ganoderma* have been interconnected with the induction of cell cycle arrest or cell death of various human cancer cells (Suarez-Arroyo et al. 2017). About 431 secondary compounds have been reported from the wide range of *Ganoderma* species of which 240 were obtained from *G. lucidum* and the rest 63, 49, 22, 19, 18, 16, 16, 15, 14, 13, 10 and 2 secondary metabolites were described from *G. applanatum/G. lipsiense*, *G. sinense, amboinense, G. colossum, G. pfeifferi, G. resinaceum, G. cochlear, G. concinna, G. australe, G. orbiforme, G. fornicatum* and *G. capense*, respectively (Baby et al. 2015). Most of the triterpenes isolated and purified from *Ganoderma* species were recognized to be potential anticancer agents (Paterson 2006; Cheng et al. 2010; Wu et al. 2012; De Silva et al. 2012). Major phytochemical observations were made on *G. lucidum* followed by *G. neo-japonicum, G. zonatum, G. applanatum* and *G. hainanense.*

Ganoderma lucidum 'the mushroom of immortality' has been reported to nourish life and provide miraculous health ease (Sanodiya et al. 2009; Batra et al. 2013). *Ganoderma lucidum* has provided a series of lanostane triterpenes which has been conventionally administered for centuries throughout Asia as a prospective cancer treatment. It has demonstrated promising anticancer efficacy singly or in adjuvant with radiotherapy and chemotherapy (Pillai et al. 2010). Chen and Zhong (2011) investigated a lanostane-type triterpenoid from *G. lucidum* as ganoderic acid T (GA-T) which came out to be a decent inhibitor of metastasis and tumour invasion. GA-T (Fig. 23.1m) acted in a dose-dependent manner to stimulate cell aggregation, suppress cell migration and inhibit cell adhesion in human colon tumour cell lines of HCT-116 p53−/− and p53+/+. GA-T also caused greater cytotoxicity towards the highly metastatic lung cancer cells by inducing apoptosis-mediated p53 expression, G1 cell cycle arrest and mitochondrial dysfunction (Tang et al. 2006). In addition, GA-T affluently inhibited metastasis in vivo as well as cancer cell invasion in vitro by restraining the nuclear translocation of NF-κB and downregulation of matrix metalloproteinase 9 (MMP-9) and inducible nitric oxide synthetase (iNOS)

(Chen et al. 2010b). However, the effectiveness of GA-T on hepatic leukaemia factor (HLF) and SMMC-7721 cell lines is quite similar. All the data signifies that GA-T has indifferent cytotoxic potency against various tumour cells (Tang et al. 2006). GA-T was also further modified on human cervical carcinoma cell lines to acquire more potent derivatives with pro-apoptotic and cytotoxic effects, with minimum consequences on non-tumour cell line (Liu et al. 2012a). GA-T observed to reveal in vivo antitumour activity in athymic mice by inhibiting the growth of solid human tumours (Tang et al. 2006).

Similarly another triterpenoid ganoderic acid Me (GA-Me) was extracted from this mushroom which had exceptional cytotoxicity in a concentration-dependent manner on human colon carcinoma (HCT-116) cells. GA-Me (Fig. 23.1n) treated tumour cells exhibited enhanced expression of antitumour protein p53 as well as Bax with significantly no change in Bcl-2, but elevated the ratio for Bax/Bcl-2. GA-Me has induced apoptotic process by disrupting the mitochondrial transmembrane potential followed by cytochrome c release and increased caspase-3 activity. Thus, GA-Me may be recommended as a novel natural apoptosis inducer for human colon carcinoma cells by mitochondrial pathway manipulation (Zhou et al. 2011). Researchers had explained another GA-Me-mediated apoptotic mechanistic pathway for breast cancer cells through inhibiting cell proliferation and downregulating the expression level of pro-survival proteins c-Myc, Bcl-2, cyclin D1 and the cell-cycle regulator protein (Li et al. 2012). Chen et al. (2008a) also investigated the influence of ganoderic acid Me on tumour invasion and explained its anti-metastasis property by hampering the cell motility and adhesion as well as suppressing the gene expression of MMP2 and MMP9. Hence, GA-Me might act as a promising apoptotic and anti-metastatic agent. The effectivity of a pair of positional isomer of ganoderic acids, namely, ganoderic acid S (GA-S) and ganoderic acid Mf (GA-Mf) has been investigated for induction of apoptosis in HeLa cells (Liu and Zhong 2010). The results revealed that both the isomers of ganoderic acids diminish the cell proliferation, and GA-S and GA-Mf induced cell cycle arrest in different phases, namely, S and G1, respectively, of human cervical cancer cells. Results obtained from flow cytometry including the phase of cell cycle arrest clearly depicted that GA-Mf was active in causing apoptosis in comparison to GA-S. Treating HeLa cells with each isomer has caused dysfunction of mitochondrial membrane potential followed by cytochrome c release from mitochondria, thereby initiating stimulation of caspase-3 and caspase-9 with an elevation in the Bax/Bcl-2 ratio. Next ganoderic acid Mk obtained from same mycelia of *G. lucidum* promoted apoptosis by diminishing the cell proliferation in concentration-dependent manner in HeLa cells (Liu et al. 2012b).

Various polyoxygenated ganoderic triterpene acids T, V, W, X, Y and Z with lanostane skeleton were obtained and purified from *G. lucidum* which manifested potent cytotoxic activity in vitro towards the hepatoma tissue culture (HTC) cell line (Ivanova et al. 2014). The triterpenoid fraction from the mature basidiocarp of *G. lucidum* also contains ganoderic acid F which showed both the anticancer and anti-metastatic activity by inhibiting angiogenesis which was produced by tumour (Kimura et al. 2002). From basidiocarps of *G. lucidum*, three more triterpene aldehydes with lanostane skeleton were derived and named as lucialdehydes A, B and C. Lucialdehydes B and C showed dynamic cytotoxicity towards sarcoma 180, sarcoma Meth-A, LLC and T-47D cell lines, but lucialdehyde C (Fig. 23.1p) revealed the better cytotoxicity towards tested cells (Gao et al. 2002).

Several lanostane-type triterpene acids were characterized from the fruiting bodies of *G. lucidum* of which 20-hydroxylucidenic acid N repressed the promotion process in mouse skin two-stage carcinogenesis (Akihisa et al. 2007). Ganoderic acid A and H inhibited the invasion of MDA-MB-231, a highly invasive breast cancer cell by hindering the transcription factors AP-1 and NF-κB and consequently diminishing the expression of CDK4 as well as suppressing the secretion of uPA (urokinase-type plasminogen activator) (Jiang et al. 2008).

Effective anti-invasive effects on hepatoma cells were displayed by the extracts of lucidenic acids A, B, C and N [14–17] which were obtained and characterized from fruiting bodies of a new strain YK - 02 of *G. lucidum* (Weng et al. 2007). Further investigations on impact of lucidenic acids (A, B, C and N) on apoptosis were made on human leukaemia cells HL-60 and the results clearly depicted that lucidenic acid B promoted apoptosis in HL-60 and had promising impact on some tumour cell lines (Hsu et al. 2008). Lucidenic acid B was found to work better than the other lucidenic acids, exhibiting both apoptotic and anti-invasive activity. It restarted the invasion of HepG2 cells via restraining TPA-induced NF-κB and AP-1 DNA-binding activities further suppressing the expression of matrix metalloproteinase 9 (Wang et al. 2008). Other two lucidenic acids (O and lucidenic lactone) were found to arrest the effect of rat DNA polymerase β, calf DNA polymerase α and human immunodeficiency virus type 1 reverse transcriptase (Mizushina et al. 1997).

A new ganoderic acid named 3α, 22β-diacetoxy-7α-hydroxy-5α-lanosta-8, 24E-dien-26-oic acid was extracted from the mycelia of *G. lucidum* with considerable cytotoxicity (Li et al. 2013). Recent studies on *Ganoderma* were attentive on the antlered form of *G. lucidum* (*G. lucidum* AF) which possess powerful pharmacological effects. Studies revealed that greater amount of triterpenes have been found in *G. lucidum* AF in contrast to normal *G. lucidum* providing potential antitumour effects (Nonaka et al. 2008; Watanabe et al. 2011).

Ganodermanontriol (GDNT) (Fig. 23.1r) is a biologically active prime *Ganoderma* alcohol which is dominant enough to supress the colony formation (anchorage-independent growth) and proliferation (anchorage-dependent growth) of metastatic, invasive and therapy-resistant MDA-MB-231 cells. Additionally, GDNT strongly inhibited the cell proliferation of MCF-7 cells while minutely preventing the growth of non-cancerous mammary epithelial cells MCF-10A. Interestingly, MDA-MB-231 cells treated with GDNT suppressed the cell cycle regulatory protein, CDC20 expression, which usually remains in highly expressed form in pre-cancerous and breast cancer cells in contrast to normal mammary epithelial cells (Jiang et al. 2011). Ganoderiol F (ganoderma alcohol) was studied which disclosed the highest in vivo antitumour effects against Lewis lung carcinoma cell (LLC)-bearing mice, and ganoderiol F has brilliantly prevented the tumour growth causing no evident side effects and toxicity (Gao et al. 2006). More ganoderic alcohols were proven to exert cytotoxic effect in some tumour cell lines which include lucidumol A [(24S)-24, 25-dihydroxylanost-8-ene-3, 7-dione] and lucidumol B [β, (24S)-lanosta-7, 9(11)-diene-3β, 24, 25-triol] (Moradali et al. 2007).

Ganoderic acid DM (GA-DM) employed anti-prostate cancer activity via retarding 5α-reductase activity (Johnson et al. 2010) and also showed cytotoxicity to other cancer cell line LnCaP (Liu et al. 2009). Wu et al. (2012) explained that GA-DM (Fig. 23.1q) showed effective anticancer property in both concentration and time-dependentmanner in MCF-7 human breast adenocarcinoma cell line by supressing cell proliferation, colony formation and inducing G1 phase cell cycle arrest. Fascinatingly, this research study disclosed that MDA-MB-231 cells require higher dose of triterpenoid extract to conciliate G1 cell cycle arrest which interpreted the greater metastatic and proliferative efficiency of these breast cancer cells. Further researchers corroborated their analysis by checking the protein expression level of the catalytic subunits of the CDK complex including CDK2 and CDK6, as well as the phosphorylation of retinoblastoma and expression level of cyclin D1 (Liu et al. 2012c; Wu et al. 2012).

Two cerebrosides obtained from *G. lucidum* which has inhibited eukaryotic DNA polymerase were (4E,8E)-ND-2′-hydroxypalmitoyl-1-O-β-D-glucopyranosyl-9-methyl-4,8-sphingadienine and (4E,8E)-N-D-2′-hydroxystearoyl-1-O-β-D-glucopyranosyl-9-methyl-4,8-sphingadienine. They selectively prevented the activities of replicative DNA polymerases, particularly the α-type, whereas they impose moderate inhibition of DNA polymerase δ and lesser effectivity on the activity of prokaryotic DNA polymerases, RNA polymerase, deoxyribonuclease I, DNA polymerase β, HIV reverse transcriptase, terminal deoxynucleotidyl transferase and ATPase (Zaidman et al. 2005). Another steroid, cerevisterol (Fig. 23.1o) studied to be a cytotoxic steroid, prevented the DNA polymerase α activity (Mizushina et al. 1997). Treating human hepatoma HuH-7 cells with ganoderic acid X produced cell apoptosis with activation of JNK nitrogen-activated protein kinases and Extracellular regulated kinase (ERK) and immediate inhibition of DNA synthesis. The apoptotic molecular mechanism was clarified by showing downregulation of Bc1-xL and dysfunction of mitochondrial membrane potential, releasing cytochrome c into the cytosol and activating caspase-3 and degrading chromosomal DNA. The ability of the compound ganoderic acid X (GA-X) is an interesting compound, which is unlikely to sensitize cancer cells to apoptosis and prevent topoisomerases activity (Li et al. 2005).

Ganoderma neo-japonicum is a polypore mushroom rarely observed breeding on decaying *Schizostachyum brachycladum* (a tropical bamboo). The Malaysian indigenous tribes—Temiars and Temuans—utilized the fruiting bodies of *G. neo-japonicum* to cure a variety of ailments including diabetes. The scientific affirmation of the medicinal properties of this mushroom has still not been explored substantially. Till date, the extracts of the mushroom have been reported to exhibit antioxidative and hepatoprotective activity in vivo (Tan et al. 2015). Two lanostanoids (ganoderal A and ganodermadiol), four ergosteroids and steroid (2β,3α,9α-trihydroxyergosta-7,22-diene) and two drimane sesquiterpenes, cryptoporic acids H and I, were isolated and characterized from the basidiocarps of *G. neo-japonicum* (Gan et al.

1998). Among the four ergosteroids, ergosta-4, 6, 8(14), 22-tetraen-3-one (Fig. 23.1s) acts as a potent cyclooxygenase (COX-1 and COX-2) inhibitor. It has been well demonstrated that COX-2, the inductive isoform of cyclooxygenase, is usually observed to be highly expressing in various solid tumours (Ferreira et al. 2010); therefore inhibiting the cyclooxygenase will prevent the tumour formation.

Ganoderma zonatum provided three lanostane-type triterpenoids (lanosta 7,9(11),24-trien-3-one 15,26-dihydroxy, ganoderic acid Y and lanosta 7,9(11),24-trien-26-oic,3-hydrox), four steroids ((22E,24R)-ergosta-7,22-dien-3β,5α.6β-triol, 5α,8α-epidiory (22E,24R)-ergosta-6,22-dien-3β-ol, ergosta-5,7,22-trien-3β-ol,7 (ergosterol) (Fig. 23.2b) and ergosta-7,22-dien-3β-ol,6) and a benzene derivative (dimethyl phthalate). Among them the remarkably oxygenated lanostane-type triterpenoid ganoderic acid Y (Fig. 23.1t) exhibited modest cytotoxic effect in response to two human tumour cell lines, A549 (lung cancer) and SMMC-7721 (liver cancer) with LD_{50} values of 29.9 and 33.5 μM, respectively (Kinge and Mih 2011).

Phytochemical effects of the extracts of *Ganoderma applanatum* could be a potential raw material for nanotechnological engineering in clinical practice. The beneficial extract from the mushroom was utilized in the treatment of African trypanosomiasis (Manasseh et al. 2012). A novel lanostanoid and six already known ergosteroids were extracted and characterized from the basidiocarp of *G. applanatum* (Gan et al. 1998). Out of six, ergosta-4,6,8 (14),22-tetraen-3-one (Fig. 23.1s) acts as a potent cyclooxygenase (COX-1 and COX-2) inhibitor, thereby preventing tumour formation (Ferreira et al. 2010).

Recently, Peng et al. have isolated 14 lanostane triterpenoids which include 9 ganoderma acids and 5 ganoderma alcohols, along with 4 known compounds from a scarce species *Ganoderma hainanense*. The study assessed the cytotoxicity for 16 compounds which include ganoderone A, lucidadiol, 4, 4, 14a-trimethyl-3,7-dioxo-5a-chol-8-en-24-oic acid and ganodermanontriol. Among them compounds 10, 12 and 17 (lucidadiol) showed preventive effect for MCF 7 cells (Peng et al. 2015).

23.2.5 *Grifola frondosa*

Grifola frondosa is a widely known edible mushroom reported to contain numerous medicinal values and potential health benefits. *G. frondosa*, commonly known as hen-of-the-woods, ram's head, sheep's head and maitake, is typically found in China, northeastern part of Japan and other Asian region and the United States. The basidiocarp of *G. frondosa* has been marketwise and widely consumed worldwide. The fruit body and some isolated bioactive substances extracted from *G. frondosa* have been frequently used to treat diabetic, diuretic, gonorrhoea and antipyretic activities, calm nerves and mind, treat haemorrhoids and stomach ailments and improve spleen (China 1986; Mizuno and Zhuang 1995; Kubo et al. 1994). Many lower-molecular-weight compounds like triacylglycerols (1-Oleoyl-2-linoleoyl-3-palmitoylglycerol) have been reported for their antitumour as well as immunomodulatory effects. Fruit bodies of *G. frondosa* are a rich source of secondary metabolites, important among these are ergosta-4, 6, 8(14), 22-tetraen-3-one (Fig. 23.1s) and ergosterol (Fig. 23.2b), a type of steroid showing potent anticancer activity. Ergosta-4, 6, 8(14), 22-tetraen-3-one and ergosterol, 1-oleoyl-2-linoleoyl-3-palmitoylglycerol can inhibit cyclooxygenase (COX) enzyme. In recent studies ergosterol established as strong protection against the bladder tumour promotion (Yazawa et al. 2000).

23.2.6 Genus *Inonotus*

Genus *Inonotus*, proposed in 1879, accommodated polypores with pigmented basidiospores and pileate habit (Sharma et al. 2013). In this genus, *Inonotus obliquus* and *I. hispidus* have been utilized as conventional therapeutics possessing excessive nutritional and remedial value, principally in treatment of malignant tumours (Kim et al. 2011; Awadh Ali et al. 2003).

A white rot fungus, *Inonotus obliquus*, broadly scattered in Europe, Asia and North American regions is typically utilized as a folk medicine for the treatment of cancer. *I. obliquus* is well familiar to local people of Japan and Russia as kabanoanatake and chaga, respectively. Kim et al. (2011) have isolated six triterpenoids from the sclerotia of *I. obliquus* (*Hymenochaetaceae*). The terpenoids were identified as 3β-hydroxylanosta-8, lanosterol, 24-dien-21-al, inotodiol (Fig. 23.1u), trametenolic acid, lanosta-8,25-dien-3,22,24-triol, ergosterolperoxide, β-sitosterol, 2,5-dihydroxy-benzaldehyde, betulin and 3,4-dihydroxybenzalacetone. Among these novel compounds 3β-hydroxylanosta-8,24-dien-21-al, ergosterol peroxide, lanosterol, inotodiol, 2,5-dihydroxy-benzaldehyde, trametenolic acid, betulin and 3,4-dihydroxybenzalacetone were examined for in vitro cytotoxic activity against A549, MCF-7 L1210, COLO 205 and HL-60 cancer cells. Of which 3β-hydroxylanosta-8,24-dien-21-al and tramentenolic acid exerted moderate cytotoxicity against L1210 cells with IC_{50} value of 62.5 and 34.4 μM, and betulin displayed frail cytotoxicity against A549 and HL-60. Additionally, 3,4-dihydroxybenzalacetone (Fig. 23.1w) manifested powerful cytotoxic effect in response to HL-60 and A549 cells. Lately the growth inhibitory effect and molecular mode of action of some major lanostane triterpenoid like tramentenolic acid, lanosterol, inotodiol and 3β-hydroxylanosta-8, 24-dien-21-al were investigated against P388 cells in the in vivo carcinogenesis test. Out of all these triterpenoids tested, inotodiol solely inhibited the proliferation, amplified DNA fragmentation and induced apoptosis via activating caspase-3/7 in P388 cells. About 10 mg/kg of inotodiol intraperitoneal administration enhanced the days of survivability of P388-bearing mice (Nomura et al. 2008). Next, a unique lanostane-type triterpenoid, spiroinonotsuoxodiol, and two other triterpenoids, inonotsuoxodiol A and inonotsudiol A, were also elucidated from the sclerotia of this species, and their cytotoxic effect was tested in L1210, P388, KB and HL-60 cells with moderate anticancer efficacy (Handa et al. 2010). Apart from these triterpenoids inonotsuoxides A (Fig. 23.1v) and B along with three known lanostane-type triterpenoids were also reported from the sclerotia of *I. obliquus*. All the compounds except for two were tested for their supressing activity on Epstein–Barr virus early antigen (EBV-EA), as an experiment for effective chemopreventive drug (Nakata et al. 2007).

From the mycelium culture broth of shaggy bracket mushroom or *Inonotus hispidus*, 6-(3,4-dihydroxystyryl)-4-hydroxy-2-pyrone (hispidin) bioactive catechols were extracted (Awadh Ali et al. 2003). Hispidin (Fig. 23.1j) interfered with the signalling pathway of various cancer cell lines, by inhibiting PKC-β. This protein kinase C-β (PKC-β) has a vital role to play in angiogenic development (Zaidman et al. 2005; Berns et al. 2000) by hampering the VEGF signal transduction pathway (Yoshiji et al. 1999). Hispidin had also markedly showed in vitro cytotoxic effects towards epithelial cancer cells like human cancerous pancreatic duct cell line (Capan-1) and human cancerous keratinocytes (SCL-1) cell line (Gonindard et al. 1997).

23.2.7 *Lentinus crinitus*

Panepoxydone (PP), an epoxy compound isolated from *Lentinus crinitus* indulged in regulating the activity of NF-κB. Panepoxydone (Fig. 23.1b) restricted the NF-κB-mediated signal transduction by preventing the phosphorylation and degradation of IκB and was found to hinder TNF-α-persuaded bracing of NF-κB in a promoter–reporter assay using COS-7 cells via inhibiting phosphorylation and degradation of IκB (Erkel et al. 1996). Additionally, panepoxydone-treated MonoMac6 cells were analysed for DNA microarray and displayed retardation in the expressivity of 33 NF-κB-dependent pro-inflammatory genes beyond prominent results on the interpretation of housekeeping genes (Erkel et al. 2007). PP possesses remarkable antitumour activity in MCF-7 cells and triple negative breast cancer cell lines, namely, MDA-MB-453, MDA-MB-468 and MDA-MB-231. PP-treated aggressive breast cancer cells portrayed notable cytotoxicity, suppressed invasion and migration and enhanced apoptosis along with significant changes in apoptosis-related protein. In PP, antitumour property seemed to be connected with its capability to prevent IκBα phosphorylation, with cytoplasmic accumulation of NF-κB (Arora et al. 2014).

23.2.8 Genus *Phellinus*

Genus *Phellinus* or wood inhabiting species belong to the family *Hymenochaetaceae*, among the remedially important mushrooms with cosmopolitan distribution. Species under this genus usually causes white rot in various types of woods of angiosperms and gymnosperms. However, different species of *Phellinus*, mycelial biomass or hymenophores were of therapeutic importance, which are used as folk remedy for dealing several disorders since ages and require vigorous scientific evaluation for the betterment of mankind (Azeem et al. 2018).

Phellinus linteus, locally called as 'Sangwhang' in Taiwan, is well approved in oriental countries like China, Japan and Korea and has been conventionally utilized as edible food and medicine. *P. linteus* carried various bioactive compounds which were familiar to boost health and treat various types of cancers (Park et al. 2009). Recently a bioactive polyphenol compound, hispolon (Fig. 23.11), was being isolated from *P. linteus*, and reports revealed that hispolon is capable of inducing apoptosis in human epidermoid KB cell following mitochondria-mediated apoptotic pathway (Chen et al. 2006). The efficiency of hispolon was further examined by Chen et al. (2008b) on human gastric cancer cells by investigating the molecular mechanism behind the cell death. They reported that hispolon has greater susceptibility towards malignant cells than normal gastric cells and potentiated ROS-mediated apoptosis in gastric cancer cell lines—SGC-7901, MGC-803 and MKN-45. Another report was yielded where hispolon was found to exert its anti-proliferative property against NB4 human leukaemia cells (Chen et al. 2013). Lu et al. (2009) proclaimed their findings on hispolon as possible inhibitor of both the breast and bladder cancer cell growth. The mechanistic pathway involved the ubiquitination and retardation of MDM2 via MDM2-recruited activated ERK1/ ERK2 on treating cells with hispolon.

Similar bioactive compound, hispolon (Fig. 23.11), had also been isolated as an active phenolic compound from the mushroom *Phellinus igniarius* which was earlier obtained from *P. linteus*. Here hispolon shows productive anticancer activity against a different cell line, hepatocellular carcinoma Hep3B cells by different-mechanistic pathway which includes ERK modulation, phosphorylation and cell cycle arrest at S phase (Huang et al. 2011).

23.2.9 *Polyporus umbellatus*

Polyporus umbellatus, one of the major mushrooms under the indexed Chinese pharmacopoeia, was widely grown and consumed in North America, Asia and Europe. More than 1000 years ago, the medicinal value of *P. umbellatus* has been documented in China and Japan (Zhao et al. 2010). One of the prime secondary metabolites in *P. umbellatus* is steroid. Among other steroids ergosta-4, 6, 8(14), 22-tetraen-3-one (ergone) is the most important steroid obtained from *Polyporus umbellatus*, which has been reported as a potential anticancer agent (Lee et al. 2005). From the previous literature survey, seven polyporusterones that isolated from this species have been reported which had dissimilar inhibitive consequences on leukaemia 1210 cells (Ohsawa et al. 1992). Zhao et al. (2010) purified and identified five known steroids, ergosta-6,22-dien-3β,5α,6β-triol, ergone, 5α,8α-epidioxy-(22E,24R)-ergosta-6,22-dien-3β-ol, (22E,24R)-ergosta-7,22-dien-3β-ol and polyporusterone B (Fig. 23.2n) from this mushroom by bioassay-guided approach. All the isolated compounds were screened to find its in vitro cytotoxic effect over HepG2 cells on time-dependent maner for 72 h. The results manifested that the cytotoxic effect of these five compounds was dose-dependent on HepG2 cells, and each compound had different ways of inhibitory effect on cancer cells. Among the five compounds, ergone represented strong inhibition of the proliferation of HepG2 cells than other four compounds. The inhibitory activity of the ergone was also compared with the other cancer cell lines HEp-2, HeLa and with normal cell HUVEC where the results showed that the inhibitory potentiality of the ergone on HeLa, HEp-2 and HepG2 was higher than HUVEC cell lines in a concentration and time-dependent manner. More analyses were made on the molecular mechanism behind the ergone extracted which was isolated from the sclerotium of *P. umbellatus*, working against human hepatocellular carcinoma HepG2 cells. Results indicated that cell proliferation is supressed due to initiation of apoptosis generated from the caspase stimulation and upon arrest at the G2/M phase (Zhao et al. 2011).

23.2.10 *Poria cocos*

The sclerotia of *Poria cocos* (belong from *Polyporaceae* family) have been mentioned as poria, and because of their diuretic efficacy, they are used as adjuvants (Haranaka et al. 1987; Yasukawa et al. 1996). Six lanostane triterpene acids were found from *P. cocos*, and their toxicity towards various cell lines like leukaemia (HL60), melanoma (CRL1579), lung (A549), breast (SK-BR-3), ovary (NIH: OVCAR-3), prostate (DU145), stomach (AZ521) and pancreas (PANC-1) was evaluated. Of all the poria-derived triterpenes, poricotriol A was observed to enhance apoptotic cell death through caspase-independent mitochondrial-mediated pathway chiefly by apoptosis-inducing factor (AIF) translocation in A549 cells, whereas poricotriol A incorporated both extrinsic and intrinsic pathways to induce cell death in HL60 cell line (Kikuchi et al. 2011). Nine lanostane-type triterpene acids were synthesized from the sclerotia of *Poria cocos*, of which only dehydroebriconic acid (Fig. 23.1k) could potentially inhibit DNA topoisomerase II activity, whereas modestly hindering the activities of DNA polymerases α, β, γ, δ, κ, ε, η, ι and λ. Lately another isolated compound, dehydrotrametenonic acid, exhibited average preventive effectivity against topoisomerase II (IC_{50} = 37.5 μM). Both dehydrotrametenonic acid and dehydroebriconic acid efficiently inhibit the proliferation as well as induced G1 phase arrest in the human gastric cancer cells (Akihisa et al. 2004; Mizushina et al. 2004).

23.3 Some Less Studied Mushrooms with Important Anticancer Secondary Metabolites

Agaricus bisporus is the most extensively cultivated species of other consumable mushrooms worldwide. A dynamic sulfhydryl reagent found from the gill tissue area of *Agaricus bisporus* mushroom termed as 490 quinone (γ-L-glutaminyl-4-hydroxy-2,5-benzoquinone) (Fig. 23.2a) has markedly supressed the L1210 murine leukaemia DNA polymerase α with little hindrance to DNA polymerase β. Inhibiting DNA synthesis by the compound has notably effected its cytotoxicity and increases the prospective of γ-L-glutaminyl-4-hydroxybenzene as a potential antitumour agent (Ferreira et al. 2010).

Agaricus blazei or almond mushroom is another well-known medicinal mushroom under the genus *Agaricus* which has been traditionally used in Brazil as an oriental medicine for the prevention of cancer. Northeastern United States, Canada, California, Britain, Hawaii, the Netherlands, Taiwan, Australia and Brazil are some of the ideal spots for growth of *Agaricus blazei*. A phytochemical ergosterol (Fig. 23.2b) has been extracted from the lipid fraction of *Agaricus blazei* which has an effective tumour growth retardation effect. Various tests suggested that the ergosterol activity as an antitumour agent might be due to its direct inhibitory effects on angiogenesis which is induced by solid tumours (Didukh et al. 2004; Takaku et al. 2001).

Albatrellus confluens, a species of polypore fungus with a beautiful pale orange cap, mainly found across various ecosystems of North America and Asia, has shown a wide range of pharmacological properties. From the fruiting bodies of this edible mushroom, two closely related bioactive phenolic compounds, grifolin (Fig. 23.2f) and neogrifolin, were isolated (Hirata and Nakanishi 1950; Vrkoc et al. 1977; Besl et al. 1977). Neogrifolin and grifolin both inhibit cancer cell growth proliferation in a wide range of cancer cell lines and trigger the induction of apoptosis by the way of dysfunction of Akt signalling pathway in the case of osteosarcoma cell lines—U2OS and MG63 (Jin et al. 2007). Grifolin mediates mitochondrial-dependent apoptosis pathway as well as death receptor signalling pathway for apoptosis induction in human nasopharyngeal carcinoma cell line CNE1 (Ye et al. 2005).

The extracts of *Amauroderma rude* had undergone various processes to yield a bioactive compound called as ergosterol. Ergosterol (Fig. 23.2b) is observed to induce cancer cell death depending on both dose and time parameter. In in vivo experiment, treating B16-injected normal mice with ergosterol has prolonged mouse survival. Further scientific reports suggested that ergosterol-arbitrated inhibition of breast cancer cell viability progresses through apoptosis via upregulating the expression of tumour suppressor Foxo3 (Li et al. 2015).

Astraeus odoratus is quite famous as superfood in Thailand but is overpriced because of the poor natural availability and difficulties in industrial production through cultivation procedure (Isaka et al. 2016). Four unique lanostane triterpenes, astraodoric acids A–D, were isolated and characterized together with other familiar compound astraodorol (artabotryols A), ergosterol, hypaphorine and nicotinic acid from *A. odoratus*. Astraodoric acids A and B went for their in vitro cytotoxic test in different human carcinoma cell lines. In the mentioned research, astraodoric acids A and B (Fig. 23.2c) exhibited better cytotoxic activity of IC_{50} 34.69 and 18.57 μg/mL against KB and 19.99 and 48.35 μg/mL against NCI-H187 cell lines (Arpha et al. 2012). Twelve more novel lanostane triterpenoids, named astraeusins A–L (1–12), were elucidated from same mushroom and subjected to investigate their cytotoxic properties against the cancer cell lines (NCI-H187, MCF-7 and KB) and also on non-cancerous Vero cells. Few of them carried a good to moderate anticancer potentiality against the tested cell lines (Isaka et al. 2016).

Three benzoquinoid macrolidic fungal bioactive metabolites were synthesized from non-toxigenic basidiomycetes of *Clitocybe clavipes* termed as clavilactones (3a, b-quinones and 3c-hydroquinone) and were recognized as efficient inhibitors of protein tyrosine kinases. Naturally synthesis 3a, b-quinones and 3c-hydroquinone showed the inhibiting property in kinase assays antagonistic against the epidermal growth factor receptor (EGF-R) tyrosine kinases and Ret/ptc1. 3b-Quinone was found to be a non-competitive inhibitor of EGF-R compared to poly [Glu (6) Ala (3) Tyr] or ATP. 3c-Hydroquinone biasedly restarted the A431 cell growth, which turns on the overexpression of active EGF-R, as a defender of IGROV-1 and SKOV-3 cells, which show low levels of the receptor over them (Zaidman et al. 2005).

Daedalea dickinsii is one of the species of brown-rot fungi belonging to the family of *Fomitopsidaceae*. Polyporenic acid C (Fig. 23.2e) is a lanostane category triterpenoid isolated from *Daedalea dickinsii* and has been reported to inhibit physiological extracellular matrix protein collagenase (MMP-1) by cleaving all structurally important α-chains from the indigenous interstitial collagens (Kawagishi et al. 1997). It has been examined that the hydroquinone derivative of polyporenic acid C is able to inhibit stromelysin (MMP-3); in effect a wide range of protein substrates, fibronectin, laminin and gelatin, and the core protein of cartilage proteoglycans were degraded. It is also a potential inhibitor of gelatinase (MMP-9) that cleaves gelatin and degrades the basement membrane (Zaidman et al. 2005). The huge MMPs mediate remodelling of physiological extracellular matrix and are involved in controlling tumour cell growth and metastasis.

Fomitella fraxinea offered some medicinally important glycans, steroids and triterpenes, which were used as a transitional medicine. Two very important triterpenoid compounds, fomitellic acids A and B, were found from the *F. fraxinea* mushroom's fruiting body that showed a potential inhibitory effect on the activities of mammalian DNA polymerases α and β (Zaidman et al. 2005). Fomitellic acid A (Fig. 23.2g) has significant effect on prevention of gastric cancer cell (NUGC) proliferation, with significant LD_{50} values of 30 and 38 μM (Mizushina et al. 2000).

Fomitopsis betulina (formerly popular as *Piptoporus betulinus*), a higher basidiomycota, generally grows on birch woods in the northern hemisphere; the fungus can also be successfully cultured both as fruiting body and as mycelium. *F. betulina* is a traditionally used medicinal mushroom for the treatment of cancer (Pleszczyńska et al. 2017). Hydroquinone (E)-2-(4-hydroxy-3-methyl-2-butenyl)-hydroquinone and polyporenic acid C (Fig. 23.2e) have been characterized from this mushroom as matrix metalloproteinase (MMPs) inhibitors. *F. betulina* has been conventionally used as a superfood for the treatment of rectal cancer in Czech Republic (Ferreira et al. 2010).

Another species from *Fomitopsis* is also famous for its medicinal importance, i.e. *Fomitopsis nigra* belonging to *Polyporaceae* family, and produces various active components, triterpenoids and triterpene glycoside complexes [1–4]. In this family, fomitoside K, which is a novel lanostane triterpene glycoside, has been isolated from methanolic extract of *Fomitopsis nigra* that is able to induce highly significant apoptotic effects on the human oral squamous YD-10B carcinoma cells (Lee et al. 2012). Fomitoside K showed significant apoptosis on various cancer cell lines including oral squamous carcinoma cell (KB), cervical cancer cell (HELA), breast cancer cell (MCF-7), human osteosarcoma epithelial cells (U2OS) and murine colon adenocarcinoma cell (C26) via mitochondrial-mediated ROS-dependent signalling pathway (Lee et al. 2012; Bhattarai et al. 2012).

Genus *Gymnopilus* is a gilled mushroom which belongs to the *Cortinariaceae* family and has been found around the world. Naturally occurring catechol-type compound hispidin [6-(3,4-dihydroxystyryl)-4-hydroxy-2-pyrone] (Fig. 23.1j) is isolated from three species of *Gymnopilus*, namely, *Gymnopilus patriae*, *Gymnopilus marginatus* and *Gymnopilus parvisporus*. Gonindard et al. (1997) synthesized hispidin isolated from *Gymnopilus marginatus* exerting effective cytotoxic effect between 1 mmol/L and 0.1 μmol/L for the cancerous SCL-1 human keratinocyte cell line, cancerous Capan-1 human pancreatic duct cell line and normal MRC-5 human fibroblasts. Amusingly treating with hispidin has enhanced the apoptosis of cancerous cells as rather than normal cells. Hispidin obtained from *Gymnopilus parvisporus* and *Gymnopilus patriae* emerged as a potent inhibitor of protein kinase C β (PKCβ) which has a vital role during angiogenesis via dysfunctioning VEGF signal transduction pathway (Zaidman et al. 2005).

Hericium erinaceus is a wild edible mushroom originating in North America, Europe and Asia and is consumed for its nutritional and health benefit value. Some reports had shown that culinary mushroom, *Hericium erinaceus*, had potential medicinal properties. Many bioactive compounds like erinacines, hericerins, resorcinols, steroids, monoterpenes and diterpenes and aromatic compounds have been isolated which were reported to have diverse biological activities including anticarcinogenic activity (Mizuno 1999a). Recently new diterpene hericerin and hericerin A have been isolated which exhibited significant reduction in the growth of human leukaemia (K562) and human prostate cancer (LANCAP) cells (Zhang et al. 2015).

Detailed exploration on the mushroom *Hexagonia speciosa* (*Polyporaceae* family) leads towards a series of oxygenated cyclohexanoids. Among cyclohexanoids, speciosin B (Fig. 23.2d) displayed a prominent cytotoxic effect on a wide range of tumour cell lines with LD_{50} values in the range of 0.23–3.30 μM (De Silva et al. 2013).

Lampteromyces japonicus is a gilled toxic mushroom originating in China and Korea but native to Japan and Eastern Asia. The fruit body contains sesquiterpene compounds illudin S and illudin M with potential antitumour and antibacterial properties. With high toxicity profile, illudin S and illudin M were studied further with various derivatives. Of these the most efficacious hemisynthetic derivatives were irofulven which showed distinct anticancer activity against solid tumour cells (McMorris et al. 1996, 2001; Poucheret et al. 2006) and other several human cancer cell lines including ovarian carcinoma, colorectal carcinoma, pancreatic carcinoma, malignant glioma and non-small cell lung cancer (Zaidman et al. 2005).

Another edible gilled mushroom belonging from the *Agaricaceae* family, *Lepiota americana* is a widespread fungus all over the world. The natural resource of sulfatase inhibitor 5, 8-epidioxy 24(*R*)-methylcholesta-6, 22-dien-3β-ol has been extracted from the basidiocarp of *Lepiota americana* (Zaidman et al. 2005). Sulfation and desulfation are the essential physiological reactions in the metabolism of most of the steroid hormones. Sulfatase inhibitor 5, 8-epidioxy

24(*R*)-methylcholesta-6,22-dien-3β-ol is important for modulation of the estrogen production for estrogen-dependent breast cancer treatment (Kim et al. 2000).

Macrolepiota procera is also popularly known as 'parasol mushroom', as it has bigger fruiting body that looks like a parasol. *Macrolepiota procera* mushroom is widespreaded over the temperate regions. Chen et al. (2018) isolated, elucidated and biologically evaluated 12 lanostane triterpenoids from the fruiting bodies of *M. procera* named as lepiotaprocerins A – L, 1–12 (Fig. 23.2i). Screening for the cytotoxicity effect of all the compounds was done on five human cancer cell lines (A-549, HL-60, SW480, SMMC-7721 and MCF-7). Among them, compounds 1–6 showed no result in the cytotoxicity assay ($IC_{50} > 40$ μM), whereas compounds 7–12 manifested significant inhibition activity against cancer cells.

Neonothopanus nambi (basidiomycete), a poisonous luminescent mushroom, is native of Malaysia, Central America, South America and Australia. Four newly isolated aristolane category sesquiterpenes named as nambinones A–C, sesquiterpene nambinone D, 1-epi-nambinone B and aurisin A and a new dimeric sesquiterpene aurisin K were obtained from the luminescent mushroom producing two specific isolates. Sesquiterpene nambione C, dimer aurisin K and aurisin A (Fig. 23.2j) exerted pronounced cytotoxic effects on NCI-H187 cancer cell lines. Additionally, aurisin A and K also proved their potential cytotoxicity against the cell lines of cholangiocarcinoma (Kanokmedhakul et al. 2012).

Omphalotus illudens commonly named as jack-o'-lantern mushroom is a large and poisonous mushroom grown mainly in Eastern North America. Similar to *Lampteromyces japonicus*, *O. illudens* also produces sesquiterpenoid toxin illudin S (Fig. 23.2k) and illudin M (Fig. 23.2l). This toxic sesquiterpenoid had already showed its antitumour effect specifically towards tumour cells than normal cells. Identical to *L. japonicus* illudin S and illudin M extracted from *O. illudens* were further tested with most potential derivatives irofulven (McMorris et al. 1996, 2001) which exerted strong anticarcinogenic effects on various human cancer cell lines (Zaidman et al. 2005) following the similar pathway as DNA and protein alkylating agent (Poucheret et al. 2006).

Epoxy compound derivative of quinones, panepoxydone (Fig. 23.1b) has been isolated from *Panus conchatus* and *Panus rudis* which was earlier been reported from *Lentinus crinitus*. This potential quinone compound was associated with regulating the action of NF-κB, which promotes tumour growth (Zaidman et al. 2005). In recent times from the cultures of *P. conchatus*, a new cadinane sesquiterpenoid was obtained and called as panutorulon. This newly obtained cadinane sesquiterpenoid was experimented for its disrupting effectivity against isozymes of 11β-hydroxysteroid dehydrogenases (11β-HSD) and for its significant activity as cytotoxic agent on five different human cancer cell lines (Ding et al. 2018).

Pleurotus cornucopiae known as Tamogi dake mushroom with a brilliant yellow pileus is observed in the field of Japan (El Bohi et al. 2005). Few low-molecular-weight bioactives such as mono- and sesquiterpenoids, ergosterol and fatty acid esters were identified in oyster mushrooms (Golak et al. 2018). Mainly four new monoterpenoids and a single sesquiterpenoid were extracted from the mycelia of *P.*

cornucopiae which fermented over rice. This sesquiterpenoid has produced its cytotoxicity towards HepG2 and HeLa cancer cells (Wang et al. 2013).

Anticancer sterol 5α,8α-epidioxy-22E-ergosta-6,9(11),22-trien-3β-ol (9,11-dehydroergosterol peroxide [9(11)- DHEP)] and 5α,8α-epidioxy-22E-ergosta-6,22-dien-3β-ol [ergosterol peroxide (EP)] were obtained from *Sarcodon aspratus*. Isolated and purified 9(11)-DHEP seems to be a potential inhibitor of HL60 leukaemia cancer cell growth and also a stronger apoptosis inducer than 5α,8α-epidioxy-22E-ergosta-6,22-dien-3β-ol [ergosterol peroxide (EP)]. Moreover 5α,8α-epidioxy-22E-ergosta-6,9(11),22-trien-3β-ol (9,11-dehydroergosterol peroxide [9(11)-DHEP)] (Fig. 23.2m) had critically restrained the growth of HT29 cancer cells, but did not show any kind of effect on WI38 normal human fibroblast cells. Examining the anticancer mechanism has revealed that aforementioned isolated sterol upregulates the expression of cyclin-dependent kinase inhibitor 1A, thereby causing apoptosis and cell cycle arrest in HT29 cancer cell line (Kobori et al. 2006).

Tricholoma terreum routinely known as dirty tricholoma or grey knight is a grey capped, large mushroom belonging to the *Tricholomataceae* family and has been found to be consumed in different countries of Europe and Australia. Four new meroterpenoids were extracted from basidiomycete of *Tricholoma terreum* named as terreumols A – D (1–4) (Fig. 23.2o), with a very rare ten-membrane ring structure system significantly showing cytotoxic effect on five different human cancer cell lines counting breast cancer MCF-7, colon cancer SW480, hepatocellular carcinoma SMMC-7721, human myeloid leukaemia HL-60 and lung cancer A-549 cells (Yin et al. 2013).

23.4 Future Perspectives

This summarized data has shown that researchers and several companies have a growing interest in the anticancer activity of different mushroom's secondary metabolites for further investigation and commercialization. Only a few compounds have passed initial phases of clinical trials: HMAF from *L. japonicus*, grifolin and neogrifolin from *Albatrellus confluens* and metabolites isolated from different species of *Ganoderma* are some examples. Mushroom-originated compounds are set to create an uprising in therapeutics to repress diverse forms of cancers. However, choosing any mushroom compounds as cancer treatment therapeutics influenced by various factors needs thorough investigation. Intensive experimentation to overcome the limitations and preclinical trials on more complex animal models that can efficiently mimic the genetic profile and the microenvironment of human tumours successively are required in this area to fulfil the growing needs for less toxic, natural origin, bioactive compounds as potent anticancer therapeutics and drugs worldwide.

23.5 Conclusions

The arch of maintaining or rebuilding a strong health defence mechanism has led researchers to search for new natural health benefit substances. As a result of constant hunt for new and efficient health-improving substances, nature, the luxurious resource of bioactive compounds, has led to develop new drugs. Nutraceuticals originated from nature are intended to play a significant role as therapeutics since ancient times. Many evidences supported that some diseases can also be treated and cured successfully by use of functional foods or through the consumption of isolated biologically active compounds as nutraceuticals or as daily dietary supplements. Mushroom shows a remarkable potentiality as immune function enhancer. Many mushrooms and their active compounds have medicinal benefits that play a crucial part in cancer treatment. Under this chapter, we have briefly covered different secondary metabolites exerting potential anticancer activity, isolated from edible as well as non-edible mushrooms. Low-molecular-weight secondary metabolites efficiently interfere with tumour initiation and progression by activating several mechanisms, like boosting immune functions by enhancing antioxidant capacity, preventing cancer by exerting specific cytotoxic effects on tumour cells, interfering with tumour angiogenesis developments and regulating different nodes of signalling pathways that are involved in tumour-suppressive mechanisms.

References

Akihisa T, Mizushina Y, Ukiya M, Oshikubo M, Kondo S, Kimura Y, Suzuki T, Tai T (2004) Dehydrotrametenonic acid and dehydroeburiconic acid from *Poria cocos* and their inhibitory effects on eukaryotic DNA polymerase alpha and beta. Biosci Biotechnol Biochem 68:448–450

Akihisa T, Nakamura Y, Tagata M, Tokuda H, Yasukawa K, Uchiyama E, Suzuki T, Kimura Y (2007) Anti-inflammatory and anti-tumor-promoting effects of triterpene acids and sterols from the fungus *Ganoderma lucidum*. Chem Biodivers 4:224–231

Arora R, Yates C, Gary BD, McClellan S, Tan M, Xi Y, Reed E, Piazza GA, Owen LB, Dean-Colomb W (2014) Panepoxydone targets NF-kB and FOXM1 to inhibit proliferation, induce apoptosis and reverse epithelial to mesenchymal transition in breast cancer. PLoS One 9(6):e98370

Arpha K, Phosri C, Suwannasai N, Mongkolthanaruk W, Sodngam S (2012) Astraodoric acids A-D: new lanostane triterpenes from edible mushroom *Astraeus odoratus* and their Mycobacterium tuberculosis H37Ra and cytotoxic activity. J Agric Food Chem 60:9834–9841

Awadh Ali NA, Mothana RA, Lesnau A, Pilgrim H, Lindequist U (2003) Antiviral activity of *Inonotus hispidus*. Fitoterapia 74:483–485

Azeem U, Dhingra GS, Shri R (2018) Pharmacological potential of wood inhabiting fungi of genus *Phellinus* Quél.: an overview. J Pharmacogn Phytochem 7:1161–1171

Baby S, Johnson AJ, Balaji G (2015) Secondary metabolites from *Ganoderma*. Phytochemistry 114:66–101

Batra P, Sharma AK, Khajuria R (2013) Probing Lingzhi or Reishi medicinal mushroom *Ganoderma lucidum* (higher Basidiomycetes): a bitter mushroom with amazing health benefits. Int J Med Mushrooms 15:127–143

Berns H, Humar R, Hengerer B, Kiefer FN, Battegay EJ (2000) RACK1 is up-regulated in angiogenesis and human carcinomas. FASEB J 14:2549–2558

Besl H, Hoefle G, Jendrny B, Jagers E, Steglich W (1977) Pilzpigmente, XXXI. Farnesylphenole aus Albatrellus-Arten (Basidiomycetes). Chem Ber 110:3770–3776

Bhattarai G, Lee YH, Lee NH, Lee IK, Yun BS, Hwang PH, Yi HK (2012) Fomitoside-K from *Fomitopsis nigra* induces apoptosis of human oral squamous cell carcinomas (YD-10B) via mitochondrial signaling pathway. Biol Pharm Bull 35:1711–1719

Bok JW, Lermer L, Chilton J, Klingeman HG, Towers GH (1999) Antitumor sterols from the mycelia of *Cordyceps sinensis*. Phytochemistry 51:891–898

Brakhage AA (2013) Regulation of fungal secondary metabolism. Nat Rev Microbiol 11:21–32

Chang KL, Kung ML, Chow NH, Su SJ (2004) Genistein arrests hepatoma cells at G2/M phase: involvement of ATM activation and upregulation of p21waf1/cip1 and Wee1. Biochem Pharmacol 67:717–726

Chen NH, Zhong JJ (2011) p53 is important for the anti-invasion of ganoderic acid T in human carcinoma cells. Phytomedicine 18:719–725

Chen W, He FY, Li YQ (2006) The apoptosis effect of hispolon from *Phellinus linteus* (Berkeley & Curtis) Teng on human epidermoid KB cells. J Ethnopharmacol 105:280–285

Chen NH, Liu JW, Zhong JJ (2008a) Ganoderic acid me inhibits tumor invasion through down-regulating matrix metalloproteinases 2/9 gene expression. J Pharmacol Sci 108:–212, 216

Chen W, Zhao Z, Li L, Wu B, Chen SF, Zhou H, Wang Y, Li YQ (2008b) Hispolon induces apoptosis in human gastric cancer cells through a ROS-mediated mitochondrial pathway. Free Radic Biol Med 45:60–72

Chen Y, Chen YC, Lin YT, Huang SH, Wang SM (2010a) Cordycepin induces apoptosis of CGTH W-2 thyroid carcinoma cells through the calcium-calpain-caspase 7-PARP pathway. J Agric Food Chem 58:11645–11652

Chen NH, Liu JW, Zhong JJ (2010b) Ganoderic acid T inhibits tumor invasion *in vitro* and *in vivo* through inhibition of MMP expression. Pharmacol Rep 62:150–163

Chen YC, Chang HY, Deng JS, Chen JJ, Huang SS, Lin IH, Kuo WL, Chao W, Huang GJ (2013) Hispolon from *Phellinus linteus* induces G0/G1 cell cycle arrest and apoptosis in NB4 human leukaemia cells. Am J Chin Med 41:1439–1457

Chen Y, Peng GF, Han XZ, Wang W, Zhang GQ, Li X (2015) Apoptosis prediction via inhibition of AKT signaling pathway by neogrifolin. Int J Clin Exp Pathol 8:1154–1164

Chen HP, Zhao ZZ, Li ZH, Huang Y, Zhang SB, Tang Y, Yao JN, Chen L, Isaka M, Feng T, Liu JK (2018) Anti-proliferative and anti-inflammatory lanostane triterpenoids from the polish tdible mushroom *Macrolepiota procera*. J Agric Food Chem 66:3146–3154

Cheng CR, Yue QX, Wu ZY, Song XY, Tao SJ, Wu XH, Xu PP, Liu X, Guan SH, Guo DA (2010) Cytotoxic triterpenoids from *Ganoderma lucidum*. Phytochemistry 71:1579–1585

China PR (1986) Dictionary of traditional Chinese medicine. Jiangsu New Medical College (eds). Shanghai Scientific, Shanghai, pp 2191–2193

Choi YH, Lee WH, Park KY, Zhang L (2000) p53-independent induction of p21 (WAF1/CIP1), reduction of cyclin B1 and G2/M arrest by the isoflavone genistein in human prostate carcinoma cells. Jpn J Cancer Res 91:164–173

De Silva DD, Rapior S, Fons F, Bahkali AH, Hyde KD (2012) Medicinal mushrooms in supportive cancer therapies: an approach to anticancer effects and putative mechanisms of action- a review. Fungal Divers 55:1–35

De Silva DD, Rapior S, Sudarman E, Stadler M, Xu J, Alias SA, Hyde KD (2013) Bioactive metabolites from macrofungi: ethnopharmacology, biological activities and chemistry. Fungal Divers 62:1–40

Debbab A, Aly AH, Proksch P (2011) Bioactive secondary metabolites from endophytes and associated marine derived fungi. Fungal Divers 49:1–12

Debbab A, Aly AH, Proksch P (2012) Endophytes and associated marine derived fungi-ecological and chemical perspectives. Fungal Divers 57:45–83

Dias DA, Urban S, Roessner U (2012) A historical overview of natural products in drug discovery. Meta 2:303–336

Didukh M, Wasser SP, Nevo E (2004) Impact of the family Agaricaceae (Fr.) Cohn on nutrition and medicine, Volz A (ed). ARG Ganter Verlag KG, Konigstein, Germany

Ding JH, Li ZH, Feng T, Liu JK (2018) A new cadinane sesquiterpenoid from cultures of the basidiomycete *Panus conchatus*. Nat Prod Res 32:2333–2337

Du YC, Wu TY, Chang FR, Lin WY, Hsu YM, Cheng FT, Lu CY, Yen MH, Tsui YT, Chen HL, Hou MF, Lu MC, Wu YC (2012) Chemical profiling of the cytotoxic triterpenoid-concentrating fraction and characterization of ergostane stereo-isomer ingredients from *Antrodia camphorata*. J Pharm Biomed Anal 58:182–192

El Bohi KM, Sabik L, Muzandu K, Shaban Z, Soliman M, Ishizuka M, Kazusaka A, Fujita S (2005) Antigenotoxic effect of *Pleurotus cornucopiae* extracts on the mutagenesis of Salmonella typhimurium TA98 elicited by benzo[a]pyrene and oxidative DNA lesions in V79 hamster lung cells. Jpn J Vet Res 52:163–172

Erkel G, Anke T, Sterner O (1996) Inhibition of NF-kappa B activation by panepoxydone. Biochem Biophys Res Commun 226:214–221

Erkel G, Wisser G, Anke T (2007) Influence of the fungal NF-kB inhibitor panepoxydone on inflammatory gene expression in MonoMac6 cells. Int Immunopharmacol 7:612–624

Evidente A, Kornienko A, Cimmino A, Andolfi A, Lefranc F, Mathieud V, Kiss R (2014) Fungal metabolites with anticancer activity. Nat Prod Rep 31:617–627

Ferreira ICFR, Vaz JA, Vasconcelos MH, Martins A (2010) Compounds from wild mushrooms with antitumor potential. Anti Cancer Agents Med Chem 10:424–436

Figueiredo L, Regis WCB (2017) Medicinal mushrooms in adjuvant cancer therapies: an approach to anticancer effects and presumed mechanisms of action. Figueiredo and Régis Nutrire 42:28

Friedman M (2015) Chemistry, nutrition, and health-promoting properties of (Lion's mane) mushroom fruiting bodies and mycelia and their bioactive compounds. J Agr Food Chem 63(32):7108–7123

Gan KH, Kuo SH, Lin C (1998) Steroidal constituents of *Ganoderma applanatum and Ganoderma neo-japonicum*. J Nat Prod 61:1421–1422

Gao JJ, Min BS, Ahn EM, Nakamura N, Lee HK, Hattori M (2002) New triterpene aldehydes, lucialdehydes A-C, from *Ganoderma lucidum* and their cytotoxicity against murine and human tumor cells. Chem Pharmacol Bull 50:837–840

Gao JJ, Hirakawa A, Min BS, Nakamura N, Hattori M (2006) *In vivo* antitumor effects of bitter principles from the antlered form of fruiting bodies of *Ganoderma lucidum*. J Nat Med 60:42–48

Golak IS, Kałużewicz A, Spiżewski T, Siwulski M, Sobieralski K (2018) Bioactive compounds and medicinal properties of oyster mushrooms (*Pleurotus* sp.). A review Folia Hort 30(2)

Gonindard C, Bergonzi C, Denier C, Sergheraert C, Klaebe A, Chavant L, Hollande E (1997) Synthetic hispidin, a PKC inhibitor, is more cytotoxic toward cancer cells than normal cells *in vitro*. Cell Biol Toxicol 13:141–153

Handa N, Yamada T, Tanaka R (2010) An unusual lanostane-type triterpenoid, spiroinonotsuoxodiol, and other triterpenoids from *Inonotus obliquus*. Phytochemistry 71:1774–1779

Haranaka K, Satomi N, Sakurai A, Yasukawa K, Takido M (1987) Antitumor activities of Zyuzentaiho-to and Cinnamomi cortex. J Med Pharmaceut Soc Wakan-Yaku 4:49–58

Harvey AL (2008) Natural products in drug discovery. Drug Discov Today 13:894–901

Hirata Y, Nakanishi K (1950) Grifolin, an antibiotic from a basidiomycete. J Biol Chem 184:135–143

Holliday J, Cleaver M (2008) Medicinal value of the cterpillar fungi species of the genus *Cordyceps* (Fr.) link (ascomycetes)-a review. Int J Med Mushrooms 10:219–234

Hsu CL, Yu YS, Yen GC (2008) Lucidenic acid B induces apoptosis in human leukemia cells via a mitochondria-mediated pathway. J Agric Food Chem 56:3973–3980

Huang GJ, Deng JS, Huang SS, Hu ML (2011) Hispolon induces apoptosis and cell cycle arrest of human hepatocellular carcinoma Hep3B cells by modulating ERK phosphorylation. J Agric Food Chem 59:7104–7113

Huang HC, Liaw CC, Yang HL, Hseu YC, Kuo HT, Tsai YC, Chien SC, Amagaya S, Chen YC, Kuo YH (2012) Lanostane triterpenoids and sterols from *Antrodia camphorata*. Phytochemistry 84:177–183

Isaka M, Palasarn S, Srikitikulchai P, Vichai V, Komwijit S (2016) Astraeusins AeL, lanostane triterpenoids from the edible mushroom *Astraeus odoratus*. Tetrahedron 72:3288–3295
Ito Y, Arita M, Adachi K, Shibata T, Sawai H, Ohno M (1981) Chirally selective synthesis of sugar moiety of nucleosides by chemicoenzymatic approach: L- and D-riboses, showdomycin, and cordycepin. Nucleic Acids Symp Ser 10:45–48
Ivanova TS, Krupodorova TA, Barshteyn VY, Artamonova AB, Shlyakhovenko VA (2014) Anticancer substances of mushroom origin. Exp Oncol 36:58–66
Jeong JW, Jin CY, Park C, Hong SH, Kim GY, Jeong YK, Lee JD, Yoo YH, Choi YH (2011) Induction of apoptosis by cordycepin via reactive oxygen species generation in human leukemia cells. Toxicol In Vitro 25:817–824
Jiang J, Grieb B, Thyagarajan A, Sliva D (2008) Ganoderic acids suppress growth and invasive behavior of breast cancer cells by modulating AP-1 and NF-κB signaling. Int J Mol Med 21:577–584
Jiang J, Jedinak A, Sliva D (2011) Ganodermanontriol (GDNT) exerts its effect on growth and invasiveness of breast cancer cells through the down-regulation of CDC20 and uPA. Biochem Biophys Res Commun 415:325–329
Jin S, Pang RP, Shen JN, Huang G, Wang J, Zhou JG (2007) Grifolin induces apoptosis via inhibition of PI3K/AKT signaling pathway in human osteosarcoma cells. Apoptosis 12:1317–1326
Johnson BM, Doonan BP, Radwan FF, Haque A (2010) Ganoderic acid DM: an alternative agent for the treatment of advanced prostate cancer. Open Prostate Cancer J 3:78–85
Joseph TP, Chanda W, Padhiar AA, Batool S, LiQun S, Zhong M, Huang M (2018) A preclinical evaluation of the antitumor activities of edible and medicinal mushrooms: a molecular insight. Integr Cancer Ther 17:200–209
Kadomatsu M, Nakajima S, Kato H, Gu L, Chi Y, Yao J, Kitamura M (2012) Cordycepin as a sensitizer to tumour necrosis factor (TNF)-α-induced apoptosis through eukaryotic translation initiation factor 2α (eIF2α)- and mammalian target of rapamycin complex 1 (mTORC1)-mediated inhibition of nuclear factor (NF)-κB. Clin Exp Immunol 168:325–332
Kanokmedhakul S, Lekphrom R, Kanomedhakul K, Hahnvajanawong C, Bua-art S, Saksirirat W, Prabpai S, Kongsaeree P (2012) Cytotoxic sesquiterpenes from luminescent mushroom *Neonothopanus nambi*. Tetrahedron 68:8261–8266
Kawagishi H, Li H, Tanno O, Inoue S, Ikeda S, Ohnishi-Kameyama M, Nagata T (1997) A lanostane-type triterpene from a mushroom *Daedalea dickinsii*. Phytochemistry 46:959–961
Kellof GJ, Sigman CC (2000) Intervention and chemoprevention of cancer. CCS Associates, Mountain View
Kikuchi T, Uchiyama E, Ukiya M, Tabata K, Kimura Y, Suzuki T, Akihisa T (2011) Cytotoxic and apoptosis-inducing activities of triterpene acids from *Poria cocos*. J Nat Prod 74:137–144
Kim H, Hall P, Smith M, Kirk M, Prasain JK, Barnes S, Grubbs C (2004) Chemoprevention by grape seed extract and Genistein in carcinogen- induced mammary cancer in rats is diet dependent. J Nutr 134:3445–3452
Kim DS, Jeong HJ, Bhat KP, Park SY, Kang SH, Yoo EH, Lee M, Lee HW, Krueger RJ (2000) Aromatse and sulfatase inhibitors from *Lepiota americana*. Planta Med 66:78–79
Kim YJ, Park J, Min BS, Shim SH (2011) Chemical constituents from the sclerotia of *Inonotus obliquus*. J Korean Soc Appl Biol Chem 54:287–294
Kimura Y, Taniguchi M, Baba K (2002) Antitumor and antimetastatic effects on liver of triterpenoid fractions of *Ganoderma lucidum*: mechanism of action and isolation of an active substance. Anticancer Res 22:3309–3318
Kinge TR, Mih AM (2011) Secondary metabolites of oil palm isolates of *Ganoderma zonatum* Murill. From Cameroon and their cytotoxicity against five human tumour cell lines. Afr J Biotechnol 10:8440–8447
Kobori M, Yoshida M, Ohnishi-Kameyama M, Takei T, Shinmoto H (2006) 5alpha, 8alpha-epidioxy-22E-ergosta-6,9(11),22-trien-3betaol from an edible mushroom suppresses growth of HL60 leukemia and HT29 colon adenocarcinoma cells. Biol Pharm Bull 29:755–759
Kozlovskii AG, Zhelifonova VP, Antipova TV (2013) Fungi of the genus Penicillium as producers of physiologically active compounds (review). Appl Biochem Microb 49:1–10

Kubo K, Aoki H, Naba N (1994) Anti-diabetic activities present in the fruit body of *Grifola frondosa* (maitake). Biol pharm bull. 17: 1106-1110.Lee IK, Jung JY, Yeom JH, Ki DW, Lee MS, yeo WH & Yun BS (2012). Fomitoside K, a new lanostane triterpene glycoside from the fruiting body of *Fomitopsis nigra*. Mycobiology 40:76–78

Lee WY, Park Y, Ahn JK, Park SY, Lee HJ (2005) Cytotoxic activity of ergosta-4,6, 8(14),22-tetraen-3-one from the sclerotia of *Polyporus umbellatus*. Bull Kor Chem Soc 26:1464–1466

Lee SJ, Moon GS, Jung KH, Kim WJ, Moon SK (2010) C-Jun N-terminal kinase 1 is required for cordycepin-mediated induction of G2/M cell-cycle arrest via p21WAF1 expression in human colon cancer cells. Food Chem Toxicol 48:277–283

Lee HJ, Burger P, Vogel M, Friese K, Bruning A (2012) The nucleoside antagonist cordycepin causes DNA double strand breaks in breast cancer cells. Investig New Drugs 30:1917–1925

Lee HH, Jeong JW, Lee JH, Kim GY, CHEONG J, Jeong YK, Yoo YH, Choi YH (2013a) Cordycepin increases sensitivity of Hep3B human hepatocellular carcinoma cells to TRAIL-mediated apoptosis by inactivating the JNK signaling pathway. Oncol Rep 30:1257–1264

Lee HH, Park C, Jeong JW, Kim MJ, Seo MJ, Kang BW, Park JU, Kim GY, Choi BT, Choi YH, Jeong YK (2013b) Apoptosis induction of human prostate carcinoma cells by cordycepin through reactive oxygen species-mediated mitochondrial death pathway. Int J Oncol 42:1036–1044

Levin RM, Juan YS, Schuler C, Leggett RE, Lin AD (2012) Medicinal properties of *Antrodia Camphorata* - a review. Curr Top Nutraceutical Res 10:53–60

Li Y, Mi C (2003) Proliferation inhibition and apoptosis onset in human ovarian carcinoma cell line SKOV3 induced by genistein. Ai Zheng 22:586–591

Li CH, Chen PY, Chang UM, Kan LS, Fang WH, Tsai KS, Lin SB (2005) Ganoderic acid X, a lanostanoid triterpene, inhibits topoisomerases and induces apoptosis of cancer cells. Life Sci 77:252–265

Li F, Wang Y, Wang X, Li J, Cui H, Niu M (2012) Ganoderic acids suppress growth and angiogenesis by modulating the Nf-kappab signaling pathway in breast cancer cells. Int J Clin Pharmacol Ther 50:712–721

Li YB, Liu RM, Zhong JJ (2013) A new ganoderic acid from *Ganoderma lucidum* mycelia and its stability. Fitoterapia 84:115–122

Li F, Zhao C, Wang L (2014) Molecular-targeted agents combination therapy for cancer: developments and potentials. Int J Cancer 134:1257–1269

Li X, Wu Q, Xie Y, Ding Y, Du WW, Sdiri M, Yang BB (2015) Ergosterol purified from medicinal mushroom *Amauroderma rude* inhibits cancer growth *in vitro* and *in vivo* by up-regulating multiple tumor suppressors. Oncotarget 6:17832–17846

Liao Z, Cox JD, Komaki R (2007) Radiochemotherapy of esophageal cancer. J Thorac Oncol 2:553–568

Liu RM, Zhong JJ (2010) Ganoderic acid Mf and S induce mitochondria mediated apoptosis in human cervical carcinoma HeLa cells. Phytomedicine 18:349–355

Liu J, Shiono J, Shimizu K, Kukita A, Kukita T, Kondo R (2009) Ganoderic acid DM: anti-androgenic osteoclastogenesis inhibitor. Bioorg Med Chem Lett 19:2154–2157

Liu RM, Li YB, Zhong JJ (2012a) Cytotoxic and pro-apoptotic effects of novel ganoderic acid derivatives on human cervical cancer cells *in vitro*. Eur J Pharmacol 68:23–33

Liu RM, Li YB, Zhong JJ (2012b) Ganoderic acid Mk from *Ganoderma lucidum* mycelia in cervical cancer HeLa cells. Lat Am J Pharm 31:43–50

Liu J, Shimizu K, Tanaka A, ShinobuW OK, Nakamura T, Kondo R (2012c) Target proteins of ganoderic acid DM provides clues to various pharmacological mechanisms. Sci Rep 2:905

Lu TL, Huang GJ, Lu TJ, Wu JB, Wu CH, Yang TC, Iizuka A, Chen YF (2009) Hispolon from *Phellinus linteus* has antiproliferative effects via MDM2-recruited ERK1/2 activity in breast and bladder cancer cells. Food Chem Toxicol 47:2013–2021

Lung MY, Hsieh CW (2011) Antioxidant property and production of exopolysaccharide from *Armillaria mellea* in submerged cultures: effect of culture aeration rate. Eng Life Sci 11:482–490

Manasseh AT, Godwin JTA, Ubleni EE, Borisde OO (2012) Phytochemical properties of Ganoderma applanatum as potential agents in the application of nanotechnology in modern day medical practice. Asian Pac J Trop Biomed 2:580–583

Mcmorris TC, Kelner MJ, Wang W, Yu J, Estes LA, Taetle R (1996) (Hydroxymethyl) acylfulvene: an illudin derivative with superior antitumor properties. J Nat Prod 59:896–899

McMorris TC, Yu J, Lira R, Dawe R, MacDonald JR, Waters SJ, Estes LA, Kelner MJ (2001) Structure-activity of antitumor agent irofulven (hydroxymethylacylfulvene) and analogues. J Org Chem 66:6158–6163

Mizuno T (1999a) Bioactive substances in *Hericium erinaceus* (bull.:Fr.) Pers. (Yamabushitake) and its medicinal utilization. Int J Med Mushrooms 1:105–119

Mizuno T (1999b) Extraction and development of antitumor-active polysaccharides from medicinal mushrooms in Japan. Int J Med Mushrooms 1:9–29

Mizuno T, Zhuang C (1995) Maitake, *Grifola frondosa*: pharmacological effects. Food Rev Int 11:135–149

Mizushina Y, Takahashi N, Hanashima L, Koshino H, Esumi Y, Uzawa J, Sugawara F, Sakaguchi K (1997) Lucidenic acid O and lactone, new terpene inhibitors of eukaryotic DNA polymerases from a basidiomycete *Ganoderma lucidum*. Bioorg Med Chem 7:2047–2052

Mizushina Y, Iida A, Ohta K, Sugawara F, Sakaguchi K (2000) Novel triterpenoids inhibit both DNA polymerase and DNA topoisomerase. Biochem J 350:757–763

Mizushina Y, Akihisa T, Ukiya M, Murakami C, Kuriyama I, Xu X, Yoshida H, Sakaguchi K (2004) A novel DNA topoisomerase inhibitor: dehydroebriconic acid, one of the lanostane-type triterpene acids from *Poria cocos*. Cancer Sci 95:354–360

Moradali MF, Mostafavi H, Ghods S, Hedjaroude GA (2007) Immunomodulating and anticancer agents in the realm of macromycetes fungi (macrofungi). Int Immunopharmacol 7:701–724

Nakata T, Yamada T, Taji S, Ohishi H, Wada S, Tokuda H, Sakuma K, Tanaka R (2007) Structure determination of inonotsuoxides a and B and *in vivo* antitumor promoting activity of inotodiol from the sclerotia of *Inonotus obliquus*. Bioorg Med Chem 15:257–264

Nomura M, Takahashi T, Uesugi A, Tanaka R, Kobayashi S (2008) Inotodiol, a lanostane triterpenoid, from *Inonotus obliquus* inhibits cell proliferation through caspase-3-dependent apoptosis. Anticancer Res 28:2691–2696

Nonaka Y, Ishibashi H, Nakai M, Shibata H, Kiso Y, Abe S (2008) Effects of the antlered form of *Ganoderma lucidum* on tumor growth and metastasis in cyclophosphamide-treated mice. Biosci Biotechnol Biochem 72:1399–1408

Ohsawa T, Yukawa M, Takao C, Murayama M, Bando H (1992) Studies on constituents of fruit body of *Polyporus umbellatus* and their cytotoxic activity. Chem Pharm Bull 40:143–147

Panda AK, Swain KC (2011) Traditional uses and medicinal potential of *Cordyceps sinensis* of Sikkim. J Ayurveda Integr Med 2:9–13

Park HG, Shim YY, Choi SO, Park WM (2009) New method development for nanoparticle extraction of water-soluble β-(1–3)-D-glucan from edible mushrooms, *Sparassis crispa* and *Phellinus linteus*. J Agric Food Chem 57:2147–2154

Paterson RRM (2001) Ganoderma- a therapeutic funfal biofactor. Phytochemistry 67:1985–2001

Paterson RRM (2006) *Ganoderma* - a therapeutic fungal biofactory. Phytochemistry 67:1985–2001

Peng X, Liu J, Xia J, Wang C, Li X, Deng Y, Bao N, Zhang Z, Qiu M (2015) Lanostane triterpenoids from *Ganoderma hainanense* J. D. Zhao. Phytochemistry 114:137–145

Petrova RD, Reznick AZ, Wasser SP, Denchev CM, Nevo E, Mahajna J (2008) Fungal metabolites modulating NF-κB activity: an approach to cancer therapy and chemoprevention (review). Oncol Rep 19:299–308

Pillai TG, Nair CKK, Janardhanan KK (2010) Enhancement of repair of radiation induced DNA strand breaks in human cells by *Ganoderma* mushroom polysaccharides. Food Chem 19:1040–1043

Pleszczyńska M, Lemieszek MK, Siwulski M, Wiater A, Rzeski W, Szczodrak J (2017) *Fomitopsis betulina* (formerly *Piptoporus betulinus*): the Iceman's polypore fungus with modern biotechnological potential. World J Microbiol Biotechnol 33:83

Popovic V, Zivkovic J, Davidovic S, Stevanovic M, Stojkovic D (2013) Mycotherapy of Cancer: an update on cytotoxic and antitumor activities of mushrooms, bioactive principles and molecular mechanisms of their action. Curr Top Med Chem 13:2791–2806

Poucheret P, Fons F, Rapior S (2006) Biological and pharmacological activity of higher fungi: 20-year retrospective analysis. Mycologie 27:311–333

Quang DN, Hashimoto T, Asakawa Y (2006) Inedible mushrooms: a good source of biologically active substances. Chem Rec 6:79–99

Rao A, Coan A, Welsh JE, Barclay WW, Koumenis C, Cramer SD (2004) Vitamin D receptor and p21/WAF1 are targets of genistein and 1,25-dihydroxyvitamin D3 in human prostate cancer cells. Cancer Res 64:2143–2147

Rao YK, Fang SH, Wu WS, Tzeng YM (2010) Constituents isolated from *Cordyceps militaris* suppress enhanced inflammatory mediator's production and human cancer cell proliferation. J Ethnopharmacol 131:363–367

Sanodiya BS, Thakur GS, Baghel RK, Prasad GBKS, Bisen PS (2009) *Ganoderma lucidum*: A Potent Pharmacological Macrofungus. Curr Pharm Biotechnol 10:717–742

Sharma JR, Das K, Mishra D (2013) The genus *Inonotus* and its related species in India. Mycosphere 4:809–818

Shervington A, Lu C (2008) Expression of multidrug resistance genes in normal and cancer stem cells. Cancer Investig 26:535–542

Silva S, Martins S, Karmali A, Rosa E (2012) Production, purification and characterisation of polysaccharides from *Pleurotus ostreatus* with antitumour activity. J Sci Food Agric 92:1826–1832

Sporn MB, Suh N (2002) Chemoprevention of cancer. Carcinogenesis 21:524–530

Suarez-Arroyo IJ, Loperena-Alvarez Y, Rosario-Acevedo R, Martínez-Montemayor MM (2017) *Ganoderma* spp.: a promising adjuvant treatment for breast cancer – a review. Molecules 4:15:1–23

Sun Y, Zhao Z, Feng Q, Xu Q, Lu L, Liu JK, Zhang L, Wu B, Li YQ (2013) Unusual spirodecane sesquiterpenes and a fumagillol analogue from *Cordyceps ophioglossoides*. Helv Chim Acta 96:76–84

Takaku T, Kimura Y, Okuda H (2001) Isolation of an antitumor compound from *Agaricus blazei* Murill and its mechanism of action. J Nutr 131:1409–1413

Tan WC, Kuppusamy UR, Phan CW, Tan YS, Raman J, Anuar AM, Sabaratnam V (2015) *Ganoderma neo-japonicum* Imazeki revisited: domestication study and antioxidant properties of its basidiocarps and mycelia. Sci Rep 5:12515

Tang W, Liu JW, Zhao WM, Wei DZ, Zhong JJ (2006) Ganoderic acid T from *Ganoderma lucidum* mycelia induces mitochondria mediated apoptosis in lung cancer cells. Life Sci 80:205–211

Thomas H, Coley HM (2003) Overcoming multidrug resistance in cancer: an update on the clinical strategy of inhibiting p-glycoprotein. Cancer Control 10:159–165

Vrkoc J, Budesinsky M, Dolejs L (1977) Phenolic meroterpenoids from the basidiomycete *Albatrellus ovinus*. Phytochemistry 16:1409–1411

Wang CJ, Chau CF, Hsich YS, Yang SF, Yen GC (2008) Lucidenic acid inhibits PMA-induced invasion of human hepatoma cells through inactivating MAPK/ERK signal transduction pathway and reducing binding activities of NF-κB and AP-1. Carcinogenesis 29:147–156

Wang S, Bao L, Zhao F, Wang Q, Li S, Ren J, Li L, Wen H, Guo L, Liu H (2013) Isolation, identification and bioactivity of monoterpenoids and sesquiterpenoids from the mycelia of edible mushroom *Pleurotus cornucopiae*. J Agric Food Chem 61:5122–5129

Wasser SP (2011) Current findings, future trends, and unsolved problems in studies of medicinal mushrooms. Appl Microbiol Biotechnol 89:1323–1332

Wasser SP (2014) Medicinal mushroom science: current perspectives, advances, evidences, and challenges. Biom J 37:345–356

Watanabe K, Shuto T, Sato M, Onuki K, Mizunoe S, Suzuki S, Sato T, Koga T, Suico MA, Kai H, Ikeda T (2011) Lucidenic acids-rich extract from antlered form of *Ganoderma lucidum* enhances TNFα induction in THP-1 monocytic cells possibly via its modulation of MAP kinases p38 and JNK. Biochem Biophys Res Commun 408:18–24

Weng CJ, Chau CF, Chen KD, Chen DH, Yen GC (2007) The antiinvasive effect of lucidenic acids isolated from a new *Ganoderma lucidum* strain. Mol Nutr Food Res 51:1472–1477

Wong YY, Moon A, Duffin R, Barthet-Barateig A, Meijer HA, Clemens MJ, de Moor CH (2010) Cordycepin inhibits protein synthesis and cell adhesion through effects on signal transduction. J Biol Chem 285:2610–2621

Wu SJ, Leu YI, Chen CH, Chao CH, Shen DY, Chan HH, Lee EJ, Wu TS, Wang YH, Shen YC, Qian K, Bastow KF, Lee KH (2010) Camphoratins A-J, potent cytotoxic and anti-inflammatory triterpenoids from fruiting body of *Taiwanofungus camphoratus*. J Nat Prod 73:1756–1762

Wu GS, Lu JJ, Guo JJ, Li YB, TanW DYY, Zhong ZF, Xu ZT, Chen XP, Wang YT (2012) Ganoderic acid DM, a natural triterpenoid, induces DNA damage, G1 cell cycle arrest and apoptosis in human breast cancer cells. Fitoterapia 83:408–414

Xu TT, Beelman RB, Lambert JD (2012) The cancer preventive effects of edible mushrooms. Anti-cancer Agent Me 12:1255–1263

Yasukawa K, Yu S-Y, Takido M (1996) Inhibitory effect of oral administration of Rikkunshito on tumor promotion in two-stage carcinogenesis in mouse skin. Journal of Traditional Medicines 13:180–184

Yazawa Y, Yokota M, Sugiyama K (2000) Antitumor promoting effect of an active component of *polyporus*, ergosterol and related compounds on rat urinary bladder carcinogenesis in a short-term test with concanavalin a. Biol Pharm Bull 23:1298–1302

Ye M, Liu JK, Lu ZX, Zhao Y, Liu SF, Li LL, Tan M, Weng XX, Li W, Cao Y (2005) Grifolin, a potential antitumor natural product from the mushroom *Albatrellus confluens* inhibits tumor cell growth by inducing apoptosis *in vitro*. FEBS Lett 579:3437–3443

Yeh CT, Rao YK, Yao CJ, Yeh CF, Li CH, Chuang SE, Luong JH, Lai GM, Tzeng YM (2009) Cytotoxic triterpenes from *Antrodia camphorata* and their mode of action in HT-29 human colon cancer cells. Cancer Lett 285:73–79

Yin X, Feng T, Li ZH, Dong ZJ, Li Y, Liu JK (2013) Highly oxygenated meroterpenoids from fruiting bodies of the mushroom *Tricholoma terreum*. J Nat Prod 76:1365–1368

Yoshiji H, Kuriyama S, Ways DK, Yoshii J, Miyamoto Y, Kawata M, Ikenaka Y, Tsujinoue H, Nakatani T, Shibuya M, Fukui H (1999) Protein kinase C lies on the signaling pathway for vascular endothelial growth factor-mediated tumor development and angiogenesis. Cancer Res 59:4413–4418

Zaidman B, Yassin M, Mahajana J, Wasser SP (2005) Medicinal mushroom modulators of molecular targets as cancer therapeutics. Appl Microbiol Biotechnol 67:453–468

Zhang Z, Liu RN, Tang QJ, Zhang JS, Yang Y, Shang XD (2015) A new diterpene from the fungal mycelia of *Hericium erinaceus*. Phytochem Lett 11:151–156

Zhao YY, Chao X, Zhang Y, Lin RC, Sun WJ (2010) Cytotoxic steroids from *Polyporus Umbellatus*. Planta Med 76:1755–1758

Zhao YY, Shen X, Chao X, Ho CC, Cheng XL, Zhang Y, Lin RC, Du KJ, Luo WJ, Chen JY (2011) Ergosta-4, 6, 8 (14), 22-tetraen-3-one induces G2/M cell cycle arrest and apoptosis in human hepatocellular carcinoma HepG2 cells. Biochim Biophys Acta 1810:384–390

Zhou L, Shi P, Chen NH, Zhong JJ (2011) Ganoderic acid Me induces apoptosis through mitochondria dysfunctions in human colon carcinoma cells. Process Biochem 46:219–225

Modulation of Fungal Metabolome by Biotic Stress

24

Geane Pereira de Oliveira, Bruna de Almeida Martins, Matheus Thomaz Nogueira Silva Lima, and Jacqueline Aparecida Takahashi

Abstract

Filamentous fungi are ubiquitous microorganisms well known as one of the major sources of pharmaceuticals and other biotechnologically useful compounds. The amazing structural diversity of fungal secondary metabolites results from years of coevolution that selected unique mechanisms of resistance and communication in specific environments or in biotic relationships. Metabolite production is controlled by transcriptional regulation, and the production of carbon-based compounds is usually restricted to opportune situations. The dynamic nature of fungal metabolism consists on a complex in vivo process regulated by interactions between gene networks, but it can be modulated by generation of in vitro biotic stress. Both in vivo and in vitro interactions trigger activation of cryptic genes leading to changes in the fungal metabolic expression such as elicitation of new metabolite production, increase or minimization of the biosynthetic pathways, or even inhibition of some metabolic routes. While biosynthesis diversification and yield improvement are interesting tools for producing bioactive compounds, metabolite suppression can be used as a biotechnologically useful tool for decreasing toxin production. This review discusses the production of secondary metabolites by filamentous fungi under different natural biotic stresses caused by algae, bacteria, plants, and other organisms and the in vitro mimetization of these interactions leading to cryptic gene expression.

G. P. de Oliveira · B. de Almeida Martins · J. A. Takahashi (✉)
Department of Chemistry, Exact Sciences Institute, Universidade Federal de Minas Gerais, Belo Horizonte-MG, Brazil
e-mail: jat@qui.ufmg.br

M. T. N. S. Lima
Department of Food Science, Faculty of Pharmacy, Universidade Federal de Minas Gerais, Belo Horizonte-MG, Brazil

T. Satyanarayana et al. (eds.), *Advancing Frontiers in Mycology & Mycotechnology*,
https://doi.org/10.1007/978-981-13-9349-5_24

Keywords
Bioactive compounds · Biotic stress · Fungal metabolites · Metabolome diversification

24.1 Introduction

24.1.1 A Molecular Overview of Metabolic Modulation in Filamentous Fungi: Importance of Coevolution Process

Filamentous fungi are ubiquitous microorganisms well known for their biotechnological potential such as for lignin degradation (Klein-Marcuschamer et al. 2012), and production of proteins (Punt et al. 2002), insecticides, enzymes such as lipases, and citric acid (Fang et al. 2012; Aguieiras et al. 2015; Niu et al. 2015), and more importantly, as one of the major sources of pharmaceuticals (Liu et al. 2010). The discovery of penicillin in 1928, the first antibiotic produced by the filamentous fungus *Penicillium notatum*, led to important developments in the areas of biology, chemistry, and medicine. After penicillin, the mechanisms for production, isolation, and identification of metabolites produced by fungal biological systems, as well as their application in the treatment of various infections, have been intensely researched (Ligon 2004).

Like penicillin, a wide range of fungi-derived pharmaceuticals called secondary metabolites of low molecular mass have been discovered, but they usually do not play a role in growth. However, these metabolites have a role in survival, communication, and competition of fungi in different environments (Calvo et al. 2002; Keller et al. 2005; Fox and Howlett 2008). The amazing structural diversity of secondary metabolites results from years of coevolution that selected unique mechanisms of resistance and communication in specific environments or in biotic relationships (Vining 1992; Brakhage et al. 2009; Brakhage 2013).

In general, genes involved in the synthesis of secondary metabolites are organized in clusters (Osbourn 2010). Different experimental strategies for modulating secondary metabolites production have been applied in fungi in order to activate cryptic genes that encode secondary metabolites (Brakhage 2013). Strategies such as nutrient variation (Hewage et al. 2014; Lima et al. 2018a), epigenetic modulation (Williams et al. 2008; Lima et al. 2018b), biotic stress (Teles et al. 2012), light incidence (Bayram et al. 2008), and temperature modifications (Nielsen et al. 2004) have been used to elicit secondary metabolite diversification. At molecular level, biosynthesis of secondary metabolites is a very complex process regulated by interactions between gene networks (Brakhage 2013). This process is controlled by transcriptional regulation in response to environmental stimulus, and the production of carbon-based compounds is usually restricted to opportune situations (Shwab and Keller 2008; Yin and Keller 2011).

The expression of some of the genes is determined at the molecular level by epigenetic factors such as the level of methylation of the gene promoter region, and by global and specific transcription factor regulators (Davies and Schwinn 2003; Keller et al. 2005). The biosynthesis of secondary metabolites usually depends on several enzymes coded by different genes in the fungal genome, as indicated by the intricate polyketide biosynthetic pathways and by genomic studies (Keller et al. 2005; Hertweck 2009).

Coevolution process influenced the development of specific strategies of communication and competitiveness between organisms, turning chemical responses intrinsic to each organism. Therefore, co-culturing is an important strategy for in vitro secondary metabolite production (Brakhage and Schroeckh 2011). This strategy, in which two or more species are grown in the mixed fermentation, aims to imitate natural fungal habitat, where they coexist with a variety of other species, inducing defense mechanisms for environmental adaptation and self-survival (Marmann et al. 2014).

This review discusses mechanisms of biotic stress and their implications in the secondary metabolites production by filamentous fungi under different co-culturing conditions. Figure 24.1 presents a graphical representation of the complex interactions of different biological pathways related to secondary metabolite production. In the omics era, discovery of interconnections between genetic complexity, protein diversity, and metabolic diversification is related to developments in genomic, transcriptomic, and metabolomic fields (Kroymann 2011). The diversity of secondary metabolites is associated with two complexity levels as indicated in Fig. 24.1: the systemic level influenced by the genetic background, environmental stimulus in response to other microorganisms, or chemical communication and, at the molecular level, its consequences that lead to activation responses of interactions over gene expression at the epigenetic level or enzymatic metabolic pathways (Brakhage and Schroeckh 2011).

24.1.2 Biotic Stress Mechanisms

The interspecies relationships may be cooperative, when the species coexist harmoniously; antagonistic, when there is competition between them, aiming at survival; or even synergistic, leading to some biological activity. Anton de Bary defined symbiosis in 1878 as "the living together of dissimilar organisms" (Sapp 1994). In this model, symbiosis is classified into three types: commensalism, when one organism

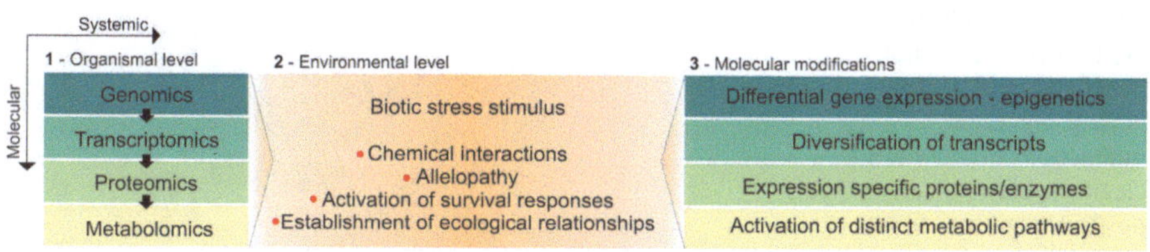

Fig. 24.1 The complexity behind the interaction of different biological pathways that elicit secondary metabolite production

has a benefit and the other one is not affected; mutualism, when there is mutual cooperation and both organisms have benefits; and parasitism, when one organism benefits at the expense of the other, on a relationship that is indispensable for the survival of the one (Dimijian 2000; Tortora et al. 2012).

Another definition of symbiosis was presented by Douglas (2010), as a "long-term association between different species from which all participating members benefit," following a line discussed and well-accepted by many biologists. In this case, symbiosis is basically considered as a mutualistic relationship.

Starting from the classical definition, mutualistic and commensal interactions can be considered cooperative relationships. In the case of parasitism, there is an antagonistic interaction, since it does not benefit one of the organisms (Roossinck 2005). However, the term antagonism came to be used in a more restricted way, as a symbiosis in which damage occurs to the organisms (Quispel 1951), leading to a new classification for symbiosis types: mutualism, commensalism, and antagonism. The latter was divided into mechanisms of competition, amensalism, and parasitism (Dighton and White 2016). Antagonistic competition refers to the fight for food and/or territory by different individuals, while amensalism is a phenomenon in which one population prevents the development of another, without obtaining any benefit (Andrade-Domínguez et al. 2014); the latter is usually carried out by the release of chemical substances, being in this case referred to as antibiosis. This last type of relationship led to coining the term antibiotic, which refers to a substance produced by one microorganism that can inhibit the development of another (Tortora et al. 2012). Synergism can be defined as a joint work between two species, in order to increase the effect that was produced individually by each one of them. This effect may be positive, generating attractive metabolic resources (Minty et al. 2013) or negative that may lead to toxin production (Xiong and Huang 2018).

The relationships herein defined are usually present when living organisms such as fungi, bacteria, lichens, plants, insects, and others have a contact in nature. Fungal relationships with several types of organisms can encompass these different interspecific interactions, as represented in Fig. 24.2. This phenomenon also occurs

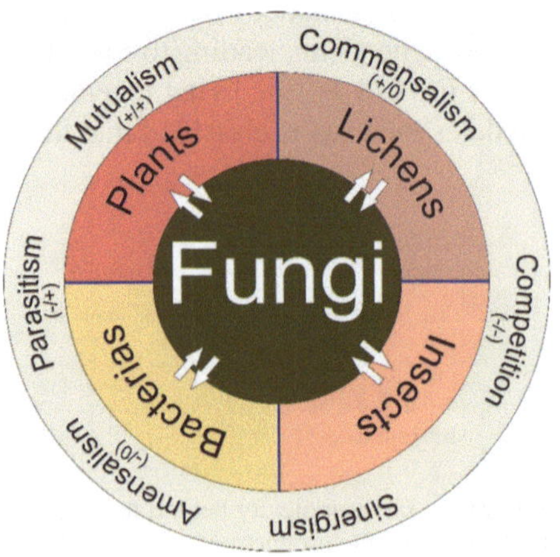

Fig. 24.2 Ecological relationships established between fungi and other organisms like plants, bacteria, insects, and lichens. Effect of each interaction is indicated by (+) for benefit, (−) not beneficial, and (0) for neutral. Synergic interactions do not have a clear association with positive/negative effects

in vitro, aiming at biosynthesis of certain secondary metabolites for industrial applications. Successful results are most often related to positive interactions (Ghosh et al. 2016).

24.2 Metabolic Modulation in Co-cultures Containing Fungi

24.2.1 Lichens, Naturally Occurring Fungi-Photobiont Biotic Interaction

Complex associations between organisms can naturally cause metabolic modulation of certain living forms toward the establishment of ecological relationships, as in lichens (Fig. 24.3). Lichens are well-known organisms which have a long evolutionary history between a mycobiont, generally a filamentous fungus from *Ascomycota* phylum, and a photobiont, a green alga or cyanobacteria (Boustie and

Fig. 24.3 Some different lichen species distributed in nature

Grube 2005; Stocker-Wörgötter 2008). The mutualistic relationship established between the photoautotrophic partner and the fungal partner (Molnár and Farkas 2010; Divakar et al. 2015) contributes to supply carbohydrates to the mycobiont, while the fungus provides essential minerals, protection from UV light, and a broad and stable surface for photosynthesis (Eisenreich et al. 2011; Wang et al. 2014; Grube et al. 2015).

The diversity of fungal species is reflected in the variability of lichen assemblages. Around 18,500 lichen species have already been identified and at least 20,000 species have been estimated to exist (Boustie and Grube 2005; Wang et al. 2014). Approximately 50% of ascomycetous fungi are able to form lichens, and most of the lichens (98%) have ascomycetes as the fungal counterpart, while a minor percentage of lichens is composed of basidiomycetes (Stocker-Wörgötter 2008; Arnold et al. 2009). Lichens have particular thallus morphology, the nonreproductive tissue, where secondary metabolites are produced. This part of the mycobiont supports the photobiont that benefits from the improved superficial area for light capture (Wang et al. 2014). The taxonomy is related predominantly to fungal species which is based on secondary metabolite profiles (Frisvad et al. 2008).

Lichens can be found in a wide range of habitats, from rainforests (Aptroot and Cáceres 2014) to glacial regions (Seppelt et al. 2010), demonstrating immense ability to adapt to different environments and the successful establishment of a symbiotic relationship. For some time, it was considered that only one fungal species and a single photobiont species were associated with a lichen complex. However, recent studies demonstrated that more than two partners can be associated (Spribille et al. 2016). This variability in species combinations led lichens to be a good source of metabolites with diverse chemical structures.

Considering that lichens are symbiotic complexes, the production of secondary metabolites might be related to communication between the partners, but the specific roles of these have not been fully elucidated (Wang et al. 2014). Research in this area is difficult, since in vitro lichen maintenance is problematic. Despite this, at least 1000 lichen-exclusive secondary metabolites are known. Most of the lichen-derived metabolites are from shikimate, polymalonate, and mevalonate biosynthetic pathways (Stocker-Wörgötter 2008; Shukla et al. 2010), like depsones, depsides, dibenzofurans, depsidones, xanthones, and chromones (Melo et al. 2011).

Secondary metabolites from lichens have been studied, and important lichen-unique molecules with potential pharmaceutical applications like antimicrobial compounds, anti-Alzheimer medicines, analgesic, anti-inflammatory, and anticancer activities have been discovered that include usnic acid (**1**), atranorin (**2**), and parietin (**3**) (Fig. 24.4) (Kowalski et al. 2011; Bačkorová et al. 2012) (Table 24.1). Secondary metabolites constitute at least 30% of lichens dry matter, and these compounds can even precipitate in their external area as reported for atranorin (**2**) and chloroatranorin (**4**) (Fig. 24.4) (Solhaug et al. 2009; Manojlović and Ranković 2012). It is interesting to note that among the metabolites (e.g., atranorin, usnic acid) identified in lichens, the bioactive ones are mainly derived from the fungal entity probably as a complementary response to the carbohydrate supply from the algal entity (Ranković and Kosanić 2015).

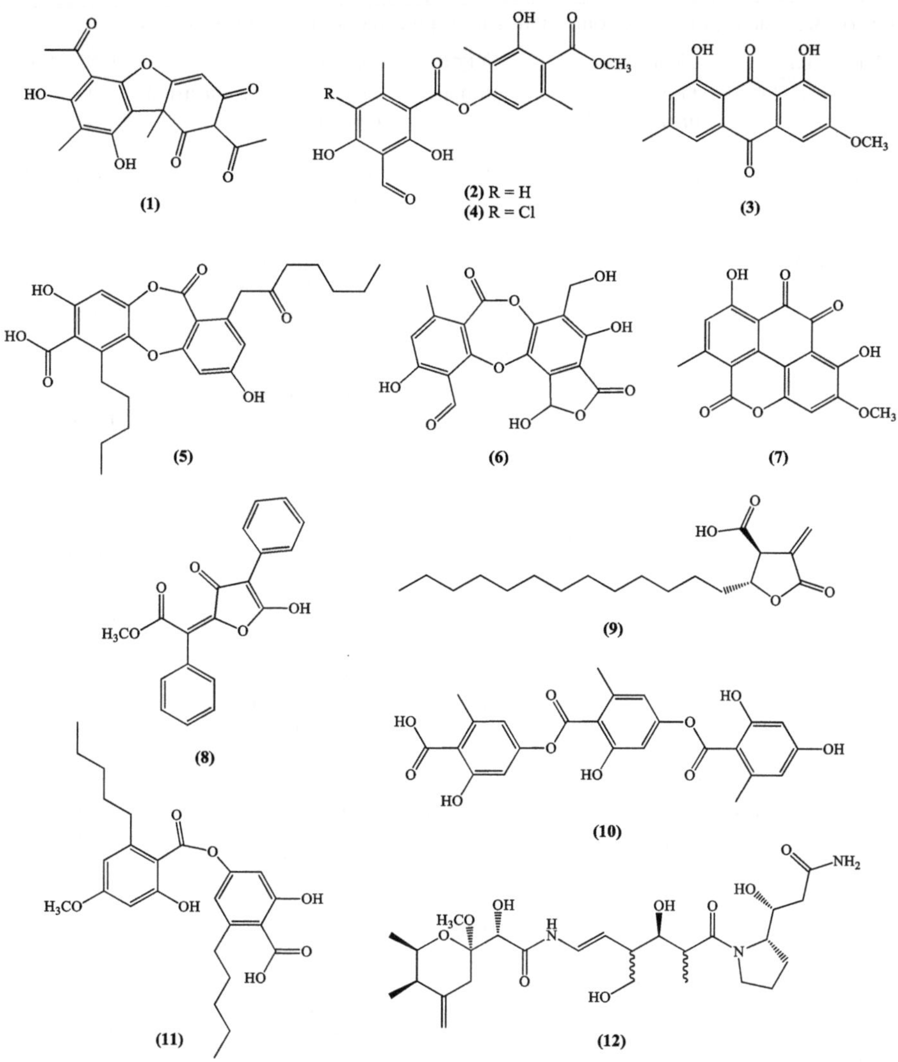

Fig. 24.4 Chemical structures of secondary metabolites from lichens

The most studied lichen-derived secondary metabolite is usnic acid (**1**), a yellow pigment vastly described as a lichen-exclusive antimicrobial compound and used in commercial products such as toothpaste and deodorants (Cocchietto et al. 2002; Ingólfsdóttir 2002). Other outstanding application of usnic acid (**1**) consists on colony growth inhibition and biofilm morphology modification, important features in hospital environments (Francolini et al. 2004). The mechanism of biofilm reduction in medical apparatus was not elucidated yet. Usnic acid (**1**) is present in different species from different lichen genera such as *Alectoria* (Einarsdóttir et al. 2010), *Flavocetraria* (Bjerke et al. 2005), and *Usnea* (Yamamoto et al. 1985). Extracts

Table 24.1 Lichen-exclusive compounds and its potential pharmaceutical activities

Compound	Lichen species	Origin	Activity	Reference
Usnic acid (**1**)	*Flavocetraria nivalis* and *Usnea florida*	Romania	Lung cancer cell invasion reduction	Yang et al. (2016)
Atranorin (**2**)	*Cladina kalbii*	Brazil	Effective antioxidant in SH-SY5Y cell lines (neuroblasts)	Melo et al. (2011)
Parietin (**3**)	*Xanthoria parietina*	Slovakia	Activation of p38 mitogen-activated protein kinase gene expression in A2780 (ovarian carcinoma) and HT-29 (colon) cancer cell lines	Bačkorová et al. (2012)
Chloroatranorin (**4**)	*Parmotrema saccatilobum*	Papua New Guinea	Analgesic and anti-inflammatory	Bugni et al. (2009)
Physodic acid (**5**)	*Cladonia portentosa* and *Pseudevernia furfuracea*	Not indicated	Neurotrophic activity	Reddy et al. (2016)
Salazinic acid (**6**)	*Parmelia sulcata*	Turkey	Antimicrobial	Candan et al. (2007)
Biruloquinone (**7**)	*Cladonia macilenta*	China	Anti-Alzheimer and antioxidant agent in β-amyloid-injured PC12 cells	Luo et al. (2013)
Vulpinic acid (**8**)	*Cladonia amaurocraea* and *Hypogymnia physodes*	Canada	Control of *Sclerotinia sclerotiorum* growth	Kowalski et al. (2011)
Protolichesterinic acid (**9**)	*Parmelia reticulata*	India	Antifungal activity over *Rhizoctonia solani*	Goel et al. (2011)
Gyrophoric acid (**10**)	*Xanthoparmelia pokornyi*	Turkey	Photoprotection of human keratinocytes	Varol et al. (2016)
Perlatolic acid (**11**)	*Cladonia portentosa* and *Pseudevernia furfuracea*	Not indicated	Neurotrophic activity and increased H3 and H4 histone acetylation	Reddy et al. (2016)

from *F. nivalis* and *U. florida* containing usnic acid (**1**) were associated with a reduction in cancer lung cell motility due to inhibition of β-catenin pathway, therefore reducing tecidual invasion by cancer cells (Yang et al. 2016). Other important biological activities indicated for usnic acid (**1**) are anti-inflammatory (Vijayakumar et al. 2000) and analgesic action (Okuyama et al. 1995).

In 2013, Pompilio et al. isolated usnic acid (**1**) and atranorin (**2**) from a Chilean lichenic species. These compounds biosynthesized from polyketide pathway, demonstrated activity against methicillin-resistant *Staphylococcus aureus* in biofilm formation related to cystic fibrosis complications. Atranorin (**2**) was able to act as an effective antioxidant in SH-SY5Y cell lines. This metabolite has been detected in a number of lichen species, including the Brazilian *Cladin kalbii* Ahti lichen (Melo et al. 2011). The antimicrobial activity of usnic acid (**1**), atranorin (**2**), a depside compound, against seven bacterial and two *Candida* species has been reported (Yilmaz et al. 2004). Among them, usnic acid (**1**) proved to be the most active compound, mainly against *Bacillus cereus*, *Streptococcus faecalis*, *Proteus vulgaris*, *Candida albicans*, and *Candida glabrata* and at lower concentrations. Antifungal activity against the important pathogenic fungus *Rhizoctonia solani* was reported for atranorin (**2**) and for protolichesterinic acid (**9**) (Fig. 24.4), being comparable to the action of a commercial fungicide, hexaconazole (Goel et al. 2011).

Atranorin (**2**) and usnic acid (**1**) demonstrated higher antitumor activity (A2780 and HT-29 cancer cell lines) than other lichenic metabolites like parietin and gyrophoric acid (**10**) (Fig. 24.4) probably as a result of mitochondrial membrane disruption (Bačkorová et al. 2012). The latter, together with vulpinic acid (**8**), presented photoprotection to human keratinocytes, demonstrating the role of these two molecules in natural protection against UV light in lichens (Varol et al. 2016).

Biruloquinone (**7**), obtained from the lichen *C. macilenta*, was reported as an efficient acetylcholinesterase inhibitor (60.5%) with possible application in Alzheimer's disease and as an antioxidant agent in P1C2 cells (Luo et al. 2013). Neurotrophic activity in the central nervous system was demonstrated for usnic acid (**1**), atranorin (**2**), physodic acid (**5**), and perlatolic acid (**11**) (Fig. 24.4) with no cytotoxic effect (Reddy et al. 2016).

Few studies regarding the production of lichen-associated secondary metabolites by cyanobacteria have been reported. In 2013, nosperin (**12**) (Fig. 24.4), a metabolite produced by polyketide pathway, was identified in *Peltigera membranacea*, originated from the cyanobacterial *Nostoc* partner (Kampa et al. 2013). Nosperin-mediated microbe-microbe communication was suggested by Kampa et al. (2013) and Scherlach and Hertweck (2018). Some studies have been published regarding *Nostoc* hepatotoxins as microcystins (Kaasalainen et al. 2012; Suzuki et al. 2016).

Besides lichen structures build between two organisms (fungus and algae), structured bacterial communities can associate with lichens turning lichenic populations more complex, promoting lichens to holobiont organisms harboring complex microbial communities (Grube and Berg 2010). In a metagenomic study published in 2015 by Grube et al., 800 bacterial species were identified in association with the lung lichen *Lobaria pulmonata*. The authors suggested that this bacterial community contributes to supply essential minerals and resistance to biotic stress and even to supplement with B12 vitamin and nitrogen. Molecular approaches revealed that a great number of species naturally co-cultured with lichens are *Alphaproteobacteria* (Grube and Berg 2010) and *Gammaproteobacteria* (Mushegian et al. 2011).

Research on the specific localization and diversity of species of bacterial communities showed that the lichen core preserves well-structured bacterial communities that vary between organisms and locations possibly related to secondary metabolite availability and lichen structure (Mushegian et al. 2011).

24.2.2 Plants and Fungi Biotic Interaction

In nature, fungi depend on other organisms or on the environment for carbon and energy. In general, the relationship between fungi and plants is mutually beneficial. At first, it consisted exclusively in the fungi survival in dead plants promoting nutrients reuse (saprophytic fungi). However, some fungi, such as mycorrhizae and endophytes, started to interact with live plants over the time. Most keep the role of mutual benefit by promoting plant growth and development, while a small portion has developed new metabolic pathways and become phytopathogenic (Grayer and Kokubun 2001).

The plant-fungi interaction also consists of an innovative source of secondary metabolites. This potential needs more attention since the accumulation of phytochemical compounds is important when in vivo chemical synthesis is economically unviable. Production of secondary metabolites generally occurs through activation of biosynthetic pathways related to plant defense for their survival (Baldi et al. 2008; Romão-Dumaresq et al. 2016). In this situation, fungi act as biotic elicitors that trigger the biosynthesis of metabolites as a plant response (Baldi et al. 2009).

The biosynthesis of phytoalexins, defense molecules produced specifically as a result of biotic stress, provides evidence for the complexity of plant-fungi interaction. Many species of microorganisms show ability to induce phytoalexin formation in plants. The term phytoalexins are restricted to antifungal substances that require the expression of enzymes involved in its biosynthetic pathway (Grayer and Kokubun 2001; Aisyah et al. 2015; Jiao et al. 2018). Isoflavonoids and sesquiterpenes are examples of classes of compounds that act as phytoalexins, synthesized by plants when they are exposed to stressful conditions, such as the attack of pathogens and insects (Dixon 2001). These attacks are generally recognized by specific receptors located in the plasma membranes of plant cells, thus initiating signal transduction network in the cytosol with subsequent activation of the plant defense system and phytoalexins biosynthesis (Zhao et al. 2005). Besides that, production of toxic radicals, such as nitric oxide and reactive oxygen species (ROS) as a defense response to pathogens is also common (Resende et al. 2003).

Another plant defense mechanism is the inhibition of fungal growth by production of antimicrobial proteins (Schlumbaum et al. 1986; van Loon et al. 2006). Most of these pathogenesis-related proteins are induced by signaling compounds (van Loon et al. 2006), in other words, compounds involved in signal transduction systems biosynthesized by plants. These compounds act as chemical elicitors, and enhancement of their biosynthesis influences directly secondary metabolite production (Zhao et al. 2005; Giri and Zaheer 2016; Jiao et al. 2018). Plant hormones, such as abscisic acid, also modulate expression of these proteins (van Loon et al. 2006).

Sperm-Coating Protein/Tpx-1/Ag5/PR-1/Sc7 (SCP/TAPS) is an important family of pathogenesis-related genes (Teixeira et al. 2012). PR-1 proteins are members of this family and are expressed by various plants, e.g., tobacco and tomato (van Loon et al. 2006). Some PR-1s homologous substances expressed by the parasitic fungus *Moniliophthora perniciosa* were detected by Teixeira et al. (2012) during interaction of this fungus with cocoa. In this case, the authors suggested that these

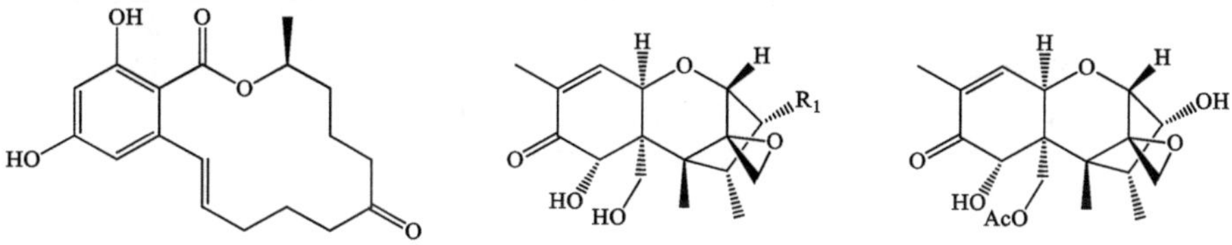

(13) **(14)** $R_1 = OH$ **(15)** $R_1 = OAc$ **(16)**

Fig. 24.5 Chemical structures of some mycotoxins produced by *Fusarium* species. (Kim and Vujanovic 2017)

substances favor colonization of *M. perniciosa* by limiting the development of microbial competitors.

In this way, plants can also stimulate the production of toxic substances by fungi, which are not produced without biotic stress. Among these substances are the so-called mycotoxins, whose production guarantees competitive advantages for the producer under stressing conditions (Magan and Aldred 2007). Some examples of mycotoxins are zearalenone **(13)**, deoxynivalenol **(14)**, 3-acetyl-deoxynivalenol **(15)**, and 15-acetyl-deoxynivalenol **(16)** (Fig. 24.5), produced by phytopathogenic species of the genus *Fusarium* spp., which affects grain fields around the world (Kim and Vujanovic 2017).

These different types of substances expressed in response to plant-fungus interaction are outlined in Fig. 24.6. In addition to this variety of responses, existence of more than 250,000 species of higher plants and 1.5 million fungal species (Grayer and Kokubun 2001) shows how plant-fungi interactions can be complex and specific, considering various possible combinations as well as the influence of abiotic factors (Baldi et al. 2009).

As already mentioned, naturally occurring interactions between plants and fungi range from beneficial to antagonist. A classical example of antagonistic interaction found in nature is the association of the phytopathogenic fungus *Botrytis cinerea* with a variety of cultivated plants and their fruits. This fungus is considered one of the most significant pathogens in terms of molecular plant pathology, being reported as an infecting agent in more than 200 species (Dean et al. 2012; Fillinger and Elad 2016). *B. cinerea* is known to be responsible for gray mold disease of grapevine (Dean et al. 2012), the decay of strawberry fruit (Abanda-Nkpwatt et al. 2006), and damages in several other cultures. Considering the economic losses promoted by this pathogen throughout the world, with a global cost of crop protection estimated at the US $ 310 million (Fillinger and Elad 2016), inhibition of this species will be very useful. Selective inhibition of *B. cinerea* growth by volatile substances like 2-phenylethanol, linalool, and nonanal present in the surface of *Fragaria x ananassa* leaves (strawberry leaves) has already been reported. These compounds also increase the population dynamics of some nonpathogenic bacteria isolated from strawberry leaves that are generally suppressed by *B. cinerea* strains (Abanda-Nkpwatt et al. 2006).

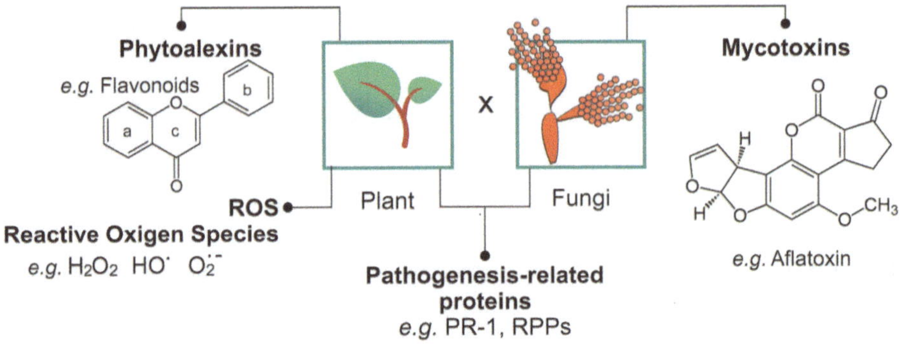

Fig. 24.6 Expression of different types of substances during plant-fungi interactions resulting in different physiological responses such as the production of ROS, secondary metabolites, and protective proteins

B. cinerea has a large number of genes that code endopolygalacturonases (endo-PG). The pattern of these genes is differentially expressed, depending on the host plant infected (Reignault et al. 2008). Polygalacturonase is the main enzyme responsible for the dissolution of the middle lamella during fruit maturation. Endo-PG randomly hydrolyzes the glycosidic bonds in the pectin molecule, leading to fruit softening during maturation (Eskin and Shahidi 2013). In 2000, Reignault et al. discussed the role of PGs in the pathogenicity of *B. cinerea* in apple, highlighting the correlation between different isoenzymes of neutral PGs with the smooth rot in the fruit.

As regards beneficial plant-fungi interactions, the presence of endophytic fungi in plant tissues, without causing apparent damage, is very important and often associated with promotion of plant resistance (Spatafora and Bushley 2015). They can be isolated by culturing surface-decontaminated plant fragments using standard protocols (Silva et al. 2010). The interaction of endophytic fungi species of *Clavicipitaceae* family with plants is an example of beneficial interaction since alkaloids produced by the endophytes protect the plant against insects and other herbivores (Marcet-Houben and Gabaldón 2016). The endophytic biodiversity of *Pachystachys lutea*, an ornamental plant, showed correlation with significant reduction of mycelial growth of the pathogens *Colletotrichum* sp. and *Fusarium oxysporum*, a feature agronomically interesting for diseases control (da Silva et al. 2018). Another example was reported by Qin et al. (2017), emphasizing the ability of some endophytic and rhizospheric fungi to help in nutrition and tolerance to stressing conditions to allow survival of the host plant in habitats under extreme conditions.

Development of co-culture approaches is a popular target nowadays for manipulation of biosynthetic pathways to intensify secondary metabolites production. *Piriformospora indica* is an example of endophytic fungus recognized as a plant growth promoter. Discovered by Varma and Franken in 1997, this species boosted the production of secondary metabolites based on plant-fungal elicitation (Baldi et al. 2009). The mechanism of interaction between *P. indica* and the plant species

Arabidopsis thaliana was investigated by Sirrenberg et al. (2007). In that study, indole-3-acetic acid (IAA) **(17)** (Fig. 24.7), a growth phytohormone, was detected, probably as a consequence of induction by compounds present in the plant root exudate. This auxin (IAA) promotes differential cell elongation, regulating plant growth and development (Zhao 2010) also benefiting *P. indica* growth in its host plant (Sirrenberg et al. 2007).

Baldi et al. (2008) established cell cultures of *P. indica* and *Sebacina vermifera* with cells of the plant species *Linum album*, in order to produce podophyllotoxins, substances widely used for the anticancer drugs synthesis and previously isolated from cell cultures of *L. album*. The increase of plant biomass in relation to control cultures was verified for both co-cultures, indicating raise in absorption of mineral nutrients by the plant cells. The successful symbiotic responses and plant cell elicitation through fungal interaction resulted in production of podophyllotoxin **(18)** and 6-methoxypodophyllotoxin **(19)** (Fig. 24.7). In that case, the hypersensitive response of the plant cells to fungal induction increased phytoalexins production as a defense mechanism.

Jiao et al. (2018) emphasized the importance of using nontoxic fungal species as elicitors, to ensure the biosafety of products formed. In their study, two GRAS (Generally Recognized as Safe) fungal species, *Aspergillus niger* and *Aspergillus oryzae*, were stabilized together with *Isatis tinctoria* L. hairy root cultures. This plant is of great interest to the pharmaceutical industry, for its proven anti-influenza efficacy and its ability to produce bioactive flavonoids. As desired, *I. tinctoria* flavonoids yield was increased in the co-culture experiments, with the superiority of *A. niger* in relation to *A. oryzae* as the fungal component in the experiment. There was an accumulation of the flavonoid aglycones liquiritigenin **(20)**, quercetin **(21)**,

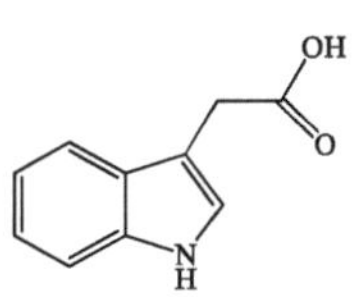

(17)

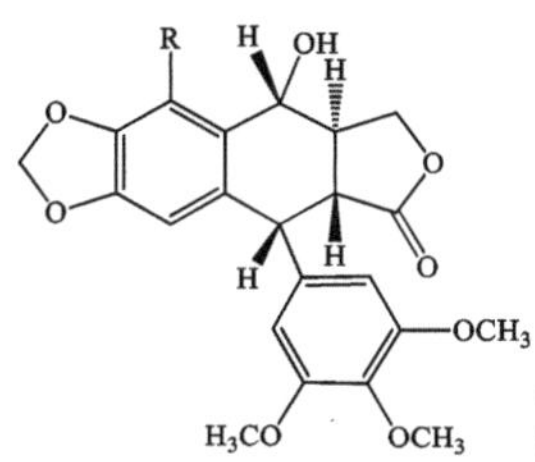

(18) R = H
(19) R = OCH_3

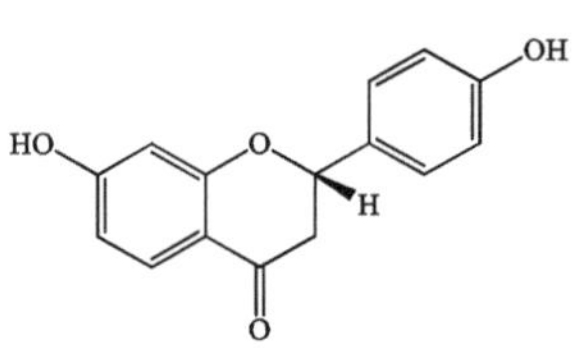

(20)

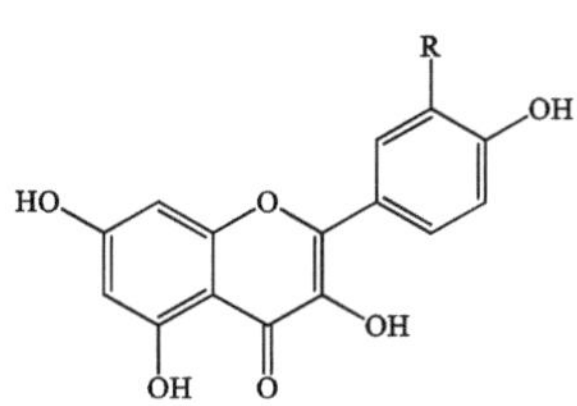

(21) R = OH
(22) R = OCH_3
(23) R = H

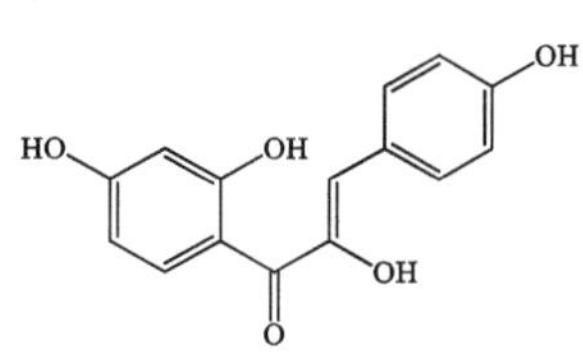

(24)

Fig. 24.7 Chemical structures of some secondary metabolites produced in fungi-plants co-cultivation experiments

isorhamnetin (**22**), kaempferol (**23**), and isoliquiritigenin (**24**) (Fig. 24.7), with the prospect of producing these bioactive compounds on an enlarged scale. The production of secondary metabolites was attributed to transcriptional activation of genes, mainly chalcone isomerase (CHI) and flavonoid 3′-hydroxylase (F3′H), involved in the biosynthetic pathway of flavonoids by regulatory systems of plant defense. In addition, intensification of production of the signal compounds nitric oxide, salicylic, and jasmonic acid was also observed. On the other hand, there was minimization in the yields of three flavonoid glycosides (rutin, neohesperidin, and buddleoside), showing another possibility of response as result of the plant-fungus interaction.

Several studies successfully addressed fungal species of *Aspergillus* genus as plant metabolites elicitors. Kümmritz et al. (2016) reported increasing oleanolic and ursolic acid production in leaves of the medicinal plant *Salvia fruticosa*, when co-cultivated with *A. niger* under optimized conditions. *A. niger* also promoted accumulation of flavonoids and glycyrrhetinic acid when cultivated with *Glycyrrhiza uralensis* Fisch adventitious root (Li et al. 2015). The same group reported the performance of *A. niger*, *A. oryzae*, and *Aspergillus flavus*, as gene expression elicitors, for the accumulation of signal molecules in *Panax ginseng* C. A. Mey (Li et al. 2016). Aisyah et al. (2015) selected *A. oryzae* to elicit induction of phytoalexins in peanuts, especially because this species does not produce aflatoxins or any other reported carcinogenic metabolite. All these studies showed the efficiency of different fungal species to increase the in vivo or in vitro production of high-value compounds by plant cultures under biotic stress.

Another filamentous fungus that stands out as a GRAS species is *Rhizopus oryzae* (Denardi-Souza et al. 2018). Aisyah et al. (2013) verified the increase in isoflavonoid phytoalexins production by soybean seedlings, as a result of an efficient fungal elicitation by *R. oryzae*. Denardi-Souza et al. (2018) showed that *R. oryzae* also triggers changes in essential nutrients contents during fermentation of agro-industrial by-products like soybean bran biomass. Using this substrate, there was achieved 58% increase in the yield of omega-3 fatty acid, 65% in proteins production, and elevation on the contents of several essential amino acids.

The composition of culture medium modifies the fungal development and, consequently, is another factor able to modulate fungal metabolome (Paranagama et al. 2007) either in single microorganism cultures or in the co-cultivation of two symbiotic organisms (Hajong et al. 2013). A comparison on the influence of soymilk dextrose agar (SDA), potato dextrose agar (PDA), potato dextrose broth (PDB), and V-8 agar in the development of soybean pathogens was performed by Xiang et al. (2014). SDA was shown to be the most effective medium, based on the radial rates of mycelial growth. Growth of *Passalora sojina* in SDA was significantly faster than in V-8 agar, while *Colletotrichum truncatum* and *Fusarium virguliforme* developed better in SDA than in PDA.

Although co-cultivation of fungi with plants is a promising technique to bring about novel biotechnologically useful metabolites, the whole process still needs intensive research to understand the role of individual parameters in directing biosynthesis. Control of these parameters will allow developing tailored industrial applications in the future based in controlled elicitation. Screening and isolation of valuable fungal species and the high cost of large-scale cultivation are also challenging (Jiao et al. 2018). In addition, as already mentioned, even if production of new metabolites or yield enhancement is the major target of plant-fungus co-cultivation, the production of metabolites may be also reduced. This is an interesting research area since processes able to decrease the biosynthesis of fungal toxins have a myriad of industrial applications. In this way, plant-fungi co-culturing proves to be a promising area of research, in which there is still much to be explored.

24.2.3 Fungi and Microorganisms in Biotic Interaction

Morphological typification derived from fungus-fungus interaction has been observed in vitro in solid media cultures due to survival strategies generated by competition between microorganisms for nutrients and space. The constant stress for survival leads to biosynthesis of enzymes, secondary metabolites, and distinct morphology for overcoming the scarcity of nutritional resources available in the natural habitat (Rayner 1988). Fungal defensive metabolites are commonly produced in response to threats imposed by competitors in order to ensure survival. The antagonistic and defense mechanisms exhibited by fungi triggering production of defensive metabolites have been used as a valuable tool in the search for bioactive compounds (Gloer 1995).

The relationship between morphology of fungal colonies and defense response to competitors has been investigated. Interestingly, physical contact is not a requirement for microbe-microbe interactions (Falconer et al. 2008), since there are also chemical-mediated interactions. Some responses demand physical proximity such as contact inhibition or colony-overgrowth, but distance inhibition and formation of zone lines, with diffusion of secondary metabolites in the medium is equally important (Bertrand et al. 2013a). Another well-established long-distance interaction occurs through the production of growth-inhibiting volatile compounds (Falconer et al. 2008).

The role of competition between fungi on secondary metabolites production has not been adequately understood yet, and furthermore, fungal metabolism has been rarely monitored during the interaction period (Chatterjee et al. 2016). However, an impressive range of natural products produced by microorganisms, belonging to diverse biosynthetic classes, was already detected using co-culturing approach (Jin et al. 2016). Although biosynthetic pathways responsible for production of defense compounds are not well known, it is clear that fungal interactions led to changes in growth, development, and adaptation patterns, in colony morphology and secondary metabolites, and proteins synthesis (Sandland et al. 2007; Wei et al. 2013).

Advances in genome sequencing have shown an unexpected genetic potential of some sequenced microorganisms for secondary metabolite production and detection of cryptic genes (Schroeckh et al. 2009; Bertrand et al. 2014a). Among cryptic gene activation strategies, biotic stress caused by co-culture is a simple, affordable, and straightforward methodology for producing compounds different from those isolated from single-microbial cultures (Cueto et al. 2001; Oh et al. 2005).

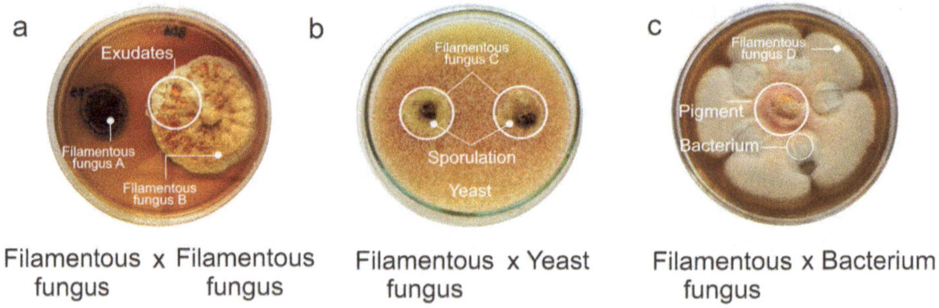

Fig. 24.8 Co-culture of filamentous fungi showing interaction with (**a**) filamentous fungus, (**b**) yeast, and (**c**) bacterium

Metabolome diversification is a direct effect of this kind of interaction (Oh et al. 2007; Nuetzmann et al. 2011), in some cases inhibition of metabolism was also observed (Chatterjee et al. 2016).

Figure 24.8 shows some typical examples of fungus-fungus (a), filamentous fungus-yeast (b), and fungus-bacterium interactions (c). In Fig. 24.8a, the metabolic response observed is overproduction of exudates by the fungal species (on the right) in the area closer to the species on the left. Figure 24.8b shows inhibition of sporulation by filamentous species in the contact zone with the yeast. In Fig. 24.8c, pigment formation by the filamentous fungus is observed in the center of the plate, as a mechanism to allow its growth through the plate, overcoming bacterium inhibition.

Under biotic stress, microorganisms can also yield compounds to act as transcriptional regulators and epigenetic modifiers as demonstrated by co-cultivation of *Aspergillus fumigatus* with the bacterial species *Streptomyces rapamycinicus*. In this situation, like in epigenetic modulation, there was the induction of a previously silent polyketide synthase pathway in the fungus, which caused the production of fumicycline A (**25**), a prenylated polyketide (Fig. 24.9) (König et al. 2013). In addition to the expression of silent gene groups, the transfer of gene fragments of full-length between microbes (horizontal gene transfer) is also a typical metabolome modulation in co-culture (Kurosawa et al. 2008).

Another example is co-cultivation of *Aspergillus nidulans* with the actinomycete *Streptomyces hygroscopicus* which resulted in specific activation of a cluster of cryptic genes that encodes polyketide archetype orsellinic acid biosynthesis (Schroeckh et al. 2009). The bacterium *S. rapamycinicus*, in turn, modifies fungal gene expression in a specific way, inducing modifications at the principal histone acetyltransferases Saga/Ada complex (Nuetzmann et al. 2011). Dialysis and electron microscopy experiments have shown that, in some cases, the close physical interaction between fungus and bacterium is necessary to provoke specific responses (Schroeckh et al. 2009; Nuetzmann et al. 2011; König et al. 2013).

Studies carried out aiming to understand the role of chemicals in microbial interactions suggest that several metabolites secreted by fungi in individual and mixed cultures are catabolized by their producers in subsequent growing stages. Other highlights are suppression and elicitation of new metabolites in mixed cultures (Chatterjee et al. 2016), showing how fungal interactions can be extraordinarily surprising. The space in which fungi are cultivated as well as nutrient contents are reported to interfere substantially in metabolite production (Bertrand et al. 2014b).

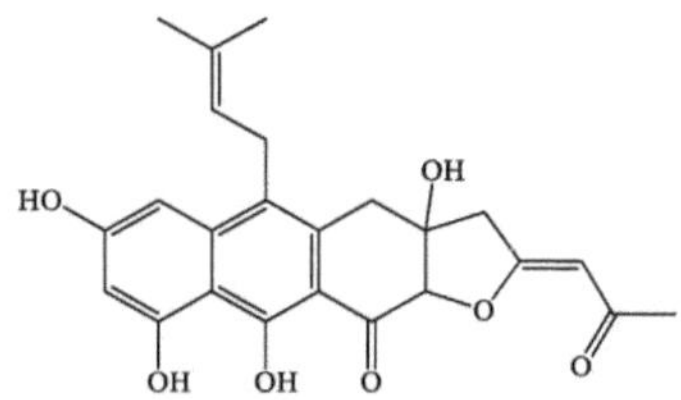

(25)

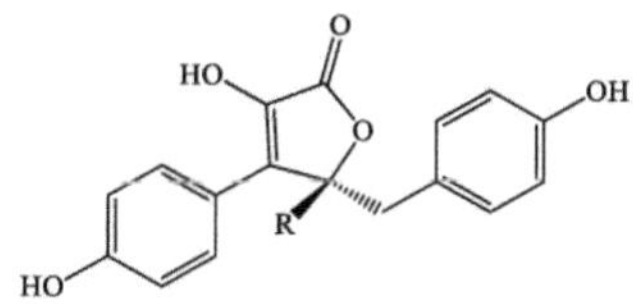

(26) R = OCH_3
(27) R = OH

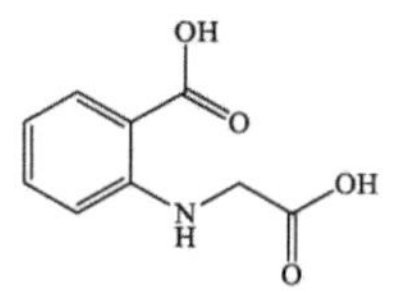

(28)

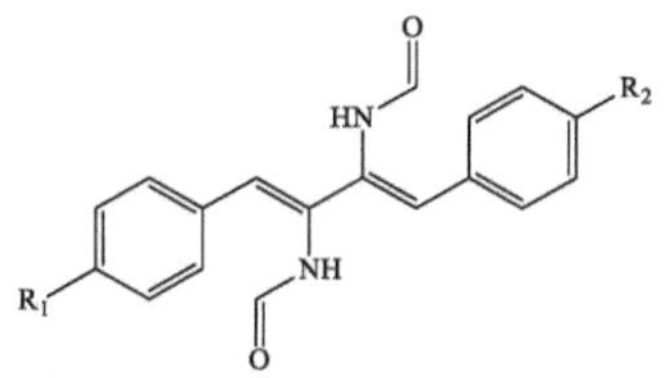

(29) $R_1 = OCH_3$; $R_2 = OSO_3H$
(30) $R_1; R_2 = OCH_3$
(31) $R_1; R_2 = OH$
(32) $R_1 = OCH_3$; $R_2 = OH$

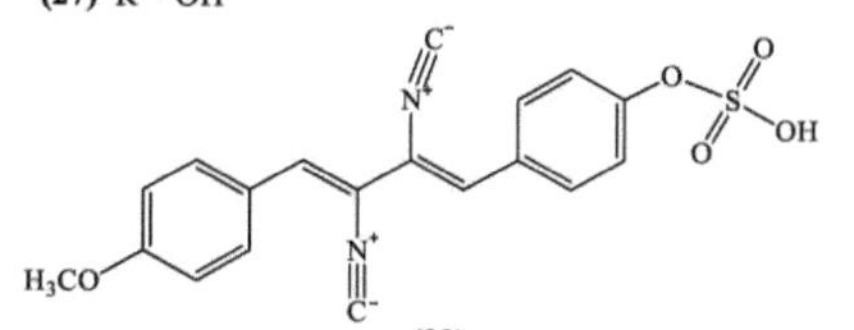

(33)

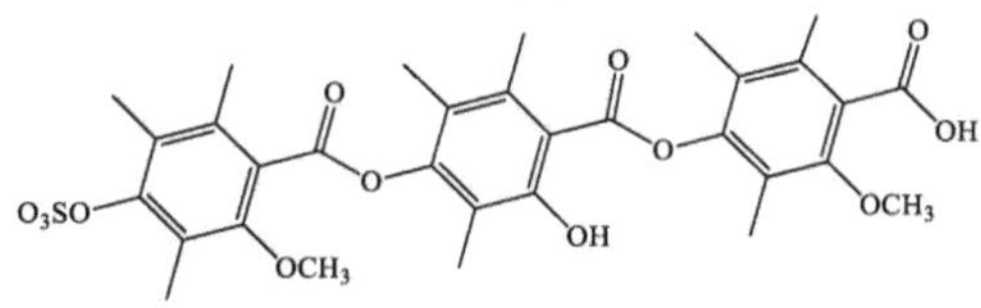

(34)

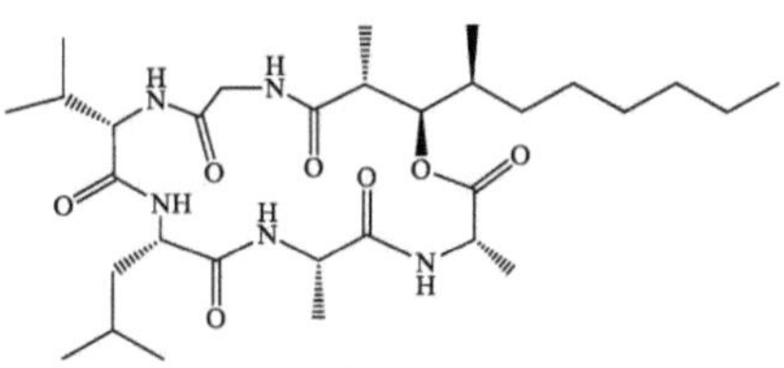

(35)

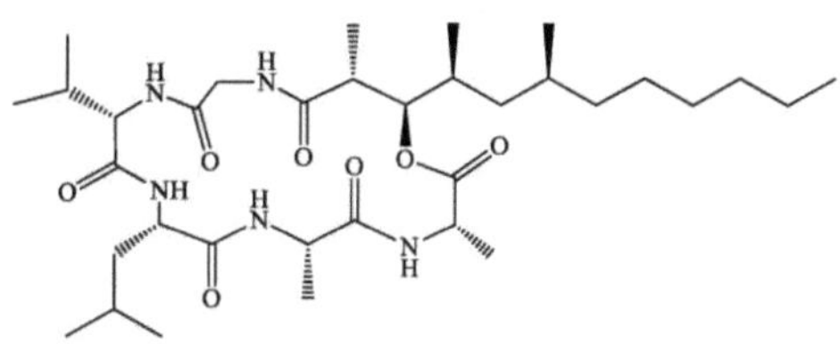

(36)

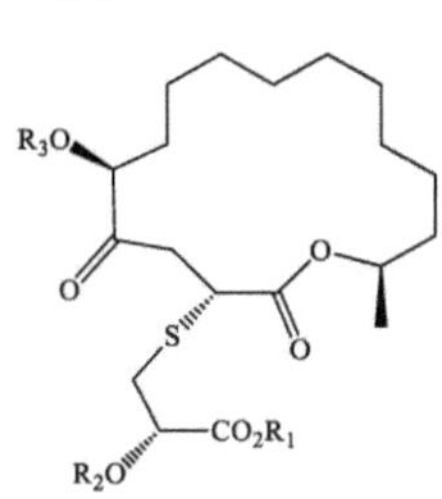

(37) $R_1; R_2; R_3 = H$
(38) $R_1; R_2 = H$; $R_3 = COCH_2CH_2CO_2H$

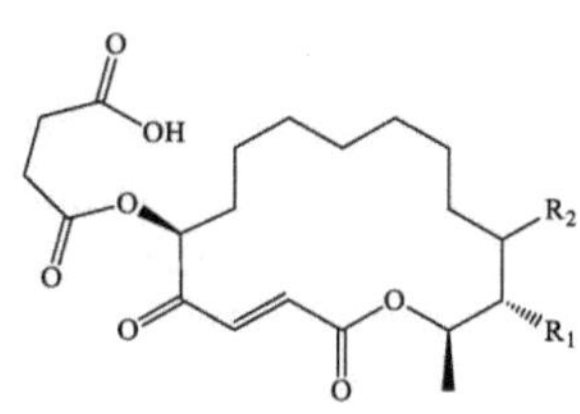

(39) $R_1 = OH$; $R_2 = H$
(40) $R_1 = H$; $R_2 = OH$

Fig. 24.9 Chemical structures of secondary metabolites produced from co-cultivation of fungi with microorganisms. (König et al. 2013 (25); Chen et al. 2015 (26–28); Zuck et al. 2011 (29–33); Bertrand et al. 2013b (34); Oh et al. 2007 (35–36); Stierle et al. 2017 (37–44); Ebrahim et al. 2016 (45–48); Shang et al. 2017 (49–51))

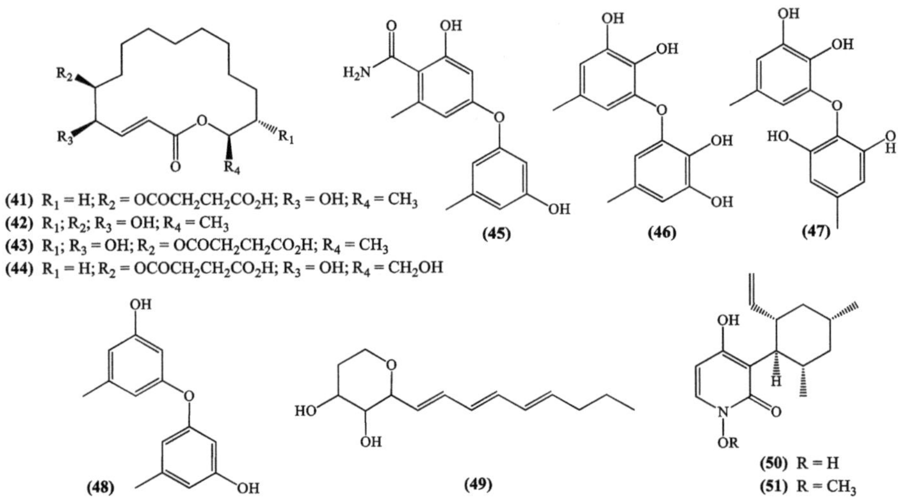

Fig. 24.9 (continued)

Biotic stress produces different degrees of inhibition between the species, as shown in experiments using mixed cultures of *Trichoderma harzianum* with other fungal isolates (Albert et al. 2011); of *Fusarium verticillioides* with *Clonostachys rosea*, where the synthesis of mycotoxins of fumonisin class by *F. verticillioides* is suppressed by *C. rosea* (Chatterjee et al. 2016); and of *A. fumigatus* with other microorganisms in agar plates (Zuck et al. 2011). Species-dependent response was described in the fermentation of *Aspergillus terreus*, a fungal species isolated from soil, with two bacteria, *Streptomyces lividans* and *Streptomyces coelicolor*, in solid medium, where the accumulation of fungal metabolites was not observed. In contrast, co-cultures of the same fungus with *Bacillus* sp. induced the production of two new butyrolactone derivatives, isobutyrolactone II **(26)** and 4-O-desmethyl-isobutyrolactone II **(27)**, along with *N*-anthranilic acid (carboxymethyl) **(28)**. Concurrently, there was up to 34 times increase in the accumulation of constitutional fungal metabolites compared to simple cultures of *A. terreus* (Chen et al. 2015). Cofermentation of *A. fumigatus* with *S. peucetius* also induced production of new metabolites, fumiformamide **(29)** and *N,N'*-((1Z,3Z)-1,4-bis(4-methoxyphenyl) buta-1,3-diene-2,3-diyl) diformamide **(30)**, along with *N*-formil derivatives **(31, 32)** and BU-4704 **(33)**, an analogue of xanthociline (Zuck et al. 2011).

Bertrand et al. (2013b) compared metabolome fingerprints resulting from simple and mixed cultivations of the fungal species *Trichophyton rubrum* and *Bionectria ochroleuca* on a solid medium. Induction of the biosynthesis of five new compounds was observed in the mixed culture, as a result of the biotic stress. Among the metabolites, 4″-hydroxysulfone-2,2″-dimetiltielavine P **(34)** was isolated. Only the non-sulfonated form was found in the extract obtained when *B. ochroleuca* was cultivated as a single culture in the medium, revealing sulfonation as a specific result of the interaction between these fungal species.

Co-culture of the marine fungus *Emericella* sp. with the bacterium *Salinispora arenicola* induced production of emericellamides A (**35**) and B (**36**), compounds unknown until then, which exhibited antibacterial activity against *S. aureus* (Oh et al. 2007). In the co-cultivation of *Penicillium fuscum* and *Penicillium camemberti/clavigerum*, the production of eight new macrolides was observed, berkeleylactones A-H (**37–44**), as well as the known antibiotic macrolide A26771B, among other metabolites. One of the isolated metabolites, berkeleylactone A (**37**), exhibited antimicrobial activity against different species of microorganisms (Stierle et al. 2017).

The same pattern of metabolic induction has been observed for endophytic fungi. For instance, under biotic stress caused by *Bacillus subtilis* or *S. lividans* in solid medium, an endophytic isolate of *Aspergillus austroafricanus* species also revealed changes in its secondary metabolism, expressed by increasing up to 29 times the production of diphenyl ethers (**45–48**) as well as production of a new austramide (**45**) (Ebrahim et al. 2016).

The complexity of molecular defenses existing in fungal co-culture was highlighted by Shang et al. (2017) by cultivating *Chaunopycnis* sp. fungal strains in the presence of *Trichoderma hamatum*. The interaction between these species resulted in transcriptional activation of a rare class of 2-acetyl-tetrahydropyran, chaunopyran A (**49**). In addition, the fungus *T. hamatum* was able to transform and inactivate the antifungal compound pyridoxantin (**50**) to methyl-pyridoxantin (**51**). Considering all these reports, it is noteworthy that the potential of biotic stress caused by co-culture in fungal metabolome modulation. The chemical structure of natural products obtained using this strategy (**25–51**) is shown in Fig. 24.9.

Co-culture also shed light in clinical studies, as in the production of fusaric acid due to the interaction between the fungi *Sarocladium strictum* and *Fusarium cf. oxysporum*, the latter was isolated from nails of onychomycosis patients (Bohni et al. 2016).

In addition to the metabolic response to biotic stress resulting from fungus-fungus and fungus-bacterium interactions, other types of co-cultures are important targets to investigate changes in microbial metabolic profiles. For instance, research on anaerobic fungal species *Piromyces* sp., isolated from goat rumen, in the presence of methanogens (anaerobic archaea) revealed citrate as one of the main metabolites produced by the fungal species. This is not a common metabolic response of this type of organism in simple cultures, where the main metabolites produced are from formate, lactate, ethanol, acetate, succinate, sugars/amino acids, and α-ketoglutarate classes. Such finding suggests the development of a new biosynthetic pathway in the co-culture and suggests that methanogens cause changes in anaerobic fungal metabolism (Cheng et al. 2013).

The microbial response in co-culture can be applied in several areas, not only for producing new bioactive secondary metabolites. Regarding environmental green applications, synergistic effects of yeasts and bacteria in co-culture were observed in the bioremediation of petroleum-contaminated soil (Zhang et al. 2014). On another application, co-culture of *Trichoderma* species with *Daedaleopsis confragosa* and *Phellinus pectinatus*, ligninolytic white rotting fungi, revealed a potential application for efficient bio-pulp production in paper and cellulose industries (Albert et al. 2011).

In summary, several studies conducted to control undesirable fungal species in the environment using other microorganisms or their metabolites support the idea that even if the dynamic nature of fungal metabolism is complex, it can be modulated by biotic stress. Current research in this field suggests that interactions between microorganisms are responsible for activation of cryptic genes leading to changes on fungal metabolome under biotic stress in nature or in vitro co-culture. Different organisms show different responses to biotic stress, such as elicitation of new metabolite production, increase or minimization of biosynthetic pathways, or even inhibition of some metabolic routes. While biosynthesis diversification and yield improvement are interesting tools for producing bioactive compounds, metabolite suppression can be used as a biotechnologically useful tool for decreasing toxins production.

24.3 Future Perspectives

Research on biotic stress caused from different types of interactions between fungal species with other organisms is another area for academic and applied research. This old natural model is shedding light on the development of new in vitro approaches to enhance secondary metabolite expression by fungi. The possibilities for metabolic diversification are in progress by investigating bacterial communities associated with lichens. This approach is adding up to a new level in terms of understanding interaction complexity and starting different models for discovering novel molecules. Increasing the knowledge of taxonomic specificity typical of lichenic associations will help to understanding the mechanisms and requirements for the association between the photobiont and the fungal species. This will result in a better knowledge of how to establish successful lichen cultures in the laboratory for the production of natural products. This will also help in improving *in situ* conditions for protection and maintenance of natural lichens considering the important role of these organisms that are sensitive to climate change and pollution.

It is also of immense interest to understand the mechanisms of infection and defense resulting from antagonistic relationships between phytopathogenic fungi and the host plants, and how these interactions directly interfere with the biochemical reactions in individuals at the transcriptional and translational levels. A number of new research areas emerged aiming at understanding and discovering effective ways to manipulate these interactions in order to ensure food and environmental security, besides reduction in agricultural losses, which has worldwide implications (George et al. 2016). For this purpose, in vitro biotic stress has been shown promising as a study alternative.

Considering the power of these interactions in the metabolic modulation in coexisting species, biotic stressing strategy is a promising resource for induction of the producing of compounds not only for the pharmaceutical industry but also for environmental and biotechnological applications. This is the case of ligninolytic enzymes that act in the degradation of agricultural residues, reducing the environmental impact caused by improper disposal of waste and even to be applied in biofuels production. Another point that deserves further investigation is toward the understanding of fungal interactions at the transcriptional level, which is extremely important to achieve better knowledge in biotic stress research.

24.4 Conclusions

The complex nature of dynamic fungal metabolism is evident either with plants, algae, or other microorganisms in all types of in vivo or in vitro co-culture systems. Biotic stress mechanisms are intrinsically linked to modulation of fungal metabolome and cryptic genes activation in response to external stimulus, resulting in the expression of not previously detected secondary metabolites, or even in the intensification in the production of those constitutively biosynthesized. Lichens are important examples of biotic interactions that bring about natural metabolic modulation. The association of photosynthetic and fungal partners promotes the biosynthesis of unique secondary metabolites mainly expressed by the fungal entity. Important compounds have been identified and biological activities with potential pharmaceutical applications have been described from lichens.

Although the interactions of fungi with other microorganisms and plants are common in nature, there is a growing interest in the in vitro mimetization of these interactions due to a high potential of fungal metabolic diversification. Thus, many studies seek to explore the biotic stress caused by co-culturing microorganisms in vitro to activate cryptic genes resulting in the production of compounds with important pharmaceutical applications. As emphasized in this review, modulation of fungal metabolome by inducing in vitro biotic stress is a successful tool to trigger unique responses and to stimulate important biosynthetic pathways.

Acknowledgements The authors gratefully acknowledge financial support from Fundação de Amparo à Pesquisa do Estado de Minas Gerais (FAPEMIG CEX APQ 02604/16), Conselho Nacional de Desenvolvimento Científico e Tecnológico (CNPq, grant # 304922/2018-8), Coordenação de Aperfeiçoamento de Pessoal de Nível Superior (CAPES), and National Institute of Science and Technology - INCT BioNat, grant # 465637/2014-0, Brazil.

References

Abanda-Nkpwatt D, Krimm U, Coiner HA, Schreiber L, Schwab W (2006) Plant volatiles can minimize the growth suppression of epiphytic bacteria by the phytopathogenic fungus *Botrytis cinerea* in co-culture experiments. Environ Exp Bot 56:108–119

Aguieiras ECG, Cavalcanti-Oliveira ED, Freire DMG (2015) Current status and new developments of biodiesel production using fungal lipases. Fuel 159:52–67

Aisyah S, Gruppen H, Madzora B, Vincken JP (2013) Modulation of isoflavonoid composition of *Rhizopus oryzae* elicited soybean (Glycine max) seedlings by light and wounding. J Agric Food Chem 61:8657–8667

Aisyah S, Gruppen H, Slager M, Helmink B, Vincken JP (2015) Modification of prenylated stilbenoids in peanut (*Arachis hypogaea*) seedlings by the same fungi that elicited them: the fungus strikes back. J Agric Food Chem 63:9260–9268

Albert S, Chauhan D, Pandya B, Padhiar A (2011) Screening of *Trichoderma* spp. as potential fungal partner in co-culturing with white rot fungi for efficient bio-pulping. Global J Biotech Biochem 6:95–101

Andrade-Domínguez A, Salazar E, Vargas-Lagunas MC, Kolter R, Encarnaci6n S (2014) Eco-evolutionary feedbacks drive species interactions. ISME J 8:1041–1054

Aptroot A, Cáceres MES (2014) New lichen species from termite nests in rainforest in Brazilian Rondônia and adjacent Amazonas. Lichenologist 46:365–372

Arnold AE, Miadlikowska J, Higgins KL, Sarvate SD, Gugger P, Way A, Hofstetter V, Kauff F, Lutzoni F (2009) A phylogenetic estimation of trophic transition networks for ascomycetous fungi: are lichens cradles of symbiotrophic fungal diversification? Syst Biol 58:283–297

Bačkorová M, Jendželovský R, Kello M, Bačkor M, Mikeš J, Fedoročko P (2012) Lichen secondary metabolites are responsible for induction of apoptosis in HT-29 and A2780 human cancer cell lines. Toxicol In Vitro 26:462–468

Baldi A, Jain A, Gupta N (2008) Co-culture of arbuscular mycorrhiza-like fungi (*Piriformospora indica* and *Sebacina vermifera*) with plant cells of *Linum album* for enhanced production of podophyllotoxins: a first report. Biotechnol Lett 30:1671–1677

Baldi A, Srivastava AK, Bisaria VS (2009) Fungal elicitors for enhanced production of secondary metabolites in plant cell suspension cultures. In: Varma A, Kharkwal AC (eds) Symbiotic Fungi. Springer, Berlin/Heidelberg, pp 373–380

Bayram O, Krappmann S, Ni M, Bok JW, Helmstaedt K, Valerius O, Braus-Stromeyer S, Kwon NJ, Keller NP, Yu JH, Braus GH (2008) VelB/VeA/LaeA complex coordinates light signal with fungal development and secondary metabolism. Science 320:1504–1506

Bertrand S, Schumpp O, Bohni N, Bujard A, Azzollini A, Monod M, Gindro K, Wolfender JL (2013a) Detection of metabolite induction in fungal co-cultures on solid media by high-throughput differential ultra-high pressure liquid chromatography-time-of-flight mass spectrometry fingerprinting. J Chromatogr A 1292:219–228

Bertrand S, Schumpp O, Bohni N, Monod M, Gindro K, Wolfender JL (2013b) De novo production of metabolites by fungal co-culture of *Trichophyton rubrum* and *Bionectria ochroleuca*. J Nat Prod 76:1157–1165

Bertrand S, Azzollini A, Schumpp O, Bohni N, Schrenzel J, Monod M, Gindro K, Wolfender JL (2014a) Multi-well fungal co-culture for de novo metabolite-induction in time-series studies based on untargeted metabolomics. Mol BioSyst 10:2289–2298

Bertrand S, Bohni N, Schnee S, Schumpp O, Gindro K, Wolfender JL (2014b) Metabolite induction *via* microorganism co-culture: a potential way to enhance chemical diversity for drug discovery. Biotechnol Adv 32:1180–1204

Bjerke JW, Elvebakk A, Domínguez E, Dahlback A (2005) Seasonal trends in usnic acid concentrations of Arctic, alpine and Patagonian populations of the lichen *Flavocetraria nivalis*. Phytochemistry 66:337–344

Bohni N, Hofstetter V, Gindro K, Buyck B, Schumpp O, Bertrand S, Monod M, Wolfender JL (2016) Production of fusaric acid by *Fusarium* spp. in pure culture and in solid medium co-cultures. Molecules 21:370

Boustie J, Grube M (2005) Lichens – a promising source of bioactive secondary metabolites. Plant Genet Resour 3:273–287

Brakhage AA (2013) Regulation of fungal secondary metabolism. Nat Rev Microbiol 11:21–32

Brakhage AA, Schroeckh V (2011) Fungal secondary metabolites – strategies to activate silent gene clusters. Fungal Genet Biol 48:15–22

Brakhage AA, Thön M, Spröte P, Scharf DH, Al-Abdallah Q, Wolke SM, Hortschansky P (2009) Aspects on evolution of fungal β-lactam biosynthesis gene clusters and recruitment of trans-acting factors. Phytochemistry 70:1801–1811

Bugni TS, Andjelic CD, Pole AR, Rai P, Ireland CM, Barrows LR (2009) Biologically active components of a Papua New Guinea analgesic and anti-inflammatory lichen preparation. Fitoterapia 80:270–273

Calvo AM, Wilson RA, Bok JW, Keller NP (2002) Relationship between secondary metabolism and fungal development. Microbiol Mol Biol Rev 66:447–459

Candan M, Yılmaz M, Tay T, Aysen A, Türk Ö (2007) Antimicrobial activity of extracts of the lichen *Parmelia sulcata* and its salazinic acid constituent. Z Naturforsch C 62:619–621

Chatterjee S, Kuang Y, Splivallo R, Chatterjee P, Karlovsky P (2016) Interactions among filamentous fungi *Aspergillus niger*, *Fusarium verticillioides* and *Clonostachys rosea*: fungal biomass, diversity of secreted metabolites and fumonisin production. BMC Microbiol 16:1–13

Chen H, Daletos G, Abdel-Aziz MS, Thomy D, Dai H, Brotz-Oesterhelt H, Lin W, Proksch P (2015) Inducing secondary metabolite production by the soil-dwelling fungus *Aspergillus terreus* through bacterial co-culture. Phytochem Lett 12:35–41

Cheng YF, Jin W, Mao SY, Zhu WY (2013) Production of citrate by anaerobic fungi in the presence of co-culture methanogens as revealed by ^{1}H NMR spectrometry. Asian-Australas J Anim Sci 26:1416–1423

Cocchietto M, Skert N, Nimis P, Sava G (2002) A review on usnic acid, an interesting natural compound. Naturwissenschaften 89:137–146

Cueto M, Jensen PR, Kauffman C, Fenical W, Lobkovsky E, Clardy J (2001) Pestalone, a new antibiotic produced by a marine fungus in response to bacterial challenge. J Nat Prod 64:1444–1446

da Silva RA, Polonio JC, Costa AT, Santos CM, Rhoden SA, Azevedo JL, Pamphile JA (2018) Bioprospection of culturable endophytic fungi associated with the ornamental plant *Pachystachys lutea*. Curr Microbiol 75:588–596

Davies KM, Schwinn KE (2003) Transcriptional regulation of secondary metabolism. Funct Plant Biol 30:913–925

Dean R, Van Kan JAL, Pretorius ZA, Hammond-Kosack KE, Pietro AD, Spanu PD, Rudd JJ, Dickman M, Kahmann R, Ellis J, Foster GD (2012) The top 10 fungal pathogens in molecular plant pathology. Mol Plant Pathol 13:414–430

Denardi-Souza T, Massarolo KC, Tralamazza SM, Badiale-Furlong E (2018) Monitoring of fungal biomass changed by *Rhizopus oryzae* in relation to amino acid and essential fatty acids profile in soybean meal, wheat and rice. CYTA – J Food 16:156–164

Dighton J, White JF (2016) The fungal community: its organization and role in the ecosystem, 4th edn. CRC Press Taylor and Francis, Boca Raton, p 597

Dimijian GG (2000) Evolving together: the biology of symbiosis, part 1. Proc (Bayl Univ Med Cent) 13:217–226

Divakar PK, Crespo A, Wedin M, Leavitt SD, Hawksworth DL, Myllys L, McCune B, Randlane T, Bjerke JW et al (2015) Evolution of complex symbiotic relationships in a morphologically derived family of lichen-forming fungi. New Phytol 208:1217–1226

Dixon RA (2001) Natural products and plant disease resistance. Nature 411:843–847

Douglas AE (2010) The symbiotic habit. Princeton University Press, Princeton, p 202

Ebrahim W, El-Neketi M, Lewald LI, Orfali RS, Lin W, Rehberg N, Kalscheuer R, Daletos G, Proksch P (2016) Metabolites from the fungal endophyte *Aspergillus austroafricanus* in axenic culture and in fungal–bacterial mixed cultures. J Nat Prod 79:914–922

Einarsdóttir E, Groeneweg J, Björnsdóttir G, Harðardottir G, Omarsdóttir S, Ingólfsdóttir K, Ögmundsdóttir H (2010) Cellular mechanisms of the anticancer effects of the lichen compound usnic acid. Planta Med 76:969–974

Eisenreich W, Knispel N, Beck A (2011) Advanced methods for the study of the chemistry and the metabolism of lichens. Phytochem Rev 10:445–456

Eskin NAM, Shahidi F (2013) Biochemistry of foods, 3rd edn. Academic, Amsterdam, p 584

Falconer RE, Bown JL, White NA, Crawford JW (2008) Modelling interactions in fungi. J R Soc Interface 5:603–615

Fang W, Azimzadeh P, Leger RJ (2012) Strain improvement of fungal insecticides for controlling insect pests and vector-borne diseases. Curr Opin Microbiol 15:232–238

Fillinger S, Elad Y (2016) Botrytis – the fungus, the pathogen and its Management in Agricultural Systems. Springer, Cham, p 486

Fox EM, Howlett BJ (2008) Secondary metabolism: regulation and role in fungal biology. Curr Opin Microbiol 11:481–487

Francolini I, Norris P, Piozzi A, Donelli G, Stoodley P (2004) Usnic acid, a natural antimicrobial agent able to inhibit bacterial biofilm formation on polymer surfaces. Antimicrob Agents Chemother 48:4360–4365

Frisvad JC, Andersen B, Thrane U (2008) The use of secondary metabolite profiling in chemotaxonomy of filamentous fungi. Mycol Res 112:231–240

George TS, Dou D, Wang X (2016) Plant–microbe interactions: manipulating signals to enhance agricultural sustainability and environmental security. Plant Growth Regul 80:1–3

Ghosh S, Chowdhury R, Bhattacharya P (2016) Mixed consortia in bioprocesses: role of microbial interactions. Appl Microbiol Biotechnol 100:4283–4295

Giri CC, Zaheer M (2016) Chemical elicitors *versus* secondary metabolite production *in vitro* using plant cell, tissue and organ cultures: recent trends and a sky eye view appraisal. Plant Cell Tissue Organ Cult 126:1–18

Gloer JB (1995) The chemistry of fungal antagonismo and defense. Can J Bot 73:1265–1274

Goel M, Dureja P, Rani A, Uniyal PL, Laatsch H (2011) Isolation, characterization and antifungal activity of major constituents of the himalayan lichen *Parmelia reticulata* Tayl. J Agric Food Chem 59:2299–2307

Grayer RJ, Kokubun T (2001) Plant-fungal interactions: the search for phytoalexins and other antifungal compounds from higher plants. Phytochemistry 56:253–263

Grube M, Berg G (2010) Microbial consortia of bacteria and fungi with focus on the lichen symbiosis. Fungal Biol Rev 23:72–85

Grube M, Cernava T, Soh J, Fuchs S, Aschenbrenner I, Lassek C, Wegner U, Becher D, Riedel K, Sensen CW, Berg G (2015) Exploring functional contexts of symbiotic sustain within lichen-associated bacteria by comparative omics. ISME J 9:412–424

Hajong S, Kumaria S, Tandon P (2013) Compatible fungi, suitable medium, and appropriate developmental stage essential for stable association of *Dendrobium chrysanthum*. J Basic Microbiol 53:1025–1033

Hertweck C (2009) The biosynthetic logic of Polyketide diversity. Angew Chem Int Ed Engl 48:4688–4716

Hewage RT, Aree T, Mahidol C, Ruchirawat S, Kittakoop P (2014) One strain-many compounds (OSMAC) method for production of polyketides, azaphilones, and an isochromanone using the endophytic fungus *Dothideomycete* sp. Phytochemistry 108:87–94

Ingólfsdóttir K (2002) Usnic acid. Phytochemistry 61:729–736

Jiao J, Gai Q-Y, Wang W, Zang Y-P, Niu L-L, Fu Y-J, Wang X (2018) Remarkable enhancement of flavonoid production in a co-cultivation system of *Isatis tinctoria* L. hairy root cultures and immobilized *Aspergillus niger*. Ind Crop Prod 112:252–261

Jin L, Quan C, Hou X, Fan S (2016) Potential pharmacological resources: natural bioactive compounds from marine-derived fungi. Mar Drugs 76:1–25

Kaasalainen U, Fewer DP, Jokela J, Wahlsten M, Sivonen K, Rikkinen J (2012) Cyanobacteria produce a high variety of hepatotoxic peptides in lichen symbiosis. Proc Natl Acad Sci U S A 109:5886–5891

Kampa A, Gagunashvili AN, Gulder TA, Morinaka BI, Daolio C, Godejohann M, Miao VP, Piel J, Andrésson Ó (2013) Metagenomic natural product discovery in lichen provides evidence for a family of biosynthetic pathways in diverse symbioses. Proc Natl Acad Sci U S A 110:129–137

Keller NP, Turner G, Bennett JW (2005) Fungal secondary metabolism - from biochemistry to genomics. Nat Rev Microbiol 3:937–947

Kim SH, Vujanovic V (2017) Biodegradation and biodetoxification of *Fusarium* mycotoxins by *Sphaerodes mycoparasitica*. AMB Express 7:1–9

Klein-Marcuschamer D, Oleskowicz-Popiel P, Simmons BA, Blanch HW (2012) The challenge of enzyme cost in the production of lignocellulosic biofuels. Biotechnol Bioeng 109:1083–1087

König CC, Scherlach K, Schroeckh V, Horn F, Nietzsche S, Brakhage AA, Hertweck C (2013) Bacterium induces cryptic meroterpenoid pathway in the pathogenic fungus *Aspergillus fumigatus*. Chembiochem 14:938–942

Kowalski M, Hausner G, Piercey-normore MD (2011) Bioactivity of secondary metabolites and thallus extracts from lichen fungi. Mycoscience 52:413–418

Kroymann J (2011) Natural diversity and adaptation in plant secondary metabolism. Curr Opin Plant Biol 14:246–251

Kümmritz S, Louis M, Haas C, Oehmichen F, Gantz S, Delenk H, Steudler S, Bley T, Steingroewer J (2016) Fungal elicitors combined with a sucrose feed significantly enhance triterpene production of a *Salvia fruticosa* cell suspension. Appl Microbiol Biotechnol 100:7071–7082

Kurosawa K, Ghiviriga I, Sambandan TG, Lessard PA, Barbara JE, Rha C, Sinskey AJ (2008) Rhodostreptomycins, antibiotics biosynthesized following horizontal gene transfer from *Streptomyces padanus* to *Rhodococcus fascians*. J Am Chem Soc 130:1126–1127

Li J, Wang J, Li J, Liu D, Li H, Gao W, Li J, Liu S (2015) *Aspergillus niger* enhance bioactive compounds biosynthesis as well as expression of functional genes in adventitious roots of *Glycyrrhiza uralensis* Fisch. Appl Biochem Biotechnol 178:576–593

Li J, Liu S, Wang J, Li J, Liu D, Li J, Gao W (2016) Fungal elicitors enhance ginsenosides biosynthesis, expression of functional genes as well as signal molecules accumulation in adventitious roots of *Panax ginseng* CA Mey. J Biotechnol 239:106–114

Ligon BL (2004) Penicillin: its discovery and early development. Semin Pediatr Infect Dis 15:52–57

Lima GS, Rocha AM, Santos GF, Silva AF, Marriel IE, Takahashi JA (2018a) Metabolic response of *Aspergillus sydowii* to OSMAC modulation produces acetylcholinesterase inhibitors. Phytochem Lett 24:39–45

Lima MTNS, dos Santos LB, Bastos RW, Nicoli JR, Takahashi JA (2018b) Antimicrobial activity and acetylcholinesterase inhibition by extracts from chromatin modulated fungi. Braz J Microbiol 49:169–176

Liu X, Ashforth E, Ren B, Song F, Dai H, Liu M, Wang J, Xie Q, Zhang L (2010) Bioprospecting microbial natural product libraries from the marine environment for drug discovery. J Antibiot 63:415–422

Luo H, Li C, Kim JC, Liu Y, Jung JS, Koh YJ, Hur J-S (2013) Biruloquinone, an acetylcholinesterase inhibitor produced by lichen-forming fungus *Cladonia macilenta*. J Microbiol Biotechnol 23:161–166

Magan N, Aldred D (2007) Why do fungi produce mycotoxins? In: Dijksterhuis J, Samson RA (eds) Food mycology: a multifaceted approach to Fungi and food. CRC Press, Boca Raton, pp 121–133

Manojlović N, Ranković B (2012) Chemical composition of three *Parmelia* lichens and antioxidant, antimicrobial and cytotoxic activities of some their major metabolites. Phytomedicine 19:1166–1172

Marcet-Houben M, Gabaldón T (2016) Horizontal acquisition of toxic alkaloid synthesis in a clade of plant associated fungi. Fungal Genet Biol 86:71–80

Marmann A, Aly A, Lin W, Wang B, Proksch P (2014) Co-cultivation – a powerful emerging tool for enhancing the chemical diversity of microorganisms. Mar Drugs 12:1043–1065

Melo MG, dos Santos JP, Serafini MR, Caregnato FF, Pasquali MA, Rabelo TK, da Rocha RF, Jr Quintans L, Araújo AA, da Silva FA, Moreira JC, Gelain DP (2011) Redox properties and cytoprotective actions of atranorin, a lichen secondary metabolite. Toxicol In Vitro 25:462–468

Minty JJ, Singer ME, Scholz SA, Bae C-H, Ahn J-H, Foster CE, Liao JC, Lin XN (2013) Design and characterization of synthetic fungal-bacterial consortia for direct production of isobutanol from cellulosic biomass. PNAS 110:14592–14597

Molnár K, Farkas E (2010) Current results on biological activities of lichen secondary metabolites: a review. Z Naturforsch C 65:157–173

Mushegian AA, Peterson CN, Baker CM, Pringle A (2011) Bacterial diversity across individual lichens. J Appl Environ Microbiol 77:4249–4252

Nielsen KF, Holm G, Uttrup LP, Nielsen PA (2004) Mould growth on building materials under low water activities. Influence of humidity and temperature on fungal growth and secondary metabolism. Int Biodeterior Biodegradation 54:325–336

Niu J, Arentshorst M, Nair PDS, Dai Z, Baker SE, Frisvad JC Ram AFJ (2015) Identification of a classical mutant in the industrial host *Aspergillus niger* by systems genetics: LaeA is required for citric acid production and regulates the formation of some secondary metabolites. G3 6:193–204

Nuetzmann HW, Reyes-Dominguez Y, Scherlach K, Schroeckh V, Horn F, Gacek A, Schümann J, Hertweck C, Strauss J, Brakhage AA (2011) Bacteria-induced natural product formation in the fungus *Aspergillus nidulans* requires Saga/Ada-mediated histone acetylation. Proc Natl Acad Sci U S A 108:14282–14287

Oh DC, Jensen PR, Kauffman CA, Fenical W (2005) Libertellenones A–D: induction of cytotoxic diterpenoid biosynthesis by marine microbial competition. Bioorg Med Chem 13:5267–5273

Oh DC, Kauffman CA, Jensen PR, Fenical W (2007) Induced production of emericellamides A and B from the marine-derived fungus *Emericella* sp. in competing co-culture. J Nat Prod 70:515–520

Okuyama E, Umeyama K, Yamazaki M, Kinoshita Y, Yamamoto Y (1995) Usnic acid and diffractaic acid as analgesic and antipyretic components of *Usnea diffracta*. Planta Med 61:113–115

Osbourn A (2010) Secondary metabolic gene clusters: evolutionary toolkits for chemical innovation. Trends Genet 26:449–457

Paranagama PA, Wijeratne EMK, Gunatilaka AAL (2007) Uncovering biosynthetic potential of plant-associated fungi: effect of culture conditions on metabolite production by *Paraphaeosphaeria quadriseptata* and *Chaetomium chiwersii*. J Nat Prod 70:1939–1945

Pompilio A, Pomponio S, Di Vincenzo V, Crocetta V, Nicoletti M, Piovano M, Garbarino JA, Di Bonaventura G (2013) Antimicrobial and antibiofilm activity of secondary metabolites of lichens against methicillin-resistant *Staphylococcus aureus* strains from cystic fibrosis patients. Future Microbiol 8:281–292

Punt PJ, Biezen N, Conesa A, Albers A, Mangnus J, Hondel C (2002) Filamentous fungi as cell factories for heterologous protein production. Trends Biotechnol 20:200–206

Qin Y, Pan X, Kubicek C, Druzhinina I, Chenthamara K, Labbé J, Yuan Z (2017) Diverse plant-associated pleosporalean fungi from saline areas: ecological tolerance and nirogen-status dependent effects on plant growth. Front Microbiol 8:1–14

Quispel A (1951) Some theoretical aspects of symbiosis. Antonie Van Leeuwenhoek 17:69–80

Ranković B, Kosanić M (2015) Lichens as a potential source of bioactive secondary metabolites. In: Ranković B (ed) Lichen secondary metabolites. Springer, Cham, pp 1–26

Rayner A (1988) Life in a collectivelessons from the fungi. New Scientist 120:49–53

Reddy RG, Veeraval L, Maitra S, Chollet-Krugler M, Tomasi S, Dévéhat FL-L, Boustie J, Chakravarty S (2016) Lichen-derived compounds show potential for central nervous system therapeutics. Phytomedicine 23:1527–1534

Reignault P, Kunz C, Delage N, Moreau E, Vedel R, Hamada W, Bompeix G, Boccara M (2000) Host and symptom-specific pectinase isozymes produced by wild-type strains and pathogenicity-altered transformants of *Botrytis cinerea*. Mycol Res 104:421–428

Reignault P, Valette-Collet O, Boccara M (2008) The importance of fungal pectinolytic enzymes in plant invasion, host adaptability and symptom type. Eur J Plant Pathol 120:1–11

Resende MLV, Salgado SML, Chaves ZM (2003) Espécies ativas de oxigênio na resposta de defesa de plantas a patógenos. Fitopatol Bras 28:123–130

Romão-Dumaresq AS, Dourado MN, Fávaro LCL, Mendes R, Ferreira A, Araújo WL (2016) Diversity of cultivated fungi associated with conventional and transgenic sugarcane and the interaction between endophytic *Trichoderma virens* and the host plant. PLoS One 11:e0158974

Roossinck MJ (2005) Symbiosis *versus* competition in plant virus evolution. Nat Rev Microbiol 3:917–924

Sandland GJ, Rodgers JK, Minchella DJ (2007) Interspecific antagonismand virulence in hosts exposed to two parasite species. J Invertebr Pathol 96:43–47

Sapp J (1994) Evolution by association: a history of Symbiosis. Oxford University Press, New York, p 272

Scherlach K, Hertweck C (2018) Mediators of mutualistic microbe–microbe interactions. Nat Prod Rep 25:303–308

Schlumbaum A, Mauch F, Vögeli U, Boller T (1986) Plant chitinases are potent inhibitors of fungal growth. Nature 324:365–367

Schroeckh V, Scherlach K, Nuetzmann HW, Shelest E, Schmid T-Heck W, Schuemann J, Martin K, Hertweck C, Brakhage AA (2009) Intimate bacterial–fungal interaction triggers biosynthesis of archetypal polyketides in *Aspergillus nidulans*. Proc Natl Acad Sci U S A 106:14558–14563

Seppelt RD, Türk R, Green TGA, Moser G, Pannewitz S, Sancho LG, Schroeter B (2010) Lichen and moss communities of Botany Bay, Granite Harbour, Ross Sea, Antarctica. Antarct Sci 22:691–702

Shang Z, Salim AA, Capon RJ (2017) Chaunopyran A: co-cultivation of marine mollusk-derived fungi activates a rare class of 2-alkenyl-tetrahydropyran. J Nat Prod 80:1167–1172

Shukla V, Joshi GP, Rawat MSM (2010) Lichens as a potential natural source of bioactive compounds: a review. Phytochem Rev 9:303–314

Shwab EK, Keller NP (2008) Regulation of secondary metabolite production in filamentous ascomycetes. Mycol Res 112:225–230

Silva GH, Oliveira CM, Teles HL, Pauletti PM, Castro-Gamboa I, Silva DHS, Bolzani VS, Young MCM, Costa-Neto CM, Pfenning LH, Berlinck RGS, Araujo AR (2010) Sesquiterpenes from *Xylaria* sp., an endophytic fungus associated with *Piper aduncum* (Piperaceae). Phytochem Lett 3:164–167

Sirrenberg A, Göbel C, Grond S, Czempinski N, Ratzinger A, Karlovsky P, Santos P, Feussner I, Pawlowski K (2007) *Piriformospora indica* affects plant growth by auxin production. Physiol Plant 131:581–589

Solhaug KA, Lind M, Nybakken L, Gauslaa Y (2009) Possible functional roles of cortical depsides and medullary depsidones in the foliose lichen *Hypogymnia physodes*. Flora – Morphol Distrib Funct Ecol Pl 204:40–48

Spatafora JW, Bushley KE (2015) Phylogenomics and evolution of secondary metabolism in plant-associated fungi. Curr Opin Plant Biol 26:37–44

Spribille T, Tuovinen V, Resl P, Vanderpool D, Wolinski H, Aime MC, Schneider K, Stabentheiner E, Toome-Heller M, Thor G, Mayrhofer H, Johannesson H, McCutcheon JP (2016) Basidiomycete yeasts in the cortex of ascomycete macrolichens. Science 353:488–492

Stierle AA, Stierle DB, Decato D, Priestley ND, Alverson JB, Hoody J, Mcgrath K, Klepacki D (2017) The berkeleylactones, antibiotic macrolides from fungal coculture. J Nat Prod 80:1150–1160

Stocker-Wörgötter E (2008) Metabolic diversity of lichen-forming ascomycetous fungi: culturing, polyketide and shikimate metabolite production, and PKS genes. Nat Prod Rep 25:188–200

Suzuki MT, Parrot D, Berg G, Grube M, Tomasi S (2016) Lichens as natural sources of biotechnologically relevant bacteria. Appl Microbiol Biotechnol 100:583–595

Teixeira PJPL, Thomazella DPT, Vidal RO, PFV d P, Reis O, Baroni RM, Franco SF, Mieczkowski P, GAG P, JMC M (2012) The fungal pathogen *Moniliophthora perniciosa* has genes similar to plant PR-1 that are highly expressed during its interaction with cacao. PLoS One 7:e45929

Teles APC, Ataliba GS, Takahashi JA (2012) Modulation of *Paecilomyces lilacinus* antimicrobial metabolite production by co-culturing with *Salmonella typhimurium*. Nat Prod Res (Print) 27:1598–1601

Tortora GJ, Funke BR, Case CL (2012) Microbiologia, 10th edn. Artmed, Porto Alegre, p 934

van Loon LC, Rep M, Pieterse CMJ (2006) Significance of inducible defense-related proteins in infected plants. Annu Rev Phytopathol 44:135–162

Varol M, Türk A, Candan M, Tay T, Koparal AT (2016) Photoprotective activity of vulpinic and gyrophoric acids toward ultraviolet b-induced damage in human keratinocytes. Phytother Res 30:9–15

Vijayakumar CS, Viswanathan S, Reddy MK, Parvathavarthini S, Kundu AB, Sukumar E (2000) Anti-inflammatory activity of (+)-usnic acid. Fitoterapia 71:564–566

Vining LC (1992) Secondary metabolism, inventive evolution and biochemical diversity – a review. Gene 115:135–140

Wang Y-Y, Liu B, Zhang X-Y, Zhou Q-M, Zhang T, Li H, Yu Y-F, Zhang X-L, Hao X-Y, Wang M, Wang L, Wei J-C (2014) Genome characteristics reveal the impact of lichenization on lichen-forming fungus *Endocarpon pusillum* Hedwig (Verrucariales, Ascomycota). BMC Genomics 15:34

Wei Q, Wang H, Chen Z, Lv Z, Xie Y, Lu F (2013) Profiling of dynamic changes in the microbial during the soy sauce fermentation process. Appl Microbiol Biotechnol 97:9111–9119

Williams RB, Henrikson JC, Hoover AR, Lee AE, Cichewicz RH (2008) Epigenetic remodeling of the fungal secondary metabolome. Org Biomol Chem 6:1895–1897

Xiang Y, Herman T, Hartman GL (2014) Utilizing soybean milk to culture soybean pathogens. Adv Microbiol 4:126–132

Xiong S, Huang C (2018) Synergistic strategies of predominant toxins in snake venoms. Toxicol Lett 287:142–154

Yamamoto Y, Mizuguchi R, Yamada Y, Yamamoto O (1985) Tissue cultures of *Usnea rubescens* and *Ramalina yasudae* and production of usnic acid in their cultures. Agric Biol Chem 49:3347–3348

Yang Y, Nguyen TT, Jeong M-H, Crişan F, Yu YH, Ha H-H, Choi KH, Jeong HG, Jeong TC, Lee KY, Kim KK, Hur JS, Kim H (2016) Inhibitory activity of (+)-usnic acid against non-small cell lung cancer cell motility. PLoS One 11:e0146575

Yılmaz M, Türk AÖ, Tay T, Kıvanç M (2004) The antimicrobial activity of extracts of the lichen cladonia foliacea and its (−)-usnic acid, atranorin, and fumarprotocetraric acid constituents. Z Naturforsch C 59:249–254

Yin W, Keller NP (2011) Transcriptional regulatory elements in fungal secondary metabolism. J Microbiol 49:329–339

Zhang C, Qi J, Cao Y (2014) Synergistic effect of yeast-bacterial co-culture on bioremediation of oil-contaminated soil. Biorem J 18:136–146

Zhao Y (2010) Auxin biosynthesis and its role in plant development. Annu Rev Plant Biol 61:49–64

Zhao J, Davis LC, Verpoorte R (2005) Elicitor signal transduction leading to production of plant secondary metabolites. Biotechnol Adv 23:283–333

Zuck KM, Shipley S, Newman DJ (2011) Induced production of N-formyl alkaloids from *Aspergillus fumigatus* by co-culture with *Streptomyces peucetius*. J Nat Prod 74:1653–1657

Marine Fungi as a Potential Source of Future Cosmeceuticals

25

Shivankar Agrawal, Sunil K. Deshmukh, and Colin J. Barrow

Abstract

Marine fungi are promising sources of bioactive metabolites. The wide array of fungal metabolites produced depends on the origin of the marine fungi, as their habitats are diverse, including living symbiotically with marine sponges and algae, or found in mangrove and bottom sediments. The metabolites discovered from marine fungi possess a range of biological activities, including antibacterial, antiviral, and anticancer activity. Some metabolites from these fungi have been used in beauty products and cosmetics and form an integral part of anti-acne, antiaging, and skin-whitening formulations. Some bioactives with novel scaffolds or new biological activities, obtained from this group of fungi, warrant further development as cosmeceuticals. The review discussed metabolites obtained from marine fungi that may be useful for beauty health products and cosmeceuticals.

S. Agrawal
TERI-Deakin Nano Biotechnology Centre, Biotechnology and Management of Bioresources Division, The Energy and Resources Institute, New Delhi, India

Centre for Chemistry and Biotechnology (CCB), School of Life and Environmental Sciences, Deakin University, Waurn Ponds, VIC, Australia

Indian Council of Medical Research (ICMR), Delhi, India

S. K. Deshmukh
Biotech & Management of Bioresources Div, The Energy and Resources Institute, New Delhi, Delhi, India

C. J. Barrow (✉)
Centre for Chemistry and Biotechnology (CCB), School of Life and Environmental Sciences, Deakin University, Waurn Ponds, VIC, Australia
e-mail: colin.barrow@deakin.edu.au

T. Satyanarayana et al. (eds.), *Advancing Frontiers in Mycology & Mycotechnology*,
https://doi.org/10.1007/978-981-13-9349-5_25

Keywords
Cosmeceutical · Marine fungi · Marine natural products · Hyperpigmentation · Acne

25.1 Introduction

Marine fungi are fungal species that predominantly exist in the marine or estuarine ecosystem, comprising ecological group with a small number of taxa. They are found associated with different sea-dwelling marine organisms like seaweeds, calcareous tubes of mollusk, mangroves, seagrass, lichens, and various other vertebrates and invertebrates (Richards et al. 2012). Broadly, marine fungi can be classified into two distinct groups, obligate and facultative marine fungi. Obligate marine fungi are predominantly sporulate and grow in marine habitat and usually develop reproductive propagules over plant or animal substrata, while facultative marine fungi flourish and grow in freshwater as well as in terrestrial region but are able to withstand and flourish in marine habitat, and frequently they have been isolated from various marine substrata (Kohlmeyer and Kohlmeyer 1980, Borse et al. 2012). Many of these groups possess specialized structures like sheaths and appendages on their propagules (spores and conidia), for anchorage on various substrata (Kohlmeyer and Kohlmeyer 1980).

Bioactive metabolites produced by these fungi belong to various chemical classes and have different biological activities (Saleem et al. 2007). Many of the fungal natural products are derived from either non-ribosomal peptides or polyketides. Some compounds belong to mixed polyketide–non-ribosomal peptides, and others are derived from other biosynthesis pathways. Over a 1000 metabolites have been reported from marine fungi with a range of biological properties and some of these are being developed as drug candidates (Deshmukh et al. 2018). Previous studies have examined taxonomy, distribution, chemistry, ecological roles, and pharmacological properties of natural products derived from marine fungi (Bugni and Ireland 2004). Although numerous compounds have been identified from marine fungi, only a few compounds have been explored for their cosmeceutical potentials, with very little information available on their biological activities or mode of action (Agrawal et al. 2018b). Marine fungi are an important source of new chemical scaffolds that can be modified to improve potential cosmeceutical activity. Here we review cosmeceutical activity of metabolites produced by marine fungi, which includes photo-protective, antiaging, antioxidant, skin-whitening, antimicrobial, and moisturizing potential. Novel cosmeceutical molecules isolated from marine fungi are depicted in Tables 25.1, 25.2, and 25.3, which are organized based on various cosmeceutical activities.

Table 25.1 Photo-protective and antiaging compounds isolated from marine fungi

	Photo-protective compounds from marine-derived fungi						
Sl. No.	Endophytic fungal strain	Source	Locality of collection	Isolated metabolite	Tested systems	Activity response*	References
1.	*Exophiala* sp.	Sponge *Halichondria panicea*	Bogil Island, Jeonnam Province, Korea	Circumdatin I (**7**), Circumdatin C (**8**), Circumdatin G (**9**)	High UVA screening activity	ED50 values of 98, 101 and 105 μM (positive control oxybenzone, ED50, 350 μM, currently used in sunscreen filter)	Zhang et al. (2008)
2.	*Annulohypoxylon stygium* (Xylariaceae family)	Marine algae *Bostrychia radicans*	Rio Escuro mangrove, Sao Paulo state, Brazil	3-Benzylidene-2-methylhexahydropyrrolo [1,2-α] pyrazine-1,4-dione (**10**), 1-(1,3-Benzodioxol-5-yl)-1,2-propanediol (**11**)	–	Exhibited no phototoxic potential and have the capability to UVB absorption and were found photo stable	Maciel et al. (2018)
3.	*Aspergillus* sp. (stock # MFA 212)	Marine red alga *Lomentaria catenata*	Golmae village, Ulsan City, Korea	Dihydroxyisoechinulin A (**12**)	Ultraviolet-A protecting activity	ED50, 130 μM, (oxybenzone, ED50, 350 μM)	Li et al. (2004a)
4.	*Aspergillus* sp.	–	–	Golmaenone (**13**), neoechinulin A (**14**)	Ultraviolet-A protecting activity	ED50 values of 90 and 170 μM	Li et al. (2004b)

(continued)

Table 25.1 (continued)

	Photo-protective compounds from marine-derived fungi						
Sl. No.	Endophytic fungal strain	Source	Locality of collection	Isolated metabolite	Tested systems	Activity response*	References
	Antiaging compounds from marine-derived fungi						
	Unidentified marine-derived fungus	*Scyphiphora hydrophyllacea*	Hainan Province, China	R-3-hydroxyundecanoic acid methylester-3-O-α-l-rhamnopyranoside **(15)**	*Staphylococcus aureus* and methicillin-resistant *S. aureus*	Modest antibacterial effect	Zeng et al. (2012)
	Paecilomyces variotii	Jellyfish *Nemopilema nomurai*	–	Paecilonic acids A **(16)** and B **(17)**	–	–	Wang et al. (2016)
	Penicillium sp.	Tunicate, *Didemnum* sp.	Mangrove forest, El-sheikh on the Egyptian Red Sea coast	Penicilloitins A **(18)** and B **(19)**	*Staphylococcus aureus*, *Escherichia coli*	Modest antibacterial effect	Mourshid et al. (2016)

Table 25.2 Antioxidant compounds isolated from marine fungi

Sl. No.	Endophytic fungal strain	Source	Locality of collection	Isolated metabolite	Tested systems	Activity response*	References
Antioxidants from mangrove-associated fungi							
1	*Phomopsis* sp. 33#	Bark/*Rhizophora stylosa*	South China Sea	Phochrodine D (**20**)	DPPH	IC_{50}, 34.0 μM	Chen et al. (2018)
2	*Pestalotiopsis* sp. HQD-6	*Rhizophora mucronata*	South China Sea	Fluturan (**21**)	DPPH	IC_{50}, 34.85 μg/mL	Zhou et al. (2018)
					ABTS	IC_{50}, 9.75 μg/mL	
3	*Ascomycota* sp. SK2YWS-L	Leaves/*Kandelia candel*	Shankou mangrove nature Reserve in Guangxi Province, China	Ascomindone A (**22**)	DPPH	IC_{50}, 18.1 μM	Tan et al. (2016)
				Ascomindone B (**23**) Ascomindone C (**24**) Ascomfurans A (**25**)	DPPH	Significant activity	
				Ascomindone A (**22**) Ascomindone B (**23**) Ascomindone C (**24**)	Hydroxyl radical scavenging assay	IC_{50} values in the range from 80 to 100 μM	
				Ascomindone A (**22**) Ascomindone B (**23**) Ascomindone C (**24**) Ascomfurans A (**25**) Ascomfurans B (**26**)	FRAP assay	Potent activity	
4	*Aspergillus* sp. 085242	Root/*Acanthus ilicifolius*	Guangxi Province, China	Asperisocoumarin A (**27**)	DPPH	EC_{50}, 125 μM	Xiao et al. (2016)
				Asperisocoumarin C (**28**)	DPPH	EC_{50}, 130 μM	

(continued)

Table 25.2 (continued)

Sl. No.	Endophytic fungal strain	Source	Locality of collection	Isolated metabolite	Tested systems	Activity response*	References
6	*Penicillium brocae* MA-192	Leaves/*Avicennia marina*	Hainan island, China	(10R,14R)-10-Hydroxydihydroresorcylide (**29**)	DPPH	IC_{50}, 14.4 μg/ml	Zhang et al. (2015)
				Brocaketone A (**30**)		IC_{50}, 5.9 μg/ml	
				Brocaketone D (**31**)		IC_{50}, 16.3 μg/ml	
7	*Alternaria*sp.R6	Root/*Myoporum bontioides*	Guangdong Province, China	(±)-(4R*,5S*,6S*)-3-amino-4,5,6-trihydroxy-2-methoxy-5-methyl-2-cyclohexen-1-one (**32**)	ABTS	EC_{50}, 8.19 μM	Wang et al. (2015)
				(±)-(4S*,5S*)-2,4,5-trihydroxy-3-methoxy-4-methoxycarbonyl-5-methyl-2-cyclopenten-1-one (**33**)		EC_{50}, 16.09 μM	
8	*Penicillium citrinum*	Stem/*Salicornia herbacea*	Jiangsu Province, China	Penicitriketo (**34**)	DPPH	IC_{50}, 85.33 μM	Wang et al. (2014)
9	*Penicillium griseofulvum* Y19–07	*Lumnitzera racemosa*	South China Sea	4-Hydroxyphenethyl methyl succinate (**35**)	DPPH	IC_{50}, 58.6 μM	Wang et al. (2009)
				4-Hydroxyphenethyl 2-(4-Hydroxyphenyl)acetate (**36**)		IC_{50}, 56.2 μM	
				4-Methylpyrocatechol (**37**)		IC_{50}, 7.1 μM	

	Antioxidants from marine sponge-associated fungi						
10	*Setosphaeria* sp. SCSIO41009	Sponge *Callyspongia* sp.	Guangdong Province, China	7-O-demethylmonocerin (**38**)	DPPH	IC_{50}, 38 μM	Pang et al. (2018)
11	*Aspergillus terreus*	*Phakellia fusca*	Xisha Islands in the South China Sea	Butyrolactone I (**39**) Butyrolactone II (**40**) 5-[(3,4-dihydro-2,2-dimethyl-2H-1-benzopyran-6-yl)- methyl]-3-hydroxy-4-(4-hydroxyphenyl)-2(5H)-furanone (**41**) Aspernolide A (**42**)	DPPH	IC_{50} in the range of 38–97 μM	Sun et al. (2018)
12	*Hypocrea koningii* PF04	*Phakellia fusca*	Yongxing Island in the South China Sea	Hypofurans A (**43**)	DPPH	IC_{50}, 27.4 μg/mL	Ding et al. (2015)
				N-isobutyl-2-phenylacetamide (**44**)	DPPH	IC_{50}, 16.8 μg/mL	
				N-(2-methylbutyl)-2-phenylacetamide (**45**)	DPPH	IC_{50}, 61.7 μg/mL	

(continued)

Table 25.2 (continued)

Sl. No.	Endophytic fungal strain	Source	Locality of collection	Isolated metabolite	Tested systems	Activity response*	References
13	*Penicillium citrinum* SpI080624G1f01	Sponge	Ishigaki Island, Okinawa prefecture, Japan	JBIR-59 (**46**)	DPPH	IC_{50}, 25 μM	Kawahara et al. (2012)
				JBIR-124 (**47**)		IC_{50}, 30 μM	
	Antioxidants from marine algae-associated fungi						
14	*Talaromyces islandicus* EN-501	*Laurencia okamurai*	Qingdao, China	8-Hydroxyconiothyrinone B (**48**)	DPPH ABTS	IC_{50},12 μM IC_{50}, 8.3 μM	Li et al. (2017)
				8,11-Dihydroxyconiothyrinone B (**49**)	DPPH ABTS	IC_{50}, 31 μM IC_{50}, 19 μM	
				4R,8-Dihydroxyconiothyrinone B (**50**)	DPPH ABTS	IC_{50}, 42 μM IC_{50}, 34 μM	
				4S,8-Dihydroxyconiothyrinone B (**51**)	DPPH ABTS	IC_{50}, 52 μM IC_{50}, 31 μM	
				4S,8-Dihydroxy-10-O-methyldendryol E (**52**)	DPPH ABTS	IC_{50}, 30 μM IC_{50}, 24 μM	

15	*Eurotium cristatum* EN-220	*Sargassum thunbergii*	Coast of Qingdao, China	Isovariecolorin I (**53**)	DPPH	IC_{50}, 20.6 μg/mL	Du et al. (2017)
				Rubrumazine B (**54**)	DPPH	IC_{50}, 28.5 μg/mL	
				Neoechinulin B (**55**)	DPPH	IC_{50}, 10.9 μg/mL	
				Neoechinulin C (**56**)	DPPH	IC_{50}, 12.1 μg/mL	
				Alkaloid E-7 (**57**)	DPPH	IC_{50}, 10.1 μg/mL	
				Didehydroechinulin (**58**)	DPPH	IC_{50}, 13.3 μg/mL	
				Echinulin (**59**)	DPPH	IC_{50}, 13.8 μg/mL	
				Dehydroechinulin (**60**)	DPPH	IC_{50}, 6.4 μg/mL	
				Variecolorin H (**61**)	DPPH	IC_{50}, 18.7 μg/mL	

(continued)

Table 25.2 (continued)

Sl. No.	Endophytic fungal strain	Source	Locality of collection	Isolated metabolite	Tested systems	Activity response*	References
16	*Aspergillus wentii* EN-48	Marine brown alga *Sargassum* sp.	–	Methyl 4-(3,4-dihydroxybenzamido) butanoate **(62)**	DPPH	IC_{50}, 5.2 μg/mL	Li et al. (2014)
				5-O-methylsulochine **(63)**	DPPH	IC_{50}, 24.7 μg/mL	
				Methyl 2-(2,6-dimethoxy-4-methylbenzoyl)-3,5-dihydroxybenzoate **(64)**	DPPH	IC_{50}, 78.2 μg/mL	
				Methyl-2-(2,6-dihydroxyl-4-methylbenzoyl)-3-hy droxy-5-methoxybenzoate **(65)**	DPPH	IC_{50}, 73.6 μg/mL	
				Physcion **(66)**	DPPH	IC_{50}, 99.4 μg/mL	
				4-(3,4-Dihydroxybenzamido) butanoic acid **(67)**	DPPH	IC_{50}, 9.6 μg/mL	
				(E)-N-(2-hydroxy-2-(4-hydroxyphenyl) ethyl)-3-(3-hydroxy-4-methoxyphenyl) acrylamide **(68)**	DPPH	IC_{50}, 82.7 μg/mL	
17	*Aspergillus* sp.	Brown alga *Ishige okamurae*	Uljin, Korea	Me 2-(6-bromo-3,4-dihydroxyphenyl)acetate **(69)**	DPPH	IC_{50}, 14.2 μM	Leutou et al. (2013)
				Me 2-(2,5-dibromo-3,4-dihydroxyphenyl)acetate **(70)**		IC_{50}, 12.1 μM	
				2-(3,4-Dihydroxyphenyl)acetic acid **(71)**		IC_{50}, 11.0 μM	

18	*Aspergillus ochraceus*	Marine red alga *Chondria crassicualis*	Yokji Island, Kyeongnam Province, Korea	(R)-(−)-5-Bromomellein (**72**)	DPPH	IC_{50}, 24 μM	Yun et al. (2013)
				(R)-(−)-Mellein (**73**)		IC_{50}, 25 μM	
				Clavatol (**74**)		IC_{50}, 30 μM,	
				Circumdatin A (**75**)		IC_{50}, 32 μM	
19	*Aspergillus ochraceus*	Brown alga *Sargassum kjellmanianum*	Dalian coastline, P. R. China,	2-Hydroxycircumdatin C (**76**)	DPPH	IC_{50} value of 9.9 μM	Cui et al. (2009)
20	Unidentified marine-derived fungus	Marine brown alga *Ishige okamurae*	Uljin, Gyeongbuk Province, Korea	N-[2-(4-hydroxyphenyl) acetyl] formamide (**77**)	DPPH	IC_{50}, 8.4 μM	Li et al. (2006)
				3,4-Dihydroxyphenyl acetic acid (**78**)		IC_{50}, 11.9 μM	
				N-[2-(4-hydroxyphenyl)ethenyl] formamide (**79**)		IC_{50}, 0.2 μM	
21	*Pseudallescheria* sp.	Marine brown alga *Agarum cribrosum*	Uljin, Gyeongbuk Province, Korea	Gliotoxin (**80**)	DPPH	IC_{50} value of 5.2 μM	Li et al. (2006)
22	*Aspergillus* sp. # MFA 212	Marine red alga *Lomentaria catenata*	Golmae village, Ulsan City, Korea	Dihydroxyisoechinulin A (**12**)	DPPH	(IC_{50}, 20 μM)	Li et al. (2004a)
23	*Aspergillus* sp. # MFA 212	Marine red alga *Lomentaria catenata*	Golmae village, Ulsan City, Korea	Golmaenone (**13**) Neoechinulin A (**14**)	DPPH	20 and 24 μM	Li et al. (2004b)
24	*Epicoccum* sp.	Marine brown alga *Fucus vesiculosus*	North Sea Costt, Tönning, Germany	Epicoccone (**81**)	DPPH	95% radical scavenging effects at 25 μg/mL	Abdel-Lateff et al. (2003a)

(continued)

Table 25.2 (continued)

Sl. No.	Endophytic fungal strain	Source	Locality of collection	Isolated metabolite	Tested systems	Activity response*	References
25	*Wardomyces anomalus*	*Enteromorpha* sp., Ulvaceae	Fehmarn Island in the Baltic Sea.	2,3,6,8-Tetrahydroxy-1-methylxanthone(**82**)	DPPH	Radical scavenging effects 94.7% at 25.0 μg/mL	Abdel-Lateff et al. (2003b)
				5-(Hydroxymethyl)-2-furancarboxylic acid (**84**)	DPPH	Radical scavenging effects 30.7% at 25.0 μg/mL	
				2,3,6,8-Tetrahydroxy-1-methylxanthone (**82**)	Inhibit peroxidation of linolenic acid	17.0% at 7.4 μg/mL	
				5-(Hydroxymethyl)-2-furancarboxylic acid (**84**)	Inhibit peroxidation of linolenic acid	2.6% at 7.4 μg/mL	
				2,3,6,8-Tetrahydroxy-1-methylxanthone (**82**), 3,6,8-trihydroxy-1-methylxanthone (**83**), 5-(hydroxymethyl)-2-furancarboxylic acid (**84**)	Inhibition of TK p56lck enzyme	100% enzyme inhibition at 200 μg/mL	
26	*Acremonium* sp.	Unidentified alga	Moraira, Mediterranean Sea	7-Isopropenylbicyclo[4.2.0] octa-1,3,5-triene-2,5-diol (**85**)	DPPH	85.5% at 25.0 μg/mL	Abdel-Lateff et al. (2002)
				2-(1-Hydroxy-1-methyl)-2,3-dihydrobenzofuran-5-ol (**86**)	DPPH	85.8% at 25.0 μg/mL)	
				2,2-Dimethylchroman-3, 6-diol (**87**)	DPPH	72.9% at 25.0 μg/mL	
				2-(3-Dihydroxy-3-methylbutyl) benzene-1,4-diol (**88**)	DPPH	90.2% at 25.0 μg/mL	

27	*Acremonium* sp.	Unidentified alga	Moraira, Mediterranean Sea	7-Isopropenylbicyclo[4.2.0] octa-1,3,5-triene-2,5-diol (**85**)	Inhibit peroxidation of linolenic acid	35.5% at 37.0 μg/mL	Abdel-Lateff et al. (2002)
				2-(1-Hydroxy-1-methyl)-2,3-dihydrobenzofuran-5-ol (**86**)	Inhibit peroxidation of linolenic acid	15.9% at 37.0 μg/mL	
				2,2-Dimethylchroman-3,6-diol (**87**)	Inhibit peroxidation of linolenic acid	9.2% at 37.0 μg/mL	
				2-(3-Dihydroxy-3-methylbutyl) benzene-1,4-diol (**88**)	Inhibit peroxidation of linolenic acid	16.6% at 37.0 μg/mL	
	Antioxidants from miscellaneous marine fungi						
28	*Aspergillus versicolor*	Deep-sea sediment	South China Sea.	Oxisterigmatocystin D (**89**)	TEAC (trolox (1.0 mmol/L) equivalent antioxidant capacity)	0.55	(Ze-Hong et al. (2018)
				Oxisterigmatocystin C (**90**)		1.16	
				Sterigmatocystine (**91**)		0.65	
				Versicolorin B (**92**)		1.03	
				UCT1072M1 (**93**)		0.97	
				Averantin (**94**)		0.89	
				Methyl-averantin (**95**)		0.86	
				Averythrin (**96**)		0.22	
				Averufanin (**97**)		0.82	
				Averufine (**98**)		0.94	
				Nidurufin (**99**)		0.62	
29	*Eurotium* sp. SCSIO F452	Sediment sample	South China Sea	(+)-Variecolortin A (**100**)	DPPH	IC_{50}, 58.4 μM	Zhong et al. (2018)
30	*Aspergillus carneus*	Sea water	South China Sea	3,4-Dihydroxy-5-(3-hydroxy-5-methylphenoxy)benzoic acid (**101**)	DPPH	IC_{50},19.3 μM	Xu et al. (2017)

(continued)

Table 25.2 (continued)

Sl. No.	Endophytic fungal strain	Source	Locality of collection	Isolated metabolite	Tested systems	Activity response*	References
31	*Aspergillus versicolor* SCSIO 41502	Marine sediment sample	South China Sea	6-Methylbenzene-1,2,4-triol (**102**), violaceol-II (**103**), cordyol C (**104**), sydowiols B (**105**), E (**106**), D (**107**)	DPPH	IC_{50} values of 18.92–52.27 μM	Huang et al. (2017)
32	*Aspergillus Niger*	Marine-mudflat-derived fungus	Suncheon Bay, Jeonnam Province, Korea	6,9-Dibromoflavasperone (**108**)		IC_{50}, 21 μM	Leutou et al. (2016a)
				Flavasperone (**109**)		IC_{50}, 25 μM	
				TMC-256A1 (**110**)		IC_{50}, 0.3 μM	
				Fonsecin (**111**)		IC_{50}, 0.02 μM	
				Aurasperone B (**112**)		IC_{50}, 0.01 μM	
33	*Thielavia* sp.	Marine-mud flat-derived fungus (bio transformation)		Thielaviazoline (**113**)	DPPH	IC_{50}, 11 μM	Leutou et al. (2016b)
34	*Thielavia* sp.	Marine-mud flat-derived fungus (bio transformation)		Thielaviazoline (**113**)	MRSA and MDRSA	MICs of 6.25 and 12.5 μg/mL	Leutou et al. (2016b)
35	Marine-derived fungus *Aspergillus* sp.			Chlorogentisyl alcohol (**114**)	DPPH	IC_{50}, 1.0 μM	Yun et al. (2011)
36	*Chrysosporium synchronum*	Brown alga *Sargassum ringgoldium*	Yokji Island of Gyeong Nam, Korea	Biotransformation of chlorogentisyl alcohol (**114**)yielded 1-O-(α-D-mannopyranosyl) chlorogentisyl alcohol (**115**)	DPPH	IC_{50}, 4.7 μM	Yun et al. (2011)
37	*Nigrospora* sp. PSU-F11	Sea fan-derived fungus	–	Nigrosporanene A (**116**)	DPPH	IC_{50}, 0.34 μg/mL	Rukachaisirikul et al. (2010)
				Nigrosporanene B (**117**)		IC_{50}, 0.24 μg/mL	

Table 25.3 Tyrosinase inhibitors and antimicrobial used for skin microbial infection from marine fungi

Sl. No.	Endophytic fungal strain	Source	Locality of collection	Isolated metabolite	Tested systems	Activity response*	References
Tyrosinase inhibitors from marine-derived fungi							
1	Marine-derived fungus *Eurotium rubrum* (chemically engineered extracts, coumarin dimerization of natural extracts)	Sea weed	Chosei-mura, Chosei-gun, Chiba prefecture, Japan	Coumarin derivative (**118**)	Tyrosinase inhibitory activity	IC_{50}, 1.7 µM	Kamauchi et al. (2018)
				Coumarin derivative (**119**)		IC_{50}, 1.2 µM	
				Coumarin dimer (**120**)		IC_{50}, 4.9 µM	
				Coumarin dimer (**121**)		IC_{50}, 1.8 µM	
				Coumarin dimer (**122**)		IC_{50}, 2.9 µM	
2	*Pestalotiopsis* sp. Z233 (in response to abiotic stress elicitation elicitation by $CuCl_2$)	*Sargassum horneri*	Wenzhou, China	1β,5α,6α,14-tetraacetoxy-9α-benzoyloxy-7βH-eudesman-2β,11-diol (**123**)	Tyrosinase inhibitory activity	IC_{50}, 14.8 µM	Wu et al. (2013)
				4α,5α-Diacetoxy-9α-benzoyloxy-7βH-eudesman-1β,2β,11,14-tetraol (**124**)		IC_{50}, 22.3 µM	

(continued)

Table 25.3 (continued)

Sl. No.	Endophytic fungal strain	Source	Locality of collection	Isolated metabolite	Tested systems	Activity response*	References
3	*Trichderma viride*	Sediment samples	Off-Izu Islands JAPAN	Homothallin II (**125**)	Tyrosinase inhibitory activity	A competitive inhibitor against the mushroom tyrosinase, and inhibit the enzyme through the binding to a copper active site of the mushroom tyrosinase	Tsuchiya et al. (2008)
4	*Botrytis* sp.	Marine red alga *Hyalosiphonia caespitose*	Collected at Dadaepo, Busan Korea	6-[(E)-hept-1-enyl]-α-pyrone, (**126**),	Tyrosinase inhibitory activity	IC_{50}, 4.5 μM	Zhang et al. (2007)
5	*Myrothecium* sp.	Marine green alga *Enteromorpha compressa*	Baegunpo, Busan Korea	Myrothenones A (**127**)	Tyrosinase inhibitory activity	IC_{50}, 0.8 μM	Li et al. (2005)
				6-n-pentyl-α-pyrone (**128**)		IC_{50}, 6.6 μM	
6	*Alternaria* sp. 581B	Marine green alga *Ulva pertusa*		Kojic acid (**129**)	Tyrosinase inhibitory activity	IC_{50}, 12.0 μM	Li et al. (2003)

	Antimicrobials from marine-derived fungi						
7	*Pleospora* sp. (PO_4)	Gut of marine isopod *Ligia oceanica*	Dongsha of Dinghai in Zhoushan, Zhejiang Province of China	Pleosporol A (**130**)	*S. epidermidis* CMCC26069	MIC, 0.5 μg/mL	Xu et al. (2018)
				Pleosporol B (**131**)		MIC, 8.0 μg/mL	
				Pleosporols C (**132**), D (**133**) (mixture of 132 and 133)		MIC, 2.0 μg/mL	
				Stemfolones (mixture of stemfolone B and 14-*epi*-stemfolone B) (**134**)		MIC, 1.0 μg/mL	
8	*Trichoderma* sp. strain MF106	–	Greenland seas	Trichodin A (**135**)	*S. epidermidis*	IC_{50}, 24 μM	Wu et al. (2014)
				Pyridoxatin (**136**)		IC_{50}, 4 μM	
9	*Spiromastix* sp.	Deep-sea sediments	South Atlantic Ocean	Spiromastixones F (**137**), G (**138**), H (**139**), I (**140**), J (**141**)	*Staphylococcus epidermidis* (both methicillin-sensitive methicillin- resistant strains (*S. epidermidis* ATCC 12228, 12–6, 12–8 methicillin-sensitive (MSSE) and *S. epidermidis* 12–8 methicillin-resistant (MRSE) strain)	MIC values in the range of 2–8 μM, (positive control levofloxacin MIC values in the range of 0.25–4 μM)	Niu et al. (2014)
10	*Aspergillus versicolor*	Gorgonian *Dichotella gemmacea*	South China Sea	Kipukasin H (**142**), Kipukasin I (**143**)	*S. epidermidis*	MIC value of 12.5 μM	Chen et al. (2014)

(continued)

Table 25.3 (continued)

Sl. No.	Endophytic fungal strain	Source	Locality of collection	Isolated metabolite	Tested systems	Activity response*	References
11	*Calcarisporium* sp. KF525	Water samples	German Wadden Sea	15G256α **(144)**	*S. epidermidis*	MIC of 12.9 μM (positive control chloramphenicol showed the MIC of 4.5 μM)	Silber et al. (2013)
				15G256β **(145)**		MIC of 16.9 μM	
				15G256π **(146)**	*P. acnes* 14.1	MIC of 4.5 μM (positive control chloramphenicol showed the MIC of 0.5 μM)	
				Calcaride A **(147)**	*S. epidermidis*	MIC of 68.8 μM	
				Calcaride B **(148)**		MIC of 52.3 μM	
				Calcaride C **(149)**		MIC of 29.6 μM	
12	Sponge-associated *Penicillium* sp.	–	–	Cis-cyclo(Leucyl-Tyrosyl) **(150)**	*S. epidermidis*	Inhibition up to 85% of biofilm formation of without interfering with bacterial growth, confirmed by SEM	Scopel et al. (2013)
13	*Penicillium chrysogenum* QEN-24S	Marine red alga *Laurencia* sp.		Conidiogenones B **(151)**	MRSA, *Pseudomonas fluorescens*, *P. aeruginosa*, and *S. epidermidis*	Each with a MIC value of 8 μg/mL	Gao et al. (2011)
					S. epidermidis	MIC 8 μg/mL	
				Conidiogenol **(152)**	*P. fluorescens* and *S. epidermidis*	Each with a MIC value of 16 μg/mL	

25.2 Cosmeceutical Application of Compounds from Marine Fungi

25.2.1 Photo-Protective Compounds

In last two decades, depletion in the ozone sphere has increased our exposure to ultraviolet radiation (UVR), particularly UVB radiation (280–315 nm) (Rozema et al. 2005). For most of the sun-dwelling aquatic or terrestrial organisms, UVB radiation imparts detrimental effect (Day and Neale 2002) including humans (Solomon 2008). Since the UVB exposure cannot be avoided by light-dependent organisms, photosynthetic organisms have developed defense mechanism to alleviate the effect of UVB, which includes accumulation of UV-absorbing substances like carotenoids, melanins, mycosporines, parietin, scytonemin, usnic acid, and various other photo-protective compounds (Dunlap and Yamamoto 1995; Rastogi and Incharoensakdi 2013). Natural photo-protective compounds have been investigated from marine and freshwater microorganisms, plants, and animals. Marine natural compounds from diverse classes have been reviewed, but photo-protectants from marine fungal sources have attracted less attention. Here, we summarize some photo-protective compounds reported from marine-derived fungus (Table 25.1).

25.2.1.1 Mycosporine and Mycosporine-Like Amino Acids (MAAs)

Mycosporines and MAAs are water-soluble molecules of low molecular weight (generally <400 Da) capable of absorbing UV of between 310 and 365 nm. Chemically, mycosporines possess either an aminocycloheximine ring with an imino alcohol or an aminocyclohexenone with nitrogen substituents (Oren and Gunde-Cimerman 2007). MAAs are derived from mycosporines bearing amino acid residues; MAAs contain amino-cyclohexenimine ring associated to an amino group, amino alcohol, or amino acid showing absorption between 320 and 360 nm (Carreto and Carignan 2011). MAAs have the capability to dissipate UV radiation without generating toxic reactive oxygen species (ROS), along with it as photo-protective it absorbs a wide range of UV spectrum (Carreto and Carignan 2011; Řezanka et al. 2004). Mycosporine–glutamicol–glucoside was detected as the major mycosporine in *Cladosporium sphaerospermum*, *C. cladosporioides*, the halophilic black yeasts *Phaeotheca triangularis* and *Hortaea werneckii*, and the halotolerant *Aureobasidium pullulans*, while only black yeasts contain a smaller amount of mycosporine–glutaminol–glucoside (Oren and Gunde-Cimerman 2007). Basidiomycetous yeast *Cryptococcus liquefaciens* also accumulates mycosporine–glutaminol–glucoside and mycosporine–glutamicol–glucoside (Kogej et al. 2006).

25.2.1.2 Carotenoids

Carotenoids, also called tetraterpenoids, are orange, red, and yellow organic pigments that are produced by plants and algae, as by well as by several bacteria, yeast, and fungi. Various marine-derived yeasts are known to produce carotenoids. The genera *Phaffia*, *Rhodotorula*, and *Xanthophyllomyces* are a rich source of astaxanthin (Ambati et al. 2014). Yeast and bacteria are poor producers of astaxanthin,

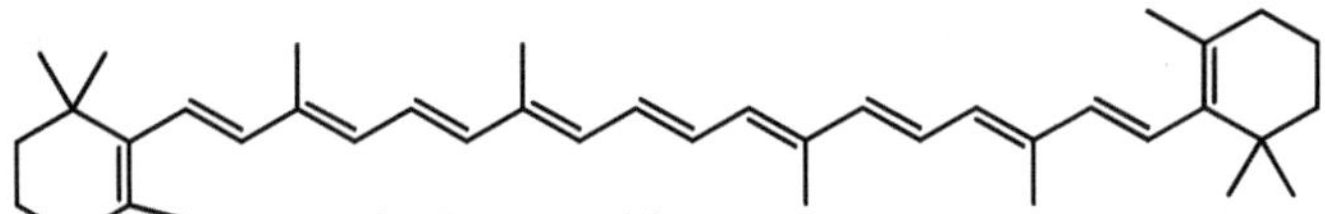

β- Carotene (1)

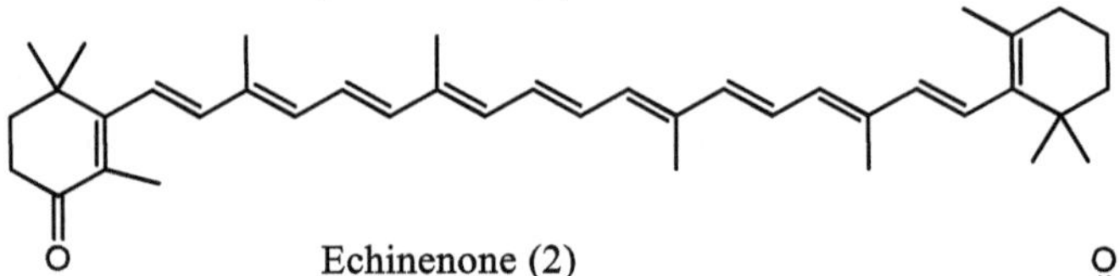

Echinenone (2)

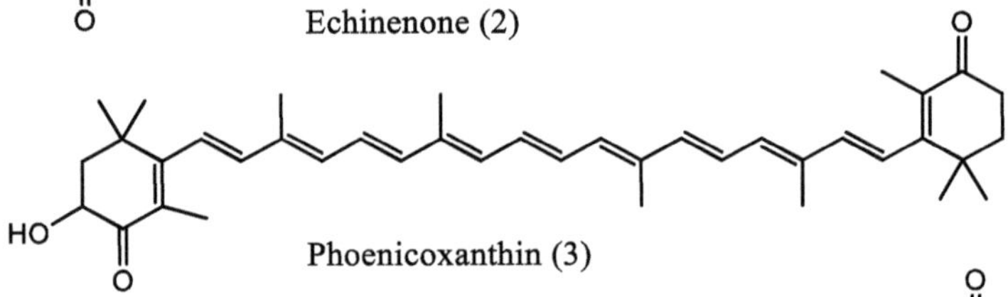

Phoenicoxanthin (3)

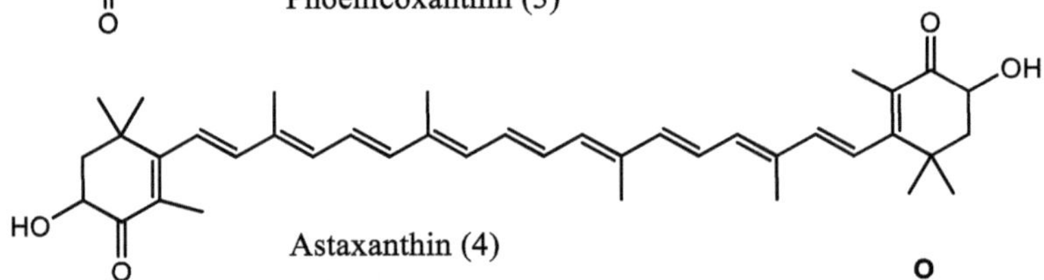

Astaxanthin (4)

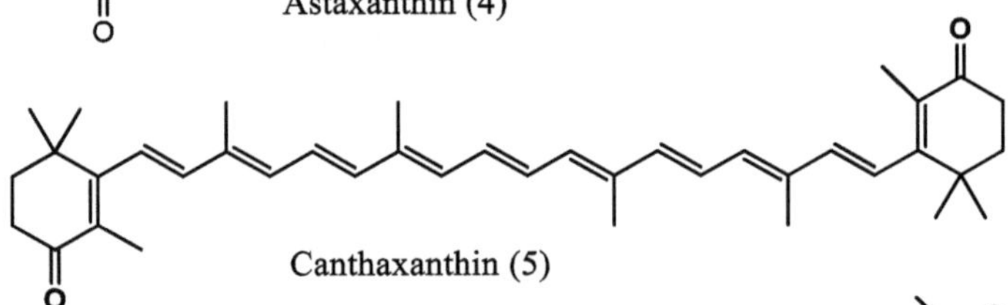

Canthaxanthin (5)

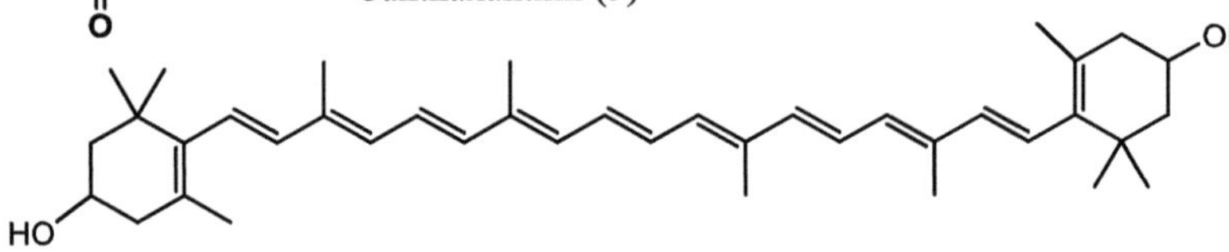

Zeaxanthin (6)

Fig. 25.1 Structures of carotenoids isolated from yeasts and fungi **(1–6)**

compared to algae, but they grow faster and have easier cultivation techniques (Bumbak et al. 2011; Mata-Gómez et al. 2014). Marine thraustochytrids are also good producers of carotenoids. They are widely distributed geographically. Thraustochytrids include planktonic and benthic forms living associated with various sea ecosystems like sediments of mangroves, estuaries, and marine ecosystem (Raghukumar 2002). Thraustochytrid species including *Thraustochytrium* strains ONC-T18 and CHN-1, *Thraustochytriidae* sp. AS4-A1 (*Ulkenia* sp.), and *Aurantiochytrium* sp. KH105 are known to be rich sources of carotenoids like of β-carotene (**1**), echinenone (**2**), phoenicoxanthin (**3**), astaxanthin (**4**), zeaxanthin (**5**), and canthaxanthin (**6**) (Fig. 25.1) [Aasen et al. 2016]. Increasing the level of carotenoids can be achieved using metabolic engineering approaches, for example, a ninefold increase for astaxanthin was achieved in the case of *Aurantiochytrium* sp. SK4 (Suen et al. 2014). Therefore, to increase the production level of carotenoids, genetic tools including CRISPR can be applied to thraustochytrids for optimizing the production of carotenoids from these and other marine microbial sources.

25.2.1.3 Other Compounds

A new benzodiazepine alkaloid, circumdatin I (**7**), together with the already known circumdatin C (**8**) and circumdatin G (**9**) (Fig. 25.2), possessing UVA protecting property, was extracted from *Exophiala* sp. fungus associated with the marine sponge *Halichondria panicea* collected from Bogil Island, Korea. All compounds displayed high UVA protecting activity with the ED_{50} values of 98, 101, and 105 μM better than oxybenzone with ED_{50} value of 350 μM presently being used as sunscreen filter (Zhang et al. 2008).

New compounds, 3-benzylidene-2-methylhexahydropyrrolo [1,2-α] pyrazine-1,4-dione (**10**) and 1-(1,3-Benzodioxol-5-yl)-1,2-propanediol (**11**) (Fig. 25.2), were extracted from *Annulohypoxylon stygium* (Xylariaceae family), the endophytic fungus associated with *Bostrychia radicans* algae isolated from the region of the Rio Escuro mangrove, Sao Paulo State, Brazil. Both compounds exhibited no phototoxic potential and have the capability to UVB absorption and were found to be photostable (Maciel et al. 2018).

Another novel indole alkaloid, dihydroxyisoechinulin A (**12**) (Fig. 25.2), was extracted from *Aspergillus* sp. associated with marine red alga *Lomentaria catenata* identified from the region of Golmae village, Ulsan City, Korea. Compound (**12**)

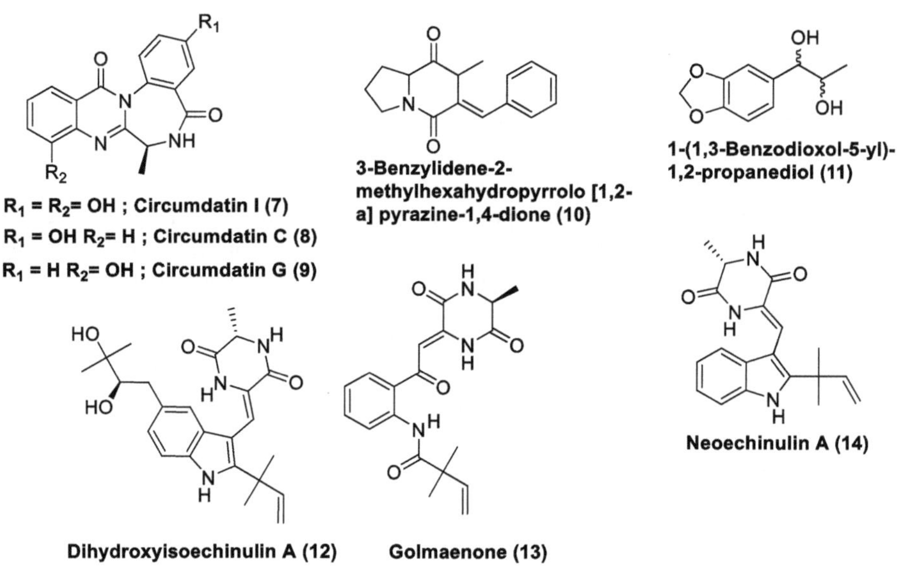

Fig. 25.2 Structures of photo-protective compounds isolated from marine fungi (**7–14**)

exhibited UVA protecting property with an ED_{50} value of 130 μM, while oxybenzone exhibited UVA protecting activity with an ED_{50} value of 350 μM (Li et al. 2004a).

The diketopiperazine alkaloid, golmaenone **(13)**, and the related alkaloid neoechinulin A **(14)** (Fig. 25.2) were extracted from the marine-derived fungus *Aspergillus* sp. Compounds **(13 and 14)** displayed a UVA quenching activity with ED50 values of 90 and 170 μM, respectively, better active than oxybenzone with the ED_{50} value of 350 μM (Li et al. 2004b).

25.2.2 Antiaging Compounds

Aging is a natural process observed in all living organisms and involves the accumulation of cellular and extracellular damage. Exposure to environmental aggressions and in particular reactive oxygen species produced during cellular metabolism generates visible signs of skin aging, primarily because of the dilapidation of the extracellular matrix present in the dermis and epidermis of skin (Giacomoni 2008; Kirkwood and Holliday 1979). Although aging is not preventable, substances applied topically may alter the aging process. Chemicals that help reduce damage are considered antiaging products (Giacomoni 2008). Numerous bioactive molecules of marine origin are already being produced at industrial scale with anti-wrinkling action, including fatty acids and exopolysaccharides (EPSs). Some of the fungal-derived antiaging compounds so far identified are summarized in Table 25.1.

25.2.2.1 Exopolysaccharides

Polysaccharides from microbial resources, especially exopolysaccharides (EPSs), are bioactive compounds that can confer anti-wrinkling action. EPSs are high molecular weight carbohydrate polymers that are an integral part of different physiological mechanisms, from attachment to intra- and inter-specific communication and competition (Poli et al. 2010). EPSs are produced by bacteria, fungi, and microalgae. However, for large-scale production, bacteria are considered better than fungi and algae (Nwodo et al. 2012). EPSs have multiple roles and can be used as an emulsifier, thickening agent, absorption, and formation of gel (Freitas et al. 2011; Suresh Kumar et al. 2007). In recent years, extreme environments such as polar, cold seeps, a benthic region of sea hydrothermal vents and hypersaline ecosystems are being explored to discover new EPSs (Cambon-Bonavita et al. 2002; Courtois et al. 2014; Jouault et al. 2001; Poli et al. 2010).

Exopolysaccharide, EPS2, was extracted from *Keissleriella* sp. YS 4108, a fungus isolated from marine sediment collected from the Yellow Sea near Shenyang Port, China, with antioxidant and free radical scavenging activity (Sun et al. 2004). EPS2 blocks the nonsite-specific strand as well Fenton reaction-mediated DNA damage at concentrations of 0.1 and 1 mg/mL and significantly inhibited human low-density lipoprotein (LDL) induced by copper-mediated oxidation concentration-dependent manner.

The marine fungus *Penicillium* sp. gxwz 446, which was isolated from the coral *Echinogorgia flora* in South China, led to the identification of two neutral exopolysaccharides, GX1–1 and GX2–1. Chemically, GX1–1 primarily consists of glucose subunit with a molecular weight of 5.0 kDa. GX1–1 was formed by monomer units of (1 → 4)-linked α-D-glucopyranose units as the central portion which was substituted by α-D-glucopyranose at C2 region repeated at every sixth sugar residues. Chemically, GX2–1 was a polymer of mannogalactoglucan with a molecular weight of 9.5 kDa. The bonds present in the backbone were composed of (1 → 4)-β-D-Glcp, (1 → 5)-β-D-Galf, (1 → 3,5)-β-D-Galf, (1 → 6)-α-D-Manp, and (1 → 2, 6)-α-D-Manp. GX1–1 was found active against RAW264.7 leading to activation of macrophage. The sulfated derivatives of GX1–1 showed to enhance the pinocytic activity of RAW264.7 cells leading to the production of NO (Sun et al. 2016).

25.2.2.2 Fatty Acids

Fatty acids have a proven role for topical application in cosmetics that aid in the repair of soft tissue, providing nourishment to the skin through stimulation of collagen production, wound healing, and anti-inflammatory properties (Ziboh et al. 2000). Among the polyunsaturated fatty acids (PUFA), the omega-3 fatty acids docosahexaenoic acid (DHA) and eicosapentaenoic acid (EPA) offer various health beneficial effects (McCusker and Grant-Kels 2010). Wild fish is the main source of omega-3 fatty acids for human consumption (Adarme-Vega et al. 2014; Covington 2004; Hutchings and Reynolds 2004). Increasing demand for DHA and EPA has resulted in research into alternative sources to fish oil, including oil plants, algae, bacteria, and fungi. Single-cell microorganisms such as yeasts and molds produce "single-cell oils" (SCOs) (Wynn and Ratledge 2005) and offer several benefits compared to fish oil. Advantages include a higher growth index for microorganisms and oil that naturally includes antioxidants such as carotenoids which partially prevent omega-3-fatty acid oxidation (Adarme-Vega et al. 2014). Thraustochytrids, fungi, and bacteria are prominent marine microorganisms that have the potential to produce these fatty acids cost-effectively at commercial scale.

A new fatty acid glucoside, *R*-3-hydroxyundecanoic acid methylester-3-*O*-α-L-rhamnopyranoside **(15)** (Fig. 25.3) was obtained from an unidentified marine-derived fungus associated with mangrove plant *Scyphiphora hydrophyllacea* collected from Hainan Province, China, and exhibited a modest inhibitory effect on *Staphylococcus aureus* and methicillin-resistant *S. aureus* [Zeng et al. 2012]. Paecilonic acids A **(16)** and B **(17)** (Fig. 25.3), two new bicyclic fatty acids, were extracted from *Paecilomyces variotii*, the marine fungus isolated from the jellyfish *Nemopilema nomurai* (Wang et al. 2016). Similarly, two new antimicrobial fatty acid esters, *penicilloitins* A **(18)** and B **(19)** (Fig. 25.3), were obtained from marine-derived fungus *Penicillium* sp. from a tunicate, *Didemnum* sp., collected from a mangrove forest located in Sharm El-Sheikh on the Egyptian Red Sea coast. Both the compounds displayed modest antimicrobial activity (Mourshid et al. 2016). It is reported that marine yeast species produce higher biomass and high lipids such as in a case of *Rhodotorula mucilaginosa* AMCQ8A (Gupta et al. 2013; Kot et al. 2016), but their potential usage in the area of cosmaceutical is still to be investigated.

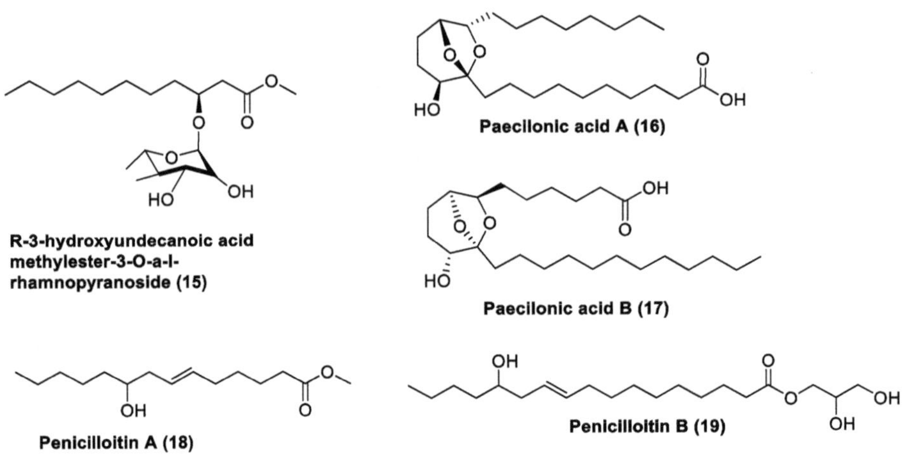

Fig. 25.3 Structures of fatty acid isolated from fungi (**15–19**)

25.2.3 Antioxidants

Oxidative damage has been implicated as the major factor in the decline in physiological function that occurs during aging. Antioxidants are the substances that protect cell membranes damaged by mitigating oxidative stress by quenching the free radicals and other reactive oxygen molecules (Wickens 2001). Currently, many synthetic antioxidants such as tertiary butyl hydroquinone, butylated hydroxytoluene, butylated hydroxyanisole, and propyl gallate are being used in cosmeceutical products, (Kim 2016). Since synthetic compounds are less preferred due to safety issues (Younes and Rinaudo 2015), antioxidants from natural origin have been widely investigated for cosmetics. Antioxidants from the marine source include carotenoids, mycosporines, MAAs, and other compounds that can be used as cosmeceutical formulation (Oren and Gunde-Cimerman 2007; Vílchez et al. 2011). Antioxidants identified from marine-derived fungi are summarized in Table 25.2.

25.2.3.1 Antioxidants from Mangrove Endophytic Fungi

A chromenopyridine derivative, phochrodine D (**20**) (Fig. 25.4), was extracted from *Phomopsis* sp. 33, an endophytic fungus associated with the bark of *Rhizophora stylosa* collected from the South China Sea. Compound (**20**) displayed moderate antioxidant activity with an IC_{50} value of 34.0 μM, while butylated hydroxytoluene (BHT), a positive control, exhibited antioxidant activity with an IC_{50} value of 25.8 μM (Chen et al. 2018).

An endophytic fungus *Pestalotiopsis* sp. HQD-6 associated with *Rhizophora mucronata* collected from the South China Sea was the source of fluturan (**21**) (Fig. 25.4), which showed significant radical scavenger activity, with IC_{50} values of 34.85 and 9.75 μg/mL, respectively, in DPPH and ABTS assays (Zhou et al. 2018).

Three novel 2,3-diaryl indone derivatives, ascomindone A (**22**), ascomindone B (**23**), and ascomindone C (**24**), and two new isobenzofuran derivatives, ascomfuran

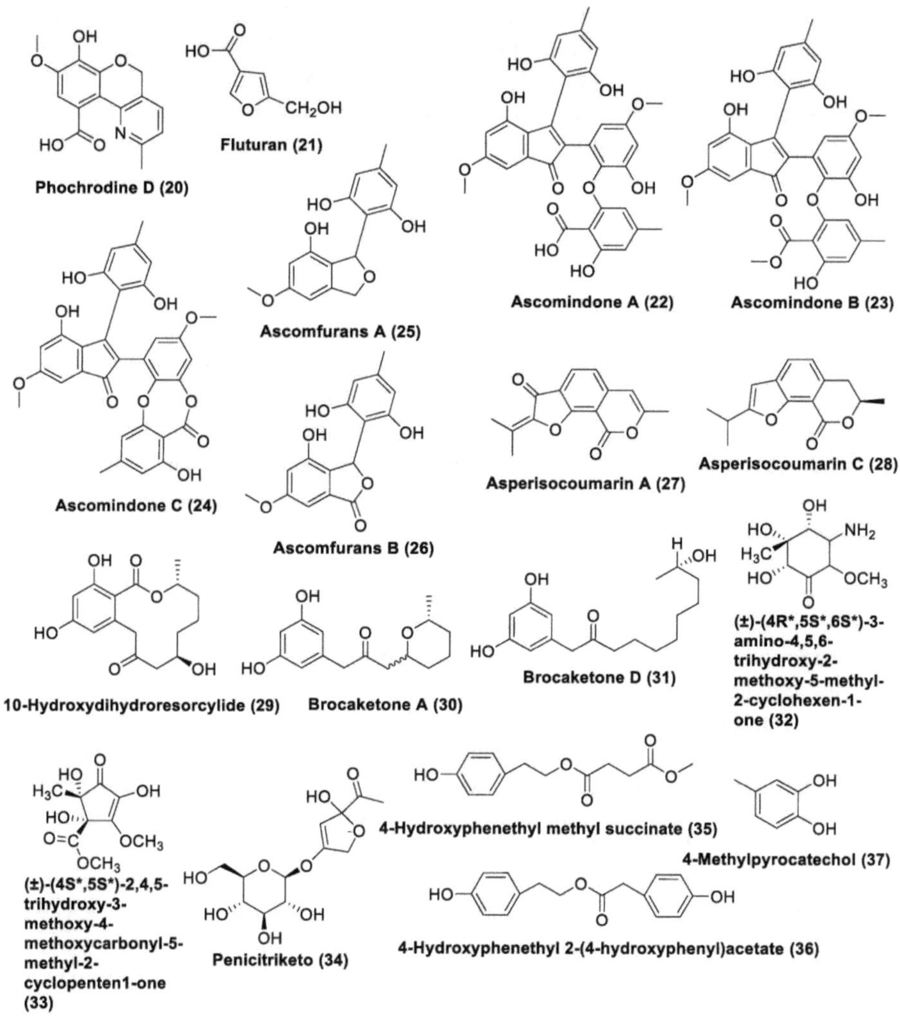

Fig. 25.4 Structures of antioxidants from mangrove endophytic fungi (**20–36**)

(**25**) and ascomfuran B (**26**) (Fig. 25.4), were extracted from *Ascomycota* sp. SK2YWS-L, an endophytic fungus residing inside the leaves of the marine mangrove *Kandelia candel*, collected from Shankou Mangrove Nature Reserve in Guangxi Province, China. Compound (**22**) exhibited the most potent scavenging capacity against DPPH radical with an IC_{50} value of 18.1 μM, while compounds (**23, 24,** and **25**) showed significant activity. In a hydroxyl radical scavenging assay, compounds (**22–24**) exhibited strong activity with the IC_{50}values in the range of 80–100 μM. In addition, all compound also displayed potent activity in FRAP assay (Tan et al. 2016).

Two isocoumarin derivatives, asperisocoumarin A (**27**) and asperisocoumarin C (**28**) (Fig. 25.4), were identified from *Aspergillus* sp. 085242, associated with the mangrove plant *Acanthus ilicifolius*, collected from Guangxi Province, China. Compounds (**27–28**) exhibited antioxidant activity with EC_{50} values of 125 and 130 μM, respectively, in DPPH scavenging assay. The positive control Vitamin C exhibited antioxidant activity with EC_{50} values of 35 μM in the same assay (Xiao et al. 2016).

Compounds (10R, 14R)-10-hydroxydihydroresorcylide (**29**), brocaketone A (**30**), and brocaketone D (**31**) (Fig. 25.4) were purified from *Penicillium brocae* MA-192, an endophytic fungus associated with the mangrove plant *Avicennia marina*, collected from Hainan Island, China. Compounds (**29–31**) showed potential DPPH radical scavenging activity with IC_{50} values of 14.4, 5.9, and 16.3 μg/ml, respectively, in comparison to BHT (IC_{50}, 18.5 μg /ml) (Zhang et al. 2015).

Racemic new cyclohexenone and cyclopentenone derivatives, (±)-(4*R*∗,5*S*∗,6*S*∗)-3-amino-4,5,6-trihydroxy-2-methoxy-5-methyl-2-cyclohexen-1-one (**32**) and (±)-(4*S*∗,5*S*∗)-2,4,5-trihydroxy-3-methoxy-4-methoxycarbonyl-5-methyl-2-cyclopenten-1-one (**33**) (Fig. 25.4), were obtained from *Alternaria* sp. R6 residing inside the roots of *Myoporum bontioides* a semi-mangrove plant collected from Guangdong Province, China. Compounds (**32**) **and** (**33**) exhibited good scavenging activities in ABTS assay with an EC_{50} value of 8.19 and 16.09 μM, respectively, compared to ascorbic acid (EC_{50}, 17.14 μM) [Wang et al. 2015].

New polyketide, penicitriketo (**34**) (Fig. 25.4) was extracted from *Penicillium citrinum*, an endophytic fungus associated with the stems of mangrove plant *Salicornia bigelovii* collected from Jiangsu Province, China. Compound (**34**) exhibited antioxidant activity in DPPH assay with an IC_{50} value of 85.33 μM. Positive control Vitamin E showed antioxidant activity with an IC_{50} value of 78.08 μM in the same assay (Wang et al. 2014).

Compounds 4-hydroxyphenethyl methyl succinate (**35**), 4-hydroxyphenethyl 2-(4-hydroxyphenyl) acetate (**36**), and 4-methylpyrocatechol (**37**) (Fig. 25.4) were purified from the marine fungus *Penicillium griseofulvum* Y19–07 associated with mangrove *Lumnitzera racemosa* collected from the South China Sea. All the compounds displayed antioxidant activity in DPPH assay with IC_{50} values of 58.6, 56.2, and 7.1 μM, respectively, while positive control vitamin E showed an IC_{50} value of 26.5 μM (Wang et al. 2009).

25.2.3.2 Antioxidants from Marine Sponge-Associated Fungi

Compound 7-*O*-demethylmonocerin (**38**) (Fig. 25.5) was purified from *Setosphaeria* sp. SCSIO41009 associated with sponge *Callyspongia* sp., collected from the sea near Xuwen County, Guangdong Province, China, and exhibited antioxidant activity with an IC_{50} value of 38 μM, in DPPH assay, while positive control ascorbic acid exhibited antioxidant activity with an IC_{50} value of 39 μM (Pang et al. 2018).

Compounds butyrolactone I (**39**), butyrolactone II (**40**), 5-[(3,4-dihydro-2,2-dimethyl-2*H*-1-benzopyran-6-yl)-methyl]-3-hydroxy-4-(4-hydroxyphenyl)-2(5H)-furanone (**41**), and aspernolide A (**42**) (Fig. 25.5) were purified from *Aspergillus terreus*, isolated from the marine sponge *Phakellia fusca* collected from the Xisha

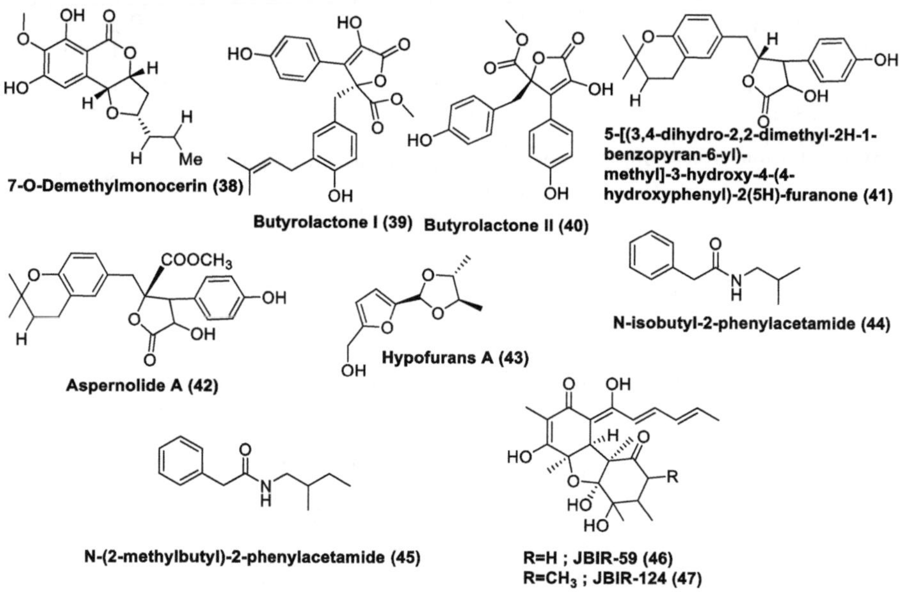

Fig. 25.5 Structures of antioxidants from sponges associated fungi **(37–47)**

Islands in the South China Sea. All compounds exhibited antioxidant activities with IC_{50} values ranging from 38 to 97 μM, while *N*-acetyl-L-cysteine exhibited the antioxidant activities with IC_{50} values 85 μM (Sun et al. 2018).

New furan derivative hypofurans A **(43)**, along with known compounds N-isobutyl-2-phenylacetamide **(44)** and N-(2-methylbutyl)-2-phenylacetamide **(45)** (Fig. 25.5), was extracted from *Hypocrea koningii* PF04 isolated from the marine sponge *Phakellia fusca* collected from Yongxing Island in the South China Sea. Compounds **(43–45)** showed moderate antioxidant activity with IC_{50} values of 27.4, 16.8, and 61.7 μg/mL, respectively, in DPPH assay (Ding et al. 2015). A new sorbicillinoid derivative designated as JBIR-59 **(46)** and its derivative designated as JBIR-124 **(47)** (Fig. 25.5) were purified from *Penicillium citrinum* SpI080624G1f01 associated with marine sponges collected from offshore of Ishigaki Island, Okinawa Prefecture, Japan. Compounds **(46)** and **(47)** exhibited antioxidant activity with an IC_{50} value of 25 and 30 μM, respectively, while α-tocopherol exhibited antioxidant activity, with an IC_{50} value of 9.0 μM in DPPH assay (Kawahara et al. 2012).

Two polysaccharides JJY-W and JJY-S were extracted from the crude exopolysaccharides of marine-derived endogenetic fungus of the sponge (*Alternaria* sp.) and were further purified by ion exchange chromatography. The chemical properties of the two polysaccharides were investigated by various chemical and spectroscopic methods. The results showed that the molecular weights of JJY-W and JJY-S were 1.4KD and 1.8KD, respectively. JJY-W was mainly composed of galactose and glucose with a small amount of mannose. JJY-S mainly consisted of mannose and glucose with a small amount of galactose. The total sugar content of JJY-W was 46%, and uronic acid was not detected. The total sugar content of JJY-S was 52%,

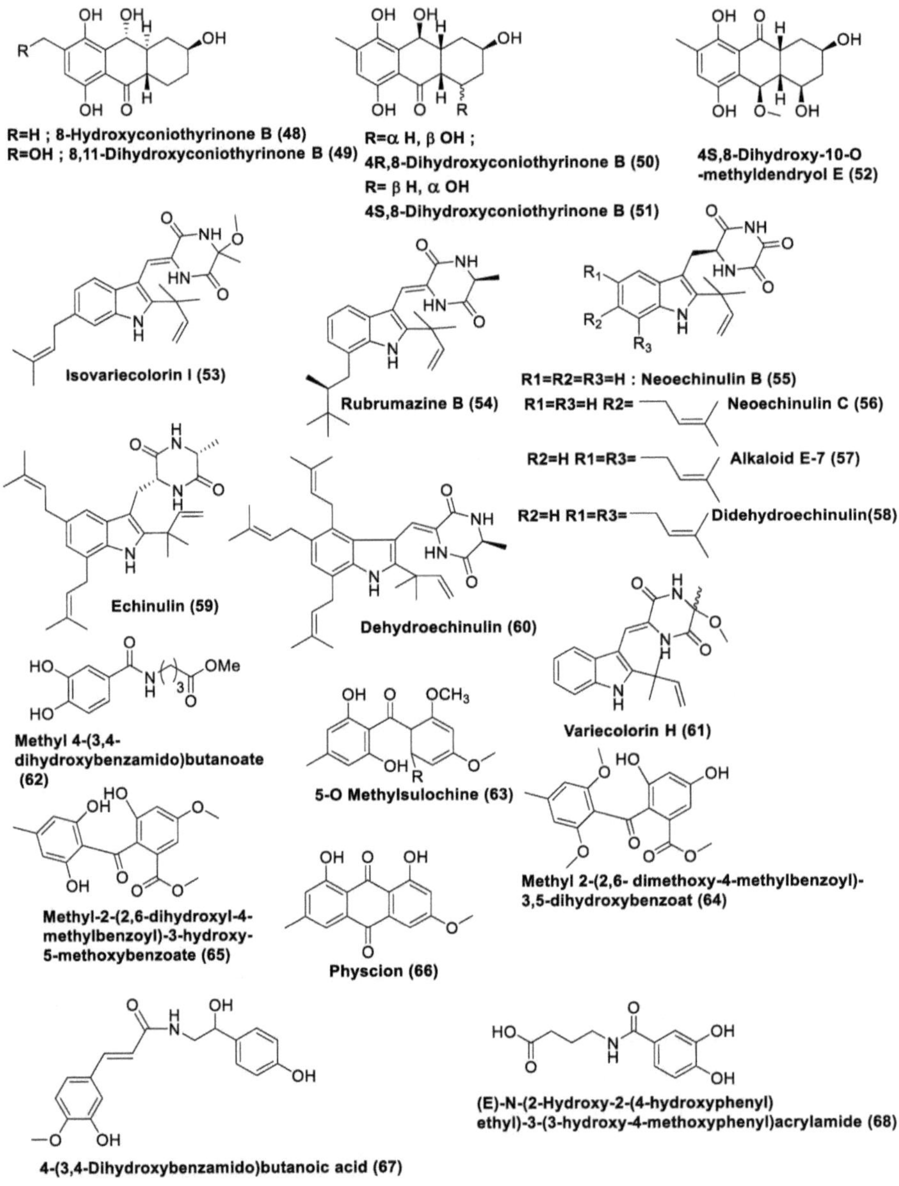

Fig. 25.6 Structures of antioxidants from marine algae-associated fungi (**48–68**)

and uronic acid content was 1.94%. JJY-W had a higher content of protein than JJY-S. Scavenging capacities of JJY-W and JJY-S on DPPH free radical and hydroxyl radical were evaluated. The results showed that two types of polysaccharides exhibited a good antioxidant activity. Moreover, JJY-W showed higher scavenging capacity on DPPH free radical than JJY-S, and JJY-S had higher hydroxyl radical scavenging capacity than JJY-W (Chen et al. 2010).

25.2.3.3 Antioxidants from Marine Algae-Associated Fungi

Polyhydroxylated hydroanthraquinone derivatives, 8-hydroxyconiothyrinone B (**48**), 8,11-dihydroxyconiothyrinone B (**49**), 4R,8-dihydroxyconiothyrinone B (**50**), 4S,8-dihydroxyconiothyrinone B (**51**), and 4S,8-dihydroxy-10-O-methyldendryol E (**52**) (Fig. 25.6), were purified from an endophytic fungus *Talaromyces islandicus* EN-501, associated with the marine red alga *Laurencia okamurai* collected from Qingdao, China. Compounds (48–**52**) displayed antioxidant activity with an IC_{50} value of 12, 31,42, 52, and 30 μM, respectively, in DPPH radical scavenging assay, while positive control BHT displayed antioxidant activity with an IC_{50} value of 61 μM. Compounds (**48–52**) also displayed moderate ABTS radical scavenging activity with IC_{50} values of 8.3, 19, 34, 31, and 24 μM, which was comparable to ascorbic acid (IC_{50}, 16 μM) [Li et al. 2017].

A new indolediketopiperazine derivative isovariecolorin I (**53**) along with known congeners including rubrumazine B (**54**), neoechinulin B (**55**), neoechinulin C (**56**), alkaloid E-7 (**57**), didehydroechinulin (**58**), echinulin (**59**), dehydroechinulin (**60**), and variecolorin H (**61**) (Fig. 25.6) was isolated from *Eurotium cristatum* EN-220, associated with the marine alga *Sargassum thunbergii* collected from the coast of Qingdao, China. Compounds (**53–61**) showed antioxidant activity in DPPH assay with IC_{50} values of 20.6, 28.5, 10.9, 12.1, 10.1, 13.3, 13.8, 6.4, and 18.7 μg/mL, while positive control ascorbic acid showed antioxidant activity with IC_{50} values of 2.0 μg/mL (Du et al. 2017).

Benzamide derivative, methyl 4-(3,4-dihydroxybenzamido)butanoate (**62**), along with six other phenolic compounds, including 5-O-methylsulochine (**63**), methyl 2-(2,6- dimethoxy-4-methylbenzoyl)-3,5-dihydroxybenzoate (**64**), methyl-2-(2,6-dihydroxyl-4-methylbenzoyl)-3-hydroxy-5-methoxybenzoate (**65**), physcion (**66**), 4-(3,4-dihydroxybenzamido)butanoic acid (**67**) and (E)-N-(2-hydroxy-2-(4-hydroxyphenyl) ethyl)-3-(3-hydroxy-4-methoxyphenyl)acrylamide (**68**) (Fig. 25.6), were extracted from *Aspergillus wentii* EN-48 associated with marine brown alga *Sargassum* sp. Compounds (**62–68**) displayed potent antioxidant activity with IC_{50} values of 5.2, 24.7, 78.2, 73.6, 99.4, 9.6, and 82.7 μg/mL, respectively, while compounds (62, 63, and 68) possess stronger activity than BHT (IC50 36.9 μg/mL) in the DPPH assay [Li et al. 2014].

Production of two new brominated dihydroxyphenylacetic acid derivatives, Me 2-(6-bromo-3,4-dihydroxyphenyl)acetate (**69**) and Me 2-(2,5-dibromo-3,4-dihydroxyphenyl) acetate (**70**), and a known compound, 2-(3,4-dihydroxyphenyl) acetic acid (**71**) (Fig. 25.7), was induced when NaBr and $CaBr_2$ used during fermentation of *Aspergillus* sp. isolated from the surface of brown alga *Ishige okamurae* collected at Uljin, Korea. Compounds (**69–71**) exhibited potent antioxidant activity in DPPH assay with IC_{50} values of 14.2, 12.1, and 11.0 μM, respectively, better than the positive control L-ascorbic acid (IC_{50}, 20.0 μM) [Leutou et al. 2013].

Similarly the addition of NaBr during fermentation of *Aspergillus ochraceus* associated with *Chondria crassicualis*, the marine red alga collected from Yokji Island, Korea, resulted in inducing the production of a new halogenated

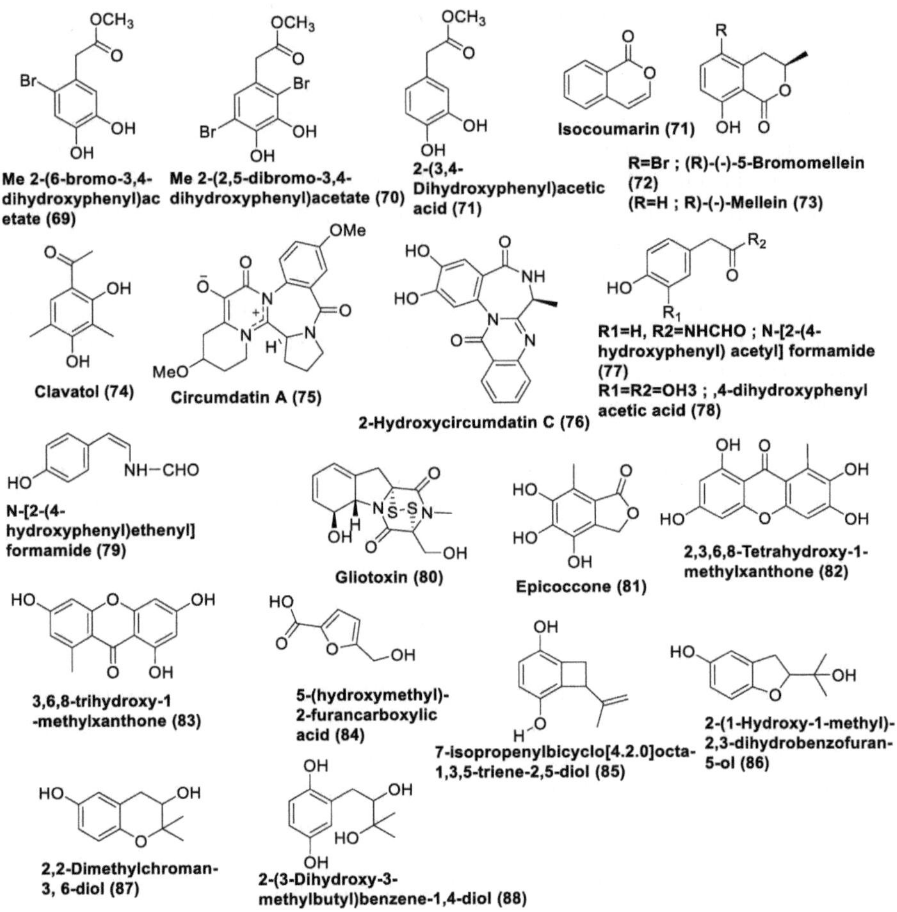

Fig. 25.7 Structures of antioxidants from marine algae-associated fungi **(69–88)**

isocoumarin, (R)-(−)-5-bromomellein **(72)**, and three known compounds, (R)-(−)-mellein **(73)**, clavatol**(74)**, and circumdatin A **(75)** (Fig. 25.7). Compounds **(72–75)** exhibited mild antioxidant activity with the IC_{50} values of each compound being 24, 25, 30, and 32 μM, respectively, in the DPPH assay, while positive control L-ascorbic acid exhibited better antioxidant activity with IC_{50} value of 20.0 μM (Yun et al. 2013).

Aspergillus ochraceus, associated with marine brown alga *Sargassum kjellmanianum* which was collected from Dalian coastline, China, was the source of benzodiazepine analog, 2-hydroxycircumdatin C **(76)** (Fig. 25.7). Compound **(76)** displayed antioxidant activity with an IC_{50} value of 9.9 μM in DPPH radical scavenging assay, which is 8.9-fold more potent than that of BHT with an IC_{50} value of 88.2 μM (Cui et al. 2009).

A new metabolite, N-[2-(4-hydroxyphenyl) acetyl]formamide **(77)**, along with known polyketides, 3,4-dihydroxyphenyl acetic acid **(78)** and

N-[2-(4-hydroxyphenyl) ethenyl]formamide **(79)** (Fig. 25.7), was isolated from an unidentified marine-derived fungus marine brown alga *Ishige okamurae* collected at Uljin, Gyeongbuk Province, Korea. Compounds **(77–79)** exhibited good antioxidant activity with IC_{50} values of 8.4, 11.9, and 0.2 μM, respectively, in DPPH assay (Li et al. 2006).

The fungal strain, *Pseudallescheria* sp., associated with marine brown alga *Agarum cribrosum* collected in the Uljin, Gyeongbuk Province, Korea, was the source of Gliotoxin **(80)** (Fig. 25.7) and showed antioxidant activity with an IC_{50} value of 5.2 μM in DPPH assay (Li et al. 2006).

A new metabolite, dihydroxyisoechinulin A **(12)** (Fig. 25.2), was extracted from *Aspergillus* sp. isolated from the marine red alga *Lomentaria catenata* collected in the Golmae village, Ulsan City, Korea. Compound **(12)** exhibited antioxidant activity with an IC_{50} value of 20 μM in DPPH assay while positive control ascorbic acid exhibited the same activity (Li et al. 2004a).

A new diketopiperazine alkaloid, golmaenone **(13)**, and related alkaloid, neoechinulin A **(14)** (Fig. 25.2), were obtained from the marine-derived fungus *Aspergillus* sp. Compounds **(13 and 14)** exhibited antioxidant activity with IC_{50} values of 20 and 24 μM, respectively, in DPPH assay, while positive control ascorbic acid exhibited antioxidant activity with IC_{50} value of 20 μM (Li et al. 2004b).

A new compound, epicoccone **(81)** (Fig. 25.7) was extracted from *Epicoccum* sp. associated with the marine alga *Fucus vesiculosus*, collected from the North Sea cost Tönning, Germany. Compound **(81)** exhibited potent activity, showing 95% activity in DPPH assay at 25 μg/mL. Compound **(81)** also inhibited the peroxidation of linolenic acid with 62% inhibition at 37 μg/mL in the TBARS assay (Abdel-Lateff et al. 2003a).

A new xanthone derivative, 2,3,6,8-tetrahydroxy-1-methylxanthone **(82)**; known xanthone derivative, 3,6,8-trihydroxy-1-methylxanthone **(83)**; and known compound, 5-(hydroxymethyl)-2-furancarboxylic acid **(84)** (Fig. 25.7) were purified from *Wardomyces anomalus*, associated with alga *Enteromorpha* sp. collected from Fehmarn Island in the Baltic Sea, Germany. Compounds **(82 and 84)** showed 94.7 and 30.7% DPPH radical scavenging activity at 25.0 μg/mL. Compounds **(82 and 84)** were also able to inhibit peroxidation of linolenic acid with 17.0 and 2.6%, respectively, at 7.4 μg/mL. All compounds **(82–84)** also displayed 100% TK p56lck enzyme inhibition activity at 200 μg/mL (Abdel-Lateff et al. 2003b).

A novel hydroquinone derivatives, 7-isopropenylbicyclo[4.2.0]octa-1,3,5-triene-2,5-diol **(85)**, and known metabolites, 2-(1-hydroxy-1-methyl)-2,3-dihydrobenzofuran-5-ol **(86)**, 2,2-dimethylchroman-3,6-diol **(87)**, and 2-(3-dihydroxy-3-methylbutyl)benzene-1,4-diol **(88)** (Fig. 25.7), were isolated from *Acremonium* sp., a fungus associated with algal material collected from Moraira, Mediterranean Sea, Spain. Compounds **(85–88)** exhibited 85.5, 85.8, 72.9, and 90.2% DPPH radical scavenging effects at 25.0 μg/mL. These compounds also inhibit 35.5, 15.9, 9.2, and 16.6% peroxidation of linolenic acid at 37.0 μg/mL (Abdel-Lateff et al. 2002).

25.2.3.4 Antioxidants from Miscellaneous Marine Fungi

Compounds oxisterigmatocystin D (**89**), oxisterigmatocystin C (**90**), sterigmatocystine (**91**), versicolorin B (**92**), UCT1072M1 (**93**), averantin (**94**), methyl-averantin (**95**), averythrin (**96**), averufanin (**97**), averufine (**98**), nidurufin (**99**) (Fig. 25.8) were purified from *Aspergillus versicolor* isolated from deep-sea sediment in South

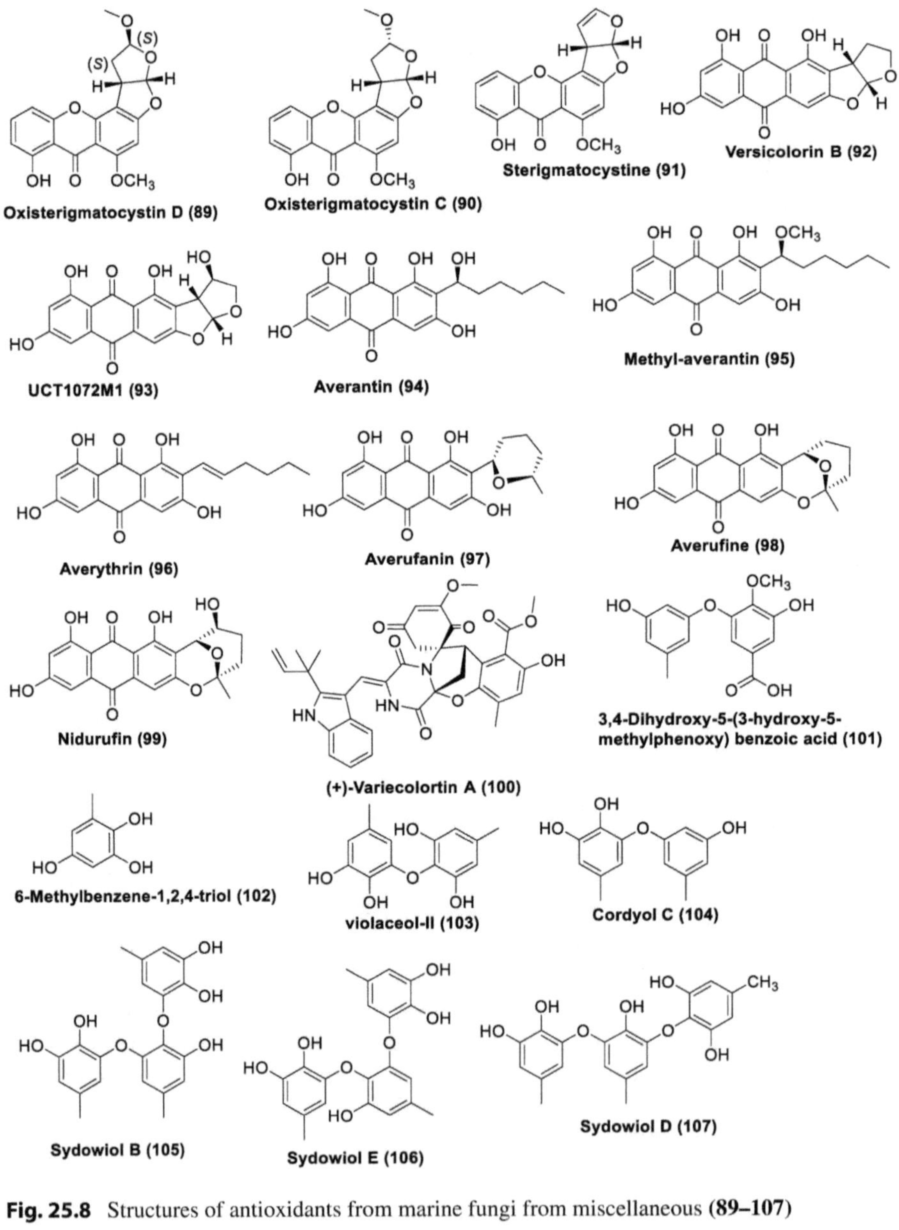

Fig. 25.8 Structures of antioxidants from marine fungi from miscellaneous (**89–107**)

China Sea. Compounds **(89–99)** exhibited antioxidant activity with 0.55, 1.16, 0.65, 1.03, 0.97, 0.89, 0.86, 0.22, 0.82, 0.94 and 0.62 TEAC (trolox (1.0 μmol/L) equivalent antioxidant capacity) values. The luciferase reporter gene assay revealed that compounds **(93, 94, 96**, and **99)** potentially activated the expression of Nrf2-regulated gene (Ze-Hong et al. 2018).

A spirocyclic diketopiperazine enantiomer, (+)-variecolortin A **(100)** (Fig. 25.8) was extracted from marine-derived fungus *Eurotium* sp. SCSIO F452 isolated from the South China Sea sediment sample. Compound **(100)** exhibited good antioxidant activity with an IC_{50} value of 58.4 μM, in DPPH assay, while the positive control ascorbic acid showed antioxidant activity with an IC_{50} value of 45.8 μM (Zhong et al. 2018).

Marine-derived fungus *Aspergillus carneus* collected from the South China Sea was the source of a known compound, 3,4-dihydroxy-5-(3-hydroxy-5-methylphenoxy)benzoic acid **(101)** (Fig. 25.8). Compound **(101)** exhibited potent antioxidant activity with an IC_{50} value of 19.3 μM in DPPH radical scavenging assay, while positive control ascorbic acid exhibited antioxidant activity with IC_{50} value of 15.3 μM (Xu et al. 2017).

Compounds 6-methylbenzene-1,2,4-triol **(102)**, violaceol-II **(103)**, cordyol C **(104)**, and sydowiols B **(105)**, E **(106)**, and D **(107)** (Fig. 25.8) were purified from *Aspergillus versicolor* SCSIO 41502 isolated from a marine sediment sample collected from the South China Sea. Compounds **(102–107)** exhibited antioxidant activity in DPPH radical scavenging assay with IC_{50} values of 18.92–52.27 μM more potent than L-ascorbic acid (Huang et al. 2017).

The supplementation of metal bromides (NaBr and CaBr2) led to production of a new brominated naphthopyranone, 6,9-dibromoflavasperone **(108)**, and three known naphtho-γ-pyranone monomers, flavasperone **(109)**, TMC-256A1**(110)**, and fonsecin **(111)**, and one naphtho-γ-pyranone dimer, aurasperone B**(112)** (Fig. 25.9), during fermentation of the marine mud flat-derived fungus *Aspergillus niger* collected at Suncheon Bay, Korea. Compounds **(108–112)** displayed potent antioxidant activity in DPPH assay, with IC_{50} values of 21, 25, 0.3, 0.02, and 0.01 μM, respectively, and compounds **(110, 111**, and **112)** were more potent than the positive control, ascorbic acid (IC_{50}, 20.0 μM) (Leutou et al. 2016a).

The microbial transformation of anthranilic acid by the marine mud flat-derived fungus *Thielavia* sp. isolated from the marine mud flat collected at Gomso Bay, Korea, produced an antibacterial polycyclic quinazoline alkaloid, thielaviazoline **(113)** (Fig. 25.9). Thielaviazoline displayed in vitro antimicrobial activity against methicillin-resistant and multidrug-resistant *Staphylococcus aureus* (MRSA and MDRSA), with minimum inhibitory concentration (MICs) of 6.25 and 12.5 μg/mL, respectively. Thielaviazoline also showed potent antioxidant activity DPPH in radical scavenging assay with an IC_{50} of 11 μM, which was more active than the positive control, L-ascorbic acid (IC_{50}, 20.0 μM) (Leutou et al. 2016b).

The biological transformation of chlorogentisyl alcohol **(114)** (Fig. 25.9), isolated from the marine-derived fungus *Aspergillus* sp., associated with marine red alga collected in Tongnyeong, Gyengnam Province, Korea, with *Chrysosporium synchronum* associated with brown alga *Sargassum ringgoldium* collected at Yokji

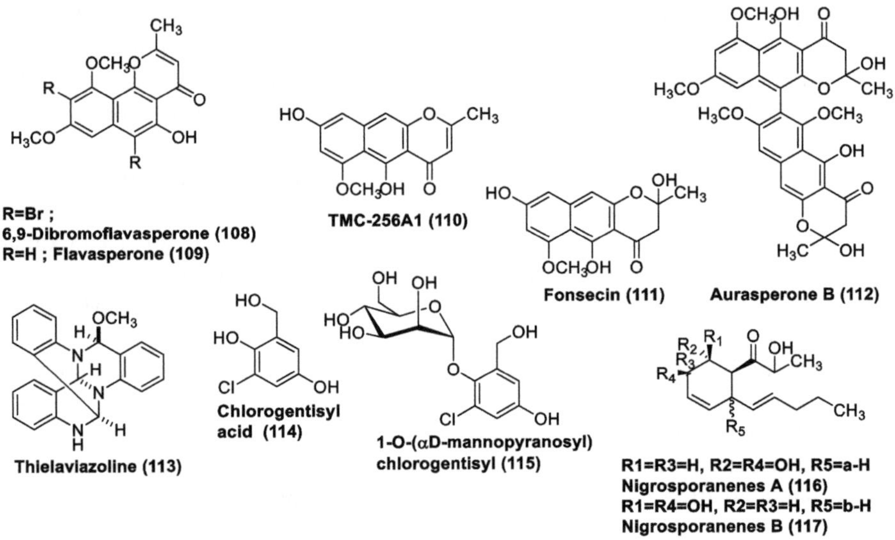

Fig. 25.9 Structures of antioxidants from marine fungi from miscellaneous (108–117)

Island of Gyeongnam, Korea, resulted in the isolation of a new glycosidic metabolite, 1-O-(α-D-mannopyranosyl)chlorogentisyl alcohol (**115**) (Fig. 25.9). Compounds (**114–115**) exhibited potent radical scavenging activity with IC_{50} values of 1.0 and 4.7 μM, respectively, in DPPH assay. Both the compounds exhibited better activity than the positive control, L-ascorbic acid with IC_{50} value of 20.0 μM (Yun et al. 2011). Nigrosporanenes A (**116**) and B (**117**) (Fig. 25.9), two new cyclohexene derivatives, were extracted from *Nigrospora* sp. PSU-F11, the sea fan-derived fungus. Both the compounds showed poor antioxidant activity with IC_{50} values of 0.34 and 0.24 μg/mL, respectively, in a DPPH assay (Rukachaisirikul et al. 2010).

25.2.4 Skin-Whitening Products

25.2.4.1 Tyrosinase Inhibitors

Tyrosinase is a copper-containing enzyme, present in microorganisms, animals, and plants, and is a key regulator of melanin biosynthesis. Its prime role involves skin and hair color determination in organisms. For the past few decades the use of tyrosinase inhibitors has become an integral component of skin-whitening creams within the cosmetic industry, with these inhibitors now being widely used in cosmeceuticals. Natural products are an important source of many tyrosinase inhibitors, but only a few of them are used for skin whitening due to safety concerns. Compounds like linoleic acid, hinokitiol, kojic acid, arbutin, naturally occurring hydroquinones, and catechols are known tyrosinase inhibitors but are also associated with side effects (Maeda and Fukuda 1991). It is estimated that the global skin-lightening

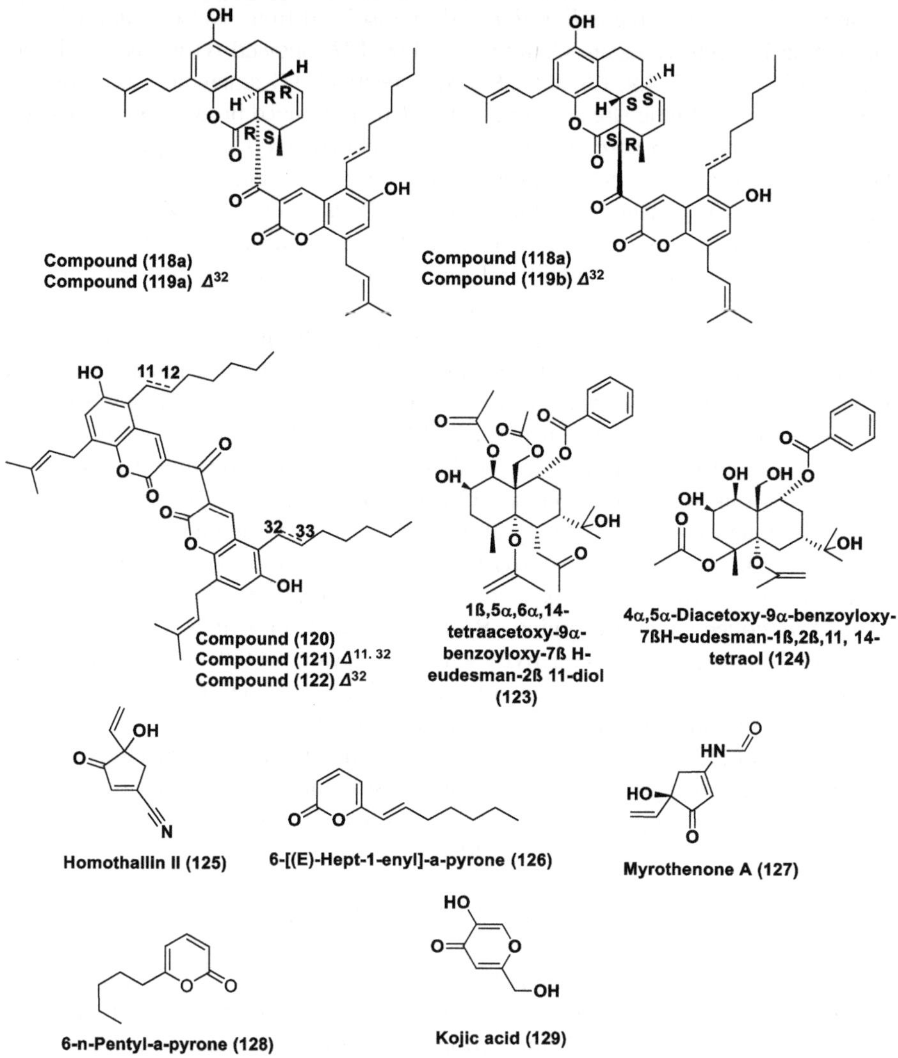

Fig. 25.10 Structures of skin-whitening compounds from marine fungi **(118–129)**

products market was valued at around USD 4075 million in 2017, and it is expected to reach approximately USD 8895 million by 2024, at a CAGR of slightly above 6.5% between 2018 and 2024 [https://globenewswire.com/news-release/2019/01/10/1685903/0/ en/Global-Skin-Lightening-Products-Market-Will-Reach-USD-8-895-Million-By-2024-Zion-Market-Research.html]. There is a need to find new tyrosinase inhibitors with minimal side effects, from marine sources including marine fungi. Some of the compounds with anti-tyrosinase inhibitory activity are presented in Table 25.3.

Two tetracyclic coumarin derivatives (**118** and **119**) and three coumarin dimers (**120–122**) (Fig. 25.10) were extracted from coumarin dimerization of natural extracts of the marine-derived fungus *Eurotium rubrum* isolated from Chosei-mura, Chosei-gun, Chiba Prefecture, Japan. Compounds **(118–122)** showed tyrosinase inhibitory activity (IC_{50}, 1.7, 1.2, 4.9, 1.8, and 2.9 μM, respectively) (Kamauchi et al. 2018).

In response to abiotic stress, elicitation by $CuCl_2$ induced production of two new sesquiterpenes, 1β,5α,6α,14-tetraacetoxy-9α-benzoyloxy-7βH-eudesman-2β,11-diol **(123)** and 4α,5α-diacetoxy-9α-benzoyloxy-7βH-eudesman-1β,2β,11,14-tetraol **(124)** (Fig. 25.10), during fermentation of the *Pestalotiopsis* sp. Z233 associated with brown alga *Sargassum horneri* collected from the seashore in Wenzhou, China. Compounds **(123 and 124)** exhibited tyrosinase inhibitory activity with an IC_{50} value of 14.8 μM and 22.3 μM (Wu et al. 2013).

Homothallin II **(125)** (Fig. 25.10) was extracted from marine-derived fungus *Trichderma viride.* Compound **(125)** inhibits the mushroom tyrosinase enzyme through the binding to a copper active site (Tsuchiya et al. 2008). A new α-pyrone derivative, 6-[(E)-hept-1-enyl]-α-pyrone **(126)** (Fig. 25.10), was obtained from *Botrytis*, isolated from the marine red alga *Hyalosiphonia caespitose* collected at Dadaepo, Busan, Korea. Compound **(126)** exhibited tyrosinase inhibitory activity with IC_{50} value of 4.5 μM, (Zhang et al. 2007). New myrothenone A **(127)** (Fig. 25.10) and a known 6-n-pentyl-α-pyrone **(128)** were extracted from the algicolous marine fungus *Myrothecium* sp. isolated from marine green alga *Enteromorpha compressa* collected at Baegunpo, Busan, Korea. Compounds **(127 and 128)** showed a tyrosinase inhibitory activity with an IC_{50} value of 0.8 and 6.6 μM, respectively, which are better active than kojic acid with IC_{50} value of 7.7 μM, currently used as a functional personal-care compound (Li et al. 2005).

Kojic acid **(129)** (Fig. 25.10) was purified from *Alternaria* sp. 581B associated with marine green alga *Ulva pertusa*, and its structure was detected by physico-chemical evidence. Kojic acid showed the significant tyrosinase inhibitory activity with IC_{50} values of 12.0 μM (Li et al. 2003).

25.2.4.2 Marine Fungi for Skin Microbial Infection

Skin ailments result from infections caused by viruses, bacteria, fungi, or parasites. The common bacterial pathogens that mostly infect skin are *Staphylococcus aureus*, *Staphylococcus epidermidis*, and group A β-hemolytic streptococci. *Propionibacterium acnes* or *Cutibacterium acnes* is an important target for the prevention and medical treatment of acne vulgaris (Agrawal et al. 2018a, b).

There is a growing emergence of resistance to commercially available antibiotics/antimycotics against clinical microbial pathogens; hence there is great need of alternative anti-infective agents.

Antibacterials identified from marine-derived fungi are summarized in Table 25.3. Four novel stemphol derivatives, pleosporols A–B **(130 and 131)** and pleosporols C–D (mixture of C and D) **(132 and 133)**, together with known stemfolones (mixture of stemfolone B and 14-*epi*-stemfolone B) **(134)** (mixture) (Fig. 25.11), were extracted from *Pleospora* sp. (PO4) isolated from the gut of marine isopod *Ligia oceanic* collected from Dongsha of Dinghai in Zhoushan, Zhejiang

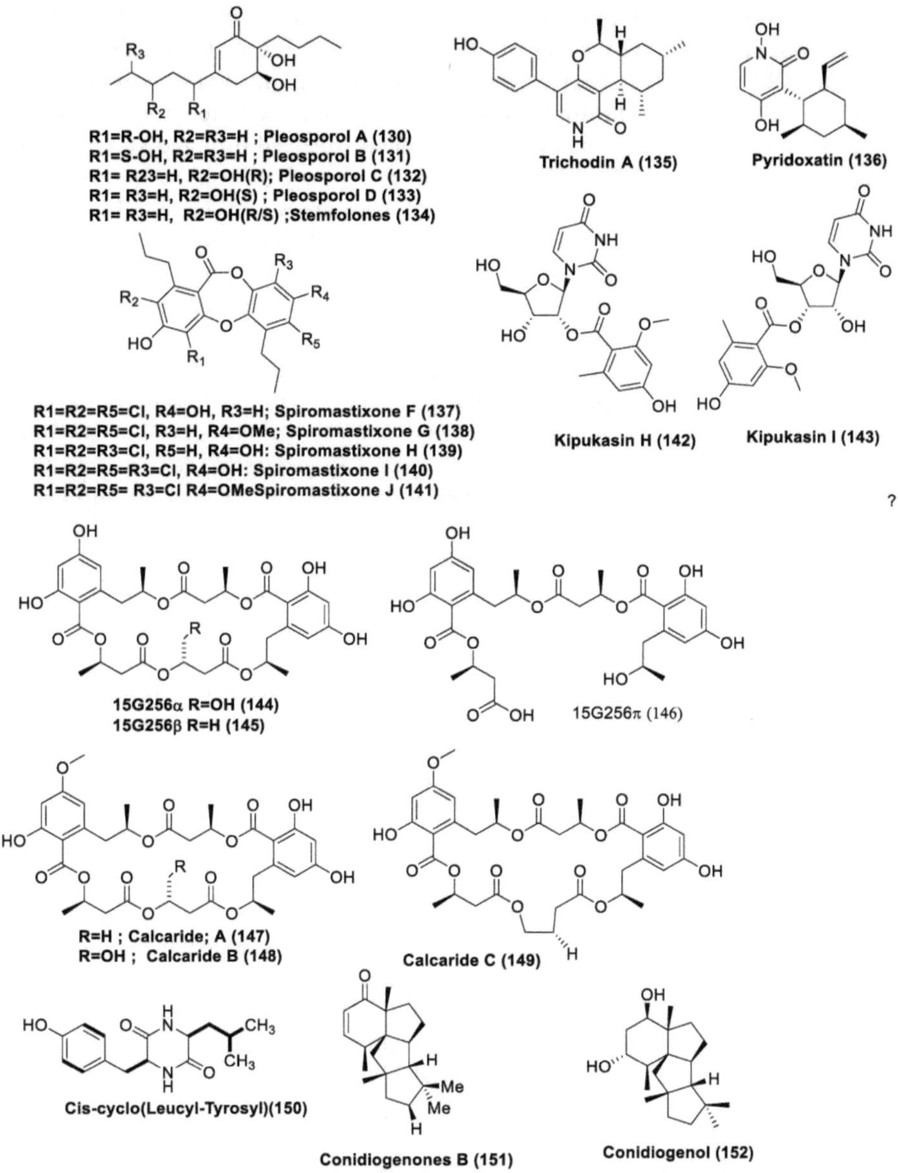

Fig. 25.11 Structures of antimicrobial compounds against skin microbial infection from marine fungi (**130–152**)

Province of China. Compounds (**130–134**) exhibited potent antimicrobial activity against *S. epidermidis* CMCC26069 with MIC values of 0.05, 8.0, 2.0, and 1.0 μg/mL (Xu et al. 2018).

Unusual pyridones, trichodin A (**135**), along with the known compound pyridoxatin (**136**) (Fig. 25.11), were obtained from marine fungus, *Trichoderma* sp.

strain MF106 isolated from the Greenland Seas. Both the compounds showed antibacterial activity with IC_{50} values of 24 μM and 4 μM, respectively, against *S. epidermidis* (Wu et al. 2014). New depsidone-based analogs, named spiromastixones F **(137)**, G **(138)**, H **(139)**, I **(140)**, J **(141)** (Fig. 25.11), were extracted from *Spiromastix* sp., a fungus isolated from the deep-sea sediment collected from the South Atlantic Ocean. Compounds **(137–141)** exhibited good antibacterial activity against *Staphylococcus epidermidis* both methicillin-sensitive and methicillin-resistant strains (*S. epidermidis* ATCC 12228, 12–6, 12–8 methicillin-sensitive (MSSE) and *S. epidermidis* 12–8 methicillin-resistant (MRSE) strain) with the MIC values in the range of 2–8 μM, while levofloxacin a positive control showed the activity in the range of 0.25–4 μM (Niu et al. 2014).

Two nucleoside derivatives Kipukasin H **(142)** and Kipukasin I **(143)** (Fig. 25.11) were extracted from *Aspergillus versicolor* derived from *Dichotella gemmacea* gorgonian collected from the South China Sea. Both compounds showed selective antibacterial activity with a MIC value of 12.5 μM against *S. epidermidis* (Chen et al. 2014).

Compounds 15G256α **(144)**, 15G256β **(145)**, and 15G256π **(146)** and known compounds calcaride A **(147)**, calcaride B **(148)**, and calcaride C **(149)** (Fig. 25.11) were extracted from *Calcarisporium* sp. KF525, isolated from German Wadden Sea water samples. Compounds **(144, 145, and 147–149)** exhibited antibacterial activity against *S. epidermidis* with MIC value of 12.9, 16.9, 68.8, 52.3, and 29.6 μM, respectively, while positive control chloramphenicol showed the MIC value of 4.5 μM against the tested strain. Compound **(146)** was found active against *P. acnes* 14.1 with the MIC value of 4.5 μM. Positive control chloramphenicol showed the MIC value of 0.5 against *P. acnes* (Silber et al. 2013).

Dipeptide cis-cyclo(Leucyl-Tyrosyl) **(150)** (Fig. 25.11) was isolated from a sponge-associated *Penicillium* sp. and exhibits up to 85% inhibition of biofilm formation of *S. epidermidis* without interfering with bacterial growth, which was confirmed by scanning electron microscopy (Scopel et al. 2013).

Conidiogenones B **(151)** and conidiogenol **(152)** (Fig. 25.11) were characterized from *Penicillium chrysogenum* QEN-24S, associated with marine red alga *Laurencia* sp. Compound **(151)** exhibited potent antibacterial activity against MRSA, *Pseudomonas fluorescens*, *P. aeruginosa*, and *S. epidermidis* with MIC value of 8 μg/mL in each case. Compound **(152)** exhibited antibacterial activity against *P. fluorescens* and *S. epidermidis* with MIC value of 16 μg/mL in each case (Gao et al. 2011).

25.3 Conclusions and Future Perspectives

Marine fungi are a unique source of novel bioactive metabolites with cosmeceutical properties. The chemical diversity of these molecules includes terpenes, steroids, polyketides, peptides, alkaloids, and polysaccharides. Some of these compounds possess good photo-protective, skin-whitening, antioxidant, antimicrobial, anti-tyrosinase, and antiaging activities. Some also have the potential for use as surfactants, emulsifiers, thickeners, stabilizers, and moisturizers. For example, the alkaloids golmaenone and neoechinulin A, extracted from the marine-derived fungus *Aspergillus* sp., exhibited an ultraviolet-A protecting activity with ED_{50} values of 90 and 170 μM, respectively, and possess better activity than oxybenzone (ED_{50}, 350 μM), which is currently being used as sunscreen (Li et al. 2004b). Most inhibitors have not been investigated for incorporation into topically applied cosmeceuticals. In addition, many identified tyrosinase inhibitors are not readily available to researchers due to limited supply, and this limits their in vivo evaluation, where relatively large amounts are needed for testing. A range of compounds with cosmeceutical potential have been identified. However, limited clinical data is currently available to substantiate the benefit of these ingredients in cosmetic products.

References

Aasen IM, Ertesvåg H, Heggeset TMB, Liu B, Brautaset T, Vadstein O, Ellingsen TE (2016) Thraustochytrids as production organisms for docosahexaenoic acid (DHA), squalene, and carotenoids. Appl Microbiol Biotechnol 100:4309–4321

Abdel-Lateff A, König GM, Fisch KM, Höller U, Jones PG, Wright AD (2002) New antioxidant hydroquinone derivatives from the algicolous marine fungus *Acremonium* sp. J Nat Prod 65:1605–1611

Abdel-Lateff A, Fisch KM, Wright AD, König GM (2003a) A new antioxidant isobenzofuranone derivative from the algicolous marine fungus *Epicoccum* sp. Planta Med 69:831–834

Abdel-Lateff A, Klemke C, König GM, Wright AD (2003b) Two new xanthone derivatives from the algicolous marine fungus *Wardomyces anomalus*. J Nat Prod 66:706–708

Adarme-Vega TC, Thomas-Hall SR, Schenk PM (2014) Towards sustainable sources for omega-3 fatty acids production. Curr Opin Biotechnol 26:14–18

Agrawal S, Adholeya A, Barrow CJ, Deshmukh SK (2018a) In-vitro evaluation of marine derived fungi against *Cutibacterium acnes*. Anaerobe 49:5–13

Agrawal S, Adholeya A, Barrow CJ, Deshmukh SK (2018b) Marine fungi: an untapped bioresource for future cosmeceuticals. Phytochem Lett 23:15–20

Ambati R, Phang S-M, Ravi S, Aswathanarayana R (2014) Astaxanthin: sources, extraction, stability, biological activities and its commercial applications—a review. Mar Drugs 12:128–152

Borse B, Bhat D, Borse K, Tuwar N, Pawar N (2012) Marine fungi of India (monograph). Broadway Publishing House, Panjim

Bugni TS, Ireland CM (2004) Marine-derived fungi: a chemically and biologically diverse group of microorganisms. Nat Prod Rep 21:143–163

Bumbak F, Cook S, Zachleder V, Hauser S, Kovar K (2011) Best practices in heterotrophic high-cell-density microalgal processes: achievements, potential and possible limitations. Appl Microbiol Biotechnol 91:31–46

Cambon-Bonavita MA, Raguenes G, Jean J, Vincent P, Guezennec J (2002) A novel polymer produced by a bacterium isolated from a deep-sea hydrothermal vent polychaete annelid. J Appl Microbiol 93:310–315

Carreto JI, Carignan MO (2011) Mycosporine-like amino acids: relevant secondary metabolites. Chemical and ecological aspects. Mar Drugs 9:387–446

Chen Y, Guo S-D, Xu J, Chen Y-L, Li H-Y, Qi X-H, Mao W-J (2010) Studies on exopolysaccharides from a marine Endogenetic fungus of the sponge (*Alternaria* sp.). Period Ocean Univ China 5:1–4

Chen M, Fu X-M, Kong C-J, Wang C-Y (2014) Nucleoside derivatives from the marine-derived fungus *Aspergillus versicolor*. Nat Prod Res 28:895–900

Chen H, Huang M, Li X, Liu L, Chen B, Wang J, Lin Y (2018) Phochrodines A-D, first naturally occurring new chromenopyridines from mangrove entophytic fungus *Phomopsis* sp. 33 Fitoterapia 124:103–107

Courtois A, Berthou C, Guézennec J, Boisset C, Bordron A (2014) Exopolysaccharides isolated from hydrothermal vent bacteria can modulate the complement system. PLoS One 9:e94965
Covington MB (2004) Omega-3 fatty acids. Am Fam Physician 70:133–140
Cui CM, Li XM, Li CS, Sun HF, Gao SS, Wang BG (2009) Benzodiazepine alkaloids from marine-derived endophytic fungus *Aspergillus ochraceus*. Helv Chim Acta 92:1366–1370
Day TA, Neale PJ (2002) Effects of UV-B radiation on terrestrial and aquatic primary producers. Annu Rev Ecol Syst 33:371–396
Deshmukh SK, Prakash V, Ranjan N (2018) Marine fungi: a source of potential anticancer compounds. Front Microbiol 8:2536
Ding LJ, Gu BB, Jiao WH, Yuan W, Li YX, Tang WZ, Yu HB, Liao XJ, Han BN, Li ZY, Xu SH (2015) New furan and cyclopentenone derivatives from the sponge-associated fungus *Hypocrea koningii* PF04. Mar Drugs 13:5579–5592
Du F-Y, Li X, Li X-M, Zhu L-W, Wang B-G (2017) Indolediketopiperazine alkaloids from *Eurotium cristatum* EN-220, an endophytic fungus isolated from the marine alga *Sargassum thunbergii*. Mar Drugs 15:24
Dunlap WC, Yamamoto Y (1995) Small-molecule antioxidants in marine organisms: antioxidant activity of mycosporine-glycine. Comp Biochem Physiol B: Biochem Mol Biol 112:105–114
Freitas F, Alves VD, Reis MA (2011) Advances in bacterial exopolysaccharides: from production to biotechnological applications. Trends Biotechnol 29:388–398
Giacomoni PU (2008) Advancement in skin aging: the future cosmeceuticals. Clin Dermatol 26:364–366
Gao SS, Li XM, Du FY, Li CS, Proksch P, Wang BG (2011) Secondary metabolites from a marine-derived endophytic fungus *Penicillium chrysogenum* QEN-24S. Mar Drugs 9(1):59–70
Gupta A, Singh D, Barrow CJ, Puri M (2013) Exploring potential use of Australian thraustochytrids for the bioconversion of glycerol to omega-3 and carotenoids production. Biochem Eng J 78:11–17
Huang Z, Nong X, Ren Z, Wang J, Zhang X, Qi S (2017) Anti-HSV-1, antioxidant and antifouling phenolic compounds from the deep-sea-derived fungus *Aspergillus versicolor* SCSIO 41502. Bioorg Med Chem Lett 27:787–791
Hutchings JA, Reynolds JD (2004) Marine fish population collapses: consequences for recovery and extinction risk. AIBS Bull 54:297–309
Jouault SC, Chevolot L, Helley D, Ratiskol J, Bros A, Sinquin C, Roger O, Fischer AM (2001) Characterization, chemical modifications and in vitro anticoagulant properties of an exopolysaccharide produced by *Alteromonas infernus*. Biochimica et Biophysica Acta (BBA)-Gen Sub 1528:141–151
Kamauchi H, Noji M, Kinoshita K, Takanami T, Koyama K (2018) Coumarins with an unprecedented tetracyclic skeleton and coumarin dimers from chemically engineered extracts of a marine-derived fungus. Tetrahedron 74:2846–2856
Kawahara T, Takagi M, Shin-ya K (2012) JBIR-124: a novel antioxidative agent from a marine sponge-derived fungus Penicillium citrinum SpI080624G1f01. J Antibiot 65:45
Kim S-K (2016) Marine cosmeceuticals: trends and prospects. CRC Press, Boca Raton, 428 pp
Kirkwood TB, Holliday R (1979) The evolution of ageing and longevity. Proc R Soc Lond B 205:531–546
Kogej T, Gostinčar C, Volkmann M, Gorbushina AA, Gunde-Cimerman N (2006) Mycosporines in extremophilic fungi—novel complementary osmolytes? Environ Chem 3:105–110
Kohlmeyer E, Kohlmeyer J (1980) Marine mycology: the higher fungi. JSTOR
Kot AM, Błażejak S, Kurcz A, Gientka I, Kieliszek M (2016) *Rhodotorula glutinis*—potential source of lipids, carotenoids, and enzymes for use in industries. Appl Microbiol Biotechnol 100:6103–6117
Leutou AS, Yun K, Kang JS, Son BW (2013) Induced production of methyl bromodihydroxyphenyl acetates by the marine-derived fungus *Aspergillus* sp. Chem Pharm Bull 61:483–485
Leutou AS, Yun K, Son BW (2016a) Induced production of 6, 9-dibromoflavasperone, a new radical scavenging naphthopyranone in the marine-mudflat-derived fungus Aspergillus niger. Arch Pharmacal Res 39:806–810

Leutou AS, Yun K, Son BW (2016b) New production of antibacterial polycyclic quinazoline alkaloid, thielaviazoline, from anthranilic acid by the marine-mudflat-derived fungus *Thielavia* sp. Nat Prod Sci 22:216–219

Li X, Jeong JH, Lee KT, Rho JR, Choi HD, Kang JS, Son BW (2003) γ-Pyrone derivatives, kojic acid methyl ethers from a marine-derived fungus *Alternaria* sp. Arch Pharm Res 26:532–534

Li Y, Li X, Kang JS, Choi HD, Son BW (2004a) New radical scavenging and ultraviolet-A protecting prenylated dioxopiperazine alkaloid related to isoechinulin A from a marine isolate of the fungus *Aspergillus*. J Antibiot 57:337–340

Li Y, Li X, Kim S-K, Kang JS, Choi HD, Rho JR, Son BW (2004b) Golmaenone, a new diketopiperazine alkaloid from the marine-derived fungus *Aspergillus* sp. Chem Charmaceut Bull 52:375–376

Li X, Kim MK, Lee U, Kim S-K, Kang JS, Choi HD, Son BW (2005) Myrothenones A and B, cyclopentenone derivatives with tyrosinase inhibitory activity from the marine-derived fungus *Myrothecium* sp. Chem Pharm Bull 53:453–455

Li X, Kim S-K, Kang JS, Choi HD, Son BW (2006) Radical scavenging hydroxyphenyl ethanoic acid derivatives from a marine-derived fungus. J Microbiol Biotechnol

Li X, Li X-M, Xu G-M, Li C-S, Wang B-G (2014) Antioxidant metabolites from marine alga-derived fungus *Aspergillus wentii* EN-48. Phytochem Lett 7:120–123

Li HL, Li XM, Li X, Wang CY, Liu H, Kassack MU, Meng LH, Wang BG (2017) Antioxidant hydroanthraquinones from the marine algal-derived endophytic fungus *Talaromyces islandicus* EN-501. J Nat Prod 80:162–168

Maciel OMC, Tavares RSN, Caluz DRE, Gaspar LR, Debonsi HM (2018) Photoprotective potential of metabolites isolated from algae-associated fungi *Annulohypoxylon stygium*. J Photochem Photobiol B Biol 178:316–322

Maeda K, Fukuda M (1991) In vitro effectiveness of several whitening cosmetic components in human melanocytes. J Soc Cosmet Chem 42:361–368

Mata-Gómez LC, Montañez JC, Méndez-Zavala A, Aguilar CN (2014) Biotechnological production of carotenoids by yeasts: an overview. Microbial Cell Factor 13:12

McCusker MM, Grant-Kels JM (2010) Healing fats of the skin: the structural and immunologic roles of the ω-6 and ω-3 fatty acids. Clin Dermatol 28:440–451

Mourshid SS, Badr JM, Risinger AL, Mooberry SL, Youssef DT (2016) Penicilloitins A and B, new antimicrobial fatty acid esters from a marine endophytic *Penicillium* species. Zeitschrift für Naturforschung C 71:387–392

Niu S, Liu D, Hu X, Proksch P, Shao Z, Lin W (2014) Spiromastixones A–O, antibacterial chlorodepsidones from a deep-sea-derived *Spiromastix* sp. fungus. J Nat Prod 77:1021–1030

Nwodo UU, Green E, Okoh AI (2012) Bacterial exopolysaccharides: functionality and prospects. Intl J Mol Sci 13:14002–14015

Oren A, Gunde-Cimerman N (2007) Mycosporines and mycosporine-like amino acids: UV protectants or multipurpose secondary metabolites? FEMS Microbiol Lett 269:1–10

Pang X, Lin X, Yang J, Zhou X, Yang B, Wang J, Liu Y (2018) Spiro-Phthalides and Isocoumarins isolated from the marine-sponge-derived fungus *Setosphaeria* sp. SCSIO41009. J Nat Prod 81:1860–1868

Poli A, Anzelmo G, Nicolaus B (2010) Bacterial exopolysaccharides from extreme marine habitats: production, characterization and biological activities. Mar Drugs 8:1779–1802

Raghukumar S (2002) Ecology of the marine protists, the Labyrinthulomycetes (Thraustochytrids and Labyrinthulids). Eur J Protistol 38:127–145

Rastogi RP, Incharoensakdi A (2013) UV radiation-induced accumulation of photoprotective compounds in the green alga *Tetraspora* sp. CU2551. Plant Physiol Biochem 70:7–13

Řezanka T, Temina M, Tolstikov A, Dembitsky V (2004) Natural microbial UV radiation filters—mycosporine-like amino acids. Folia Microbiol 49:339–352

Richards TA, Jones MD, Leonard G, Bass D (2012) Marine fungi: their ecology and molecular diversity. Annu Rev Mar Sci 4:495–522

Rozema J, Boelen P, Blokker P (2005) Depletion of stratospheric ozone over the Antarctic and Arctic: responses of plants of polar terrestrial ecosystems to enhanced UV-B, an overview. Environ Pollut 137:428–442

Rukachaisirikul V, Khamthong N, Sukpondma Y, Phongpaichit S, Hutadilok-Towatana N, Graidist P, Sakayaroj J, Kirtikara K (2010) Cyclohexene, diketopiperazine, lactone and phenol derivatives from the sea fan-derived fungi *Nigrospora* sp. PSU-F11 and PSU-F12. Arch Pharm Res 33:375–380

Saleem M, Ali MS, Hussain S, Jabbar A, Ashraf M, Lee YS (2007) Marine natural products of fungal origin. Nat Prod Rep 24:1142–1152

Scopel M, Abraham W-R, Henriques AT, Macedo AJ (2013) Dipeptide cis-cyclo (Leucyl-Tyrosyl) produced by sponge associated *Penicillium* sp. F37 inhibits biofilm formation of the pathogenic *Staphylococcus epidermidis*. Bioorg Med Chem Lett 23:624–626

Silber J, Ohlendorf B, Labes A, Erhard A, Imhoff J (2013) Calcarides A–E, antibacterial macrocyclic and linear polyesters from a *Calcarisporium* strain. Marine Ddrugs 11:3309–3323

Solomon KR (2008) Effects of ozone depletion and UV-B radiation on humans and the environment. Atmosphere-Ocean 46:185–202

Suen YL, Tang H, Huang J, Chen F (2014) Enhanced production of fatty acids and astaxanthin in *Aurantiochytrium* sp. by the expression of *Vitreoscilla* hemoglobin. J Agric Food Chem 62:12392–12398

Sun C, Wang J-W, Fang L, Gao X-D, Tan R-X (2004) Free radical scavenging and antioxidant activities of EPS2, an exopolysaccharide produced by a marine filamentous fungus *Keissleriella* sp. YS 4108. Life Sci 75:1063–1073

Sun K, Chen Y, Niu Q, Zhu W, Wang B, Li P, Ge X (2016) An exopolysaccharide isolated from a coral-associated fungus and its sulfated derivative activates macrophages. Int J Biol Macromol 82:387–394

Sun Y, Liu J, Li L, Gong C, Wang S, Yang F, Hua H, Lin H (2018) New butenolide derivatives from the marine sponge-derived fungus *Aspergillus terreus*. Bioorg Med Chem Lett 28:315–318

Suresh Kumar A, Mody K, Jha B (2007) Bacterial exopolysaccharides–a perception. J Basic Microbiol 47:103–117

Tan C, Liu Z, Chen S, Huang X, Cui H, Long Y, Lu Y, She Z (2016) Antioxidative polyketones from the mangrove-derived fungus ascomycota sp. SK2YWS-L. Sci Rep 6:36609

Tsuchiya T, Yamada K, Minoura K, Miyamoto K, Usami Y, Kobayashi T, Hamada-Sato N, Imada C, Tsujibo H (2008) Purification and determination of the chemical structure of the tyrosinase inhibitor produced by *Trichoderma viride* strain H1-7 from a marine environment. Biol Pharm Bull 31:1618–1620

Vílchez C, Forján E, Cuaresma M, Bédmar F, Garbayo I, Vega JM (2011) Marine carotenoids: biological functions and commercial applications. Mar Drugs 9:319–333

Wang Y-N, Tian L, Hua H-M, Lu X, Sun S, Wu H-H, Pei Y-H (2009) Two new compounds from the broth of the marine fungus *Penicillium griseofulvum* Y19-07. J Asian Nat Prod Res 11:912–917

Wang X, Wang H, Liu T, Xin Z (2014) A PKS I gene-based screening approach for the discovery of a new polyketide from *Penicillium citrinum* Salicorn 46. Appl Microbiol Biotechnol 98:4875–4885. https://doi.org/10.1007/s00253-014-5572-3

Wang J, Ding W, Wang R, Du Y, Liu H, Kong X, Li C (2015) Identification and bioactivity of compounds from the mangrove endophytic fungus *Alternaria* sp. Mar Drugs 13:4492–4504. https://doi.org/10.3390/md13074492

Wang H, Hong J, Yin J, Liu J, Liu Y, Choi JS, Jung JH (2016) Paecilonic acids A and B, bicyclic fatty acids from the jellyfish-derived fungus *Paecilomyces variotii* J08NF-1. Bioorg Med Chem Lett 26:2220–2223

Wickens AP (2001) Ageing and the free radical theory. Respir Physiol 128:379–391

Wu B, Wu X, Sun M, Li M (2013) Two novel tyrosinase inhibitory sesquiterpenes induced by CuCl2 from a marine-derived fungus *Pestalotiopsis* sp. Z233. Marine Drugs 11:2713–2721

Wu B, Oesker V, Wiese J, Schmaljohann R, Imhoff J (2014) Two new antibiotic pyridones produced by a marine fungus, *Trichoderma* sp. strain MF106. Mar Drugs 12:1208–1219

Wynn JP, Ratledge C (2005) Microbial production of oils and fats. In: Food biotechnology. CRC Press, Boca Raton, pp 459–488

Xiao Z, Chen S, Cai R, Lin S, Hong K, She Z (2016) New furoisocoumarins and isocoumarins from the mangrove endophytic fungus *Aspergillus* sp. 085242. Beilstein J Org Chem 12:2077–2085. https://doi.org/10.3762/bjoc.12.196

Xu L-L, Zhang C-C, Zhu X-Y, Cao F, Zhu H-J (2017) Bioactive phenyl ether derivatives from the marine-derived fungus *Aspergillus carneus*. Nat Prod Res 31:1875–1879

Xu J, Liu P, Li X, Gan L, Wang P (2018) Novel stemphol derivatives from a marine fungus *Pleospora* sp. Nat Product Res:1–7

Younes I, Rinaudo M (2015) Chitin and chitosan preparation from marine sources. Structure, properties and applications. Mar Drugs 13:1133–1174

Yun K, Kondempudi CM, Choi HD, Kang JS, Son BW (2011) Microbial mannosidation of bioactive chlorogentisyl alcohol by the marine-derived fungus *Chrysosporium synchronum*. Chem Pharm Bull 59:499–501

Yun K, Feng Z, Choi HD, Kang JS, Son BW (2013) New production of (R)-(−)-5-bromomellein, a dihydroisocoumarin derivative from the marine-derived fungus *Aspergillus ochraceus*. Chem Nat Compd 49:24–26

Ze-Hong W, Dong L, Ying X, Jian-Liang C, Wen-Han L (2018) Antioxidant xanthones and anthraquinones isolated from a marine-derived fungus *Aspergillus versicolor*. Chin J Nat Med 16:219–224

Zeng Y-B, Wang H, Zuo W-J, Zheng B, Yang T, Dai H-F, Mei W-L (2012) A fatty acid glycoside from a marine-derived fungus isolated from mangrove plant *Scyphiphora hydrophyllacea*. Mar Drugs 10:598–603

Zhang D, Li X, Kang JS, Choi HD, Son BW (2007) A new α-Pyrone derivative, 6-[(E)-Hept-1-enyl]-α-pyrone, with Tyrosinase inhibitor activity from a marine isolate of the fungus *Botrytis*. Bull Kor Chem Soc 28:887

Zhang D, Yang X, Kang JS, Choi HD, Son BW (2008) Circumdatin I, a new ultraviolet-A protecting benzodiazepine alkaloid from a marine isolate of the fungus *Exophiala*. J Antibiot 61:40–42

Zhang P, Meng L-H, Mándi A, Li X-M, Kurtán T, Wang B-G (2015) Structure, absolute configuration, and conformational study of resorcylic acid derivatives and related congeners from the fungus *Penicillium brocae*. RSC Adv 5:39870–39877. https://doi.org/10.1039/c5ra02203g

Zhong W, Wang J, Wei X, Chen Y, Fu T, Xiang Y, HuangX TX, Xiao Z, Zhang W, Zhang S, Long L, Wang F (2018) Variecolortins A–C, three pairs of spirocyclic diketopiperazine enantiomers from the marine-derived fungus *Eurotium* sp. SCSIO F452. Org Lett 20:4593–4596

Zhou J, Diao X, Wang T, Chen G, Lin Q, Yang X, Xu J (2018) Phylogenetic diversity and antioxidant activities of culturable fungal endophytes associated with the mangrove species *Rhizophora stylosa* and R. mucronata in the South China Sea. PLoS One 13:e0197359

Ziboh VA, Miller CC, Cho Y (2000). Metabolism of polyunsaturated fatty acids by skin epidermal enzymes: generation of antiinflammatory and antiproliferative metabolites. Am J Clin Nutr 71:361s–366s

Index

T. Satyanarayana et al. (eds.), *Advancing Frontiers in Mycology & Mycotechnology*,
https://doi.org/10.1007/978-981-13-9349-5